Safety Evaluation Report

Related to the License Renewal of Palo Verde Nuclear Generating Station, Units 1, 2, and 3

Docket Numbers 50-528, 50-529, and 50-530

Arizona Public Service Company

Manuscript Completed: March 2011
Date Published: April 2011

Office of Nuclear Reactor Regulation

ABSTRACT

This safety evaluation report (SER) documents the technical review of the Palo Verde Nuclear Generating Station, Units 1, 2, and 3 (PVNGS), license renewal application (LRA) by the U.S. Nuclear Regulatory Commission (NRC) staff (the staff). By letter dated December 11, 2008, and supplemented by letter dated April 14, 2009, Arizona Public Service Company (APS) (the applicant) submitted the LRA in accordance with Title 10 of the *Code of Federal Regulations*, Part 54 "Requirements for Renewal of Operating Licenses for Nuclear Power Plants." APS requests renewal of the PVNGS operating licenses (facility operating license numbers NPF-41, NPF-51, and NPF-74) for a period of 20 years beyond the current expiration dates of midnight on June 1, 2025, for Unit 1; April 24, 2026, for Unit 2; and November 25, 2027, for Unit 3.

PVNGS is a three-unit, nuclear-powered, steam electric generating facility located in Maricopa County, AZ, approximately 26 miles west of the Phoenix metropolitan area boundary. The NRC issued the construction permits on May 25, 1976, for all three units, and it issued the operating licenses on June 1, 1985, for Unit 1; April 24, 1986, for Unit 2; and November 25, 1987, for Unit 3. PVNGS employs a pressurized water reactor (PWR) design with a dry ambient containment. Each of the units uses a System 80 PWR nuclear steam supply system provided by Combustion Engineering, Incorporated. Bechtel Power Corporation is responsible for the engineering and construction of the station and designed the balance of the plant. The licensed power output is 3,990 megawatts-thermal per unit with a net electrical output of approximately 1,346 megawatts-electric per unit.

On August 6, 2010, the staff issued an SER with Open Item Related to the License Renewal of Palo Verde Nuclear Generating Station, Units 1, 2, and 3, in which the staff identified one open item and five confirmatory items necessitating further review. This SER presents the status of the staff's review of information submitted through March 17, 2011, the cutoff date for consideration in the SER. The open and confirmatory items identified in the SER with Open Item were resolved before the staff made a final determination. SER Sections 1.5 and 1.6 summarize these open and confirmatory items. SER Section 6.0 provides the staff's final conclusion of the LRA review.

TABLE OF CONTENTS

Abstract iii

Table of Contents v

List of Tables xii

Abbreviations xiii

1.0 Introduction and General Discussion 1-1
    1.1 Introduction 1-1
    1.2 License Renewal Background 1-2
        1.2.1 Safety Review 1-3
        1.2.2 Environmental Review 1-4
    1.3 Principal Review Matters 1-5
    1.4 Interim Staff Guidance 1-6
    1.5 Summary of Open Items 1-6
    1.6 Summary of Confirmatory Items 1-10
    1.7 Summary of Additional Items 1-12
    1.8 Summary of Proposed License Conditions 1-13

2.0 Scoping and Screening Methodology 2-1
    2.1 Scoping and Screening Methodology 2-1
        2.1.1 Introduction 2-1
        2.1.2 Summary of Technical Information in the Application 2-1
        2.1.3 Scoping and Screening Program Review 2-2
            2.1.3.1 Implementation Procedures and Documentation Sources
                      for Scoping and Screening 2-3
            2.1.3.2 Quality Controls Applied to License Review Application
                      Development 2-6
            2.1.3.3 Training 2-7
            2.1.3.4 Conclusion of Scoping and Screening Program Review 2-8
        2.1.4 Plant Systems, Structures, and Components Scoping Methodology .. 2-8
            2.1.4.1 Application of the Scoping Criteria in 10 CFR 54.4(a)(1) 2-8
            2.1.4.2 Application of the Scoping Criteria in 10 CFR 54.4(a)(2) .. 2-10
            2.1.4.3 Application of the Scoping Criteria in 10 CFR 54.4(a)(3) .. 2-18
            2.1.4.4 Plant-Level Scoping of Systems and Structures 2-22
            2.1.4.5 Mechanical Scoping 2-23
            2.1.4.6 Structural Scoping 2-25
            2.1.4.7 Electrical Component Scoping 2-26
            2.1.4.8 Scoping Methodology Conclusion 2-27
        2.1.5 Screening Methodology 2-27
            2.1.5.1 General Screening Methodology 2-27
            2.1.5.2 Mechanical Component Screening 2-29
            2.1.5.3 Structural Component Screening 2-30
            2.1.5.4 Electrical Component Screening 2-31
            2.1.5.5 Screening Methodology Conclusion 2-33
        2.1.6 Summary of Evaluation Findings 2-33
    2.2 Plant-Level Scoping Results 2-33
        2.2.1 Introduction 2-33
        2.2.2 Summary of Technical Information in the Application 2-33

Table of Contents

		2.2.3	Staff Evaluation	2-33

- 2.2.3 Staff Evaluation 2-33
- 2.2.4 Conclusion 2-35
- 2.3 Scoping and Screening Results: Mechanical Systems 2-35
  - 2.3.1 Reactor Vessel, Internals, and Reactor Coolant System 2-35
    - 2.3.1.1 Reactor Vessel and Internals 2-36
    - 2.3.1.2 Reactor Coolant System 2-37
    - 2.3.1.3 Pressurizer 2-38
    - 2.3.1.4 Steam Generators 2-39
    - 2.3.1.5 Reactor Core 2-40
  - 2.3.2 Engineered Safety Features 2-41
    - 2.3.2.1 Containment Leak Test System 2-41
    - 2.3.2.2 Containment Purge System 2-41
    - 2.3.2.3 Containment Hydrogen Control System 2-42
    - 2.3.2.4 Safety Injection and Shutdown Cooling System 2-42
  - 2.3.3 Auxiliary Systems 2-43
    - 2.3.3.1 Fuel Handling and Storage System 2-45
    - 2.3.3.2 Spent Fuel Pool Cooling and Cleanup System 2-45
    - 2.3.3.3 Essential Cooling Water System 2-47
    - 2.3.3.4 Essential Chilled Water System 2-48
    - 2.3.3.5 Normal Chilled Water System 2-49
    - 2.3.3.6 Nuclear Cooling Water System 2-50
    - 2.3.3.7 Essential Spray Pond System 2-52
    - 2.3.3.8 Nuclear Sampling System 2-53
    - 2.3.3.9 Compressed Air System 2-54
    - 2.3.3.10 Chemical Volume and Control System 2-55
    - 2.3.3.11 Control Building Heating, Ventilation, and Air Conditioning System 2-57
    - 2.3.3.12 Auxiliary Building Heating, Ventilation, and Air Conditioning System 2-58
    - 2.3.3.13 Fuel Building Heating, Ventilation, and Air Conditioning System 2-59
    - 2.3.3.14 Containment Building Heating, Ventilation, and Air Conditioning System 2-59
    - 2.3.3.15 Diesel Generator Building Heating, Ventilation, and Air Conditioning System 2-60
    - 2.3.3.16 Radwaste Building Heating, Ventilation, and Air Conditioning System 2-61
    - 2.3.3.17 Turbine Building Heating, Ventilation, and Air Conditioning System 2-61
    - 2.3.3.18 Miscellaneous Site Structures and Spray Pond Pump House Heating, Ventilation, and Air Conditioning System 2-62
    - 2.3.3.19 Fire Protection System 2-63
    - 2.3.3.20 Diesel Generator Fuel Oil Storage and Transfer System 2-70
    - 2.3.3.21 Diesel Generator 2-70
    - 2.3.3.22 Domestic Water System 2-72
    - 2.3.3.23 Demineralized Water System 2-73
    - 2.3.3.24 Water Reclamation Facility Fuel System 2-74
    - 2.3.3.25 Service Gases (Nitrogen and Hydrogen) System 2-75
    - 2.3.3.26 Gaseous Radwaste System 2-76
    - 2.3.3.27 Radioactive Waste Drains System 2-76

Table of Contents

		2.3.3.28 Station Blackout Generator System	2-78
		2.3.3.29 Cranes, Hoists, and Elevators	2-78
		2.3.3.30 Miscellaneous Auxiliary Systems In-Scope ONLY for Criterion 10 CFR 54.4(a)(2)	2-79
	2.3.4	Steam and Power Conversion Systems	2-81
		2.3.4.1 Main Steam System	2-81
		2.3.4.2 Condensate Storage and Transfer System	2-83
		2.3.4.3 Auxiliary Feedwater System	2-84
		2.3.4.4 Condensate System	2-85
		2.3.4.5 Feedwater System	2-85
		2.3.4.6 Main Turbine System	2-86
		2.3.4.7 Steam Generator Feedwater Pump Turbine System	2-86
		2.3.4.8 Feedwater Heater Extraction, Drains, and Vents System	2-87
2.4	Scoping and Screening Results: Structures		2-87
	2.4.1	Containment Building	2-88
		2.4.1.1 Summary of Technical Information in the Application	2-88
		2.4.1.2 Staff Evaluation	2-89
		2.4.1.3 Conclusion	2-89
	2.4.2	Control Building	2-90
		2.4.2.1 Summary of Technical Information in the Application	2-90
		2.4.2.2 Staff Evaluation and Conclusion	2-90
	2.4.3	Diesel Generator Building	2-90
		2.4.3.1 Summary of Technical Information in the Application	2-90
		2.4.3.2 Staff Evaluation and Conclusion	2-90
	2.4.4	Turbine Building	2-91
		2.4.4.1 Summary of Technical Information in the Application	2-91
		2.4.4.2 Staff Evaluation	2-91
		2.4.4.3 Conclusion	2-91
	2.4.5	Auxiliary Building	2-92
		2.4.5.1 Summary of Technical Information in the Application	2-92
		2.4.5.2 Staff Evaluation	2-92
		2.4.5.3 Conclusion	2-92
	2.4.6	Radwaste Building	2-92
		2.4.6.1 Summary of Technical Information in the Application	2-92
		2.4.6.2 Staff Evaluation and Conclusion	2-93
	2.4.7	Main Steam Support Structure	2-93
		2.4.7.1 Summary of Technical Information in the Application	2-93
		2.4.7.2 Staff Evaluation and Conclusion	2-93
	2.4.8	Station Blackout Generator Structures	2-94
		2.4.8.1 Summary of Technical Information in the Application	2-94
		2.4.8.2 Staff Evaluation and Conclusion	2-94
	2.4.9	Fuel Building	2-94
		2.4.9.1 Summary of Technical Information in the Application	2-94
		2.4.9.2 Staff Evaluation	2-95
		2.4.9.3 Conclusion	2-95
	2.4.10	Spray Pond and Associated Water Control Structures	2-95
		2.4.10.1 Summary of Technical Information in the Application	2-95
		2.4.10.2 Staff Evaluation and Conclusion	2-96
	2.4.11	Tank Foundations and Shells	2-96
		2.4.11.1 Summary of Technical Information in the Application	2-96
		2.4.11.2 Staff Evaluation and Conclusion	2-96

Table of Contents

		2.4.12	Transformer Foundations and Electrical Structures 2-96
			2.4.12.1 Summary of Technical Information in the Application 2-96
			2.4.12.2 Staff Evaluation 2-97
			2.4.12.3 Conclusion 2-97
		2.4.13	Yard Structures (In-Scope) 2-97
			2.4.13.1 Summary of Technical Information in the Application 2-97
			2.4.13.2 Staff Evaluation 2-98
			2.4.13.3 Conclusion 2-98
		2.4.14	Supports 2-99
			2.4.14.1 Summary of Technical Information in the Application 2-99
			2.4.14.2 Staff Evaluation and Conclusion 2-99
		2.4.15	Fire Barriers 2-99
			2.4.15.1 Summary of Technical Information in the Application 2-99
			2.4.15.2 Staff Evaluation 2-100
			2.4.15.3 Conclusion 2-101
	2.5	Scoping and Screening Results: Electrical and Instrumentation and Control Systems 2-101	
		2.5.1	Electrical and Instrumentation and Control Systems Component Groups 2-102
			2.5.1.1 Summary of Technical Information in the Application 2-102
			2.5.1.2 Staff Evaluation 2-103
			2.5.1.3 Conclusion 2-104
	2.6	Conclusion for Scoping and Screening 2-104	
3.0	Aging Management Review Results 3-1		
	3.0	Applicant's Use of the Generic Aging Lessons Learned Report 3-1	
		3.0.1	Format of the License Renewal Application 3-2
			3.0.1.1 Overview of Table 1s 3-2
			3.0.1.2 Overview of Table 2s 3-2
		3.0.2	Staff's Review Process 3-3
			3.0.2.1 Review of Aging Management Programs 3-4
			3.0.2.2 Review of Aging Management Review Results 3-5
			3.0.2.3 Updated Final Safety Analysis Report Supplement 3-6
			3.0.2.4 Documents Reviewed 3-6
		3.0.3	Aging Management Programs 3-6
			3.0.3.1 Aging Management Programs Consistent with the Generic Aging Lessons Learned Report 3-9
			3.0.3.2 Aging Management Programs Consistent with the Generic Aging Lessons Learned Report, with Exceptions or Enhancements 3-44
			3.0.3.3 Aging Management Programs Not Consistent with or Not Addressed in the Generic Aging Lessons Learned Report 3-128
		3.0.4	Quality Assurance Program Attributes Integral to Aging Management Programs 3-134
			3.0.4.1 Summary of Technical Information in the Application 3-134
			3.0.4.2 Staff Evaluation 3-134
			3.0.4.3 Conclusion 3-135
	3.1	Aging Management of Reactor Vessel, Internals and Reactor Coolant System 3-136	
		3.1.1	Summary of Technical Information in the Application 3-136
		3.1.2	Staff Evaluation 3-136

		3.1.2.1	Aging Management Review Results Consistent with the Generic Aging Lessons Learned Report 3-151
		3.1.2.2	Aging Management Review Results Consistent with the Generic Aging Lessons Learned Report for Which Further Evaluation Is Recommended 3-158
		3.1.2.3	Aging Management Review Results Not Consistent with or Not Addressed in the Generic Aging Lessons Learned Report 3-179
	3.1.3	Conclusion	 3-181
3.2	Aging Management of Engineered Safety Features Systems 3-181
	3.2.1	Summary of Technical Information in the Application 3-181
	3.2.2	Staff Evaluation	 3-181
		3.2.2.1	Aging Management Review Results Consistent with the Generic Aging Lessons Learned Report 3-189
		3.2.2.2	Aging Management Review Results Consistent with the Generic Aging Lessons Learned Report for Which Further Evaluation Is Recommended 3-196
		3.2.2.3	Aging Management Review Results Not Consistent with or Not Addressed in the Generic Aging Lessons Learned Report 3-203
	3.2.3	Conclusion	 3-206
3.3	Aging Management of Auxiliary Systems 3-206
	3.3.1	Summary of Technical Information in the Application 3-207
	3.3.2	Staff Evaluation	 3-207
		3.3.2.1	Aging Management Review Results Consistent with the Generic Aging Lessons Learned Report 3-221
		3.3.2.2	Aging Management Review Results Consistent with the Generic Aging Lessons Learned Report for Which Further Evaluation Is Recommended 3-230
		3.3.2.3	Aging Management Review Results Not Consistent with or Not Addressed in the Generic Aging Lessons Learned Report 3-250
	3.3.3	Conclusion	 3-269
3.4	Aging Management of Steam and Power Conversion Systems 3-269
	3.4.1	Summary of Technical Information in the Application 3-269
	3.4.2	Staff Evaluation	 3-270
		3.4.2.1	Aging Management Review Results Consistent with the Generic Aging Lessons Learned Report 3-275
		3.4.2.2	Aging Management Review Results Consistent with the Generic Aging Lessons Learned Report for Which Further Evaluation Is Recommended 3-279
		3.4.2.3	Aging Management Review Results Not Consistent with or Not Addressed in the Generic Aging Lessons Learned Report 3-288
	3.4.3	Conclusion	 3-291
3.5	Aging Management of Structures and Component Supports 3-291
	3.5.1	Summary of Technical Information in the Application 3-291
	3.5.2	Staff Evaluation	 3-292
		3.5.2.1	Aging Management Review Results Consistent with the Generic Aging Lessons Learned Report 3-303

Table of Contents

		3.5.2.2	Aging Management Review Results Consistent with the Generic Aging Lessons Learned Report for Which Further Evaluation Is Recommended............................3-306

                3.5.2.3   Aging Management Review Results Not Consistent with or Not Addressed in the Generic Aging Lessons Learned Report 3-321

       3.5.3   Conclusion...3-327

  3.6   Aging Management of Electrical and Instrumentation and Controls..............3-328
       3.6.1   Summary of Technical Information in the Application....................3-328
       3.6.2   Staff Evaluation 3-328
                3.6.2.1   Aging Management Review Results Consistent with the Generic Aging Lessons Learned Report 3-331
                3.6.2.2   Aging Management Review Results Consistent with the Generic Aging Lessons Learned Report for Which Further Evaluation Is Recommended............................3-332
                3.6.2.3   Aging Management Review Results Not Consistent with or Not Addressed in the Generic Aging Lessons Learned Report 3-338
       3.6.3   Conclusion..3-338

  3.7   Conclusion for Aging Management Review Results...................................3-338

4.0 Time Limited Aging Analyses..4-1
  4.1   Time Limited Aging Analyses 4-1
       4.1.1   Identification of Time Limited Aging Analyses 4-1
       4.1.2   Summary of Technical Information in the Application.......................4-1
       4.1.3   Staff Evaluation 4-1
       4.1.4   Conclusion..4-12
  4.2   Reactor Vessel Neutron Embrittlement 4-12
       4.2.1   Neutron Fluence, Upper Shelf Energy and Adjusted Reference Temperature 4-13
       4.2.2   Pressurized Thermal Shock 4-16
       4.2.3   Pressure Temperature Limits 4-17
       4.2.4   Low Temperature Overpressure Protection.............................4-19
  4.3   Metal Fatigue Analysis 4-19
       4.3.1   Enhanced Fatigue Aging Management Program 4-21
       4.3.2   American Society of Mechanical Engineers III Fatigue Analysis of Class 1 Vessels, Piping, and Components...................................4-42
       4.3.3   Fatigue and Cycle Based Time Limited Aging Analyses of American Society of Mechanical Engineers III, Subsection NG, Reactor Pressure Vessel Internals 4-75
       4.3.4   Effects of the Reactor Coolant System Environment on Fatigue Life of Piping and Components (Generic Safety Issue 190).......................4-76
       4.3.5   Assumed Thermal Cycle Count for Allowable Secondary Stress Range Reduction Factor in American National Standards Institute B31.1 and American Society of Mechanical Engineers III Class 2 and 3 Piping 4-82
  4.4   Environmental Qualification of Electrical Equipment.................................4-85
       4.4.1   Summary of Technical Information in the Application....................4-85
       4.4.2   Staff Evaluation 4-86
       4.4.3   Updated Final Safety Analysis Report Supplement.........................4-86
       4.4.4   Conclusion..4-87
  4.5   Concrete Containment Tendon Prestress Analyses....................................4-87
       4.5.1   Summary of Technical Information in the Application....................4-87

		4.5.2	Staff Evaluation 4-88
		4.5.3	Updated Final Safety Analysis Report Supplement..........................4-89
		4.5.4	Conclusion..4-89

    4.6   Containment Liner Plate, Equipment Hatch and Personnel Air Locks, Penetrations, and Polar Crane Brackets 4-90

        4.6.1   Absence of a Time Limited Aging Analysis for Containment Liner Plate, Polar Crane Brackets, Equipment Hatch and Personnel Air Locks, and Containment Penetrations (Except Main Steam, Main Feedwater, and Recirculation Sump Suction Penetrations) 4-90

        4.6.2   Design Cycles for the Main Steam and Main Feedwater Penetrations 4-92

        4.6.3   Design Cycles for the Recirculation Sump Suction Line Penetrations 4-93

    4.7   Other Plant Specific Time Limited Aging Analyses 4-94

        4.7.1   Load Cycle Limits of Cranes, Lifts, and Fuel Handling Equipment Designed to Crane Manufacturers Association of America Standard 70 4-94

        4.7.2   Absence of Time Limited Aging Analyses for Metal Corrosion Allowances and Corrosion Effects..4-97

        4.7.3   Inservice Flaw Growth Analyses that Demonstrate Structural Stability for 40 Years 4-97

        4.7.4   Fatigue Crack Growth and Fracture Mechanics Stability Analyses of Half Nozzle Repairs to Alloy 600 Material in Reactor Coolant Hot Legs and Supporting Corrosion Analyses 4-98

        4.7.5   Corrosion Analyses of Pressurizer Ferritic Materials Exposed to Reactor Coolant by Half Nozzle Repairs of Pressurizer Heater Sleeve Alloy 600 Nozzles...4-103

        4.7.6   Absence of a Time Limited Aging Analysis for Reactor Vessel Underclad Cracking Analyses 4-104

        4.7.7   Absence of a Time Limited Aging Analysis for a Reactor Coolant Pump Flywheel Fatigue Crack Growth Analysis............................4-104

        4.7.8   Building Absolute or Differential Heave or Settlement, Including Possible Effects of Changes in Perched Groundwater Lens...........4-104

    4.8   Absence of Time Limited Aging Analyses Supporting Title 10, Part 50.12, Exemptions, of the Code of Federal Regulations..4-108

    4.9   Conclusion for Time Limited Aging Analyses 4-108

5.0   Review by the Advisory Committee on Reactor Safeguards 5-1

6.0   Conclusion 6-1

Appendix A: Palo Verde Nuclear Generating Station Units 1, 2, and 3 License Renewal Commitments A-1

Appendix B: Chronology B-1

Appendix C: Principal Contributors C-1

Appendix D: References D-1

LIST OF TABLES

Table 1.4-1	Current Interim Staff Guidance	1-6
Table 2.2-1	Missing Systems or Structures in Table 2.2 1 of the LRA	2-34
Table 3.0-1	Aging Management Programs	3-6
Table 3.1-1	Staff Evaluation for Reactor Vessel, Reactor Vessel Internals and Reactor Coolant System Components in the GALL Report	3-137
Table 3.2-1	Staff Evaluation for Engineered Safety Features Systems Components in the GALL Report	3-182
Table 3.3-1	Staff Evaluation for Auxiliary System Components in the GALL Report	3-208
Table 3.4-1	Staff Evaluation for Steam and Power Conversion Systems Components in the GALL Report	3-270
Table 3.5-1	Staff Evaluation for Structures and Component Supports Components in the GALL Report	3-293
Table 3.6-1	Staff Evaluation for Electrical and Instrumentation and Controls in the GALL Report	3-329
Table 4.1-1	Alloy 82/182 Dissimilar Metal Welds Mitigation	4-9
Table 4.3-1	Alloy 600 Small-Bore Hot Leg Nozzle Repairs	4-62
Table A-1	Palo Verde Nuclear Generating Station License Renewal Commitments	A-1

ABBREVIATIONS

ACI	American Concrete Institute
ACRS	Advisory Committee on Reactor Safeguards
ACSR	aluminum conductor steel reinforced
ACU	air conditioning unit
ADAMS	Agencywide Document Access and Management System
AERM	aging effect requiring management
AFW	auxiliary feedwater
AHU	air handling unit
AMP	aging management program
AMR	aging management review
ANSI	American National Standards Institute
APS	Arizona Public Service Company
ART	adjusted reference temperature
ASM	American Society for Metals
ASME	American Society of Mechanical Engineers
ASTM	American Society for Testing and Materials
ATWS	anticipated transient without scram
BTP	branch technical position
BWR	boiling water reactor
CASS	cast austenitic stainless steel
CBF	cycle-based fatigue
CC	cycle counting
CE	Combustion Engineering, Incorporated
CEA	control element assembly
CEDM	control element drive mechanism
CEOG	Combustion Engineering Owners Group
CFR	Code of Federal Regulations
CLB	current licensing basis
CMMA	Crane Manufacturer's Association of America
CSS	core support structure
CST	condensate storage tank
CTL	Construction Technology Laboratories
CUF	cumulative usage factor
CVCS	chemical and volume control system
DBA	design basis accident
DBE	design basis event
DG	diesel generator

Abbreviations

ECCS	emergency core cooling system
EDG	emergency diesel generator
EFPY	effective full-power year
EOL	end of life
EPRI	Electric Power Research Institute
EQ	environmental qualification
ESF	engineered safety feature
ESP	essential spray pond
ETL	electro-thermal link
FAC	flow-accelerated corrosion
FR	*Federal Register*
FSAR	final safety analysis report
GALL	Generic Aging Lessons Learned
GEIS	Generic Environmental Impact Statement
GL	generic letter
GSI	generic safety issue
HELB	high-energy line break
HPSI	high-pressure safety injection
HVAC	heating, ventilation, and air conditioning
I&C	instrumentation and controls
IASCC	irradiation assisted stress corrosion cracking
IE	Office of Inspection and Enforcement
IGSCC	intergranular stress corrosion cracking
IN	information notice
IPA	integrated plant assessment
ISG	interim staff guidance
ISI	inservice inspection
ISO	International Standardization Organization
LBB	leak-before-break
LOCA	loss of coolant accident
LR	license renewal
LRA	license renewal application
LRDMT	license renewal data management tool
LTOP	low temperature over pressure protection
MEB	metal enclosed bus

Abbreviations

MIC	microbiologically-induced corrosion
MNSA	mechanical nozzle seal assembly
MRV	minimum required value
MSSS	main steam support structure
NAS	National Academy of Sciences
NATM	Nuclear Administrative Technical Manual
NDE	nondestructive examination
NEI	Nuclear Energy Institute
NESC	National Electrical Safety Code
NFPA	National Fire Protection Association
NPF	Nuclear Power Facility
NRC	U.S. Nuclear Regulatory Commission
NSAC	Nuclear Safety Analysis Center
NSSS	nuclear steam supply system
NUREG	NRC Technical Report Designation
OBE	operating-basis earthquake
P&ID	piping and instrumentation drawing
PDT	pressure differential transmitter
PEC	pulsed-eddy current
PLL	predicted lower limit
P-T	pressure-temperature
PTS	pressurized thermal shock
PVC	polyvinyl chloride
PVNGS	Palo Verde Nuclear Generating Station
PWR	pressurized water reactor
PWSCC	primary water stress corrosion cracking
QA	quality assurance
RAI	request for additional information
RCP	reactor coolant pump
RCPB	reactor coolant pressure boundary
RCS	reactor coolant system
RFO	refuel outage
RG	regulatory guide
RIS	regulatory information summary
RMWT	reactor makeup water tank
RPV	reactor pressure vessel

Abbreviations

RR	relief request
RTD	resistance temperature detector
RT_{NDT}	resistance temperature nil ductility transition
RV	reactor vessel
RVID	reactor vessel integrity database
RWT	refueling water tank
SBF	stress-based fatigue
SBO	station blackout
SBOG	station blackout generator
SC	structure and component
SCC	stress-corrosion cracking
SE	safety evaluation
SER	safety evaluation report
SG	steam generator
SRP-LR	Standard Review Plan-License Renewal
SSC	system, structure, and component
SSE	safe-shutdown earthquake
SWMS	site work management system
TLAA	time-limited aging analysis
TR	topical report
TS	technical specifications
TSR	Technical Requirements Manual Surveillance Requirement
UFSAR	Updated Final Safety Analysis Report
UGS	upper guide structure
USE	upper-shelf energy
UT	ultrasonic testing
WCAP	Westinghouse Commercial Atomic Power Report
WRF	water reclamation facility

1.0 INTRODUCTION AND GENERAL DISCUSSION

1.1 Introduction

This document is a safety evaluation report (SER) on the license renewal application (LRA) for Palo Verde Nuclear Generating Station, Units 1, 2, and 3 (PVNGS), as filed by Arizona Public Service Company (APS) (the applicant). By letter dated December 11, 2008, as supplemented by letter dated April 14, 2009, APS submitted its application to the U.S. Nuclear Regulatory Commission (NRC) for renewal of the PVNGS operating license for an additional 20 years. The NRC staff (the staff) prepared this report to summarize the results of its safety review of the LRA for compliance with Title 10, Part 54, "Requirements for Renewal of Operating Licenses for Nuclear Power Plants," of the *Code of Federal Regulations* (10 CFR Part 54). The NRC project manager for the license renewal review is Ms. Lisa Regner. Ms. Regner may be contacted by telephone at 301-415-1906, or by electronic mail at Lisa.Regner@NRC.gov. Alternatively, written correspondence may be sent to the following address:

Division of License Renewal
U.S. Nuclear Regulatory Commission
Washington, D.C., 20555-0001
Attention: Lisa Regner, Mail Stop O11-F1

In its December 11, 2008, submission letter, the applicant requested renewal of the operating licenses issued under Section 103 (Operating License Nos. NPF-41, NPF-51, and NPF-74) of the Atomic Energy Act of 1954, as amended, for PVNGS for a period of 20 years beyond the current expiration at midnight on June 1, 2025 (Unit 1), April 24, 2026 (Unit 2), and November 25, 2027 (Unit 3).

PVNGS is located approximately 26 miles west of Phoenix, AZ. The NRC issued the construction permits on May 25, 1976, for all three units, and issued the operating licenses on June 1, 1985, for Unit 1; April 24, 1986, for Unit 2; and November 25, 1987, for Unit 3. PVNGS uses a pressurized water reactor (PWR) design with a dry ambient containment. Each of the PVNGS units uses a System 80 PWR nuclear steam supply system provided by Combustion Engineering, Incorporated (CE). Bechtel Power Corporation was responsible for the engineering and construction of the station and designed the balance of the plant. The licensed power output is 3,990 megawatts-thermal per unit with a net electrical output of approximately 1,346 megawatts-electric per unit. The updated final safety analysis report (UFSAR) contains details of the plant and the site.

The license renewal process consists of two concurrent reviews, a technical review of safety issues and an environmental review. The NRC regulations in 10 CFR Part 54, "Requirements for Renewal of Operating Licenses for Nuclear Power Plants" and 10 CFR Part 51, "Environmental Protection Regulations for Domestic Licensing and Related Regulatory Functions," respectively, set forth requirements for these reviews. The safety review for the PVNGS license renewal is based on the applicant's LRA and on its responses to the staff's requests for additional information (RAIs). The applicant supplemented the LRA and provided clarifications through its responses to the staff's RAIs in audits, meetings, and docketed correspondence. On August 8, 2010, the staff issued an SER with Open Items related to the License Renewal of Palo Verde Nuclear Generating Station, Units 1, 2, and 3, in which the staff identified one open item for further review. Subsequently, the applicant amended the LRA and provided responses to the staff's RAIs and docketed correspondence. Unless otherwise noted, the staff reviewed and considered information submitted through March 17, 2011. The staff reviewed information received after that date depending on the stage of the safety review and the volume and complexity of the information.

Introduction and General Discussion

The public may view the LRA and all pertinent information and materials at the NRC Public Document Room, located on the first floor of One White Flint North, 11555 Rockville Pike, Rockville, MD 20852-2738 (301-415-4737 or 800-397-4209), the Litchfield Park Branch Library, West Wigwam Boulevard, Litchfield Park, AZ 85340, and the Sam Garcia Western Avenue Library, 495 East Western Avenue, Avondale, AZ 85323. In addition, the public may find the LRA, as well as materials related to the license renewal review, on the NRC Web site at http://www.nrc.gov/reactors/operating/licensing/renewal.html.

This SER summarizes the results of the staff's safety review of the LRA and describes the technical details considered in evaluating the safety aspects of the unit's proposed operation for an additional 20 years beyond the term of the current operating licenses. The staff reviewed the LRA in accordance with NRC regulations and the guidance in NUREG-1800, Revision 1, "Standard Review Plan for Review of License Renewal Applications for Nuclear Power Plants" (SRP-LR), dated September 2005.

SER Sections 2-4 address the staff's evaluation of license renewal issues considered during the review of the application. SER Section 5 is reserved for the report of the Advisory Committee on Reactor Safeguards (ACRS). The conclusions of this SER are in Section 6.

SER Appendix A is a table showing the applicant's regulatory commitments for renewal of the operating licenses. SER Appendix B is a chronology of the principal correspondence between the staff and the applicant regarding the LRA review. SER Appendix C is a list of principal contributors to the SER, and Appendix D is a bibliography of the references in support of the staff's review.

In accordance with 10 CFR Part 51, the staff prepared a draft plant-specific supplement to NUREG-1437, "Generic Environmental Impact Statement for License Renewal of Nuclear Plants (GEIS)." This supplement discusses the environmental considerations for license renewal for PVNGS. The staff issued the draft, plant-specific GEIS, NUREG-1437, Supplement 44, "Generic Environmental Impact Statement for License Renewal of Nuclear Plants Regarding the Palo Verde Nuclear Generating Station, Units 1, 2, and 3 – Draft Report" in August of 2010. The final, plant-specific GEIS Supplement is scheduled to be issued in January 2011.

1.2 License Renewal Background

Pursuant to the Atomic Energy Act of 1954, as amended, and NRC regulations, operating licenses for commercial power reactors are issued for 40 years and can be renewed for up to 20 additional years. The original 40-year license term was selected based on economic and antitrust considerations rather than on technical limitations; however, some individual plant and equipment designs may have been engineered for an expected 40-year service life.

In 1982, the staff anticipated interest in license renewal and held a workshop on nuclear power plant aging. This workshop led the NRC to establish a comprehensive program plan for nuclear plant aging research. From the results of that research, a technical review group concluded that many aging phenomena are readily manageable and pose no technical issues precluding life extension for nuclear power plants. In 1986, the staff published a request for comment on a policy statement that would address major policy, technical, and procedural issues related to license renewal for nuclear power plants.

In 1991, the staff published 10 CFR Part 54, the License Renewal Rule (Volume 56, page 64943, of the *Federal Register* (56 FR 64943), dated December 13, 1991). The staff participated in an industry-sponsored demonstration program to apply 10 CFR Part 54 to a pilot plant and to gain the experience necessary to develop implementation guidance. To establish a scope of review for license renewal, 10 CFR Part 54 defined age-related degradation unique to license renewal. During the demonstration program, the staff found that adverse aging effects on plant systems and components are managed during the period of initial license and that the scope of the review did not allow sufficient credit for management programs. This was particularly true for the implementation of 10 CFR 50.65, "Requirements for Monitoring the Effectiveness of Maintenance at Nuclear Power Plants," which regulates management of plant-aging phenomena. As a result of this finding, the staff amended 10 CFR Part 54 in 1995.

As published in May 8, 1995 (60 FR 22461), the amended 10 CFR Part 54 establishes a regulatory process that is simpler, more stable, and more predictable than the previous 10 CFR Part 54. In particular, as amended, 10 CFR Part 54 focuses on the management of adverse aging effects rather than on the identification of age-related degradation unique to license renewal. The staff made these rule changes to ensure that important systems, structures, and components (SSCs) will continue to perform their intended functions during the period of extended operation. In addition, the amended 10 CFR Part 54 clarifies and simplifies the integrated plant assessment process to be consistent with the revised focus on passive, long-lived structures and components (SCs).

Concurrent with these initiatives, the staff pursued a separate rulemaking effort (61 FR 28467, June 5, 1996) and amended 10 CFR Part 51 to focus the scope of the review of environmental impacts of license renewal in order to fulfill NRC responsibilities under the National Environmental Policy Act of 1969.

1.2.1 Safety Review

License renewal requirements for power reactors are based on two key principles:

(1) The regulatory process is adequate to ensure that the licensing bases of all currently operating plants maintain an acceptable level of safety with the possible exceptions of the detrimental aging effects on the functions of certain SSCs, as well as a few other safety-related issues, during the period of extended operation.

(2) The plant-specific licensing basis must be maintained during the renewal term in the same manner and to the same extent as during the original licensing term.

In implementing these two principles, 10 CFR 54.4, "Scope," defines the scope of license renewal as including those SSCs (1) that are safety-related, (2) whose failure could affect safety-related functions, or (3) that are relied on to demonstrate compliance with the NRC's regulations for fire protection, environmental qualification (EQ), pressurized thermal shock (PTS), anticipated transient without scram (ATWS), and station blackout (SBO).

Pursuant to 10 CFR 54.21(a), a license renewal applicant must review all SSCs within the scope of 10 CFR Part 54 to identify SCs subject to an aging management review (AMR). Those SCs subject to an AMR perform an intended function without moving parts or without change in configuration or properties and are not subject to replacement based on a qualified life or specified time period. Pursuant to 10 CFR 54.21(a), a license renewal applicant must demonstrate that the aging effects will be managed such that the intended function(s) of those SCs will be maintained consistent with the current licensing basis (CLB) for the period of extended operation. However, active equipment is considered to be adequately monitored and maintained by existing programs. In other words, detrimental aging effects that may affect active equipment can be readily identified and corrected through routine surveillance, performance monitoring, and maintenance. Surveillance and maintenance programs for active equipment, as well as other maintenance aspects of plant design and licensing basis, are required throughout the period of extended operation.

Introduction and General Discussion

Pursuant to 10 CFR 54.21(d), the LRA is required to include an UFSAR supplement with a summary description of the applicant's programs and activities for managing aging effects and an evaluation of time-limited aging analyses (TLAAs) for the period of extended operation.

License renewal also requires TLAA identification and updating. During the plant design phase, certain assumptions about the length of time the plant can operate are incorporated into design calculations for several plant SSCs. In accordance with 10 CFR 54.21(c)(1), the applicant must either show that these calculations will remain valid for the period of extended operation, project the analyses to the end of the period of extended operation, or demonstrate that the aging effects on these SSCs will be adequately managed for the period of extended operation.

In 2005, the NRC revised Regulatory Guide (RG) 1.188, "Standard Format and Content for Applications to Renew Nuclear Power Plant Operating Licenses." This RG endorses Nuclear Energy Institute (NEI) 95-10, Revision 6, "Industry Guideline for Implementing the Requirements of 10 CFR Part 54 - The License Renewal Rule," issued in June 2005. NEI 95-10 details an acceptable method of implementing 10 CFR Part 54. The staff also used the SRP-LR to review the LRA.

In the LRA, the applicant fully utilized the process defined in NUREG-1801, Revision 1, "Generic Aging Lessons Learned (GALL) Report," dated September 2005. The GALL Report summarizes staff-approved aging management programs (AMPs) for many SCs subject to an AMR. If an applicant commits to implementing these staff-approved AMPs, they can greatly reduce the time, effort, and resources for LRA review, thus improving the efficiency and effectiveness of the license renewal review process. The GALL Report summarizes the aging management evaluations, programs, and activities credited for managing aging for most of the SCs used throughout the industry. The report is also a quick reference for both applicants and staff reviewers to AMPs and activities that can manage aging adequately during the period of extended operation.

1.2.2 Environmental Review

Part 51 of 10 CFR contains regulations on environmental protection regulations. In December 1996, the staff revised the environmental protection regulations to facilitate the environmental review for license renewal. The staff prepared the GEIS to document its evaluation of possible environmental impacts associated with nuclear power plant license renewals. For certain types of environmental impacts, the GEIS contains generic findings that apply to all nuclear power plants and are codified in Appendix B, "Environmental Effect of Renewing the Operating License of a Nuclear Power Plant," to Subpart A, "National Environmental Policy Act - Regulations Implementing Section 102(2)," of 10 CFR Part 51. Pursuant to 10 CFR 51.53(c)(3)(i), a license renewal applicant may incorporate these generic findings in its environmental report. In accordance with 10 CFR 51.53(c)(3)(ii), an environmental report also must include analyses of environmental impacts that must be evaluated on a plant-specific basis (i.e., Category 2 issues).

In accordance with the National Environmental Policy Act of 1969 and 10 CFR Part 51, the staff reviewed the plant-specific environmental impacts of license renewal, including whether there was new and significant information not considered in the GEIS. As part of its scoping process, the staff held public meetings on June 25, 2009. One meeting was in the afternoon at Tonopah Valley High School, in Tonopah, AZ. The other was in the evening at Estrella Mountain Community College in Avondale, AZ. The purpose of these meetings was to seek comments from local stakeholders on plant-specific environmental issues. The draft, plant-specific GEIS, Supplement 44, documents the results of the environmental review and makes a preliminary recommendation as to the license renewal action. The staff held public meetings similar to the scoping meetings discussed above on September 15, 2010, in Tonopah and Avondale, AZ, to discuss the draft, plant-specific GEIS, Supplement 44. After considering all comments on the

Introduction and General Discussion

draft GEIS received from stakeholders, the staff will publish the final, plant-specific GEIS currently scheduled for January 2011.

1.3 Principal Review Matters

Part 54 of 10 CFR describes the requirements for the renewal of operating licenses for nuclear power plants. The staff's technical review of the LRA was in accordance with NRC guidance and 10 CFR Part 54 requirements. Section 54.29, "Standards for Issuance of a Renewed License," of 10 CFR sets forth the license renewal standards. This SER describes the results of the staff's safety review.

Pursuant to 10 CFR 54.19(a), the NRC requires a license renewal applicant to submit general information, which the applicant provided in LRA Section 1. The staff reviewed LRA Section 1 and finds that the applicant has submitted the required information.

Pursuant to 10 CFR 54.19(b), the NRC requires that the LRA include "conforming changes to the standard indemnity agreement, 10 CFR 140.92, Appendix B, to account for the expiration term of the proposed renewed license." On this issue, the applicant stated in the LRA:

> The current indemnity agreement for Palo Verde states in Article VII that the agreement shall terminate "at the time of expiration of that license specified in Item 3 of the Attachment to the agreement". Item 3 of the Attachment to the indemnity agreement, as amended, lists license numbers NPF-41, NPF-51, and NPF-74.

> APS requests that conforming changes be made to the indemnity agreement, and/or the Attachment to the agreement, as required, to ensure that the indemnity agreement continues to apply during both the terms of the current licenses and the terms of the renewed licenses. APS understands that no changes may be necessary for this purpose if the current license numbers are retained.

The staff intends to maintain the original license numbers upon issuance of the renewed licenses, if approved. Therefore, no conforming changes need to be made to the indemnity agreement, and the 10 CFR 54.19(b) requirements have been met.

Pursuant to 10 CFR 54.21, "Contents of Application - Technical Information," the NRC requires that the LRA contain, (a) an integrated plant assessment, (b) a description of any CLB changes during the staff's review of the LRA, (c) an evaluation of TLAAs, and (d) a final safety analysis report supplement. LRA Sections 3 and 4 and Appendix B address the license renewal requirements of 10 CFR 54.21(a), (b), and (c). LRA Appendix A satisfies the license renewal requirements of 10 CFR 54.21(d).

Pursuant to 10 CFR 54.21(b), the NRC requires that each year following submission of the LRA and at least three months before the scheduled completion of the staff's review, the applicant submit an LRA amendment identifying any CLB changes to the facility that affect the contents of the LRA, including the UFSAR supplement. By letter dated December 7, 2009, the applicant submitted an LRA update that summarizes the CLB changes that have occurred during the staff's review of the LRA. The LRA was accepted for review on May 15, 2009 (74 FR 22978). This submission satisfies 10 CFR 54.21(b) requirements for the SER with Open Items.

Pursuant to 10 CFR 54.22, "Contents of Application - Technical Specifications," the NRC requires that the LRA include changes or additions to the technical specifications (TS) that are necessary to manage aging effects during the period of extended operation. In LRA Appendix D, the applicant stated that no changes to the TS are required to support the LRA. This statement adequately addresses the 10 CFR 54.22 requirement.

Introduction and General Discussion

The staff evaluated the technical information required by 10 CFR 54.21 and 10 CFR 54.22 in accordance with NRC regulations and SRP-LR guidance. SER Sections 2, 3, and 4 document the staff's evaluation of the LRA technical information.

As required by 10 CFR 54.25, "Report of the Advisory Committee on Reactor Safeguards," the ACRS will issue a report documenting its evaluation of the staff's LRA review and SER. SER Section 5 is reserved for the ACRS report, when it is issued. SER Section 6 documents the findings required by 10 CFR 54.29.

1.4 Interim Staff Guidance

License renewal is a living program. The staff, industry, and other interested stakeholders gain experience and develop lessons learned with each renewed license. The lessons learned address the staff's performance goals of maintaining safety, improving effectiveness and efficiency, reducing regulatory burden, and increasing public confidence. Interim staff guidance (ISG) is documented for use by the staff, industry, and other interested stakeholders until incorporated into such license renewal guidance documents as the SRP-LR and GALL Report.

Table 1.4-1 shows the current set of ISGs, as well as the SER sections in which the staff addresses them.

Table 1.4-1. Current Interim Staff Guidance

ISG Issue (Approved ISG Number)	Purpose	SER Section
Nickel-alloy Components in the Reactor Coolant Pressure Boundary (LR-ISG-19B)	To address the cracking of nickel-alloy components in the reactor pressure boundary (ISG is under development. NEI and the Electric Power Research Institute-modification/rework package will develop an augmented inspection program for GALL AMP XI.M11-B. This AMP will not be completed until the NRC approves an augmented inspection program for nickel-alloy base metal components and welds as proposed by Electric Power Research Institute-modification/rework package).	3.0.3.3.1
Corrosion of Drywell Shell in Mark I Containments (LR-ISG-2006-01)	To address concerns related to corrosion of drywell shell in Mark I containments	Not applicable
Changes to GALL AMP XI.E6, "Electrical Cable Connections Not Subject to 10 CFR 50.49 Environmental Qualification Requirements" (LR-ISG-2007-02)	To address the frequency of inspection of electrical cable connections not subject to 10 CFR 50.49 before the period of extended operation. The staff has addressed industry comments and a notice of availability of the Final LR-ISG-2007-02 was published in the *Federal Register*. See 74 FR 68287, dated December 23, 2009.	3.0.3.1.8
Aging Management of Spent Fuel Pool Neutron-Absorbing Materials Other than Boraflex (LR-ISG-2009-01)	To provide guidance as to one acceptable approach for managing the effects of aging during the period of extended operation for certain neutron-absorbing spent fuel pool components within the scope of the License Renewal Rule (10 CFR Part 54, "Requirements for Renewal of Operating Licenses for Nuclear Power Plants")	Not applicable

1.5 Summary of Open Items

As a result of its review of the LRA, the staff identified the following open item. An item is considered open if, based on the applicant's submittals, it does not meet all applicable

Introduction and General Discussion

regulatory requirements at the time of the issuance of this SER. The staff has assigned a unique identifying number to this open item.

Open Item 4.3-1

The staff's review of LRA Section 4.3 "Metal Fatigue," is inconclusive regarding the applicant's compliance with regulatory requirements. The staff requires further information to reach a conclusion on the proposed metal fatigue analysis as discussed in detail in SER Section 4.3. The following presents a brief description of the RAIs that the staff has issued to the applicant:

- Clarify which transients required a 25-percent occurrences assumption and justify why this assumption yields a conservative 60-year cycle basis (RAI 4.3-1).

  Response: In a June 29, 2010, letter the applicant provided information including the transients for which it applied a 25-percent occurrences assumption. The applicant also provided justification that this assumption yields a conservative basis because the applicant performed a detailed review of plant documentation for occurrences of these transients. The staff's detailed review is found in SER Section 4.3.1.4.2.

- Explain why the cumulative usage factors (CUFs) for the instrument nozzles at Unit 1 are 5 times greater than at Units 2 and 3 (RAI 4.3-2).

  Response: In letters dated June 29, and October 13, 2010, the applicant provided information concerning dissimilarities in the CUFs for the instrument nozzles in Unit 1 versus Units 2 and 3. The applicant stated the differences were due to variations in modeling and analysis methods and assumptions. The staff's detailed review is found in SER Section 4.3.1.4.2.

- Provide the allowable stress limits and stress ranges from the revised design analyses for the reactor coolant hot leg sample line piping and the steam generator (SG) downcomer and feedwater recirculation line piping (RAI 4.3-3).

  Response: In a June 29, 2010, response, the applicant provided the requested information related to the reactor coolant hot leg sample line piping and the SG downcomer and feedwater recirculation line piping and the ASME Code allowable stress limits and stress range reduction factors. The staff's detailed review is found in SER Section 4.3.5.2.

- Demonstrate that the environmental factors used in analyzing the reactor pressure vessel components are the maximums for a given material and provide a basis and justification for any assumptions (RAI 4.3-4).

  Response: In a June 29, 2010, response, the applicant stated that the "maximum applicable" F_{en} factors for the low alloy steel RPV shell and lower head, RPV inlet and outlet nozzles, and safety injection nozzle (forging knuckle) were all computed using the guidance in NUREG/CR-6583, "Effects of LWR Coolant Environments on Fatigue Design Curves of Carbon and Low-Alloy Steels." The applicant also provided the basis and justification for assumptions used in the analysis. The staff's detailed review is found in SER Section 4.3.4.2.

- Describe the methodology used for the environmental factor calculation of the charging system nozzle and the safety injection nozzle and justify assumptions (RAI 4.3-5).

  Response: In a June 29, 2010, response, the applicant provided the details of the methodology used to calculate a more accurate F_{en} value for the charging nozzle and safety injection nozzle. The staff's detailed review is found in SER Section 4.3.4.2.

Introduction and General Discussion

- Justify using an environmental factor value of 1.49 for the nickel alloy pressurizer heater penetrations and describe plans to update the CUF calculation methodology consistent with NUREG/CR-6909 (RAI 4.3-6).

  Response: In a June 29, 2010, response, the applicant committed to confirm the conservatism of its use of an F_{en} value of 1.49 or perform a reanalysis of the pressurizer heater penetrations using an F_{en} value calculated using the methodology in NUREG/CR-6909. The staff's detailed review is found in SER Section 4.3.4.2.

- Provide the monitoring basis for the pressurizer spray nozzles (RAI 4.3-7).

  Response: In an August 12, 2010, response, the applicant clarified that the higher F_{en} value for the surge line elbow will result in a higher environmentally-assisted fatigue usage factor compared to the pressurizer spray nozzle. Further, since the stratification effects on the surge line are only associated with the surge line elbow, this results in the surge line elbow as the bounding component compared to the pressurizer spray nozzle. The staff's detailed review is found in SER Section 4.3.1.2.2.

- Clarify whether Transient 17 (LRA Table 4.3-3) refers to initiation of the pressurizer spray system or containment spray system and provide the basis for correlating the tracking of this transient to Transient 12 (RAI 4.3-8).

  Response: In an August 12, 2010, response, the applicant clarified that Transient 17 refers to the initiation of auxiliary pressurizer spray and provided the relationship between Transient 17 and 12. The applicant explained the sequence of events to initiate auxiliary spray during cooldown which provided the basis for the correlation between the two transients. The staff's detailed review is found in SER Section 4.3.1.4.2.

- Clarify if the transient cycle counting procedure has been updated to include Transient 25 (LRA Table 4.3-3); if not, specify when it will be updated (RAI 4.3-9).

  Response: In an August 12, 2010, response, the applicant stated that the cycle-counting surveillance procedure was updated to include Transient 25 (in Amendment 14 dated April 28, 2010) and the applicant's enhanced Metal Fatigue Program will monitor this transient during the period of extended operation to ensure that it does not exceed the design limit. The staff's detailed review is found in SER Section 4.3.1.4.2.

- Clarify whether Transient 79 (LRA Table 4.3-3) is the American Society of Mechanical Engineers (ASME) Code Section XI required system leak test; if so, justify the number of occurrences stated in the LRA (RAI 4.3-10).

  Response: In an August 12, 2010, letter, the applicant clarified that its operating practice is to perform the ASME Code Section XI leak test concurrently with a plant heatup without a separate thermal transient and, therefore, the fatigue effects are appropriately accounted for as a plant heatup transient and not as a separate leak test transient. Further, the applicant clarified that the ASME Code fatigue analyses account for the fatigue effects of plant heatup, cooldown, and ASME Code Section XI leak test as separate transients. The staff's detailed review is found in SER Section 4.3.1.4.2.

- Clarify the meaning of "significant contributors to usage factor" and how this is associated with the corrective action limits in the metal fatigue AMP (RAI 4.3-11).

  Response: In an August 12, 2010, response, the applicant clarified that the "significant contributors" to fatigue include all transients listed in the UFSAR tables. The applicant also stated that the cycle counting corrective action limits associated with all transients listed in LRA Table 4.3-2 will be tracked by the enhanced Metal Fatigue Program. This

Introduction and General Discussion

will ensure that the assumptions made in the analyses of record and design limits are not exceeded. The staff's detailed review is found in SER Section 4.3.1.5.2.

- Clarify whether the scope of corrective actions for CUF monitoring includes both ASME Code Class 1 components and ASME Code Class 2 components analyzed to Class 1 requirements; if not, provide justification (RAI 4.3-12).

    Response: In letters dated July 21 and August 12, 2010, the applicant clarified that the scope of the enhanced Metal Fatigue Program includes all components, including Class 2 and 3 components, with a CUF analysis. Further, the applicant's program will ensure that the design limit of 1.0 is not exceeded or corrective actions will be taken to reanalyze, repair, or replace the component before the design limit is exceeded. The staff's detailed review is found in SER Section 4.3.1.4.2.

- Justify why the SG tubes CUF calculation is not a TLAA and provide the basis for the CUF value of zero; justify omitting AMR items for pressurizer components cumulative fatigue damage; justify omitting AMR items for ASME Code Class 2 and 3 or American National Standards Institute (ANSI) B31.1 components cumulative fatigue damage (RAI 4.3-13).

    Response: In an August 12, 2010, response, the applicant stated that the SG tube CUF value was taken from the applicable design report for each unit. The applicant further clarified that the zero value for the SG tube CUF is based on the cyclic stress range being below the endurance limit. Also, the applicant amended LRA Table 3.1.2-4 to include the ANSI B31.1 component AMR items. The staff's detailed review is found in SER Sections 3.1.2.2.1, 3.2.2.2.1, 3.3.2.2.1, and 3.4.2.2.1.

- Clarify whether cycle-based monitoring will be performed on reactor pressure vessel studs only (not lugs); if so, provide the transients contributing to fatigue usage and quantify the contribution for both studs and lugs (RAI 4.3-14).

    Response: In an August 12, 2010, response, the applicant stated that both RPV studs and RPV external bottom head shear lugs will be monitored individually by cycle counting, and action limits for the enhanced Metal Fatigue Program will be established to allow for corrective actions before the design basis number of events is exceeded. The staff's detailed review is found in SER Section 4.3.2.1.2.

- Identify the type of low cycle fatigue analysis referred to in LRA Sections 4.3.2.7 and 4.3.2.8 and justify why the low-cycle fatigue analysis is not a TLAA (RAI 4.3-15).

    Response: In an August 12, 2010, response, the applicant stated that the low cycle fatigue analysis referred to in LRA Sections 4.3.2.7 and 4.3.2.8 is not a TLAA and does not contain an implicit fatigue analysis, cycle-based fatigue flaw growth, or cycle-based fracture mechanics analysis. The staff's detailed review is found in SER Section 4.1.3.1.5.

- Clarify the current design basis CUF values and limits for the transients evaluated for the regenerative and letdown heat exchangers (RAI 4.3-16).

    Response: In an August 12, 2010, response, the applicant provided the design basis CUFs for the components. In the response, the applicant also explained the design basis transients analyzed and associated limits for the transients. During the review of the analyses of record for the heat exchangers, the applicant noted that the analysis assumed a higher number of cycles for significant design transients and a lower number of cycles for less significant transients than those stated in the UFSAR for several transients. The applicant stated that none of the transient limits have been challenged by current operating history. The applicant also stated that the inconsistency between

Introduction and General Discussion

the transient assumptions in the UFSAR and those in the analyses are in the applicant's corrective action process for evaluation. The staff's detailed review is found in SER Section 4.3.2.10.2.

- Identify which reactor vessel internal components are designed to ASME Section III NG requirements and which require a CUF design calculation; for those required, provide the design basis CUF values and limits for the transients evaluated; justify the use of cycle-based monitoring for those with high CUF values (RAI 4.3-17).

    Response: In an August 12, 2010, response, the applicant identified the ASME Code Section III, Subsection NG reactor vessel internal components and clarified the design basis CUFs and transients for those components. Further, the applicant stated it will use cycle counting in its enhanced Metal Fatigue Program to track these transients to ensure that when action limits are reached that corrective actions are taken to maintain fatigue usage below the design limit of 1.0. The staff's detailed review is found in SER Section 4.3.3.2.

- Identify the source documents for the equation references in LRA Section 4.3.5 and identify the analysis of record for the recirculating SG downcomer and feedwater recirculation lines (RAI 4.3-18).

    Response: In a August 12, 2010, response, the applicant clarified that the 7000-thermal cycles fatigue analysis is the analysis of record for the recirculating SG downcomer and feedwater recirculation lines and has been identified as a TLAA consistent with 10 CFR 54.21(c)(1). The staff's detailed review is found in SER Section 4.3.5.2.

1.6 Summary of Confirmatory Items

Based on its review of the LRA, including additional information submitted through July 9, 2010, the staff identified the following confirmatory items. An item is considered confirmatory if the staff and the applicant have reached a satisfactory resolution, but the applicant has not yet formally submitted the resolution. The staff has assigned a unique identifying number to each confirmatory item.

Confirmatory Item 2.1.4.2-1
SER Section 2.1.4.2 Application of the Scoping Criteria in 10 CFR 54.4(a)(2)

During the scoping and screening methodology audit, performed on-site October 19-22, 2009, the staff determined that the nonsafety-related, abandoned containment spray chemical addition tanks in Units 1 and 3 had not been included in the scope of license renewal. The associated piping was cut and capped for these tanks, but they had not been verified to be drained. The tanks were found to contain liquid and, thus, are in the scope of license renewal under 10 CFR 54.4(a)(2). The applicant responded to the RAI by providing a commitment to drain the tanks by August 30, 2010.

By letter dated November 10, 2010, the applicant stated that the containment spray chemical addition tanks and associated piping components have been drained. The staff's detailed review is found in SER Section 2.1.4.2.

Confirmatory Item 3.0.3.2.2-1
SER Section 3.0.3.2.2 Flow-Accelerated Corrosion Program

PVNGS experienced a through-wall leak in stainless steel high pressure safety injection (HPSI) system piping which was determined by the applicant to be caused by erosion from cavitation. The applicant resolved this issue by periodic replacement of the affected sections of the HPSI system for all three units at approximately 7.5-year intervals. In its review of the apparent cause evaluation, the staff noted that the extent of condition analysis identified components in other

Introduction and General Discussion

safety-related systems that were potentially susceptible to the same aging effect. The staff was unclear whether the applicant evaluated all identified in-scope line items susceptible to this aging effect. The applicant agreed to submit information regarding the resolution of the extent of condition for the HPSI cavitation erosion issue.

The applicant responded by letters dated July 30, and September 3, 2010, stating that the extent of condition evaluation was completed. The applicant also provided a commitment (Commitment No. 59) stating that it would complete inspections of other potentially susceptible piping locations by June 30, 2012, and would incorporate any remaining components found to exhibit flow-related degradation into a comparable periodic replacement plan. The staff's detailed review is found in SER Section 3.0.3.2.2.

Confirmatory Item 3.0.3.2.13-1
SER Section 3.0.3.2.13 One-Time Inspection of ASME Code Class 1 Small-Bore Piping

The applicant has experienced two failures of ASME Code Class 1 small-bore piping socket welds that the applicant attributed to a design defect. The applicant stated that the design issue has been resolved, but the staff considers this an aging-related failure requiring management. The staff needs assurance that a sufficient number of samples, as recommend by the GALL Report, will be selected to assure identification of small-bore piping socket weld inside-diameter cracking. The applicant stated that it would modify the One-Time Inspection Program (discussed in SER Section 3.0.3.1.6) to volumetrically inspect 10 percent of the socket weld population for each unit.

In its response to Confirmatory Item 3.0.3.2.13-1, dated July 30, 2010, and supplemented by letter dated December 3, 2010, the applicant revised its AMP to volumetrically inspect at least 10 percent of Class 1 socket welds per unit with a maximum of 25 welds and that weld selection will be based on risk insight and the potential for aging degradation. The staff's detailed review is found in SER Section 3.0.3.2.13.

Confirmatory Item 3.1.2.2.14-1
SER Section 3.1.2.2.14 Wall Thinning Due to Flow-Accelerated Corrosion

The GALL Report identifies that wall thinning due to flow-accelerated corrosion (FAC) can occur in steel SG feedrings and supports. The applicant stated that feedring wall thinning, as addressed in Information Notice (IN) 91-19, is not applicable to PVNGS due to the model of SGs in use, and that no action is required. However, the staff does not consider IN 91-19 to be limited to CE SGs. During a conference call on July 9, 2010, the applicant clarified that the material of the SG feedring is FAC resistant. The applicant also explained that the scope of the SG degradation assessment done before every outage includes secondary side SG internals.

The applicant clarified that the material of the SG feedring is fabricated from P11 steel and, therefore, is FAC resistant. The applicant also explained that the Steam Generator Tube Integrity Program considers wall thinning of the SG feedring and applicable operating experience as part of the secondary side SG Degradation Assessment, performed before every outage. The staff's detailed review is found in SER Section 3.1.2.2.14.

Confirmatory Item 3.3.2.1.1-1
SER Section 3.3.2.1 AMR Results Consistent with the GALL Report

The staff needs further information on how the applicant manages elastomer components exposed to raw water in auxiliary systems for aging during the period of extended operation. The applicant stated it would provide information demonstrating that the polyvinyl chloride and polyethylene components are not susceptible to loss of material due to erosion and that the AMPs proposed to manage aging of the elastomer-lined carbon steel piping are appropriate.

Introduction and General Discussion

The applicant provided information demonstrating that the polyvinyl chloride components are not susceptible to loss of material due to erosion since they are used in low-velocity systems and because the material has good resistance to abrasion and erosion. The submittal also contained information describing why the AMPs proposed to manage aging of elastomer-lined carbon steel piping are appropriate. The staff's detailed review is found in SER Section 3.3.2.1.1.

1.7 Summary of Additional Items

Since issuance of the SER with Open Items on August 6, 2010, the staff identified and resolved the following additional items prior to issuance of this SER. The items are summarized below and the appropriate sections are identified which provide details of the staff's review.

Inaccessible Medium Voltage Cables Not Subject to 10 CFR 50.49 Environmental Qualification Requirements Program (Section 3.0.3.1.9)

The staff determined that the applicant's Inaccessible Medium Voltage Cables AMP did not consider the most recent industry operating experience and the staff's position on the scope of the program and inspection frequency. The applicant responded by letter dated October 13, 2010, providing additional information and a revised commitment to expand the scope of the program to include inaccessible low-voltage cables (480 volts to 2 kilo-volts) and to increase the manhole inspection and cable test frequencies. The staff's detailed review is found in SER Section 3.0.3.1.9.

Buried Piping and Tanks Inspection Program (Section 3.0.3.2.12)

The staff determined that the applicant's Buried Piping and Tanks AMP did not consider the most recent industry operating experience and the staff's position on the scope and inspection frequency. In its response dated October 13, 2010, as supplemented by letter dated November 10, 2010, the applicant provided additional information and a revised commitment to describe its inspections of buried piping and tanks. See SER Section 3.0.3.2.12 for additional details.

NUREG/CR-6260 Limiting Locations (Section 4.3.4.2)

The staff requested the applicant to confirm and justify that the plant-specific components or locations listed in LRA Table 4.3-11 (except the pressurizer surge line pressurizer elbow) are bounding for the generic NUREG/CR-6260 locations and the additional location (pressurizer heater penetrations). The staff also requested the applicant to confirm and justify that the LRA Table 4.3-11 locations selected for environmentally-assisted fatigue analyses consists of the most limiting locations for the plant. If these locations are not bounding, the applicant was requested to clarify the locations that require an environmentally-assisted fatigue analysis and the actions that will be taken for these additional locations.

By letter dated December 3, 2010, the applicant provided additional information to address the staff's concern. The applicant committed to confirm and justify that the LRA locations were limiting. The staff's detailed review is found in SER Section 4.3.4.2.

Selective Leaching AMP (Section 3.0.3.2.11)

The staff determined that the applicant's inspection methodology was not defined in accordance with the staff's position on sample size. The applicant responded by letter dated December 3, 2010, providing additional information and a revised commitment to further describe its inspection parameters. The staff's detailed review is found in SER Section 3.0.3.2.11.

Introduction and General Discussion

Steam Generator Tube Denting and Welds Susceptible to Primary Water Stress Corrosion Cracking (Sections 3.1.2.1.1, item 3.1.1-79, and Section 3.1.2.1.2, item 3.1.1-81)

The staff determined that the applicant did not consider the most recent operating experience and the staff's position on primary water stress corrosion cracking of SG tube-to-tubesheets welds and divider plate bar welds. In addition the staff noted a discrepancy in the LRA associated with the SG tube denting aging mechanism. The applicant amended the LRA to correct the aging effect for SG tubes to be consistent with the GALL Report. In addition, the applicant addressed the staff's concerns about tube-to-tubesheet and divider plate bar welds by committing to address the aging management of these welds. The staff's detailed review is found in SER Sections 3.1.2.1.1 and 3.1.2.1.2.

1.8 Summary of Proposed License Conditions

Following the staff's review of the LRA, including subsequent information and clarifications from the applicant, the staff identified three proposed license conditions.

The first license condition requires the applicant to include the UFSAR supplement required by 10 CFR 54.21(d) in the next UFSAR update required by 10 CFR 50.71(e) following the issuance of the renewed licenses. The applicant may make changes to the programs and activities described in the UFSAR supplement provided changes are evaluated in accordance with the criteria set forth in 10 CFR 50.59.

The second license condition requires the applicant to complete future activities described in the UFSAR supplement before the period of extended operation.

The third license condition requires that all capsules in the reactor vessel that are removed and tested meet the requirements of American Society for Testing and Materials E 185-82 to the extent practicable for the configuration of the specimens in the capsule. The staff must approve any changes to the capsule withdrawal schedule before implementation, including spare capsules. All capsules placed in storage must be maintained for future insertion. The staff must approve any changes to storage requirements.

Introduction and General Discussion

2.0 Scoping and Screening Methodology

2.1 Scoping and Screening Methodology

2.1.1 Introduction

Title 10 of the *Code of Federal Regulations (CFR)*, Section 54.21, "Contents of Application - Technical Information," (10 CFR 54.21) requires an integrated plant assessment (IPA) for each license renewal application (LRA). The IPA must list and identify all of the systems, structures and components (SSCs) within the scope of license renewal in accordance with 10 CFR 54.4(a) and all structures and components (SCs) subject to an aging management review (AMR) in accordance with 10 CFR 54.21(a).

LRA Section 2.1, "Scoping and Screening Methodology," describes the scoping and screening methodology used to identify the SSCs at the Palo Verde Nuclear Generating Station (PVNGS) within the scope of license renewal and the SCs subject to an AMR. The U.S. Nuclear Regulatory Commission (NRC) staff (the staff) reviewed the scoping and screening methodology of the Arizona Public Service Company (APS) (the applicant) to determine if it meets the scoping requirements of 10 CFR 54.4(a) and the screening requirements of 10 CFR 54.21.

In developing the scoping and screening methodology for the LRA, the applicant stated that it considered the requirements of 10 CFR Part 54, "Requirements for Renewal of Operating Licenses for Nuclear Power Plants," (the Rule) as well as statements of consideration related for the Rule and the guidance of Nuclear Energy Institute (NEI) 95-10, Revision 6, "Industry Guideline for Implementing the Requirements of 10 CFR Part 54 - The License Renewal Rule," dated June 2005 (NEI 95-10). Additionally, in developing this methodology, the applicant stated that it considered the correspondence between the NRC, other applicants, and the NEI.

2.1.2 Summary of Technical Information in the Application

In LRA Sections 2 and 3, the applicant provides the technical information required by 10 CFR 54.4, "Scope," and 10 CFR 54.21(a). This safety evaluation report (SER) contains sections entitled "Summary of Technical Information in the Application," which provide information taken directly from the LRA.

In LRA Section 2.1, the applicant describes the process used to identify the SSCs that meet the license renewal scoping criteria under 10 CFR 54.4(a) and the process used to identify the SCs that are subject to an AMR, as required by 10 CFR 54.21(a)(1). The applicant provided the results of the process used for identifying the SCs subject to an AMR in the following LRA sections:

- Section 2.2, "Plant Level Scoping Results"
- Section 2.3, "Scoping and Screening Results: Mechanical Systems"
- Section 2.4, "Scoping and Screening Results: Structures"
- Section 2.5, "Scoping and Screening Results: Electrical and Instrumentation and Control Systems"

In LRA Section 3.0, "Aging Management Review Results," the applicant describes its aging management results as follows:

- Section 3.1, "Aging Management of Reactor Vessel, Internals, and Reactor Coolant System"

Scoping and Screening Methodology

- Section 3.2, "Aging Management of Engineered Safety Features"
- Section 3.3, "Aging Management of Auxiliary Systems"
- Section 3.4, "Aging Management of Steam and Power Conversion Systems"
- Section 3.5, "Aging Management of Containments, Structures, and Component Supports"
- Section 3.6, "Aging Management of Electrical and Instrumentation and Controls"

LRA Section 4, "Time-Limited Aging Analyses," states the applicant's evaluation of time-limited aging analyses (TLAAs).

2.1.3 Scoping and Screening Program Review

The staff evaluated the LRA scoping and screening methodology in accordance with the guidance contained in NUREG-1800, Revision 1, "Standard Review Plan for Review of License Renewal Applications for Nuclear Power Plants," (SRP-LR), Section 2.1, "Scoping and Screening Methodology." The following regulations form the basis for the acceptance criteria for the scoping and screening methodology review:

- 10 CFR 54.4(a), as it relates to the identification of plant SSCs within the scope of the Rule
- 10 CFR 54.4(b), as it relates to the identification of the intended functions of SSCs within the scope of the Rule
- 10 CFR 54.21(a)(1) and (a)(2), as they relate to the methods used by the applicant to identify plant SCs subject to an AMR

As part of the review of the applicant's scoping and screening methodology, the staff reviewed the activities described in the following sections of the LRA using the guidance contained in the SRP-LR:

- Section 2.1, to ensure that the applicant described a process for identifying SSCs that are within the scope of license renewal in accordance with the requirements of 10 CFR 54.4(a)
- Section 2.2, to ensure that the applicant described a process for determining the SCs that are subject to an AMR in accordance with the requirements of 10 CFR 54.21(a)(1) and (a)(2)

In addition, the staff conducted a scoping and screening methodology audit at PVNGS, located 26 miles west of the Phoenix, AZ metropolitan area, during the week of October 19-22, 2009. The audit focused on ensuring that the applicant had developed and implemented adequate guidance to conduct the scoping and screening of SSCs in accordance with the methodologies described in the LRA and the requirements of the Rule. The staff reviewed implementation of the project-level guidelines and topical reports describing the applicant's scoping and screening methodology. The staff conducted detailed discussions with the applicant on the implementation and control of the license renewal program and reviewed the administrative control documentation used by the applicant during the scoping and screening process, the quality practices used by the applicant to develop the LRA, and the training and qualification of the LRA development team.

The staff evaluated the quality attributes of the applicant's aging management program activities described in Appendix A, "Updated Final Safety Analysis Report Supplement," and Appendix B, "Aging Management Programs" of the LRA. On a sampling basis, the staff performed a system review of the safety injection and shutdown cooling, diesel generator fuel oil storage and

Scoping and Screening Methodology

transfer, auxiliary feedwater, and turbine building. This assessment included a review of the scoping and screening results reports and supporting design documentation used to develop the reports. The purpose of the staff's review was to ensure that the applicant had appropriately implemented the methodology outlined in the administrative controls and to verify that the results are consistent with the current licensing basis (CLB) documentation.

2.1.3.1 Implementation Procedures and Documentation Sources for Scoping and Screening

The staff reviewed the applicant's scoping and screening implementing procedures as documented in the Scoping and Screening Methodology Audit Trip Report, dated April 7, 2010, to verify that the process used to identify SCs subject to an AMR was consistent with the SRP-LR. Additionally, the staff reviewed the scope of CLB documentation sources and the process used by the applicant to ensure that applicant's commitments, as documented in the CLB and relative to the requirements of 10 CFR 54.4 and 10 CFR 54.21, were appropriately considered and that the applicant adequately implemented its procedural guidance during the scoping and screening process.

2.1.3.1.1 Summary of Technical Information in the Application

In LRA Section 2.1, "Scoping and Screening Methodology," the applicant addressed the following information sources for the license renewal scoping and screening process:

- Updated Final Safety Analysis Report (UFSAR)
- Safety Evaluation Reports
- Technical Specifications
- Licensing correspondence
- Engineering drawings
- License renewal position papers
- Plant equipment database

The applicant stated that it used a variety of documents, including those listed above, to apply scoping criteria in determining and confirming SSC functions.

2.1.3.1.2 Staff Evaluation

Scoping and Screening Implementing Procedures. The staff reviewed the applicant's scoping and screening methodology implementing procedures, including license renewal guidelines, documents, and reports, as documented in the audit report, to ensure the guidance is consistent with the requirements of the Rule, the SRP-LR, and NEI 95-10. The staff finds that the overall process used to meet the 10 CFR Part 54 requirements and described in the implementing procedures and AMRs is consistent with the Rule, the SRP-LR, and industry guidance.

The applicant's implementing procedures contain guidance for determining plant SSCs within the scope of the Rule and for determining which SCs, within the scope of license renewal, are subject to an AMR. During the review of the implementing procedures, the staff focused on the consistency of the detailed procedural guidance with information in the LRA. This information included the implementation of NRC staff positions documented in the SRP-LR and the information in the applicant's responses, dated February 5, April 1, and April 2, 2010, to the staff's requests for additional information (RAIs) dated December 23, 2009.

After reviewing the LRA and supporting documentation, the staff determined that the scoping and screening methodology instructions are consistent with the methodology description provided in LRA Section 2.1. The applicant's methodology is sufficiently detailed to provide concise guidance on the scoping and screening implementation process to be followed during the LRA activities.

Scoping and Screening Methodology

Sources of Current Licensing Basis Information. The staff reviewed the scope and depth of the applicant's CLB review to verify that the methodology is sufficiently comprehensive to identify SSCs within the scope of license renewal as well as SCs requiring an AMR. Under 10 CFR 54.3(a), the CLB is the set of NRC requirements applicable to a specific plant and a licensee's written commitments for ensuring compliance with, and operation within, applicable NRC requirements. The CLB also includes the plant-specific design bases that are docketed and in effect and applicable NRC regulations, orders, license conditions, exemptions, technical specifications, and design-basis information (documented in the most recent UFSAR). The CLB also includes licensee commitments, remaining in effect, that the applicant made in docketed licensing correspondence such as licensee responses to NRC bulletins, generic letters and enforcement actions, and licensee commitments documented in NRC safety evaluations or licensee event reports.

During the audit, the staff reviewed pertinent information sources used by the applicant including the UFSAR, design-basis information, and license renewal drawings. In addition, the applicant's license renewal process identified additional sources of plant information pertinent to the scoping and screening process, including the plant equipment database, controlled drawings, technical correspondence, analyses, and reports. The staff confirmed that the applicant's detailed license renewal program guidelines specified the use of the CLB source information in developing scoping evaluations.

The plant equipment database, UFSAR, design basis information, and plant drawings were the applicant's primary repository for system identification and component safety classification information. During the audit, the staff reviewed the applicant's administrative controls for the plant equipment database, design-basis information, and other information sources used to verify system information. Plant administrative procedures describe these controls and govern their implementation. Based on a review of the administrative controls and a sample of the system classification information contained in the applicable documentation, the staff concludes that the applicant has established adequate measures to control the integrity and reliability of system identification and safety classification data. Therefore, the staff concludes that the information sources used by the applicant during the scoping and screening process provide a sufficiently controlled source of system and component data to support scoping and screening evaluations.

During the staff's review of the applicant's CLB evaluation process, the applicant explained the incorporation of updates to the CLB and the process used to ensure that it adequately incorporates those updates into the license renewal process. The staff determined that LRA Section 2.1 provided a description of the CLB and related documents used during the scoping and screening process that is consistent with the guidance contained in the SRP-LR.

In addition, the staff reviewed the implementing procedures and results reports used to support the identification of SSCs that the applicant relied on to demonstrate compliance with the safety-related criteria, nonsafety-related criteria, and the regulated events criteria detailed in 10 CFR 54.4(a). The applicant's license renewal program guidelines provided a listing of documents used to support scoping and screening evaluations. The staff finds these design documentation sources to be useful for ensuring that the initial scope of SSCs identified by the applicant was consistent with the plant's CLB.

The staff determined that additional information would be required to complete the review of the applicant's scoping methodology. RAI 2.1-1, dated December 23, 2009, states that during review of the LRA and the performance of the scoping and screening methodology audit performed onsite October 19-22, 2009, the staff determined that although differences exist between the three units, the applicant had provided a single set of license renewal drawings to assist the staff in its review. In addition, the staff's review of license renewal drawing LR-PVNGS-CT-01-M-CTP-001 identified a vent and drain valve, which is present only in

Scoping and Screening Methodology

PVNGS Unit 2, but is not identified or described in the LRA. As a result, the staff asked that the applicant provide the following information:

- A description of the process used to identify and document the differences in system, structure, or component configurations and the material and environments between similar structures or components between the three PVNGS units

- A list of any differences of systems, structures, or components included within the scope of license renewal and any structures or component subject to AMR between the three PVNGS units

The applicant responded to RAI 2.1-1 by letter dated February 5, 2010, which stated the following:

> PVNGS Piping and Instrumentation Drawings (P&ID) were used with the LRDMT [license renewal data management tool - containing information derived from the plant equipment database] for scoping and screening of mechanical components. For each mechanical system, P&IDs for units 1, 2, and 3 were reviewed for obvious differences. The most representative P&ID was selected for development of the license renewal boundary drawing and was used as the working drawing for scoping and screening of components. After in-scope license renewal boundaries were established on a plant system P&ID, each in-scope component on the P&ID was checked off, and scoping and screening information was entered into the LRDMT (component by component). After all in-scope components were checked off on the working P&ID, any unevaluated components in the LRDMT were reconciled. Some of these components were shown on other interfacing drawings and had to be evaluated accordingly for being in-scope of license renewal. Components that were clearly out of scope based on the P&ID in-scope boundaries and [plant equipment database] research were not included in-scope in the LRDMT. Any remaining LRDMT system components were evaluated and determined whether or not to be within the scope of license renewal based on [plant equipment database] and current licensing basis (CLB) research. Some of these components were minor unit differences. Each of these unit difference components were then evaluated for intended function, material type, and internal and external environments and documented in the LRDMT on a component by component basis. Thus, all mechanical components in the LRDMT (i.e., SWMS [site equipment database]) were accounted for and evaluated for license renewal whether the component was applicable to one, two, or three units.

> In cases where a mechanical system had unit differences for in-scope components, system boundary drawings were sometimes modified to show the unit-specific components with a note explaining additional unit-specific components are in-scope, as was the case for the valve noted in this RAI on boundary drawing LR-PVNGS-CT-01-MCTP-001. Whether unit-specific components have been shown on the boundary drawings or not, the unit-specific components have been appropriately scoped and screened as described above.

> Unit differences are identified and documented as described in the general discussion above. The differences are contained in the LRDMT, and affected components scoped and screened as discussed above. There is no separate reporting or documenting of these minor unit differences since it is inherent in the scoping and screening methodology that mechanical system components are

Scoping and Screening Methodology

> evaluated individually and determined to be, or not to be, within the scope of license renewal.
>
> Unit-difference components were found based on the P&ID review and [plant equipment database]/LRDMT reconciliation. These components were then evaluated on an individual component basis. Each component was researched through SWMS (Site Work Management System), P&IDs, component drawings, component specifications, etc., to determine the material type and appropriate environments on a one for one component basis. The results were documented in the LRDMT for each of these components.
>
> The review of the scoping and screening methodology concluded that the methodology did not preclude identification of SSCs which should have been included in the scope of license renewal. PVNGS Units 1, 2, and 3 components have been included in the appropriate component type/material/environment groups for aging evaluation management. No additional SSCs were added to the scope of license renewal based on this review.
>
> [The response included] a list of differences of systems, structures, or components included within the scope of license renewal and any structures or component subject to aging management review, between the three PVNGS units.
>
> The scoping and screening methodology was reviewed with respect to the issue identified. It was determined that the methodology did not preclude identification of SSCs which should have been included in the scope of license renewal. The specific valve in question, although not on the Unit 1 boundary drawing used for PVNGS, was noted on the boundary drawing as existing in Unit 2 only. The valve was included in the PVNGS LRDMT and was evaluated for aging management.

The staff reviewed the applicant's response to RAI 2.1-1 and determined that the applicant had provided documentation that explained and clarified how differences in SSCs between the three PVNGS units were identified, documented, and dispositioned and provided a list of the identified differences. The staff determined that the methodology as explained, provided an adequate method to ensure that differences between units were evaluated and identified on the single set of license renewal drawings. RAI 2.1-1 is resolved.

2.1.3.1.3   Conclusion

Based on its review of LRA Section 2.1, the detailed scoping and screening implementing procedures, the results from the scoping and screening audit, and the response to the RAI, the staff concludes that the applicant's scoping and screening methodology consider CLB information in a manner consistent with the Rule, the SRP-LR, and NEI 95-10 guidance and, therefore, is acceptable.

2.1.3.2   Quality Controls Applied to License Review Application Development

2.1.3.2.1   Staff Evaluation

The staff reviewed the quality assurance controls used by the applicant to ensure that it adequately carried out the scoping and screening methodologies used in the LRA. The applicant applied the following quality assurance processes during the LRA development:

- Written procedures and guidelines governed the scoping and screening methodology.
- The applicant's team reviewed the LRA in a structured self-assessment.

Scoping and Screening Methodology

- Internal assessment teams reviewed the LRA. These teams included the license renewal team, system engineers, subject matter experts, quality assurance audit team, plant review board, and off-site review committee. Each of these teams included different levels of plant and organizational management.

- External assessment teams, including industry peers, reviewed the LRA. Recent license renewal applicants also participated and provided additional benchmarking.

- The applicant addressed and managed comments received through the assessment process using peer and management review.

2.1.3.2.2 Conclusion

On the basis of its review of pertinent LRA development guidance, discussion with the applicant's license renewal staff, review of the applicant's documentation of the activities performed to assess the quality of the LRA, and review of the information provided in response to RAI 2.1-1, the staff concludes that the applicant's quality assurance activities meet current regulatory requirements and provide assurance that LRA development activities were performed in accordance with the applicant's license renewal program requirements.

2.1.3.3 Training

2.1.3.3.1 Staff Evaluation

The staff reviewed the applicant's training process to ensure that it applied the guidelines and methodology for the scoping and screening activities in a consistent and appropriate manner. As outlined in the implementing procedures, the applicant requires training for all personnel participating in the development of the LRA and uses only trained and qualified personnel to prepare the scoping and screening implementing procedures. The training included the following activities:

- Initial qualification was completed before the project started and included the review of the license renewal processes, license renewal project guidelines, and relevant industry documents such as 10 CFR Part 54 regulations, NEI 95-10, SRP-LR, and NUREG-1801, "Generic Aging Lessons Learned Report," Revision 1.

- Training was required for the license renewal project personnel, which followed documented and written guidance.

- License renewal project personnel completed training in general license renewal requirements, license renewal project procedures, and discipline-specific areas. Project personnel with license renewal project experience also supplied mentoring services.

- Plant personnel received information systems training and attended aging management program and TLAA workshops.

The staff reviewed the applicant's written procedures and, on a sampling basis, reviewed completed qualification and training records, and completed training checklists for some of the applicant's license renewal personnel. The staff determined that the applicant had developed and implemented adequate procedures to control the training of personnel performing LRA activities.

2.1.3.3.2 Conclusion

On the basis of discussions with the applicant's license renewal project personnel, responsible for the scoping and screening process, and its review of selected documentation in support of the process, the staff concludes that the applicant's personnel are adequately trained to implement the scoping and screening methodology described in the applicant's implementing procedures and the LRA.

Scoping and Screening Methodology

2.1.3.4 Conclusion of Scoping and Screening Program Review

On the basis of a review of information provided in LRA Section 2.1, a review of the applicant's detailed scoping and screening implementing procedures, discussions with the applicant's license renewal personnel, and the results from the scoping and screening methodology audit, the staff concludes that the applicant's scoping and screening program is consistent with the SRP-LR and the requirements of 10 CFR 54 and, therefore, is acceptable.

2.1.4 Plant Systems, Structures, and Components Scoping Methodology

LRA Section 2.1 describes the applicant's methodology used to scope SSCs per the requirements of the 10 CFR 54.4(a) criteria. The LRA states that the scoping process categorized the entire plant in terms of systems and structures with respect to license renewal. According to the LRA, systems and structures were evaluated against criteria provided in 10 CFR 54.4 (a)(1), (2), and (3) to determine if the item should be considered within the scope of license renewal. The LRA states that the scoping process identified the SSCs that are safety-related and perform or support an intended function for responding to a design basis event (DBE); are nonsafety-related but their failure could prevent accomplishment of a safety-related function; or support a specific requirement for one of the five regulated events applicable to license renewal. Section 2.1.1 "Introduction," states that the scoping methodology utilized is consistent with 10 CFR 54 and with the industry guidance contained in NEI 95-10.

2.1.4.1 Application of the Scoping Criteria in 10 CFR 54.4(a)(1)

2.1.4.1.1 Summary of Technical Information in the Application

As required by 10 CFR 54.4(a)(1), plant SSCs within the scope of license renewal must include safety-related SSCs, which are those SSCs that are relied upon to remain functional during and following DBEs (as defined in 10 CFR 50.49 (b)(1)), to ensure the following functions:

(i)   The integrity of the reactor coolant pressure boundary

(ii)   The capability to shut down the reactor and maintain it in a safe shutdown condition

(iii)   The capability to prevent or mitigate the consequences of accidents, which could result in potential off-site exposures comparable to those referred to in 10 CFR 50.34(a)(1), 10 CFR 50.67(b)(2), or 10 CFR 100.11, as applicable

LRA Section 2.1.2.1, "Title 10 CFR 54.4(a)(1) - Safety-Related," states the following:

> Safety-related classifications for systems and structures at PVNGS are reported in the UFSAR or in design basis documents such as engineering drawings, evaluations, or calculations. Safety-related classifications for components are documented on engineering drawings and in a plant equipment database. The safety-related classification as reported in these source documents has been relied upon to identify SSCs satisfying one or more of the criteria of 10 CFR 54.4(a)(1). These SSCs have been identified as within the scope of license renewal.

The LRA points to the UFSAR Appendix 17.2C *"Terms and Definitions,"* which defines "safety-related" as the equipment and SCs that are relied upon to remain functional, during and following design bases events, to ensure the following:

- The integrity of the reactor coolant boundary

- The capability to shut down the reactor and maintain it in a safe condition

- The capability to prevent or mitigate the consequences of accident, which could result in potential offsite exposures comparable to the guideline exposures of 10 CFR 100

Scoping and Screening Methodology

The UFSAR review identified the set of DBEs and confirmed that the license renewal process had evaluated the associated SSCs consistent with the criteria of the Rule. The exposure guidelines used for license renewal are the same as 10 CFR 54.4 with the exception of the guidelines cited for off-site exposures. In addition to the guidelines of 10 CFR 100, 10 CFR 54.4(a)(1)(iii) references the dose guidelines of 10 CFR 50.34(a)(1) and 10 CFR 50.67(b)(2). These different exposure guidelines appear in three different code sections to address similar accident analyses performed by licensees for different reasons. The guidelines of 10 CFR 50.34(a)(1) are applicable to facilities seeking a construction permit and are, therefore, not applicable to license renewal. The exposure guidelines of 10 CFR 50.67(b)(2) address the use of alternate source terms, which are not applicable under the PVNGS CLB. Therefore, the applicant stated that use of the safety-related classification designators is consistent with 10-CFR 54.4(a)(1) scoping criteria.

2.1.4.1.2 Staff Evaluation

As stated above, under 10 CFR 54.4(a)(1), the applicant must consider all safety-related SSCs relied upon to remain functional, during and following a DBE, to ensure the integrity of the reactor coolant pressure boundary, the ability to shut down the reactor and maintain it in a safe shutdown condition, or the capability to prevent or mitigate the consequences of accidents that could result in potential offsite exposures.

With regard to identification of DBEs, Section 2.1.3, "Review Procedures," of the SRP-LR states the following:

> The set of DBEs as defined in the Rule is not limited to Chapter 15 (or equivalent) of the UFSAR. Examples of DBEs that may not be described in this chapter include external events, such as floods, storms, earthquakes, tornadoes, or hurricanes, and internal events, such as a high energy line break. Information regarding DBEs as defined in 10 CFR 50.49(b)(1) may be found in any chapter of the facility UFSAR, the Commission's regulations, NRC orders, exemptions, or license conditions within the CLB. These sources should also be reviewed to identify SSCs relied upon to remain functional during and following DBEs (as defined in 10 CFR 50.49(b)(1)) to ensure the functions described in 10 CFR 54.4(a)(1).

During the audit, the applicant stated that it evaluated the types of events listed in NEI 95-10 (i.e., anticipated operational occurrences, DBEs, external events, and natural phenomena) that were applicable to PVNGS. The staff reviewed the applicant's basis documents, which described all design basis conditions in the CLB and addressed all events defined by 10 CFR 50.49(b)(1) and 10 CFR 54.4(a)(1). The PVNGS UFSAR and basis documents discussed events such as internal and external flooding, tornados, and missiles. The staff concludes that the applicant's evaluation of DBEs was consistent with the SRP-LR.

The applicant scoped SSCs for the 10 CFR 54.4(a)(1) criterion in accordance with the license renewal implementing procedures, which provide guidance for the preparation, review, verification, and approval of the scoping evaluations to ensure the adequacy of the results of the scoping process. The staff reviewed the implementing procedures governing the applicant's evaluation of safety-related SSCs and sampled the applicant's reports of the scoping results to ensure that the methodology was applied in accordance with the implementing procedures. In addition, the staff discussed the methodology and results with applicant personnel who were responsible for these evaluations.

The staff reviewed the applicant's evaluation of the Rule and CLB definitions pertaining to 10 CFR 54.4(a)(1) and determined that the CLB definition of "safety-related," met the definition of "safety-related" specified in 10 CFR 54.4(a)(1). The staff reviewed a sample of the license renewal scoping results for the safety injection and shutdown cooling, diesel generator fuel oil storage and transfer, auxiliary feedwater, and turbine building to provide additional assurance that the applicant adequately implemented its scoping methodology in accordance with 10 CFR 54.4(a)(1). The staff verified that the applicant developed the scoping results for each of the sampled systems consistent with the methodology, identified the SSCs credited for performing intended functions, and adequately described the basis for the results as well as the intended functions. The staff also confirmed that the applicant had identified and used pertinent engineering and licensing information to identify the SSCs required to be within the scope of license renewal in accordance with 10 CFR 54.4(a)(1) criteria.

2.1.4.1.3 Conclusion

On the basis of its review of systems (on a sampling basis), discussions with the applicant, and review of the applicant's scoping process, the staff concludes that the applicant's methodology for identifying systems and structures that are within the scope of license renewal is consistent with the SRP-LR and the requirements of 10 CFR 54.4(a)(1), and, therefore, is acceptable.

2.1.4.2 Application of the Scoping Criteria in 10 CFR 54.4(a)(2)

2.1.4.2.1 Summary of Technical Information in the Application

LRA Section 2.1.2.2, "Title 10 CFR 54.4(a)(2) - Nonsafety-Related Affecting Safety-Related," states the following:

> 10 CFR 54.4(a)(2) requires that plant SSCs within the scope of license renewal include all nonsafety-related SSCs whose failure could prevent satisfactory accomplishment of any of the functions identified for safety-related SSCs. The guidance provided in NEI 95-10, Appendix F, was used to develop the methodology for scoping to the criterion of 10 CFR 54.4(a)(2).

Section 2.1.2.2 of the LRA goes on to state that the methodology includes identification of nonsafety-related SSCs that are connected to safety-related SSCs and nonsafety-related SSCs that could spatially interact with safety-related SSCs. Determination and identification of any other SSCs satisfying criterion 10 CFR 54.4(a)(2) was completed as described below, based on review of applicable CLB documents, plant specific and industry operating experience, and by system and structure functional evaluations.

LRA Section 2.1.2.2 states the following in relation to nonsafety-related SSCs providing functional support to safety-related SSCs:

> The PVNGS UFSAR and other current licensing basis documents were reviewed for every nonsafety-related plant system or structure, to determine whether the system or structure was credited with supporting satisfactory accomplishment of a safety-related function. Nonsafety-related systems or structures credited in CLB documents with supporting accomplishment of a safety-related function were classified as satisfying criterion 10 CFR 54.4(a)(2) and were included within the scope of license renewal.

LRA Section 2.1.2.2 states the following in relation to nonsafety-related SSCs directly connected to safety-related SSCs:

> Nonsafety-related SSCs were included within the scope of license renewal, as applicable, up to the first seismic anchor past the safety/nonsafety interface for those nonsafety-related mechanical SSCs that are connected to a safety-related SSC and must provide structural integrity. In most cases, an actual seismic anchor exists to serve as the boundary for the nonsafety structural integrity feature. In cases where seismic anchors were not available to serve as the license renewal boundary, other methods as provided for in NEI 95-10, including

Scoping and Screening Methodology

equivalent anchors were utilized to establish the license renewal boundary. Other methods included:

- A combination of restraints or supports such that the nonsafety-related piping and associated structures and components attached to safety-related piping is included in-scope up to a boundary point that encompasses two (2) supports in each of three (3) orthogonal directions.

- A base-mounted component (e.g., pump, heat exchanger, tank, etc.) that is a rugged component and is designed not to impose loads on connecting piping is included in scope as it has a support function for the safety-related piping.

- A flexible connection that is considered a pipe stress analysis model end point when the flexible connection effectively decouples the piping system (i.e., does not support loads or transfer loads across it to connected piping).

- A free end of nonsafety-related piping, such as a drain pipe that ends at an open floor drain.

- A point where buried piping exits the ground. The buried portion of the piping should be included in the scope of license renewal. A determination that the buried piping is well founded on compacted soil that is not susceptible to liquification must be documented.

- Nonsafety-related piping runs that are connected at both ends to safety-related piping include the entire run of nonsafety-related piping.

- LRA Section 2.1.2.2 states, in relation to nonsafety-related SSCs with the potential for spatial interaction with safety-related SSCs:

Nonsafety-related SSCs which are not connected to safety-related piping and/or which are not required for structural integrity, but have a spatial relationship such that their failure could adversely impact the performance of a safety-related SSC intended function, were included in the scope of license renewal per NEI 95-10, Appendix F. PVNGS applied the preventative option for 10 CFR 54.4(a)(2) scoping. The preventative option as implemented at PVNGS is based on a "spaces" approach for scoping of nonsafety-related systems with potential spatial interaction with safety-related SSCs. Potential spatial interaction is assumed in any structure that contains active or passive safety-related SSCs. The structures of concern for potential spatial interaction were identified based on the review of the CLB to determine which structures contained safety-related SSCs. Plant walkdowns were performed as required to confirm that all structures containing safety-related SSCs had been identified. For structures that contain safety-related SSCs, there may be selected rooms within the structure that do not contain any safety-related SSCs. CLB document reviews and plant walkdowns were utilized as appropriate to confirm that these rooms did not contain safety-related SSCs, thereby eliminating spatial interactions concerns from these rooms.

Nonsafety-related systems and components that contain water, oil, or steam, and are located inside structures that contain safety-related SSCs are included in-scope for potential spatial interaction under criterion 10 CFR 54.4(a)(2), unless located in an excluded room. All high-energy lines located inside primary containment are included within the scope of license renewal. High-energy lines

Scoping and Screening Methodology

located outside primary containment are included within the scope of license renewal if their failure could adversely impact any safety-related SSC's. Safety-related high-energy lines are in-scope under 10 CFR 54.4(a)(1), and nonsafety-related high-energy lines are in-scope under 10 CFR 54.4(a)(2). The potential effects of flooding as a consequence of a pipe break or critical crack were analyzed on a case-by-case basis to ensure that the operability of safety-related equipment would not be impaired. Floor drains and curbs required for flood mitigation are within the scope of license renewal under 10 CFR 54.4(a)(2). Supports for all nonsafety-related SSCs within these structures are included within the scope of license renewal.

2.1.4.2.2 Staff Evaluation

Under 10 CFR 54.4(a)(2), the applicant must consider all nonsafety-related SSCs whose failure could prevent the satisfactory accomplishment of functions for safety-related SSCs, relied on during and following a DBE, to ensure: (1) the integrity of the reactor coolant pressure boundary, (2) the ability to shut down the reactor and maintain it in a safe shutdown condition, or (3) the capability to prevent or mitigate the consequences of accidents that could result in potential offsite exposures comparable to those referred to in 10 CFR 50.34(a)(1), 10 CFR 50.67(b)(2), or 10 CFR 100.11.

Regulatory Guide (RG) 1.188, Revision 1, endorses the use of NEI 95-10, Revision 6. NEI 95-10 discusses the staff's position on 10 CFR 54.4(a)(2) scoping criteria; including nonsafety-related SSCs typically identified in the CLB; consideration of missiles, cranes, flooding, and high-energy line breaks; nonsafety-related SSCs connected to safety-related SSCs; nonsafety-related SSCs in proximity to safety-related SSCs; and mitigative and preventative options related to nonsafety-related and safety-related SSCs interactions.

In addition, the staff's position (as discussed in NEI 95-10, Revision 6) is that applicants should not consider hypothetical failures, but rather should base evaluations on the plant's CLB, engineering judgment and analyses, and relevant operating experience. NEI 95-10 further describes operating experience as all documented plant-specific and industry-wide experience that can be used to determine the plausibility of a failure. Documentation would include NRC generic communications and event reports, plant-specific condition reports, industry reports such as safety operational event reports, and engineering evaluations. The staff reviewed LRA Section 2.1.2.2, in which the applicant described the scoping methodology for nonsafety-related SSCs under 10 CFR 54.4(a)(2). In addition, the staff reviewed the applicant's implementing document and results report, which documents the guidance and corresponding results of the applicant's scoping review pursuant to 10 CFR 54.4(a)(2). The applicant stated in Section 2.1.2.2 that it performed the review in accordance with the guidance contained in NEI 95-10, Revision 6, Appendix F.

<u>Nonsafety-Related SSCs Required to Perform a Function that Supports a Safety-Related SSC</u>. The staff determined that nonsafety-related SSCs required to remain functional to support a safety-related function had been reviewed by the applicant for inclusion within the scope of license renewal in accordance with 10 CFR 54.4(a)(2). The staff reviewed the evaluating criteria discussed in LRA Section 2.1.2.2 and the applicant's 10 CFR 54.4(a)(2) implementing document. The staff confirmed that the applicant had reviewed the UFSAR, plant drawings, plant equipment database, and other CLB documents to identify the nonsafety-related systems and structures that function to support a safety-related system whose failure could prevent the performance of a safety-related intended function. The applicant also considered missiles, overhead handling systems, internal and external flooding, and high-energy line breaks. Accordingly, the staff finds that the applicant implemented an acceptable method for including nonsafety-related systems that perform functions that support safety-related intended functions, within the scope of license renewal as required by 10 CFR 54.4(a)(2).

Nonsafety-Related SSCs Directly Connected to Safety-Related SSCs. The staff confirmed that nonsafety-related SSCs, directly connected to SSCs, had been reviewed by the applicant for inclusion within the scope of license renewal in accordance with 10 CFR 54.4(a)(2). The staff reviewed the evaluating criteria discussed in LRA Section 2.1.2.2 and the applicant's 10 CFR 54.4(a)(2) implementing document. The applicant had reviewed the safety-related to nonsafety-related interfaces for each mechanical system in order to identify the nonsafety-related components located between the safety to nonsafety-related interface and license renewal structural boundary.

The staff determined that in order to identify the nonsafety-related SSCs connected to safety-related SSCs and required to be structurally sound to maintain the integrity of the safety-related SSCs, the applicant used a combination of the following to identify the portion of nonsafety-related piping systems to include within the scope of license renewal:

- Seismic anchors
- Equivalent anchors
- Bounding conditions described in NEI 95-10 Revision 6, Appendix F (i.e., base-mounted component, flexible connection, inclusion to the free end of nonsafety-related piping, or inclusion of the entire piping run)

The staff finds that the applicant implemented an acceptable method for including nonsafety-related SSCs directly connected to safety-related SSCs, within the scope of license renewal as required by 10 CFR 54.4(a)(2).

Nonsafety-Related SSCs with the Potential for Spatial Interaction with Safety-Related SSCs. The staff confirmed that nonsafety-related SSCs with the potential for spatial interaction with safety-related SSCs had been reviewed by the applicant for inclusion within the scope of license renewal in accordance with 10 CFR 54.4(a)(2). The staff reviewed the evaluating criteria discussed in the LRA Section 2.1.2.2 and the applicant's 10 CFR 54.4(a)(2) implementing procedure. The applicant considered physical impacts (i.e., pipe whip, jet impingement), harsh environments, flooding, spray, and leakage when evaluating the potential for spatial interactions between nonsafety-related systems and safety-related SSCs. The staff further confirmed that the applicant used a "spaces" approach to identify the portions of nonsafety-related systems with the potential for spatial interaction with safety-related SSCs. The "spaces" approach focuses on the interaction between nonsafety-related and safety-related SSCs that are located in the same space, which was defined for the purposes of the review as a structure containing active or passive safety-related SSCs.

LRA Section 2.1.2.2 and the applicant's implementing document state that the applicant had included mitigative features when considering the affect of nonsafety-related SSCs on safety-related SSCs for occurrences discussed in the CLB. The staff reviewed the applicant's CLB information, primarily contained in the UFSAR, related to missiles, crane load drops, flooding, and high-energy line breaks. The staff determined that the applicant also considered the features designed to protect safety-related SSCs from the effects of these occurrences through the use of mitigating features, such as floor drains and curbs. The staff confirmed that the applicant had included the mitigating features within the scope of license renewal in accordance with 10 CFR 54.4(a)(2).

LRA Section 2.1.2.2 and the applicant's implementing document state that the applicant had used a preventive approach, which considered the affect of nonsafety-related SSCs contained in the same space as safety-related SSCs. The staff determined that the applicant had evaluated all nonsafety-related SSCs containing liquid or steam and located in spaces containing safety-related SSCs. The applicant used a spaces approach to identify the nonsafety-related SSCs that were located within the same space as safety-related SSCs.

Scoping and Screening Methodology

As described in the LRA and for the purpose of the scoping review, a space was defined as a structure containing active or passive safety-related SSCs. In addition, the staff determined that following the identification of the applicable mechanical systems, the applicant identified its corresponding structures for potential spatial interaction, based on a review of the CLB and plant walkdowns. Nonsafety-related systems and components that contain liquid or steam and are located inside structures that contain safety-related SSCs were included within the scope of license renewal, unless it was in an evaluated area and determined not to contain safety-related SSCs. The staff also determined that, based on plant and industry operating experience, the applicant excluded the nonsafety-related SSCs containing air or gas from the scope of license renewal, with the exception of portions that are attached to safety-related SSCs and required for structural support. The staff confirmed that those nonsafety-related SSCs determined to contain liquid or steam and located within a space containing safety-related SSCs were included within the scope of license renewal in accordance with 10 CFR 54.4(a)(2).

The staff determined that additional information would be required to complete the review of the applicant's scoping methodology. RAI 2.1-2, dated December 23, 2009, states that during the scoping and screening methodology audit, performed onsite October 19-22, 2009, the staff determined that the applicant had not included certain fluid-filled, nonsafety-related SSCs, adjacent to safety-related SSCs, within the scope of license renewal. The applicant's basis for not including the nonsafety-related SSCs was information contained in the applicant's "Moderate Energy Crack Evaluation" document.

The staff asked that the applicant perform a review of the issue and provide the basis for the determination that the Moderate Energy Crack Evaluation is part of the CLB. The staff also asked for a description and analysis of the pertinent information contained in the Moderate Energy Crack Evaluation that gives the basis for the conclusion that failure of the nonsafety-related, fluid filled SSCs could not prevent the satisfactory accomplishment of safety-related functions for SSCs relied on to remain functional during and following a DBE. This issue was also reviewed during a NRC license renewal inspection, done the week of February 22, 2010 (Inspection Report dated April 29, 2010), and subsequently documented in Palo Verde Action Request 3440560.

The applicant responded to RAI 2.1-2 and issues identified during the NRC license renewal inspection the week of February 22, 2010, by letter dated April 1, 2010, which states the following:

> PVNGS UFSAR Table 3.6-3 provides the methods of protection of safety-related systems from the effects of high and moderate energy line breaks. The methods specified in the table are layout, enclosure, and redundancy. The Moderate Energy Crack Evaluation was prepared to verify, in part, the protection methods specified in PVNGS UFSAR were met. Therefore, Moderate Energy Crack Evaluation, 13-MC-ZZ- 642, is part of the current licensing basis (CLB) as defined in 10 CFR 54.3.

> Palo Verde has revised its 10 CFR 54.4(a)(2) implementing document to delete the use of information contained in the Moderate Energy Crack Evaluation. The Moderate Energy Crack Evaluation will not be used to provide the basis for determination that failure of the non-safety-related fluid-filled structures, systems, and components (SSCs) could not prevent the satisfactory accomplishment of safety-related functions for SSCs relied on to remain functional during and following a design basis event.

> The information contained in the Moderate Energy Crack Evaluation had been used to evaluate spatial interaction of nonsafety-related SSCs in the Auxiliary Building, Control Building, Diesel Generator Building, and Fuel Building. As a

Scoping and Screening Methodology

result of the revision to the 10 CFR 54.4(a)(2) implementing document, the nonsafety-related SSCs that were previously excluded from the scope of license renewal using information contained in the Moderate Energy Crack Evaluation and could interact with safety-related SSCs in the Auxiliary Building, Control Building, Diesel Generator Building, and Fuel Building were evaluated and included within the scope of license renewal in accordance with criterion 10 CFR 54.4(a)(2).

The staff reviewed the applicant's response to RAI 2.1-2 and determined that the applicant had revised its process and evaluated nonsafety-related SSCs in the auxiliary building, control building, diesel generator building, fuel building, and anywhere near safety-related SSCs. This evaluation determined if there is a potential for spatial interaction without consideration of the previously cited Moderate Energy Crack Evaluation. As a result of the process revision and the evaluation subsequently performed, the applicant has included additional nonsafety-related SSCs within the scope of license renewal in accordance with 10 CFR 54.4(a)(2). The staff determined that the revised process is in accordance with the requirements of 10 CFR 54.4(a)(2) and is, therefore, acceptable. RAI 2.1-2 is closed.

RAI 2.1-3, dated December 23, 2009, states that during the scoping and screening methodology audit, performed onsite October 19-22, 2009, the staff determined that the following nonsafety-related SSCs had not been included within the scope of license renewal. The staff asked that the applicant perform a review and provide the basis for not including the following nonsafety-related SSCs, attached or adjacent to safety-related SSCs, within the scope of license renewal in accordance with 10 CFR 54.4(a)(2):

(1) Nonsafety-related pipe attached to the safety-related penetration of the condensate storage tank

(2) Nonsafety-related, abandoned containment spray chemical addition tanks, located in the auxiliary building along with safety-related SSCs, for which the associated piping had been cut and capped but the tanks had not been verified to be dry (PVNGS Units 1 and 3)

(3) Nonsafety-related, fluid-filled SSCs located within the turbine building and adjacent to a penetration into the safety-related main steam support structure (MSSS)

(4) Nonsafety-related, fluid-filled SSCs located on the auxiliary building roof and adjacent to opening in the safety-related MSSS

The applicant responded to RAI 2.1-3(1) by a letter dated February 5, 2010, which states the following:

> The non-highlighted nonsafety-related piping attached to the condensate storage tank (CST) has been reviewed, and two of the six lines were added to the scope of license renewal based on criterion 10 CFR 54.4(a)(3). The scoping methodology did not preclude the identification of these two lines. These lines were identified as a result of correcting the tank level during the review. License Renewal boundary drawing PVNGS-CT-01-M-CTP-001 has been revised (Revision 3) to add the following LR Note 1 that provides the bases for the conclusion that the other four nonsafety-related piping lines attached to the CST are not within the scope of license renewal:
>
>> The tank penetrations are above the minimum required tank level, and therefore the piping and components are not required to maintain the tank pressure boundary. The pipe sizes are much smaller than the tank, and consequently impose no structural impact because they do not have a structural integrity function nor

Scoping and Screening Methodology

> do the lines have a spatial interaction with safety-related components, and are not within the scope of license renewal based on criterion 10 CFR 54.4(a)(2). The tank penetrations are not associated with venting. Consequently, the piping and components are not within the scope of license renewal for SBO based on 10 CFR 54.4(a)(3).

The staff reviewed the applicant's response to RAI 2.1-3(1) and determined that the applicant had provided documentation of an evaluation and basis for not including within the scope of license renewal four of the six nonsafety-related pipes attached to the CST. The staff determined that the applicant had provided documentation of the basis that two of the six nonsafety-related pipes attached to the CST had been re-evaluated and determined to have a 10 CFR 54.4(a)(3) intended function and appropriately included within the scope of license renewal. RAI 2.1-3(1) is resolved.

The applicant responded to RAI 2.1-3(2) by letter dated April 2, 2010, which states the following:

> The containment spray chemical addition tanks were taken out of service and abandoned in Units 1, 2, and 3. However, a residual amount of fluid containing a diluted solution of the original chemical was identified to be captured in the Units 1 and 3 tanks, and suspected in the associated piping in all three units. The following commitment is being added to LRA Table A4-1, License Renewal Commitments.
>
> > By August 30, 2010, APS will ensure that that the abandoned containment spray chemical addition tanks and associated piping components in PVNGS Units 1, 2, and 3 are drained to preclude any spatial interactions with safety-related components.
>
> When these actions are completed, the Units 1, 2, and 3 abandoned containment spray chemical addition tanks and the associated piping components will not be within the scope of license renewal.

The staff determined that, in order to resolve the concern in RAI 2.1-3(2), it needs confirmation that the abandoned containment spray chemical addition tanks and associated piping components in PVNGS Units 1, 2, and 3 are drained to prevent any spatial interactions with safety-related components. This was identified as <u>Confirmatory Item 2.1.4.2-1</u>.

By letter dated November 10, 2010, the applicant stated that the containment spray chemical addition tanks and associated piping components have been drained. The concern in RAI 2.1-3(2) is resolved and Confirmatory Item 2.1.4.2-1 is closed.

The applicant responded to RAI 2.1-3(3) by letter dated February 5, 2010, which states the following:

> During a site audit conducted for initial plant licensing, the NRC staff questioned the ability of a fire to spread through the unprotected wall opening between elevation 120 feet of the main steam support structure (MSSS) and the turbine building. The NRC question and APS response is documented in UFSAR Section 9A, Question 9A.121, and the response was accepted by the NRC in Section 9.5.1.3 of Supplement No. 6 to the PVNGS Safety Evaluation Report (NUREG-0857). In the response, APS indicated that the wall openings between the MSSS and the turbine building are unsealed to allow cooling of the hot piping anchor/support attachments at the concrete structure. A compartment devoid of in situ combustibles is located between zones 74A and 74B (formerly zone 74) of the MSSS and the turbine building. Ventilation exhaust fans use this compartment as a supply plenum to pull cooling air flow over the pipe

> support/anchors from the turbine building and the MSSS. The air flow is away from the safety-related equipment in zones 74A and 74B.
>
> Based on this response, a failure of nonsafety-related, fluid-filled SSCs located within the turbine building and adjacent to a penetration will not result in any spray effect to the components in main steam support structure since the ventilation exhaust fans will pull air away from both the main steam support structure and the turbine building. Also, the safety-related components inside the MSSS are environmentally qualified to maintain intended functions during a single area line break inside the MSSS (UFSAR Section 10.3). Therefore, there are no fluid filled SSCs located within the turbine building whose failure could prevent satisfactory accomplishment of safety-related functions as described in 10 CFR 54.4(a)(1).

The staff reviewed the applicant's response to RAI 2.1-3(3), which provided a discussion on the purpose of the wall openings between the turbine building and the MSSS. The openings direct airflow and the environmental qualification (EQ) of safety-related SSCs located within the MSSS. The staff determined that the applicant had provided an evaluation and basis for not including nonsafety-related SSCs, located in the turbine building, within the scope of license renewal. The potential for spatial interaction with safety-related SSCs in the MSSS was prevented by building design and mitigated by EQ of safety-related SSCs located in the MSSS. RAI 2.1-3(3) is resolved.

The applicant responded to RAI 2.1-3(4) by letter dated February 5, 2010, which states the following:

> The MSSS is a safety-related Category I structure that provides shelter, protection, and support for the license renewal intended functions of the safety-related SSCs located inside the MSSS. The roofing membrane and the concrete and structural steel of the external walls, including any openings in the walls, are within scope of license renewal with license renewal intended functions of shelter, protection and support. The above grade portion of the MSSS is designed to be open to natural circulation of outside air (UFSAR 3.11.4) including the adjacent auxiliary building roof. The nonsafety-related fluid-filled SSCs located on the auxiliary building roof adjacent to the MSSS are evaluated as not within scope of license renewal based on criteria of 10 CFR 54.4(a)(2) leakage barrier considerations. This is because the walls of the MSSS, including any openings in the walls, are designed to be open to natural circulation of outside air and are evaluated as providing shelter, protection and support to any safety-related components inside the MSSS from rain and water spray arising from the failure of nonsafety-related fluid-filled SSCs on the auxiliary building roof.

The staff reviewed the applicant's response to RAI 2.1-3(4) and determined that the applicant had provided a discussion and evaluation on the purpose of the building openings; they are designed to allow airflow, but also, to provide protection to safety-related SSCs from outside moisture such as rain. The staff determined that the applicant had provided an evaluation and basis for not including nonsafety-related SSCs, located adjacent to the MSSS, within the scope of license renewal in that the potential for spatial interaction with safety-related SSCs was prevented by building design. RAI 2.1-3(4) is resolved.

2.1.4.2.3 Conclusion

On the basis of its review of the applicant's scoping process, discussions with the applicant, and review of the information provided in the response to RAIs 2.1-2 and 2.1-3, the staff concludes that the applicant's methodology for identifying and including nonsafety-related SSCs that could

Scoping and Screening Methodology

affect the performance of safety-related SSCs within the scope of license renewal is consistent with the scoping criteria of 10 CFR 54.4(a)(2), and, therefore, is acceptable.

2.1.4.3 Application of the Scoping Criteria in 10 CFR 54.4(a)(3)

2.1.4.3.1 Summary of Technical Information in the Application

LRA Section 2.1.2.3, "Title 10 CFR 54.4(a)(3) - Regulated Events," states the following:

> 10 CFR 54.4(a)(3) requires that plant SSCs within the scope of license renewal include all SSCs relied on in safety analyses or plant evaluations to perform a function that demonstrates compliance with the regulations for fire protection (10 CFR 50.48), environmental qualification (10 CFR 50.49), pressurized thermal shock (10 CFR 50.61), anticipated transients without scram (10 CFR 50.62), and station blackout (10 CFR 50.63). Position papers were prepared to provide input to the SSC scoping process. The purpose of these position papers was to evaluate the PVNGS CLB relative to the regulated events, identify the systems and structures that are relied upon to demonstrate compliance with each of these regulations, and document the results of this review. Guidance provided by the position papers was used during system and structure scoping to identify system and structure intended functions for Criterion (a)(3), and again during component scoping as necessary to determine which components are credited in the regulated events. SSCs credited in the regulated events have been classified as satisfying criterion 10 CFR 54.4(a)(3) and have been identified as within the scope of license renewal.

Fire Protection. LRA Section 2.1.2.3.1, "Fire Protection," describes the scoping of systems and structures relied on in safety analyses or plant evaluations to perform a function that demonstrates compliance with the fire protection criterion. LRA Section 2.1.2.3.1 states the following:

> Criterion 10 CFR 54.4(a)(3) requires that plant SSCs within the scope of license renewal include all SSCs relied on in safety analyses or plant evaluations to perform a function that demonstrates compliance with the regulations for fire protection (10 CFR 50.48). 10 CFR 50.48 requires each operating nuclear power plant to have a fire protection plan that satisfies the requirement of Criterion 3 of 10 CFR 50 Appendix A, and further requires all nuclear power plants licensed to operate prior to January 1, 1979, to comply with Sections III.G, III.J and III.O of Appendix R to 10 CFR 50.

> The PVNGS Fire Protection Program licensing basis is based on Appendix A to Branch Technical Position (BTP) APCSB 9.5-1, "Guidelines for Fire Protection for Nuclear Power Plants Docketed Prior to July 1, 1976" and 10 CFR 50, Appendix R, "Fire Protection Program for Nuclear Power Facilities Operating Prior to January 1, 1979." The primary CLB document for PVNGS is UFSAR Section 9.5.1, "Fire Protection System." Appendices 9A and 9B of the UFSAR provide the fire hazards analysis and other information concerning the design and license bases, including comparisons to Appendix A to BTP APCSB 9.5-1 and to 10 CFR 50 Appendix R. PVNGS Fire Hazards Analysis is presented in UFSAR Appendix 9B.2. The Fire Hazards Analysis shows that redundant safety systems required to achieve and maintain hot standby and cold shutdown are adequately protected against fire damage.

The applicant included CLB SSCs classified as satisfying criterion 10 CFR 54.4(a)(3) related to fire protection that were identified for PVNGS as within the scope of license renewal.

Scoping and Screening Methodology

Environmental Qualification. LRA Section 2.1.2.3.2, "Environmental Qualification (EQ)," describes the scoping of systems and structures relied on in safety analyses or plant evaluations to perform a function in compliance with the EQ criterion. LRA Section 2.1.2.3.2 states the following:

> Criterion 10 CFR 54.4(a)(3) requires that plant SSCs within the scope of license renewal include all SSCs relied on in safety analyses or plant evaluations to perform a function that demonstrates compliance with the regulations for environmental qualification (10 CFR 50.49). The PVNGS environmental qualification (EQ) program applies to electrical equipment important to safety that is located in a harsh environment. The UFSAR, Section 3.11, states that environmental design criteria for PVNGS conform to 10 CFR 50, Appendix A, General Design Criterion 4, "Environmental and Missile Design Bases." The safety-related systems and components required to mitigate the consequences of a design basis accident (DBA), or to attain a safe shutdown of the reactor, are designed to remain functional during and after exposure to normal operation environmental conditions and following the specific DBA which they are intended to mitigate. ... All components within the scope of the PVNGS EQ program which demonstrate compliance with 10 CFR 50.49 and the systems containing those components were classified as satisfying criterion 10 CFR 54.4(a)(3) and were identified as within the scope of license renewal.

Pressurized Thermal Shock. LRA Section 2.1.2.3.3, "Pressurized Thermal Shock (PTS)," states the following:

> Criterion 10 CFR 54.4(a)(3) requires that plant SSCs within the scope of license renewal include all SSCs relied on in safety analyses or plant evaluations to perform a function that demonstrates compliance with the regulations for pressurized thermal shock (10 CFR 50.61).

The applicant determined that the only component within the scope of the license renewal rule for PTS is the reactor pressure vessel. The applicant stated that the calculation of nil-ductility transition reference temperature is a TLAA, as defined by 10 CFR 54.3(a). Chapter 4 of the applicant's LRA addresses this separately.

Anticipated Transient Without Scram. LRA Section 2.1.2.3.4, "Anticipated Transients Without Scram," (ATWS) describes the scoping of systems and structures relied on in safety analyses or plant evaluations to perform a function in compliance with the ATWS criterion. LRA Section 2.1.2.3, Subsection 2.1.2.3.4, states the following:

> Criterion 10 CFR 54.4(a)(3) requires that plant SSCs within the scope of license renewal include all SSCs relied upon in safety analyses or plant evaluations to perform a function that demonstrates compliance with the regulations for anticipated transients without scram (10 CFR 50.62). An anticipated transient without scram (ATWS) is a postulated operational transient that generates an automatic scram signal accompanied by a failure of the reactor protection system to shutdown the reactor. The ATWS Rule required improvements in the design to reduce the probability of failure to shutdown the reactor following anticipated transients, and to mitigate the consequences of an ATWS event. ... The following equipment is required by the ATWS Rule for reduction of risk from an ATWS event at PVNGS:
>
> - Supplementary protection system which includes the diverse scram system
>
> - Diverse auxiliary feedwater actuation system

2-19

Scoping and Screening Methodology

- Diverse turbine trip circuitry

The applicant further stated that components designed to satisfy the ATWS rule which demonstrate compliance with 10 CFR 50.62, were classified as satisfying criterion 10 CFR 54.4(a)(3) and were identified as within the scope of license renewal.

Station Blackout. LRA Section 2.1.2.3, Subsection 2.1.2.3.5, "Station Blackout," (SBO) describes the scoping of systems and structures relied on in safety analyses or plant evaluations to perform a function in compliance with the SBO criterion. LRA Section 2.1.2.3.5, states the following:

> Criterion 10 CFR 54.4(a)(3) requires that plant SSCs within the scope of license renewal include all SSCs relied on in safety analyses or plant evaluations to perform a function that demonstrates compliance with the regulations for station blackout (10 CFR 50.63). The SBO rule (10 CFR 50.63) requires that nuclear power plants have the capability to withstand and recover from the loss of offsite and onsite AC power of a specified duration (the coping duration). Regulatory Guide 1.155 provides guidance on selecting the time period for which a licensee must cope with the SBO. PVNGS used RG 1.155 to calculate a plant-specific coping time period. A "sixteen hour" coping duration was determined for PVNGS based on expected frequency of loss of offsite power and the probable time needed for its restoration. Redundancy and reliability in onsite emergency AC power source (emergency diesel generators) was also factored in the evaluation.

The applicant developed a position paper to summarize the results of a detailed review of the SBO documentation. The position paper identifies the SSCs credited with coping and recovering from a SBO. The SSCs identified in the SBO position paper were used in scoping evaluations to identify SSCs that demonstrate compliance with 10 CFR 50.63. The LRA included all SSCs classified as satisfying criterion 10 CFR 54.4(a)(3) related to SBO as within the scope of license renewal.

2.1.4.3.2 Staff Evaluation

The staff reviewed the applicant's approach in identifying SSCs relied on to perform functions meeting the requirements of the regulations on fire protection, EQ, ATWS, PTS, and SBO. The staff reviewed the applicant's methodology, boundary scoping drawings, position papers, and LRA to assess the scoping process for these regulated safety systems. The staff also evaluated, on a sampling basis, the SSCs included within the scope of license renewal under 10 CFR 54.4(a)(3).

The staff confirmed that the applicant's implementing procedures, as described in LRA Section 2.1.1.1.3, "Technical Position Papers," were used for identifying SSCs within the scope of license renewal under 10 CFR 54.4(a)(3). The applicant evaluated the PVNGS CLB to identify all SSCs that perform functions addressed in 10 CFR 54.4(a)(3), "Regulated Events," and then included these SSCs within the scope of license renewal as documented in the specific regulated event(s) position papers. The staff determined that these position paper results reference the information sources used for determining the SSCs credited for compliance with the events listed in the specified regulations for the applicable license renewal regulated events.

Fire Protection. The staff determined that the applicant's fire protection scoping documents identified SSCs in the scope of license renewal required for fire protection. The applicant used documents, such as the UFSAR, fire protection design basis calculations, and the Fire Protection Design Basis Manual, to identify the SSCs within the scope of license renewal for fire protection. The staff reviewed the source documents used by the applicant to identify SSCs within the scope of license renewal, primarily the UFSAR, Section 9.5.1, "Fire Protection Systems."

The staff also reviewed, on a sampling basis, the scoping results in conjunction with the LRA and the CLB information to validate the methodology for including SSCs within the scope of license renewal. The staff determined that the applicant's scoping included SSCs that perform intended functions to meet the requirements of 10 CFR 50.48. Based on its review of the CLB documents and the sample review, the staff determined that the applicant's scoping methodology is adequate for including SSCs credited in performing fire protection functions within the scope of license renewal.

Environmental Qualification. The staff confirmed that the applicant's EQ scoping document required the inclusion of safety-related electrical equipment, nonsafety-related electrical equipment whose failure under postulated environmental conditions could prevent satisfactory accomplishments of safety functions of the safety-related equipment, and certain post-accident monitoring equipment, as defined in 10 CFR 50.49(b)(1), (b)(2), and (b)(3). The staff determined that the applicant used the CLB, as described in the UFSAR Section 3.11 and its EQ program manual, to identify SSCs necessary to meet the requirements of 10 CFR 50.49. SWMS, the site's equipment database, contains the EQ identifications for specific components. The staff reviewed the LRA, implementing procedures, scoping results reports, and master EQ component equipment list in the equipment database to verify that the applicant identified SSCs within the scope of license renewal that meet EQ requirements. Based on that review, the staff determined that the applicant's scoping methodology is adequate for identifying EQ SSCs within the scope of license renewal.

Pressurized Thermal Shock. The staff confirmed that the applicant's PTS scoping document included the scoping methodology used to review the CLB information to identify SSCs within the scope of license renewal to meet 10 CFR 50.61, "Fracture Toughness Requirements for Protection Against Pressurized Thermal Shock Events." The applicant's scoping methodology resulted in a determination that the reactor vessel is within the scope of license renewal. The staff reviewed the basis document and position paper and determined that the methodology was appropriate for identifying SSCs with functions credited for complying with the PTS regulation. The staff finds that the scoping results included SSCs that perform intended functions to meet the requirements of 10 CFR 50.61. The staff determined that the applicant's scoping methodology is adequate for including SSCs within the scope of license renewal which are credited in meeting PTS requirements.

Anticipated Transient Without Scram. The staff confirmed that the applicant's ATWS scoping document included the scoping methodology used to review the CLB and the site equipment database to identify plant SSCs within the scope of license renewal which are credited for ATWS mitigation. The staff reviewed these documents, and the LRA, in conjunction with the scoping results to validate the methodology for identifying ATWS SSCs within the scope of license renewal. The staff determined that the scoping results included systems and structures that perform intended functions to meet 10 CFR 50.62 requirements. The staff determined that the applicant's scoping methodology is adequate for identifying SSCs within the scope of license renewal which are credited in meeting the ATWS regulation.

Station Blackout. The staff confirmed that the applicant's SBO scoping document included the scoping methodology used to review CLB documents to identify SSCs within the scope of license renewal which are credited for SBO mitigation. The CLB documents used by the applicant include the plant-specific SBO calculations, UFSAR, drawings, modifications, the site equipment database, and plant procedures. On a sampling basis, the staff reviewed these documents and the LRA, in conjunction with the scoping results, to validate the applicant's methodology. The staff finds that the scoping results included SSCs that perform intended functions meeting 10 CFR 50.63 requirements. The staff determined that the applicant's scoping methodology is adequate for identifying SSCs within the scope of license renewal which are credited in complying with the SBO regulation.

2.1.4.3.3   Conclusion

On the basis of the sample reviews, discussions with the applicant, review of the LRA, and review of the implementing procedures and reports, the staff concludes that the applicant's methodology for identifying systems and structures meets the scoping criteria pursuant to 10 CFR 54.4(a)(3) and, therefore, is acceptable.

2.1.4.4   Plant-Level Scoping of Systems and Structures

2.1.4.4.1   Summary of Technical Information in the Application

System and Structure Level Scoping. LRA Section 2.1, "Scoping and Screening Methodology," states the following:

> For systems, structures and components (SSCs) within the scope of license renewal, 10 CFR 54.21(a)(1) requires the license renewal applicant to identify and list the structures and components subject to an aging management review (AMR). 10 CFR 54.21(a)(2) further requires that the methods used to implement the requirements of 10 CFR 54.21(a)(1) be described and justified. This section of the application provides a description and justification of the methodology used to identify and list structures and components that are within the scope of license renewal and subject to an AMR. PVNGS Unit 1, Unit 2, and Unit 3 are constructed of similar materials with similar environments. Unless otherwise noted throughout this application, plant systems and structures discussed in this application apply to PVNGS Units 1, 2 and 3.

LRA Section 2.1.1, "Introduction," states the following:

> The first step in the integrated plant assessment (IPA) process identified the plant SSCs within the scope of 10 CFR 54. This step is called scoping. For those SSCs identified to be within the scope of the license renewal rule, the second step of the IPA process then identified and listed the structures and components that are subject to an AMR. This step of the process is called screening. The scoping and screening steps have been performed consistent with the requirements of 10 CFR 54, the Statements of Consideration supporting the license renewal rule, and the guidance provided in NEI 95-10, "Industry Guideline for Implementing the Requirements of 10 CFR Part 54 - The License Renewal Rule." Section 2.1.1.1 provides a discussion of the documentation used to perform scoping and screening.

LRA Section 2.1.3, "Scoping Methodology," states the following:

> Scoping of the SSCs was performed to the criteria of 10 CFR 54.4(a) to identify those SSCs within the scope of the license renewal rule. The scoping evaluation results have been retained in the license renewal database. The following sections describe the methodology used for scoping. Separate discussions of mechanical system scoping methodology, structures scoping methodology, and electrical and I&C system scoping methodology are provided.

2.1.4.4.2   Staff Evaluation

The staff reviewed the applicant's methodology for the scoping of plant systems and components to ensure consistency with 10 CFR 54.4. The applicant documented the methodology used to determine the systems and components within the scope of license renewal in implementing procedures and scoping result reports for each system. The scoping process defined the plant in terms of systems and structures. Specifically, the implementing procedures identify the systems and structures that are subject to 10 CFR 54.4 review, describe the processes for capturing the results of the review, and determine if the system or structure performs intended functions consistent with the criteria of 10 CFR 54.4(a). The applicant completed the scoping process for all systems and structures to ensure that it addressed the entire plant.

The applicant documented the results of the plant-level scoping process in accordance with the implementing documents and provided the results in the systems and structures documents and reports. These results included information such as a description of the structure or system, a listing of functions performed by the system or structure, description of intended functions, 10 CFR 54.4(a) scoping criteria met by the system or structure, references, and the basis for the intended function classification of the system or structure. During the audit, the staff reviewed a sampling of the documents and reports and concluded that the applicant's scoping results contained an appropriate level of detail to document the scoping process.

2.1.4.4.3 Conclusion

Based on its review of the LRA, site guidance documents, and a sampling of system scoping results reviewed during the audit, the staff concludes that the applicant's methodology for identifying SSCs, and their intended functions, within the scope of license renewal is consistent with the requirements of 10 CFR 54.4 and, therefore, is acceptable.

2.1.4.5 Mechanical Scoping

2.1.4.5.1 Summary of Technical Information in the Application

LRA Section 2.1.3.1 states the following:

> A list of all mechanical systems was developed using the plant equipment database and system plant numbering procedure and is documented in the Plant Systems and Aging Management Programs Position Paper. These mechanical systems were evaluated to each of the criteria of 10 CFR 54.4(a).

LRA Section 2.1.3.1 further states that for each system, the applicant performed the following:

> Identification of the System Purpose and Functions. A description was prepared for each mechanical system that included the purpose and summarized the functions that the system was designed to perform. This summary description was prepared using information obtained from the UFSAR system descriptions, current licensing basis documents, design basis documents (including piping and instrumentation drawings), and system operating descriptions.
>
> Determination of the System Evaluation Boundary. After the system functions were identified, the system evaluation boundary was determined and marked-up on piping and instrumentation drawings. All of the components needed for the system to perform its intended functions are included within the license renewal boundary.
>
> Comparison of System Functions Against 10 CFR 54.4(a)(1-3). All system functions were compared against the criteria of 10 CFR 54.4(a)(1), (a)(2) and (a)(3). The system functions were identified from the information sources previously described. Each of the system functions satisfying the scoping criteria in 10 CFR 54.4(a) was identified as a system intended function. Functions performed by safety-related portions of the evaluated system were identified as satisfying criterion (a)(1) and were classified as intended functions. Functions performed by nonsafety-related systems or parts of systems that are required to ensure success of a safety-related function were identified as satisfying criterion

Scoping and Screening Methodology

(a)(2) and classified as intended functions. Functions that were credited in one of the regulated events were identified as satisfying criterion (a)(3) and classified as intended functions. A function may have been classified as an intended function under more than one of the three criteria in 10 CFR 54.4. Any system that performed one or more intended functions (i.e., satisfying criterion (a)(1), (a)(2), or (a)(3)) was classified as a system within the scope of the license renewal rule.

Identification of Supporting Systems. After a system was determined to be in the scope of the Rule for criteria (a)(1) or (a)(3), an evaluation was performed to identify all of its supporting systems. Each of the supporting systems was then reviewed to determine if its failure could prevent satisfactory accomplishment of any intended functions of the in-scope system. When it was determined that a supporting system was needed to maintain an intended function of the in-scope system, the supporting system was determined to be in-scope. When a supporting system was identified as being in-scope, the scoping evaluation for the supporting system was reviewed and revised as necessary. This step in the scoping process ensured that all supporting systems' intended functions were identified.

Component Level Scoping. A component was determined to be in-scope if it was determined that the component was needed to fulfill a system intended function meeting the safety-related criteria of 10 CFR 54.4(a)(1), the nonsafety-related affecting safety-related criterion of 10 CFR 54.4(a)(2), and/or if the component was needed to support the criteria of 10 CFR 54.4(a)(3) for regulated events. Components meeting one of these three criteria were identified in the license renewal database as within the scope of the Rule. Components not meeting one of these three criteria were identified in the license renewal database as out of scope.

2.1.4.5.2 Staff Evaluation

The staff evaluated LRA Section 2.1.3.1 and the guidance in the implementing procedures and reports to review the mechanical scoping process. The project documents and reports provided instructions for identifying the evaluation boundaries. The staff reviewed the implementing documents and the CLB documents associated with mechanical system scoping, and found that the guidance and CLB source information noted above are acceptable to identify mechanical components and support structures in mechanical systems that are within the scope of license renewal. The staff had detailed discussions with the applicant's license renewal project personnel and reviewed documentation pertinent to the scoping process. The staff assessed if the applicant had appropriately applied the scoping methodology outlined in the LRA and implementing procedures and if the scoping results were consistent with CLB requirements. The staff determined that the applicant's procedure is consistent with the description provided in LRA Section 2.1.3.1 and the guidance contained in the SRP-LR, Section 2.1, and that the applicant adequately implemented this procedure.

On a sampling basis, the staff reviewed the applicant's scoping reports for the safety injection and shutdown cooling system, diesel generator fuel oil storage and transfer system, and auxiliary feedwater system mechanical component types that met the scoping criteria of 10 CFR 54.4. The staff also reviewed the implementing procedures and discussed the methodology and results with the applicant. The staff verified that the applicant had identified and used pertinent engineering and licensing information in determining the safety injection and shutdown cooling system, diesel generator fuel oil storage and transfer system, and auxiliary feedwater system mechanical component types required to be within the scope of license renewal. As part of the review process, the staff evaluated each system's intended function, the basis for inclusion of the intended function, and the process used to identify each component type.

The staff verified that the applicant had identified and highlighted system P&IDs to develop the license renewal boundaries in accordance with the procedural guidance. Additionally, the staff determined that the applicant had independently verified the results in accordance with the governing procedures. The staff confirmed that knowledgeable personnel, familiar with plant systems and license renewal, were utilized by the applicant to perform independent reviews of the marked-up drawings to ensure accurate identification of system intended functions. The staff also confirmed that the applicant had performed additional cross-discipline verification and independent reviews of the resultant highlighted drawings before final approval of the scoping effort.

2.1.4.5.3   Conclusion

On the basis of its review of the LRA, scoping implementing procedures, and the sampling system review of mechanical scoping results, the staff concludes that the applicant's methodology for identifying mechanical SSCs within the scope of license renewal is in accordance with the requirements of 10 CFR 54.4 and, therefore, is acceptable.

2.1.4.6   Structural Scoping

2.1.4.6.1   Summary of Technical Information in the Application

LRA Section 2.1.3.2, "Structure Scoping Methodology," states the following:

> A list of all structures was developed that included buildings, tank foundations, and other miscellaneous structures. The list of structures used for scoping was developed through review of site plot drawings in conjunction with a walkdown of the property. The UFSAR was relied upon to identify the safety classifications of structures and structural components. Category I structures and structural components were considered safety-related.
>
> Structure descriptions were prepared, including the structure purpose and all functions. Structure evaluation boundaries were determined, including examination of structure interfaces. This information was included in the license renewal database. All structure functions were evaluated against the criteria of 10 CFR 54.4(a)(1), (a)(2) and (a)(3) and the results of this evaluation were documented in the license renewal database. In those instances where the structure intended functions required support from other structures or systems, the supporting systems or structures were identified and evaluated against the criteria in 10 CFR 54.4(a)(2). A list of references supporting the evaluation of each structure was documented in the license renewal database.
>
> For structures determined to be within the scope of license renewal, structural drawings were reviewed to identify structural elements (such as steel structures, foundations, floors, walls, ceilings, penetrations, stairways or curbs). For in-scope structures, all structural components that are required to support the intended functions of the structure were entered into the license renewal database and were identified as in-scope of license renewal. Some individual structural components fabricated from the same material and exposed to the same environment were replaced in the database with a generic component, such as "Structural Steel" to represent all of the carbon steel beams and columns in a given building. For each in-scope structure, all of the structural components listed in the license renewal database were evaluated and a determination was made as to whether the structural component was required to support the intended functions of the structure. Structural components that support the

intended functions of the structure were identified in the license renewal database as within the scope of license renewal.

2.1.4.6.2 Staff Evaluation

The staff evaluated LRA Section 2.1.3.2 and the guidance in the implementing procedures and reports to review the structural scoping process. The project documents and reports provided instructions for identifying the evaluation boundaries. The staff reviewed the applicant's approach for identifying structures relied upon to perform the functions described in 10 CFR 54.4(a). As part of this review, the staff discussed the methodology with the applicant, reviewed the documentation developed to support the review, and evaluated the scoping results for a sample of structures (e.g., turbine building) that were identified within the scope of license renewal. The staff determined that the applicant had identified and developed a list of plant structures and the structures intended functions through a review of the site equipment database, UFSAR, drawings, system notebook, and walkdowns. Each structure the applicant identified was evaluated against the criteria of 10 CFR 54.4(a)(1), (a)(2), and (a)(3).

The staff reviewed selected portions of the plant equipment database, CLB information, drawings, and implementing procedures to verify the adequacy of the methodology. The staff reviewed the applicant's methodology for identifying structures meeting the scoping criteria as defined in 10 CFR 54.4(a). The staff also reviewed the scoping methodology implementing procedures and discussed the methodology and results with the applicant. In addition, on a sampling basis, the staff reviewed the applicant's scoping reports, including information contained in the source documentation, for the turbine building, to verify that the application of the methodology would provide the results as documented in the LRA.

The staff verified that the applicant identified and used pertinent engineering and licensing information in order to determine that the turbine building was required to be included in the scope of license renewal. As part of the review process, the staff evaluated the intended functions identified for the turbine building, the structural components within the building, the basis for inclusion of the intended function, and the process used to identify each of the component types.

2.1.4.6.3 Conclusion

On the basis of its review of information in the LRA, scoping implementation procedures, and a sampling review of structural scoping results for the turbine building, the staff concludes that the applicant's methodology for identification of the structural SSCs within the scope of license renewal is in accordance with the requirements of 10 CFR 54.4 and, therefore, is acceptable.

2.1.4.7 Electrical Component Scoping

2.1.4.7.1 Summary of Technical Information in the Application

LRA Section 2.1.3.3, "Electrical and I&C System Scoping Methodology" states the following:

> Scoping process for electrical and instrumentation and control (I&C) systems was performed by a system level scoping and then a component level scoping. System level scoping methodology was performed similar to the mechanical system-level scoping, which utilized the UFSAR descriptions, database records, current licensing basis documents and design basis documents that were applicable to the system. Furthermore, the system safety classification and all systems functions were determined by reviewing these documents and were evaluated against the criteria of 10 CFR 54.4(a)(1), (a)(2) and (a)(3). The component level scoping methodology included all electrical and I&C components that perform an intended function as described in 10 CFR 54.4 for systems within the scope of license renewal. Furthermore, the scoping of

installed electrical components were identified by reviewing plant drawings and databases and a list of typical electrical components provided by NEI 95-10 was utilized to determine which electrical component types are installed. Any electrical component types that are installed but were not listed in the plant equipment database were added into the license renewal database for evaluation during component screening.

2.1.4.7.2   Staff Evaluation

The staff evaluated LRA Section 2.1.3.3 and the guidance contained in the implementing procedures and reports to review the electrical scoping process. The staff reviewed the applicant's approach to identifying electrical and I&C SSCs relied upon to perform the functions described in 10 CFR 54.4(a). The staff reviewed portions of the documentation used by the applicant to perform the electrical scoping process including the UFSAR, plant equipment database, CLB documentation, documents, procedures, drawings, specifications, codes, and standards.

The staff noted that after the applicant completed the electrical and I&C component scoping, it categorized in-scope electrical components into component types. These component types include electrical and I&C components with common characteristics and functions, such as cable, connections, fuse holders, terminal blocks, high-voltage transmission conductor, connections and insulators, metal enclosed bus, switchyard bus, and connections.

As part of this review, the staff discussed the methodology with the applicant, reviewed the implementing procedures developed to support the review, and evaluated a sampling of the scoping results of SSCs that were identified as within the scope of license renewal. The staff determined that the applicant had included electrical and I&C components, including electrical and I&C components contained in mechanical or structural systems, within the scope of license renewal on a commodity basis.

2.1.4.7.3   Conclusion

On the basis of its review of information contained in the LRA, scoping implementing procedures, scoping bases documents, and review of a sample of electrical scoping results, the staff concludes that the applicant's methodology for the scoping of electrical components within the scope of license renewal is in accordance with the requirements of 10 CFR 54.4, and therefore, is acceptable.

2.1.4.8   Scoping Methodology Conclusion

On the basis of its review of the LRA, implementing procedures, and a review of a sample of scoping results, the staff concludes that the applicant's scoping methodology is consistent with the guidance contained in the SRP-LR, and identified those SSCs, (1) that are safety-related, (2) whose failure could affect safety-related functions, and (3) that are necessary to demonstrate compliance with the NRC's regulations for fire protection, EQ, PTS, ATWS, and SBO. The staff concluded that the applicant's methodology is consistent with the requirements of 10 CFR 54.4(a), and, therefore, is acceptable.

2.1.5   Screening Methodology

2.1.5.1   General Screening Methodology

2.1.5.1.1   Summary of Technical Information in the Application

LRA Section 2.1.4, "Screening Methodology," and accompanying subsections, describe the screening process that identifies the SCs within the scope of license renewal that are subject to an AMR. Section 2.1.4 states the following:

Scoping and Screening Methodology

> Screening is the process of identifying, and listing the structures and components that are subject to an aging management review. This section, and the accompanying subsections for mechanical systems, electrical and instrument and control systems, and structures, describes the process used to perform screening. All SSCs listed in the license renewal database were scoped to the criteria of 10 CFR 54.4(a). All of the structures and components categorized as within the scope of license renewal were screened against the criteria of 10 CFR 54.21(a)(1)(i) and (1)(ii) to determine whether they are subject to aging management review. The screening methodology utilized is described in this section of the application. The word "passive" is used in the screening process for all components that perform intended functions without moving parts, or a change in configuration or properties. All components that are not "passive" are known as "active". The word "long-lived" is used in the screening process for all components that are not subject to replacement based on qualified life or specific time period. Components that are not "long-lived" are known as "short-lived."
>
> NEI 95-10, Appendix B, *"Typical Structure, Component and Commodity Groupings and Active/Passive Determinations for the Integrated Plant Assessment,"* provides industry guidance for screening structures and components. The guidance provided in NEI 95-10, Appendix B, has been incorporated into the license renewal screening process. Slightly differing screening methodologies have been applied for mechanical systems, electrical and instrument and control systems, and structures.

2.1.5.1.2   Staff Evaluation

Under 10 CFR 54.21, each LRA must contain an IPA identifying SCs within the scope of license renewal that are subject to an AMR. The IPA must identify components that perform an intended function without moving parts or a change in configuration or properties (passive), as well as components that are not subject to periodic replacement based on a qualified life or specified period (long-lived). In addition, the IPA must include a description and justification of the methodology used to determine the passive and long-lived SCs and a demonstration that the effects of aging on those SCs will be adequately managed so that the intended functions will be maintained under all design conditions imposed by the plant-specific CLB for the period of extended operation.

The staff reviewed the methodology used by the applicant to identify the mechanical and structural components and electrical commodity groups within the scope of license renewal that should be subject to an AMR. The applicant implemented a process for determining which SCs are subject to an AMR in accordance with the requirements of 10 CFR 54.21(a)(1). In LRA Section 2.1.4, the applicant discusses these screening activities as they relate to the component types and commodity groups within the scope of license renewal.

The staff determined that the screening process used by the applicant evaluated the component types and commodity groups included within the scope of license renewal to determine which ones were long-lived and passive and, therefore subject to an AMR. The staff reviewed LRA Section 2.3, "Scoping and Screening Results: Mechanical Systems," Section 2.4, "Scoping and Screening Results: Structures," and Section 2.5, "Scoping and Screening Results: Electrical and Instrumentation and Controls Systems." These sections provide the results of the process used to identify component types and commodity groups subject to an AMR. The staff also reviewed, on a sampling basis, the screening results reports for safety injection and shutdown cooling, diesel generator fuel oil storage and transfer, auxiliary feedwater, and turbine building.

The applicant provided the staff with a detailed discussion of the processes used for each discipline and provided administrative documentation that described the screening

Scoping and Screening Methodology

methodology. SER Section 2.1.5.2, "Mechanical Component Screening," Section 2.1.5.3, "Structural Component Screening," and Section 2.1.5.4, "Electrical Component Screening," below, discuss specific methods for mechanical, electrical, and structural component screening.

2.1.5.1.3 Conclusion

On the basis of its review of the LRA, the implementing procedures, and a review of a sample of screening results, the staff concludes that the applicant's screening methodology is consistent with the guidance contained in the SRP-LR and identifies passive and long-lived components in the scope of license renewal that are subject to an AMR. The staff concludes that the applicant's process for determining which component types and commodity groups subject to an AMR is consistent with the requirements of 10 CFR 54.21 and, therefore, is acceptable.

2.1.5.2 Mechanical Component Screening

2.1.5.2.1 Summary of Technical Information in the Application

LRA Section 2.1.4.1 states the following:

> In mechanical systems, component screening was a continuation of the component scoping activity. After a mechanical system component was categorized in the license renewal database as in scope, the classification as an active or passive component was determined based on evaluation of the component description and type. The active/passive component determinations documented in NEI 95-10, revision 6, Appendix B, provided guidance for this activity. In-scope components that were determined to be passive and long-lived were identified in the license renewal database as subject to aging management review.
>
> Each component that was identified as subject to an aging management review was evaluated to determine its component intended function(s). The component intended function(s) was identified based on an evaluation of the component type and the way(s) in which the component supports the system intended functions. Most in-scope passive components perform only one intended function. However, a few in-scope component types may perform more than one function. The results of the component screening were recorded in the license renewal database. The list of component intended functions utilized in the screening of mechanical system components can be found in Table 2.1-1, "Intended Functions Abbreviations and Definitions."
>
> During the screening process, a few in-scope passive components were identified in the screening process as short-lived components. Components that were identified during screening as short-lived were eliminated from the aging management review process and the basis for the classification as short-lived was documented in the license renewal database. All other in-scope passive components were identified in the license renewal database as subject to an aging management review. During the aging management review process, if detailed review of maintenance procedures and requirements determined that a component previously categorized as long-lived was subject to replacement based on a qualified life or specified time period; the component was re-categorized as short-lived and eliminated from the aging management review evaluation process.

2.1.5.2.2 Staff Evaluation

The staff reviewed the mechanical screening methodology discussed in LRA Section 2.1.4.1, the implementing documents, the scoping and screening reports, and the license renewal

Scoping and Screening Methodology

drawings. The staff determined that the mechanical system screening process began with the results from the scoping process and that the applicant reviewed each system evaluation boundary as depicted on the P&IDs to identify passive and long-lived components. Additionally, the staff determined that the applicant had identified all passive and long-lived components that perform or support an intended function within the system evaluation boundaries and determined that those components are subject to an AMR. The results of the review are documented in the scoping and screening reports, which contain information such as the information sources reviewed and the intended functions of the component.

The staff verified that the applicant established mechanical system evaluation boundaries for each system within the scope of license renewal and that it determined these boundaries by mapping the system boundary onto P&IDs. The staff confirmed that the applicant reviewed the components within the system boundary to determine whether the component supported the system intended function. The staff further confirmed that the applicant reviewed those components that supported the system's intended function to determine whether the component was passive and long-lived and, therefore, subject to an AMR.

The staff reviewed selected portions of the UFSAR, plant equipment database, CLB documentation, databases and documents, procedures, drawings, specifications, and selected scoping and screening reports. The staff conducted detailed discussions with the applicant's license renewal team and reviewed documentation pertinent to the screening process. The staff also performed a walkdown of portions of the selected systems with plant engineers to verify documentation. The staff assessed whether the mechanical screening methodology outlined in the LRA and the associated procedures were appropriately implemented. The staff also assessed if the scoping results were consistent with CLB requirements. During the scoping and screening methodology audit, the staff discussed the screening methodology with the applicant and, on a sampling basis, reviewed the applicant's screening reports for the safety injection and shutdown cooling system, diesel generator fuel oil storage and transfer system, and auxiliary feedwater system to verify proper implementation of the screening process. Based on these audit activities, the staff did not identify any discrepancies between the methodology documented and the implementation results.

2.1.5.2.3   Conclusion

On the basis of its review of the LRA, discussions with the applicant, screening implementation procedures, UFSAR, plant equipment database, CLB documentation, procedures, drawings, specifications, selected scoping and screening reports, and review of a sample of the safety injection and shutdown cooling system, diesel generator fuel oil storage and transfer system, and auxiliary feedwater system screening reports, the staff concludes that the applicant's methodology for identification of mechanical components within the scope of license renewal and subject to an AMR is in accordance with the requirements of 10 CFR 54.21(a)(1) is acceptable.

2.1.5.3   Structural Component Screening

2.1.5.3.1   Summary of Technical Information in the Application

LRA Section 2.1.4.2 states the following:

> Structures and structural components typically perform their functions without moving parts and without a change in configuration or properties. When a structure or structural component was determined to be in-scope of license renewal by the scoping process described in Section 2.1.3.2, the structure screening methodology classified the component as passive. This is consistent with guidance found in NEI 95-10, Revision 6, Appendix B. During the structural screening process, the intended function(s) of structural components were

2-30

Scoping and Screening Methodology

determined and recorded in the license renewal database. In the structure screening process, an evaluation was made to determine whether in-scope structural components were subject to replacement based on a qualified time period. If an in-scope structural component was determined to be subject to replacement based on a qualified time period, the component was identified as short-lived and was excluded from an AMR. In such a case, the basis for determining that the structural component was short-lived was documented in the license renewal database.

2.1.5.3.2   Staff Evaluation

The staff reviewed the screening methodology used to identify structural components that are subject to an AMR as required in 10 CFR 54.21(a)(1) as discussed in LRA Sections 2.1.4.2, implementing procedures, scoping and screening reports, and license renewal drawings. The staff confirmed that the applicant had reviewed the structures included within the scope of license renewal, identified the passive, long-lived components with component level intended functions, and determined that those components are subject to an AMR.

The staff reviewed selected portions of the UFSAR, structure system information, the plant equipment database, and scoping and screening reports that the applicant used to perform the structural scoping and screening. The staff also reviewed screening activities on a sampling basis that documented the SCs within the scope of license renewal. The staff had detailed discussions with the applicant's license renewal team and reviewed documentation pertinent to the screening process to assess if the screening methodology outlined in the LRA and associated procedures were appropriately implemented and if the scoping results are consistent with CLB requirements.

During the scoping and screening methodology audit, the staff reviewed, on a sampling basis, the applicant's screening reports for the turbine building to verify proper implementation of the screening process. Based on these onsite review activities, the staff did not identify any discrepancies between the methodology documented and the implementation results.

2.1.5.3.3   Conclusion

Based on the staff's review of information contained in the LRA, implementing procedures, plant equipment database, and a sampling of the turbine building screening results, the staff concludes that the applicant's methodology for identifying structural components within the scope of license renewal and subject to an AMR is in accordance with the requirements of 10 CFR 54.21(a)(1) and, therefore, is acceptable.

2.1.5.4   Electrical Component Screening

2.1.5.4.1   Summary of Technical Information in the Application

LRA Section 2.1.4.3, "Electrical and I&C System Component Screening Methodology," states that the screening of electrical and I&C components used a "spaces" approach that is consistent with the guidance in NEI 95-10. This approach is based on identifying areas where the bounding environmental conditions exist. It also states that the bounding environmental conditions are applied during the AMR in order to evaluate the aging effects that are associated with electrical component types that are located within the bounding area. The LRA states that the use of the "spaces" approach for AMR of electrical components types eliminates the need to associate electrical and I&C components with specific systems that are within the scope of license renewal.

The LRA also states that the applicant categorized in-scope electrical components as "active" or "passive" based on the determinations documented in NEI 95-10, Appendix B. Furthermore, the LRA states that passive long-lived electrical and I&C components that perform an intended

Scoping and Screening Methodology

function without moving parts, or without change in configuration or properties, were grouped into component types such as cable, connections, fuse holders, terminal blocks, high-voltage transmission conductor, connections and insulators, metal enclosed bus, switchyard bus, and connections. The LRA states that component-level intended function(s) were determined for each in-scope passive electrical component type, identified in the license renewal database as being subject to AMR, and recorded in the license renewal database.

LRA Table 2.5-1, "Electrical and I&C Component Groups Requiring Aging Management Review," lists that the resulting AMR electrical commodity groups of long-lived passive components subject to an AMR are as follows:

- Cable connections (metallic parts)
- Connector
- High voltage insulator
- Insulated cables and connections
- Metal enclosed bus (bus or connections)
- Metal enclosed bus (enclosure)
- Metal enclosed bus (insulation and insulators)
- Penetrations electrical
- Switchyard bus and connections
- Terminal block
- Transmission conductors and connections

2.1.5.4.2   Staff Evaluation

The staff reviewed the applicant's methodology used for electrical component screening in LRA Section 2.1.4.3, "Electrical and I&C System Component Screening Methodology," and Section 2.5, "Scoping and Screening Results: Electrical and Instrumentation and Control Systems," the applicant's implementing procedures, bases documents, and electrical AMR reports. The staff confirmed that the applicant used the screening process described in these documents, along with the information contained in NEI 95-10, Appendix B and the SRP-LR, to identify the electrical and I&C components subject to an AMR.

The staff determined that the applicant identified commodity groups found to meet the passive criteria described in NEI 95-10. In addition, the staff determined that the applicant evaluated the identified passive commodities to identify if they were subject to replacement based on a qualified life or specified period (short-lived) or not subject to replacement based on a qualified life or specified period (long-lived). The applicant determined that the remaining passive, long-lived components were subject to an AMR.

The staff performed a review to determine if the screening methodology outlined in the LRA and implementing procedures were appropriately implemented and if the scoping results are consistent with CLB requirements. During the scoping and screening methodology audit, the staff reviewed selected screening reports and discussed the reports with the applicant to verify proper implementation of the screening process. Based on these onsite review activities, the staff did not identify any discrepancies between the methodology documented and the implementation results.

2.1.5.4.3   Conclusion

On the basis of its review of the LRA, the screening implementation procedures, selected portions of the UFSAR, plant equipment database, CLB documentation, procedures, drawings, specifications and selected scoping and screening reports, a sample of the results of the screening methodology, and discussion with the applicant, the staff concludes that the applicant's methodology for identification of electrical components within the scope of license renewal and subject to an AMR is in accordance with the requirements of 10 CFR 54.21(a)(1) and, therefore, is acceptable.

2.1.5.5 Screening Methodology Conclusion

On the basis of its review of the LRA, the screening implementing procedures, a sample review of screening results, and discussions with the applicant's staff, the staff concludes that the applicant's screening methodology is consistent with the guidance contained in the SRP-LR and identifies those passive, long-lived components within the scope of license renewal that are subject to an AMR. The staff concludes that the applicant's methodology is consistent with the requirements of 10 CFR 54.21(a)(1) and, therefore, is acceptable.

2.1.6 Summary of Evaluation Findings

On the basis of its review of the information presented in LRA Section 2.1, the supporting information in the scoping and screening implementing procedures and reports, the information presented during the scoping and screening methodology audit, the applicant's responses to the staff's RAIs, sample system reviews, and discussions with the applicant, the staff confirms that the applicant's scoping and screening methodology is consistent with the requirements of 10 CFR 54.4 and 10 CFR 54.21(a)(1). The staff also concludes that the applicant's description and justification of its scoping and screening methodology are adequate to meet the requirements of 10 CFR 54.21(a)(1). Based on this review, the staff concludes that the applicant's methodology for identifying systems and structures within the scope of license renewal and SCs requiring an AMR, is acceptable.

2.2 Plant-Level Scoping Results

2.2.1 Introduction

In LRA Section 2.1, the applicant described the methodology for identifying SSCs within the scope of license renewal. In LRA Section 2.2, the applicant describes the results of the application of its scoping methodology to determine which SSCs must be included within the scope of license renewal. The staff reviewed the plant-level scoping results to determine whether the applicant has properly identified all systems and structures relied upon to mitigate DBEs, as required by 10 CFR 54.4(a)(1). In addition, the staff ensured that the applicant noted all systems and structures that, if they failed, could prevent satisfactory accomplishment of any safety-related functions, as required by 10 CFR 54.4(a)(2) and systems and structures relied on in safety analyses or plant evaluations to perform functions required by regulations referenced in 10 CFR 54.4(a)(3).

2.2.2 Summary of Technical Information in the Application

In LRA Table 2.2-1, the applicant listed the plant mechanical systems, structures, and electrical and I&C systems within the scope of license renewal. Based on the DBEs considered in the plant's CLB, other CLB information relating to nonsafety-related systems and structures, and certain regulated events, the applicant identified plant-level systems and structures within the scope of license renewal as defined by 10 CFR 54.4.

2.2.3 Staff Evaluation

In LRA Section 2.1, the applicant described its methodology for identifying systems and structures within the scope of license renewal and subject to an AMR. The staff's evaluation is in SER Section 2.1. To verify that the applicant properly implemented its methodology, the staff's review focused on the implementation results shown in LRA Table 2.2-1 to confirm that there were no omissions of plant-level systems or structures from the scope of license renewal.

Scoping and Screening Methodology

The staff determined if the applicant properly identified the systems and structures within the scope of license renewal in accordance with 10 CFR 54.4. The staff reviewed selected systems and structures that the applicant did not identify as within the scope of license renewal to verify if the systems and structures have any intended functions requiring their inclusion within the scope of license renewal. The staff's review of the applicant's implementation was conducted in accordance with the guidance in SRP-LR Section 2.2, "Plant-Level Scoping Results."

The staff's review of LRA Section 2.2 identified an area where additional information was necessary to complete the review of the applicant's scoping and screening results. The applicant responded to the staff's RAI as discussed below.

In RAI 2.2-1, dated December 3, 2009, the staff noted the following UFSAR systems and structures (see Table 2.2-1) could not be located in LRA Table 2.2-1.

Table 2.2-1 Missing Systems or Structures in Table 2.2-1 of the LRA

UFSAR Section	System or Structures
Section 12.2, "Radiation Sources," Table 12.2-11, "Systems Used In Post-Accident Shielding Review"	Post-Accident Sampling System
Section 14.2.12, "Individual Test Descriptions"	Post-Accident Monitoring System
Section 7.2.5, "Supplementary Protection System"	Supplementary Protection System
Section 3.2, "Classification of Structures, Components, and Systems," Table 3.2-1, "Quality Classification of Structures, Systems, and Components", item 25. Structures	Equipment building

The staff asked the applicant to provide the reasoning for not including the above systems and structures in LRA Table 2.2-1.

In its response, by letter dated January 18, 2010, the applicant stated the following:

> The Post Accident Sampling System has been removed from the licensing and technical basis of the Palo Verde plant.
>
> The Post Accident Monitoring System refers to the post accident monitoring instrumentation. UFSAR Table 1.8-1 identifies Post Accident Monitoring instrumentation in 30 plant systems which perform the Post Accident Monitoring function. These 30 systems are included in LRA Table 2.2-1, Scoping Results.
>
> Supplementary Protection System
>
> UFSAR Section 7.2.5 identifies the Supplementary Protection System as part of the Reactor Protection System. The Reactor Protection System is identified in LRA Table 2.2-1 as a system within the scope of license renewal.
>
> Equipment Building
>
> UFSAR Table 3.2-1, "Quality Classification of Structures, Systems, and Components," identifies the Equipment Building as part of the Containment Building. The Containment Building is identified in LRA Table 2.2-1 as a structure within the scope of license renewal.

Based on its review, the staff finds the applicant's response to RAI 2.2-1 acceptable because the applicant clarified why the post-accident sampling system, post-accident monitoring, supplementary protection system, and equipment building are not included in Table 2.2-1. Therefore, the staff's concern described in RAI 2.2-1 is resolved.

2.2.4 Conclusion

The staff reviewed LRA Section 2.2, the RAI response, and the UFSAR supporting information to determine if the applicant failed to identify any systems and structures within the scope of license renewal. The staff finds no such omissions. Based on its review, the staff concludes that the applicant adequately identified, in accordance with 10 CFR 54.4, the systems and structures within the scope of license renewal.

2.3 Scoping and Screening Results: Mechanical Systems

This section documents the staff's review of the applicant's scoping and screening results for mechanical systems. Specifically, this section discusses the following:

- Reactor vessel, internals, and reactor coolant system (RCS)
- Engineered safety features (ESFs)
- Auxiliary systems
- Steam and power conversion systems

In accordance with the requirements of 10 CFR 54.21(a)(1), the applicant must list passive, long-lived SCs within the scope of license renewal and subject to an AMR. To verify that the applicant properly implemented its methodology, the staff's review focused on the implementation results. This focus allowed the staff to verify that the applicant identified the mechanical system SCs that meet the scoping criteria and are subject to an AMR, confirming that there were no omissions.

The staff evaluated mechanical systems using the evaluation methods described here (Section 2.3), in the guidance in SRP-LR Section 2.3, and, where applicable, the system functions as described in the UFSAR. The objective was to determine if the applicant has identified, in accordance with 10 CFR 54.4, components and supporting structures for mechanical systems that meet the license renewal scoping criteria. Similarly, the staff evaluated the applicant's screening results to verify that all passive, long-lived components are subject to an AMR, as required by 10 CFR 54.21(a)(1).

In its scoping evaluation, the staff reviewed the LRA, applicable sections of the UFSAR, and license renewal boundary drawings, and other licensing basis documents, as appropriate, for each mechanical system within the scope of license renewal. The staff reviewed relevant licensing basis documents for each mechanical system to confirm that the LRA specified all intended functions defined by 10 CFR 54.4(a). The staff's review then focused on identifying any components with intended functions defined by 10 CFR 54.4(a) that the applicant may have omitted from the scope of license renewal.

After reviewing the scoping results, the staff evaluated the applicant's screening results. For those SCs with intended functions delineated under 10 CFR 54.4(a), the staff verified the applicant properly screened out only SCs that have functions performed with moving parts or a change in configuration or properties or SCs that are subject to replacement after a qualified life or specified time period, as described in 10 CFR 54.21(a)(1). The staff confirmed the remaining SCs received an AMR, as required by 10 CFR 54.21(a)(1). The staff asked for additional information to resolve any omissions or discrepancies noted.

2.3.1 Reactor Vessel, Internals, and Reactor Coolant System

LRA Section 2.3.1 identifies the reactor vessel, reactor vessel internals, and RCS SCs subject to an AMR for license renewal. The applicant described the supporting SCs of the reactor vessel, internals, and RCS in the following LRA sections:

- 2.3.1.1, "Reactor Vessel and Internals"

Scoping and Screening Methodology

- 2.3.1.2, "Reactor Coolant System"
- 2.3.1.3, "Pressurizer"
- 2.3.1.4, "Steam Generators"
- 2.3.1.5, "Reactor Core"

2.3.1.1 Reactor Vessel and Internals

2.3.1.1.1 Summary of Technical Information in the Application

LRA Section 2.3.1.1 states that the reactor is a PWR with two reactor coolant loops. It states that the reactor vessel is a vertically mounted cylindrical vessel with a hemispherical lower head welded to the vessel and a removable hemispherical upper closure head. The applicant describes the reactor internals as comprised of the following component groups: core support structure (CSS), upper guide structure (UGS), flow skirt, in-core instrumentation support structures that support and orient the fuel assemblies, and control element assemblies (CEA) and in-core instrumentation that guide the reactor coolant through the vessel. The applicant further states that upper flange of the core support barrel, which rests on a ledge in the reactor vessel, supports the CSS at its upper end. It states the CSS consists of the core support barrel assembly, the lower support structure assembly, and the core shroud assembly. Further, the UGS aligns and laterally supports the upper end of the fuel assemblies, maintaining the control element spacing, holding down the fuel assemblies during operation, preventing fuel assemblies from being lifted out of position during a severe accident condition, and protecting the control elements from the effects of coolant cross flow in the upper plenum. The UGS consists of the UGS support barrel assembly, the UGS CEA shroud assembly, and the UGS holddown ring. The flow skirt is a right circular cylinder, perforated with flow holes and reinforced with two stiffening rings. It reduces inequalities in core inlet flow distributions and prevents the formation of large vortices in the lower plenum. The in-core support system begins outside the pressure vessel, penetrates the bottom of the vessel boundary, and ends in the upper end of the fuel assembly.

The intended function of the reactor vessel is to support the reactor core and control rod drive mechanisms and to provide a pressure boundary for reactor coolant. The reactor internals support the core, maintain fuel alignment, direct coolant flow, and provide gamma and neutron shielding. Portions of the reactor vessel and internals support fire protection, PTS, and SBO requirements.

LRA Table 2.3.1-1 lists the component types that require an AMR as follows.

The following components are in the reactor vessel:

- Control element drive mechanism (CEDM) housing (upper and lower)
- CEDM nozzles
- Closure bolting
- Closure head bolts
- Flange leak monitoring tube
- Head vent penetration
- In-core instrument guide tube
- In-core instrument nozzle
- Nozzle safe ends and welds
- Nozzles
- Shell
- Shell bottom head
- Support pads and shear keys

The following components are in the reactor vessel internals:

- Core stop lug and surveillance capsule holder
- CSS
- Flow skirt
- In-core instrument support structures
- UGS

2.3.1.1.2   Staff Evaluation and Conclusion

The staff reviewed LRA Section 2.3.1.1 and UFSAR Sections 3.9.5, 4.1, and 5.3 to determine whether the applicant failed to identify any components within the scope of license renewal. In addition, the staff's review determined if the applicant failed to identify any components subject to an AMR. The staff found no such omissions. On the basis of its review, the staff concludes the applicant has appropriately identified the reactor vessel system mechanical components within the scope of license renewal, as required by 10 CFR 54.4(a), and that the applicant has adequately identified the mechanical components subject to an AMR in accordance with the requirements stated in 10 CFR 54.21(a)(1).

2.3.1.2   Reactor Coolant System

2.3.1.2.1   Summary of Technical Information in the Application

LRA Section 2.3.1.2 states the following:

> The reactor is a pressurized water reactor with two coolant loops. The reactor coolant system circulates water in a closed cycle, removing heat from the reactor core and internals and transferring it to a secondary (steam generating) system. The steam generators provide the interface between the reactor coolant (primary) system and the main steam (secondary) system. Reactor coolant is prevented from mixing with the secondary steam by the steam generator tubes and the steam generator tube sheet.
>
> System pressure is controlled by the pressurizer, where steam and water are maintained in thermal equilibrium. Steam is formed by energizing immersion heaters in the pressurizer, or is condensed by the pressurizer spray to limit pressure variations caused by contraction or expansion of the reactor coolant.
>
> Reactor coolant loop penetrations include a charging and a letdown nozzle; the pressurizer surge line in one reactor vessel outlet pipe; the four safety injection inlet nozzles, one in each reactor vessel inlet pipe; two outlet nozzles to the shutdown cooling system, one in each reactor vessel outlet pipe; pressurizer spray nozzle; vent and drain connections; and sample and instrument connections.

LRA Section 2.3.1.2 goes on to state that the major components of the RCS are the reactor vessel and internals; two parallel coolant loops; a pressurizer connected to one of the reactor vessel outlet pipes; and associated piping, valves, and instrumentation. Each loop contains one steam generator (SG) and two reactor coolant pumps (RCPs). All components are located inside the containment building.

The LRA states that the intended functions of the RCS are to maintain RCS pressure during normal operation, maintain system integrity, transfer heat from the reactor to other systems during certain DBEs, act as a heat sink, allow for reactivity control, allow for removal of non-condensable gases, and provide a barrier against release of radioactivity generated within the reactor. Portions of the RCS are within the scope of license renewal as nonsafety-related affecting safety-related components. Portions of the RCS support fire protection, EQ, ATWS, and SBO requirements.

Scoping and Screening Methodology

LRA Table 2.3.1-2 lists the component types that require an AMR as follows:

- Class 1 piping (greater than or equal to 4 inches)
- Closure bolting
- Filter
- Flame arrestor
- Flexible hoses
- Heat exchanger (RCP high-pressure cooler)
- Heat exchanger (RCP seal cooler)
- Orifice
- Piping
- Pump
- Sight gauge
- Tank
- Thermo well
- Tubing
- Valve

2.3.1.2.2   Staff Evaluation and Conclusion

The staff reviewed LRA Section 2.3.1.2, UFSAR Sections 5.1 and 5.2, and the license renewal boundary drawings to determine if the applicant failed to identify any components within the scope of license renewal. In addition, the staff's review determined if the applicant failed to identify any components subject to an AMR. The staff found no such omissions. Based on its review, the staff concludes the applicant has appropriately identified the RCS mechanical components within the scope of license renewal, as required by 10 CFR 54.4(a), and that the applicant has adequately identified the mechanical components subject to an AMR in accordance with the requirements stated in 10 CFR 54.21(a)(1).

2.3.1.3   Pressurizer

2.3.1.3.1   Summary of Technical Information in the Application

LRA Section 2.3.1.3 states the following:

> The purpose of the pressurizer is to maintain the reactor coolant system operating pressure within acceptable limits. The pressurizer includes one pressurizer vessel connected to one of the primary coolant loops for each unit, and is part of the reactor coolant pressure boundary. The pressurizer contains components for maintaining reactor coolant system pressure, which consist of electric heaters to increase reactor coolant system pressure and an internal spray nozzle to reduce reactor coolant system pressure.
>
> The reactor coolant system contains the piping system components associated with the pressurizer, excluding the pressurizer vessel and its internals. The pressurizer is located in the containment building.

The LRA goes on to state that the intended functions of the pressurizer are to maintain the RCS operating pressure within acceptable limits to mitigate the consequences of accidents by regulating the temperature and pressure of the coolant. The pressurizer holds steam and water in thermal equilibrium. The pressurizer is part of the reactor coolant pressure boundary and supports fire protection and SBO requirements.

LRA Table 2.3.1-3 lists the component types that require an AMR as follows:

- Closure bolting

- Pressurizer heater bundle diaphragm plate
- Pressurizer heater sheaths and sleeves
- Pressurizer instrument penetrations
- Pressurizer integral support
- Pressurizer lower head
- Pressurizer manways and covers
- Pressurizer nozzle thermal sleeves
- Pressurizer nozzles
- Pressurizer safe ends
- Pressurizer shell and upper head

2.3.1.3.2 Staff Evaluation and Conclusion

The staff reviewed LRA Section 2.3.1.3, UFSAR Section 5.1, and UFSAR Section 5.4.10 to determine if the applicant failed to identify any components within the scope of license renewal. In addition, the staff's review determined if the applicant failed to identify any components subject to an AMR. The staff found no such omissions. On the basis of its review, the staff concludes the applicant has appropriately identified the pressurizer system mechanical components within the scope of license renewal, as required by 10 CFR 54.4(a), and that the applicant has adequately identified the mechanical components subject to an AMR in accordance with the requirements stated in 10 CFR 54.21(a)(1).

2.3.1.4 Steam Generators

2.3.1.4.1 Summary of Technical Information in the Application

LRA Section 2.3.1.4 states the following:

> The purpose of the steam generator system is to provide heat removal from the reactor coolant system through the generation of steam and also to act as an assured source of steam to the steam driven auxiliary feedwater pump. The system consists of the primary and secondary pressure boundaries of the steam generators including all pieces and parts within the pressure boundary and all penetrations out to the safe ends of the penetration nozzles.

The LRA goes on to state that the intended function of the SG is to provide heat removal from the coolant by the generation of steam for DBE mitigation, SBO, and fire safe shutdown requirements. The SG provides a source of steam to the turbine driven auxiliary feedwater pump. The SG primary channel head and tubes form part of the reactor coolant pressure boundary. The SG outlet nozzles restrict main steam flow in the event of a main steam line break. The SG system supports fire protection, ATWS, and SBO requirements.

LRA Table 2.3.1-4 lists the component types that require an AMR as follows:

- SG closure bolting
- SG feedring
- SG flow distribution baffle
- SG internal structures
- SG plugs and stakes
- SG primary head
- SG primary head divider plate
- SG primary manways and flanges
- SG primary nozzle dam retention ring
- SG primary nozzles and safe ends
- SG secondary manways and flanges

- SG secondary nozzles and safe ends
- SG secondary shell
- SG tubes
- SG tubesheet
- Tubing

2.3.1.4.2 Staff Evaluation and Conclusion

The staff reviewed LRA Section 2.3.1.4 and UFSAR Sections 5.4.2, 5.4.4, and 10.3 to determine if the applicant failed to identify any SSC within the scope of license renewal. In addition, the staff's review determined if the applicant failed to identify any components subject to an AMR. The staff found no such omissions. On the basis of its review, the staff concludes the applicant has appropriately identified the SG system mechanical components within the scope of license renewal, as required by 10 CFR 54.4(a) and that the applicant has adequately identified the mechanical components subject to an AMR in accordance with the requirements stated in 10 CFR 54.21(a)(1).

2.3.1.5 Reactor Core

2.3.1.5.1 Summary of Technical Information in the Application

LRA Section 2.3.1.5 states the following:

> The reactor core is composed of 241 fuel assemblies and 89 Control Element Assemblies (CEAs). The fuel assembly, which provides for 236 fuel rod and 20 guide tube positions (16 x 16 array), consists of 5 guide tubes welded to 11 fuel rod spacer grids and is closed at the top and bottom by end fittings. Each of the 5 guide tubes displace four fuel rod positions and provides guidance channels for the CEAs over their entire length of travel with in-core instrumentation inserted in the central guide tube of selected fuel assemblies. The in-core instrumentation is routed into the bottom of the fuel assemblies through the bottom head of the reactor vessel.
>
> Each fuel rod consists of slightly enriched uranium in the form of sintered uranium dioxide pellets, enclosed in a pressurized zircaloy or ZIRLO tube that forms a hermetic enclosure.

The LRA goes on to state that the intended system function of the reactor core is for each fuel rod to transfer heat to the coolant and cladding. The CEAs and guide tubes control short-term reactivity changes and are used for reactor shutdown. The initial reactor reactivity control relies on the CEAs that are inserted when the reactor trip breaker is de-energized. However, the CEA is short-lived, with a lifetime of about five cycles due to accumulative neutron burn-up. The fuel assemblies are also short-lived, since they are replaced at regular intervals based on plant fuel cycle. Therefore, the applicant stated that it found no components in the reactor core system subject to AMR.

2.3.1.5.2 Staff Evaluation and Conclusion

The staff reviewed LRA Section 2.3.1.5 and UFSAR Sections 4.1, 4.2, and 4.3 to determine if the applicant failed to identify any components within the scope of license renewal. In addition, the staff's review determined if the applicant failed to identify any components subject to an AMR. The staff found no such omissions. On the basis of its review, the staff concludes the applicant has appropriately identified that no mechanical components in the reactor core system are within the scope of license renewal, as required by 10 CFR 54.4(a). In addition, the staff finds that the applicant has adequately identified that there are no mechanical components subject to an AMR in accordance with the requirements stated in 10 CFR 54.21(a)(1).

2.3.2 Engineered Safety Features

LRA Section 2.3.2 identifies the ESF SCs subject to an AMR for license renewal. ESFs in nuclear plants mitigate the consequences of design-basis or loss-of-coolant accidents.

The applicant described the supporting SCs of the ESFs in the following LRA sections:

- 2.3.2.1, "Containment Leak Test System"
- 2.3.2.2, "Containment Purge System"
- 2.3.2.3, "Containment Hydrogen Control System"
- 2.3.2.4, "Safety Injection and Shutdown Cooling System"

The staff's findings, based on the review of LRA Sections 2.3.2.1–2.3.2.4, are in SER Sections 2.3.2.1–2.3.2.4, respectively.

2.3.2.1 Containment Leak Test System

2.3.2.1.1 Summary of Technical Information in the Application

LRA Section 2.3.2.1.1 describes the containment leak test system, which is comprised of filters, dryers, instrumentation, piping, and valves associated with delivering compressed air to the containment for conducting the integrated leak rate test. The LRA also states that the purpose of the system is to provide a means for periodic testing of containment leakage by pressurizing the containment building and monitoring leakage to the atmosphere. During normal plant operation, the system is isolated, and containment penetrations are sealed with blank flanges. These blank flanges form part of the containment boundary.

LRA Table 2.3.2-1 identifies the components subject to an AMR for the containment leak test system by component type and intended function. Portions of the system contain safety-related components relied upon to remain functional during and following DBEs.

2.3.2.1.2 Staff Evaluation and Conclusion

The staff reviewed the LRA, UFSAR Sections 6.2.1, 6.2.4, and 6.2.6, and license renewal boundary drawings to determine if the applicant failed to identify any components within the scope of license renewal. In addition, the staff's review determined if the applicant failed to identify any components subject to an AMR. The staff found no such omissions. On the basis of its review, the staff concludes the applicant has appropriately identified the containment leak test system mechanical components within the scope of license renewal, as required by 10 CFR 54.4(a), and that the applicant has adequately identified the mechanical components subject to an AMR in accordance with the requirements stated in 10 CFR 54.21(a)(1).

2.3.2.2 Containment Purge System

2.3.2.2.1 Summary of Technical Information in the Application

LRA Section 2.3.2.2.1 describes the containment purge system that consists of a refueling purge and a power access purge. The refueling purge train consists of a supply air handling unit (AHU) and an exhaust fan. It is used for high flow rate purge during refueling and is closed during normal power generation. The power access purge is comprised of a supply AHU and charcoal exhaust filtration unit. It is used for low flow rate purge before and during power access periods.

LRA Table 2.3.2-2 identifies the components subject to an AMR for the containment purge system by component type and intended function. The containment purge system contains safety-related components relied upon to remain functional during and following DBEs. The failure of nonsafety-related SSCs in the containment purge system potentially could prevent the satisfactory accomplishment of a safety-related function. In addition, the system performs functions that support EQ requirements.

Scoping and Screening Methodology

2.3.2.2.2 Staff Evaluation and Conclusion

The staff reviewed the LRA, UFSAR Sections 6.2.4.2.3, 9.4.6.2.2, and 7.3.1.1.10.1, and license renewal boundary drawings to determine if the applicant failed to identify any components within the scope of license renewal. In addition, the staff's review determined if the applicant failed to identify any components subject to an AMR. The staff found no such omissions. On the basis of its review, the staff concludes the applicant has appropriately identified the containment purge system mechanical components within the scope of license renewal, as required by 10 CFR 54.4(a), and that the applicant has adequately identified the mechanical components subject to an AMR in accordance with the requirements stated in 10 CFR 54.21(a)(1).

2.3.2.3 Containment Hydrogen Control System

2.3.2.3.1 Summary of Technical Information in the Application

LRA Section 2.3.2.3.1 describes the containment hydrogen control system, which is comprised of two hydrogen recombiners and associated control cabinets and one hydrogen purge exhaust air filtration unit. The system is staged in Unit 1 and shared by all three units at the site. The system has the necessary electrical and mechanical connections to accommodate installation in any unit within 72 hours of a loss of coolant accident (LOCA). The containment hydrogen control system monitors the hydrogen concentration in the containment building and maintains the hydrogen concentration inside the containment below the lower combustible limit of 4 percent by volume in air following a LOCA.

LRA Table 2.3.2-3 identifies the components subject to an AMR for the containment hydrogen control system by component type and intended function. The containment hydrogen control system contains safety-related components relied upon to remain functional during and following DBEs. In addition, the system performs functions that support EQ requirements.

2.3.2.3.2 Staff Evaluation and Conclusion

The staff reviewed the LRA, UFSAR Sections 1.2.4, 6.2.4, 6.2.5, Table 3.9-25, Table 3.9-27, and Table 6.2.4-1, and license renewal boundary drawings to determine if the applicant failed to identify any components within the scope of license renewal. In addition, the staff's review determined if the applicant failed to identify any components subject to an AMR. The staff found no such omissions. On the basis of its review, the staff concludes the applicant has appropriately identified the containment hydrogen control system mechanical components within the scope of license renewal, as required by 10 CFR 54.4(a), and that the applicant has adequately identified the mechanical components subject to an AMR in accordance with the requirements stated in 10 CFR 54.21(a)(1).

2.3.2.4 Safety Injection and Shutdown Cooling System

2.3.2.4.1 Summary of Technical Information in the Application

LRA Section 2.3.2.4.1 describes the safety injection and shutdown cooling system, which provides the high-pressure and low-pressure safety injection functions, the shutdown cooling function, and the containment spray function. Each unit has two safety injection trains comprised of the high-pressure pump, low-pressure pump, containment spray pump, heat exchanger, and safety injection tanks. The system also includes the trisodium phosphate baskets, which maintain post-LOCA sump fluid pH levels within acceptable limits; containment sumps, including screens and liners; piping that penetrates containment, including the necessary containment isolation valves; and the refueling water tank (RWT).

LRA Table 2.3.2-4 identifies the components subject to an AMR for the safety injection and shutdown cooling system by component type and intended function. The safety injection and shutdown cooling system contains safety-related components relied upon to remain functional during and following DBEs. The failure of nonsafety-related SSCs in the safety injection and shutdown cooling system potentially could prevent the satisfactory accomplishment of a safety-related function. In addition, the system performs functions that support fire protection, EQ, and SBO requirements.

2.3.2.4.2 Staff Evaluation and Conclusion

The staff reviewed the LRA; UFSAR Sections 5.4.7, 6.2.2, 6.2.4, 6.3, 6.5.2, and 8.3.1.1.10; and license renewal boundary drawings to determine if the applicant failed to identify any components within the scope of license renewal. In addition, the staff's review determined if the applicant failed to identify any components subject to an AMR. The staff found no such omissions. On the basis of its review, the staff concludes the applicant has appropriately identified the safety injection and shutdown cooling system mechanical components within the scope of license renewal, as required by 10 CFR 54.4(a). In addition, the staff finds that the applicant has adequately identified the mechanical components subject to an AMR in accordance with the requirements stated in 10 CFR 54.21(a)(1).

2.3.3 Auxiliary Systems

LRA Section 2.3.3 identifies the auxiliary systems SCs subject to an AMR for license renewal. The applicant described the supporting SCs of the auxiliary systems in the following LRA sections:

- 2.3.3.1, "Fuel Handling and Storage System"
- 2.3.3.2, "Spent Fuel Pool Cooling and Cleanup System"
- 2.3.3.3, "Essential Cooling Water System"
- 2.3.3.4, "Essential Chilled Water System"
- 2.3.3.5, "Normal Chilled Water System"
- 2.3.3.6, "Nuclear Cooling Water System"
- 2.3.3.7, "Essential Spray Pond System"
- 2.3.3.8, "Nuclear Sampling System"
- 2.3.3.9, "Compressed Air System"
- 2.3.3.10, "Chemical Volume and Control System"
- 2.3.3.11, "Control Building HVAC [heating, ventilating and air-conditioning] System"
- 2.3.3.12, "Auxiliary Building HVAC System"
- 2.3.3.13, "Fuel Building HVAC System"
- 2.3.3.14, "Containment Building HVAC System"
- 2.3.3.15, "Diesel Generator Building HVAC System"
- 2.3.3.16, "Radwaste Building HVAC System"
- 2.3.3.17, "Turbine Building HVAC System"
- 2.3.3.18, "Miscellaneous Site Structures/Spray Pond Pump House HVAC System"
- 2.3.3.19, "Fire Protection System"
- 2.3.3.20, "Diesel Generator Fuel Oil Storage and Transfer System"
- 2.3.3.21, "Diesel Generator"

Scoping and Screening Methodology

- 2.3.3.22, "Domestic Water System"
- 2.3.3.23, "Demineralized Water System"
- 2.3.3.24, "WRF [water reclamation facility] Fuel System"
- 2.3.3.25, "Service Gases (N_2 and H_2 [nitrogen and hydrogen]) System"
- 2.3.3.26, "Gaseous Radwaste System"
- 2.3.3.27, "Radioactive Waste Drains System"
- 2.3.3.28, "Station Blackout Generator System"
- 2.3.3.29, "Cranes, Hoists, and Elevators"
- 2.3.3.30, "Miscellaneous Auxiliary Systems In-Scope ONLY for Criterion 10 CFR 54.4(a)(2)"

Auxiliary Systems Generic Requests for Additional Information. In RAI 2.3-1, dated December 3, 2009, the staff noted portions of several systems have spatial interaction as nonsafety affecting safety-related components in the fuel building and in the auxiliary building and are within the scope of license renewal as nonsafety affecting safety-related components based on the criterion of 10 CFR 54.4(a)(2). Many spatial interaction terminations are shown on license renewal drawings as license renewal boundaries for 10 CFR 54.4(a)(2) piping. However, the basis for the spatial interaction termination cannot be determined (e.g., entering a building or room with no safety-related components, becoming buried pipe). The staff asked the applicant to provide the bases for the spatial interaction terminations. During the scoping and screening audit, the staff verified that 19 of the identified spatial interaction terminations complied with the criteria for spatial interaction boundaries.

In its response, dated January 18, 2010, the applicant clarified that spatial interaction terminations are associated only with the following two situations: (1) piping exits an area with safety-related components to an area with no safety-related components, and (2) piping has an open end to atmosphere. The applicant verified that all of the above identified spatial interaction terminations met these two criteria or made corrections to the drawings to meet these criteria.

Based on its review, the staff finds the applicant's response to RAI 2.3-1 for those spatial interaction terminations to be acceptable because the applicant verified or made corrections to the drawings to meet the clarified spatial interaction termination criteria. Therefore, the staff's concern described in RAI 2.3-1 is resolved.

As part of the staff's review, the following RAI identified instances of boundary drawing errors where the continuation notation for piping from one boundary drawing to another boundary drawing could not be found or was incorrect.

In RAI 2.3-2, dated December 3, 2009, the staff noted drawings where the staff was unable to identify the license renewal boundary because: (1) continuations were not provided or were incorrect, or (2) the continuation drawing was not provided. The staff asked the applicant to provide additional information to locate the continuations.

In its response, dated January 18, 2010, the applicant provided sufficient information, in response to RAIs supplied in the individual system sections, to locate the license renewal boundaries. Based on its review, the staff finds the applicant's response to RAI 2.3-2 acceptable because the applicant provided the continuation locations. Therefore, the staff's concern described in RAI 2.3-2 is resolved.

Scoping and Screening Methodology

2.3.3.1 Fuel Handling and Storage System

2.3.3.1.1 Summary of Technical Information

LRA Section 2.3.3.1 describes the fuel handling and storage system, which consists of cranes, elevators, fuel storage racks, lift rigs, machines, transfer systems, and trolleys. The purpose of the fuel handling and storage system is to provide onsite storage and manipulation capability for fuel assemblies and CEAs, to provide for the servicing of the reactor vessel closure head and internals and to provide radiation shielding for spent fuel.

LRA Table 2.3.3-1 identifies the components subject to an AMR for the fuel handling and storage system by component type and intended function. The fuel handling and storage system contains safety-related components relied upon to remain functional during and following DBEs. The failure of nonsafety-related SSCs in the system potentially could prevent the satisfactory accomplishment of a safety-related function.

2.3.3.1.2 Staff Evaluation

The staff reviewed LRA Section 2.3.3.1; UFSAR Sections 9.1.1, 9.1.2, and 9.1.4; and the license renewal boundary drawings using the evaluation methodology described in SER Section 2.3 and the guidance in SRP-LR Section 2.3. The staff's review identified an area in which additional information was necessary to complete the review of the applicant's scoping and screening results. The applicant responded to the staff's RAI as discussed below.

In RAI 2.3.3.1-1, dated December 3, 2009, the staff noted in LRA Section 2.3.3.1, "System Description," the CEA change platform was listed as a component that is within the scope of license renewal. The CEA change platform was not included as a component subject to AMR in Table 2.3.3-1 for the fuel handling and storage system. The staff asked the applicant to provide additional information explaining why the CEA change platform is not included as a component subject to an AMR in LRA Table 2.3.3-1.

In its response, dated January 18, 2010, the applicant stated that the CEA change platform consists of several major components—some are passive and subject to AMR and some are active and not subject to AMR. The applicant then described the major components and listed where these components were included in LRA Table 2.3.3-1.

Based on its review, the staff finds the applicant's response to RAI 2.3.3.1-1 acceptable because the applicant clarified which CEA change platform components are subject to AMR and included in LRA Table 2.3.3-1. Therefore, the staff's concern described in RAI 2.3.3.1-1 is resolved.

2.3.3.1.3 Conclusion

The staff reviewed the LRA Section 2.3.3.1; UFSAR Sections 9.1.1, 9.1.2, and 9.1.4; and the RAI response to determine if the applicant failed to identify any components within the scope of license renewal. In addition, the staff's review determined if the applicant failed to identify any components subject to an AMR. The staff found no such omissions. On the basis of its review, the staff concludes the applicant has appropriately identified the fuel handling and storage system mechanical components within the scope of license renewal, as required by 10 CFR 54.4(a). The staff also finds that the applicant has adequately identified the mechanical components subject to an AMR in accordance with the requirements stated in 10 CFR 54.21(a)(1).

2.3.3.2 Spent Fuel Pool Cooling and Cleanup System

2.3.3.2.1 Summary of Technical Information

LRA Section 2.3.3.2 describes the spent fuel pool cooling and cleanup system, which consists of two sub-systems—one for removal of decay heat from the spent fuel and one for maintaining

Scoping and Screening Methodology

pool clarity and reduction of radiation at the pool's surface. There are two independent trains for each cooling and cleanup subsystem. Each cooling subsystem train includes a spent fuel cooling pump, a fuel pool heat exchanger and related piping, valves, and instrumentation. The spent fuel pool cooling pumps circulate fuel pool water through the two fuel pool heat exchangers. Each cleanup subsystem train includes a strainer, pump, filter, ion exchanger, and related piping and instrumentation. During normal operation, one or both trains may be lined up to continuously clean the water in the spent fuel pool or the RWT. During refueling, the system can be aligned to the refueling pool. Additionally, the system provides a backup source of borated water to the chemical and volume control system (CVCS) via the spent fuel pool to achieve safe shutdown.

LRA Table 2.3.3-2 identifies the components subject to an AMR for the spent fuel pool cooling and cleanup system by component type and intended function. The spent fuel pool cooling and cleanup system contains safety-related components relied upon to remain functional during and following DBEs. The failure of nonsafety-related SSCs in the spent fuel pool cooling and cleanup system potentially could prevent the satisfactory accomplishment of a safety-related function. In addition, portions of the system are necessary to support SBO requirements.

2.3.3.2.2 Staff Evaluation

The staff reviewed LRA Section 2.3.3.2, UFSAR Sections 9.1.3 and 9.3.4.5, and the license renewal boundary drawings using the evaluation methodology described in SER Section 2.3 and the guidance in SRP-LR Section 2.3. The staff's review identified an area in which additional information was necessary to complete the review of the applicant's scoping and screening results. The applicant responded to the staff's RAI as discussed below.

In RAI 2.3.3.2-1, dated December 3, 2009, the staff noted that included on LRA drawing LR-PVNGS-PC-01-M-PCP-001 (G-2 and C-2) and in LRA Section 2.3.3.2 is a component described as being an ion-exchanger. However, it appears in LRA Table 2.3.3-2 that the ion-exchanger is a demineralizer since an ion-exchanger is not listed. The staff asked the applicant to provide additional information explaining why the component described as an ion-exchanger on the LRA drawing and in LRA Section 2.3.3.2 appears to be identified as a demineralizer in LRA Table 2.3.3-2.

In its response, dated January 18, 2010, the applicant stated that individual component names and component types in the LRA are consistent with the component names and component types as they appear in the plant equipment database. It listed the ion-exchangers depicted on drawing LR-PVNGS-PC-01-M-PCP-001 as component type "demineralizer" consistent with the component type assigned within the plant equipment database.

Based on its review, the staff finds the applicant's response to RAI 2.3.3.2-1 acceptable because the applicant clarified why the component described as an ion-exchanger on the LRA drawing and in LRA Section 2.3.3.2 is identified as a demineralizer in LRA Table 2.3.3-2. Therefore, the staff's concern described in RAI 2.3.3.2-1 is resolved.

2.3.3.2.3 Conclusion

The staff reviewed the LRA Section 2.3.3.2, UFSAR Sections 9.1.3 and 9.3.4.5, RAI response, and boundary drawings to determine if the applicant failed to identify any components within the scope of license renewal. In addition, the staff's review determined if the applicant failed to identify any components subject to an AMR. The staff found no such omissions. On the basis of its review, the staff concludes the applicant has appropriately identified the spent fuel pool cooling and cleanup system mechanical components within the scope of license renewal, as required by 10 CFR 54.4(a), and that the applicant has adequately identified the mechanical components subject to an AMR in accordance with the requirements stated in 10 CFR 54.21(a)(1).

2.3.3.3 Essential Cooling Water System

2.3.3.3.1 Summary of Technical Information

LRA Section 2.3.3.3 describes the essential cooling water system, which is comprised of two separate, independent, and redundant trains including a heat exchanger, surge tank, pump, chemical addition tank, piping, valves, and associated I&Cs. The purpose of the system is to remove heat from all essential components required for normal and emergency shutdown of the plant, except the diesel generators, and reject the heat to the essential spray ponds (ESPs) through the essential cooling water heat exchanger. The system also provides a back-up source of cooling water for the fuel pool cooling heat exchangers, RCPs, CEDM normal air cooling units, nuclear sample coolers, and normal chillers. The system also provides an intermediate barrier between the RCS and the ESP system to reduce the possibility of radioactive leakage to the environment.

LRA Table 2.3.3-3 notes the components subject to an AMR for the essential cooling water system by component type and intended function. The essential cooling water system contains safety-related components relied upon to remain functional during and following DBEs. The failure of nonsafety-related SSCs in the system potentially could prevent the satisfactory accomplishment of a safety-related function. In addition, the system performs functions that support fire protection, EQ, and SBO requirements.

2.3.3.3.2 Staff Evaluation

The staff reviewed LRA Section 2.3.3.3, UFSAR Section 9.2.2.1, and the license renewal boundary drawings using the evaluation methodology described in SER Section 2.3 and the guidance in SRP-LR Section 2.3. The staff's review identified an area in which additional information was necessary to complete the review of the applicant's scoping and screening results. The applicant responded to the staff's RAI as discussed below.

In RAI 2.3.3.3-1, dated December 3, 2009, the staff noted an essential cooling water system drawing showed certain 1-inch lines within the scope of license renewal as nonsafety affecting safety-related components based on the criterion of 10 CFR 54.4(a)(2). However, the applicant showed parts of these lines continuing to the drain as not within the scope of license renewal. The staff asked the applicant to provide additional information explaining why the lines to the drains are not within the scope of license renewal and justify the boundary locations with respect to the applicable requirements of 10 CFR 54.4(a).

In its response, dated January 18, 2010, the applicant stated that the relief valve drain lines have been added to the scope of license renewal based on 10 CFR 54.4(a)(2) criterion. The applicant provided a revised drawing to show that it added the relief valve drain lines to the scope of license renewal. In addition, the applicant provided a revision to LRA Table 3.3.2-3 to add the leakage boundary function for these lines. Based on its review, the staff finds the applicant's response to RAI 2.3.3.3-1 acceptable because the applicant has revised the drawings to show the relief valve drain lines within scope for license renewal and added the leakage boundary function to LRA Table 3.3.2-3. Therefore, the staff's concern described in RAI 2.3.3.3-1 is resolved.

2.3.3.3.3 Conclusion

The staff reviewed the LRA Section 2.3.3.3, UFSAR Section 9.2.2.1, the RAI response, and original and revised boundary drawings to determine if the applicant failed to identify any components within the scope of license renewal. In addition, the staff's review determined if the applicant failed to identify any components subject to an AMR. On the basis of its review, the staff concludes the applicant has appropriately identified the essential cooling water system mechanical components within the scope of license renewal, as required by 10 CFR 54.4(a), and that the applicant has adequately identified the mechanical components subject to an AMR in accordance with the requirements stated in 10 CFR 54.21(a)(1).

2.3.3.4 Essential Chilled Water System

2.3.3.4.1 Summary of Technical Information

LRA Section 2.3.3.4 describes the essential chilled water system, which is a closed loop system with two independent trains. Each train consists of a chilled water refrigeration unit, a chilled water circulation pump, an expansion tank, control valves, instrumentation, and insulated piping. The purpose of the system is to cool all ESF air handling equipment so that a suitable environment can be maintained for personnel and equipment during a transient or DBE. The system does not operate under normal operations, but the applicant starts it upon actuation of an ESF signal.

LRA Table 2.3.3-4 identifies the components subject to an AMR for the essential chilled water system by component type and intended function. The essential chilled water system contains safety-related components relied upon to remain functional during and following DBEs. The failure of nonsafety-related SSCs in the system potentially could prevent the satisfactory accomplishment of a safety-related function. In addition, the system performs functions that support fire protection and SBO requirements.

2.3.3.4.2 Staff Evaluation

The staff reviewed LRA Section 2.3.3.4; UFSAR Sections 6.4, 9.2.9.2 and 9.5.1; and the license renewal boundary drawings using the evaluation methodology described in SER Section 2.3 and the guidance in SRP-LR Section 2.3. The staff's review identified areas in which additional information was necessary to complete the review of the applicant's scoping and screening results. The applicant responded to the staff's RAIs as discussed below.

In RAI 2.3.3.4-1, dated December 3, 2009, the staff noted an essential chilled water system drawing showed several lines in and out of the air conditioning units (ACUs) within the scope of license renewal based on the criteria of 10 CFR 54.4(a)(2). However, the applicant showed 18 ACUs as not within the scope of license renewal. The staff asked the applicant to provide additional information explaining why the ACUs are not within the scope of license renewal and to justify the boundary locations with respect to the applicable requirements of 10 CFR 54.4(a).

In its response, dated January 18, 2010, the applicant confirmed that the ACUs are within scope of license renewal. The applicant showed the ACUs as dashed lines on the drawing, indicating that the units are in other plant systems, and it correctly highlighted them in those plant systems. Based on its review, the staff finds the applicant's response to RAI 2.3.3.4-1 acceptable because the applicant clarified the scoping classification of the ACUs in question, and the staff verified that the applicant properly highlighted the ACUs on the essential chilled water system boundary drawings. Therefore, the staff's concern described in RAI 2.3.3.4-1 is resolved.

In RAI 2.3.3.4-2, dated December 3, 2009, the staff noted an essential chilled water system drawing showed 1½- and 2-inch lines within the scope of license renewal as nonsafety affecting safety-related components based on the criteria of 10 CFR 54.4(a)(2). However, the applicant showed portions of these lines, downstream of seismic anchors, as not within the scope of license renewal for spatial interaction. The staff asked the applicant to give additional information explaining why these sections of pipe are not within the scope of license renewal.

In its response, dated January 18, 2010, the applicant stated that the pipe lines are not in scope as they are open-ended gas lines that pose no potential for spatial interaction. Based on its review, the staff finds the applicant's response to RAI 2.3.3.4-2 acceptable because the applicant clarified the scoping classification of the pipe lines in question, and the staff agrees that there is no potential for spatial interaction. Therefore, the staff's concern described in RAI 2.3.3.4-2 is resolved.

In RAI 2.3.3.4-3, dated December 3, 2009, the staff noted an essential chilled water system drawing showed a valve and the capped end upstream of the valve as within the scope of license renewal for 10 CFR 54.4(a)(1) and 10 CFR 54.4(a)(2), respectively. However, the applicant showed a small portion of the line, in between the valve and capped end, as out of scope for license renewal. The staff asked the applicant to provide additional information explaining why this section of pipe is not within the scope of license renewal and to justify the boundary locations with respect to the applicable requirements of 10 CFR 54.4(a).

In its response, dated January 18, 2010, the applicant stated that it inadvertently omitted the highlighting of the pipe between the valve and the capped end of the pipe. The applicant provided a revised drawing to show the pipe segment as within scope. Based on its review, the staff finds the applicant's response to RAI 2.3.3.4-3 acceptable because the applicant clarified the scoping classification for the pipe segment in question and provided a revised drawing. Therefore, the staff's concern described in RAI 2.3.3.4-3 is resolved.

In RAI 2.3.3.4-4, dated December 3, 2009, the staff noted an essential chilled water system drawing shows a 1½-inch line as within the scope of license renewal for 10 CFR 54.4(a)(2). The applicant showed the continuation of this line on another license renewal drawing not within the scope of license renewal. The staff asked the applicant to provide additional information explaining the discrepancy. In its response, dated January 18, 2010, the applicant stated the drawing has been revised to include the 1½-inch line as within the scope of license renewal based on the criteria of 10 CFR 54.4(a)(2). Based on its review, the staff finds the applicant's response to RAI 2.3.3.4-4 acceptable because the applicant clarified the scoping classification for the line in question, and the staff received updated drawings. Therefore, the staff's concern described in RAI 2.3.3.4-4 is resolved.

2.3.3.4.3 Conclusion

The staff reviewed the LRA Section 2.3.3.4; UFSAR Sections 6.4, 9.2.9.2, and 9.5.1; RAI responses; and original and revised boundary drawings to determine if the applicant failed to identify any components within the scope of license renewal. In addition, the staff's review determined if the applicant failed to identify any components subject to an AMR. On the basis of its review, the staff concludes the applicant has appropriately identified the essential chilled water system mechanical components within the scope of license renewal, as required by 10 CFR 54.4(a), and that the applicant has adequately identified the mechanical components subject to an AMR in accordance with the requirements stated in 10 CFR 54.21(a)(1).

2.3.3.5 Normal Chilled Water System

2.3.3.5.1 Summary of Technical Information

LRA Section 2.3.3.5 describes the normal chilled water system, which is a closed-loop system consisting of chilled water refrigeration units, chilled water circulation pumps, an expansion tank, control valves, instrumentation, and insulated piping. The applicant states the purpose of the system is to supply cooling to air handling equipment so that plant ventilation can maintain a suitable environment for personnel and equipment. Further, the system operates during normal plant operations, during hot standby, and during scheduled refueling or maintenance shutdown periods.

LRA Table 2.3.3-5 identifies the components subject to an AMR for the normal chilled water system by component type and intended function. The LRA states that the normal chilled water system contains safety-related components relied upon to remain functional during and following DBEs. It further states that the failure of nonsafety-related SSCs in the system potentially could prevent the satisfactory accomplishment of a safety-related function. In addition, portions of the system perform functions that support EQ requirements.

2.3.3.5.2   Staff Evaluation

The staff reviewed LRA Section 2.3.3.5, UFSAR Section 9.2.9.1 and Table 6.2.4-1, and the license renewal boundary drawings using the evaluation methodology described in SER Section 2.3 and the guidance in SRP-LR Section 2.3. The staff's review identified areas in which additional information was necessary to complete the review of the applicant's scoping and screening results. The applicant responded to the staff's RAIs as discussed below.

In RAI 2.3.3.5-1, dated December 3, 2009, the staff noted a license renewal normal chilled water system drawing showed several lines in and out of ACUs and AHUs within the scope of license renewal under 10 CFR 54.4(a)(2). However, the applicant showed 12 ACUs and AHUs as not within the scope of license renewal. The staff asked the applicant to give additional information explaining why these components are not within the scope of license renewal and to justify the boundary locations with respect to the applicable requirements of 10 CFR 54.4(a).

In its response, dated January 18, 2010, the applicant confirmed that the 12 ACUs and AHUs are within the scope of license renewal. The applicant showed the ACUs and AHUs as dashed lines on the drawing, indicating that the units are in other plant systems, and it correctly highlighted them in those plant systems. Based on its review, the staff finds the applicant's response to RAI 2.3.3.5-1 acceptable because the applicant clarified the scoping classification of the ACUs and AHUs in question. The staff verified that the applicant highlighted the ACUs and AHUs as in scope for license renewal on the identified drawings and included them in Section 2.3.3.5. Therefore, the staff's concern described in RAI 2.3.3.5-1 is resolved.

In RAI 2.3.3.5-2, dated December 3, 2009, the staff noted on a license renewal normal chilled water system drawing that 11 lines attached to 10 CFR 54.4(a)(2) lines are shown as not within scope for license renewal. The staff asked the applicant to provide additional information explaining why these pipe sections are not within the scope of license renewal and to justify the boundary locations with respect to the applicable requirements of 10 CFR 54.4(a).

In its response, dated January 18, 2010, the applicant submitted a revised drawing showing the 11 relief valve drain and AHU pan drain lines within the scope of license renewal under 10 CFR 54.4 (a)(2). Based on its review, the staff finds the applicant's response to RAI 2.3.3.5-2 acceptable because the staff confirmed that the applicant revised the drawing to show the relief valve drain and AHU pan drain lines in question as within scope of license renewal and subject to AMR. Therefore, the staff's concern described in RAI 2.3.3.5-2 is resolved.

2.3.3.5.3   Conclusion

The staff reviewed the LRA Section 2.3.3.5, UFSAR Section 9.2.9.1 and Table 6.2.4-1, RAI responses, and original and revised boundary drawings to determine if the applicant failed to identify any components within the scope of license renewal. In addition, the staff's review determined if the applicant failed to identify any components subject to an AMR. On the basis of its review, the staff concludes the applicant has appropriately identified the normal chilled water system mechanical components within the scope of license renewal, as required by 10 CFR 54.4(a), and that the applicant has adequately identified the mechanical components subject to an AMR in accordance with the requirements stated in 10 CFR 54.21(a)(1).

2.3.3.6   Nuclear Cooling Water System

2.3.3.6.1   Summary of Technical Information in the Application

LRA Section 2.3.3.6 describes the nuclear cooling water system, which consists of one closed-loop train including two, full-capacity pumps, redundant heat exchangers, an

Scoping and Screening Methodology

expansion tank, heat exchangers associated with nonsafety-related plant auxiliary systems and components, instrumentation, and piping. The purpose of the system is to provide cooling to auxiliary systems and components such as the RCPs, the boric acid concentrator, the waste gas compressor, the radwaste evaporator, the normal chilled water chillers, the letdown heat exchanger, the fuel pool heat exchangers, the CEDMs, the auxiliary steam vent condenser, and various sample coolers.

LRA Table 2.3.3-6 identifies the components subject to an AMR for the nuclear cooling water system by component type and intended function. The applicant states that the nuclear cooling water system contains safety-related components relied upon to remain functional during and following DBEs. The applicant also identifies that the failure of nonsafety-related SSCs in the system potentially could prevent the satisfactory accomplishment of a safety-related function. In addition, the system performs functions that support fire protection and EQ requirements.

2.3.3.6.2   Staff Evaluation

The staff reviewed LRA Section 2.3.3.6; UFSAR Sections 6.2.4, 8.3.1.1.3, 9.2.2.2, 9.5.1, and 15.6.5; and the license renewal boundary drawings using the evaluation methodology described in SER Section 2.3 and the guidance in SRP-LR Section 2.3. The staff's review identified areas in which additional information was necessary to complete the review of the applicant's scoping and screening results. The applicant responded to the staff's RAIs as discussed below.

In RAI 2.3.3.6-1, dated December 3, 2009, the staff noted several portions of the nuclear cooling water system are within the scope of license renewal as nonsafety affecting safety-related components based on the criterion of 10 CFR 54.4(a)(2). However, the applicant showed 26 lines attached to 10 CFR 54.4(a)(2) lines as not within the scope of license renewal. The staff asked the applicant to provide additional information explaining why these sections of pipe are not within the scope of license renewal and to justify the boundary locations with respect to the applicable requirements of 10 CFR 54.4(a).

In its response, dated January 18, 2010, the applicant provided revised license renewal boundary drawings showing the relief valve drain lines and drain lines within the scope of license renewal under 10 CFR 54.4(a)(2). The applicant also provided a revised Table 3.3.2-6 to include a leakage boundary spatial function for the relief valve drain lines and to include drain lines in a wetted gas environment. Based on its review, the staff finds the applicant's response to RAI 2.3.3.6-1 acceptable because the applicant clarified the scoping classification of the relief valve drain lines in question and included appropriate components as subject to AMR. Therefore, the staff's concern described in RAI 2.3.3.6-1 is resolved.

In RAI 2.3.3.6-2, dated December 3, 2009, the staff noted a license renewal nuclear cooling water system drawing showed several lines in and out of the ACUs within the scope of license renewal for 10 CFR 54.4(a)(2). However, the applicant showed two ACUs as not within the scope of license renewal. The staff asked the applicant to provide additional information explaining why the units are not within the scope of license renewal and to justify the boundary locations with respect to the applicable requirements of 10 CFR 54.4(a).

In its response, dated January 18, 2010, the applicant confirmed that the ACUs are in scope for license renewal. The applicant showed the ACUs as dashed lines on the drawing, indicating that the units are in other plant systems. The applicant correctly highlighted the ACUs as in scope of license renewal on the corresponding system drawing. Based on its review, the staff finds the applicant's response to RAI 2.3.3.6-2 acceptable because the applicant clarified the scoping classification of the ACUs in question, and the staff verified that the applicant properly highlighted the ACUs on the corresponding system drawing. Further, the staff verified that these components have been included in Section 2.3.3.6 as subject to AMR. Therefore, the staff's concern described in RAI 2.3.3.6-2 is resolved.

Scoping and Screening Methodology

In RAI 2.3.3.6-3, dated December 3, 2009, the staff noted a license renewal nuclear cooling water system drawing showed an 8-inch line as within the scope of license renewal for 10 CFR 54.4(a)(3) whereas a small portion of the same line is shown as not within the scope of license renewal for 10 CFR 54.4(a)(2). The staff asked the applicant to provide additional information to clarify the scoping classification for this pipe section. In its response, dated January 18, 2010, the applicant stated the drawing has been revised to indicate the line is within scope of license renewal based on 10 CFR 54.4(a)(3). The applicant provided the revised drawing. Based on its review, the staff finds the applicant's response to RAI 2.3.3.6-3 acceptable because the applicant clarified the scoping classification of the pipe line in question, and the staff verified the change on the revised drawing. Therefore, the staff's concern described in RAI 2.3.3.6-3 is resolved.

2.3.3.6.3 Conclusion

The staff reviewed the LRA Section 2.3.3.6; UFSAR Sections 6.2.4, 8.3.1.1.3, 9.2.2.2, 9.5.1, and 15.6.5; RAI responses; and original and revised boundary drawings to determine if the applicant failed to identify any components within the scope of license renewal. In addition, the staff's review determined if the applicant failed to identify any components subject to an AMR. On the basis of its review, the staff concludes the applicant has appropriately identified the nuclear cooling water system mechanical components within the scope of license renewal, as required by 10 CFR 54.4(a), and that the applicant has adequately identified the mechanical components subject to an AMR in accordance with the requirements stated in 10 CFR 54.21(a)(1).

2.3.3.7 Essential Spray Pond System

2.3.3.7.1 Summary of Technical Information in the Application

LRA Section 2.3.3.7 describes the ESP system, which is comprised of two separate, redundant trains including a pump, ESP, piping, valves, and I&Cs. The applicant states that the purpose of the system is to provide cooling water to nuclear safety-related components and dissipate heat to the atmosphere by the ESPs (ultimate heat sink) under normal and accident conditions. Also, the LRA states that each train alone has a 100-percent heat dissipation capacity for safe shutdown during a loss of offsite power.

LRA Table 2.3.3-7 lists the components subject to an AMR for the ESP system by component type and intended function. The applicant describes the ESP pond system as containing safety-related components relied upon to remain functional during and following DBEs. The failure of nonsafety-related SSCs in the system potentially could prevent the satisfactory accomplishment of a safety-related function. In addition, the system performs functions that support fire protection and SBO requirements.

2.3.3.7.2 Staff Evaluation

The staff reviewed LRA Section 2.3.3.7, UFSAR Sections 9.2.1 and 9.2.5, and the license renewal boundary drawings using the evaluation methodology described in SER Section 2.3 and the guidance in SRP-LR Section 2.3. The staff's review identified an area in which additional information was necessary to complete the review of the applicant's scoping and screening results. The applicant responded to the staff's RAI as discussed below.

In RAI 2.3.3.7-1, dated December 3, 2009, the staff noted a license renewal ESP system drawing showed two 1-inch lines as within the scope of license renewal for 10 CFR 54.4(a)(2). However, the applicant showed the continuation of these lines, after the seismic anchor to the drains, as not within the scope of license renewal. The staff asked the applicant to provide additional information explaining why portions of these lines are not within the scope of license renewal and to justify the boundary locations with respect to the applicable requirements of 10 CFR 54.4(a).

In its response, dated January 18, 2010, the applicant stated the relief valve drain lines have been added to the scope of license renewal based on the criteria of 10 CFR 54.4(a)(2). The applicant revised the drawing to reflect the drain lines as within the scope of license renewal. The applicant also revised LRA Table 3.3.2-7 to include drain lines in a wetted gas environment. Based on its review, the staff finds the applicant's response to RAI 2.3.3.7-1 acceptable because the applicant clarified that the drain lines are in scope of license renewal, and the staff verified the revised drawings show the relief drain valve lines within scope for license renewal. Further, the staff confirmed that these components were included in the LRA as subject to AMR. Therefore, the staff's concern described in RAI 2.3.3.7-1 is resolved.

2.3.3.7.3 Conclusion

The staff reviewed the LRA Section 2.3.3.7, UFSAR Sections 9.2.1, and 9.2.5, RAI response, and original and revised boundary drawings to determine if the applicant failed to identify any components within the scope of license renewal. In addition, the staff's review determined if the applicant failed to identify any components subject to an AMR. On the basis of its review, the staff concludes the applicant has appropriately identified the ESP system mechanical components within the scope of license renewal, as required by 10 CFR 54.4(a), and that the applicant has adequately identified the mechanical components subject to an AMR in accordance with the requirements stated in 10 CFR 54.21(a)(1).

2.3.3.8 Nuclear Sampling System

2.3.3.8.1 Summary of Technical Information in the Application

LRA Section 2.3.3.8 describes the nuclear sampling system, which allows collection of samples from the RCS and auxiliary systems for analysis during normal and post-accident conditions without requiring access to containment. Sample points include the RCS hot leg, pressurizer surge line, pressurizer steam space, safety injection and shutdown cooling system, and CVCSs. The LRA states that the nuclear sampling system consists of sampling lines, heat exchangers, sample vessels, sample sinks or racks, analysis equipment, and instrumentation.

LRA Table 2.3.3-8 lists the components subject to an AMR for the nuclear sampling system by component type and intended function. Portions of the nuclear sampling system contain safety-related components relied upon to remain functional during and following DBEs. The failure of portions of the nonsafety-related SSCs in the system potentially could prevent the satisfactory accomplishment of a safety-related function. In addition, the system performs functions that support fire protection and EQ requirements.

2.3.3.8.2 Staff Evaluation

The staff reviewed LRA Section 2.3.3.8; UFSAR Sections 3.11, 6.2.4, 8.3.2.1, 9.3.2, 9.5.1, and 15.6.5; and the license renewal boundary drawings using the evaluation methodology described in SER Section 2.3 and the guidance in SRP-LR Section 2.3. The staff's review identified areas in which additional information was necessary to complete the review of the applicant's scoping and screening results. The applicant responded to the staff's RAIs as discussed below.

In RAI 2.3.3.8-1, dated December 3, 2009, the staff noted that a license renewal drawing for the nuclear sampling system showed the continuation of a 1-inch pipe section, in scope for criteria 10 CFR 54.4(a)(2), to another drawing where the continuation was not within the scope of license renewal. The staff asked the applicant to provide additional information to justify the discrepancy.

In its response, dated January 18, 2010, the applicant stated that the 1-inch line is not within the scope of license renewal because it is not safety-related and neither connects to safety-related equipment nor passes through areas occupied by safety-related equipment for spatial interaction. The applicant also stated that it inadvertently colored the origin of 1-inch line as within the scope of license renewal. The applicant revised this drawing and submitted it to the staff. Based on its review, the staff finds the applicant's response to RAI 2.3.3.8-1 acceptable because the applicant clarified why the 1-inch pipe was not within the scope of license renewal. The staff verified the corrected highlighting on revised drawing. Therefore, the staff's concern described in RAI 2.3.3.8-1 is resolved.

In RAI 2.3.3.8-2, dated December 3, 2009, the staff noted a license renewal drawing for the nuclear sampling system showed a continuation of a 2-inch pipe section in scope for license renewal, to a "hot lab sink drain" on an associated drawing; however, the continuation was not shown as within the scope of license renewal. The staff asked the applicant to provide additional information to justify the discrepancy.

In its response, dated January 18, 2010, the applicant stated that the 2-inch line is within the scope of license renewal in the nuclear sampling system but not within the scope in the chemical waste system. The applicant also stated that the drawing has been revised to add a spatial interaction termination flag on the 2-inch line and the highlighting downstream of the spatial interaction termination flag before the continuation to the license renewal boundary drawing was removed. The applicant submitted the revised drawing to the staff. Based on its review, the staff finds the applicant's response to RAI 2.3.3.8-2 acceptable because the applicant justified the reason for not including the continuation on the drawing within scope. The staff confirmed changes to the license renewal boundary drawing. Therefore, the staff's concern described in RAI 2.3.3.8-2 is resolved.

In RAI 2.3.3.8-3, dated December 3, 2009, the staff noted a license renewal drawing for the nuclear sampling system showed a continuation of a ½-inch pipe section to "equipment drain tank" on an associated license renewal drawing. The associated license renewal drawing referred to a third drawing, but this drawing was not consistent with the other two. The staff asked the applicant to provide additional information to locate the continuation of the ½-inch pipe section.

In its response, dated January 18, 2010, the applicant stated that the ½-inch pipe section continues from both drawings and is shown on both drawings. The inconsistency was noted in the continuation grid coordinates, which was caused by using a Unit 3 P&ID with a continuation to a Unit 1 drawing. Based on its review, the staff finds the applicant's response to RAI 2.3.3.8-3 acceptable because the applicant clarified the inconsistency in grid locations. Therefore, the staff's concern described in RAI 2.3.3.8-3 is resolved.

2.3.3.8.3   Conclusion

The staff reviewed the LRA Section 2.3.3.8; UFSAR Sections 3.11, 6.2.4, 8.3.2.1, 9.3.2, 9.5.1 and 15.6.5; RAI responses; and original and revised boundary drawings to determine if the applicant failed to identify any components within the scope of license renewal. In addition, the staff's review determined if the applicant failed to identify any components subject to an AMR. On the basis of its review, the staff concludes the applicant has appropriately identified the nuclear sampling system mechanical components within the scope of license renewal, as required by 10 CFR 54.4(a), and that the applicant has adequately identified the mechanical components subject to an AMR in accordance with the requirements stated in 10 CFR 54.21(a)(1).

2.3.3.9   Compressed Air System

2.3.3.9.1   Summary of Technical Information in the Application

LRA Section 2.3.3.9 describes the compressed air system as comprised of two subsystems: the instrument air system and the service and breathing air system. The applicant stated that the instrument air subsystem provides filtered, dry, oil-free air for pneumatic instrument operation and the control of pneumatic actuators using three air compressors, three air

Scoping and Screening Methodology

receivers, and six air dryer units. The applicant also stated the instrument air system also has nitrogen back-up capability. The service and breathing air subsystem supplies oil-free breathable air using one air compressor, two air receivers, and one refrigerated air dryer to service air stations and breathing air stations throughout the plant. The instrument air subsystem is required for normal plant operation but is not required for safe shutdown of the plant.

LRA Table 2.3.3-9 identifies the components subject to an AMR for the compressed air system by component type and intended function. This section states that the compressed air system contains safety-related components relied upon to remain functional during and following DBEs. The applicant stated the failure of nonsafety-related SSCs in the system potentially could prevent the satisfactory accomplishment of a safety-related function. In addition, portions of the system perform functions that support fire protection, EQ, and SBO requirements.

2.3.3.9.2   Staff Evaluation and Conclusion

The staff reviewed the LRA Section 2.3.3.9, UFSAR Section 9.3.1, and license renewal boundary drawings to determine if the applicant failed to identify any components within the scope of license renewal. In addition, the staff's review determined if the applicant failed to identify any components subject to an AMR. The staff found no such omissions. On the basis of its review, the staff concludes the applicant has appropriately identified the compressed air system mechanical components within the scope of license renewal, as required by 10 CFR 54.4(a), and that the applicant has adequately identified the mechanical components subject to an AMR in accordance with the requirements stated in 10 CFR 54.21(a)(1).

2.3.3.10   Chemical Volume and Control System

2.3.3.10.1   Summary of Technical Information in the Application

LRA Section 2.3.3.10 describes the chemical volume and control system, which adjusts the purity, volume, and boric acid concentration of the reactor coolant. The system's major components are the RWT, the reactor drain tank, the equipment drain tank, the gas stripper, the boric acid concentrator, heat exchangers, filters, ion exchangers, piping, valves, and various pumps, including the charging pumps. The LRA states that the system has many functions including those listed below:

- Maintain the chemistry and purity of the reactor coolant
- Maintain volume in the RCS
- Receive, store, separate, and process reactor grade, borated waste
- Provide borated water to the emergency core cooling system for injection to the RCS
- Control the boron concentration in the RCS
- Provide auxiliary pressurizer spray for control of pressure and cooling
- Provide and receive injection water to and from the RCP seals
- Supply demineralized reactor makeup water to various auxiliary equipment
- Provide a means for continuous removal of noble gases from the reactor coolant
- Provide makeup to the spent fuel pool
- Provide purification of shutdown cooling flow
- Provide makeup to the reactor coolant system for losses from small leaks
- Provide water to the auxiliary feedwater system as makeup for reactor heat removal

LRA Table 2.3.3-10 identifies the components subject to an AMR for the chemical volume and control system by component type and intended function. The chemical volume and control system contains safety-related components relied upon to remain functional during and following DBEs. The failure of nonsafety-related SSCs in the system potentially could prevent

Scoping and Screening Methodology

the satisfactory accomplishment of a safety-related function. In addition, the system performs functions that support fire protection, EQ, and SBO requirements.

2.3.3.10.2 Staff Evaluation

The staff reviewed LRA Section 2.3.3.10; UFSAR Sections 1.2.10.2, 3.11, 6.2.4, 8.3.1.1.10, and 9.3.4; and the license renewal boundary drawings using the evaluation methodology described in SER Section 2.3 and the guidance in SRP-LR Section 2.3. The staff's review identified areas in which additional information was necessary to complete the review of the applicant's scoping and screening results. The applicant responded to the staff's RAIs as discussed below.

In RAI 2.3.3.10-1, dated December 3, 2009, the staff noted portions of the CVCS are within the scope of license renewal as nonsafety affecting safety-related components based on the criterion of 10 CFR 54.4(a)(2). However, the applicant showed 22 lines attached to 10 CFR 54.4(a)(2) lines as not within the scope of license renewal. The staff asked the applicant to provide additional information explaining why these sections of pipe are not within the scope of license renewal and to justify the boundary locations with respect to the applicable requirements of 10 CFR 54.4(a).

In its response, dated January 18, 2010, the applicant provided additional information for each of the 22 locations on their scoping status and, in all cases, submitted revised license renewal boundary drawings to clarify the boundary locations. The applicant also revised Table 3.3.2-10 to include components in a wetted gas environment. Based on its review, the staff finds the applicant's response to RAI 2.3.3.10-1 acceptable because the applicant clarified boundary locations, and the staff verified the revised drawings. Therefore, the staff's concern described in RAI 2.3.3.10-1 is resolved.

In RAI 2.3.3.10-2, dated December 3, 2009, the staff noted that the applicant did not provide a continuation drawing in the license renewal package for certain license renewal boundary drawings. The staff asked the applicant to provide additional information to locate the license renewal boundaries.

In its response, dated January 18, 2010, the applicant provided additional information to clarify the boundary locations. Based on its review, the staff finds the applicant's response to RAI 2.3.3.10-2 acceptable because the boundary locations were provided. Therefore, the staff's concern described in RAI 2.3.3.10-2 is resolved.

In RAI 2.3.3.10-3, dated December 3, 2009, the staff noted that license renewal drawings showed 16 lines as not within the scope of license renewal, but the lines are connected to piping and tanks, which are shown as within the scope of license renewal. The staff asked the applicant to provide additional information to justify why these lines are not within the scope of license renewal.

In its response, dated January 18, 2010, the applicant provided additional information for each of the 16 locations on their scoping status and, in many cases, submitted revised license renewal boundary drawings to clarify the boundary locations. The applicant also revised Table 3.3.2-10 to include components in a wetted gas environment. Based on its review, the staff finds the applicant's response to RAI 2.3.3.10-3 acceptable because the boundary locations were provided. The staff reviewed and verified the revised boundary drawings. Therefore, the staff's concern described in RAI 2.3.3.10-3 is resolved.

2.3.3.10.3 Conclusion

The staff reviewed the LRA Section 2.3.3.10; UFSAR Sections 1.2.10.2, 3.11, 6.2.4, 8.3.1.1.10, and 9.3.4; RAI responses; and original and revised boundary drawings to determine if the applicant failed to identify any components within the scope of license renewal. In addition, the staff's review determined if the applicant failed to identify any components subject to an AMR.

Scoping and Screening Methodology

On the basis of its review, the staff concludes the applicant has appropriately identified the chemical volume and control system mechanical components within the scope of license renewal, as required by 10 CFR 54.4(a), and that the applicant has adequately identified the mechanical components subject to an AMR in accordance with the requirements stated in 10 CFR 54.21(a)(1).

2.3.3.11 Control Building Heating, Ventilation, and Air Conditioning System

2.3.3.11.1 Summary of Technical Information in the Application

LRA Section 2.3.3.11 describes the control building HVAC system, which is comprised of four subsections: (1) control room normal HVAC, (2) control building normal HVAC, (3) control room essential HVAC, and (4) control building essential HVAC. The LRA states that the functions of the control building HVAC system are to maintain an environment in the control room complex, suitable for prolonged occupancy throughout the duration of postulated accidents; to provide control room isolation to prevent intrusion of poisonous gases, smoke, or airborne radioactivity; to maintain a suitable environment for the ESF switchgear, ESF equipment rooms, and battery rooms during postulated accidents; and to ventilate and exhaust battery rooms to maintain hydrogen below flammable concentrations.

LRA Table 2.3.3-11 identifies the components subject to an AMR for the control building HVAC system by component type and intended function. The applicant further stated that the control building HVAC system contains safety-related components relied upon to remain functional during and following DBEs. The failure of portions of the nonsafety-related SSCs in the system potentially could prevent the satisfactory accomplishment of a safety-related function. In addition, portions of the system perform functions that support fire protection and SBO requirements.

2.3.3.11.2 Staff Evaluation

The staff reviewed LRA Section 2.3.3.11, UFSAR Sections 6.4 and 9.4.1, and the license renewal boundary drawings using the evaluation methodology described in SER Section 2.3 and the guidance in SRP-LR Section 2.3. The staff's review of LRA Section 2.3.3.11 identified an area in which additional information was necessary to complete the review of the applicant's scoping and screening results. The applicant responded to the staff's RAI as discussed below.

In RAI 2.3.3.11-1, dated November 13, 2009, the staff requested that the applicant explain the scoping status of the fixed louvers in the air inlet and outlet. The applicant did not highlight the louvers as being in scope on the license renewal boundary drawings.

In its response, dated December 11, 2009, the applicant stated that the fixed louvers installed in an exterior wall of the control building are structural components within scope of license renewal. The applicant stated that these louvers were included in the component types of "barrier" and "structural steel" in LRA Section 2.4.2. The applicant explained that the license renewal drawings did not have the louvers highlighted because those drawings are mechanical boundary drawings, and only mechanical components within the license renewal scope are highlighted on those drawings. The single license renewal drawing for structures was based on the site plan. Based on its review, the staff finds the applicant's response to RAI 2.3.3.11-1 acceptable because the applicant clarified that the items (louvers) identified in the RAI are within scope and are addressed in LRA Section 2.4.2. Therefore, the staff's concern described in RAI 2.3.3.11-1 is resolved.

2.3.3.11.3 Conclusion

The staff reviewed the LRA Section 2.3.3.11, UFSAR Sections 6.4 and 9.4.1, RAI response, and license renewal boundary drawings to determine if the applicant failed to identify any components within the scope of license renewal. In addition, the staff's review determined if the applicant failed to identify any components subject to an AMR. On the basis of its review, the staff concludes the applicant has appropriately identified the control building HVAC system mechanical components within the scope of license renewal, as required by 10 CFR 54.4(a), and that the applicant has adequately identified the mechanical components subject to an AMR in accordance with the requirements stated in 10 CFR 54.21(a)(1).

2.3.3.12 Auxiliary Building Heating, Ventilation, and Air Conditioning System

2.3.3.12.1 Summary of Technical Information in the Application

LRA Section 2.3.3.12 describes the auxiliary building HVAC system, which consists of two subsystems: (1) the auxiliary building normal HVAC and (2) the auxiliary building essential HVAC. The auxiliary building normal HVAC subsystem maintains environmental conditions suitable for personnel comfort and safe operation of equipment during normal plant operation. The auxiliary building essential HVAC subsystem maintains the required thermal environment for the ESF equipment rooms and auxiliary feedwater pump rooms during accident conditions.

LRA Table 2.3.3-12 identifies the components subject to an AMR for the auxiliary building HVAC system by component type and intended function. The auxiliary building HVAC system contains safety-related components relied upon to remain functional during and following DBEs. The failure of portions of the nonsafety-related SSCs in the system potentially could prevent the satisfactory accomplishment of a safety-related function. In addition, portions of the system perform functions that support fire protection, EQ, and SBO requirements.

2.3.3.12.2 Staff Evaluation

The staff reviewed LRA Section 2.3.3.12, UFSAR Section 9.4.2, and the license renewal boundary drawings using the evaluation methodology described in SER Section 2.3 and the guidance in SRP-LR Section 2.3. The staff's review of LRA Section 2.3.3.12 identified an area in which additional information was necessary to complete the review of the applicant's scoping and screening results. The applicant responded to the staff's RAI as discussed below.

In RAI 2.3.3.12-1, dated November 13, 2009, the staff requested that the applicant explain the scoping status of two fire dampers on a license renewal drawing for auxiliary building HVAC system. The applicant did not highlight these fire dampers as being in scope on the drawing.

In its response, dated December 11, 2009, the applicant stated that it should have highlighted one fire damper as being in scope as it has fire barrier and non-safety-related structural support functions. The applicant described this as an apparent drawing preparation oversight and corrected it. The applicant stated that the other fire damper was not in scope; thus, was correctly not highlighted on the drawing. This damper is not mounted in a fire barrier wall. Although not depicted on the drawing, the air that passes through the damper comes from the open areas of the elevations above via stairwells. Based on its review, the staff finds the applicant's response to RAI 2.3.3.12-1 acceptable because the applicant clarified that one fire damper was in scope and that the other fire damper was not in scope, as the wall in which it was mounted was not a fire barrier. Therefore, the staff's concern described in RAI 2.3.3.12-1 is resolved.

2.3.3.12.3 Conclusion

The staff reviewed LRA Section 2.3.3.12, UFSAR Section 9.4.2, license renewal boundary drawings, and RAI response to determine if the applicant failed to properly identify any components within the scope of license renewal. In addition, the staff's review determined whether the applicant failed to identify any components subject to an AMR. On the basis of its review, the staff concludes the applicant has appropriately identified the auxiliary building HVAC system mechanical components within the scope of license renewal, as required by

10 CFR 54.4(a), and that the applicant has adequately identified the mechanical components subject to an AMR in accordance with the requirements stated in 10 CFR 54.21(a)(1).

2.3.3.13 Fuel Building Heating, Ventilation, and Air Conditioning System

2.3.3.13.1 Summary of Technical Information in the Application

LRA Section 2.3.3.13 describes the fuel building HVAC system, which consists of the fuel building normal and essential HVAC subsystems. The LRA states that the fuel building normal HVAC subsystem operates during normal modes of operation and distributes tempered outside air throughout the building. The subsystem maintains environmental conditions suitable for personnel comfort and safe operation of equipment during normal operation. The fuel building essential HVAC subsystem operates only in the event of a fuel handling accident or LOCA and directs filtered exhaust to the fuel building vents to minimize airborne radiation releases. If radiation monitors detect high radiation levels, the fuel building is isolated, the essential air filtration units start, the normal AHUs are secured, and negative pressure is established in the building.

LRA Table 2.3.3-13 identifies the components subject to an AMR for the fuel building HVAC system by component type and intended function. The fuel building HVAC system contains safety-related components relied upon to remain functional during and following DBEs. The failure of portions of nonsafety-related SSCs in the system potentially could prevent the satisfactory accomplishment of a safety-related function. In addition, portions of the system perform functions that support fire protection and EQ requirements.

2.3.3.13.2 Staff Evaluation and Conclusion

The staff reviewed LRA Section 2.3.3.13, UFSAR Sections 6.5.1 and 9.4.5, and the license renewal boundary drawings using the evaluation methodology described in SER Section 2.3 and the guidance in SRP-LR Section 2.3 to determine if the applicant failed to identify any components within the scope of license renewal. In addition, the staff's review determined if the applicant failed to identify any components subject to an AMR. The staff found no such omissions. On the basis of its review, the staff concludes the applicant has appropriately identified the fuel building HVAC system mechanical components within the scope of license renewal, as required by 10 CFR 54.4(a), and that the applicant has adequately identified the mechanical components subject to an AMR in accordance with the requirements stated in 10 CFR 54.21(a)(1).

2.3.3.14 Containment Building Heating, Ventilation, and Air Conditioning System

2.3.3.14.1 Summary of Technical Information in the Application

LRA Section 2.3.3.14 describes the containment building HVAC system, which controls air temperature to ensure operability of containment building equipment, provide filtration to maintain airborne radioactivity levels below permissible limits, and provide an environment suitable for maintenance and refueling activities.

The LRA also states that containment building HVAC system functions during normal plant operations, containment pre-access periods, or during extended shutdowns. It is comprised of the following subsystems:

- Containment building normal cooling subsystem
- Containment building normal cleanup subsystem
- CEDM cooling subsystem
- Reactor cavity cooling subsystem
- Pressurizer cooling subsystem
- Tendon gallery ventilation subsystem

Scoping and Screening Methodology

- Main steam support structure ventilation subsystem

LRA Table 2.3.3-14 identifies the components subject to an AMR for the containment building HVAC system by component type and intended function. The containment building HVAC system contains safety-related components relied upon to remain functional during and following DBEs. The failure of portions of the nonsafety-related SSCs in the system potentially could prevent the satisfactory accomplishment of a safety-related function. In addition, portions of the system perform functions that support fire protection and EQ requirements.

2.3.3.14.2 Staff Evaluation and Conclusion

The staff reviewed LRA Section 2.3.3.14, UFSAR Sections 6.2.4 and 9.4.6, and the license renewal boundary drawings using the evaluation methodology described in SER Section 2.3 and the guidance in SRP-LR Section 2.3 to determine if the applicant failed to identify any components within the scope of license renewal. In addition, the staff's review determined if the applicant failed to identify any components subject to an AMR. The staff found no such omissions. On the basis of its review, the staff concludes the applicant has appropriately identified the containment building HVAC system mechanical components within the scope of license renewal, as required by 10 CFR 54.4(a), and that the applicant has adequately identified the mechanical components subject to an AMR in accordance with the requirements stated in 10 CFR 54.21(a)(1).

2.3.3.15 Diesel Generator Building Heating, Ventilation, and Air Conditioning System

2.3.3.15.1 Summary of Technical Information in the Application

LRA Section 2.3.3.15 describes the diesel generator building HVAC system, which is comprised of two separate and independent HVAC trains, one for each of the diesel generator compartments in each diesel generator building. Each train consists of the diesel generator building normal and essential HVAC subsystems. The normal subsystem maintains environmental conditions during normal operation suitable for personnel comfort and safe operation of equipment when the diesel generator is not running. The essential HVAC subsystem maintains the appropriate environment for the diesel generator and its auxiliaries during emergency conditions when the diesel generator is required to operate.

LRA Table 2.3.3-15 identifies the components subject to an AMR for the diesel generator building HVAC system by component type and intended function. The diesel generator building HVAC system contains safety-related components relied upon to remain functional during and following DBEs. The failure of portions of the nonsafety-related SSCs in the system potentially could prevent the satisfactory accomplishment of a safety-related function. In addition, portions of the system perform functions that support fire protection and SBO requirements.

2.3.3.15.2 Staff Evaluation

The staff reviewed LRA Section 2.3.3.15, UFSAR Section 9.4.7, and the license renewal boundary drawings using the evaluation methodology described in SER Section 2.3 and the guidance in SRP-LR Section 2.3. The staff's review of LRA Section 2.3.3.15 identified an area in which additional information was necessary to complete the review of the applicant's scoping and screening results. The applicant responded to the staff's RAI as discussed below.

In RAI 2.3.3.15-1, dated November 13, 2009, the staff asked that the applicant explain the scoping status of the fixed ventilation louvers shown on the license renewal drawing for the diesel generator building HVAC system. The staff noted that the applicant did not highlight these louvers as being in scope on the drawing.

In its response, dated December 11, 2009, the applicant stated that these fixed louvers are installed in an external wall and interior wall of the building and are included in scope as described in LRA Section 2.4.3 as part of the structural steel component type providing

Scoping and Screening Methodology

structural support and shelter and protection functions. The drawing did not have the louvers highlighted because that drawing is a mechanical boundary drawing and only mechanical components within the license renewal scope are highlighted on those drawings. The single license renewal drawing for structures was based on the site plan. Based on its review, the staff finds the applicant's response to RAI 2.3.3.15-1 acceptable because the applicant clarified that the items (louvers) identified in the RAI were in scope and were addressed in LRA Section 2.4.3. Therefore, the staff's concern described in RAI 2.3.3.15-1 is resolved.

2.3.3.15.3 Conclusion

The staff reviewed the LRA Section 2.3.3.15, UFSAR Section 9.4.7, license renewal boundary drawings, and RAI response to determine if the applicant failed to identify any components within the scope of license renewal. In addition, the staff's review determined if the applicant failed to identify any components subject to an AMR. On the basis of its review, the staff concludes the applicant has appropriately identified the diesel generator building HVAC system mechanical components within the scope of license renewal, as required by 10 CFR 54.4(a), and that the applicant has adequately identified the mechanical components subject to an AMR in accordance with the requirements stated in 10 CFR 54.21(a)(1).

2.3.3.16 Radwaste Building Heating, Ventilation, and Air Conditioning System

2.3.3.16.1 Summary of License Renewal Application Technical Information

LRA Section 2.3.3.16 describes the radwaste building HVAC system, which is a once-through ventilation system with no recirculation, except for the radwaste control room, which has a recirculation AHU. The system provides a suitable environment for personnel comfort and safe operation of equipment. Further, the building airflow patterns inhibit the spread of airborne radioactivity and maintain a slight negative pressure in the building.

LRA Table 2.3.3-16 identifies the components subject to an AMR for the radwaste building HVAC system by component type and intended function. The failure of portions of the nonsafety-related SSCs in the radwaste building HVAC system potentially could prevent the satisfactory accomplishment of a safety-related function. In addition, portions of the system perform functions that support fire protection requirements.

2.3.3.16.2 Staff Evaluation and Conclusion

The staff reviewed LRA Section 2.3.3.16, UFSAR Section 9.4.3 and Appendix 9B.2.10, and the license renewal boundary drawings using the evaluation methodology described in SER Section 2.3 and the guidance in SRP-LR Section 2.3 to determine if the applicant failed to identify any components within the scope of license renewal. In addition, the staff's review determined if the applicant failed to identify any components subject to an AMR. The staff found no such omissions. On the basis of its review, the staff concludes the applicant has appropriately identified the radwaste building HVAC system mechanical components within the scope of license renewal, as required by 10 CFR 54.4(a). The staff also finds that the applicant has adequately identified the mechanical components subject to an AMR in accordance with the requirements stated in 10 CFR 54.21(a)(1).

2.3.3.17 Turbine Building Heating, Ventilation, and Air Conditioning System

2.3.3.17.1 Summary of License Renewal Application Technical Information in the Application

LRA Section 2.3.3.17 describes the turbine building HVAC system, which is comprised of three major subsystems. The general area HVAC subsystem maintains environmental conditions suitable for equipment operation during normal plant operations and shutdown periods. The battery and switchgear room HVAC subsystem prevents the accumulation of hydrogen gas in the battery room during normal plant operation and shutdown periods. The lube oil room HVAC

Scoping and Screening Methodology

subsystem removes combustible gases and heat from the lube oil room during normal plant operation and shutdown periods.

LRA Table 2.3.3-17 identifies the components subject to an AMR for the turbine building HVAC system by component type and intended function. Portions of the turbine building HVAC system perform functions that support fire protection and SBO requirements.

2.3.3.17.2 Staff Evaluation

The staff reviewed LRA Section 2.3.3.17, UFSAR Section 9.4.4 and Appendix 9B.2.20.1, and the license renewal boundary drawings using the evaluation methodology described in SER Section 2.3 and the guidance in SRP-LR Section 2.3. The staff's review of LRA Section 2.3.3.17 identified an area in which additional information was necessary to complete the review of the applicant's scoping and screening results. The applicant responded to the staff's RAI as discussed below.

In RAI 2.3.3.17-1, dated November 13, 2009, the staff requested that the applicant explain the scoping status of the backdraft dampers and ducting downstream of certain fire dampers because the applicant highlighted the fire dampers as in the scope of license renewal on the license renewal drawings.

In its response, dated December 11, 2009, the applicant stated that these fire dampers are installed in a two-hour fire barrier wall, but are not credited for a fire protection function. The fire dampers are credited for a SBO function; therefore, they are in the scope of license renewal and are highlighted on the drawing. The downstream ducting and backdraft dampers do not have a fire protection or a SBO function and are not in scope; thus, the applicant did not highlight them on the drawing.

Based on its review, the staff finds the applicant's response to RAI 2.3.3.17-1 acceptable because the applicant clarified that fire dampers were in scope for a SBO function, and the downstream ducting and backdraft dampers were not in scope. Therefore, the staff's concern described in RAI 2.3.3.15-1 is resolved.

2.3.3.17.3 Conclusion

The staff reviewed the LRA, UFSAR Section 9.4.4 and Appendix 9B.2.20.1, license renewal boundary drawings, and the RAI response to determine if the applicant failed to identify any components within the scope of license renewal. In addition, the staff's review determined if the applicant failed to identify any components subject to an AMR. On the basis of its review, the staff concludes the applicant has appropriately identified the turbine building HVAC system mechanical components within the scope of license renewal, as required by 10 CFR 54.4(a), and that the applicant has adequately identified the mechanical components subject to an AMR in accordance with the requirements stated in 10 CFR 54.21(a)(1).

2.3.3.18 Miscellaneous Site Structures and Spray Pond Pump House Heating, Ventilation, and Air Conditioning System

2.3.3.18.1 Summary of Technical Information in the Application

LRA Section 2.3.3.18 describes the miscellaneous site structures HVAC systems, including the spray pond pump house HVAC system. The spray pond pump house HVAC system is the only miscellaneous site structures HVAC system that is within the scope of license renewal. Each unit has two redundant ESP pump houses located next to the ESPs. Each house is equipped with one essential ventilation exhaust fan, which maintains room temperature at or below the spray pond qualification temperature during emergency or post accident operation of the ESP system.

Scoping and Screening Methodology

LRA Table 2.3.3-18 identifies the components subject to an AMR for the miscellaneous site structures and the spray pond pump house HVAC system by component type and intended function. The spray pond pump house HVAC system contains safety-related components relied upon to remain functional during and following DBEs. Additionally, portions of the spray pond pump house HVAC system perform functions that support fire protection and SBO requirements.

2.3.3.18.2 Staff Evaluation and Conclusion

The staff reviewed LRA Section 2.3.3.18, UFSAR Section 9.4.8, Appendix 9B.2.7, and Appendix 9B.2.8, and the license renewal boundary drawings using the evaluation methodology described in SER Section 2.3 and the guidance in SRP-LR Section 2.3 to determine if the applicant failed to identify any components within the scope of license renewal. In addition, the staff's review determined if the applicant failed to identify any components subject to an AMR. The staff found no such omissions. On the basis of its review, the staff concludes the applicant has appropriately identified the ESP pump house HVAC system mechanical components within the scope of license renewal, as required by 10 CFR 54.4(a), and that the applicant has adequately identified the mechanical components subject to an AMR in accordance with the requirements stated in 10 CFR 54.21(a)(1).

2.3.3.19 Fire Protection System

2.3.3.19.1 Summary of Technical Information in the Application

LRA Section 2.3.3.19 describes the fire protection system as comprised of the following:

- Two 50-percent diesel-driven fire water pumps, one 50-percent motor-driven fire pump, fire water pump drivers, fire water tanks, and underground distribution system including outside loop, hydrants, sectional control valves, and isolation valves

- Hose stations, standpipes, halon, CO_2, deluge, and preaction systems within the power block, including control valves, spray nozzles, and sprinkler heads

- Diesel fuel oil supply to the motor-driven fire pumps

- A jockey pump with associated piping

The LRA states that the purpose of the fire protection system is to minimize the effects of fire on plant SSCs important to safety, such that a fire will not compromise the ability to achieve safe shutdown of the plant. Further, the LRA states that the safety-related components at the containment penetrations are included in this system.

LRA Table 2.3.3-19 identifies the components subject to an AMR for the fire protection system by component type and intended function. According to the LRA, the fire protection system has intended functions under 10 CFR 54.4(a)(1), 10 CFR 54.4(a)(2), and 10 CFR 54.4(a)(3).

2.3.3.19.2 Staff Evaluation

The staff reviewed LRA Section 2.3.3.19, UFSAR Section 9.5.1 and Appendix 9B, and license renewal drawings using the evaluation methodology described in SER Section 2.3 and guidance in SRP-LR, Section 2.3. During its review, the staff evaluated the system functions described in the LRA and UFSAR to verify that the applicant had not omitted from the scope of license renewal any components with intended functions under 10 CFR 54.4(a). The staff then reviewed those components that the applicant identified as within the scope of license renewal to verify that the applicant had not omitted any passive or long-lived components subject to an AMR in accordance with 10 CFR 54.21(a)(1).

The staff also reviewed the fire protection CLB documents listed in the Operating License Conditions (2.C(7), 2.C(6), and 2.F) and NUREG-0857, "Safety Evaluation Report related to the

Scoping and Screening Methodology

operation of Palo Verde Nuclear Generating Station Units 1, 2 and 3," dated November 1981 through Supplement 11.

This review included the applicant's commitments to 10 CFR 50.48, "Fire protection" (i.e., approved fire protection program), as provided in the responses to Appendix A to the BTP Auxiliary Systems Branch (APS) 9.5-1, documented in PVNGS, Units 1, 2, and 3, SER, and NUREG-0857, dated November 1981 through Supplement 11.

During its review of LRA Section 2.3.3.19, the staff identified areas in which additional information was necessary to complete its review of the applicant's scoping and screening results. The applicant responded to the staff's RAIs as discussed below.

In RAI 2.3.3.19-1, dated November 12, 2009, the staff questioned why a license renewal drawing for the fire protection system shows the fire water system and associated components (outside transformers) as outside the scope of license renewal (i.e., not highlighted). In the RAI, the staff requested that the applicant verify if these fire water systems and associated components are within the scope of license renewal, in accordance with 10 CFR 54.4(a) and subject to an AMR, in accordance with 10 CFR 54.21(a)(1). The staff asked that the applicant justify excluding these components from the scope of license renewal and an AMR.

In its response, dated December 23, 2009, the applicant provided scoping and screening results for the fire protection system components in question. The applicant stated the following:

> Transformers shown on drawing LR-PVNGS-FP-01-M-FPP-002 at coordinates B-2 through B-7 are electrical components within the scope of license renewal. These transformers are excluded from aging management review in accordance with 10 CFR 54.21(a)(1). NEI 95-10 Appendix B item 104 identifies that transformers are excluded from an AMR under 10 CFR 54.21(a)(1).

> These structures, systems, and components (SSCs) were not highlighted on drawing LR-PVNGS-FP-01-M-FPP-002 because those drawings are mechanical boundary drawings and only mechanical components within scope of license renewal are highlighted on mechanical boundary drawings. The scoping and screening methodology for mechanical systems is further detailed in the PVNGS LRA Section 2.1.3.1, and the scoping and screening methodology for electrical systems is further detailed in PVNGS LRA Section 2.1.3.3. A single license renewal drawing, LR-PVNGS-ELEC-E-MAA-001, was created for electrical [staff] based on the switchyard one-line diagram.

In evaluating this response, the staff found that it was incomplete and that review of LRA Section 2.3.3.19 could not be completed. The applicant did not explain why fire water systems installed on outdoor transformers are not within the scope of license renewal and subject to an AMR. The applicant indicated that electrical components are within the scope of license renewal, but transformers are excluded from an AMR in accordance with 10 CFR 54.21(a)(1). The staff's question concerned scoping and AMR of fire water systems installed on outdoor transformers. This resulted in the staff holding a telephone conference with the applicant on January 18, 2010, to discuss information necessary to resolve the concern in RAI 2.3.3.19-1. During the call, the applicant explained that the fire water systems were determined to have no license renewal intended function. The applicant stated that PVNGS complies with Appendix A to BTP APCSB 9.5-1 by locating all oil-filled transformers 50-feet from any building containing safety-related systems.

The staff reviewed the commitment to 10 CFR 50.48, "Fire protection" (i.e., approved fire protection program) and performed a point-by-point comparison with BTP APCSP 9.5-1 as documented in the UFSAR, Table 9B.3-1. Section D.1(h) of the BTP APCSP 9.5-1 recommends that buildings containing safety-related systems should be protected from

Scoping and Screening Methodology

exposure or spill fires involving oil filled transformers by location of the transformers at least 50 feet away or verifying that building walls within 50-feet of oil filled transformers are without openings and have a fire resistance rating of at least 3-hours. UFSAR Table 9B.3-1 states that oil filled transformers are located at least 50-feet from any building containing safety-related systems with the exception of the west ESF transformer, which is located approximately 48 feet from the 3-hour rated auxiliary building exterior wall. Based on the applicant's information in UFSAR Table 9B.3-1 and the applicant's compliance statements, the staff finds that the fire protection systems for the subject outdoor oil filled transformers are correctly excluded from the scope of the license renewal. Therefore, the staff's concern described in RAI 2.3.3.19-1 is resolved.

In RAI 2.3.3.19-2, dated November 12, 2009, the staff questioned why the license renewal fire protection drawing identified the fuel tank dikes and fire protection system valves and drains in the cable spreading room and radwaste building as out of the scope of license renewal.

In the RAI, the staff requested that the applicant verify if these fire protection system components are within the scope of license renewal, in accordance with 10 CFR 54.4(a) and subject to an AMR, in accordance with 10 CFR 54.21(a)(1). The staff requested that the applicant justify excluding these components from the scope of license renewal and an AMR.

In its response, dated December 23, 2009, the applicant provided scoping and screening results for the fire protection system components in question in the license renewal drawing. For the upper and lower cable spreading room and radwaste building, the applicant stated the following:

> LRA drawing LR-PVNGS-FP-01-M-FPP-003 has been revised to reflect that the fire water system valves and drains in the Upper and Lower Cable Spreading Room and Radwaste Building are within the scope of license renewal. The additional components have been highlighted in green on Revision 1 of LRA drawing LR-PVNGS- FP-01-M-FPP-003.

Based on its review, the staff finds the applicant's response to RAI 2.3.3.19-2 acceptable because it indicated that the fire protection system and the components in question are within scope of license renewal and subject to an AMR.

For fire water system valves and drains, the applicant stated the following:

> LRA drawing LR-PVNGS-FP-01-M-FPP-006 has been revised to reflect that the fire water system valves and drains are within the scope of license renewal, with the exception of the valves noted below. The additional components are highlighted in green on Revision 1 of LRA drawing LR-PVNGS-FP-01-M-FPP-006.

> The following valves shown on LRA drawing LR-PVNGS-FP-01-M-FPP-006 are not within the scope of license renewal because they are not part of the criteria (a)(3) pressure boundary and do not have a criterion (a)(2) function. Criterion (a)(2) components are non-safety-related and are in rooms where there is potential for spatial interaction with safety-related equipment. These valves are not located in rooms with the potential for spatial interaction.

> Fire water system valves not within the scope of license renewal:

> - V407 & V408, Preaction Valve V711 vent & drain
> - V419 & V420, Preaction Valve V729 vent & drain
> - V421 & V422, Preaction Valve V726 vent & drain
> - V423 & V424, Preaction Valve V723 vent & drain

2-65

Scoping and Screening Methodology

- V425 & V426, Preaction Valve V732 vent & drain

Based on its review, the staff finds the applicant's response to RAI 2.3.3.19-2 acceptable because it indicated that the water system valves and drains in question are within the scope of license renewal and subject to an AMR, with the exception of the valves mentioned above. These valves are drain valves and can be isolated from the rest of the system to maintain the fire water system pressure boundary. Therefore, these valves are not relied on to perform a pressure boundary intended function, and are not subject to an AMR. The applicant stated that the failure of the drain valves will not prevent satisfactory accomplishment of a safety function and are not required to demonstrate compliance with 10 CFR 54.4(a)(3). The staff agrees with the applicant's exclusion of the valves from the scope of license renewal since these valves can be isolated and do not perform a 10 CFR 54.4(a)(1) or (3) function; therefore, the staff's concern is resolved.

For fire water system valves, drains, and fuel tank dikes on the license renewal drawing for fire protection, the applicant stated the following:

> LRA drawing LR-PVNGS-FP-AO-M-FPP-001 has been revised to show fire water system valves and drains within the scope of license renewal, with the exception of the items noted below. The additional components are highlighted in green on Revision 1 of LRA drawing LR-PVNGS-FP-AO-M-FPP-001.
>
> Portions of the fire water system shown on drawing LR-PVNGS-FP-AO-M-FPP-001 and listed below do not have a criteria (a)(3) function or a criterion (a)(2) function. Criterion (a)(2) components are non-safety-related and are in rooms where there is potential for spatial interaction with safety-related equipment. These valves are not located in rooms with the potential for spatial interaction.
>
> Portions of the Fire Water System not within the scope of license renewal:
>
> - Caustic Injection Pump & associated piping and valves
> - Sulfite Injection Pump & associated piping and valves
> - Motor Driven Recirculation Pump & associated piping and valves (for chemical addition)
> - Fire Water Pump test/recirculation piping and valves
> - Water Reclamation fire suppression piping and valves
> - Chemical Storage Building fire suppression piping and valves
> - Vehicle Maintenance Facility fire suppression piping and valves
> - Station Blackout Gas Turbine General Area fire suppression piping and valves
> - Low Level Radioactive Waste Material Storage Facility fire suppression piping and valves
>
> The fuel oil tank dikes are in the scope of license renewal and screened in as generic concrete components in LRA Section 2.4.13, "Yard Structures." The fuel oil tank dikes are not highlighted in green on LRA drawing LR-PVNGS-FP-AO-M-FPP-001 because they are not mechanical components. Only mechanical components within scope of license renewal are highlighted on mechanical boundary drawings. A single license renewal drawing, LR-PVNGS-STR-OOB-001, was created for structures based on the site plan.

As a result of these changes, LRA Tables 2.3.3-19 and 3.3.2-19 have been revised and are included in Enclosure 2 as LRA Amendment 6. A cast iron orifice for the branch line associated with the jockey pump in the Fire Pump House has been added to LRA Tables 2.3.3-19 and 3.3.2-19. The material for the small diameter system air check valves for preaction valves was corrected from cast iron to copper alloy and resulted in an additional line on LRA Table 2.3.3-19 for the interior environment associated with the copper alloy check valves.

Based on its review, the staff finds the applicant's response to RAI 2.3.3.19-2 acceptable because it indicated that the water system valves and drains in question are within the scope of license renewal and subject to an AMR, with exception of the certain fire protection components. The components do not have a license renewal intended function and are, therefore, outside the scope of license renewal and are not subject to an AMR. Further, the applicant identified that the fuel oil tank dikes are within the scope of license renewal and included in LRA Section 2.4.13, "Yard Structures." In addition, the applicant identified which portions of the fire protection system and components are excluded from the scope of license renewal in accordance with 10 CFR 54.21(a)(2) and 10 CFR 54.21(a)(3). The staff's concern is resolved.

In RAI 2.3.3.19-3, dated November 12, 2009, the staff questioned why a license renewal fire protection drawing identified certain components associated with the carbon dioxide (CO_2) fire suppression system in the Auxiliary Building as outside the scope of license renewal and not subject to an AMR.

In the RAI, the staff requested that the applicant verify whether these CO_2 fire suppression system components are in the scope of license renewal, in accordance with 10 CFR 54.4(a) and subject to an AMR, in accordance with 10 CFR 54.21(a)(1). The staff requested that the applicant justify excluding these components from the scope of license renewal and an AMR.

In its response, dated December 23, 2009, the applicant provided the scoping and screening results for the CO_2 fire suppression system components in question in license renewal drawing LR-PVNGS-FP-01-M-FPP-004. For the CO_2 fire suppression system components, the applicant stated the following:

> LRA drawing LR-PVNGS-FP-01-M-FPP-004 has been revised (Revision 1) to show the following additional components within scope of license renewal (highlighted in green):
>
> - Drain valve V380 at coordinate B-5
>
> - Switchgear Building CO_2 hose stations
>
> - CO_2 Storage Unit components VE052, VW001, V743, UV116 and associated piping
>
> The Electric Vaporizer and CO_2 supply line for the Main Generator purge do not have a license renewal intended function and are not within the scope of license renewal.

Based on its review, the staff finds applicant's response to RAI 2.3.3.19-3 acceptable because it indicated that the CO_2 fire suppression system components in question are within the scope of license renewal and subject to an AMR. Further, the applicant determined that the electric vaporizer and CO_2 supply line for the main generator have no license renewal intended function and, therefore, are not within the scope of license renewal and subject to an AMR. The staff finds this acceptable, and the staff's concern is resolved.

Scoping and Screening Methodology

In RAI 2.3.3.19-4, dated November 12, 2009, the staff stated that LRA Tables 2.3.3-19 and 3.3.2-19 exclude several types of fire protection components that appear in NUREG-0857, "Safety Evaluation Report Related to the Operation of Palo Verde Nuclear Generating Station, Units 1, 2, and 3," and license renewal drawings. These components include the following:

- Pipe fittings, pipe supports, hangers, and couplings
- Fire hose stations, connections, and racks
- Filter housings
- Halon 1301 storage bottles
- Dikes for oil spill confinement
- Floor drains and curbs for fire-fighting water
- Passive components in the diesel fuel fire pump

The staff requested that the applicant verify whether LRA Tables 2.3.3-19 and 3.3.2-19 should include the components listed above. If they are excluded from the scope of license renewal and not subject to an AMR, the staff requested that the applicant justify their exclusion.

In its response, dated December 23, 2009, the applicant stated that it evaluated the list of components and determined no changes to the LRA are required as described below:

> Pipe fittings and couplings. Pipe fittings and couplings are evaluated as the component type "piping." This is consistent with the definition of piping, piping components, and piping elements noted in the Generic Aging Lessons Learned Report, Chapter IX.B. The component type "piping" is identified in LRA Table 2.4-14 as a component within the scope of license renewal and subject to an AMR.

> Pipe supports and hangers. Fire protection pipe supports and hangers are evaluated as structural component type "supports non-ASME" and are identified in LRA Table 2.4-14 as components within the scope of license renewal and subject to an AMR.

> Fire hose stations, connections and racks. Fire hose stations, connections and racks are evaluated as component types "piping" and "valve" and are identified in LRA Table 2.3.3-19 as components within the scope of license renewal and subject to an AMR.

> Filter housings. Filter housings are evaluated as part of the component type "Filter" in LRA Table 2.3.3-19 as components within the scope of license renewal and subject to an AMR.

> Halon 1301 storage bottles. The component type of "tank" specified in LRA Table 2.3.3-19 includes Halon 1301 storage bottles within the scope of license renewal and subject to an AMR. This is consistent with the definition of tanks noted in the Generic Aging Lessons Learned Report, Chapter IX.B.

> Dikes for oil spill confinement. The response to question 9A.86(c) documents that in the diesel generator building, the day tank room door curbs are sized to a height to contain the full volume of the day tank and its associated piping. These curbs are evaluated as component type "concrete element" and are identified in LRA Table 2.4-3 as components within the scope of license renewal and subject to an AMR for fire protection. There are no oil containment dikes in the outside areas that are within the scope of license renewal and subject to an AMR. The UFSAR, Table 9B.3-1, Section D.1(h), documents that all oil filled transformers are located at least 50 feet from any building containing safety-related systems with the exception of the west ESF transformer, which is located approximately

Scoping and Screening Methodology

48 feet from the 3-hour-rated auxiliary building exterior wall. Therefore, dikes for oil spill confinement are not within the scope of license renewal. The UFSAR, Section 3.8.4.1 documents that the diesel generator fuel oil storage tanks are located underground with approximately 10 feet of earth cover.

Floor drains and curbs for fire-fighting water. Floor drains are evaluated as component type "piping" and are identified in LRA Table 2.3.3-30 as components within the scope of license renewal and subject to an AMR. Curbs for containing fire-fighting water are evaluated as component type "concrete element" and are identified for each building in LRA Section 2.4 as components within the scope of license renewal and subject to an AMR.

Passive components in the diesel fuel fire pump engine. These components do not have unique component identification numbers and are integral to the diesel engine, and are evaluated as part of the engine. The fire pump diesel engine is an active component and is excluded from an AMR in accordance with 10 CFR 54.21(a)(1) as further detailed in NEI 95-10 Appendix B, which states that fire pump diesel engines are excluded from an AMR under 10 CFR 54.21(a)(1).

In reviewing the applicant's response to the RAI, the staff found it had resolved each item in the RAI. Although the description of the "piping" line item, provided in LRA Tables 2.3.3-19 and 3.3.2-19, does not list these components specifically, the applicant stated that it considers this line item to include the pipe fittings and couplings. In addition, the applicant addressed pipe support and hangers in LRA Table 2.4-14, "Supports," under component type "support non-ASME." The applicant said that it evaluated the fire hose station, connection, and racks under the component types "piping" and "valves" in LRA Table 2.3.3.19, with AMR results provided in Table 3.3.2-19. The applicant addressed the filter housing under the component type "filter" in LRA Table 2.3.3.19, with AMR results in Table 3.3.2.19. Similarly, the applicant addressed Halon 1301 storage bottles under the component type "tank" in LRA Table 2.3.3.19, with AMR results in Table 3.3.2.19. The applicant addressed the structural AMR of dikes for oil spill confinement and curbs for containing fire-fighting water under the component type "concrete element" in LRA Table 2.4-3, "Diesel Generator Building." Floor drains for fire-fighting water are considered under component type "piping" in LRA Table 2.3.3-30, "Miscellaneous Auxiliary Systems In-Scope only based on Criterion 10 CFR 54.4(a)(2)."

In its response, the applicant also confirmed that it evaluated passive components in the diesel fuel fire pump engine as part of the engine because these components do not have unique component identification numbers and are integral to the diesel engine. Further, the staff agrees with the applicant's determination that the fire pump diesel engines are excluded from an AMR under 10 CFR 54.21(a)(1).

Based on its review, the staff finds the applicant's response to RAI 2.3.3.19-4 acceptable because it confirms that the components in question are within the scope of license renewal and subject to an AMR. The staff's concern is resolved.

2.3.3.19.3 Conclusion

The staff reviewed the LRA Section 2.3.3.19, UFSAR Section 9.5.1 and Appendix 9B, RAI responses, and license renewal drawings to determine if the applicant failed to identify any fire protection systems and components within the scope of license renewal. In addition, the staff's review determined if the applicant failed to identify any components subject to an AMR. On the basis of its review, the staff concludes that the applicant has adequately identified the fire protection system components within the scope of license renewal, as required by 10 CFR 54.4(a). The staff also finds that the applicant has adequately identified the fire

Scoping and Screening Methodology

protection system components subject to an AMR in accordance with the requirements stated in 10 CFR 54.21(a)(1).

2.3.3.20 Diesel Generator Fuel Oil Storage and Transfer System

2.3.3.20.1 Summary of Technical Information in the Application

LRA Section 2.3.3.20 describes the diesel generator fuel oil storage and transfer system, which is comprised of an underground diesel fuel oil storage tank, diesel fuel oil transfer pump, diesel fuel oil day tank, piping, valves, and instrumentation for each diesel generator. The purpose of the system is to provide fuel oil for the emergency diesel generators (EDGs).

LRA Table 2.3.3-20 identifies the components subject to an AMR for the diesel generator fuel oil storage and transfer system by component type and intended function. The diesel generator fuel oil storage and transfer system contains safety-related components relied upon to remain functional during and following DBEs. The failure of portions of the nonsafety-related SSCs in the system potentially could prevent the satisfactory accomplishment of a safety-related function. In addition, portions of the system perform functions that support fire protection requirements.

2.3.3.20.2 Staff Evaluation

The staff reviewed LRA Section 2.3.3.20, UFSAR Section 9.5.4 and Appendix 9B.2.1, and the license renewal boundary drawings using the evaluation methodology described in SER Section 2.3 and the guidance in SRP-LR Section 2.3. The staff's review identified an area in which additional information was necessary to complete the review of the applicant's scoping and screening results. The applicant responded to the staff's RAI as discussed below.

In RAI 2.3.3.20-1, dated December 3, 2009, the staff noted that a drawing specified in LRA Section 2.3.3.20, could not be found in the package. The staff asked the applicant to provide additional information to verify which drawing was the correct drawing to use during the scoping and screening review.

In its response, dated January 18, 2010, the applicant stated that the drawing number in LRA Section 2.3.3.20 is incorrect, but the provided drawing is the correct drawing. The applicant submitted a correction to the LRA.

Based on its review, the staff finds the applicant's response to RAI 2.3.3.20-1 acceptable because the staff confirmed that the applicant corrected LRA Section 2.3.3.20 to include the appropriate drawing. Therefore, the staff's concern described in RAI 2.3.3.20-1 is resolved.

2.3.3.20.3 Conclusion

The staff reviewed the LRA Section 2.3.3.20, UFSAR Section 9.5.4 and Appendix 9B.2.1, RAI response, and boundary drawings to determine if the applicant failed to identify any components within the scope of license renewal. In addition, the staff's review determined if the applicant failed to identify any components subject to an AMR. On the basis of its review, the staff concludes the applicant has appropriately identified the diesel generator fuel oil storage and transfer system mechanical components within the scope of license renewal, as required by 10 CFR 54.4(a). The staff also finds that the applicant has adequately identified the mechanical components subject to an AMR in accordance with the requirements stated in 10 CFR 54.21(a)(1).

2.3.3.21 Diesel Generator

2.3.3.21.1 Summary of Technical Information in the Application

LRA Section 2.3.3.21 describes the diesel generator system, which is comprised of two diesel generators per unit, each driven by a four-cycle, 20-cylinder diesel engine. The purpose of the

Scoping and Screening Methodology

system is to provide a reliable onsite power source capable of starting and supplying the essential loads necessary to shut down the plant safely and maintain it in a safe shutdown condition if a loss of offsite power should occur.

The diesel generator system consists of the following subsystems:

- Diesel generator cooling water system
- Diesel generator starting system
- Diesel generator lubrication system
- Diesel generator combustion air intake and exhaust system
- Fuel oil system

LRA Table 2.3.3-21 identifies the components subject to an AMR for the diesel generator system by component type and intended function. The diesel generator system contains safety-related components relied upon to remain functional during and following DBEs. The failure of portions of the nonsafety-related SSCs in the system potentially could prevent the satisfactory accomplishment of a safety-related function. In addition, portions of the system perform functions that support fire protection and SBO requirements.

2.3.3.21.2 Staff Evaluation

The staff reviewed LRA Section 2.3.3.21; UFSAR Sections 7.4.1.1.1, 8.3.1.1.4, 9.5.4, 9.5.5, 9.5.6, 9.5.7, and 9.5.8 and the license renewal boundary drawings using the evaluation methodology described in SER Section 2.3 and the guidance in SRP-LR Section 2.3. The staff's review identified areas in which additional information was necessary to complete the review of the applicant's scoping and screening results. The applicant responded to the staff's RAIs as discussed below.

In RAI 2.3.3.21-1, dated December 3, 2009, the staff noted components that are within the scope of license renewal for criteria 10 CFR 54.4(a)(1) or 10 CFR 54.4(a)(2) but are not included in Table 2.3.3-21. The staff asked the applicant to provide additional information explaining why specified components are not included as component types subject to an AMR in LRA Table 2.3.3-21.

In its response, dated January 18, 2010, the applicant stated that individual component names and component types in the LRA are consistent with the component names and component types as they appear in the plant equipment database as follows:

- Diesel air intake silencers are included in LRA Table 2.3.3-21 with the component type "filter."

- Local observation glasses, LG 344 and LG 343, are included in LRA Table 2.3.3-21 as component type "sight gauge."

- Turbocharger housings are included in LRA Table 2.3.3-21 as component type "blower."

- Diesel generator air intake manifolds are evaluated as within scope of license renewal as integral parts of the EDG.

- Starting air headers are included in LRA Table 2.3.3-21 and subject to AMR with the component type "piping."

In a follow-up response, dated March 1, 2010, the applicant stated that it evaluated injector housings as within the scope of license renewal as integral parts of the EDG.

Based on its review, the staff finds the applicant's response to RAI 2.3.3.21-1 acceptable because the applicant clarified how each of the components listed above that are subject to an

Scoping and Screening Methodology

AMR are presented in Table 2.3.3-21 and provided the rationale for those components not subject to an AMR. Therefore, the staff's concern described in RAI 2.3.3.21-1 is resolved.

In RAI 2.3.3.21-2, dated December 3, 2009, the staff noted that the three pipelines lack drawing continuation information at the end locations of criterion 10 CFR 54.4(a)(2) pipe. The staff asked the applicant to provide additional information to locate the license renewal boundary.

In its response, dated January 18, 2010, the applicant stated that the appropriate license renewal terminal component symbol has been added to the end locations of each of the three 10 CFR 54.4(a)(2) pipes. The applicant provided revised drawings to the staff.

Based on its review, the staff finds the applicant's response to RAI 2.3.3.21-2 acceptable because the appropriate license renewal terminal component symbol has been added to the drawing continuation locations. The staff verified the additions by reviewing the revised drawings. Therefore, the staff's concern described in RAI 2.3.3.21-2 is resolved.

2.3.3.21.3 Conclusion

The staff reviewed the LRA Section 2.3.3.21; UFSAR Sections 7.4.1.1.1, 8.3.1.1.4, 9.5.4, 9.5.5, 9.5.6, 9.5.7, and 9.5.8; RAI responses; and original and revised boundary drawings to determine if the applicant failed to identify any components within the scope of license renewal. In addition, the staff's review determined if the applicant failed to identify any components subject to an AMR. On the basis of its review, the staff concludes the applicant has appropriately identified the diesel generator mechanical components within the scope of license renewal, as required by 10 CFR 54.4(a), and that the applicant has adequately identified the mechanical components subject to an AMR in accordance with the requirements stated in 10 CFR 54.21(a)(1).

2.3.3.22 Domestic Water System

2.3.3.22.1 Summary of Technical Information in the Application

LRA Section 2.3.3.22 describes the domestic water system, which is comprised of a well water supply subsystem, a water treatment subsystem, and a storage and transfer subsystem, which are all shared facilities. The LRA states that the purpose of the system is to process local onsite well water to remove suspended solids and part of the dissolved solids, chlorinate and neutralize the processed water, and store and transfer the domestic water to each unit and common facilities at the site.

LRA Table 2.3.3-22 identifies the components subject to an AMR for the domestic water system by component type and intended function. The failure of portions of the nonsafety-related SSCs in the domestic water system potentially could prevent the satisfactory accomplishment of a safety-related function. In addition, portions of the system perform functions that support fire protection requirements.

2.3.3.22.2 Staff Evaluation

The staff reviewed LRA Section 2.3.3.22, UFSAR Section 9.2.4, and the license renewal boundary drawings using the evaluation methodology described in SER Section 2.3 and the guidance in SRP-LR Section 2.3. The staff's review identified areas in which additional information was necessary to complete the review of the applicant's scoping and screening results. The applicant responded to the staff's RAIs as discussed below.

In RAI 2.3.3.22-1, dated December 3, 2009, the staff noted that a license renewal drawing for the domestic water system shows the continuation of two 1½-inch lines as within the scope of license renewal. The continuations on the associated drawing are not within the scope of license renewal. The staff asked the applicant to provide additional information to clarify the scoping classification for these pipe sections.

Scoping and Screening Methodology

In its response, dated January 18, 2010, the applicant stated the highlighting of the continuation of the 1½-inch lines was inadvertently omitted on the drawing. The applicant revised the drawing to highlight these continuations and provided it to the staff. Based on its review, the staff finds the applicant's response to RAI 2.3.3.22-1 acceptable because the applicant has revised the drawing to highlight these lines as within scope. The staff verified the additions by reviewing the revised drawings. Therefore, the staff's concern described in RAI 2.3.3.22-1 is resolved.

In RAI 2.3.3.22-2, dated December 3, 2009, the staff noted that a license renewal drawing for the domestic water system shows a continuation of line that could not be located. The applicant was requested to provide additional information to locate the license renewal boundary.

In its response, dated January 18, 2010, the applicant stated that the line has been cut and capped and no longer continues to the associated drawing. The drawing was revised to show that the line is cut and capped and does not continue. The applicant provided the staff with a revised drawing. Based on its review, the staff finds the applicant's response to RAI 2.3.3.22-2 acceptable because the applicant stated the line is currently cut and capped and does not continue. The staff verified the additions by reviewing the revised drawings. Therefore, the staff's concern described in RAI 2.3.3.22-2 is resolved.

In RAI 2.3.3.22-3, dated December 3, 2009, the staff noted license renewal drawings for the domestic water system show components and associated relief valves and drain lines as not within the scope of license renewal. The staff noted that similar components on other domestic water system drawings are within the scope of license renewal. The staff asked the applicant to provide additional information to explain why the components and associated relief valves and drain lines are not within the scope of license renewal.

In its response, dated January 18, 2010, the applicant stated that it revised the drawing to reflect that the components and associated relief and drain lines have been added to the scope of license renewal based on the criterion of 10 CFR 54.4(a)(2). The applicant provided the staff with revised drawings. Additionally, the applicant revised LRA Section 3.3.2.1.22 and LRA Table 3.3.2-22 to add components in a wetted gas environment.

Based on its review, the staff finds the applicant's response to RAI 2.3.3.22-3 acceptable because the applicant added the components with its associated relief valves and drain lines as within the scope of license renewal. The staff verified the additions by reviewing the revised drawings. Therefore, the staff's concern described in RAI 2.3.3.22-3 is resolved.

2.3.3.22.3 Conclusion

The staff reviewed the LRA Section 2.3.3.22, UFSAR Section 9.2.4, RAI responses, and original and revised boundary drawings to determine if the applicant failed to identify any components within the scope of license renewal. In addition, the staff's review determined if the applicant failed to identify any components subject to an AMR. On the basis of its review, the staff concludes the applicant has appropriately identified the domestic water system mechanical components within the scope of license renewal, as required by 10 CFR 54.4(a), and that the applicant has adequately identified the mechanical components subject to an AMR in accordance with the requirements stated in 10 CFR 54.21(a)(1).

2.3.3.23 Demineralized Water System

2.3.3.23.1 Summary of Technical Information in the Application

LRA Section 2.3.3.23 describes the demineralized water system, which is comprised of piping components, pumps, tanks and demineralizers. It removes dissolved gas and solids from the water processed from the reverse osmosis units of the domestic water system. The system also stores and transfers demineralized water to multiple systems in each unit and to common facilities in the chemical production system.

LRA Table 2.3.3-23 identifies the components subject to an AMR for the demineralized water system by component type and intended function. The demineralized water system contains safety-related components relied upon to remain functional during and following DBEs. The failure of portions of the nonsafety-related SSCs in the system potentially could prevent the satisfactory accomplishment of a safety-related function.

2.3.3.23.2 Staff Evaluation and Conclusion

The staff reviewed LRA Section 2.3.3.23, UFSAR Section 9.2.3, and the license renewal boundary drawings using the evaluation methodology described in SER Section 2.3 and the guidance in SRP-LR Section 2.3 to determine if the applicant failed to identify any components within the scope of license renewal. In addition, the staff's review determined if the applicant failed to identify any components subject to an AMR. The staff found no such omissions. On the basis of its review, the staff concludes the applicant has appropriately identified the demineralized water system mechanical components within the scope of license renewal, as required by 10 CFR 54.4(a), and that the applicant has adequately identified the mechanical components subject to an AMR in accordance with the requirements stated in 10 CFR 54.21(a)(1).

2.3.3.24 Water Reclamation Facility Fuel System

2.3.3.24.1 Summary of Technical Information in the Application

LRA Section 2.3.3.24 describes the WRF fuel system that receives, stores, and supplies fuel oil for the lime re-calcining furnaces, the auxiliary boilers, and the SBO generators.

LRA Table 2.3.3-24 identifies the component types subject to an AMR for the WRF fuel system by component type and intended function. Portions of the WRF fuel system perform functions that support SBO.

2.3.3.24.2 Staff Evaluation

The staff reviewed LRA Section 2.3.3.24, UFSAR Sections 1.2.10.3.9 and 8.3.1.1.10, and the license renewal boundary drawings using the evaluation methodology described in SER Section 2.3 and the guidance in SRP-LR Section 2.3. The staff's review identified areas in which additional information was necessary to complete the review of the applicant's scoping and screening results. The applicant responded to the staff's RAIs as discussed below.

In RAI 2.3.3.24-1, dated December 3, 2009, the staff noted the license renewal drawing for the WRF fuel system shows two lines as not within the scope of license renewal that are connected to tanks which are shown as within the scope of license renewal. There is also no indication as to where these lines go. The staff asked the applicant to provide additional information as to why these lines are not in the scope of license renewal and to identify the line continuations.

In its response, dated January 18, 2010, the applicant provided additional information and submitted a revised license renewal boundary drawing to locate the license renewal boundary.

Based on its review, the staff finds the applicant's response to RAI 2.3.3.24-1 acceptable because the applicant provided boundary locations. This was verified by staff review of the revised drawing. Therefore, the staff's concern described in RAI 2.3.3.24-1 is resolved.

In RAI 2.3.3.24-2, dated December 3, 2009, the staff noted on the license renewal drawing for the WRF fuel system that certain pipe lines attached to the fuel oil storage tanks lack drawing continuation information at the end locations of pipe. The staff asked the applicant to provide additional information to locate the license renewal boundaries.

Scoping and Screening Methodology

In its response, dated January 18, 2010, the applicant provided additional information for each of the locations on the drawing including the license renewal boundaries. Based on its review, the staff finds the applicant's response to RAI 2.3.3.24-2 acceptable because the boundary locations were provided. Therefore, the staff's concern described in RAI 2.3.3.24-2 is resolved.

2.3.3.24.3 Conclusion

The staff reviewed the LRA Section 2.3.3.24, UFSAR Sections 1.2.10.3.9 and 8.3.1.1.10, RAI responses, and original and revised boundary drawings to determine if the applicant failed to identify any components within the scope of license renewal. In addition, the staff's review determined if the applicant failed to identify any components subject to an AMR. On the basis of its review, the staff concludes the applicant has appropriately identified the WRF fuel system mechanical components within the scope of license renewal, as required by 10 CFR 54.4(a), and that the applicant has adequately identified the mechanical components subject to an AMR in accordance with the requirements stated in 10 CFR 54.21(a)(1).

2.3.3.25 Service Gases (Nitrogen and Hydrogen) System

2.3.3.25.1 Summary of Technical Information in the Application

LRA Section 2.3.3.25 describes the service gases system, which consists of two subsystems that supply nitrogen and hydrogen gas to various systems.

LRA Table 2.3.3-25 identifies the components subject to an AMR for the service gases system by component type and intended function. The applicant stated that the service gases system contains safety-related components relied upon for containment isolation. The failure of portions of the nonsafety-related SCs in the system potentially could prevent the satisfactory accomplishment of a safety-related function. In addition, portions of the system perform functions that support EQ and SBO requirements.

2.3.3.25.2 Staff Evaluation

The staff reviewed LRA Section 2.3.3.25; UFSAR Sections 6.2.4, 9.3.6, and Table 6.4.2-1; and the license renewal boundary drawings using the evaluation methodology described in SER Section 2.3 and the guidance in SRP-LR Section 2.3. The staff's review identified an area in which additional information was necessary to complete the review of the applicant's scoping and screening results. The applicant responded to the staff's RAI as discussed below.

In RAI 2.3.3.25-1, dated December 3, 2009, the staff noted the license renewal service gases system drawing shows continuation of a pipe section in scope for license renewal from the nitrogen accumulators to another drawing. Review of the associated drawing shows no apparent continuation. The staff asked the applicant to provide additional information to locate the license renewal boundary.

In its response, dated January 18, 2010, the applicant stated that the continuation of the pipelines is within the scope of license renewal and specified the drawing. Based on its review, the staff finds the applicant's response to RAI 2.3.3.25-1 acceptable because the applicant provided information to locate the license renewal boundaries. The staff reviewed the drawing and verified the license renewal boundaries. Therefore, the staff's concern described in RAI 2.3.3.25-1 is resolved.

2.3.3.25.3 Conclusion

The staff reviewed LRA Section 2.3.3.25, UFSAR Sections 6.2.4, 9.3.6, and Table 6.4.2-1, RAI response, and original and revised boundary drawings to determine if the applicant failed to identify any components within the scope of license renewal. In addition, the staff's review determined if the applicant failed to identify any components subject to an AMR. On the basis of its review, the staff concludes the applicant has appropriately identified the service gases system mechanical components within the scope of license renewal, as required by 10 CFR 54.4(a), and that the applicant has adequately identified the mechanical components subject to an AMR in accordance with the requirements stated in 10 CFR 54.21(a)(1).

2.3.3.26 Gaseous Radwaste System

2.3.3.26.1 Summary of Technical Information in the Application

LRA Section 2.3.3.26 describes the gaseous radwaste system, which is comprised of a waste gas surge tank and three waste gas decay tanks, waste gas compressors, piping, and valves. The system collects and processes radioactive and potentially radioactive waste gas and limits the release of gaseous activity.

LRA Table 2.3.3-26 identifies the components subject to an AMR for the gaseous radwaste system by component type and intended function. The gaseous radwaste system contains safety-related components relied upon for containment isolation. The failure of portions of the nonsafety-related SCs in the system potentially could prevent the satisfactory accomplishment of a safety-related function. In addition, portions of the system perform functions that support EQ requirements.

2.3.3.26.2 Staff Evaluation

The staff reviewed LRA Section 2.3.3.26, UFSAR Section 11.3 and Table 6.2.4-1, and the license renewal boundary drawings using the evaluation methodology described in SER Section 2.3 and the guidance in SRP-LR Section 2.3. The staff's review identified an area in which additional information was necessary to complete the review of the applicant's scoping and screening results. The applicant responded to the staff's RAI as discussed below.

In RAI 2.3.3.26-1, dated December 3, 2009, the staff noted the license renewal drawing for the gaseous radwaste system shows continuation of a 1-inch pipe section that is shown as within the scope of license renewal from the volume control tank relief to another drawing. The staff noted that the pipe section on the associated drawing was not within the scope of license renewal. The staff asked the applicant to provide additional information to clarify the scoping classification for this pipe section.

In its response, dated January 18, 2010, the applicant stated the drawing was not correct. The 1-inch line is not within scope of license renewal based on the scoping criteria of 10 CFR 54.4(a). The applicant provided the staff with a revised drawing. Based on its review, the staff finds the applicant's response to RAI 2.3.3.26-1 acceptable because the applicant clarified that the 1-inch line was not scoped correctly. The staff reviewed the drawing and verified the changes. Therefore, the staff's concern described in RAI 2.3.3.26-1 is resolved.

2.3.3.26.3 Conclusion

The staff reviewed the LRA Section 2.3.3.26, UFSAR Section 11.3 and Table 6.2.4-1, RAI response, and original and revised boundary drawings to determine if the applicant failed to identify any components within the scope of license renewal. In addition, the staff's review determined if the applicant failed to identify any components subject to an AMR. On the basis of its review, the staff concludes the applicant has appropriately identified the gaseous radwaste system mechanical components within the scope of license renewal, as required by 10 CFR 54.4(a), and that the applicant has adequately identified the mechanical components subject to an AMR in accordance with the requirements stated in 10 CFR 54.21(a)(1).

2.3.3.27 Radioactive Waste Drains System

2.3.3.27.1 Summary of Technical Information in the Application

LRA Section 2.3.3.27 describes the radioactive waste drains system, which consists of piping components, valves, filters, drains and pumps. The radioactive waste drains system collects non-corrosive, radioactive or potentially radioactive liquid wastes from equipment and floor drains of various buildings and pumps the collected liquid wastes to the liquid radwaste system for processing. In addition, the radioactive waste drains system provides for indication of flooding of watertight rooms and leakage detection for the refueling pool and fuel pool liner plate.

LRA Table 2.3.3-27 identifies the components subject to an AMR for the radioactive waste drains system by component type and intended function. The radioactive waste drains system contains safety-related components relied upon for containment isolation. The failure of portions of the nonsafety-related SCs in the system potentially could prevent the satisfactory accomplishment of a safety-related function. In addition, portions of the system perform functions that support EQ requirements.

2.3.3.27.2 Staff Evaluation

The staff reviewed LRA Section 2.3.3.27, UFSAR Section 9.3.3 and Table 6.2.4-1, and the license renewal boundary drawings using the evaluation methodology described in SER Section 2.3 and the guidance in SRP-LR Section 2.3. The staff's review identified areas in which additional information was necessary to complete the review of the applicant's scoping and screening results. The applicant responded to the staff's RAIs as discussed below.

In RAI 2.3.3.27-1, dated December 3, 2009, the staff noted the license renewal drawing for the radioactive waste drains system shows a section of piping continuing from another drawing (essential ACU) as not within the scope of license renewal. The essential ACU piping section is included within scope for 10 CFR 54.4(a)(2). The staff asked the applicant to provide additional information to justify why this section of piping is included within the scope of license renewal on one drawing (essential ACU drawing) and not within the scope of license renewal on the radioactive waste drains system drawing.

In its response, dated January 18, 2010, the applicant stated the portion of the line extending onto the radioactive waste drains system license renewal boundary drawing has been revised to include the line within the scope of license renewal based on criterion 10 CFR 54.4(a)(2). The applicant provided the staff with a revised drawing. Based on its review, the staff finds the applicant's response to RAI 2.3.3.27-1 acceptable because the applicant indicated the continuation of the line as within scope. The staff reviewed the drawing and verified the changes. Therefore, the staff's concern described in RAI 2.3.3.27-1 is resolved.

In RAI 2.3.3.27-2 and RAI 2.3.3.27-3, dated December 3, 2009, the staff noted the license renewal radioactive waste drains system drawing shows continuations of pipe sections, in scope for license renewal, to drawings that show the lines as not within the scope of license renewal. The staff asked the applicant to provide additional information to explain the discrepancy.

In its response, dated January 18, 2010, the applicant stated the continuations of the pipe sections are within scope of license renewal and should have been shown as within scope for 10 CFR 54.4(a)(2). The applicant provided the staff with revised drawings for both RAIs. Based on its review, the staff finds the applicant's response to RAI 2.3.3.27-2 and RAI 2.3.3.27-3 acceptable because the applicant included the continuations in scope for 10 CFR 54.4(a)(2) on the revised drawings. The staff reviewed the drawings and verified the changes. Therefore, the staff's concerns, described in RAI 2.3.3.27-2 and RAI 2.3.3.27-3, are resolved.

2.3.3.27.3 Conclusion

The staff reviewed the LRA Section 2.3.3.27, UFSAR Section 9.3.3 and Table 6.2.4-1, RAI responses, and original and revised boundary drawings to determine if the applicant failed to identify any components within the scope of license renewal. In addition, the staff's review determined if the applicant failed to identify any components subject to an AMR. On the basis of its review, the staff concludes the applicant has appropriately identified the radioactive waste drains system mechanical components within the scope of license renewal, as required by 10 CFR 54.4(a), and that the applicant has adequately identified the mechanical components subject to an AMR in accordance with the requirements stated in 10 CFR 54.21(a)(1).

2.3.3.28 Station Blackout Generator System

2.3.3.28.1 Summary of Technical Information in the Application

LRA Section 2.3.3.28 describes the SBO generator system, which consists of two 100-percent capacity turbine generators. The SBO generator system provides alternate AC power to necessary station loads via the safety-related 4.16 kV buses during an SBO event in any one unit.

LRA Table 2.3.3-28 identifies the components subject to an AMR for the SBO generator system by component type and intended function. Portions of the SBO generator system perform functions that support SBO requirements.

2.3.3.28.2 Staff Evaluation and Conclusion

The staff reviewed LRA Section 2.3.3.28, UFSAR Section 1.2.10.3.9, and the license renewal boundary drawings using the evaluation methodology described in SER Section 2.3 and the guidance in SRP-LR Section 2.3 to determine if the applicant failed to identify any components within the scope of license renewal. In addition, the staff's review determined if the applicant failed to identify any components subject to an AMR. The staff found no such omissions. On the basis of its review, the staff concludes the applicant has appropriately identified the SBO generator system mechanical components within the scope of license renewal, as required by 10 CFR 54.4(a). In addition, the staff finds that the applicant has adequately identified the mechanical components subject to an AMR in accordance with the requirements stated in 10 CFR 54.21(a)(1).

2.3.3.29 Cranes, Hoists, and Elevators

2.3.3.29.1 Summary of Technical Information in the Application

LRA Section 2.3.3.29 describes the cranes, hoists, and elevators group, which provide lifting and maneuvering capacity in various buildings. The applicant identified the following cranes and trolleys in the LRA as within the scope of license renewal:

- Auxiliary building 4-ton trolleys
- Main steam supply system south room 140-foot elevation hoist assemblies
- Diesel generator building train A, 5-ton bridge crane
- Diesel generator building train B, 5-ton bridge crane
- Diesel generator building room 1, 25-ton trolley
- Diesel generator building room 2, 25-ton trolley

LRA Table 2.3.3-29 identifies the components subject to an AMR for the cranes, hoists, and elevators group by component type and intended function. The cranes, hoists, and elevators group contains nonsafety-related components, the failure of which potentially could prevent the satisfactory accomplishment of a safety-related function.

2.3.3.29.2 Staff Evaluation and Conclusion

The staff reviewed LRA Section 2.3.3.29 and the license renewal boundary drawings using the evaluation methodology described in SER Section 2.3 and the guidance in SRP-LR Section 2.3 to determine if the applicant failed to identify any components within the scope of license renewal. In addition, the staff's review determined if the applicant failed to identify any components subject to an AMR. The staff found no such omissions. On the basis of its review, the staff concludes the applicant has appropriately identified the cranes, hoists and elevators

Scoping and Screening Methodology

group mechanical components within the scope of license renewal, as required by 10 CFR 54.4(a), and that the applicant has adequately identified the mechanical components subject to an AMR in accordance with the requirements stated in 10 CFR 54.21(a)(1).

2.3.3.30 Miscellaneous Auxiliary Systems In-Scope ONLY for Criterion 10 CFR 54.4(a)(2)

2.3.3.30.1 Summary of Technical Information in the Application

LRA Section 2.3.3.30 describes those nonsafety-related auxiliary systems or nonsafety-related portions of auxiliary systems with the potential for adverse spatial interaction with safety-related systems or components. The applicant identified the following auxiliary systems in the LRA as within the scope of license renewal based only on the criterion of 10 CFR 54.4(a)(2):

Auxiliary Steam. The auxiliary steam system consists of an auxiliary boiler, transfer pumps, receivers, tanks, piping, and valves, and provides a source of steam for various nonsafety-related functions during plant startup, shutdown, normal operations, and testing evolutions.

Chemical Waste. The chemical waste system consists of the following five sub-systems: (1) the radioactive chemical waste sub-system that collects the corrosive radioactive waste from the chemical laboratory and decontamination stations; (2) the cooling water waste sub-system that collects the chemically treated cooling water from the auxiliary and radwaste buildings for reuse or disposal; (3) the condensate polisher regeneration waste sub-system that collects the rinse washes from the condensate polisher demineralizers and neutralizes the waste for disposal; (4) the spent regenerate waste sub-system that collects and neutralizes the rinse washes from the makeup demineralizers for disposal; and (5) the yard areas chemical tank drains sub-system.

Liquid Radwaste. The liquid radwaste system collects, processes, monitors, and recycles or disposes of liquid radwaste. The liquid radwaste system consists of instrumentation and process components such as piping, filters, pumps, tanks, and an evaporator.

Oily Waste and Non-Radioactive Waste. The oily waste and non-radioactive waste system collects and transports liquid waste from equipment and floor drains of the turbine building, the control building, the diesel generator buildings, the fire pump house, and the yard area. The system removes entrained oil from the wastewater for disposal and conveys the oil-free water to the evaporation pond.

Solid Radwaste. The solid radwaste system consists of the following four subsystems: (1) spent resin transfer subsystem, (2) wet waste processing subsystem, (3) dry waste disposal subsystem, and (4) the filter handling and disposal subsystem. The system, comprised of piping, valves, tanks and pumps, provides processing and packaging capability for concentrated waste solutions and spent resins.

Sanitary Sewage and Treatment. The sanitary sewage and treatment system collects the sanitary wastewater from facilities throughout the plant through drain piping and transports it through one wet well, one sewage lift station, and one surge tank to the three package sewage treatment units, where the waste is treated and clarified.

Secondary Chemical Control. The secondary chemical control system is an integrated system that operates concurrently to maintain the required operating water chemistry of the condensate and feedwater under all normal operating and upset or abnormal conditions. It is comprised of the following four subsystems: (1) the condensate demineralizer subsystem that maintains required water chemistry of the condensate and feedwater loop during upset or abnormal conditions; (2) the SG blowdown processing subsystem that compensates for the concentrating effect of the SGs by continuous blowdown and processing; (3) the chemical monitoring and addition subsystems that establish and maintain the proper chemistry within the condensate,

Scoping and Screening Methodology

feedwater, and SG secondary side water and provide continuous indication of significant chemical parameters in the secondary system; and (4) the online process sampling subsystem that takes continuous samples from the main condenser, condensate demineralizers, main feedwater lines, SG blowdown lines and downcomer, and circulating water lines.

LRA Table 2.3.3-30 identifies the components subject to an AMR for the miscellaneous systems within the scope of license renewal only for criterion 10 CFR 54.4(a)(2) by component type and intended function. The miscellaneous systems described above contain nonsafety-related SCs that potentially could prevent the satisfactory accomplishment of a safety-related function. Portions of the system (oily waste and non-radioactive waste system) perform functions that support fire protection requirements.

2.3.3.30.2 Staff Evaluation

The staff reviewed LRA Section 2.3.3.30, UFSAR Sections 3.6, 9.3.2, 9.3.3, 10.4.6, 11.2.2.3, and 11.4, and the license renewal boundary drawings using the evaluation methodology described in SER Section 2.3 and the guidance in SRP-LR Section 2.3. The staff's review identified areas in which additional information was necessary to complete the review of the applicant's scoping and screening results. The applicant responded to the staff's RAIs as discussed below.

In RAI 2.3.3.30-1, dated December 3, 2009, the staff noted the license renewal drawing for the non-radioactive waste system shows sections of 4-inch lines as not within the scope of license renewal. The two lines are connected to lines which are shown within the scope of license renewal based on 10 CFR 54.4(a)(2). The staff asked the applicant to justify why the 4-inch lines, attached to lines that are within scope for license renewal, are shown as not within the scope for license renewal.

In its response, dated January 18, 2010, the applicant stated that the 4-inch lines are drain lines from the fuel oil day tank floors, and they are within the scope of license renewal. The drawing was revised to add this piping and associated components within the scope of license renewal for 10 CFR 54.4(a)(3). In addition, the applicant revised LRA Table 2.2-1 and Section 2.3.3.30 to delete the oily waste and non-radioactive waste system due to the addition of the fire protection intended functions. Various LRA sections were created to identify that the oily waste and non-radioactive waste system also supports fire protection requirements based on the criteria of 10 CFR 54.4(a)(3).

Based on its review, the staff finds the applicant's response to RAI 2.3.3.30-1 acceptable because the applicant reanalyzed the components in question and found them to be within scope for license renewal per 10 CFR 54.4(a)(3), and it revised the LRA accordingly. The staff verified the changes by reviewing the revised drawing and LRA sections. Therefore, the staff's concern described in RAI 2.3.3.30-1 is resolved.

In RAI 2.3.3.30-2, dated December 3, 2009, the staff noted that 4-inch piping lines on the license renewal drawing for oily waste and non-radioactive waste system are shown as within scope for 10 CFR 54.4(a)(2). Several lines attached to these 4-inch lines, however, are shown as not within the scope of license renewal for 10 CFR 54.4(a)(2). The staff asked the applicant to provide additional information to justify why the attached lines are not within the scope of license renewal.

In its response, dated January 18, 2010, the applicant stated that all vent and drain lines attached to the 4-inch drain lines are within the scope of license renewal in accordance with criterion of 10 CFR 54.4(a)(2). The applicant provided a revised drawing. In addition, the applicant revised LRA Section 3.3.2.1.31 and LRA Table 3.3.2-31 to add the oily waste and non-radioactive system vent lines to the scope of license renewal. This resulted in revisions to

2-80

LRA Table 3.3.2-31 and the addition of a poly-vinyl chloride (PVC) material to the list of oily waste and non-radioactive system materials in LRA Section 3.3.2.1.31.

Based on its review, the staff finds the applicant's response to RAI 2.3.3.30-2 acceptable because the applicant reanalyzed the components in question and justified the associated license renewal boundaries. The staff reviewed the revised drawings and verified the changes. Therefore, the staff's concern described in RAI 2.3.3.30-2 is resolved.

In RAI 2.3.3.30-3, dated December 3, 2009, the staff noted that a 4-inch piping line on the license renewal drawing for oily waste and non-radioactive waste system was shown as not within scope for 10 CFR 54.4(a)(2). Several adjacent lines are shown as within the scope of license renewal. The staff asked the applicant to provide additional information to justify why the attached lines are not within the scope of license renewal.

In its response, dated January 18, 2010, the applicant stated that highlighting of the continuation of the 4-inch pipe line was inadvertently omitted on the license renewal boundary drawing. The applicant revised the drawing to show the 4-inch pipe line continuation as highlighted, including the 6-inch pipe line where the piping connects to the sump cover. Based on its review, the staff finds the applicant's response to RAI 2.3.3.30-3 acceptable because the applicant acknowledged the inadvertent omission of highlighted piping, clarified the license renewal boundary, and revised the LRA drawings accordingly. The staff reviewed the revised drawings and verified the changes. Therefore, the staff's concern described in RAI 2.3.3.30-3 is resolved.

2.3.3.30.3 Conclusion

The staff reviewed the LRA Section 2.3.3.30; UFSAR Sections 3.6, 9.3.2, 9.3.3, 10.4.6, 11.2.2.3, and 11.4; RAI responses; and original and revised boundary drawings to determine if the applicant failed to identify any components within the scope of license renewal. In addition, the staff's review determined if the applicant failed to identify any components subject to an AMR. On the basis of its review, the staff concludes the applicant has appropriately identified the miscellaneous auxiliary systems in-scope only for criterion 10 CFR 54.4(a)(2) mechanical components within the scope of license renewal, as required by 10 CFR 54.4(a). In addition, the staff finds that the applicant has adequately identified the mechanical components subject to an AMR in accordance with the requirements stated in 10 CFR 54.21(a)(1).

2.3.4 Steam and Power Conversion Systems

LRA Section 2.3.4 identifies the steam and power conversion systems SCs within the scope of license renewal and subject to an AMR. The applicant described the supporting SCs of the steam and power conversion systems in the following LRA sections:

- 2.3.4.1, "Main Steam System"
- 2.3.4.2, "Condensate Storage and Transfer System"
- 2.3.4.3, "Auxiliary Feedwater System"
- 2.3.4.4, "Condensate System"
- 2.3.4.5, "Feedwater System"
- 2.3.4.6, "Main turbine System"
- 2.3.4.7, "SG Feedwater Pump Turbine System"
- 2.3.4.8, "Feedwater Heater Extraction, Drains and Vents System"

2.3.4.1 Main Steam System

2.3.4.1.1 Summary of Technical Information in the Application

LRA Section 2.3.4.1 describes the main steam system, which is comprised of the main steam supply system, turbine bypass system, portions of the feedwater system, and portions of the SG blowdown system. The main steam system delivers steam from the SGs to the high-pressure

Scoping and Screening Methodology

turbine for a range of operating conditions. Each main steam line contains a pneumatically-operated atmospheric dump valve, five spring-loaded safety valves, one main steam isolation valve, a cross-tie header, and associated vent and drain valves. The turbine bypass system contains eight air-operated valves and has the capability to remove the heat from the SGs to minimize transient effects on the RCS during startup, hot shutdown, cooldown, and load rejection.

LRA Table 2.3.4-1 identifies the components subject to an AMR for the main steam system by component type and intended function. The main steam system contains safety-related components relied upon to remain functional during and following DBEs. The failure of portions of the nonsafety-related SSCs in the system potentially could prevent the satisfactory accomplishment of a safety-related function. In addition, portions of the system perform functions that support fire protection, EQ, SBO, and ATWS requirements.

2.3.4.1.2   Staff Evaluation

The staff reviewed LRA Section 2.3.4.1; UFSAR Sections 10.3, 10.4.4, 10.4.7, and 10.4.9; and the license renewal boundary drawings using the evaluation methodology described in SER Section 2.3 and the guidance in SRP-LR Section 2.3. The staff's review identified areas in which additional information was necessary to complete the review of the applicant's scoping and screening results. The applicant responded to the staff's RAIs as discussed below.

In RAI 2.3.4.1-1, dated December 3, 2009, the staff noted license renewal drawings show several locations where the license renewal spatial interaction termination cannot be determined (listed below). The staff asked the applicant to provide additional information to locate the license renewal spatial interaction terminations.

> Drawing LR-PVNGS-SG-01-M-SGP-002:
>
> Piping N-007-DCDA-8" upstream of valve UV-172 (G-13)
> Piping N-010-DBDB-8" upstream of valve UV-175 (C-13)
> Piping E-039-DABA-6" downstream of valve UV-5000 (E-2)
> Piping E-048-DABA-6" downstream of valve UV-5008 (B-2)
>
> Drawing LR-PVNGS-SG-01-M-SGP-001-02:
>
> Piping N-335-HDDA-1" upstream of valve V346 (G-4)
> Piping N-335-HDDA-1" upstream of valve V348 (G-4)
> Piping N-321-HDDA-1" upstream of valve V358 (D-4)
> Piping N-321-HDDA-1" upstream of valve V357 (D-4)

In its response, dated January 18, 2010, the applicant provided additional information for each of the locations described above and submitted a revised license renewal boundary drawing to clarify the spatial interaction terminations. Based on its review, the staff finds the applicant's response to RAI 2.3.4.1-1 acceptable because it clarified the spatial interaction terminations. The staff reviewed the revised drawing and verified the spatial interaction terminations. Therefore, the staff's concern described in RAI 2.3.4.1-1 is resolved.

In RAI 2.3.4.1-2, dated December 3, 2009, the staff noted license renewal drawing LR-PVNGS-SG-01-M-SGP-002 (D-8 and G-8) shows two flow nozzles out of each of the SGs as not within the scope of license renewal. The staff asked the applicant to provide additional information explaining why the flow nozzles out of the SGs are not within the scope of license renewal.

In its response, dated January 18, 2010, the applicant stated the flow nozzles are in scope and submitted a revised license renewal boundary drawing. Based on its review, the staff finds the applicant's response to RAI 2.3.4.1-2 acceptable because the applicant stated the flow nozzles are within scope. The staff verified the modifications on the revised license renewal boundary

Scoping and Screening Methodology

drawing and found them acceptable. Therefore, the staff's concern described in RAI 2.3.4.1-2 is resolved.

In RAI 2.3.4.1-3, dated December 3, 2009, the staff noted license renewal drawing LR-PVNGS-SG-01-M-SGP-001-01 shows eight boxes on the main steam piping downstream of the main steam isolation valves that the applicant did not define (listed below). The staff asked the applicant to provide additional information explaining this box symbol and if this component type is subject to an AMR.

- Two boxes on piping E-206-DLBB-28 inches downstream of valve UV-170 (G-9)
- Two boxes on piping E-207-DLBB-28 inches downstream of valve UV-180 (E-9)
- Two boxes on piping E-208-DLBB-28 inches downstream of valve UV-171 (D-9)
- Two boxes on piping E-209-DLBB-28 inches downstream of valve UV-181 (B-9)

In its response, dated January 18, 2010, the applicant stated the box symbols represent pipe whip restraints that are within scope and subject to an AMR.

Based on its review, the staff finds the applicant's response to RAI 2.3.4.1-3 acceptable because the applicant stated the boxes represent pipe whip restraints and are within scope and subject to an AMR. Therefore, the staff's concern described in RAI 2.3.4.1-3 is resolved.

In RAI 2.3.4.1-4, dated December 3, 2009, the staff noted license renewal drawing LR-PVNGS-SG-01-M-SGP-001-02 (D-8 and G-8) shows two drag resistors N-299-HBDB-54" and N-300-HBDB-54" as not within the scope of license renewal. However, the inlet piping E-059-DLBB-12" and E-084-DLBB-12" as well as the outlet piping N-306-GBDB-1" and N-312-GBDB-1" are within the scope of license renewal. In addition, the applicant showed FX-178 and FX-179 as within the scope of license renewal inside of the drag resistors. The staff asked the applicant to provide additional information explaining why the drag resistors are not within the scope of license renewal.

In its response, dated January 18, 2010, the applicant stated the pipe shrouds and drag resistors have been added to the scope of license renewal for 10 CFR 54.4(a)(2). The applicant submitted a revised drawing. Based on its review, the staff finds the applicant's response to RAI 2.3.4.1-4 acceptable because the pipe shrouds and drag resistors to the scope of license renewal. The staff reviewed revised drawing LR-PVNGS-SG-01-M-SGP-001-02 and verified that the drawing shows the pipe shroud and drag resistors as within scope of license renewal. Therefore, the staff's concern described in RAI 2.3.4.1-4 is resolved.

2.3.4.1.3 Conclusion

The staff reviewed LRA Section 2.3.4.1; UFSAR Sections 10.3, 10.4.4, 10.4.7, and 10.4.9; RAI responses; and original and revised boundary drawings to determine if the applicant failed to identify any components within the scope of license renewal. In addition, the staff's review determined if the applicant failed to identify any components subject to an AMR. On the basis of its review, the staff concludes the applicant has appropriately identified the main steam system mechanical components within the scope of license renewal, as required by 10 CFR 54.4(a), and that the applicant has adequately identified the mechanical components subject to an AMR in accordance with the requirements stated in 10 CFR 54.21(a)(1).

2.3.4.2 Condensate Storage and Transfer System

2.3.4.2.1 Summary of Technical Information in the Application

LRA Section 2.3.4.2 describes the condensate storage and transfer system, which is comprised of one CST, two condensate transfer pumps, and associated piping, valves, and I&Cs. The condensate storage and transfer system provides the source of feedwater for the auxiliary feedwater system for reactor decay heat removal during hot standby conditions and for cooling the reactor. The system also maintains feedwater inventory in the secondary system during startup, shutdown, hot standby, and normal power operations. The condensate storage and transfer system also provides makeup water for the essential cooling water system, essential chilled water system, diesel generator system, and the spent fuel pool.

LRA Table 2.3.4-2 identifies the components subject to an AMR for the condensate storage and transfer system by component type and intended function. The condensate storage and transfer system contains safety-related components relied upon to remain functional during and following DBEs. The failure of portions of the nonsafety-related SSCs in the system potentially could prevent the satisfactory accomplishment of a safety-related function. In addition, portions of the system perform functions that support fire protection and SBO requirements.

2.3.4.2.2 Staff Evaluation

The staff reviewed LRA Section 2.3.4.2, UFSAR Sections 3.8.4.1.7 and 9.2.6, and the license renewal boundary drawings using the evaluation methodology described in SER Section 2.3 and the guidance in SRP-LR Section 2.3. The staff's review identified an area in which additional information was necessary to complete the review of the applicant's scoping and screening results. The applicant responded to the staff's RAI as discussed below.

In RAI 2.3.4.2-1, dated December 3, 2009, the staff noted license renewal drawing LR-PVNGS-CT-01-M-CTP-001 (C-2) shows line N-031-HCDA-3" as not within the scope of license renewal for 10 CFR 54.4(a)(2). However, the continuation of this 3-inch line on drawing LR-PVNGS-PC-01-M-PCP-001 (H-11) shows this line is within the scope of license renewal for 10 CFR 54.4(a)(2). The staff asked the applicant to provide additional information explaining why there is a difference in the scope classification between the drawing LR-PVNGS-CT-01-M-CTP-001 and the continuation drawing.

In its response, dated January 18, 2010, the applicant stated that it inadvertently omitted highlighting of the continuation of pipe line N-031-HCDA-3". The applicant revised the drawing to show the pipe line N-031-HCDA-3" continuation as within the scope of license renewal. The revised drawing was provided to the staff. Based on its review, the staff finds the applicant's response to RAI 2.3.4.2-1 acceptable because it corrected the drawing continuation discrepancy. The staff reviewed the revised drawing and verified the continuation of the line is highlighted as within scope for license renewal. Therefore, the staff's concern described in RAI 2.3.4.2-1 is resolved.

2.3.4.2.3 Conclusion

The staff reviewed LRA Section 2.3.4.2, UFSAR Sections 3.8.4.1.7 and 9.2.6, RAI response, and original and revised boundary drawings to determine if the applicant failed to identify any components within the scope of license renewal. In addition, the staff's review determined if the applicant failed to identify any components subject to an AMR. On the basis of its review, the staff concludes the applicant has appropriately identified the condensate storage and transfer system mechanical components within the scope of license renewal, as required by 10 CFR 54.4(a). The staff also finds that the applicant has adequately identified the mechanical components subject to an AMR in accordance with the requirements stated in 10 CFR 54.21(a)(1).

2.3.4.3 Auxiliary Feedwater System

2.3.4.3.1 Summary of Technical Information in the Application

LRA Section 2.3.4.3 describes the auxiliary feedwater system, which is comprised of two motor-driven pumps, one turbine-driven pump, and associated piping and valves. The auxiliary feedwater system supplies feedwater from the CST to the SGs during fill, startup, hot standby, normal shutdown, and emergency conditions.

Scoping and Screening Methodology

LRA Table 2.3.4-3 identifies the components subject to an AMR for the auxiliary feedwater system by component type and intended function. The auxiliary feedwater system contains safety-related components relied upon to remain functional during and following DBEs. The failure of portions of the nonsafety-related SSCs in the system potentially could prevent the satisfactory accomplishment of a safety-related function. In addition, portions of the system perform functions that support fire protection, EQ, SBO, and ATWS requirements.

2.3.4.3.2 Staff Evaluation and Conclusion

The staff reviewed LRA Section 2.3.4.3; UFSAR Sections 7.3.5, 9.2.6, and 10.4.9; and the license renewal boundary drawings to determine if the applicant failed to identify any components within the scope of license renewal. In addition, the staff's review determined if the applicant failed to identify any components subject to an AMR. The staff found no such omissions. On the basis of its review, the staff concludes the applicant has appropriately identified the auxiliary feedwater system mechanical components within the scope of license renewal, as required by 10 CFR 54.4(a), and that the applicant has adequately identified the mechanical components subject to an AMR in accordance with the requirements stated in 10 CFR 54.21(a)(1).

2.3.4.4 Condensate System

2.3.4.4.1 Summary of Technical Information in the Application

LRA Section 2.3.4.4 describes the condensate system, which is comprised of the main condenser, condenser hotwell, condensate pumps, and associated piping and valves. The condensate system collects condensate from the exhaust steam of the main turbines, feedwater pump turbines, and steam cycle drains in the main condenser hotwell and delivers de-aerated water to the suction of the main feedwater pumps.

LRA Table 2.3.4-4 identifies the components subject to an AMR for the condensate system by component type and intended function. The condensate system does not contain safety-related components relied upon to remain functional during and following DBEs. Portions of the system perform functions that support fire protection requirements. The function to meet these requirements is performed by electrical components; therefore, there are no mechanical components in the condensate system requiring an AMR.

2.3.4.4.2 Staff Evaluation and Conclusion

The staff reviewed LRA Section 2.3.4.4, UFSAR Sections 10.4.1 and 10.4.7, and applicable boundary drawings to determine if the applicant failed to identify any components within the scope of license renewal. In addition, the staff's review determined if the applicant failed to identify any components subject to an AMR. The staff found no such omissions. On the basis of its review, the staff concludes that the applicant has appropriately identified the condensate system mechanical components within the scope of license renewal, as required by 10 CFR 54.4(a), and that the applicant has adequately identified the mechanical system components subject to an AMR in accordance with the requirements stated in 10 CFR 54.21(a)(1).

2.3.4.5 Feedwater System

2.3.4.5.1 Summary of Technical Information in the Application

LRA Section 2.3.4.5 describes the feedwater system, which is comprised of two interconnected trains with turbine-driven feedwater pumps, three stages of high-pressure feedwater heaters, and associated piping, valves, and components. The feedwater system receives condensate from the condensate system and delivers feedwater, at required pressure and temperature, to the SGs.

Scoping and Screening Methodology

LRA Table 2.3.4-5 identifies the components subject to an AMR for the feedwater system by component type and intended function. The feedwater system does not contain safety-related components relied upon to remain functional during and following DBEs. Portions of the system perform functions that support fire protection requirements. The function to meet these requirements is performed by electrical components; therefore, there are no mechanical components in the feedwater system requiring an AMR.

2.3.4.5.2 Staff Evaluation and Conclusion

The staff reviewed LRA Section 2.3.4.5, UFSAR Section 10.4.7, and applicable boundary drawings to determine if the applicant failed to identify any components within the scope of license renewal. In addition, the staff's review determined if the applicant failed to identify any components subject to an AMR. The staff found no such omissions. On the basis of its review, the staff concludes the applicant has appropriately identified the feedwater system mechanical components within the scope of license renewal, as required by 10 CFR 54.4(a), and that the applicant has adequately identified the mechanical components subject to an AMR in accordance with the requirements stated in 10 CFR 54.21(a)(1).

2.3.4.6 Main Turbine System

2.3.4.6.1 Summary of Technical Information in the Application

LRA Section 2.3.4.6 describes the main turbine system, which is comprised of one double-flow high-pressure turbine, three double-flow low-pressure turbines, four moisture separator-re-heaters, and the associated piping, valves, and instrumentation. The purpose of the main turbine system is to convert steam from the main steam system to mechanical energy to drive the main generator. In addition, the system supplies extraction steam for feedwater heating and hot reheat steam for the main feedwater pump turbines.

LRA Table 2.3.4-6 identifies the components subject to an AMR for the main turbine system by component type and intended function. The main turbine system does not contain safety-related components relied upon to remain functional during and following DBEs. Portions of the system perform functions that support ATWS requirements. The function to meet these requirements is performed by electrical components; therefore, there are no mechanical components in the main turbine system requiring an AMR.

2.3.4.6.2 Staff Evaluation and Conclusion

The staff reviewed LRA Section 2.3.4.6, UFSAR Section 10.2, and applicable boundary drawings to determine if the applicant failed to identify any components within the scope of license renewal. In addition, the staff's review determined if the applicant failed to identify any components subject to an AMR. The staff found no such omissions. On the basis of its review, the staff concludes the applicant has appropriately identified the main turbine system mechanical components within the scope of license renewal, as required by 10 CFR 54.4(a), and that the applicant has adequately identified the mechanical components subject to an AMR in accordance with the requirements stated in 10 CFR 54.21(a)(1).

2.3.4.7 Steam Generator Feedwater Pump Turbine System

2.3.4.7.1 Summary of Technical Information in the Application

LRA Section 2.3.4.7 describes the SG feedwater pump turbine system, which is comprised of pump turbines, hydraulic actuator systems, and associated piping systems. The SG feedwater pump turbines provide the motive force to drive the SG turbine-driven feedwater pumps.

LRA Table 2.3.4-7 identifies the components subject to an AMR for the SG feedwater turbine system by component type and intended function. The SG feedwater turbine system does not contain safety-related components relied upon to remain functional during and following DBEs.

Portions of the system perform functions that support fire protection requirements. The function to meet these requirements is performed by electrical components and active portions of the main feedwater pump turbine stop valves. No passive mechanical components are relied upon to perform this function. Therefore, there are no mechanical components in the SG feedwater turbine system requiring an AMR.

2.3.4.7.2 Staff Evaluation and Conclusion

The staff reviewed LRA Section 2.3.4.7, UFSAR Section 10.4.7, and applicable boundary drawings to determine if the applicant failed to identify any components within the scope of license renewal. In addition, the staff's review determined if the applicant failed to identify any components subject to an AMR. The staff found no such omissions. On the basis of its review, the staff concludes the applicant has appropriately identified the SG feedwater pump turbine system mechanical components within the scope of license renewal, as required by 10 CFR 54.4(a). The staff also finds that the applicant has adequately identified the mechanical components subject to an AMR in accordance with the requirements stated in 10 CFR 54.21(a)(1).

2.3.4.8 Feedwater Heater Extraction, Drains, and Vents System

2.3.4.8.1 Summary of Technical Information in the Application

LRA Section 2.3.4.8 describes the feedwater heater extraction, drains, and vents system, which is comprised of three trains of four-stage low-pressure heaters, two trains of three-stage high-pressure heaters, and associated piping, drains, and vents. The feedwater heater extraction, drains, and vents system supplies preheated feedwater to the SGs to improve cycle efficiency and minimize thermal stress on the feedwater piping and SG feedwater nozzles.

LRA Table 2.3.4-8 identifies the components subject to an AMR for the feedwater heater extraction, drains, and vents system by component type and intended function. The feedwater heater extraction, drains, and vents system does not contain safety-related components relied upon to remain functional during and following DBEs. Portions of the system perform functions that support fire protection requirements. The function to meet these requirements is performed by electrical components; therefore, there are no mechanical components in the feedwater heater extraction, drains, and vents system requiring an AMR.

2.3.4.8.2 Staff Evaluation and Conclusion

The staff reviewed LRA Section 2.3.4.8, UFSAR Sections 10.2.2 and 10.4.7, and applicable boundary drawings to determine if the applicant failed to identify any components within the scope of license renewal. In addition, the staff's review determined if the applicant failed to identify any components subject to an AMR. The staff found no such omissions. On the basis of its review, the staff concludes the applicant has appropriately identified the feedwater heater extraction, drains, and vents system mechanical components within the scope of license renewal, as required by 10 CFR 54.4(a). In addition, the staff finds that the applicant has adequately identified the mechanical components subject to an AMR in accordance with the requirements stated in 10 CFR 54.21(a)(1).

2.4 Scoping and Screening Results: Structures

This section documents the staff's review of the applicant's scoping and screening results for structures. Specifically, this section discusses the following structures:

- 2.4.1, "Containment Building"
- 2.4.2, "Control Building"
- 2.4.3, "Diesel Generator Building"

Scoping and Screening Methodology

- 2.4.4, "Turbine Building"
- 2.4.5, "Auxiliary Building"
- 2.4.6, "Radwaste Building"
- 2.4.7, "Main Steam Support Structure"
- 2.4.8, "Station Blackout Generator Structures"
- 2.4.9, "Fuel Building"
- 2.4.10, "Spray Pond and Associated Water Control Structures"
- 2.4.11, "Tank Foundations and Shells"
- 2.4.12, "Transformer Foundations and Electrical Structures"
- 2.4.13, "Yard Structures (In-scope)"
- 2.4.14, "Supports"
- 2.4.15, "Fire Barriers"

In accordance with the requirements of 10 CFR 54.21(a)(1), the applicant listed passive, long-lived SCs that are within the scope of license renewal and subject to an AMR. To verify that the applicant properly implemented its methodology, the staff focused its review on the implementation results. The staff confirmed that there were no omissions of structural components that meet the scoping criteria and are subject to an AMR.

The staff evaluated the information provided in the LRA in the same manner for all structures. The objective of the review was to determine if the structural components that appeared to meet the scoping criteria specified in 10 CFR 54.4 were identified by the applicant as within the scope of license renewal, in accordance with 10 CFR 54.4. Similarly, the staff evaluated the applicant's screening results to verify that all long-lived, passive SCs were subject to an AMR in accordance with 10 CFR 54.21(a)(1).

To perform its evaluation, the staff reviewed the applicable LRA sections, focusing its review on SCs that had not been identified as within the scope of license renewal. The staff reviewed relevant licensing basis documents, including the UFSAR, for each structure to determine if the applicant had omitted components with intended functions delineated under 10 CFR 54.4(a) from the scope of license renewal. The staff also reviewed licensing basis documents to determine if the applicant specified all intended functions delineated under 10 CFR 54.4(a) in the LRA. If the staff found omissions, it requested additional information to resolve the discrepancies.

Once the staff completed its review of the scoping results, the staff evaluated the applicant's screening results. For those SCs with intended functions, the staff sought to determine if the functions are performed with moving parts or a change in configuration or properties or if they are subject to replacement based on a qualified life or specified time period, as described in 10 CFR 54.21(a)(1). For those that did not meet either of these criteria, the staff sought to confirm that these structural components were subject to an AMR as required by 10 CFR 54.21(a)(1). If the staff found discrepancies, it requested additional information to resolve them.

2.4.1 Containment Building

2.4.1.1 Summary of Technical Information in the Application

In LRA Section 2.4.1, the applicant stated that the containment building is a seismic Category I structure. The shell of the building is a pre-stressed, reinforced concrete, cylindrical structure with a hemispherical dome roof. The applicant also stated that the containment building foundation is a conventionally reinforced concrete mat, circular in plan, constructed separately from other structures. The applicant further stated that the interior of the containment building shell is lined with carbon steel plates welded together to form a barrier which is essentially leak tight. The liner is thickened locally around the penetrations, large brackets, and major attachments that transfer loads through the liner plate to the concrete structure. Attachments to the shell wall are brackets for support of the polar crane, electrical conduit and cable tray, spray piping, lighting, and ventilation. The major structural components of the containment building are as follows:

- Post-tensioning system
- Steel liner plate
- Penetrations
- Containment building internal structures
- Emergency sumps

LRA Table 2.4-1 identifies the components subject to an AMR for the containment building by component type and intended function.

2.4.1.2 Staff Evaluation

The staff reviewed LRA Section 2.4.1 using the evaluation methodology described in SER Section 2.4 and the guidance in SRP-LR Section 2.4. During its review of the LRA Section 2.4.1, the staff identified areas in which additional information was necessary to complete the evaluation of the applicant's scoping and screening results for the containment building.

In RAI 2.4.1-1, dated November 2, 2009, the staff requested that the applicant provide additional information to confirm that the hatches and plug, caulking, and sealant are to be included in the scope of license renewal and subject to an AMR or to justify their exclusion.

In its response to the RAI, dated December 17, 2009, the applicant stated that the containment access hatches and associated components are identified as component types "hatch-emergency airlock," "hatch-equipment," and "hatch-personnel airlock" in LRA Table 2.4-1. Component type "hatch-equipment" includes the concrete missile barrier for the equipment hatch. The applicant also stated that it noted these components as within the scope of license renewal and subject to an AMR. The applicant further stated that the seals in the containment building are identified as component types "compressible joints and seals" or "fire barrier seals" in LRA Table 2.4-1. LRA Table 2.4-1 lists these components as components within the scope of license renewal and subject to an AMR.

Based on its review, the staff finds the response to RAI 2.4.1-1 acceptable because the hatches and plug, caulking, and sealant, as described by the applicant, do perform a 10 CFR 54.4(a) intended function for license renewal and are subject to an AMR. The staff's concern described in RAI 2.4.1-1 is resolved.

2.4.1.3 Conclusion

The staff reviewed LRA Section 2.4.1; UFSAR Sections 2.5.4.5, 3.8.1, 6.2.2, 6.2.6.2 and Appendix 9B.2.11; RAI response; and applicable boundary drawings to determine if the applicant failed to identify any components within the scope of license renewal. In addition, the staff's review determined if the applicant failed to identify any components subject to an AMR. The staff found no such omissions. On the basis of its review, the staff concludes the applicant has appropriately identified the containment building structural components within the scope of license renewal, as required by 10 CFR 54.4(a), and that the applicant has adequately identified the structural components subject to an AMR in accordance with the requirements stated in 10 CFR 54.21(a)(1).

Scoping and Screening Methodology

2.4.2 Control Building

2.4.2.1 *Summary of Technical Information in the Application*

In LRA Section 2.4.2, the applicant stated that the control building is a safety-related, seismic Category I structure that provides support, shelter, and protection to ESF and nuclear auxiliary systems. The applicant also stated that the control building is within the scope of license renewal based on the criteria of 10 CFR 54.4(a)(1). The applicant further stated that the control building shelters and protects nonsafety-related SSCs whose failure could prevent performance of a safety-related function. Therefore, it is within the scope of license renewal based on the criterion of 10 CFR 54.4(a)(2). The applicant further stated that the portions of the control and corridor buildings support fire protection, ATWS, and SBO requirements based on the criteria of 10 CFR 54.4(a)(3).

LRA Table 2.4-2 lists the components subject to an AMR for the control building by component type and intended function.

2.4.2.2 *Staff Evaluation and Conclusion*

The staff reviewed LRA Section 2.4.2; UFSAR Sections 2.5.4.8.1, 3.8.4.1.3, and 3.8.4; and applicable boundary drawings to determine if the applicant failed to identify any components within the scope of license renewal. In addition, the staff's review determined if the applicant failed to identify any components subject to an AMR. The staff found no such omissions. On the basis of its review, the staff concludes the applicant has appropriately identified the control building structural components within the scope of license renewal, as required by 10 CFR 54.4(a), and that the applicant has adequately identified the structural components subject to an AMR in accordance with the requirements stated in 10 CFR 54.21(a)(1).

2.4.3 Diesel Generator Building

2.4.3.1 *Summary of Technical Information in the Application*

In LRA Section 2.4.3, the applicant stated that the diesel generator building is a seismic Category I, multi-story, box-type, structural steel and reinforced concrete structure, which houses the EDGs, fuel oil day tanks, exhaust silencers, and exhaust stacks. The applicant stated that the diesel generator building provides structural support and protection of components relied upon to perform a safe shutdown and maintain the plant in a safe shutdown condition; therefore, it is with the scope of license renewal based on the criteria of 10 CFR 54.4(a)(1). The applicant also stated that the diesel generator building shelters and protects nonsafety-related SSCs whose failure could prevent performance of a safety-related function. Therefore, it is within the scope of license renewal based on the criteria of 10 CFR 54.4(a)(2). The applicant further stated that the portions of the diesel generator building support SBO requirements and provide support, shelter, and protection for components necessary to demonstrate compliance with fire protection requirements. The diesel generator building is within the scope of license renewal based on the criteria of 10 CFR 54.4(a)(3).

LRA Table 2.4-3 identifies the components subject to an AMR for the diesel generator building by component type and intended function.

2.4.3.2 *Staff Evaluation and Conclusion*

The staff reviewed LRA Section 2.4.3.2, UFSAR Section 3.8.4.1.4 and Appendix 9B2.4, and applicable boundary drawings to determine if the applicant failed to identify any components within the scope of license renewal. In addition, the staff's review determined if the applicant failed to identify any components subject to an AMR. The staff found no such omissions. On the basis of its review, the staff concludes the applicant has appropriately identified the diesel generator building structural components within the scope of license renewal, as required by

2.4.4 Turbine Building

2.4.4.1 Summary of Technical Information in the Application

In LRA Section 2.4.4, the applicant stated that the turbine building is a seismic Category II structure, whose behavior was analyzed under the extreme environmental loads (e.g., tornado, safe-shutdown earthquake) to verify that a collapse would not occur. The applicant also stated that the turbine building is within the scope of license renewal based on the criteria of 10 CFR 54.4(a)(2). The applicant further stated that the turbine building physically supports and protects systems and components that are required for fire protection and SBO based on the criteria of 10 CFR 54.4(a)(3).

LRA Table 2.4-4 identifies the components subject to an AMR for the turbine building by component type and intended function.

2.4.4.2 Staff Evaluation

The staff reviewed LRA Section 2.4.4 and UFSAR Sections 3.3.2.3, 3.8.4.4, and Appendix 9B.2.20.1 using the evaluation methodology described in SER Section 2.4 and the guidance in SRP-LR Section 2.4.

During its review of the LRA Section 2.4.4, the staff identified areas in which additional information was necessary to complete the evaluation of the applicant's scoping and screening results for the turbine building.

In RAI 2.4.4-1, dated November 2, 2009, the staff requested that the applicant provide additional information to confirm the inclusion or justify the exclusion of the compressible joints and seals, since it is not clear if it was included in the LRA Table 2.4-4 as being within the scope of license renewal and subsequently evaluated for an AMR.

In its response to the RAI, dated December 17, 2009, the applicant stated that the structural seals in the turbine building are identified as component types "fire barrier seals" or "roofing membrane" in LRA Table 2.4-4. The applicant stated that these components are identified as within the scope of license renewal and subject to an AMR.

Based on its review, the staff finds the response to RAI 2.4.4-1 acceptable because the compressible joints and seals in the turbine building, as described by the applicant, do perform a 10 CFR 54.4(a) intended function for license renewal and are subject to an AMR. The staff's concern described in RAI 2.4.4-1 is resolved.

2.4.4.3 Conclusion

The staff reviewed LRA Section 2.4.4; UFSAR Sections 3.3.2.3, 3.8.4.4, and Appendix 9B.2.20.1; RAI response; and applicable boundary drawings to determine if the applicant failed to identify any components within the scope of license renewal. In addition, the staff's review determined if the applicant failed to identify any components subject to an AMR. The staff found no such omissions. On the basis of its review, the staff concludes the applicant has appropriately identified the turbine building structural components within the scope of license renewal, as required by 10 CFR 54.4(a), and that the applicant has adequately identified the structural components subject to an AMR in accordance with the requirements stated in 10 CFR 54.21(a)(1).

Scoping and Screening Methodology

2.4.5 Auxiliary Building

2.4.5.1 Summary of Technical Information in the Application

In LRA Section 2.4.5, the applicant stated that the auxiliary building is a multi-story structural steel and reinforced concrete seismic Category I structure, housing the safety injection system, containment spray system, containment combustible gas control system, CVCS, and containment isolation system. The applicant also stated that the auxiliary building provides support, shelter, and protection to the ESF and nuclear auxiliary systems. In addition, the auxiliary building performs functions that support fire protection and SBO.

LRA Table 2.4-5 identifies the components subject to an AMR for the auxiliary building by component type and intended function.

2.4.5.2 Staff Evaluation

The staff reviewed LRA Section 2.4.5 using the evaluation methodology described in SER Section 2.4 and the guidance in SRP-LR Section 2.4.

During its review of the LRA Section 2.4.5, the staff identified areas in which additional information was necessary to complete the evaluation of the applicant's scoping and screening results for the auxiliary building.

In RAI 2.4.5-1, dated November 2, 2009, the staff requested that the applicant provide additional information to confirm that the boot seal penetrations are to be included in the scope of license renewal and subject to an AMR or to justify their exclusion.

In its response to the RAI, dated December 17, 2009, the applicant stated that the structural seals in the auxiliary building are identified as component types in the LRA as, "roofing membrane," "fire barrier seals," "compressible joints or seals," or "caulking or sealant." The component type "caulking or sealant" includes boot seal penetrations. LRA Table 2.4-5 identifies these components as within the scope of license renewal and subject to an AMR.

Based on its review, the staff finds the response to RAI 2.4.5-1 acceptable because the boot seal penetrations in the auxiliary building, as described by the applicant, do perform a 10 CFR 54.4(a) intended function for license renewal and are subject to an AMR. The staff's concern described in RAI 2.4.5-1 is resolved.

2.4.5.3 Conclusion

The staff reviewed LRA Section 2.4.5, UFSAR Sections 2.5.4.8.1 and 3.8.4.1, RAI response, and applicable boundary drawings to determine if the applicant failed to identify any components within the scope of license renewal. In addition, the staff's review determined if the applicant failed to identify any components subject to an AMR. The staff found no such omissions. On the basis of its review, the staff concludes the applicant has appropriately identified the auxiliary building structural components within the scope of license renewal, as required by 10 CFR 54.4(a), and that the applicant has adequately identified the structural components subject to an AMR in accordance with the requirements stated in 10 CFR 54.21(a)(1).

2.4.6 Radwaste Building

2.4.6.1 Summary of Technical Information in the Application

In LRA Section 2.4.6, the applicant stated that the radwaste building is a rectangular, multistory, reinforced concrete structure that houses radioactive waste treatment facilities, tanks, filters, and other miscellaneous equipment. The applicant also stated that the radwaste building is a seismic Category II structure, whose behavior was checked under the extreme environmental (e.g., tornado, safe-shutdown earthquake) loads to verify that a collapse would not occur. This ensures that external safety-related SSCs would not be damaged by the radwaste building during a DBE; therefore, the radwaste building is within the scope of license renewal under the criteria of 10 CFR 54.4(a)(2). The applicant also stated that the radwaste building physically supports and protects systems and components that are required for fire protection based on the criteria of 10 CFR 54.4(a)(3).

LRA Table 2.4-6 identifies the components subject to an AMR for the radwaste building by component type and intended function.

2.4.6.2 Staff Evaluation and Conclusion

The staff reviewed LRA Section 2.4.6, UFSAR Sections 3.8.4.4 and Table 9B3-1, and applicable boundary drawings to determine if the applicant failed to identify any components within the scope of license renewal. In addition, the staff's review determined if the applicant failed to identify any components subject to an AMR. The staff found no such omissions. On the basis of its review, the staff concludes the applicant has appropriately identified the radwaste building structural components within the scope of license renewal, as required by 10 CFR 54.4(a), and that the applicant has adequately identified the structural components subject to an AMR in accordance with the requirements stated in 10 CFR 54.21(a)(1).

2.4.7 Main Steam Support Structure

2.4.7.1 Summary of Technical Information in the Application

In LRA Section 2.4.7, the applicant stated that the MSSS is a box-type reinforced concrete seismic Category I structure supported by a reinforced concrete basemat founded on granular backfill. It houses the atmospheric dump valves, main steam isolation valves, feedwater isolation valves, and essential auxiliary feedwater pumps and their equipment.

The applicant stated that MSSS is safety-related since it provides support, shelter, and protection to ESFs and nuclear auxiliary systems; therefore, it is within the scope of license renewal based on the criteria of 10 CFR 54.4(a)(1). The applicant also stated that the MSSS shelters and protects nonsafety-related SSCs whose failure could prevent performance of a safety-related function; therefore, it is within the scope of license renewal based on the criterion of 10 CFR 54.4(a)(2).

The applicant finally stated that the portions of the MSSS support fire protection and SBO requirements based on the criteria of 10 CFR 54.4(a)(3).

LRA Table 2.4-7 identifies the components subject to an AMR for the MSSS by component type and intended function.

2.4.7.2 Staff Evaluation and Conclusion

The staff reviewed LRA Section 2.4.7, UFSAR Sections 2.5.4.8.1 and 3.8.4.1.5, and applicable boundary drawings to determine if the applicant failed to identify any components within the scope of license renewal. In addition, the staff's review determined if the applicant failed to identify any components subject to an AMR. The staff found no such omissions. On the basis of its review, the staff concludes the applicant has appropriately identified the MSSS structural components within the scope of license renewal, as required by 10 CFR 54.4(a), and that the applicant has adequately identified the structural components subject to an AMR in accordance with the requirements stated in 10 CFR 54.21(a)(1).

Scoping and Screening Methodology

2.4.8 Station Blackout Generator Structures

2.4.8.1 Summary of Technical Information in the Application

In LRA Section 2.4.8, the applicant stated that the SBO generator structures consist of the SBO generator concrete foundation and the turbine control building, which is a steel structure with metal siding and concrete foundation. The applicant also stated that the fuel oil tanks, which are located plant north of the SBO generators, are founded directly on compacted backfill and there are no structural components that require aging management. The applicant further stated that the SBO generator structures are nonsafety-related structures that provide support, shelter, and protection for components required to demonstrate compliance with SBO requirements based on the criteria of 10 CFR 54.4(a)(3).

LRA Table 2.4-8 identifies the components subject to an AMR for the SBO generator structures by component type and intended function.

2.4.8.2 Staff Evaluation and Conclusion

The staff reviewed LRA Section 2.4.8, UFSAR Section 8.3.1.1.10, and applicable boundary drawings to determine if the applicant failed to identify any components within the scope of license renewal. In addition, the staff's review determined if the applicant failed to identify any components subject to an AMR. The staff found no such omissions. On the basis of its review, the staff concludes the applicant has appropriately identified the SBO generator structures structural components within the scope of license renewal, as required by 10 CFR 54.4(a), and that the applicant has adequately identified the structural components subject to an AMR in accordance with the requirements stated in 10 CFR 54.21(a)(1).

2.4.9 Fuel Building

2.4.9.1 Summary of Technical Information in the Application

In LRA Section 2.4.9, the applicant stated that the fuel building is a seismic Category I rectangular reinforced concrete structure supported on a reinforced concrete base slab, founded on granular backfill. The elevated floors and roof are reinforced concrete, supported by reinforced concrete bearing walls. The applicant also stated that the fuel building contains the spent fuel pool, new fuel storage area, the dry spent fuel storage system loading and transfer equipment, the spent fuel pool cooling heat exchangers and pumps, and other miscellaneous equipment.

The applicant further stated that the spent fuel pool, including the transfer canal, cask loading pit, and cask wash down area consist of reinforced concrete walls and floors lined with stainless steel plates. The applicant also stated that the cask loading pit and cask wash down area gate seals are designed as seismic Category I. These seals are designed to remain functional during and after accident conditions under the criteria of 10 CFR 54.4(a)(1). The applicant stated that the fuel building shelters and protects nonsafety-related SSCs whose failure could prevent performance of a safety-related function; therefore, it is within the scope of license renewal based on the criterion of 10 CFR 54.4(a)(2). The applicant also stated that portions of the fuel building support fire protection requirements based on the criteria of 10 CFR 54(a)(3).

The fuel transfer canal gate seals are designed as seismic Category II. The applicant further stated that during accident conditions, the water elevation would remain above the pool cooling system suction piping and more than 10 feet of water coverage would be available to shield the spent fuel assemblies.

LRA Table 2.4-9 identifies the components subject to an AMR for the fuel building by component type and intended function.

2.4.9.2 Staff Evaluation

The staff reviewed LRA Section 2.4.9 using the evaluation methodology described in SER Section 2.4 and the guidance in SRP-LR Section 2.4.

During its review of the LRA Section 2.4.9, the staff identified areas in which additional information was necessary to complete the evaluation of the applicant's scoping and screening results for the fuel building. In RAI 2.4.9-1, dated November 2, 2009, the staff requested that the applicant provide additional information to confirm whether the fire barrier coatings and wraps are to be included in the scope of license renewal and subject to an AMR or to justify their exclusion.

In its response to the RAI, dated December 17, 2009, the applicant stated that there are no fire barrier coatings or wraps within the scope of license renewal and subject to an AMR in the fuel building. The UFSAR, Appendix 9B.2, "Fire Hazards Analysis," describes the fire protection evaluation for the fuel building in UFSAR Section 9B.2.6. The applicant indicated that this evaluation documents that no fire barrier coatings or wraps are credited for protection of structural members or raceways in the fuel building.

Based on its review, the staff finds the response to RAI 2.4.9-1 acceptable because it states that no fire barrier coatings or wraps are credited for protection of structural members or raceways in the fuel building. Therefore, the staff's concern described in RAI 2.4.9-1 is resolved.

2.4.9.3 Conclusion

The staff reviewed LRA Section 2.4.9; UFSAR Sections 1.2.12.4, 3.8.4.1.2, and 9.1; the RAI response; and applicable boundary drawings to determine if the applicant failed to identify any components within the scope of license renewal. In addition, the staff's review determined if the applicant failed to identify any components subject to an AMR. The staff found no such omissions. On the basis of its review, the staff concludes the applicant has appropriately identified the fuel building structural components within the scope of license renewal, as required by 10 CFR 54.4(a), and that the applicant has adequately identified the structural components subject to an AMR in accordance with the requirements stated in 10 CFR 54.21(a)(1).

2.4.10 Spray Pond and Associated Water Control Structures

2.4.10.1 Summary of Technical Information in the Application

In LRA Section 2.4.10, the applicant stated that the spray pond and associated water control structures include two ESPs per unit. Each pond has an intake structure to feed the cooling loop and a pond inlet for the return line. The applicant also stated that the ESPs, the pump houses, the intake structures, and the sumps are safety-related, seismic Category I, reinforced concrete structures, and founded on natural sands. Each pond serves one train of the ESP system to provide the ultimate heat sink for cooling auxiliary systems required for safe reactor shutdown; therefore, they are within the scope of license renewal based on the criteria of 10 CFR 54.4(a)(1). The applicant stated that the spray pond and associated water control structures shelter and protect nonsafety-related SSCs whose failure could prevent performance of a safety-related function; therefore, they are within the scope of license renewal based on the criterion of 10 CFR 54.4(a)(2). Finally, the applicant stated that the ESPs, the pump houses, and the intake structures provide structural support fire protection and SBO requirements and are within the scope of license renewal based on the criteria of 10 CFR 54.4(a)(3).

LRA Table 2.4-10 identifies the components subject to an AMR for the spray pond and associated water control structures by component type and intended function.

Scoping and Screening Methodology

2.4.10.2 Staff Evaluation and Conclusion

The staff reviewed LRA Section 2.4.10, UFSAR Sections 2.4.11.6 and 3.8.4.1.6, and applicable boundary drawings to determine if the applicant failed to identify any components within the scope of license renewal. In addition, the staff's review determined if the applicant failed to identify any components subject to an AMR. The staff found no such omissions. On the basis of its review, the staff concludes the applicant has appropriately identified the spray pond and associated water control structures structural components within the scope of license renewal, as required by 10 CFR 54.4(a). In addition, the staff finds that the applicant has adequately identified the structural components subject to an AMR in accordance with the requirements stated in 10 CFR 54.21(a)(1).

2.4.11 Tank Foundations and Shells

2.4.11.1 Summary of Technical Information in the Application

In LRA Section 2.4.11, the applicant stated that the tank foundations and shells are reinforced concrete structures that provide structural support for the CST, RWT, RWT valve pit, and reactor makeup water tank (RMWT). The CST and RWT are seismic Category I structures and have concrete shells, built-up roofs, and stainless steel liners. The applicant stated that the RWT is safety-related and supplies the required volume of borated water for safety injection following a LOCA; therefore, the tank foundations and shells are within the scope of license renewal based on the criteria of 10 CFR 54.4(a)(1). The applicant stated that the tank foundations and shells shelter and protect nonsafety-related SSCs whose failure could prevent performance of a safety-related function; therefore, they are within the scope of license renewal based on the criterion of 10 CFR 54.4(a)(2). The applicant also stated that the CST, RWT, and RMWT foundations and shells provide structural support and protection for SSCs required for fire protection and SBO; therefore, they are within the scope of license renewal based on the criteria of 10 CFR 54.4(a)(3).

LRA Table 2.4-11 identifies the components subject to an AMR for the tank foundations and shells by component type and intended function.

2.4.11.2 Staff Evaluation and Conclusion

The staff reviewed LRA Section 2.4.11; UFSAR Sections 3.5D, 3.8.4.1.7, 3.8.4.1.8 and Appendices 9B.2.20.3, 9B.2.20.5, and 9B.2.9.3; and applicable boundary drawings to determine if the applicant failed to identify any components within the scope of license renewal. In addition, the staff's review determined if the applicant failed to identify any components subject to an AMR. The staff found no such omissions. On the basis of its review, the staff concludes the applicant has appropriately identified the tank foundations and shells structural components within the scope of license renewal, as required by 10 CFR 54.4(a), and that the applicant has adequately identified the structural components subject to an AMR in accordance with the requirements stated in 10 CFR 54.21(a)(1).

2.4.12 Transformer Foundations and Electrical Structures

2.4.12.1 Summary of Technical Information in the Application

In LRA Section 2.4.12, the applicant stated that reinforced concrete pads, founded on granular backfill, support the transformer foundations and electrical structures. The applicant also stated that all of the transmission towers to the first breakers in the SRP 500 kV switchyard and the towers supporting the transmission lines to the ESF and startup transformers are steel towers with reinforced concrete drilled caisson foundations. Electrical cables from the transformers are installed in buried concrete duct banks, and manholes are provided along these duct banks for cable installation and access. The applicant further stated that the concrete duct banks, and manholes provide structural support, shelter, and protection for the electrical cables.

The applicant stated that the concrete duct banks and the manholes provide shelter and protection for SSCs required for SBO recovery and are within the scope of license renewal based on the criteria of 10 CFR 54.4(a)(3). Additionally, the applicant stated that the concrete fire barrier walls separating the ESF, main, normal, and auxiliary transformers provide spatial separation and fire barriers to meet the requirements for fire protection and are, therefore, within scope based on the criteria of 10 CFR 54.4(a)(3).

LRA Table 2.4-12 identifies the components subject to an AMR for the transformer foundations and electrical structures by component type and intended function.

2.4.12.2 Staff Evaluation

The staff reviewed LRA Section 2.4.12 using the evaluation methodology described in SER Section 2.4 and the guidance in SRP-LR Section 2.4.

During its review of the LRA Section 2.4.12, the staff identified areas in which additional information was necessary to complete the evaluation of the applicant's scoping and screening results for the transformer foundations and electrical structures. In RAI 2.4.12-1, dated November 2, 2009, the staff requested that the applicant provide additional information to confirm that the electrical penetrations are to be included in the scope of license renewal and subject to an AMR or to justify their exclusion.

In its response to the RAI, dated December 17, 2009, the applicant stated that there are no electrical penetrations within the scope of license renewal and subject to an AMR in LRA Section 2.4.12, "Transformer Foundations and Electrical Structures." Electrical components that connect these structures with other buildings are routed through electrical penetrations that are evaluated with those buildings in their respective subsections in LRA Section 2.4.

Based on its review, the staff finds the response to RAI 2.4.12-1 acceptable because the applicant clarified that the components are within the scope of license renewal and are evaluated in their respective subsections in LRA Section 2.4. Therefore, the staff's concern described in RAI 2.4.12-1 is resolved.

2.4.12.3 Conclusion

The staff reviewed LRA Section 2.4.12, UFSAR Sections 2.5.4.8.1 and Appendix 9B.2.21, the RAI response, and applicable boundary drawings to determine if the applicant failed to identify any components within the scope of license renewal. In addition, the staff's review determined if the applicant failed to identify any components subject to an AMR. The staff found no such omissions. On the basis of its review, the staff concludes the applicant has appropriately identified the transformer foundations and electrical structures structural components within the scope of license renewal, as required by 10 CFR 54.4(a), and that the applicant has adequately identified the structural components subject to an AMR in accordance with the requirements stated in 10 CFR 54.21(a)(1).

2.4.13 Yard Structures (In-Scope)

2.4.13.1 Summary of Technical Information in the Application

In LRA Section 2.4.13, the applicant stated that the yard structures (in-scope) consist of the condensate and essential pipe tunnels, CST pump house, and the diesel fuel oil tank vault. The yard structures are seismic Category I, reinforced concrete structures, which provide structural support and shelter and protection for safety-related components. The applicant also stated that the condensate and essential pipe tunnels, CST pump house, diesel fuel oil tank vault, and fire pump house provide spatial fire barriers and structural support for the fire suppression components. The applicant further stated that the essential pipe tunnels, condensate tunnel, and CST pump house provide shelter and protection for SSCs required for SBO recovery.

The applicant stated that the condensate and essential pipe tunnels, the CST pump house, and the diesel fuel oil tank vault provide structural support and protection for safety-related components that they rely on to shutdown the reactor and maintain it in a safe shutdown condition. Therefore, these structures are within the scope of license renewal based on the criteria of 10 CFR 54.4(a)(1). The applicant also stated that these yard structures shelter and protect nonsafety-related SSCs whose failure could prevent performance of a safety-related function; therefore, they are within the scope of license renewal based on the criterion of 10 CFR 54.4(a)(2). The applicant stated that these structures provide spatial fire barriers and structural support for fire suppression components and are, therefore, within the scope of license renewal based on the criteria of 10 CFR 54.4(a)(3). The LRA states that the essential pipe tunnel, the condensate tunnel, and the CST tank pump house provide shelter and protection for SSCs required for SBO recovery; therefore, these structures are also within the scope of license renewal based on the criteria of 10 CFR 54.4(a)(3).

LRA Table 2.4-13 identifies the in-scope components subject to an AMR for the yard structures by component type and intended function.

2.4.13.2 Staff Evaluation

The staff reviewed LRA Section 2.4.13 using the evaluation methodology described in SER Section 2.4 and the guidance in SRP-LR Section 2.4.

During its review of the LRA Section 2.4.13, the staff identified areas in which additional information was necessary to complete the evaluation of the applicant's scoping and screening results for the yard structures. In RAI 2.4.13-1, dated November 2, 2009, the staff requested that the applicant provide additional information to confirm the "duct banks and manholes" and "transmission towers" are to be included in the scope of license renewal and subject to an AMR or justify their exclusion.

In its response to the RAI, dated December 17, 2009, the applicant stated that it evaluated "duct banks and manholes" and "transmission towers" in LRA Section 2.4.12, "Transformer Foundations and Electrical Structures," and LRA Table 2.4-12.

The applicant also stated that, to clarify the LRA, it added the following sentence at the end of the "Structure Description," and just before the "Structure Intended Function," in Section 2.4.13, "Yard Structures (In-scope)":

> Duct banks and manholes and transmission towers are evaluated in Section 2.4.12, Transformer Foundations and Electrical Structures, and Table 2.4-12.

Based on its review, the staff finds the response to RAI 2.4.13-1 acceptable. For clarification, the applicant added a sentence, "Duct banks and manholes and transmission towers are evaluated in Section 2.4.12, Transformer Foundations and Electrical Structures, and Table 2.4-12." Therefore, the staff's concern described in RAI 2.4.13-1 is resolved.

2.4.13.3 Conclusion

The staff reviewed LRA Section 2.4.13; UFSAR Appendix 9B and Tables 3.5-9, 9.5.2, and 9B.3-1; the RAI response; and applicable boundary drawings to determine if the applicant failed to identify any components within the scope of license renewal. In addition, the staff's review determined if the applicant failed to identify any components subject to an AMR. The staff found no such omissions. On the basis of its review, the staff concludes the applicant has appropriately identified yard structures structural components within the scope of license renewal, as required by 10 CFR 54.4(a), and that the applicant has adequately identified the structural components subject to an AMR in accordance with the requirements stated in 10 CFR 54.21(a)(1).

2.4.14 Supports

2.4.14.1 Summary of Technical Information in the Application

In LRA Section 2.4.14, the applicant stated that the supports are within the scope of license renewal because they support and protect components within the scope of license renewal. Supports are integral parts of all systems, and many of these supports are not uniquely identified with component identification numbers. However, the applicant stated that support characteristics such as design, materials of construction, environments, and anticipated stressors are similar. Therefore, the applicant evaluated structural supports for mechanical and electrical components as commodities across system boundaries.

The applicant's commodity evaluation applies to structural supports for structures within the scope of license renewal. The applicant addressed the following structural supports for mechanical components in the LRA: supports for American Society of Mechanical Engineers (ASME) Class 1 piping and components; supports for ASME Class 2 and Class 3 piping and components; and supports for HVAC ducts, tube track, instrument tubing, instruments, and non-ASME piping and components. The applicant stated that the LRA addressed the following electrical components and supports: cable trays and supports, conduit and supports, and electrical panels and enclosures. The applicant further stated that the ASME Class 1 piping and component commodity group includes the supports for the following RCS components: reactor vessel, pressurizer, SG, and RCP.

The applicant also stated that supports are safety-related components relied upon to remain functional during and following DBEs and are, therefore, within the scope of license renewal under the criteria of 10 CFR 54.4(a)(1). The failure of nonsafety-related SSCs in the supports could prevent the satisfactory accomplishment of a safety-related function and are, therefore, within the scope of license renewal under the criteria of 10 CFR 54.4(a)(2). In addition, the supports perform functions that support fire protection, PTS, and SBO and are, therefore, within the scope of license renewal under the criteria of 10 CFR 54.4(a)(3).

LRA Table 2.4-14 identifies the components subject to an AMR for the supports by component type and intended function.

2.4.14.2 Staff Evaluation and Conclusion

The staff reviewed LRA Section 2.4.14, UFSAR Sections 3.8.3.1 and 5.4.14, and applicable boundary drawings to determine if the applicant failed to identify any components within the scope of license renewal. In addition, the staff's review determined whether the applicant failed to identify any components subject to an AMR. The staff found no such omissions. On the basis of its review, the staff concludes the applicant has appropriately identified supports components within the scope of license renewal, as required by 10 CFR 54.4(a), and that the applicant has adequately identified the structural components subject to an AMR in accordance with the requirements stated in 10 CFR 54.21(a)(1).

2.4.15 Fire Barriers

2.4.15.1 Summary of Technical Information in the Application

The tables in LRA Section 2.4 show the function of "fire barriers" for one of the component types that are credited as part of the fire protection system. For example, LRA Tables 2.4-1 through 2.4-9 and Table 2.4-13 list the fire barrier function for concrete block (masonry walls), concrete elements, fire barrier coatings and wraps, fire barrier doors, and fire barrier seals. Therefore, fire barrier elements are within the scope of license renewal and subject to an AMR.

Scoping and Screening Methodology

LRA Table 2.1-1 identifies the intended function for the fire barriers and states that the rated fire barriers confine or retard a fire from spreading to or from adjacent areas of the plant. According to the LRA, fire barriers have an intended function under 10 CFR 50.54(a)(1) and 10 CFR 50.54(a)(3).

2.4.15.2 Staff Evaluation

The staff reviewed the LRA Section 2.4 (fire barrier portion only), UFSAR, and license renewal drawings using the evaluation methodology described in the SER Section 2.4 and guidance in SRP-LR, Section 2.4. During its review, the staff evaluated the system functions described in the LRA and UFSAR to verify that the applicant had not omitted from the scope of license renewal any components with intended functions under 10 CFR 54.4(a). The staff then reviewed those components that the applicant identified as within the scope of license renewal to verify that the applicant had not omitted any passive or long-lived components subject to an AMR in accordance with 10 CFR 54.21(a)(1).

The staff also reviewed the fire protection CLB documents listed in the Operating License, (license conditions 2.C(7), 2.C(6), and 2.F) and NUREG-0857, "Safety Evaluation Report related to the operation of Nuclear Generating Station Units 1, 2 and 3," dated November 1981 through Supplement 11.

This review of the fire protection CLB documents included applicant commitments to 10 CFR 50.48, "Fire protection" (i.e., approved fire protection program), as provided in the responses to Appendix A to BTP APCSB 9.5-1, "Guidelines for Fire Protection for Nuclear Power Plants," and Appendix A to the BTP APS 9.5-1, documented in PVNGS Units 1, 2, and 3 SERs and NUREG-0857, dated November 1981 through Supplement 11.

During its review of LRA Section 2.4, the staff identified areas in which additional information was necessary to complete its review of the applicant's scoping and screening results. The applicant responded to the staff's RAIs as discussed below.

In RAI 2.4-1, dated November 12, 2009, the staff asked why the applicant excluded several types of fire barrier assemblies and components from the scope of license renewal. These fire barrier components include the following:

- Table 2.4-1, "Fire Barrier Concrete Block (masonry walls) and Fire Barrier Doors"
- Table 2.4-3, "Fire Barrier Concrete Block (masonry walls) and Fire Barrier Coatings and Wraps"
- Table 2.4-4, "Fire Barrier Concrete Elements"
- Table 2.4-6, "Fire Barrier Concrete Block (masonry walls) and Fire Barrier Coatings and Wraps"
- Table 2.4-7, "Fire Barrier Concrete Block (masonry walls), Fire Barrier Doors, and Fire Barrier Coatings and Wraps"
- Table 2.4-8, "Fire Barrier Concrete Elements, Fire Barrier Concrete Block (masonry walls), Fire Barrier doors, Fire Barrier seals, and Fire Barrier Coatings and Wraps"
- Table 2.4-9, "Fire Barrier Concrete Block (masonry walls) and Fire Barrier Coatings and Wraps"
- Table 2.4-13, "Fire Barrier Coatings and Wraps"

The staff asked that the applicant verify whether LRA Section 2.4 should include the fire barrier assembles and components listed above. If the applicant excluded these from the scope of license renewal, the staff asked that the applicant justify this exclusion.

Scoping and Screening Methodology

In its response, dated December 23, 2009, the applicant provided the results of the scoping and screening process for the listed fire barrier assembles and components. In reviewing its response to RAI 2.4-1, the staff found that the applicant had addressed and resolved each item in the RAI. The applicant confirmed that none of the following assembles and components are credited for performing a fire barrier function:

- Concrete blocks (masonry walls) and doors in the containment building (LRA Table 2.4-1)

- Concrete blocks (masonry walls) and coatings or wraps in the diesel generator building (LRA Table 2.4-3)

- Concrete elements in the turbine building (LRA Table 2.4-4)

- Concrete blocks (masonry walls) and coatings or wraps in the radwaste building (LRA Table 2.4-6)

- Concrete blocks (masonry walls) in the MSSS (LRA Table 2.4-7)

- Concrete elements, concrete blocks (masonry walls), doors, seals, and coatings or wraps in the SBO generator structures (LRA Table 2.4-8)

- Concrete blocks (masonry walls) and coatings or wraps in the fuel building (LRA Table 2.4-9)

- Coatings or wraps in yard structures (LRA Table 2.4-13)

However, the applicant credited fire barrier coatings and wraps for performing a fire barrier function in the MSSS, and it added fire barrier coatings and wraps to LRA Table 2.4-7, Section 3.5.2.1.7, and LRA Table 3.5.2-7.

Based on its review, the staff finds the applicant's response to RAI 2.4-1 acceptable, because it clarified that fire barrier assemblies and components in question are not required to perform a fire barrier function. The applicant noted that only fire barrier coatings and wraps in the MSSS are within the scope of license renewal and subject to an AMR. The staff's concern is resolved.

2.4.15.3 Conclusion

The staff reviewed the LRA, UFSAR, RAI responses, and license renewal drawings to determine if the applicant failed to identify any fire protection systems and components within the scope of license renewal. In addition, the staff sought to determine if the applicant failed to identify any components subject to an AMR. On the basis of its review, the staff concludes that the applicant has adequately identified the fire barriers within the scope of license renewal, as required by 10 CFR 54.4(a), and those subject to an AMR, as required by 10 CFR 54.21(a)(1).

2.5 Scoping and Screening Results: Electrical and Instrumentation and Control Systems

This section documents the staff's review of the applicant's scoping and screening results for electrical and I&C systems. Specifically, this section discusses electrical and I&C component commodity groups.

In accordance with the requirements of 10 CFR 54.21(a)(1), the applicant must list passive, long-lived SSCs within the scope of license renewal and subject to an AMR. To verify that the applicant properly implemented its methodology, the staff's review focused on the implementation results. This focus allowed the staff to confirm that there were no omissions of electrical and I&C system components that meet the scoping criteria and are subject to an AMR.

Scoping and Screening Methodology

The staff's evaluation of the information in the LRA was the same for all electrical and I&C systems. The objective was to determine whether the applicant has identified, in accordance with 10 CFR 54.4, components and supporting structures for electrical and I&C systems that appear to meet the license renewal scoping criteria. Similarly, the staff evaluated the applicant's screening results to verify that all passive, long-lived components were subject to an AMR in accordance with 10 CFR 54.21(a)(1).

In its scoping evaluation, the staff reviewed the applicable LRA sections, focusing on components that the applicant did not identify as within the scope of license renewal. The staff reviewed the UFSAR for each electrical and I&C system to determine whether the applicant has omitted from the scope of license renewal components with intended functions delineated under 10 CFR 54.4(a).

After its review of the scoping results, the staff evaluated the applicant's screening results. For those SSCs with intended functions, the staff sought to determine whether the functions are performed with moving parts or a change in configuration or properties or the SSCs are subject to replacement after a qualified life or specified time period, as described in 10 CFR 54.21(a)(1). For those meeting neither of these criteria, the staff sought to confirm that these SSCs were subject to an AMR, as required by 10 CFR 54.21(a)(1).

2.5.1 Electrical and Instrumentation and Control Systems Component Groups

2.5.1.1 Summary of Technical Information in the Application

LRA Section 2.5 describes the electrical and I&C systems. The applicant's scoping method includes all plant electrical and I&C components. Evaluation of electrical systems includes electrical and I&C components in mechanical systems. The applicant states that the plant-wide basis approach for the review of plant equipment eliminates the need to identify each unique component and its specific location and prevents improper exclusion of components from an AMR.

For the electrical and I&C components identified as within the scope of license renewal, the applicant grouped them into component groups regardless of their system association. The applicant applied the screening criteria in 10 CFR 54.21(a)(1)(i) and 10 CFR 54.21(a)(1)(ii) to this list of component groups to identify those that perform an intended function without moving parts or without a change in configuration or properties and to remove the component groups that are subject to replacement based on a qualified life or specified time period. The following list identifies the component groups that the applicant determined to be subject to an AMR and their intended functions:

- Connections (metallic parts)—electrical continuity
- Connectors—electrical continuity
- High-voltage insulators—insulation and structural support
- Insulated cables and connections—electrical continuity and insulation
- Metal enclosed bus and connections—electrical continuity
- Bus bar and connections—electrical continuity
- Bus enclosure—expansion, separation, and structural support
- Bus insulation and insulators—insulation
- Penetrations electrical—electrical continuity and insulation
- Switchyard bus and connections—electrical continuity

Scoping and Screening Methodology

- Terminal Blocks - electrical continuity

- Transmission conductors and connections—electrical continuity

- Electrical equipment subject to 10 CFR 50.49 EQ requirements (TLAA, see LRA Section 4)

- Grounding conductors—ground metal structures and equipment

- Cable tie wraps—installation aid and cable spacing

2.5.1.2 *Staff Evaluation*

The staff reviewed LRA Section 2.5, UFSAR Sections 7 and 8, and the license renewal boundary drawings using the evaluation methodology described in SER Section 2.5 and the guidance in SRP-LR Section 2.5, "Scoping and Screening Results: Electrical and Instrumentation and Controls Systems."

During its review, the staff evaluated the system functions described in the LRA and UFSAR to verify that the applicant had not omitted from the scope of license renewal any components with intended functions delineated under 10 CFR 54.4(a). The staff then reviewed those components that the applicant identified as within the scope of license renewal to verify that the applicant had not omitted any passive and long-lived components subject to an AMR in accordance with the requirements of 10 CFR 54.21(a)(1).

General Design Criterion 17 of 10 CFR Part 50, Appendix A, requires two physically independent circuits supply electric power from the transmission network to the onsite electric distribution system to minimize the likelihood of simultaneous failure. Additionally, there is guidance provided in SRP-LR Section 2.5.2.1.1. For purposes of the license renewal rule, the staff has determined that the following is true:

> The plant system portion of the offsite power system that is used to connect the plant to the offsite power source meeting the requirements of 10 CFR 54.4(a)(3). This path typically includes the switchyard circuit breakers that connect to the offsite system power transformers (startup transformers), the transformers themselves, the intervening overhead or underground circuits between circuit breaker and transformer and transformer and onsite electrical system, and the associated control circuits and structures.

According to this guidance, ensuring that the appropriate offsite power system long-lived passive SSCs that are part of this circuit path are subject to an AMR will assure that the bases underlying the SBO requirements are maintained over the period of the extended operation.

The applicant included the complete circuits between the onsite circuits and up to and including switchyard breakers (which includes the associated controls and structures) supplying the AE-NAN-X01, AE-NAN-X02, and AE-NAN-X03 startup transformers within the scope of license renewal. Switchyard breakers 925, 928, 945, 948, 995, and 998 are the scoping boundary for the primary and backup sources of offsite power. For PVNGS Unit 1, the primary offsite power source is from startup transformers AE-NAN-X03 and AE-NAN-X02, while the backup offsite power source is from startup transformer AE-NAN-X01. For PVNGS Unit 2, the primary offsite power source is from startup transformers AE-NAN-X01 and AE-NAN-X03, while the backup offsite power source is from startup transformer AE-NAN-X02. For PVNGS Unit 3, the primary offsite power source is from startup transformers AE-NAN-X01 and AE-NAN-X02, while the backup offsite power source is from startup transformer AE-NAN-X03. Consequently, the staff concludes that the scoping is consistent with NRC guidance.

The staff noted that the applicant had not included cable tie-wraps in any component group. In the LRA, the applicant states that it did a review to determine if cable tie-wraps meet the scoping criteria of 10 CFR 54.4. The applicant states that it uses cable tie-wraps as an aid during cable installation, but it does not perform any license renewal functions, and seismic qualification of cable trays does not credit the use of electrical cable tie-wraps. Furthermore, the applicant considered the failure of plastic cable tie-wraps and concluded that such failure would not affect safety-related equipment or any design-basis event. The staff reviewed the UFSAR and found that cable tie-wraps are not credited in the design basis. Based on this review and the information provided in the LRA, the staff finds the applicant's exclusion of cable tie-wraps from SSCs subject to an AMR acceptable.

2.5.1.3 Conclusion

The staff reviewed the LRA, UFSAR, and the license renewal boundary drawing to determine if the applicant failed to identify any SSCs within the scope of license renewal. The staff has found no such omissions. In addition, the staff's review determined if the applicant failed to identify any components subject to an AMR. The staff found no such omissions. On the basis of its review, the staff concludes that the applicant has adequately identified the electrical and I&C systems components within the scope of license renewal, as required by 10 CFR 54.4(a), and those subject to an AMR, as required by 10 CFR 54.21(a)(1).

2.6 Conclusion for Scoping and Screening

The staff reviewed the information in LRA Section 2, "Scoping and Screening Methodology for Identifying Structures and Components Subject to Aging Management Review and Implementation Results" and determines that the applicant's scoping and screening methodology is consistent with 10 CFR 54.21(a)(1). Further, the applicant's methodology is consistent with the staff's positions on the treatment of safety-related and nonsafety-related SSCs within the scope of license renewal and on SCs subject to an AMR, and is consistent with the requirements of 10 CFR 54.4 and 10 CFR 54.21(a)(1).

On the basis of its review, the staff concludes that the applicant has adequately identified those systems and components within the scope of license renewal, as required by 10 CFR 54.4(a), and those subject to an AMR, as required by 10 CFR 54.21(a)(1).

The staff concludes that there is reasonable assurance that the applicant will continue to conduct the activities authorized by the renewed licenses in accordance with the CLB, and any changes to the CLB, in order to comply with 10 CFR 54.21(a)(1), the Atomic Energy Act of 1954, as amended, and NRC regulations.

3.0 AGING MANAGEMENT REVIEW RESULTS

This section of the safety evaluation report (SER) evaluates aging management programs (AMPs) and aging management reviews (AMRs) for Palo Verde Nuclear Generating Station, Units 1, 2, and 3 (PVNGS), by the U.S. Nuclear Regulatory Commission (NRC) staff (the staff). In Appendix B of its license renewal application (LRA), Arizona Public Service Company (APS) (the applicant) describes the 40 AMPs that it relies on to manage or monitor the aging of passive, long-lived structures and components (SCs).

In LRA Section 3, the applicant provided the results of the AMRs for those SCs identified in LRA Section 2 as within the scope of license renewal and subject to an AMR.

3.0 Applicant's Use of the Generic Aging Lessons Learned Report

In preparing its LRA, the applicant credited NUREG-1801, Revision 1, "Generic Aging Lessons Learned (GALL) Report," dated September 2005. The GALL Report contains the staff's generic evaluation of the existing plant programs and documents the technical basis for determining where existing programs are adequate without modification and where existing programs should be augmented for the period of extended operation. The evaluation results, documented in the GALL Report, note that many of the existing programs are adequate to manage the aging effects for particular license renewal SCs. The GALL Report also contains recommendations on specific areas for which existing programs should be augmented for license renewal. An applicant may reference the GALL Report in its LRA to demonstrate that its programs correspond to those reviewed and approved in the report.

The GALL Report provides a summary of staff-approved AMPs to manage or monitor the aging of SCs subject to an AMR. If an applicant commits to implementing these staff-approved AMPs, they will greatly reduce the time, effort, and resources for LRA review and improve the efficiency and effectiveness of the license renewal review process. The GALL Report also serves as a quick reference for applicants and staff reviewers to the AMPs and activities that the staff has determined will adequately manage or monitor aging during the period of extended operation.

The GALL Report lists the systems, structures, and components (SSCs); the SC materials; environments to which the SCs are exposed; the aging effects of the materials and environments; the AMPs credited with managing or monitoring the aging effects; and recommendations for further applicant evaluations of aging management for certain component types.

To determine whether use of the GALL Report would improve the efficiency of the LRA review, the staff conducted a demonstration of the GALL Report process in order to model the format and content of safety evaluations based on it. The results of the demonstration project confirmed that the GALL Report process will improve the efficiency and effectiveness of the LRA review while maintaining the staff's focus on public health and safety. NUREG-1800, Revision 1, "Standard Review Plan for Review of License Renewal Applications for Nuclear Power Plants" (SRP-LR), dated September 2005, was prepared based on both the GALL Report model and lessons learned from the demonstration project.

The staff's review of the LRA was in accordance with Title 10, Part 54, of the *Code of Federal Regulations* (10 CFR Part 54), "Requirements for Renewal of Operating Licenses for Nuclear Power Plants," as well as the guidance of the SRP-LR and the GALL Report.

In addition to its review of the LRA, the staff conducted an onsite audit of AMPs during the week of December 7, 2009. The staff designed the onsite audit for maximum efficiency of the LRA review. The applicant can respond to questions and the staff can readily evaluate the applicant's responses. This process reduces the need for formal correspondence between the staff and the applicant, and the result is an improvement in review efficiency.

3.0.1 Format of the License Renewal Application

The applicant submitted an application that follows the standard LRA format agreed upon by the staff and the Nuclear Energy Institute (NEI) by letter dated April 7, 2003. This revised LRA format incorporated lessons learned from the staff's reviews of the previous five LRAs, which used a format developed from information gained during a demonstration project conducted by the staff and NEI to evaluate the use of the GALL Report in the LRA review process.

The organization of LRA Section 3 parallels that of SRP-LR Chapter 3. LRA Section 3 presents AMR results information in the following two table types:

(1)   Table 1s: Table 3.x.1 – where "3" indicates the LRA section number, "x" indicates the subsection number from the GALL Report, and "1" indicates that this table type is the first in LRA Section 3

(2)   Table 2s: Table 3.x.2-y – where "3" indicates the LRA section number, "x" indicates the subsection number from the GALL Report, "2" indicates that this table type is the second in LRA Section 3, and "y" indicates the system table number

The content of the previous LRAs and of the PVNGS application is essentially the same. The intent of the revised format of the PVNGS LRA was to modify the tables in LRA Section 3 to provide additional information that would help in the staff's review. In its Table 1s, the applicant summarized the portions of the application that it considered consistent with the GALL Report. In its Table 2s, the applicant noted the linkage between the scoping and screening results in LRA Section 2 and the AMRs in LRA Section 3.

3.0.1.1   Overview of Table 1s

Each Table 1 compares, in summary, how the facility aligns with the corresponding tables in the GALL Report. The tables are essentially the same as Tables 1–6 in the GALL Report, except that an "Item Number" column has replaced the "Type" column, and a "Discussion" column replaced by the "Discussion Item Number in the GALL" column. The "Item Number" column is a means for the staff reviewer to cross-reference Table 2s with Table 1s. In the "Discussion" column, the applicant provided clarifying information. The following are examples of information that might be contained within this column:

- further evaluation recommended—information or reference to where that information is located
- the name of a plant-specific program
- exceptions to GALL Report assumptions
- discussion of how the line is consistent with the corresponding line item in the GALL Report when the consistency may not be obvious
- discussion of how the item is different from the corresponding line item in the GALL Report (e.g., when an exception is taken to a GALL Report AMP)

The format of each Table 1 allows the staff to align a specific row in the table with the corresponding GALL Report table row to easily check for consistency.

3.0.1.2   Overview of Table 2s

Each Table 2 provides the detailed results of the AMRs for components subject to an AMR, as noted in LRA Section 2. The LRA has a Table 2 for each of the systems or structures within a

Aging Management Review Results

specific system grouping (e.g., reactor coolant system (RCS), engineered safety features (ESFs), auxiliary systems, etc.). For example, the ESF group has tables that are specific to the containment spray system, containment isolation system, and emergency core cooling system (ECCS). Each Table 2 consists of the following nine columns:

(1) **Component Type** – The first column lists LRA Section 2 component types subject to an AMR in alphabetical order.

(2) **Intended Function** – The second column identifies the license renewal intended functions including abbreviations, where applicable, for the listed component types. Definitions and abbreviations of intended functions are in LRA Table 2.0-1.

(3) **Material** – The third column lists the particular construction material(s) for the component type.

(4) **Environment** – The fourth column lists the environments to which the component types are exposed. This column notes the internal and external service environments. A list of these environments is provided in LRA Tables 3.0-1, 3.0-2, and 3.0-3.

(5) **Aging Effect Requiring Management (AERM)** – The fifth column lists the AERMs. As part of the AMR process, the applicant determined AERMs for each combination of material and environment.

(6) **Aging Management Programs** – The sixth column lists the AMPs that the applicant uses to manage the identified aging effects.

(7) **NUREG-1801 Vol. 2 Item** – The seventh column lists the GALL Report item(s) that are similar to the AMR results, as noted in the LRA. The applicant compared each combination of component type, material, environment, AERM, and AMP in LRA Table 2 with the GALL Report items. If there are no corresponding items in the GALL Report, the applicant leaves the column blank.

(8) **Table 1 Item** – The eighth column lists the corresponding summary item number from LRA Table 1. If the applicant notes AMR results consistent with the GALL Report in each LRA Table 2, the Table 1 line item summary number should be listed in LRA Table 2. If there is no corresponding item in the GALL Report, the applicant leaves column eight blank.

(9) **Notes** – The ninth column lists the corresponding notes used to show how the information in each Table 2 aligns with the information in the GALL Report. An NEI working group developed the notes, which are identified by letters. SER Section 3.0.2.2 describes the generic notes A–E, which indicate that AMR items are essentially consistent with the GALL Report. Each AMR section that provides the staff's review of items that are not consistent with the GALL Report (3.x.2.3) contains an explanation of notes F–J. Any plant-specific notes identified by numbers provide additional information about the consistency of the line item with the GALL Report.

3.0.2 Staff's Review Process

The staff conducted three types of evaluations of the AMRs and AMPs:

(1) For items that the applicant stated were consistent with the GALL Report, the staff conducted either an audit or a technical review to determine consistency.

(2) For items that the applicant stated were consistent with the GALL Report with exceptions, enhancements, or both, the staff conducted either an audit or a technical review of the item to determine consistency. In addition, the staff conducted either an audit or a technical review of the applicant's technical justifications for the exceptions or

Aging Management Review Results

the adequacy of the enhancements. The SRP-LR states that an applicant may take one or more exceptions specific to the GALL Report AMP elements; however, any deviation from, or exception to, the GALL Report AMP should be described and justified. Therefore, the staff considers exceptions as being portions of the GALL Report AMP that the applicant does not intend to implement. In some cases, an applicant may choose an existing plant program that does not meet all the program elements defined in the GALL Report AMP. However, the applicant may make a commitment to augment the existing program to satisfy the GALL Report AMP before the period of extended operation. Therefore, the staff considers these augmentations or additions to be enhancements. Enhancements include, but are not limited to, activities needed to ensure consistency with the GALL Report recommendations. Enhancements may expand, but not reduce, the scope of an AMP.

(3) For other items, the staff conducted a technical review to verify conformance with 10 CFR 54.21(a)(3) requirements.

Staff audits and technical reviews of the applicant's AMPs and AMRs determine whether the aging effects on SCs can be adequately managed to maintain their intended function(s), consistent with the plant's current licensing basis (CLB), for the period of extended operation, as required by 10 CFR Part 54.

3.0.2.1 Review of Aging Management Programs

For AMPs for which the applicant claimed consistency with the GALL Report AMPs, the staff conducted either an audit or a technical review to verify the claim. For each AMP with one or more deviations, the staff evaluated each deviation to determine whether the deviation was acceptable and whether the modified AMP would adequately manage the aging effect(s) for which it was credited. For AMPs not evaluated in the GALL Report, the staff performed a full review to determine their adequacy. The staff evaluated the AMPs against the following 10 program elements defined in SRP-LR, Appendix A.

(1) <u>Scope of the Program</u> – Scope of the program should include the specific SCs subject to an AMR for license renewal.

(2) <u>Preventive Actions</u> – Preventive actions should prevent or mitigate aging degradation.

(3) <u>Parameters Monitored or Inspected</u> – Parameters monitored or inspected should be linked to the degradation of the particular structure or component's intended function(s).

(4) <u>Detection of Aging Effects</u> – Detection of aging effects should occur before there is a loss of structure or component intended function(s). This includes aspects such as method or technique (i.e., visual, volumetric, surface inspection), frequency, sample size, data collection, and timing of new or one-time inspections to ensure the timely detection of aging effects.

(5) <u>Monitoring and Trending</u> – Monitoring and trending should provide predictability of the extent of degradation as well as timely corrective or mitigative actions.

(6) <u>Acceptance Criteria</u> – Acceptance criteria, against which the need for corrective action will be evaluated, should ensure that the structure or component's intended function(s) are maintained under all CLB design conditions during the period of extended operation.

(7) <u>Corrective Actions</u> – Corrective actions, including root cause determination and prevention of recurrence, should be timely.

(8) <u>Confirmation Process</u> – Confirmation process should ensure that preventive actions are adequate and that corrective actions are completed and are effective.

(9) <u>Administrative Controls</u> – Administrative controls should provide for a formal review and approval process.

(10) <u>Operating Experience</u> – Operating experience of the AMP, including past corrective actions resulting in program enhancements or additional programs, should provide objective evidence to support the conclusion that the effects of aging will be adequately managed to maintain the SC intended function(s) during the period of extended operation.

SER Section 3.0.3 documents details of the staff's audit evaluation of program elements 1–6.

The staff reviewed the applicant's quality assurance (QA) program and documented its evaluations in SER Section 3.0.4. The staff's evaluation of the QA program included assessment of the "corrective actions," "confirmation process," and "administrative controls" program elements (elements 7, 8, and 9).

The staff also reviewed the information on the "operating experience" program element (element 10) for each program and documented its evaluation in SER Section 3.0.3.

3.0.2.2 Review of Aging Management Review Results

Each LRA Table 2 contains information concerning whether or not the AMRs noted by the applicant align with the GALL Report AMRs. For a given AMR in a Table 2, the staff reviewed the intended function, material, environment, AERM, and AMP combination for a particular system component type. Item numbers in column seven of the LRA, "NUREG-1801 Vol. 2 Item," correlate to an AMR combination as identified in the GALL Report. The staff also conducted onsite audits to verify these correlations. A blank in column seven indicates that the applicant was unable to identify an appropriate correlation in the GALL Report. The staff also conducted a technical review of combinations not consistent with the GALL Report. The next column, "Table 1 Item," refers to a number indicating the correlating row in Table 1.

For component groups evaluated in the GALL Report for which the applicant claimed consistency with the report and for which the applicant does not recommend further evaluation, the staff's audit and review determined whether the plant-specific components of these GALL Report component groups were bounded by the GALL Report evaluation.

The applicant noted, for each AMR line item, how the information in the tables aligns with the information in the GALL Report. The staff audited those AMRs, with notes A–E showing how the AMR is consistent with the GALL Report.

Note A indicates that the AMR line item is consistent with the GALL Report for component, material, environment, and aging effect. In addition, the AMP is consistent with the GALL Report AMP. The staff audited these line items to verify consistency with the GALL Report and validity of the AMR for the site-specific conditions.

Note B indicates that the AMR line item is consistent with the GALL Report for component, material, environment, and aging effect. In addition, the AMP takes some exceptions to the GALL Report AMP. The staff audited these line items to verify consistency with the GALL Report, and verified that the exceptions to the GALL Report AMPs have been reviewed and accepted. The staff also determined whether the applicant's AMP was consistent with the GALL Report AMP and whether the AMR was valid for the site-specific conditions.

Note C indicates that the component for the AMR line item is consistent with, although different from, the GALL Report for material, environment, and aging effect. In addition, the AMP is consistent with the GALL Report AMP. This note shows that the applicant was unable to find a listing of some system components in the GALL Report; however, the applicant found in the GALL Report a different component with the same material, environment, aging effect, and AMP as the component under review. The staff audited these line items to verify consistency with the

GALL Report. The staff also determined whether the AMR line item of the different component was applicable to the component under review and whether the AMR was valid for the site-specific conditions.

Note D indicates that the component for the AMR line item is consistent with, although different from, the GALL Report for material, environment, and aging effect. In addition, the AMP takes some exceptions to the GALL Report AMP. The staff audited these line items to verify consistency with the GALL Report. The staff verified whether the AMR line item of the different component was applicable to the component under review and whether the noted exceptions to the GALL Report AMPs have been reviewed and accepted. The staff also determined whether the applicant's AMP was consistent with the GALL Report AMP and whether the AMR was valid for the site-specific conditions.

Note E indicates that the AMR line item is consistent with the GALL Report for material, environment, and aging effect, but credits a different AMP. The staff audited these line items to verify consistency with the GALL Report. The staff also determined whether the credited AMP would manage the aging effect consistently with the GALL Report AMP and whether the AMR was valid for the site-specific conditions.

3.0.2.3 *Updated Final Safety Analysis Report Supplement*

The staff also reviewed the Updated Final Safety Analysis Report (UFSAR) supplement, which summarizes the applicant's programs and activities for managing aging effects for the period of extended operation, as required by 10 CFR 54.21(d).

3.0.2.4 *Documents Reviewed*

In its review, the staff used the LRA, including supplements and amendments, the SRP-LR, and the GALL Report. During the onsite audit, the staff examined documentation to verify the applicant's justifications that the activities and programs will adequately manage the effects of aging on the SCs. The staff also conducted detailed discussions and interviews with the applicant's license renewal project personnel and others with technical expertise relevant to aging management. The staff's Audit Report can be found in a letter to the applicant dated April 7, 2010.

3.0.3 Aging Management Programs

SER Table 3.0-1 presents the AMPs credited by the applicant and described in LRA Appendix B. The table also identifies which GALL Report AMPs that the applicant claims are consistent with its analyses, a comparison to the Gall Report AMP, and the section of this SER that documents the staff's evaluation of the program.

Table 3.0-1. Aging Management Programs

Applicant Aging Management Program	LRA Sections	New or Existing Program	Applicant Comparison to the GALL Report	GALL Report Aging Management Programs	SER Section
American Society of Mechanical Engineers (ASME) Section XI In-service Inspections (ISIs), Subsections IWB, IWC, and IWD Program	A1.1 B2.1.1	Existing	Consistent	XI.M1, "ASME Section XI Inservice Inspection, Subsections IWB, IWC, and IWD"	3.0.3.1.1
Water Chemistry Program	A1.2 B2.1.2	Existing	Consistent with enhancement[a]	XI.M2, "Water Chemistry"	3.0.3.2.1
Reactor Head Closure Studs Program	A1.3 B2.1.3	Existing	Consistent	XI.M3, "Reactor Head Closure Studs"	3.0.3.1.2

Aging Management Review Results

Applicant Aging Management Program	LRA Sections	New or Existing Program	Applicant Comparison to the GALL Report	GALL Report Aging Management Programs	SER Section
Boric Acid Corrosion	A1.4 B2.1.4	Existing	Consistent	XI.M10, "Boric Acid Corrosion"	3.0.3.1.3
Nickel-Alloy Penetration Nozzles Welded to the Upper RV Closure Heads of Pressurized Water Reactors (PWRs)	A1.5 B2.1.5	Existing	Consistent	XI.M11A, "Nickel-Alloy Penetration Nozzles Welded to the Upper Reactor Vessel Closure Heads of Pressurized Water Reactors"	3.0.3.1.4
Flow-Accelerated Corrosion (FAC) Program	A1.6 B2.1.6	Existing	Consistent with exceptions	XI.M17, "Flow-Accelerated Corrosion"	3.0.3.2.2
Bolting Integrity Program	A1.7 B2.1.7	Existing	Consistent with exceptions	XI.M18, "Bolting Integrity"	3.0.3.2.3
Steam Generator (SG) Tube Integrity Program	A1.8 B2.1.8	Existing	Consistent	XI.M19, "Steam Generator Tube Integrity"	3.0.3.1.5
Open-Cycle Cooling Water System Program	A1.9 B2.1.9	Existing	Consistent with enhancements	XI.M20, "Open-Cycle Cooling Water System"	3.0.3.2.4
Closed-Cycle Cooling Water System Program	A1.10 B2.1.10	Existing	Consistent with enhancements and exceptions	XI.M21, "Closed-Cycle Cooling Water System"	3.0.3.2.5
Inspection of Overhead Heavy Load and Light Load (Related to Refueling) Handling Systems Program	A1.11 B2.1.11	Existing	Consistent with enhancement	XI.M23, "Inspection of Overhead Heavy Load and Light Load (Related to Refueling) Handling Systems"	3.0.3.2.6
Fire Protection Program	A1.12 B2.1.12	Existing	Consistent with enhancements and exceptions	XI.M26, "Fire Protection"	3.0.3.2.7
Fire Water System Program	A1.13 B2.1.13	Existing	Consistent with enhancements and exceptions	XI.M27, "Fire Water System"	3.0.3.2.8
Fuel Oil Chemistry Program	A1.14 B2.1.14	Existing	Consistent with enhancements and exceptions	XI.M30, Fuel Oil Chemistry	3.0.3.2.9
RV Surveillance Program	A1.15 B2.1.15	Existing	Consistent with enhancements	XI.M31, "Reactor Vessel Surveillance"	3.0.3.2.10
One-Time Inspection Program	A1.16 B2.1.16	New	Consistent	XI.M32, "One-Time Inspection"	3.0.3.1.6
Selective Leaching of Materials Program	A1.17 B2.1.17	New	Consistent with exceptions	XI.M33, "Selective Leaching of Materials"	3.0.3.2.11
Buried Piping and Tanks Inspection Program	A1.18 B2.1.18	New	Consistent with exceptions	XI.M34, "Buried Piping and Tanks Inspection"	3.0.3.2.12
One-Time Inspection of ASME Code Class 1 Small-Bore Piping	A1.19 B2.1.19	Existing	Consistent with exceptions	XI.M35, "One-Time Inspection of ASME Code Class 1 Small-Bore Piping"	3.0.3.2.13
External Surfaces Monitoring Program	A1.20 B2.1.20	New	Consistent with exception	XI.M36, "External Surfaces Monitoring Program"	3.0.3.2.14

Aging Management Review Results

Applicant Aging Management Program	LRA Sections	New or Existing Program	Applicant Comparison to the GALL Report	GALL Report Aging Management Programs	SER Section
Reactor Coolant System Supplement	A1.21 B2.1.21	Not applicable	Not applicable	XI.M11A, " Nickel-Alloy Penetration Nozzles Welded to the Upper Reactor Vessel Closure Heads of Pressurized Water Reactors" and XI.M16 "PWR Vessel Internals"	3.0.3.1.7
Inspection of Internal Surfaces in Miscellaneous Piping and Ducting Components Program	A1.22 B2.1.22	New	Consistent with exceptions	XI.M38, "Inspection of Internal Surfaces in Miscellaneous Piping and Ducting Components"	3.0.3.2.15
Lubricating Oil Analysis Program	A1.23 B2.1.23	Existing	Consistent with exceptions	XI.M39, "Lubricating Oil Analysis"	3.0.3.2.16
Electrical Cables and Connections Not Subject to 10 CFR 50.49 Environmental Qualification (EQ) Requirements Program	A1.24 B2.1.24	New	Consistent	XI.E1, "Electrical Cables and Connections Not Subject to 10 CFR 50.49 Environmental Qualification Requirements"	3.0.3.1.8
Electrical Cables and Connections Not Subject to 10 CFR 50.49 Environmental Qualification Requirements Used in Instrumentation Circuits Program	A1.25 B2.1.25	Existing	Consistent with enhancements	XI.E2, "Electrical Cables and Connections Not Subject to 10 CFR 50.49 Environmental Qualification Requirements Used in Instrumentation Circuits"	3.0.3.2.17
Inaccessible Medium Voltage Cables Not Subject to 10 CFR 50.49 Environmental Qualification Requirements Program	A1.26 B2.1.26	New	Consistent	XI.E3, "Inaccessible Medium Voltage Cables Not Subject to 10 CFR 50.49 Environmental Qualification Requirements"	3.0.3.1.9
ASME Section XI, Subsection IWE Program	A1.27 B2.1.27	Existing	Consistent with exceptions	XI.S1, "ASME Section XI, Subsection IWE"	3.0.3.2.18
ASME Section XI, Subsection IWL Program	A1.28 B2.1.28	Existing	Consistent	XI.S2, "ASME Section XI, Subsection IWL"	3.0.3.1.10
ASME Section XI, Subsection IWF Program	A1.29 B2.1.29	Existing	Consistent	XI.S3, "ASME Section XI, Subsection IWF"	3.0.3.1.11
10 CFR Part 50 Appendix J Program	A1.30 B2.1.30	Existing	Consistent	XI.S4, "10 CFR 50, Appendix J"	3.0.3.1.12
Masonry Wall Program	A1.31 B2.1.31	Existing	Consistent with enhancement	X.S5, "Masonry Wall Program"	3.0.3.2.19
Structures Monitoring Program	A1.32 B2.1.32	Existing	Consistent with enhancement	XI.S6, "Structures Monitoring Program"	3.0.3.2.20
RG 1.127, Inspection of Water-Control Structures Associated with Nuclear Power Plants	A1.33 B2.1.33	Existing	Consistent with enhancement	XI.S7, "RG 1.127, Inspection of Water-Control Structures Associated with Nuclear Power Plants"	3.0.3.2.21
Nickel-Alloy Aging Management Program	A1.34 B2.1.34	Existing	Plant Specific	Not Applicable	3.0.3.3.1

Aging Management Review Results

Applicant Aging Management Program	LRA Sections	New or Existing Program	Applicant Comparison to the GALL Report	GALL Report Aging Management Programs	SER Section
Electrical Cable Connections Not Subject to 10 CFR 50.49 Environmental Qualification Requirements Program	A1.35 B2.1.35	New	Consistent	XI.E6, "Electrical Cable connections Not Subject to 10 CFR 50.49 Environmental Qualification Requirements"	3.0.3.1.13
Metal Enclosed Bus Program	A1.36 B2.1.36	New	Consistent	XI.E4, "Metal Enclosed Bus"	3.0.3.1.14
Fuse Holders Program	A1.37 B2.1.37	New	Consistent	XI.E5, "Fuse Holders"	3.0.3.1.15
Metal Fatigue of Reactor Coolant Pressure Boundary (RCPB) Program	A2.1 B3.1	Existing	Consistent with enhancements	X.M1, "Metal Fatigue of Reactor Coolant Pressure Boundary"	3.0.3.2.22
Environmental Qualification (EQ) of Electric Components Program	A2.2 B3.2	Existing	Consistent	X.E1, "Environmental Qualification (EQ) of Electric Components"	3.0.3.1.16
Concrete Containment Tendon Prestress Program	A2.3 B3.3	Existing	Consistent with enhancements	X.S1, "Concrete Containment Tendon Prestress"	3.0.3.2.23

[a] The enhancement was implemented during staff review, and the AMP is now consistent.

3.0.3.1 Aging Management Programs Consistent with the Generic Aging Lessons Learned Report

In LRA Appendix B, the applicant identified the following AMPs as consistent with the GALL Report:

- ASME Section XI ISI, Subsections IWB, IWC, and IWD Program
- Reactor Head Closure Studs Program
- Boric Acid Corrosion Program
- Nickel-Alloy Penetration Nozzles Welded to the Upper Reactor Vessel (RV) Closure heads of Pressurized Water Reactors Program
- Steam Generator Tube Integrity Program
- One-Time Inspection Program
- Reactor Coolant System Supplement
- Electrical Cables and Connections Not Subject to 10 CFR 50.49 Environmental Qualification Requirements Program
- Inaccessible Medium Voltage Cables Not Subject to 10 CFR 50.49 Environmental Qualification Requirements Program
- ASME Section XI, Subsection IWL Program
- ASME Section XI, Subsection IWF Program
- 10 CFR Part 50, Appendix J Program

Aging Management Review Results

- Electrical Cable Connections Not Subject to 10 CFR 50.49 Environmental Qualification Requirements Program
- Metal Enclosed Bus Program
- Fuse Holders Program
- Environmental Qualification (EQ) of Electrical Components Program

3.0.3.1.1 American Society of Mechanical Engineers Section XI In-Service Inspection, Subsections IWB, IWC, and IWD Program

Summary of Technical Information in the Application. LRA Section B2.1.1 describes the existing ASME Section XI ISI, Subsections IWB, IWC, and IWD Program as consistent with the GALL Report AMP XI.M1, "ASME Section XI Inservice Inspection, Subsections IWB, IWC, and IWD." The applicant stated that the ASME Section XI ISI, Subsections IWB, IWC, and IWD Program manages cracking, loss of fracture toughness, and loss of material in Class 1, 2, and 3 piping and components within the scope of license renewal. The applicant stated that the program includes periodic visual, surface, and volumetric examinations and leakage tests of Class 1, 2, and 3 pressure-retaining components, including welds, pump casings, valve bodies, integral attachments, and pressure-retaining bolting, as identified in ASME Code, Section XI, Tables IWB-2500-1, IWC-2500-1, and IWD-2500-1.

The applicant stated that Units 1, 2, and 3 are in their third 10-year ISI intervals. In conformance with 10 CFR 50.55a(g)(4)(ii), the applicant updates its program at the end of each successive 120-month period to comply with the requirements of the latest edition of the code (in effect, twelve months before the start of the inspection interval). The applicant further stated that the current ASME Code Section XI ISI Program is consistent with ASME Code Section XI, 2001 edition–2003 addenda. In addition, during the period of extended operation, the applicant will use the latest ASME Code edition at that time, consistent with provisions of 10 CFR 50.55a. The applicant stated that the program provides measures for monitoring to detect aging effects before a loss of intended function and provides measures for repair and replacement of components in which aging effects are detected.

Staff Evaluation. During its audit, the staff reviewed the applicant's claim of consistency with the GALL Report. The staff also reviewed and confirmed that the plant's conditions are bounded by the conditions for which the GALL Report was evaluated.

The staff noted that the applicant's current, third-interval ASME Section XI ISI, Subsections IWB, IWC, and IWD Program does not use risk-informed ISI methodology. The staff also noted that the applicant's program includes CLB requirements under 10 CFR 50.55a that require code cases N-722 and N-729-1 be used to provide additional examinations. These examinations are for pressurized water reactors (PWR) pressure retaining welds in Class 1 components fabricated with Alloy 600/82/182 materials (code case N-722), and alternative examination requirements for PWR RV upper heads with nozzles having pressure-retaining partial-penetration welds (code case N-729-1). The staff also confirmed that all components required by the ASME Code to be examined in each preceding 10-year ISI interval, have been examined. The staff finds these features of the applicant's program to be acceptable because they are consistent with the applicant's CLB and with the GALL Report's recommendations.

The staff compared elements one through six of the applicant's program to the corresponding elements of the GALL Report AMP XI.M1. As discussed in the Audit Report, the staff confirmed that these elements are consistent with the corresponding elements of the GALL Report AMP XI.M1. Based on its audit, the staff finds that elements one through six of the applicant's ASME Section XI ISI, Subsections IWB, IWC, and IWD Program are consistent with the corresponding program elements of the GALL Report AMP XI.M1 and, therefore, acceptable.

Aging Management Review Results

Operating Experience. LRA Section B2.1.1 summarizes operating experience related to the ASME Section XI ISI, Subsections IWB, IWC, and IWD Program. The applicant stated that review of its plant-specific operating experience for its program has not revealed any program adequacy issues or implementation issues with the ASME Section XI ISI, Subsections IWB, IWC, and IWD Program. The applicant stated that it evaluates industry operating experience for relevancy to its ASME Section XI ISI, Subsections IWB, IWC, and IWD Program and that it takes and documents appropriate actions. The applicant stated that, based on these results, the ASME Section XI ISI, Subsections IWB, IWC, and IWD Program is effective in monitoring ASME Class 1, 2, and 3 components and detecting aging effects before any loss of intended function.

The applicant stated that its review of the Second 10-Year ISI Interval Summary Reports for Units 1, 2, and 3 found that there were no code repairs or code replacements required for continued service of ASME Code Section XI, IWB, IWC, and IWD components during the 10-year period. The applicant further stated that the summary reports did not identify any implementation issues with the ASME Section XI ISI, Subsections IWB, IWC, and IWD Program for ASME Code Section XI, IWB, IWC, and IWD components.

The applicant stated that it updates the ASME Section XI ISI, Subsections IWB, IWC, and IWD Program to account for industry operating experience, and annual updates of ASME Code Section XI allow the applicant to update the code to reflect industry operating experience. The applicant also stated that the requirement to update the ASME Section XI ISI, Subsections IWB, IWC, and IWD Program, to reference more recent editions of ASME Code Section XI, at the end of each 10-year inspection interval ensures that the program reflects enhancements due to operating experience that have been incorporated into ASME Code Section XI.

The staff reviewed operating experience information, in the application and during the audit, to determine whether the applicant reviewed the applicable aging effects and industry and plant-specific operating experience. As discussed in the Audit Report, the staff conducted an independent search of the plant operating experience information to determine if the applicant adequately incorporated and evaluated operating experience related to this program. The staff found that the applicant adequately identified and incorporated industry and applicable plant-specific operating experience.

Based on its audit and review of the application, the staff finds that operating experience related to the applicant's program demonstrates that it can adequately manage the detrimental effects of aging on SSCs within the scope of the program. The staff confirmed that the "operating experience" program element satisfies the criterion of SRP-LR Section A.1.2.3.10 and, therefore, the staff finds it acceptable.

UFSAR Supplement. LRA Section A1.1 provides the UFSAR supplement for the ASME Section XI ISI, Subsections IWB, IWC, and IWD Program. The staff reviewed this UFSAR supplement's description of the program and notes that it conforms to the recommended description for this type of program as described in SRP-LR Table 3.1-2. The staff also notes that the applicant committed (Commitment No. 3) to the ongoing implementation of the existing ASME Section XI ISI, Subsections IWB, IWC, and IWD Program to manage the aging of applicable components during the period of extended operation. The staff determines that the information in the UFSAR supplement is an adequate summary description of the program, as required by 10 CFR 54.21(d).

Conclusion. On the basis of its review of the applicant's ASME Section XI ISI, Subsections IWB, IWC, and IWD Program, the staff finds all program elements consistent with the GALL Report. The staff concludes that the applicant has demonstrated that the effects of aging will be adequately managed so that the intended function(s) will be maintained consistent with the CLB for the period of extended operation, as required by 10 CFR 54.21(a)(3). In addition, the staff reviewed the UFSAR supplement for this AMP, and concludes that it provides an adequate summary description of the program, as required by 10 CFR 54.21(d).

3.0.3.1.2 Reactor Head Closure Studs Program

Summary of Technical Information in the Application. LRA Section B2.1.3 describes the existing Reactor Head Closure Studs Program as consistent with the GALL Report AMP XI.M3, "Reactor Head Closure Studs." The applicant stated that the Reactor Head Closure Studs Program conducts ASME Code Section XI inspections of RV flange stud hole threads and of reactor head closure studs, nuts, and washers to manage the aging effects of cracking and loss of materials in these components. The applicant also stated that its program follows the preventive measures described in NRC Regulatory Guide (RG) 1.65, "Materials and Inspections for Reactor Closure Studs," and that it uses a lubricant on RV flange stud hole threads, reactor head closure stud and nut threads, and washer faces after reactor head closure stud, nut, and washer cleaning and examination.

The applicant stated that it detects potential cracking or loss of material in RV flange stud hole threads and in reactor head closure studs, nuts, and washers through visual or volumetric examinations, in accordance with ASME Code Section XI requirements. The applicant stated that these examinations are conducted during refueling outages, after the studs are removed from the vessel flange holes. The applicant stated that during flood-up of the reactor cavity, in support of refueling activities, the RV flange holes are plugged with water tight plugs to assure that the holes, studs, nuts, and washers are protected from borated water.

The applicant stated that its program currently implements the requirements of ASME Code Section XI, Subsection IWB, 2001 edition through the 2003 addenda and that its program is updated during each successive 120-month inspection interval to comply with the requirements of the latest ASME Code Section XI edition in accordance with the provisions of 10 CFR 50.55a.

Staff Evaluation. During its audit, the staff reviewed the applicant's claim of consistency with the GALL Report. The staff also reviewed and confirmed that the plant's conditions are bounded by the conditions for which the GALL Report was evaluated.

The staff compared elements one through six of the applicant's program to the corresponding elements of the GALL Report AMP. As discussed in the Audit Report, the staff confirmed that each element of the applicant's program is consistent with the corresponding element of the GALL Report AMP Reactor Head Closure Studs, with the exception of the "preventive actions" and "detection of aging effects" program elements. For these elements, the staff determined the need for additional clarification, which resulted in the issuance of requests for additional information (RAIs).

In the GALL Report AMP Reactor Head Closure Studs, the "preventive actions" program element states that preventive measures include the use of acceptable surface treatments and stable lubricants and that implementation of this measure can reduce the potential for stress corrosion cracking (SCC) or intergranular stress corrosion cracking (IGSCC). In its review of the applicant's "preventive actions" program element, the staff noted that the applicant uses the thread lubricant Lubrikol L1G6M5, which contains 1-percent molybdenum disulphide on the reactor head bolt closure studs. The staff further noted that, although it is not explicitly referenced in the GALL Report AMP, NUREG-1339, "Resolution of Generic Safety Issue 29: Bolting Degradation or Failure in Nuclear Power Plants," includes specific recommendations against the use of thread lubricants containing molybdenum disulfide.

By letter dated December 29, 2010, the staff issued RAI B2.1.7-3 requesting the applicant to explain how it will manage the aging effects of concern in NUREG-1339, related to use of a thread lubricant containing molybdenum disulfide, to ensure that the preventive actions

Aging Management Review Results

described in the GALL Report AMP Reactor Head Closure Studs are implemented during the period of extended operation.

In its response, dated March 24, 2010, the applicant stated that it follows the preventive measures in RG 1.65 to prevent aging effects due to corrosion or hydrogen embrittlement. The applicant stated that its RV closure studs are not metal-plated and that it currently uses "Super Molly 402-40 or equivalent (e.g., Lubrikol L1G6M5)" as lubricant on RV flange stud hole threads, reactor head closure stud and nut threads, and washer faces after reactor head closure stud, nut and washer cleaning and examination are complete. The applicant further stated that administrative controls limit the use of Super Molly 402-40 to only these applications and that, while Super Molly 402-40 is compatible with the RV flange, stud, nut, and washer materials, it is in the process of a phased withdrawal of lubricants containing molybdenum disulfide associated with these applications from the site. In a conference call (January 11, 2011), the applicant confirmed that this would be completed before the period of extended operation. The applicant further stated that it has had minimal issues with galling of reactor head closure studs and that it has not identified any cases of SCC or IGSCC with its RV studs, nuts, flange stud holes, or washers.

Based on the its review, the staff finds the applicant's response to RAI B2.1.7-03 acceptable because the applicant has administrative controls in place to limit the use of Super Molly 402-40 or other lubricants containing molybdenum disulfide, and the applicant is in the process of a phased withdrawal of lubricants containing molybdenum disulfide from the site. The applicant resolved the staff's concern, described in RAI B2.1.7-03.

In the GALL Report AMP Reactor Head Closure Studs, the "detection of aging effects" program element states that Examination Category B-G-1, for pressure-retaining bolting greater than 2-inch diameter in the RV, specifies surface and volumetric examination of studs when removed from the RV. In its review of the applicant's "detection of aging effects" program element, the staff noted that the applicant performs volumetric (not volumetric and surface) examination of reactor head closure studs when removed from the RV flange.

By letter dated December 29, 2009, the staff issued RAI B2.1.3-01 requesting the applicant to explain why implementation of only volumetric examinations, rather than volumetric and surface examinations, for removed closure studs is not identified as an exception to recommendations in the GALL Report. Additionally, the staff asked the applicant to justify how the use of only volumetric inspections for these components will provide adequate detection of aging effects during the period of extended operation.

In its response, dated February 19, 2010, the applicant stated that it determined that no exception to the GALL Report AMP Reactor Head Closure Studs, "detection of aging effects" program element was necessary because the volumetric examination of reactor head closure studs, when removed, meets the requirements of ASME Code Section XI, 2001 edition, including the 2002 and 2003 addenda. The applicant further stated that it appears that the phrase "surface and volumetric examination of studs when removed" should have been changed to "surface or volumetric examination of studs when removed" when the ASME code version cited in the GALL Report was changed.

The staff confirmed that ASME Code Section XI, 2001 edition, including 2002 and 2003 addenda, Examination Category B-G-1 for pressure retaining bolting greater than 2-inches diameter in RVs no longer includes a specific examination requirement for reactor head closure studs "when removed." In the 2001 code edition, the requirement for item B6.30 ("closure studs, when removed") was changed to specify either volumetric or surface examination; and in the 2002 addenda, item B6.30 was deleted entirely, which resulted in volumetric examination being normally specified, with surface examination permitted to be substituted for volumetric examination when studs are removed. The staff further confirmed that there are no statements

3-13

Aging Management Review Results

in the GALL Report indicating that additional surface examination is recommended for license renewal, and there are no additional requirements in the 10 CFR 50.55a endorsements of ASME Code Section XI, 2001 edition, with 2002 and 2003 addenda, to indicate that augmentation of the Examination Category B-G-1 specifications is required.

Based on its review, the staff finds the applicant's response to RAI B2.1.3-01 acceptable because (1) the applicant's volumetric examination of reactor head closure studs is consistent with specifications if ASME Code Section XI, 2001 edition, including the 2002 and 2003 addenda, which is referenced in the GALL Report, (2) neither the GALL Report nor 10 CFR 50.55a specify augmentation of the ASME Code Section XI, Examination Category B-G-1 requirements, and (3) either surface or volumetric examinations are capable of detecting cracking in reactor head closure studs prior to the aging mechanism affecting the CLB function of the studs. The applicant, therefore, resolved the staff's concern described in RAI B2.1.3-01.

Based on its audit and review of the applicant's responses to RAIs B2.1.7-03 and B2.1.3-01, the staff finds that elements one through six of the applicant's Reactor Head Closure Studs Program are consistent with the corresponding program elements of the GALL Report AMP XI.M3 and, therefore, acceptable.

Operating Experience. LRA Section B2.1.3 summarizes operating experience related to the Reactor Head Closure Studs Program. The applicant stated that its review of its plant-specific operating experience has not revealed any implementation issues with the Reactor Head Closure Studs Program and that it has not identified any instances of cracking, due to SCC or IGSCC, with RV studs, nuts, flange stud holes, or washers.

The applicant stated that review of its Refueling Outage ISI Reports for the second 10-year ISI interval found there were no repair or replacement items associated with the RV closure studs, nuts, washers or flange thread holes. The applicant also stated that it updated its ISI Program, which is the basis for its Reactor Head Closure Studs Program, to account for industry operating experience. Annual updates of ASME Code Section XI allow the applicant to update the code to reflect industry operating experience.

The staff reviewed operating experience information in the application and during the audit to determine whether the applicant reviewed the applicable aging effects and industry and plant-specific operating experience. As discussed in the Audit Report, the staff conducted an independent search of the plant operating experience information to determine if the applicant adequately incorporated and evaluated operating experience related to this program. The staff found that the applicant adequately identified and incorporated industry and applicable plant-specific operating experience.

Based on its audit and review of the application, the staff finds that operating experience related to the applicant's program demonstrates that it can adequately manage the detrimental effects of aging on SSCs within the scope of the program. The staff confirmed that the "operating experience" program element satisfies the criterion of SRP-LR Section A.1.2.3.10 and, therefore, the staff finds it acceptable.

UFSAR Supplement. LRA Section A1.3 provides the UFSAR supplement for the Reactor Head Closure Stud Program. The staff reviewed this UFSAR supplement description of the program and notes that it conforms to the recommended description for this type of program as described in SRP-LR Table 3.1-2. The staff also notes that the applicant committed (Commitment No. 5) to the ongoing implementation of the existing Reactor Head Closure Studs Program to manage the aging of applicable components during the period of extended operation. The staff determines that the information in the UFSAR supplement is an adequate summary description of the program, as required by 10 CFR 54.21(d).

Aging Management Review Results

Conclusion. On the basis of its review of the applicant's Reactor Head Closure Studs Program, the staff finds all program elements consistent with the GALL Report. The staff concludes that the applicant has demonstrated that the effects of aging will be adequately managed so that the intended function(s) will be maintained consistent with the CLB for the period of extended operation, as required by 10 CFR 54.21(a)(3). The staff also reviewed the UFSAR supplement for this AMP, and concludes that it provides an adequate summary description of the program, as required by 10 CFR 54.21(d).

3.0.3.1.3   Boric Acid Corrosion Program

Summary of Technical Information in the Application. LRA Section B2.1.4 describes the existing Boric Acid Corrosion Program as consistent with the GALL Report AMP XI.M10, "Boric Acid Corrosion." The applicant stated that this program manages loss of material in components with materials susceptible to boric acid corrosion. The applicant also stated that the program includes provisions to identify leakage, inspect and examine for evidence of leakage, evaluate leakage, and initiate corrective actions. The program also includes the tracking and trending of boric acid leakage from plant components.

Staff Evaluation. During its audit, the staff reviewed the applicant's claim of consistency with the GALL Report. The staff also reviewed and confirmed that the plant's conditions are bounded by the conditions for which the GALL Report was evaluated.

The staff compared elements one through six of the applicant's program to the corresponding element of the GALL Report AMP Boric Acid Corrosion. As discussed in the Audit Report, the staff confirmed that these elements are consistent with the corresponding elements of the GALL Report AMP Boric Acid Corrosion. Based on its audit, the staff finds that elements one through six of the applicant's Boric Acid Corrosion Program are consistent with the corresponding program elements of the GALL Report AMP Boric Acid Corrosion and, therefore, acceptable.

Operating Experience. LRA Section B2.1.4 summarizes operating experience related to the Boric Acid Corrosion Program. The applicant stated that, in response to recent NRC generic communications, it revised the RCS pressure boundary integrity walkdowns to include periodic visual inspection of the RCS for indications of leakage. The applicant also stated that there were several instances where it made containment entries to investigate increased RCS leakage in which it detected boric acid leakage through the visual observation of active leaks or residual deposits. The applicant stated that these cases illustrate that the program is effective in identifying boric acid corrosion before the loss of component intended function.

The staff reviewed operating experience information, in the application and during the audit, to determine whether the applicant reviewed applicable aging effects and industry and plant-specific operating experience. In addition to the information provided in the LRA, the staff reviewed the relevant operating experience recorded in onsite records, including condition reports, disposition requests, and corrective maintenance work orders. The following two examples of site-specific operating experience also demonstrate the effectiveness of the program:

(1)   Based on a visual inspection, the applicant found a check valve with boric acid crystals, indicating that leakage had occurred followed by evaporation and the subsequent appearance of crystals. Following this observation, the applicant cleaned the area and conducting a surveillance, which confirmed the presence of a small leak at the suspected location. The applicant issued a work order for valve seal replacement. The applicant attributed the leaking valve seal to access difficulty, which led to the application of unsatisfactory torque on the associated fixture during the prior valve maintenance cycle.

(2) Inspection reports include cases where the applicant observed small, recurring leaks on valves in the boric acid piping system. In these cases, the applicant generated appropriate work orders for repairs and replacements, as indicated by the circumstances. In addition, the applicant conducted engineering evaluations to determine that the correct components were used for the associated pressures and boric acid environments.

Based on its audit and review of the application, the staff finds that operating experience related to the applicant's program demonstrates that it can adequately manage the detrimental effects of aging on SSCs within the scope of the program, and that implementation of the program has resulted in the applicant taking corrective actions. The staff confirmed that the "operating experience" program element satisfies the criterion in SRP-LR Section A.1.2.3.10 and, therefore, the staff finds it acceptable.

UFSAR Supplement. LRA Section A1.4 provides the UFSAR supplement for the Boric Acid Corrosion Program. The staff reviewed the UFSAR supplement description of the program and notes that it conforms to the recommended description for this type of program, as described in SRP-LR Tables 3.1-2, 3.2-2, 3.3-2, 3.4-2, 3.5-2, and 3.6-2. The staff also notes that the applicant committed (Commitment No. 6) to the ongoing implementation of the existing Boric Acid Corrosion Program to manage the aging of applicable components during the period of extended operation. The staff determines that the information in the UFSAR supplement is an adequate summary description of the program, as required by 10 CFR 54.21(d).

Conclusion. On the basis of its review of the applicant's Boric Acid Corrosion program, the staff finds all program elements consistent with the GALL Report. The staff concludes that the applicant has demonstrated that the effects of aging will be adequately managed so that the intended function(s) will be maintained consistent with the CLB for the period of extended operation, as required by 10 CFR 54.21(a)(3). The staff also reviewed the UFSAR supplement for this AMP and concludes that it provides an adequate summary description of the program, as required by 10 CFR 54.21(d).

3.0.3.1.4 Nickel-Alloy Penetration Nozzles Welded to the Upper Reactor Vessel Closure Heads of Pressurized Water Reactors

Summary of Technical Information in the Application. LRA Section B2.1.5 describes the existing Nickel-Alloy Penetration Nozzles Welded to the Upper RV Closure Heads of PWR Program (Nickel-Alloy Head Penetration Program) as consistent with the GALL Report AMP XI.M11A, "Nickel-Alloy Penetration Nozzles Welded to the Upper Reactor Vessel Closure Heads of Pressurized Water Reactors." The applicant stated that the program manages cracking, due to primary water stress corrosion cracking (PWSCC), and loss of material, due to boric acid wastage in nickel-alloy pressure vessel head penetration nozzles. The applicant stated that the program includes the RV closure head, the upper vessel head penetration nozzles, and associated welds. The applicant also stated that it mitigated cracking through the control of water chemistry. The applicant further stated that it managed the aging effects of cracking and loss of material through a combination of visual and non-visual inspection techniques, as described in ASME Code Case N-729-1, as modified by 10 CFR 50.55a(g)(6)(ii)(D)(2)–(6).

Staff Evaluation. During its audit, the staff reviewed the applicant's claim of consistency with the GALL Report. The staff also reviewed and confirmed that the plant's conditions are bounded by the conditions for which the GALL Report was evaluated.

The staff compared elements one through six of the applicant's program to the corresponding elements of the GALL Report AMP. In making this comparison, the staff noted that, in its original submission, the applicant proposed this AMP to be consistent with the GALL Report AMP with one enhancement. The proposed enhancement was to add the inspection

Aging Management Review Results

requirements described in ASME Code Case N-729-1 as modified by 10 CFR 50.55a(g)(6)(ii)(D)(2)–(6). The staff also noted that in Supplement 1 to the LRA, dated April 10, 2009, the inspection requirements described in the code case were incorporated into the AMP, thereby eliminating the need for enhancement. The staff further noted that the governing document discussed in the GALL Report AMP is NRC Order EA-03-009 or any subsequent NRC requirements that may be established to supersede the requirements of Order EA-03-009. The staff finally noted that the governing document in the LRA AMP is Code Case N-729-1 (the Code Case resulted from the incorporation of the inspection requirements of Order EA-03-009 into 10 CFR 50.55a and ASME Code Section XI). The staff finds the incorporation of the originally proposed enhancement into the AMP and the use of Code Case N-729-1 acceptable because 10 CFR 50.55a(g)(6)(ii)(D)(1) states that NRC Order EA-03-009 is withdrawn in favor of ASME Code Case N-729-1, as modified by 10 CFR 50.55a(g)(6)(ii)(D)(2)-(6). The staff finds that elements one through six of the applicant's Nickel-Alloy Head Penetration Program are consistent with the corresponding program elements of the GALL Report AMP XI.M11A and, therefore, are acceptable.

Operating Experience. LRA Section B2.1.5 summarizes operating experience related to the Nickel-Alloy Head Penetration Program. The applicant stated that it has conducted all inspections as required by NRC order EA-03-009 or Code Case N-729-1. For Units 1 and 3, the applicant reported that it observed no evidence of cracking or metal wastage. For Unit 2, the applicant observed indications on the vent line on one occasion and removed these indications by machining. The applicant subsequently applied a weld overlay over this area. The applicant also stated that the head for Unit 2 was replaced in the fall of 2009, Unit 1 head was replaced in spring of 2010, and the head for Unit 3 is scheduled for replacement in the fall of 2010. Based on the use of different alloys for the penetration nozzles in the replacement heads, the staff considers operating experience accumulated using the old heads to be of little value for predicting the performance of the new heads.

The staff reviewed all available industry and plant specific operating experience to determine if the applicant incorporated this experience in the AMP. If not, the staff determined if the failure of the applicant to incorporate this experience in the AMP would adversely affect the ability of the applicant to adequately manage the aging of the components under consideration, through the use of the AMP. The staff found that the applicant adequately identified and incorporated industry and applicable plant-specific operating experience.

Based on its audit and review of the application, the staff finds that operating experience related to the applicant's program demonstrates that it can adequately manage the detrimental effects of aging on SSCs within the scope of the program, and that implementation of the program has resulted in the applicant taking corrective actions. The staff confirmed that the "operating experience" program element satisfies the criterion in SRP-LR Section A.1.2.3.10 and, therefore, the staff finds it acceptable.

UFSAR Supplement. LRA Section A1.5 provides the UFSAR supplement for the Nickel-Alloy Head Penetration Program. The staff reviewed this UFSAR supplement description of the program and notes that it contains the critical aspects of the recommended description for this type of program as described in SRP-LR Table 3.1-2. The staff also notes that the applicant committed (Commitment No. 7) to the ongoing implementation of the existing Nickel-Alloy Head Penetration Program to manage the aging of applicable components during the period of extended operation. The staff determines that the information in the UFSAR supplement is an adequate summary description of the program, as required by 10 CFR 54.21(d).

Conclusion. On the basis of its review of the applicant's Nickel-Alloy Head Penetration Program, the staff finds all program elements consistent with the GALL Report. The staff concludes that the applicant has demonstrated that the effects of aging will be adequately managed so that the intended function(s) will be maintained consistent with the CLB for the

Aging Management Review Results

period of extended operation, as required by 10 CFR 54.21(a)(3). The staff also reviewed the UFSAR supplement for this AMP and concludes that it provides an adequate summary description of the program, as required by 10 CFR 54.21(d).

3.0.3.1.5   Steam Generator Tube Integrity Program

Summary of Technical Information in the Application. LRA Section B2.1.8 describes the existing SG Tube Integrity Program as consistent with the GALL Report AMP XI.M19, "Steam Generator Tube Integrity." The applicant stated that the SG Tube Integrity Program includes measures to prevent, inspect, and assess degradation; to assess structural and leakage integrity; to maintain primary and secondary chemistry controls; and to conduct required maintenance and repair activities necessary to manage aging mechanisms. The applicant further stated that the SG Tube Integrity Program is consistent with NEI 97-06, "Steam Generator Program Guidelines," and with the requirements of the technical specifications (TS), which encompass and exceed the requirements of RG 1.121, "Bases for Plugging Degraded Steam Generator Tubes."

The applicant stated that each unit has two identical recirculating SGs, designed by Combustion Engineering, Inc. (CE), which are considered a modified CE System 80 design. The original SGs were replaced in Units 1, 2, and 3 during the fall of 2005, 2003, and 2007, respectively. Each replacement SG has 12,580 Alloy 690 thermally-treated tubes. The tubes are hydraulically expanded into the tubesheet for the entire tubesheet thickness. The tube support system is similar to the original design and, like the original design, is fabricated from type 409 ferritic stainless steel. In addition to the tubing material change, the U-bend region in the first 17 rows were stress relieved after bending, to minimize the potential for SCC.

Staff Evaluation. During its audit, the staff reviewed the applicant's claim of consistency with the GALL Report. The staff also reviewed and confirmed that the plant's conditions are bounded by the conditions for which the GALL Report was evaluated.

The staff compared elements one through six of the applicant's program to the corresponding elements of the GALL Report AMP, "Steam Generator Tube Integrity." As discussed in the Audit Report, the staff confirmed that these elements are consistent with the corresponding elements of the GALL Report AMP, "Steam Generator Tube Integrity." Based on its audit, the staff finds that elements one through six of the applicant's SG Tube Integrity AMP are consistent with the corresponding program elements of the GALL Report AMP, "Steam Generator Tube Integrity" and, therefore, are acceptable.

Operating Experience. LRA Section B2.1.8 summarizes operating experience related to the SG Tube Integrity program. The applicant stated that wear is the only active damage mechanism in the SGs, specifically, wear as the result of tubing interaction with the tube supports. Most of the wear indications observed were in the region around the stay cylinder and at either the diagonal supports or vertical support 3. The applicant also provided the following operational experience:

> Due to certain historically observed wear phenomenon, PVNGS has employed conservative administrative plugging criteria related to support wear mechanisms. For example, support wear indications are removed for wear rate greater than or equal to 35% for a normal operating cycle if no previous wear is identified. This plugging criterion is designed to ensure that the structural and accident leakage performance criteria specified in the PVNGS Technical Specifications are not exceeded in the subsequent operating cycle. It was expected, based on RSG [replacement steam generators] redesign, that the conditions necessary to generate high wear rates in the Batwing Stay Cylinder (BWSC) and Cold Leg Corner (CLC) regions were eliminated. While this was clearly the case for CLC wear, the RSG inspection results during the initial inspection in Unit 2 (U2R12) indicated that the RSGs continued to exhibit similar

wear conditions within the BWSC region. As a result of these findings, a decision was made prior to Unit 1 and Unit 3 RSG installation to plug and stake all of the "frontline" BWSC tubes. The subsequent inspections during U2R13, U1R13, U2R14, U1R14, U3R14, and U2R15 have indicated that the BWSC wear issue exists in the RSGs of all three PVNGS units.

On February 19, 2004, Unit 2 was operating at full power when radiation monitors displayed indications of a low level primary to secondary leak. Shortly thereafter the leak rate was calculated to be 11.8 gallons per day, even though grab samples indicated 3 gallons per day, and the decision was made to shut the unit down to find and repair the leak.

After cooling the plant down and performing tests, one RSG tube was found to be leaking and was plugged. Further analysis showed that the cause of the leak was from a puncture received from a wood screw that was used in the construction of the shipping crates for the tubes when the RSGs were being manufactured. The tubes were placed in the crate and the crate assembled around them. One screw that was used near the outer diameter of the top of the tube bend protruded through the wood and began to puncture the tube material. The screw did not completely penetrate the tube and the unit was operated from its post-outage startup to this date when the tube finally began leaking. Contamination to the secondary plant was minimal and the unit entered Mode 1 on March 9, 2004. Corrective actions put in place after the event prevented recurrence in the Unit 1 and 3 RSGs.

The staff reviewed operating experience information in the application and during the audit to determine if the applicant reviewed aging effects and industry and plant-specific operating experience. The staff conducted an independent search of the plant operating experience information to determine if the applicant had adequately incorporated and evaluated operating experience related to this program. The staff found that the applicant had adequately identified and incorporated industry and applicable plant-specific operating experience.

Based on its audit and review of the application, the staff finds that operating experience related to the applicant's program demonstrates that it can adequately manage the detrimental effects of aging on SSCs within the scope of the program, and that implementation of the program has resulted in the applicant taking corrective actions. The staff confirmed that the "operating experience" program element satisfies the criterion in SRP-LR Section A.1.2.3.10 and, therefore, the staff finds it acceptable.

UFSAR Supplement. LRA Section A1.8 provides the UFSAR supplement for the SG Tube Integrity Program. The staff reviewed this UFSAR supplement description of the program and notes that it conforms to the recommended description for this type of program, as described in SRP-LR Table 3.1-2. The staff also notes that the applicant committed (Commitment No. 10) to the ongoing implementation of the existing SG Tube Integrity Program to manage the aging of applicable components during the period of extended operation.

The staff finds that the information in the UFSAR supplement is an adequate summary description of the program, as required by 10 CFR 54.21(d).

Conclusion. On the basis of its review of the applicant's SG Tube Integrity Program, the staff finds all program elements consistent with the GALL Report. The staff concludes that the applicant has demonstrated that the effects of aging will be adequately managed so that the intended function(s) will be maintained consistent with the CLB for the period of extended operation, as required by 10 CFR 54.21(a)(3). The staff also reviewed the UFSAR supplement for this AMP, and concludes that it provides an adequate summary description of the program, as required by 10 CFR 54.21(d).

Aging Management Review Results

3.0.3.1.6   One-Time Inspection Program

Summary of Technical Information in the Application. LRA Section B2.1.16 describes the One-Time Inspection Program as a new program, consistent with the GALL Report AMP XI.M32, "One-Time Inspection." The applicant stated that its program performs one-time inspections of plant piping and components to verify the effectiveness of the Water Chemistry, Fuel Oil Chemistry, and Lubricating Oil Analysis Programs. The applicant also stated that the aging effects evaluated by the One-Time Inspection Program are loss of material, cracking, and reduction of heat transfer. The applicant further stated that the One-Time Inspection Program includes the identification and statistical sampling of the piping and component population for the selected material and environment combinations, as well as the inspection of the selected components per ASME Code Sections V or XI, as applicable.

Staff Evaluation. During its audit, the staff reviewed the applicant's claim of consistency with the GALL Report. The staff also reviewed and confirmed that the plant's conditions are bounded by the conditions for which the GALL Report was evaluated.

The staff compared elements one through six of the applicant's program to the corresponding elements of the GALL Report, "One-Time Inspection AMP." As discussed in the Audit Report, the staff confirmed that each element of the applicant's program is consistent with the corresponding element of the GALL Report AMP, with the exception of the "detection of aging effects" program element. For this element, the staff determined the need for additional clarification, which resulted in the issuance of an RAI, discussed below.

During the audit, the applicant stated that it would perform surface examination techniques to detect cracking in addition to the enhanced visual and volumetric inspection techniques recommended by the GALL Report AMP, "One-Time Inspection." However, it was not clear to the staff how the applicant would use surface examinations to detect cracking. By letter dated December 29, 2009, the staff issued RAI B2.1.16-2 requesting that the applicant clarify if the surface examination techniques used to detect cracking will replace the enhanced visual or volumetric inspections recommended by the GALL Report or if the surface inspections will supplement the enhanced visual or volumetric inspections. The staff also asked that the applicant clarify if it will perform the surface examinations on wetted surfaces.

In its response, dated February 19, 2010, the applicant stated that depending on each component's environment and accessibility, a proper non-destructive examination (NDE) technique selected from the One-Time Inspection Program procedure will be used to identify aging effects in piping and components, including those that are exposed to water, lube oil, and fuel oil environments. The applicant also stated that the NDE techniques discussed in the One-Time Inspection Program are consistent with ASME Code NDE techniques. The staff finds the applicant's response to RAI B2.1.16-2 acceptable because the program will perform NDE techniques that are capable of detecting loss of material or cracking, are appropriate for the given material and environment combinations, and are consistent with industry standards. The applicant's response resolved the staff's concern, as described in RAI B2.1.16-2.

Based on its audit and review of the applicant's response to RAI B2.1.16-2, the staff finds that elements one through six of the applicant's One-Time Inspection Program are consistent with the corresponding program elements of the GALL Report AMP, "One-Time Inspection" and, therefore, are acceptable.

Operating Experience. LRA Section B2.1.16 summarizes the operating history related to the One-Time Inspection Program. The applicant stated that it will perform one-time inspections during the 10-year period before the period of extended operation to identify possible aging effects. Since this is a new program, there is no plant-specific operating experience to review. However, the applicant also stated that ASME Code Section XI ISI inspection techniques have proven effective in detecting aging effects before the loss of intended function. In addition, a

Aging Management Review Results

review of plant-specific operating experience associated with the ISI Program has not revealed any ISI Program adequacy issues.

The staff reviewed the operating experience information, in the application and during the audit, to determine if the applicant reviewed applicable aging effects and industry and plant-specific operating experience associated with the ISI Program. As discussed in the Audit Report, the staff conducted an independent search of the plant operating experience information to determine if the applicant had adequately incorporated and evaluated the operating history related to its program. During its review, the staff noted the applicant's operating experience review may not have been comprehensive and determined the need for additional clarification, which resulted in the issuance of the following RAI.

The applicant stated in the LRA that the plant-specific operating experience associated with the ISI Program had not revealed any inadequacies or implementation issues. The staff noted that, although there is no plant specific operating experience associated with the One-Time Inspection Program application, any operating experience resulting from maintenance activities of systems and components that will be within the scope of the ISI Program should be included in the review. By letter dated December 29, 2009, the staff issued RAI B2.1.16-1 requesting that the applicant provide a summary of operating experience resulting from observations of loss of material, cracking, and loss of heat transfer identified during maintenance as well as the associated corrective action activities taken.

In its response, dated February 19, 2010, the applicant stated that an examination of plant-specific aging effect related operating experience from 1996 through October 2009 identified 16,000 maintenance and corrective action entries. The applicant also stated that these entries discussed, "...mechanical joint leakage, out of specification chemistry, bearing wear products, presence of water in oil, manufacturing defects, and construction flaws," and that the items were more closely related to the Water Chemistry, Fuel Oil Chemistry, and Lubricating Oil Analysis Programs than the One-Time Inspection Program. The applicant also stated that it did not find any operating experience related to loss of material, cracking, and loss of heat transfer for components that the One-Time Inspection Program will manage. The staff finds the plant-specific operating experience associated with the applicant's One-Time Inspection Program is bounded by the GALL Report and, therefore, is acceptable. The applicant resolved the staff's concern, as described in RAI B2.1.16-1.

Based on its audit, review of the application, and review of the applicant's response to RAI B2.1.16-1, the staff finds that operating experience related to the applicant's program demonstrates that the applicant can adequately manage the detrimental effects of aging on SSCs within the scope of the program, and that implementation of the program has resulted in the applicant taking corrective actions. The staff confirmed that the "operating experience" program element satisfies the criterion in SRP-LR Section A.1.2.3.10 and, therefore, is acceptable.

UFSAR Supplement. LRA Section A1.16 provides the UFSAR supplement for the One-Time Inspection Program. The staff reviewed this UFSAR supplement description of the program and notes that it conforms to the recommended description for this type of program, as described in SRP-LR Tables 3.1-2, 3.2-2, 3.3-2 and 3.4-2. The staff also notes that the applicant committed (Commitment No. 18) to implement the new One-time Inspection Program before entering the period of extended operation to manage the aging of applicable components and to inspect fuel oil tank bottoms (Commitment No. 16) for thickness with a one-time ultrasonic or pulsed eddy current examination. The staff determines that the information in the UFSAR supplement is an adequate summary description of the program, as required by 10 CFR 54.21(d).

Conclusion. On the basis of its audit and review of the applicant's One Time Inspection Program, the staff determines that those program elements for which the applicant claimed

Aging Management Review Results

consistency with the GALL Report are consistent. The staff concludes that the applicant has demonstrated that the effects of aging will be adequately managed so that the intended function(s) will be maintained consistent with the CLB for the period of extended operation, as required by 10 CFR 54.21(a)(3). The staff also reviewed the UFSAR supplement for this AMP, and concludes that it provides an adequate summary description of the program, as required by 10 CFR 54.21(d).

3.0.3.1.7   Reactor Coolant System Supplement

Summary of Technical Information in the Application. In LRA Section B2.1.21, the applicant described its RCS Supplement Program, stating that, in accordance with the SRP-LR, Section 3.1, "Aging Management of Reactor Vessel, Internals, and Coolant System," this program supplements the AMPs for the RCS components with the following additional requirements.

- *For Reactor Coolant System Nickel-Alloy Pressure Boundary Components*. The applicant will implement applicable, (1) NRC orders, bulletins and generic letters associated with nickel alloys, (2) staff-accepted industry guidelines, (3) participate in the industry initiatives, such as owners group programs and the Electric Power Research Institute (EPRI) Materials Reliability Program, to manage the aging effects associated with nickel alloys, (4) upon completion of these programs, but not less than 24 months before entering the period of extended operation, the applicant will submit an inspection plan for RCS nickel-alloy pressure boundary components to the NRC for review and approval.

- *For Reactor Vessel Internals*. The applicant will, (1) participate in industry programs for investigating and managing aging effects on reactor internals, (2) evaluate and implement the results of the industry programs as applicable to the reactor internals, and (3) upon completion of these programs, but not less than 24 months before entering the period of extended operation, the applicant will submit an inspection plan for reactor internals to the NRC for review and approval.

Staff Evaluation. The staff reviewed the applicant's proposed RCS Supplement Program to confirm whether this AMP is consistent with the GALL Report. For RCS nickel-alloy pressure boundary components, the GALL Report AMP XI.M11A, "Nickel-Alloy Nozzles and Penetrations," states that, "[g]uidance for the aging management of other nickel-alloy nozzles and penetrations is provided in the AMR line items of Chapter IV, as appropriate." The staff noted that the GALL Report Table IV.A2 recommends specific AMPs for the management of postulated aging effects that may occur in the nickel-alloy control rod drive mechanism penetration pressure housings (IV.A2-11), core support pads and guide lugs (IV.A2-12), and bottom head penetration instrument tubes (IV.A2-19) covered by the following LRA Sections:

- Section 3.1.2.2.13, "Cracking due to Primary Water Stress Corrosion Cracking (PWSCC)"

- Section 3.1.2.2.16, "Cracking due to Stress Corrosion Cracking and Primary Water Stress Corrosion Cracking"

The GALL Report does not recommend any further evaluation if the applicant commitment, specified under the Table IV.A2, column heading "Aging Management Program (AMP)," for these components (or line items) is confirmed. The applicant must meet the following commitment, as required in SRP-LR Sections 3.1.2.2.13 and 3.1.2.2.16:

> Comply with applicable NRC Orders and provide a commitment in the FSAR supplement to submit a plant-specific AMP to implement applicable (1) Bulletins and Generic Letters and (2) staff-accepted industry guidelines.

Similarly, for reactor pressure vessel (RPV) internals, the GALL Report AMP XI.M16, "Reactor Vessel Internals," states that, "[g]uidance for the aging management of PWR Vessel Internals is provided in the AMR line items of Chapter IV, as appropriate." The following LRA sections cover the management of postulated aging effects that may occur for PWRs:

- Section 3.1.2.2.6, "Loss of Fracture Toughness Due to Neutron Irradiation Embrittlement and Void Swelling"

- Section 3.1.2.2.9, "Loss of Preload Due to Stress Relaxation"

- Section 3.1.2.2.12, "Cracking Due to Stress Corrosion Cracking and Irradiation-Assisted Stress Corrosion Cracking (IASCC)"

- Section 3.1.2.2.15, "Changes in Dimensions Due to Void Swelling"

- Section 3.1.2.2.17, "Cracking Due to Stress Corrosion Cracking, Primary Water Stress Corrosion Cracking, and Irradiation-Assisted Stress Corrosion Cracking (IASCC)"

The GALL Report does not recommend any further evaluation if the applicant commitment, specified under the Table IV.B3 column heading "Aging Management Program (AMP)," for these RPV internals (or line items) is confirmed. The applicant must meet the following commitment, as required in SRP-LR Sections 3.1.2.2.6, 3.1.2.2.9, 3.1.2.2.12, 3.1.2.2.15, and 3.1.2.2.17:

> No further aging management review is necessary if the applicant provides a commitment in the FSAR supplement to (1) participate in the industry programs for investigating and managing aging effects on reactor internals; (2) evaluate and implement the results of the industry programs as applicable to the reactor internals; and (3) upon completion of these programs, but not less than 24 months before entering the period of extended operation, submit an inspection plan for reactor internals to the NRC for review and approval.

By comparing the contents of the RCS Supplement Program with Commitment No. 23 in LRA Appendix A as well as with the commitments specified in the SRP-LR and GALL Report Tables IV.A2 and IV.B3, the staff concludes that the RCS Supplement Program is equivalent to Commitment No. 23, which exceeds the SRP-LR required commitment for certain RCS nickel-alloy components. Essentially, the RCS Supplement Program is the same as the SRP-LR required commitment for certain RPV internals. Hence, the staff considers the applicant's RCS Supplement Program as a means of fulfilling Commitment No. 23, designed to meet a key aging management guideline provided in SRP-LR Sections 3.1.2.2.13 and 3.1.2.2.16 for certain RCS nickel-alloy components and SRP-LR Sections 3.1.2.2.6, 3.1.2.2.9, 3.1.2.2.12, 3.1.2.2.15, and 3.1.2.2.17 for certain RPV internals. Therefore, the staff determines that the 10 program elements for a typical GALL Report AMP do not apply to the applicant's RCS Supplement Program.

For the RCS nickel-alloy components, in addition to the RCS Supplement Program (or equivalently, implementation of Commitment No. 23), the staff verified that LRA Sections 3.1.2.2.13 and 3.1.2.2.16 also require the ASME Code Section XI ISI (IWB, IWC, and IWD) Program and Water Chemistry Program to mitigate the specific aging mechanisms. For RPV internals, in addition to the RCS Supplement Program, the staff verified that LRA Sections 3.1.2.2.12 and 3.1.2.2.17 also require control of water chemistry to mitigate the specific aging mechanisms. Staff evaluation of the ASME Code Section XI ISI Program and the Water Chemistry Program can be found in SER Sections 3.0.3.1.1 and 3.0.3.2.1.

The staff noted that the lists of components for the RPV internals, in LRA Table 3.1.2-1 under the aging effects of LRA Sections 3.1.2.2.6, 3.1.2.2.9, 3.1.2.2.12, and 3.1.2.2.15, do not seem to be consistent with the lists of components in GALL Report Table IV.B3, for which the RCS

Supplement Program is credited for part or all of the aging management. These seeming inconsistencies are largely due to the plant-specific features of the RPV internals, which consist of components not identical to those listed in GALL Report Table IV.B3. Sections 3.1.2.2.6, 3.1.2.2.9, 3.1.2.2.12, and 3.1.2.2.15 of this SER contain the staff's resolution of the RAIs related to these inconsistencies.

Based on the staff's review above and the staff's resolution of RAIs related to inconsistency of component listing between the LRA and the GALL Report, the staff concludes that the RCS Supplement Program is equivalent to Commitment No. 23. Commitment No. 23 is designed to meet the SRP-LR and GALL Report Tables IV.A2 and IV.B3 requirements for RCS nickel-alloy components and the RPV internals under the aging mechanisms identified earlier. Hence, working with appropriate AMP(s), as specified in GALL Report Tables IV.A2 and IV.B3, the RCS Supplement Program is acceptable for management of aging effects listed above for the RCS nickel-alloy components and the RPV internals.

UFSAR Supplement. LRA Section A1.21 provides the UFSAR supplement for the Reactor Coolant System Supplement. The staff reviewed this UFSAR supplement description of the program and notes that it conforms to the recommendations described in SRP-LR Table 3.1-2. The staff also notes that the applicant committed (Commitment No. 23) to implement the Reactor Coolant System Supplement before entering the period of extended operation. The staff determines that the information in the UFSAR supplement is an adequate summary description of the program, as required by 10 CFR 54.21(d).

Conclusion. On the basis of its review of the applicant's RCS Supplement Program, the staff determines that this AMP is a program designed to fulfill Commitment 23. This AMP is consistent with the GALL Report because GALL Report Tables IV.A2 and IV.B3 provide guidance for the management of the relevant RCS nickel-alloy pressure boundary components and RPV internals, which require the commitments as stated in Commitment No. 23. The staff concludes that, combined with other specific AMPs, the applicant has demonstrated that the effects of aging for certain RCS nickel-alloy components and the RPV internals will be adequately managed so that the intended functions will be maintained consistent with the CLB for the period of extended operation, as required by 10 CFR 54.21(a)(3). The staff also reviewed the UFSAR Supplement for this AMP, and concludes that it provides an adequate summary description of the program, as required by 10 CFR 54.21(d).

3.0.3.1.8 Electrical Cables and Connections Not Subject to the Environmental Qualification Requirements Program Under Title 10, Part 50.49 of the Code of Federal Regulations

Summary of Technical Information in the Application. LRA Section B2.1.24 describes the new Electrical Cables and Connections Not Subject to 10 CFR 50.49 EQ Requirements Program as consistent with the GALL Report AMP XI.E1 "Electrical Cables and Connections Not Subject to 10 CFR 50.49 EQ Requirements." The applicant stated that the Electrical Cables and Connections Program manages embrittlement, melting, cracking, swelling, surface contamination, or discoloration to ensure that electrical cables, connections and terminal blocks, not subject to the EQ requirements of 10 CFR 50.49 and within the scope of license renewal, are capable of performing their intended functions.

Staff Evaluation. During its audit, the staff reviewed the applicant's claim of consistency with the GALL Report. The staff also reviewed and confirmed that the plant's conditions are bounded by the conditions for which the GALL Report was evaluated.

The staff compared elements 1 through 6 of the applicant's program to the corresponding elements of the GALL Report AMP, "Electrical Cables and Connections Not Subject to 10 CFR 50.49 EQ Requirements Program." As discussed in the Audit Report, the staff confirmed that each element of the applicant's program is consistent with the corresponding

Aging Management Review Results

element of the GALL Report AMP, with the exception of the "scope of program" element. In order to verify whether the "scope of program" element is consistent with the corresponding element of the GALL Report AMP, the staff issued RAI B2.1.24-1 to the applicant in a letter dated December 29, 2009. This RAI requested the applicant to explain why electrical containment penetrations were not included in the scope of the program.

In its response, dated February 19, 2010, the applicant stated that LRA Table 3.6.2-1 shows the penetrations electrical line being managed using the AMP Electrical Cables and Connections Not Subject to 10 CFR 50.49 EQ Requirements. The applicant also stated that the Electrical Cables and Connections Not Subject to 10 CFR 50.49 EQ Requirements AMP was revised to clarify that the non-EQ electrical containment penetrations are included in the program. The applicant revised the LRA AMP section to include the following clarification, "[c]onnection insulation material includes termination kits and tape used to insulate splices that are normally located within junction boxes, terminal blocks located within terminal boxes, and non-EQ electrical containment penetrations."

The staff finds the applicant's response acceptable because the applicant included non-EQ electrical containment penetrations in the scope of Electrical Cables and Connections Not Subject to 10 CFR 50.49 EQ Requirement AMP. Including electrical containment penetrations makes the applicant's AMP consistent with the recommendations in the GALL Report AMP. The staff's concern in RAI B2.1.24-1 is resolved.

Based on its audit, review of the application, and review of the applicant's response to RAI B2.1.24-1, the staff finds that elements one through six of the applicant's Electrical Cables and Connections Not Subject to 10 CFR 50.49 EQ Requirements Program are consistent with the corresponding program elements of the GALL Report AMP and, therefore, are acceptable.

Operating Experience. The staff also reviewed the "operating experience" program element described in LRA Section B2.1.24. The applicant stated that the Electrical Cables and Connections Not Subject to 10 CFR 50.49 EQ Requirements Program is a new program; therefore, there is no plant-specific operating experience. The applicant stated, however, that it reviewed the plant operating history and found three minor cases of cable aging due to adverse environments. It found a lighting power cable with degraded insulation, and it replaced the cable. In the second case, it found conduits were run too close to a steam line, and it relocated the conduits and mega-ohm tested the cables. No cable degradation was found. In the third case, the applicant found water leaking from a pull box. It abandoned the cable and sealed the conduit. The applicant further stated that it will evaluate industry and plant-specific operating experience in the development and implementation of this program.

The staff conducted an independent search of the applicant's condition report database for operating experience relevant to the Electrical Cables and Connections Not Subject to 10 CFR 50.49 EQ Requirements Program. The staff verified that the operating experience described in the applicant's basis document adequately addresses the operating history relative to this AMP.

Based on its audit and review of the application, the staff finds that operating experience related to the applicant's program demonstrates that it can adequately manage the detrimental effects of aging on SSCs within the scope of the program, and that implementation of the program has resulted in the applicant taking corrective actions. The staff confirmed that the "operating experience" program element satisfies the criterion in SRP-LR Section A.1.2.3.10 and, therefore, is acceptable.

UFSAR Supplement Review. LRA Section A1.24 provides the UFSAR supplement for the Electrical Cables and Connections Not Subject to 10 CFR 50.49 EQ Requirements Program. The staff reviewed this UFSAR supplement description of the program and notes that it conforms to the recommended description for this type of program, as described in SRP-LR

Aging Management Review Results

Table 3.6-2. The staff also notes that the applicant committed (Commitment No. 26) to implement the Electrical Cables and Connections Not Subject to 10 CFR 50.49 EQ Requirements Program before entering the period of extended operation. The staff determines that the information in the UFSAR supplement is an adequate summary description of the program, as required by 10 CFR 54.21(d).

Conclusion. On the basis of its audit and review of the applicant's Electrical Cables and Connections Not Subject to 10 CFR 50.49 EQ Requirements Program, the staff finds that all program elements are consistent with the GALL Report. The staff concludes that the applicant has demonstrated that the effects of aging will be adequately managed so that the intended function(s) will be maintained consistent with the CLB for the period of extended operation, as required by 10 CFR 54.21(a)(3). The staff also reviewed the UFSAR supplement for this AMP, and concludes that it provides an adequate summary description of the program, as required by 10 CFR 54.21(d).

3.0.3.1.9 Inaccessible Medium Voltage Cables Not Subject to the Environmental Qualification Requirements Program Under Title 10, Part 50.49 of the Code of Federal Regulations

Summary of Technical Information in the Application. LRA Section B2.1.26 describes the Inaccessible Medium Voltage Cables Not Subject to 10 CFR 50.49 EQ Requirements Program as a new program that is consistent with the GALL Report AMP XI.E3, "Inaccessible Medium Voltage Cables Not Subject To 10 CFR 50.49 Environmental Qualification Requirements." The applicant stated that the Inaccessible Medium Voltage Cables Not Subject to 10 CFR 50.49 EQ Requirements Program manages localized damage and breakdown of insulation. This damage and breakdown leads to electrical failure in inaccessible medium voltage cables that are exposed to adverse localized environments caused by significant moisture and significant voltage. Therefore, the program is designed to ensure that inaccessible medium voltage cables not subject to the EQ requirements of 10 CFR 50.49 and within the scope of license renewal, are capable of performing their intended function.

The applicant also stated that it will inspect all cable manholes that contain in-scope, non-EQ inaccessible medium voltage cables for water collection and remove the collected water as required. The applicant further stated that it will perform this inspection and water removal based on actual plant experience, but at least every two years. In addition, the applicant stated that it will test all in-scope, non-EQ inaccessible medium voltage cables routed through manholes to provide an indication of conductor insulation condition. The applicant stated that it will perform testing at least once every 10 years, and the first test will be completed before the period of extended operation.

Staff Evaluation. During its audit, the staff reviewed the applicant's claim of consistency with the GALL Report. The staff also reviewed and confirmed that the plant's conditions are bounded by the conditions for which the GALL Report was evaluated.

The staff compared elements one through six of the applicant's program to the corresponding elements of the GALL Report AMP Inaccessible Medium Voltage Cables Not Subject To 10 CFR 50.49 EQ Requirements. As discussed in the Audit Report, the staff confirmed that these elements are consistent with the corresponding elements of the GALL Report AMP. Based on its audit, the staff finds that elements one through six of the applicant's Inaccessible Medium Voltage Cables Not Subject to 10 CFR 50.49 EQ Requirements Program are consistent with the corresponding program elements of the GALL Report AMP and, therefore, are acceptable.

Operating Experience. LRA Section B2.1.26 summarizes operating experience related to the Inaccessible Medium Voltage Cables Not Subject to 10 CFR 50.49 EQ Requirements Program. Since this is a new program, there is no plant-specific operating experience. However, the

Aging Management Review Results

applicant stated that a review of relevant operating history records did not reveal any failure of inaccessible medium voltage cables. The applicant stated that it has experienced cases where medium voltage cable splices have been subjected to water intrusion resulting in low megger readings. During manhole walkdowns in 2009, the applicant identified one manhole containing water that submerged the cables and a subsequent inspection of connected manholes found additional water. The applicant also stated that a review of manhole walkdowns, including connected manholes, revealed recurring instances of water intrusion.

A review of the PVNGS response to GL 2007-01, "Inaccessible or Underground Power Cable Failures That Disable Accident Mitigation Systems or Causes Plant Transients," stated that, for inaccessible or underground cables within the scope of 10 CFR 50.56 (the Maintenance Rule), no in-service power cable failures were found, but two power cable dielectric strength failures were identified. The applicant reworked the cables identified. A subsequent design change from motor-operated valves to manual-actuation valves changed the cables' classification to "spare." The applicant attributed the cable failure to water intrusion. The applicant's response to GL 2007-01 also indicated that PVNGS has experienced degraded cable splices on power cables installed in manholes. The applicant repaired the identified splices, per its corrective action program. The applicant attributed these failures to water intrusion.

In response to these findings, the applicant stated that it is evaluating changes to its existing manhole inspection, de-watering program, and preventive maintenance basis documents to improve program effectiveness. The applicant further, in LRA B2.1.26, that it will evaluate industry and plant-specific operating experience in the development and implementation of this program.

The staff reviewed relevant operating history, in the application and during the audit, to determine if the applicant reviewed the applicable aging effects and industry operating experience and plant-specific operating history documents. As discussed in the Audit Report, the staff conducted an independent search of the plant operating history information to determine if the applicant had adequately incorporated and evaluated operating history related to this program. The staff also confirmed that the applicant addressed inaccessible medium voltage cable operating experience identified after issuance of the GALL Report.

During its review, the staff identified operating experience that warranted additional information to ensure the program would be effective in adequately managing aging effects during the period of extended operation. This resulted in the issuance of an RAI. By letter dated December 29, 2009, the staff issued RAI B2.1.26-1. This RAI requested the applicant to explain how PVNGS meets the GALL Report AMP, "Inaccessible Medium Voltage Cables Not Subject To 10 CFR 50.49 EQ Requirements" program element, "scope of program," applicability based on plant operating experience that shows that inaccessible medium voltage cables are exposed to significant moisture for more than a few days. In addition, the staff requested that the applicant describe how it will use plant operating experience and condition reports documentation to develop the Inaccessible Medium Voltage Cables Not Subject to 10 CFR 50.49 EQ Requirements Program to minimize the potential for inaccessible medium voltage cable to be exposed to significant moisture (e.g., inspection frequency determination based on periodic and event -driven significant moisture exposure and appropriate corrective action).

The applicant responded by letters, dated February 19, 2010, and May 21, 2010, and stated that the Inaccessible Medium Voltage Cables Not Subject to 10 CFR 50.49 EQ Requirements program element for "scope of program" states, "[t]he scope of this program includes all in-scope inaccessible medium voltage cables not subject to the EQ requirements of 10 CFR 50.49 that are exposed to significant moisture (lasts more than a few days) simultaneously with significant voltage (energized greater than 25 percent of the time)." The applicant also stated that this is consistent with the GALL Report AMP program element for "scope of program." The

Aging Management Review Results

applicant revised LRA Sections A1.26 and B2.1.26 to reflect this response as consistent with GALL Report AMP definitions of significant voltage and moisture and inspection frequency.

The applicant also stated that plant operating experience was used in the development of the inspection frequency for the AMP Inaccessible Medium Voltage Cables Not Subject to 10 CFR 50.49 EQ Requirements Program. The LRA AMP element for "preventive action" states the following:

> Inspection for water collection within the cable manholes is being performed based on plant experience with water accumulation. The inspection frequencies will be established to be at least once every 2 years for all manholes within the scope of license renewal. If any of the manholes are found to contain water, the manholes are pumped dry, the source of the water is investigated, and the inspection frequency will be increased based on past experience.

The applicant further stated that it changed the manhole inspection frequency from a maximum five-year interval to two years maximum. The applicant also stated that the preventive maintenance program groups manholes into three frequencies of inspections based on their history of water intrusion: two weeks, six months, and two years. The two-week preventive maintenance task requires inspection of manholes if it has rained 0.3 inches or more within a 24-hour period since the last time the preventive maintenance was performed. The applicant also inspects manholes, grouped into this preventive maintenance task, on a six-month frequency to ensure they are always inspected, even during dry periods. The applicant stated that under the preventive maintenance program, a manhole will not be moved to the two-year reduced frequency inspection interval until it has been found dry for two years. The applicant revised LRA Sections A1.26 and B2.1.26 to reflect this response, including operating-experience based inspection schedules.

With the information provided by the applicant's RAI response, the staff finds the Inaccessible Medium Voltage Cables Not Subject to 10 CFR 50.49 EQ Requirements Program acceptable because the applicant revised LRA Sections A.1.26 and B2.1.26 to be consistent with the guidance of the GALL Report AMP such that there is reasonable assurance that it will adequately manage inaccessible medium voltage cable exposure to significant moisture during the period of extended operation. The staff's concern described in RAI B2.1.26-1 is resolved.

The application of AMP XI.E3 to inaccessible medium voltage power cables was based on the operating experience available at the time Revision 1 of the GALL Report was developed. However, recently -identified industry operating experience indicates that the presence of water or moisture can be a contributing factor in inaccessible power cables failures at lower service voltages (480V to 2kV). Applicable operating experience was identified in licensee responses to Generic Letter (GL) 2007-01, "Inaccessible or Underground Power Cable Failures that Disable Accident Mitigation Systems or Cause Plant Transients," which included failures of power cable operating at service voltages of less than 2kV where water was considered a contributing factor. The staff has concluded, based on recently -identified industry operating experience concerning the failure of inaccessible low voltage power cables (400v to 2kV) in the presence of significant moisture, that these cables should be addressed in an AMP. The staff notes that the applicant's AMP does not address these inaccessible low voltage power cables.

By letter dated September 27, 2010, the staff issued RAI B2.1.26-3 requesting the applicant to:

1. Provide a summary of the evaluation of recently-identified industry operating experience and any plant-specific operating experience concerning inaccessible low voltage power cable failures within the scope of license renewal (not subject to 10 CFR 50.49 environmental qualification requirements), and how this operating experience applies to the need for additional aging management activities at PVNGS.

Aging Management Review Results

2. Discuss how PVNGS will manage the effects of aging on inaccessible low voltage power cables within the scope of license renewal and subject to aging management review; with consideration of recently-identified industry operating experience and any plant-specific operating experience. The discussion should include the AMP description, program elements (i.e., scope of program, parameters monitored or inspected, detection of aging effects, and corrective actions), and UFSAR summary description to demonstrate reasonable assurance that the intended functions of inaccessible low voltage power cables subject to adverse localized environments will be maintained consistent with the CLB through the period of extended operation.

3. Evaluate whether the Inaccessible Medium Voltage Program test and inspection frequencies, including event-driven inspections, incorporate recent industry and plant-specific operating experience for both inaccessible low and medium voltage cables. Discuss how the Inaccessible Medium Voltage Program will ensure that future industry and plant-specific operating experience will be incorporated into the program.

The applicant responded by letter dated October 13, 2010, and stated that a review of PVNGS operating experience identified two low voltage power cable testing failures reported in response to GL 2007-01. The applicant also stated that the two low voltage cables did not meet the acceptance criteria for insulation resistance tests and the failures were attributed to water intrusion. The applicant further stated that the two low voltage cables have been abandoned. These failures are previously discussed in more detail in the GL 2007-01 discussion above. The applicant finally stated that no subsequent low voltage power cable failures have been identified. The applicant also stated that a review of industry operating experience is part of the PVNGS corrective action program.

As stated above, the applicant's AMP requires an event-driven inspection program and a 2-year periodic inspection frequency as part of the Inaccessible Medium Voltage Cables Not Subject to 10 CFR 50.49 EQ Requirements Program. The applicant stated that the LRA has been revised to increase the periodic inspection to at least annually. The applicant also stated that based on PVNGS and industry operating experience, the LRA has been revised to increase the periodic cable testing to at least once every 6 years. The applicant further stated that the "energized greater than 25%" significant voltage criterion was removed such that all in-scope inaccessible power cables whether de-energized or energized will be tested. In addition, based on plant-specific and industry operating experience, the applicant also revised the LRA to add low voltage (480V and above) non-EQ inaccessible power cables and associated manholes into the scope of the Inaccessible Medium Voltage Cables Not Subject to 10 CFR 50.49 EQ Requirements Program.

Based on the applicant's responses to RAIs B2.1.26-1 and B2.1.26-3, the staff finds that the applicant has appropriately expanded the program scope to include inaccessible low voltage power cables (480V to 2kv) and clarified that no inaccessible power cable was excluded based on the "significant voltage" criterion. Further, the staff finds that cable insulation testing is appropriate because it considers plant-specific and industry operating experience, and the actual frequency of testing is based on safety significance and testing history. This approach is consistent with the discussion of operating experience in the SRP-LR Section A.1.2.3.10, which states that applicants should consider plant-specific and applicable industry operating experience for its AMPs. Finally, the staff finds that the applicant's inspection frequency for cable manholes containing inaccessible in-scope power cables is appropriate since it considers applicable industry and plant-specific operating experience including cable manhole water accumulation.

The actual periodic frequency of inspection will be established based on inspection results. Collected water in manholes is removed to ensure cables are not submerged and inspections are performed based on event-driven occurrences such as rain. Given that plant-specific

Aging Management Review Results

operating experience has shown water accumulation in cable manholes within the scope of this AMP, an inspection frequency determined through inspection results and additional inspections based on event-driven occurrences is acceptable because the applicant's existing inspections will continue to inform the program's inspection frequencies (i.e., to provide feedback for changes of the inspection periodicity as appropriate).

Based on its audit and review of the application, and review of the applicant's response to RAI B2.1.26-1 and B2.1.26-3, the staff finds that operating experience related to the applicant's program demonstrates that it can adequately manage the detrimental effects of aging on SSCs within the scope of the program, and that implementation of the program has resulted in the applicant taking corrective actions. The staff confirmed that the "operating experience" program element satisfies the criterion in SRP-LR Section A.1.2.3.10 and, therefore, is acceptable.

UFSAR Supplement. LRA Section A1.26 provides the UFSAR supplement for the Inaccessible Medium Voltage Cables Not Subject to 10 CFR 50.49 EQ Requirements Program. The staff reviewed this UFSAR supplement description of the program against the recommended description for this type of program, as described in SRP-LR Table 3.6-2. The staff found that it required additional information for its review as discussed below.

By letter dated December 29, 2009, the staff issued RAI B2.1.26-2 to request that the applicant justify why the LRA Appendix A, Section A1.26, "Inaccessible Medium Voltage Cables Not Subject To 10 CFR 50.49 Environmental Qualification Requirements" summary description does not include definitions of significant moisture, significant voltage, and minimum electrical manhole inspection frequencies, consistent with SRP-LR Table 3.6-2. The applicant responded by letter dated February 19, 2010, and stated that it has revised LRA Sections A1.26 and B2.1.26 to include the definitions of significant moisture, voltage, and inspection frequency.

With the information provided by the applicant's RAI response, the staff finds the UFSAR supplement acceptable because the applicant revised LRA Section A1.26 to include definitions of significant moisture, significant voltage, and inspection frequency to be consistent with the guidance of SRP-LR Table 3.6-2. The staff's concern described in RAI B2.1.26-2 is resolved.

In addition, as part of its response to RAI B2.1.26-3, the applicant revised LRA Section A1.26 to add low voltage power cable (480V or greater) to the scope of its Inaccessible Medium Voltage Cables Not Subject to 10 CFR 50.49 EQ Requirements Program. The applicant also clarified that no inaccessible power cables were excluded from the program due to the "significant voltage" criterion. UFSAR Section A1.26 was further revised by the applicant to include revised cable test and manhole inspection frequencies of 6 years and 1 year, respectively. The applicant also revised Commitment No. 28 to reflect the changes in inspection and test frequencies and increase the scope of the program to include inaccessible power cables of 480V and above. With the resolution of RAIs B2.1.26-2 and B2.1.26-3, the staff determines that the applicant's UFSAR supplement provides an adequate summary description consistent with guidance of SRP-LR Table 3.6.

Based on the information provided by the RAI responses, the staff finds the UFSAR supplement acceptable because the applicant revised LRA Section A1.26 to include definitions of significant moisture, significant voltage, inspection and test frequencies, and increased program scope to be consistent with the guidance of SRP-LR Table 3.6-2.

The staff also notes that the applicant committed (Commitment No. 28) to implement the new Inaccessible Medium Voltage Cables Not Subject to 10 CFR 50.49 EQ Requirements Program before entering the period of extended operation to manage the aging of applicable components.

The staff determines that the information in the UFSAR supplement, as amended, is an adequate summary description of the program, as required by 10 CFR 54.21(d).

Conclusion. On the basis of its review of the applicant's Inaccessible Medium Voltage Cables Not Subject to 10 CFR 50.49 EQ Requirements Program and RAI responses, the staff finds all program elements consistent with the GALL Report. The staff concludes that the applicant has demonstrated that the effects of aging will be adequately managed so that the intended function(s) will be maintained consistent with the CLB for the period of extended operation, as required by 10 CFR 54.21(a)(3). The staff also reviewed the UFSAR supplement and RAI response for this AMP, and concludes that it provides an adequate summary description of the program, as required by 10 CFR 54.21(d).

3.0.3.1.10 American Society of Mechanical Engineers Section XI, Subsection IWL Program

Summary of Technical Information in the Application. LRA Section B2.1.28 describes the existing ASME Section XI, IWL Program as consistent with the GALL Report AMP XI.S2, "ASME Section XI, Subsection IWL." The LRA states that the program manages cracking, loss of material, and increase in porosity and permeability of the concrete containment building and post-tensioned system. Included in this inspection program are the concrete containment structure (includes all accessible areas of the concrete dome, cylinder walls, and buttresses) and the post-tensioning system (includes tendons, end anchorages, and concrete surfaces around the end anchorages). The applicant visually examines concrete surface areas for indications of distress or deterioration. It measures tendon prestress forces by lift-off and examines tendon wires for corrosion or mechanical damage.

For the inspection interval from August 1, 2001–July 31, 2011, the applicant performs IWL ISIs in accordance with the 1992 Edition of ASME Section XI (with 1992 Addendum), Subsection IWL, supplemented with the applicable requirements of 10 CFR 50.55a(b)(2) and additional commitments. The applicant stated that the IWL ISI Program is consistent with the 2001 edition of ASME Section XI, Subsection IWL, including the 2002 and 2003 Addenda. In conformance with 10 CFR 50.55a(g)(4)(ii), the applicant updates the IWL ISI Program during each successive 120-month inspection interval to comply with the requirements of the latest edition of the Code, specified twelve months before the start of the inspection interval, as required by 10 CFR 50.55a(g)(4)(ii).

Staff Evaluation. During its audit, the staff reviewed the applicant's claim of consistency with the GALL Report. The staff also reviewed and confirmed that the plant's conditions are bounded by the conditions for which the GALL Report was evaluated.

The staff compared elements one through six of the applicant's program to the corresponding elements of the GALL Report AMP, "ASME Section XI, Subsection IWL." As discussed in the Audit Report, the staff confirmed that each element of the applicant's program is consistent with the corresponding elements of the GALL Report AMP except for the "detection of aging effects" element. In order to obtain the information necessary to verify if the "detection of aging effects" element is consistent with the corresponding elements of the GALL Report AMP, the staff issued RAI B2.1.28-2 in a letter dated December 29, 2009. This RAI requested the applicant to provide justification for increasing the frequency of concrete surface examination to 10 years, plus or minus one year. The GALL AMP ASME Section XI, Subsection IWL Program, "detection of aging effects" element and ASME Section IWL-2410 require that the inspections of concrete surfaces be performed at one, three, and five years following the structural integrity test. Thereafter, the applicant performs inspections at five-year intervals.

In its response to RAI B2.1.28-2, by letter dated February 19, 2010, the applicant stated that the frequency of inspection for the ASME Section XI, Subsection IWL AMP is consistent with ASME Section XI Subsection IWL paragraph IWL-2421, "Sites with Multiple Plants." This paragraph allows inspection intervals of every ten years to be staggered, so that the applicant inspects at least one unit every five years. Therefore, the existing program is consistent with ASME Section XI Subsection IWL and GALL, and no exception is required.

Aging Management Review Results

The staff found the applicant's response to B2.1.28-2, concerning the inspection of concrete surfaces, not acceptable because IWL-2421 only allows the increase in inspection frequency to 10 years for examinations required by IWL-2524 and IWL-2525, which are requirements for examination of tendon anchorage areas and corrosion protection medium and free water in the post-tensioning system, respectively. IWL-2410 describes the requirements for concrete surface examinations. Therefore, the staff issued a follow-up RAI B2.1.28, in a letter dated April 28, 2010, asking the applicant to provide additional information on this issue.

In its response to RAI B2.1.28-3, in a letter dated May 21, 2010, the applicant stated that during the current inspection interval of 10 years, Relief Request (RR)-L3—approved by the NRC in a letter to APS dated October 6, 2000—allows a frequency of 10 years for performing Section XI, Subsection IWL inspections of the concrete containment exterior surfaces. Subsequent intervals will be in accordance with the requirements of ASME Section XI, Subsection IWL (five-year frequency), supplemented with the applicable requirements of 10 CFR 50.55a, unless an RR is approved to alter these frequencies. The staff finds the applicant's response to RAI B2.1.28-3 acceptable because it bases the containment concrete surface examination frequency during the current 10-year interval on NRC approved RR-L3. In addition, after the current 10-year interval, the applicant will perform containment exterior surface examination at a frequency of five years, as required by IWL-2410. The staff's concern as described in RAI B2.1.28-3 is resolved.

Operating Experience. LRA Section B2.1.28 summarizes operating experience related to the ASME Section XI ISI, IWL Program. The applicant stated that the ASME Section XI, Subsection IWL Program inspects the post-tensioned concrete containment in accordance with 10 CFR 50.55a(b)(2)(viii). When observed degradation could indicate the presence of degradation in inaccessible areas, or when the conditions described in 10 CFR 50.55a(b)(2)(viii)(C or D) are detected, the ISI Program Engineer is notified, and the ISI Summary Report includes the conditions. The applicant further stated that its review of operating experience has identified only two instances where observed degradation was significant enough to warrant inclusion in a Summary Report. The grease spots identified in these cases are located on the containment exterior concrete surface. An engineering evaluation determined that these are cosmetic conditions and that there are no detrimental affects to the structure.

The staff reviewed operating experience information, in the application and during the audit, to determine whether the applicant reviewed the applicable aging effects and industry and plant-specific operating experience. As discussed in the Audit Report, the staff conducted an independent search of the plant operating experience information to determine whether the applicant had adequately incorporated and evaluated operating experience related to this program.

The LRA Section B2.1.28 states that the applicant has reviewed the information documented in Information Notice (IN) 99-10 concerning degrading of prestress tendon systems in prestressed concrete containments for applicability. The LRA further states that the existing tendon integrity surveillance procedures are based on the guidance of RG 1.35. The applicant monitors and evaluates degradation and conditions, discussed in IN 99-10, PVNGS tendon integrity surveillance procedures. A trend of degradation, described in IN 99-10, has not occurred.

The operating experience section of the LRA Section B2.1.28 states that tendon integrity surveillance procedures are based on the guidance of RG 1.35. This statement is not consistent with GALL Report, AMP XI-S2, "ASME Section XI, Subsection IWL Program," element 10 (operating experience), which states that implementation of ASME Section XI, Subsection IWL, in accordance with 10 CFR 50.55a, is a necessary element of aging management for concrete containments through the period of extended operation. Therefore, in a letter dated December 29, 2009, the staff issued RAI B2.1.28-1 asking the applicant to explain why tendon integrity surveillance procedures conform to RG 1.35 instead of 10 CFR 50.55a.

Aging Management Review Results

In response to the RAI B2.1.28-1, in a letter dated February 19, 2010, the applicant stated that LRA Section B2.1.28 addresses the ASME Section XI, Subsection IWL inspections performed in accordance with 10 CFR 50.55a. LRA Section B3.3 addresses the Concrete Containment Tendon Prestress Program, which manages the loss of tendon prestress and is consistent with RG 1.35.1, Proposed Revision 0. The applicant revised LRA Section B2.1.28 to delete the last paragraph in the operating experience discussion and replace it with the statement, "[f]or discussion of IN 99-10, see Appendix B3.3, Concrete Containment Tendon Prestress."

The staff finds the applicant's response to RAI B2.1.28-1 acceptable because it removed the reference to RG 1.35 from LRA Section B2.1.28. In addition, the program description section of the LRA ASME Section XI, Subsection IWL Program makes reference to the appropriate sections of 10 CFR 50.55a that provide regulations on the scope and frequency of inspections for concrete containments. LRA Section B3.3, "Concrete Containment Tendon Prestress Program," addresses the issues reported in IN 99-10, including comparing the regression analysis trend lines of the individual lift-off values of prestressing tendons with minimum required value (MRV) and predicted lower limit (PLL) for each tendon group. The staff's concerns, identified in RAI B2.1.28-1, are resolved.

Based on its audit and review of the application, the staff finds that operating experience related to the applicant's program demonstrates that it can adequately manage the detrimental effects of aging on SSCs within the scope of the program, and that implementation of the program has resulted in the applicant taking corrective actions. The staff confirmed that the "operating experience" program element satisfies the criterion in SRP-LR Section A.1.2.3.10 and, therefore, is acceptable.

UFSAR Supplement. LRA Section A1.28 provides the UFSAR supplement for the ASME Section XI, Subsection IWL Program. The staff reviewed this UFSAR supplement description of the program and notes that it conforms to the recommended description for this type of program, as described in SRP-LR Table 3.5-2. The staff also notes that the applicant committed (Commitment No. 30) to the ongoing implementation of the existing ASME Section XI, Subsection IWL Program to manage the aging of concrete containment and post-tensioning system during the period of extended operation. The staff determines that the information in the UFSAR supplement is an adequate summary description of the program, as required by 10 CFR 54.21(d).

Conclusion. On the basis of its review of the applicant's ASME Section XI, Subsection IWL Program, the staff finds all program elements consistent with the GALL Report. The staff concludes that the applicant has demonstrated that the effects of aging will be adequately managed so that the intended function(s) will be maintained consistent with the CLB for the period of extended operation, as required by 10 CFR 54.21(a)(3). The staff also reviewed the UFSAR supplement for this AMP, and concludes that it provides an adequate summary description of the program, as required by 10 CFR 54.21(d).

3.0.3.1.11 American Society of Mechanical Engineers Section XI, Subsection IWF Program

Summary of Technical Information in the Application. LRA Section B2.1.29 describes the existing ASME Section XI, IWF Program as consistent with the GALL Report AMP XI.S3, "ASME Section XI, Subsection IWF." The applicant stated that the program provides a systematic method for periodic NDE, visual examination, and testing of SSCs to ensure the integrity of component pressure boundaries and supports. The applicant stated that the ASME Section XI, Subsection IWF program manages loss of material, cracking, and loss of mechanical function that could result in loss of intended function for Class 1, 2, and 3 component supports. There are no ASME Section XI, Subsection IWF, Class MC supports. In conformance with 10 CFR 50.55a(g)(4)(ii), the ISI Program is updated during each successive

Aging Management Review Results

120-month inspection interval to comply with the requirements of the latest edition of the ASME Code specified by 10 CFR 50.55a(g)(4)(ii).

Staff Evaluation. During its audit, the staff reviewed the applicant's claim of consistency with the GALL Report. The staff also reviewed and confirmed that the plant's conditions are bounded by the conditions for which the GALL Report was evaluated.

The staff compared elements one through six of the applicant's program to the corresponding elements of the GALL Report AMP ASME Section XI, Subsection IWF. As discussed in the Audit Report, the staff confirmed that these elements are consistent with the corresponding elements of the GALL Report AMP. Based on its audit, the staff finds that elements one through six of the applicant's ASME Section XI ISI, IWF Program are consistent with the corresponding program elements of the GALL Report AMP and are, therefore, acceptable.

Operating Experience. Program element 10, "operating experience" of the LRA ASME Section XI, Subsection IWF Program summarizes operating experience related to the ASME Section XI ISI, IWF Program. The applicant stated that the plant-specific operating experience for the ISI Program has not revealed any implementation issues with the ASME Section XI, Subsection IWF Program. The applicant updates the ASME Section XI, Subsection IWF Program to account for industry operating experience and periodically revises the program to reflect operating experience. The requirement to update the ASME Section XI, Subsection IWF Program to reference more recent editions of ASME Section XI at the end of each inspection interval ensures that the program reflects enhancements, due to operating experience, that have been incorporated into ASME Section XI Code. The applicant further stated that an evaluation of the results of the latest ISI examinations indicated that the integrity of the IWF support systems has been maintained. The applicant corrected all discrepancies or determined "use-as-is" in accordance with work control practices and ASME Section XI Code.

The staff reviewed operating experience information, in the application and during the audit, to determine if the applicant reviewed the applicable aging effects and industry and plant-specific operating experience. As discussed in the Audit Report, the staff conducted an independent search of the plant operating experience information to determine whether the applicant had adequately incorporated and evaluated operating experience related to this program.

During the audit, the staff noted that the license renewal aging management industry operating experience report for the ASME Section XI, Subsection IWF Program states that the degradation identified in IN 2009-04 for constant (spring) supports is not age-related. However, the staff noted that the IN discusses possible age-related degradation of mechanical constant supports that may adversely affect the analyzed stresses of connected piping systems. To address this concern, the staff issued RAI B2.1.29-1 in a letter dated December 29, 2009. In this RAI, the staff requested the applicant to explain the basis of the statement made in its basis document that the constant supports degradation, identified in IN 2009-04, is not age-related.

In response to the RAI B2.1.29-1, dated February 19, 2010, the applicant stated that the results of the evaluation to date indicate that the degradation of the constant spring supports, as identified in IN 2009-04, was not age-related. The apparent cause was a design issue involving the configuration of the supporting structural members. Therefore, the AMP does not require changes. The applicant further stated that when the ASME Section XI, Subsection IWF inspection identified the degradation addressed by IN 2009-04, the scope of inspections was expanded in accordance with the requirements of ASME Section XI, Subsection IWF. No other constant spring supports were found to have the same condition. It was determined that the extent of condition was limited to four supports in each unit—one constant on each main steam line in each of the three units. The applicant plans to take corrective actions after the root cause evaluation is completed in accordance with the Corrective Action Program.

Aging Management Review Results

The staff finds the applicant's response to RAI B2.1.29-1 concerning IN 2009-04 acceptable because the preliminary results of the root cause evaluation indicate that degradation of the four constant steam line supports is not age-related. Apparently, the degradation is due to improper design and support selection. Other constant supports of similar design installed in the plant did not experience degradation. In addition, the applicant plans to take corrective actions after the root cause evaluation is completed. The staff's concern in RAI B2.1.29-1 is resolved.

Based on its audit and review of the application, the staff finds that operating experience related to the applicant's program demonstrates that it can adequately manage the detrimental effects of aging on SSCs within the scope of the program, and that implementation of the program has resulted in the applicant taking corrective actions. The staff confirmed that the "operating experience" program element satisfies the criterion in SRP-LR Section A.1.2.3.10 and, therefore, is acceptable.

UFSAR Supplement. LRA Section A1.29 provides the UFSAR supplement for the ASME Section XI, Subsection IWF Program. The staff reviewed this UFSAR supplement description of the program and notes that it conforms to the recommended description for this type of program, as described in SRP-LR Table 3.5-2. The staff also notes that the applicant committed (Commitment No. 31) to the ongoing implementation of the existing ASME Section XI, Subsection IWF Program to manage aging of applicable component supports during the period of extended operation. The staff determines that the information in the UFSAR supplement is an adequate summary description of the program, as required by 10 CFR 54.21(d).

Conclusion. On the basis of its review of the applicant's ASME Section XI, Subsection IWF Program, the staff finds all program elements consistent with the GALL Report. The staff concludes that the applicant has demonstrated that the effects of aging will be adequately managed so that the intended function(s) will be maintained consistent with the CLB for the period of extended operation, as required by 10 CFR 54.21(a)(3). The staff also reviewed the UFSAR supplement for this AMP, and concludes that it provides an adequate summary description of the program, as required by 10 CFR 54.21(d).

3.0.3.1.12 Title 10 of the Code of Federal Regulations, Part 50, Appendix J Program

Summary of Technical Information in the Application. LRA Section B2.1.30 describes the existing 10 CFR Part 50, Appendix J Program as consistent with the GALL Report AMP XI.S4, "10 CFR Part 50, Appendix J." The LRA further states that the program assures that leakage, through the primary containment and systems and components penetrating containment, does not exceed allowable leakage rate limits in the TS. The applicant further states that the program does not prevent degradation but, instead, provides measures for monitoring to detect degradation before a loss of intended function. The applicant is implementing Option B of the program, which allows the testing intervals to be performance-based.

Staff Evaluation. During its audit, the staff reviewed the applicant's claim of consistency with the GALL Report. The staff also reviewed and confirmed that the plant's conditions are bounded by the conditions for which the GALL Report was evaluated.

The staff compared elements one through six of the applicant's program to the corresponding elements of the GALL Report AMP, "10 CFR Part 50, Appendix J." As discussed in the Audit Report, the staff confirmed that these elements are consistent with the corresponding elements of the GALL Report AMP. Based on its audit, the staff finds that elements one through six of the applicant's 10 CFR Part 50, Appendix J Program are consistent with the corresponding program elements of the GALL Report AMP, "10 CFR Part 50, Appendix J" and are, therefore, acceptable.

Aging Management Review Results

Operating Experience. LRA Section B2.1.30 summarizes operating experience related to the 10 CFR Part 50, Appendix J Program. The applicant provided the results of the most recent Type A, Integrated Leak Rate Tests for all three units. All three units successfully passed the last Type A tests. The applicant discussed leakage through a containment isolation valve during the Type A test, which led to hardware modifications, procedural changes, and preventative maintenance tasks. The applicant further explained that since it has carried out these corrective actions, it has not identified any further problems for this or similar valves. The applicant also stated that it noted Type B and C test failures due to debris, corrosion products, and general degradation of valve seating surfaces, which have been corrected by cleaning or adjusting the connecting components.

The staff reviewed operating experience information, in the application and during the audit, to determine if the applicant reviewed the applicable aging effects and industry and plant-specific operating experience. As discussed in the Audit Report, the staff conducted an independent search of the plant operating experience information to determine if the applicant had adequately incorporated and evaluated operating experience related to this program. The staff found that the applicant had adequately identified and incorporated industry and applicable plant-specific operating experience.

Based on its audit and review of the application, the staff finds that operating experience related to the applicant's program demonstrates that it can adequately manage the detrimental effects of aging on SSCs within the scope of the program, and that implementation of the program has resulted in the applicant taking corrective actions. The staff confirmed that the "operating experience" program element satisfies the criterion in SRP-LR Section A.1.2.3.10 and is, therefore, acceptable.

UFSAR Supplement. LRA Section A1.30 provides the UFSAR supplement for the 10 CFR Part 50, Appendix J Program. The staff reviewed this UFSAR supplement description of the program and notes that it conforms to the recommended description for this type of program, as described in SRP-LR Table 3.5-2. The staff also notes that the applicant committed (Commitment No. 32) to the ongoing implementation of the existing 10 CFR Part 50, Appendix J Program to manage the aging of applicable components during the period of extended operation. The staff determines that the information in the UFSAR supplement is an adequate summary description of the program, as required by 10 CFR 54.21(d).

Conclusion. On the basis of its review of the applicant's 10 CFR Part 50, Appendix J Program, the staff finds all program elements consistent with the GALL Report. The staff concludes that the applicant has demonstrated that the effects of aging will be adequately managed so that the intended function(s) will be maintained consistent with the CLB for the period of extended operation, as required by 10 CFR 54.21(a)(3). The staff also reviewed the UFSAR supplement for this AMP, and concludes that it provides an adequate summary description of the program, as required by 10 CFR 54.21(d).

3.0.3.1.13 Electrical Cable Connections Not Subject to 10 CFR 50.49 Environmental Qualification Requirements Program

Summary of Technical Information in the Application. LRA Section B2.1.35 describes the new Electrical Cable Connections Not Subject to 10 CFR 50.49 EQ Requirements Program that, when implemented, will be consistent with license renewal interim staff guidance, LR-ISG-2007-02, *Changes to Generic Aging Lessons Learned (GALL) Report Aging Management Program (AMP) XI.E6, "Electrical Cable Connections Not Subject to 10 CFR 50.49 Environmental Qualification Requirements."* This program will also be consistent with the GALL Report AMP XI.E6, "Electrical Cable Connections Not Subject to 10 CFR 50.49 Environmental Qualification Requirements." The applicant stated that this AMP manages the effects of loosening of bolted connections due to thermal cycling, ohmic heating, electrical transients,

Aging Management Review Results

vibration, chemical contamination, corrosion, and oxidation to ensure that electrical cable connections, not subject to the EQ requirements of 10 CFR 50.49 and within the scope of license renewal, are capable of performing their intended function. The applicant also stated that it performs infrared thermography testing on non-EQ electrical cable connections associated with active and passive components within the scope of license renewal. The applicant further stated that it will test a representative sample of external connections at least once before the period of extended operation using infrared thermography to confirm that there are no AERM. The applicant stated that selection of the sample to be tested is based on application (medium and low voltage), circuit loading, and environment. The applicant also stated that it documents the technical basis for the sample selection. Finally, the applicant stated that the one-time testing of a sample of non-EQ electrical cable connectors is representative of other non-EQ electrical cable connectors with similar application, circuit loading conditions, and environments that are bounded by the testing.

Staff Evaluation. During its audit, the staff reviewed the applicant's claim of consistency with the GALL Report. The staff also reviewed and confirmed that the plant's conditions are bounded by the conditions for which the GALL Report was evaluated.

The staff compared elements one through six of the applicant's program to the corresponding elements of the GALL Report AMP, "Electrical Cable Connections Not Subject to 10 CFR 50.49 EQ Requirements." As discussed in the Audit Report, the staff confirmed that each element of the applicant's program is consistent with the corresponding element of the GALL Report AMP, with the exception of elements "scope of program," "parameters monitored or inspected," and "detection of aging effects." The staff's concern and resolution are discussed below.

During the audit, the staff determined that the LRA AMP and the associated UFSAR Supplement (LRA Appendix A, Section A1.35) were not consistent with the GALL Report AMP, "Electrical Cable Connections Not Subject to 10 CFR 50.49 EQ Requirements." However, the LRA AMP and UFSAR Supplement were consistent with the program description and program elements of the (at that time) proposed LR-ISG-2007-02 issued for public comment, by letter dated August 29, 2007. During the audit, the staff determined that it needed clarification, and the staff considered issuing an RAI on the subject. Subsequent to the audit, a notice of availability of the final LR-ISG-2007-02 was published in the Federal Register on December 23, 2009. The staff, therefore, re-evaluated the LRA AMP and UFSAR Supplement based on the changes to the aging management recommendations provided by ISG 2007-02. Based on its audit and re-evaluation, the staff finds that elements one through six of the applicant's program are consistent with the corresponding elements of the GALL Report AMP, as modified by LR-ISG-2007-02; therefore, the staff's concern is resolved.

Operating Experience. LRA Section B2.1.35 summarizes operating experience related to the Electrical Cable Connections Not Subject to 10 CFR 50.49 EQ Requirements Program. The applicant stated that it routinely performs infrared thermography on electrical components and connections. The applicant also stated that a review of plant operating experience identified scans where electrical cable connections showed thermal anomalies. The applicant further stated that it cleaned and re-tightened these connections. The applicant also stated that these thermal anomalies did not cause any loss of equipment intended function. The applicant concluded that there is sufficient confidence that the implementation of the Electrical Cable Connections Not Subject to 10 CFR 50.49 EQ Requirements Program will provide confirmation that supports industry operating experience in that electrical connections have not experienced a high degree of failures.

The staff reviewed relevant operating experience information, in the application and during the audit, to determine whether the applicant reviewed the applicable aging effects and industry and plant-specific operating experience. As discussed in the Audit Report, the staff conducted walkdowns, interviewed the applicant's staff, and reviewed onsite documentation provided by

Aging Management Review Results

the applicant. The staff also conducted an independent search of the applicant's operating experience information to determine whether the applicant had adequately incorporated and evaluated operating experience related to this program. The staff found that the applicant had adequately identified and incorporated industry and applicable plant-specific operating experience.

Based on its audit and review of the application, the staff finds that operating experience related to the applicant's program demonstrates that it can adequately manage the detrimental effects of aging on SSCs within the scope of the program, and that implementation of the program has resulted in the applicant taking corrective actions. The staff confirmed that the "operating experience" program element satisfies the criterion in SRP-LR Section A.1.2.3.10 and is, therefore, acceptable.

UFSAR Supplement. LRA Section A1.35 provides the UFSAR supplement for the Electrical Cable Connections Not Subject to 10 CFR 50.49 EQ Requirements Program. The staff reviewed this UFSAR supplement description of the program and notes that it conforms to the recommended description for this type of program as described in SRP-LR Table 3.6-2 and as modified by the applicant's implementation of LR-ISG-2007-02. The staff also noted that the applicant committed (Commitment No. 37) to implement the new Electrical Cable Connections Not Subject to 10 CFR 50.49 EQ Requirements Program before entering the period of extended operation to manage the aging of applicable components. The staff determines that the information in the UFSAR supplement, as amended, is an adequate summary description of the program, as required by 10 CFR 54.21(d).

Conclusion. On the basis of its review of the applicant's Electrical Cable Connections Not Subject to 10 CFR 50.49 EQ Requirements Program, the staff finds all program elements consistent with the GALL Report and ISG 2007-02. The staff concludes that the applicant has demonstrated that the effects of aging will be adequately managed so that the intended function(s) will be maintained consistent with the CLB for the period of extended operation, as required by 10 CFR 54.21(a)(3). The staff also reviewed the UFSAR supplement for this AMP, and concludes that it provides an adequate summary description of the program, as required by 10 CFR 54.21(d).

3.0.3.1.14 Metal Enclosed Bus Program

Summary of Technical Information in the Application. LRA Section B2.1.36 describes the new Metal Enclosed Bus (MEB) Program as consistent with the GALL Report AMP XI.E4, "Metal Enclosed Bus." The applicant stated that the MEB Program manages the effects of loose connections, embrittlement, cracking, melting, swelling or discoloration of insulation, loss of material of bus enclosure assemblies, hardening of boots and gaskets, and cracking of internal bus supports to ensure that MEBs within the scope of license renewal are capable of performing their intended function. The applicant also stated that MEBs within the scope of this program are the buses used during recovery from station blackout (SBO). The applicant further stated that it will inspect a sample of the MEB accessible bolted connections will for evidence of overheating. It will perform contact resistance testing on a sample of accessible splice plates to check for loose connections. It will inspect each bus section for cracks, corrosion, foreign debris, excessive dust buildup, and evidence of water intrusion. The applicant will inspect the bus insulation for signs of embrittlement, cracking, melting, swelling, or discoloration, which may indicate overheating or aging degradation. It will inspect the internal bus supports for structural integrity and signs of cracks, and it will inspect the bus enclosure assemblies for loss of material due to corrosion and hardening of boots and gaskets. The applicant further stated that it will complete the MEB program before the period of extended operation and once every 10 years thereafter.

Aging Management Review Results

Staff Evaluation. During its audit, the staff reviewed the applicant's claim of consistency with the GALL Report. The staff also reviewed and confirmed that the plant's conditions are bounded by the conditions for which the GALL Report was evaluated.

The staff compared elements one through six of the applicant's program to the corresponding elements of the GALL Report AMP, "Metal Enclosed Bus." As discussed in the Audit Report, the staff confirmed that each element of the applicant's program is consistent with the corresponding element of the GALL Report AMP, with the exception of "acceptance criteria" discussed below. For this area, the staff determined the need for additional clarification, which resulted in the issuance of an RAI.

The GALL Report uses XI.S6, "Structures Monitoring Program," for inspection of the exterior of MEBs and accessible gaskets and sealant associated with the exterior of MEBs. In the GALL Report, "Structures Monitoring Program," under "acceptance criteria," it states that, for each structure/aging effect combination, the acceptance criteria are selected to ensure that the need for corrective actions will be identified before loss of intended functions. The staff reviewed the applicant's plant basis document and noted that, under the "acceptance criteria" element, the applicant did not specify the acceptance criteria for inspecting the exterior of MEBs, including gasket and sealants. In a letter dated December 29, 2009, the staff issued RAI B2.1.36-1, which asked the applicant to provide acceptance criteria for inspecting the exterior of MEBs. In response to the staff's request, the applicant submitted a letter dated February 19, 2010, and stated that the acceptance criteria of the LRA MEB AMP has been revised to include acceptance criteria for inspecting the exterior of MEBs. The following paragraph has been added:

> Visual inspection is the primary method for detecting external corrosion and material aging degradation. The exterior of MEBs will be inspected for general corrosion. No unacceptable indication of corrosion will be allowed to exist. For boots and gaskets discoloration, checkering, and cracking are indications of hardening. Physical manipulation during the visual inspection can also be used to verify the absence of hardening or cracking. No unacceptable indications of cracking will be allowed to exist. The program depends on the judgment and experience of the inspector to assess material condition. All unacceptable indications as a result of the inspection will be entered into the corrective action process.

The staff finds the applicant's response acceptable because the applicant revised element six of the MEB Program to include acceptable criteria for inspecting the exterior of MEBs. The staff's concern in RAI B2.1.36-1 is resolved.

Based on its audit and review, the staff finds that elements one through six of the applicant's MEB Program are consistent with the corresponding program elements of the GALL Report AMPs and are, therefore, acceptable.

Operating Experience. LRA Section B2.1.36 summarizes operating experience related to the MEB Program. The applicant stated that the MEB Program is a new program; therefore, plant-specific operating experience to verify the effectiveness of the program is not available. The applicant stated, however, that industry experience has shown that failures have occurred on MEBs caused by cracked insulation and moisture or debris buildup internal to the bus. The applicant also stated that operating experience has shown that bus connections exposed to appreciable ohmic heating during operation may experience loosening due to repeated cycling of connected loads. IN 2000-14 "Non Vital Bus Fault Leads to Fire and Loss of Offsite Power" and IN 89-64 "Electrical Bus Bar Failures" are examples of non-segregated bus duct failures.

The applicant stated that a review of relevant plant-specific operating experience has determined that there have been no problems resulting in the loss of intended function of the

Aging Management Review Results

MEBs. Sections of the MEBs are inspected every other outage, and thermography is performed on the bus at the transformer connections once every six months. The inspection results for the MEBs during the last 10 years have revealed that only one splice plate required rework and repairs to cracked Noryl sleeving. The applicant also stated that it will evaluate industry and plant-specific operating experience in the development and implementation of this program.

The staff reviewed relevant operating history information, in the application and during the audit, to determine if the applicant reviewed the applicable aging effects and industry and plant-specific operating experience. As discussed in the Audit Report, the staff conducted an independent search of the plant operating experience information to determine whether the applicant had adequately incorporated and evaluated operating experience relevant to this program. The staff found that the applicant had adequately identified and incorporated industry and applicable plant-specific operating experience.

Based on its audit and review of the application, the staff finds that operating experience related to the applicant's program demonstrates that it can adequately manage the detrimental effects of aging on SSCs within the scope of the program. The staff confirmed that the "operating experience" program element satisfies the criterion in SRP-LR Section A.1.2.3.10 and is, therefore, acceptable.

UFSAR Supplement. LRA Section A1.36 provides the UFSAR supplement for the MEB Program, as amended. The staff reviewed this UFSAR supplement description of the program and notes that it conforms to the recommended description for this type of program, as described in SRP-LR Table 3.6-2. The staff also notes that the applicant committed (Commitment No. 38) to implement the new MEB Program before entering the period of extended operation to manage the aging of applicable components. The staff determines that the information in the UFSAR supplement is an adequate summary description of the program, as required by 10 CFR 54.21(d).

Conclusion. On the basis of its review of the applicant's MEB Program, the staff finds all program elements consistent with the GALL Report. The staff concludes that the applicant has demonstrated that the effects of aging will be adequately managed so that the intended function(s) will be maintained consistent with the CLB for the period of extended operation, as required by 10 CFR 54.21(a)(3). The staff also reviewed the UFSAR supplement for this AMP, and concludes that it provides an adequate summary description of the program, as required by 10 CFR 54.21(d).

3.0.3.1.15 Fuse Holders Program

Summary of Technical Information in the Application. LRA Section B2.1.37 describes the new Fuse Holders Program as consistent with the GALL Report AMP XI.E5, "Fuse Holders." The applicant stated that the Fuse Holders Program manages thermal fatigue, mechanical fatigue, vibration, chemical contamination, and corrosion of the metallic portions of the fuse holders to ensure that fuse holders within the scope of license renewal are capable of performing their intended function. The applicant also stated that it will test fuse holders that perform a license renewal function located outside of active devices for deterioration of the metallic clamps by using thermography. The applicant stated that it will perform fuse holder testing at least once every 10 years, with the first test completed before the period of extended operation. The applicant further stated that it will base the acceptance criteria on temperature rise above the reference temperature.

Staff Evaluation. During its audit, the staff reviewed the applicant's claim of consistency with the GALL Report. The staff also reviewed and confirmed that the plant's conditions are bounded by the conditions for which the GALL Report was evaluated. The staff compared elements one through six of the applicant's program to the corresponding elements of the GALL Report AMP. As discussed in the Audit Report, the staff confirmed that these elements are consistent with the corresponding elements of the GALL Report AMP, "Fuse Holders." Based on its audit and review of the application, the staff finds that elements one through six of the applicant's Fuse Holders Program are consistent with the corresponding program elements of the GALL Report AMP and are, therefore, acceptable.

Operating Experience. LRA Section B2.1.37 summarizes operating experience related to the Fuse Holders Program. The applicant stated that the Fuse Holders Program is a new program; therefore, plant-specific operating experience to verify the effectiveness of the program is not available. The applicant stated, however, that industry operating experience forms the basis for this program and is included in the operating experience element of the corresponding GALL Report AMP description. The relative operating experience shows that loosening of fuse holders and corrosion of fuse clips are aging mechanisms that, if left unmanaged, can lead to a loss of electrical continuity function. The applicant stated that, as additional industry and applicable plant-specific operating experience become available, it will evaluate the operating experience and appropriately incorporate it into the program. The applicant concluded that implementation of the Fuse Holders Program provides reasonable assurance that aging effects will be managed such that systems and components within the scope of this program will continue to perform their intended functions consistent with the CLB for the period of extended operation.

The staff reviewed operating experience information, in the application and during the audit, to determine if the applicant reviewed the applicable aging effects and industry and plant-specific operating history. As discussed in the Audit Report, the staff conducted an independent search of the plant operating experience information to determine if the applicant had adequately incorporated and evaluated operating experience related to this program. In addition, the staff confirmed that the applicant addressed operating experience identified after issuance of the GALL Report. The staff found that the applicant had adequately identified and incorporated industry and applicable plant-specific operating experience.

Based on its audit and review of the application, the staff finds that operating experience related to the applicant's program demonstrates that it can adequately manage the detrimental effects of aging on SSCs within the scope of the program. The staff finds that the applicant's program will appropriately address operating experience related to Fuse Holders and will adequately manage the detrimental effects of aging on SSCs within the scope of the program. The staff confirmed that the "operating experience" program element satisfies the criterion in SRP-LR Section A.1.2.3.10 and is, therefore, acceptable.

UFSAR Supplement. LRA Section A1.37 provides the UFSAR supplement for the Fuse Holders Program. The staff reviewed this UFSAR supplement description of the program and notes that it conforms to the recommended description for this type of program as described in SRP-LR Table 3.6-2. The staff also notes that the applicant committed (Commitment No. 50) to implement the new Fuse Holders Program prior to entering the period of extended operation to manage the aging of applicable components. The staff determines that the information in the UFSAR supplement is an adequate summary description of the program, as required by 10 CFR 54.21(d).

Conclusion. On the basis of its review of the applicant's Fuse Holders Program, the staff finds all program elements consistent with the GALL Report. The staff concludes that the applicant has demonstrated that the effects of aging will be adequately managed so that the intended function(s) will be maintained consistent with the CLB for the period of extended operation, as required by 10 CFR 54.21(a)(3). The staff also reviewed the UFSAR supplement for this AMP, and concludes that it provides an adequate summary description of the program, as required by 10 CFR 54.21(d).

Aging Management Review Results

3.0.3.1.16 Environmental Qualification of Electrical Components Program

Summary of Technical Information in the Application. LRA Section B3.2 describes the existing EQ of Electrical Components Program as consistent with the GALL Report AMP X.E1, "Environmental Qualification (EQ) of Electric Components." The applicant stated that the EQ Program manages the effects of component thermal, radiation, and cyclic aging effects using 10 CFR 50.49(f) methods. The applicant also states that electrical equipment within the scope of the EQ Program is environmentally qualified, in accordance with NUREG-0588, "Interim Staff Position on Equipment Qualification of Safety-Related Electrical Equipment," Category 1 requirements, as supplemented by 10 CFR 50.49. The applicant further stated that, as required by 10 CFR 50.49, EQ components not qualified for the current license term are to be refurbished or replaced or have their qualification extended before reaching the aging limits established in the evaluation. The applicant also stated that it considers aging evaluations for EQ components that specify a qualification of at least 40 years to be time-limited aging analysis (TLAAs) for license renewal. SER Section 4.4 provides the staff's review of these TLAAs.

Staff Evaluation. During its audit, the staff reviewed the applicant's claim of consistency with the GALL Report. The staff also reviewed and confirmed that the plant's conditions are bounded by the conditions for which the GALL Report was evaluated. The staff compared elements one through six of the applicant's program to the corresponding elements of the GALL Report AMP. As discussed in the Audit Report, the staff confirmed that these elements are consistent with the corresponding elements of the GALL Report AMP. Based on its audit, the staff finds the that elements one through six of the applicant's EQ of Electrical Components Program are consistent with the corresponding program elements of the GALL Report AMP, "EQ of Electrical Components" and are, therefore, acceptable.

Operating Experience. LRA Section B3.2 summarizes operating experience related to the EQ of Electrical Components Program. The applicant stated in LRA Section B3.2 that the EQ Program complies with 10 CFR 50.49. The applicant also stated that the program includes consideration of operating experience and NRC correspondence. Specifically, the applicant stated that it evaluates operating experience; system, equipment or component related information, as reported through NRC bulletins, notices, circulars, GLs; and Part 21 Notifications for applicability to the EQ Program under the Regulatory Interaction and Correspondence Control Procedure. The applicant further stated that, when it identifies an emerging industry aging issue that affects the qualification of an EQ component, it evaluates the affected component and takes appropriate corrective actions.

During the audit, the staff asked that the applicant provide the results of recent industry or applicant EQ self-assessments. The applicant referenced two EQ Program self-assessments, dated 2007 and 2008, and an EQ of electrical equipment benchmarking report, also dated 2008. The 2007 self-assessment concluded that the EQ Program complied with the requirements of 10 CFR 50.49. The report noted some improvements, but also some declining trends, since a 1999 EQ assessment. The report identified recommendations for improvement in EQ training for interfacing personnel, procedural interfaces, work backlog, and tracking methods. The 2007 self-assessment also recommended improvements for file documentation. The 2008 benchmarking concluded that there were no conditions adverse to quality and no areas of improvement with "high" significance. The benchmark did identify one area of improvement as "medium," and it classified 47 areas as "low" significance. The assessment noted each area of improvement and proposed recommendations, as appropriate. The applicant generated condition reports to address the benchmark findings.

The 2008 self-assessment concluded that the EQ Program generally met 10 CFR 50.49 requirements, noting that certain aspects of the program were not consistent with industry guidance or procedures. The report noted a lack of specific training, examples of incomplete evaluations and procedure use, and limited industry participation by the program engineers.

Aging Management Review Results

Again, the applicant noted areas of improvement and generated condition reports to address the self-assessment findings and recommendations. During the audit, the applicant provided the disposition of the recommended and corrective actions identified by the self-assessment and benchmarking reports. The applicant has completed the majority of the self-assessment and benchmarking recommended and corrective actions, with the remaining actions scheduled for completion by July 31, 2010.

The staff reviewed the operating experience information, in the application and during the audit, to determine if the applicant reviewed the applicable aging effects and industry and plant-specific operating experience. The staff interviewed the applicant's technical staff to confirm that plant-specific operating experience did not reveal any degradation not bounded by industry experience. The operating experience identified in the applicant's basis documents demonstrated that it identified and dispositioned corrective actions to ensure EQ Program effectiveness. The staff confirmed that the applicant addressed operating experience identified after issuance of the GALL Report.

Based on its audit and review of the LRA, the staff finds that operating experience related to the applicant's program demonstrates that it can adequately manage the detrimental effects of aging on SSCs within the scope of the program, and that implementation of this program has resulted in the applicant taking corrective actions. The staff confirmed that the "operating experience" program element satisfies the criterion in SRP-LR Section A.1.2.3.10 and is, therefore, acceptable.

UFSAR Supplement. LRA Appendix A, Section A2.2 provides the UFSAR supplement for the LRA Section B3.2, EQ of Electrical Components Program. The staff reviewed the UFSAR supplement description of the program against the recommended description for this type of program, as described in SRP-LR Tables 4.4-1 and 4.4-2, and noted that it did not include reanalysis attributes consistent with the description of the AMP in LRA Section B3.2 or the TLAA in LRA Section 4.4. The GALL Report, "EQ of Electrical Components Program" states that reanalysis of an aging evaluation is normally performed to extend the qualification by reducing excess conservatism incorporated in the prior evaluation. Important attributes of a reanalysis include analytical methods, data collection and reduction methods, underlying assumptions, acceptance criteria, and corrective actions (if acceptance criteria are not met).

By letter dated December 29, 2009, the staff issued RAI 4.4-1 to request that the applicant provide justification for not including reanalysis attributes in accordance with 10 CFR 54.21(c)(1)(iii) in the UFSAR supplement. The applicant responded by letter dated February 19, 2010, and stated LRA Sections A2.2 and A3.3 have been revised to include the following:

> Reanalysis of aging evaluations to extend the qualifications of components is performed on a routine basis as part of the EQ program. Important attributes for the reanalysis of aging evaluations include analytical methods, data collection and reduction methods, underlying assumptions, acceptance criteria and corrective actions (if acceptance criteria are not met).

With the information provided by the applicant's RAI response, the staff finds the UFSAR supplement acceptable because the applicant revised LRA Sections A2.2 and A3.3 UFSAR supplement descriptions to be consistent with the guidance of SRP-LR Table 4.4.2. The staff's concern described in RAI 4.4-1 is resolved.

The staff also noted that the applicant committed (Commitment No. 40) to re-evaluate existing EQ evaluations before the period of extended operation. The staff noted that the applicant's commitment is inconsistent with the license renewal commitments for existing programs in that the existing EQ Program is considered an AMP but is not included in the applicant's commitment. The applicant's commitment requires that existing EQ evaluations be re-evaluated

3-43

Aging Management Review Results

before the period of extended operation but not as an ongoing program, consistent with the applicant's Table A4-1 for existing programs.

By letter dated December 29, 2009, the staff issued RAI 3.2-1 to request that the applicant provide justification for not referencing the existing EQ Program or noting that it is to be implemented on an ongoing basis. The applicant responded by letter dated February 19, 2010, and stated that it revised LRA Table A4-1, Commitment No. 40, to credit the EQ Program for license renewal. The applicant also stated that maintaining qualification through the period of extended operation requires that it re-evaluate existing EQ evaluations before the period of extended operation. With the information provided by the applicant's RAI response, the staff finds the UFSAR supplement commitment acceptable because the applicant revised Table A4-1, "License Renewal Commitments" to be consistent with the guidance of SRP-LR Table 4.4.2. The staff's concern described in RAI 4.4-1 is resolved.

The staff determines that the information in the UFSAR supplement, as amended, is an adequate summary description of the program, as required by 10 CFR 54.21(d).

Conclusion. On the basis of its review of the applicant's EQ of Electrical Components Program, the staff finds all program elements consistent with the GALL Report. The staff concludes that the applicant has demonstrated that the effects of aging will be adequately managed so that the intended function(s) will be maintained consistent with the CLB for the period of extended operation, as required by 10 CFR 54.21(a)(3). The staff also reviewed the UFSAR supplement for this AMP, and concludes that it provides an adequate summary description of the program, as required by 10 CFR 54.21(d).

3.0.3.2 Aging Management Programs Consistent with the Generic Aging Lessons Learned Report, with Exceptions or Enhancements

In LRA Appendix B, the applicant stated that the following AMPs are, or will be, consistent with the GALL Report, with exceptions or enhancements:

- Water Chemistry Program
- Flow-Accelerated Corrosion Program
- Bolting Integrity Program
- Open-Cycle Cooling Water System Program
- Closed-Cycle Cooling Water System Program
- Inspection of Overhead Heavy Load and Light Load Handling Systems Program
- Fire Protection Program
- Fire Water System Program
- Fuel Oil Chemistry Program
- RV Surveillance Program
- Selective Leaching of Materials Program
- Buried Piping and Tanks Inspection Program
- One-Time Inspection of ASME Code Class 1 Small-Bore Piping Program
- External Surfaces Monitoring Program
- Inspection of Internal Surfaces in Miscellaneous Piping and Ducting Components Program

Aging Management Review Results

- Lubricating Oil Analysis Program

- Electrical Cables and Connections Not Subject to 10 CFR 50.49 Environmental Qualification Requirements Used in Instrumentation Circuits Program

- ASME Section XI, Subsection IWE Program

- Masonry Wall Program

- Structures Monitoring Program

- RG 1.127, "Inspection of Water-Control Structures Associated with Nuclear Power Plants Program"

- Metal Fatigue of Reactor Coolant Pressure Boundary Program

- Concrete Containment Tendon Prestress Program

For AMPs that the applicant claimed are consistent with the GALL Report, with exception(s) or enhancement(s), the staff performed an audit and review to confirm that those attributes or features of the program, for which the applicant claimed consistency with the GALL Report, were indeed consistent. The staff also reviewed the exception(s) or enhancement(s) to the GALL Report to determine if they were acceptable and adequate. The following sections document the results of the staff's audits and reviews.

3.0.3.2.1 Water Chemistry Program

Summary of Technical Information in the Application. LRA Section B2.1.2 describes the existing Water Chemistry Program as consistent, with an enhancement, with the GALL Report AMP XI.M2, "Water Chemistry." The applicant stated that the program manages cracking, denting, loss of material, reduction of heat transfer, and wall thinning in primary and secondary water systems. The applicant also stated that the program manages hardening and loss of strength; however, by letter dated February 19, 2010, the applicant amended its LRA to delete these aging effects from the scope of the Water Chemistry Program because they are not included in the GALL Report. In the LRA, the applicant also stated that the scope of the primary Water Chemistry Control Program includes the RCS and related auxiliary systems containing treated borated water and that the scope of the secondary Water Chemistry Control Program is the SG secondary side and the secondary cycle systems. The applicant also stated that the methods used to manage water chemistry rely on limiting the concentration of chemistry species known to cause corrosion as well as adding chemical species known to inhibit degradation by their influence on pH and dissolved oxygen levels. The applicant further stated that in low-flow areas or stagnant portions of the systems, where chemical sampling may not be as effective in determining local environmental conditions, it will use a one-time inspection of a representative group of components to verify the effectiveness of the Water Chemistry Program.

Staff Evaluation. During its audit, the staff reviewed the applicant's claim of consistency with the GALL Report. The staff also reviewed and confirmed that the plant's conditions are bounded by the conditions for which the GALL Report was evaluated.

The staff compared elements one through six of the applicant's program to the corresponding elements of the GALL Report, "Water Chemistry Program." As discussed in the Audit Report, the staff confirmed that each element of the applicant's program is consistent with the corresponding element of the GALL Report AMP, with the exception of the "monitoring and trending" program element. The "monitoring and trending" program element of the GALL Report, "Water Chemistry Program" recommends the use of increased sampling to verify the effectiveness of corrective actions whenever corrective actions are taken to address an abnormal chemistry condition. However, during the audit, the staff found that the applicant's Water Chemistry Program does not indicate that an increased sampling frequency or other

Aging Management Review Results

measures will be taken when an abnormal chemistry condition occurs. For this element, the staff determined the need for additional clarification, which resulted in the issuance of an RAI.

By letter dated December 29, 2009, the staff issued RAI B2.1.2-1 requesting that the applicant provide information on its evaluation of the effectiveness of corrective actions when an abnormal chemistry condition occurs. In its response, dated February 19, 2010, the applicant stated that it revised its program implementing procedure to add guidance on increasing the sampling rate due to an abnormal chemistry condition. The staff finds the applicant's response acceptable because the change to the implementing procedure makes the applicant's Water Chemistry Program consistent with the recommendations in the GALL Report, "Water Chemistry Program." The staff's concern described in RAI B2.1.2-1 is resolved.

The staff also reviewed the portions of the "scope of program" and "preventive actions" program elements associated with the enhancement to determine if the program will be adequate to manage the aging effects for which it is credited. The following section describes the staff's evaluation of this enhancement.

Enhancement. LRA Section B2.1.2 identifies an enhancement to the "scope of program" and "preventive actions" program elements. The applicant stated that it will enhance plant procedures to address sampling of effluents from new secondary system cation resins for purgeable and non-purgeable organic carbon. However, by letter dated February 19, 2010, the applicant amended the LRA to remove this enhancement and to remove the commitment to revise the plant procedures. In its letter, the applicant stated that it does not test for purgeable and non-purgeable organic carbon from the effluent of new resins used in the secondary system because an alternate analysis accomplishes the same result. The applicant also stated that it has incorporated this alternative analysis into the plant procedure containing the specifications for bulk chemistry. The staff finds removal of the enhancement acceptable because the applicant revised its procedures to include the alternate method of testing the cation resin effluent, and the applicant's preventive actions are consistent with the GALL Report's recommendations. The applicant samples water chemistry in accordance with the guidance provided in the Electric Power Research Institute (EPRI) water chemistry guidelines.

Based on its audit and review of the applicant's response to RAI B2.1.2-1, the staff finds that elements one through six of the applicant's Water Chemistry Program are consistent with the corresponding program elements of the GALL Report, "Water Chemistry Program" and are, therefore, acceptable.

Operating Experience. LRA Section B2.1.2 summarizes operating experience related to the Water Chemistry Program. The applicant stated that the program is consistent with the EPRI water chemistry guidelines and, therefore, benefits from the operating experience captured in these guidelines. Concerning primary water chemistry control, the applicant also stated that all three units experienced high fluoride concentrations. These high fluoride concentrations occurred following several refueling outages, and the applicant determined them to be the result of degradation of an eddy current probe conduit that has since been repaired. There have been no subsequent discernable releases of fluoride. Regarding secondary water chemistry control, the applicant stated that condenser tube plugs were installed in incorrect locations, which led to the ingress of chloride, sodium, and sulfate. The applicant took corrective actions to include the creation of official tube sheet maps and implementation of administrative controls on installation and verification of tube plugs.

The staff reviewed operating experience information, in the application and during the audit, to determine if the applicant reviewed the applicable aging effects and industry and plant-specific operating experience. As discussed in the Audit Report, the staff conducted an independent search of the plant operating experience information to determine if the applicant had adequately incorporated and evaluated operating experience related to this program. The staff

Aging Management Review Results

found that the applicant had adequately identified and incorporated industry and applicable plant-specific operating experience.

Based on its audit and review of the application, the staff finds that operating experience related to the applicant's program demonstrates that it can adequately manage the detrimental effects of aging on SSCs within the scope of the program, and that implementation of the program has resulted in the applicant taking corrective actions. The staff confirmed that the "operating experience" program element satisfies the criterion in SRP-LR Section A.1.2.3.10 and is, therefore, acceptable.

UFSAR Supplement. LRA Section A1.2, as amended, provides the UFSAR supplement for the Water Chemistry Program. By letter dated February 19, 2010, the applicant amended this supplement to remove hardening and loss of strength as aging effects managed by the Water Chemistry Program. The staff reviewed this UFSAR supplement description of the program, as amended, and notes that it conforms to the recommended description for this type of program as described in SRP-LR Tables 3.1-2, 3.2-2, 3.3-2, 3.4-2, and 3.5-2. The staff also notes that the applicant committed (Commitment No. 4) to the ongoing implementation of the existing Water Chemistry Program to manage the aging of applicable components during the period of extended operation. The staff determines that the information in the UFSAR supplement, as amended, is an adequate summary description of the program, as required by 10 CFR 54.21(d).

Conclusion. On the basis of its audit and review of the applicant's Water Chemistry Program, the staff determines that those program elements for which the applicant claimed consistency with the GALL Report are consistent. The staff concludes that the applicant has demonstrated that the effects of aging will be adequately managed so that the intended function(s) will be maintained consistent with the CLB for the period of extended operation, as required by 10 CFR 54.21(a)(3). The staff also reviewed the UFSAR supplement for this AMP, and concludes that it provides an adequate summary description of the program, as required by 10 CFR 54.21(d).

3.0.3.2.2   Flow-Accelerated Corrosion Program

Summary of Technical Information in the Application. LRA Section B2.1.6 describes the existing Flow-Accelerated Corrosion (FAC) Program as consistent, with an exception, to the GALL Report AMP XI.M17, "Flow-Accelerated Corrosion." The applicant stated that the FAC Program manages wall thinning due to FAC on the internal surfaces of carbon or low alloy steel piping, elbows, reducers, expanders, and valve bodies that contain both single-phase and two-phase high energy fluids. The applicant further stated that the program uses the EPRI computer program CHECKWORKS, along with the implementing guidelines contained in Nuclear Safety Analysis Center (NSAC) 202L-R3, "Recommendations for an Effective Flow Accelerated Corrosion Program" (NSAC-202L-R3).

Staff Evaluation. During its audit, the staff reviewed the applicant's claim of consistency with the GALL Report. The staff also reviewed and confirmed that the plant's conditions are bounded by the conditions for which the GALL Report was evaluated. The staff compared elements one through six of the applicant's program to the corresponding elements of the GALL Report, "Flow-Accelerated Corrosion Program." The staff confirmed that these elements are consistent with the corresponding elements of the GALL Report AMP, with an exception to the "scope of program" and "detection of aging effects" program elements.

Exception. LRA Section B2.1.6 states that there is an exception to the "scope of program" and "detection of aging effects" program elements. The GALL Report AMP states that in the "scope of program" and "detection of aging effects" program elements, the FAC Program relies on implementation of EPRI guidelines in NSAC-202L-R2; however, the guidelines provided in the PVNGS governing procedure were based on the recommendations in EPRI Guideline NSAC-202L-R3. The applicant stated that the new revision of the EPRI guidelines incorporate

Aging Management Review Results

lessons learned and improvements to detection, modeling, and mitigation technologies that became available since the publication of NSAC-202L-R2. The staff previously reviewed NSAC-202L-R3 (NUREG-1929, "Safety Evaluation Report Related to the License Renewal of Beaver Valley Power Station, Units 1 and 2") and determined that it is equivalent to NSAC-202L-R2. In addition, NSAC-202L-R3 allows the use of the averaged band method, which is another method for determining the wear of piping components from ultrasonic testing (UT) inspection. The staff notes that EPRI documents are created using industry experience over several years and finds that the averaged band method provides another method to determine the wear of piping components from UT inspections. The staff finds this method to be more accurate, thereby resulting in better prediction of remaining life as well as less rework. The staff finds the use of EPRI NSAC-202L-R3 acceptable because it will continue to allow the applicant to manage wall thinning due to FAC on the internal surfaces of carbon and low alloy steel piping and components that contain both single-phase and two-phase high-energy fluids.

Based on its review, the staff finds that elements one through six of the applicant's FAC Program, with an acceptable exception, are consistent with the corresponding program elements of the GALL Report AMP and are, therefore, acceptable.

Operating Experience. LRA Section B2.1.6 summarizes operating experience related to the FAC Program. The applicant stated that their review of work orders from 1996–2009 revealed no reported FAC-related leak or rupture from components within the scope of license renewal. FAC Program inspections identified wall thinning on most of the work orders, and there were cases where the allowable wall thickness was reached and more rigorous stress analyses were performed to justify continued service and postpone replacement. The applicant also provided the following operational experience:

> For previous refueling outages from R10 through R14 of all three units, 66 to 166 locations of large-bore systems were originally selected for inspection before the outage. The scope was expanded if necessary based on UT findings. An inspection location included the subject component (such as an elbow) and its adjacent area (such as upstream and downstream piping). For small-bore systems, 16 to 52 inspections were selected before the outage. The scope was also expanded if necessary based on UT findings. The replacements for each outage are scheduled on proactive basis, determined by the projected remaining service life based on FAC analyses and by programmatic strategy based on industry experience and cost comparison to further inspections. The selections of FAC-resistance materials are stainless steel, chrome-moly alloy, or carbon steel with trace chromium content > 0.1%. Baseline inspections were performed for selected replacement locations of chrome-moly alloy and carbon steel with trace chromium content > 0.1%.

Further review of operating experience by the staff identified a through-wall leak in stainless steel piping downstream of a valve in the high-pressure safety injection (HPSI) recirculation line to the refueling water tank. The apparent cause evaluation for this issue, completed in January 2007, stated that the failure was due to erosion by damaging cavitation, which occurs when the system is aligned to perform surveillance testing and to fill the safety injection tanks. The apparent cause also identified several corrective actions in the FAC Program to prevent recurrence. These initial corrective actions included examining additional components in other systems that were identified as susceptible to cavitation erosion, revising the program to inspect portions of all three unit's HPSI recirculation piping every 18 months in order to assess degradation rates, and reviewing the current program for identification of erosion damage in carbon steel systems due to cavitation.

The applicant ultimately resolved this issue by implementing a periodic replacement program for the affected sections of the HPSI system for all three units at approximately 7.5-year intervals.

Aging Management Review Results

This action removed the associated HPSI components from the scope of license renewal, in accordance with 10 CFR 54.21(a)(1)(ii), since they were no longer considered long-lived components. In addition, the extent of condition analysis addressed stainless piping and components in systems associated with heat removal and implied that the FAC Program would address the cavitation erosion issue in carbon steel piping systems. The staff noted, however, that the FAC Program's computer program, CHECWORKS, specifically excludes cavitation erosion in its evaluations.

The staff held conference calls with the applicant on June 17, July 8, and August 27, 2010, to discuss the above concern regarding the resolution of the extent of condition of the cavitation erosion issue. This issue was previously identified as Confirmatory Item 3.0.3.2.2-1, in the staff's "Safety Evaluation Report With Open Items Related to the License Renewal of the PVNGS."

In its response dated July 30, as supplemented on September 3, 2010, the applicant stated that its extent of condition evaluation identified 26 components and associated piping segments that were potentially susceptible to cavitation in each unit. The applicant also stated that, of the locations inspected to date by ultrasonic testing, only the HPSI recirculation piping discussed above exhibited cavitation erosion. The applicant also added Commitment No. 59 to LRA Table A4-1, stating that it would complete the inspections of other potentially-susceptible piping locations by June 30, 2012, and would incorporate any remaining components found to exhibit flow-related degradation into a comparable periodic replacement plan. The applicant further stated that a review of stainless steel and carbon steel in-scope components and piping in infrequently-operated systems did not identify any other locations potentially susceptible to cavitation erosion. This review was initially excluded from the extent of condition evaluation. In addition, the applicant provided the bases for determining the replacement interval for the HPSI recirculation piping segments.

The staff finds the applicant's responses acceptable because the extent of condition reviews now account for both stainless steel and carbon steel piping for infrequently-operated, in-scope systems, and the applicant committed to complete inspections for other locations susceptible to cavitation erosion. Based on the above, the staff finds that past corrective actions have resulted in appropriate enhancements and, therefore, the concern described in RAI B2.1.6-1 is resolved, and Confirmatory Item 3.0.3.2.2-1 is closed.

The staff reviewed operating experience information in the application to determine if the applicant reviewed the applicable aging effects and industry and plant-specific operating experience. The staff conducted an independent search of the plant operating experience information to determine if the applicant had adequately incorporated and evaluated operating experience related to this program. The staff found that the applicant had adequately identified and incorporated industry and applicable plant-specific operating experience.

Based on its review of the application, and review of the applicant's responses to RAI B2.1.6-1, the staff finds that operating experience related to the applicant's program demonstrates that it can adequately manage the detrimental effects of aging on SSCs within the scope of the program and that implementation of the program has resulted in the applicant taking corrective actions. The staff confirmed that the "operating experience" program element satisfies the criterion in SRP-LR Section A.1.2.3.10 and is, therefore, acceptable.

UFSAR Supplement. LRA Section A1.6 provides the UFSAR supplement for the FAC Program. The staff reviewed this UFSAR supplement description of the program and notes that the applicant amended Section A1.6, by letter dated June 21, 2010, to include the following statement, "[t]he program relies on implementation of the EPRI guidelines of NSAC-202L, *Recommendations for an Effective Flow Accelerated Corrosion Program*." As amended, the UFSAR supplement description of the program conforms to the recommended description for

Aging Management Review Results

this type of program as described in SRP-LR Tables 3.3-2 and 3.4-2. The staff also notes that the applicant committed (Commitment No. 8) to the ongoing implementation of the existing FAC Program to manage the aging of applicable components with implementation of the enhancement before entering the period of extended operation. The staff finds that the information in the UFSAR supplement is an adequate summary description of the program, as required by 10 CFR 54.21(d).

Conclusion. On the basis of its audit and review of the applicant's FAC Program, the staff finds that those program elements for which the applicant claimed consistency with the GALL Report are consistent. In addition, the staff reviewed the exception and justification and finds that the AMP, with the exception, is adequate to manage the aging effects for which the LRA credits it. The staff concludes that the applicant has demonstrated that the effects of aging will be adequately managed so that the intended function(s) will be maintained consistent with the CLB for the period of extended operation, as required by 10 CFR 54.21(a)(3). The staff also reviewed the UFSAR supplement for this AMP, and concludes that it provides an adequate summary description of the program, as required by 10 CFR 54.21(d).

3.0.3.2.3 Bolting Integrity Program

Summary of Technical Information in the Application. LRA Section B2.1.7 describes the existing Bolting Integrity Program as consistent, with exceptions, with the GALL Report AMP XI.M18, "Bolting Integrity." The applicant stated that the Bolting Integrity Program manages cracking, loss of material, and loss of preload for pressure retaining bolting and ASME component support bolting. The applicant also stated that the program includes preload control, selection of bolting material, use of lubricants and sealants consistent with EPRI good bolting practices, and periodic inspections for indication of aging effects. The applicant further stated that three other AMPs supplement the Bolting Integrity Program for managing loss of preload, cracking, and loss of material. These three AMPs are: (1) ASME Section XI ISI, Subsections IWB, IWC and IWD Program, discussed in LRA Section B2.1.1; (2) ASME Section XI, Subsection IWF Program, discussed in LRA Section B2.1.29; and (3) External Surfaces Monitoring Program, discussed in LRA Section B2.1.20.

In addition, the applicant stated that the general practices established in the Bolting Integrity Program are consistent with EPRI NP-5067, "Good Bolting Practices," Volumes 1 and 2, and the recommendations in NUREG-1339, "Resolution of Generic Safety Issue 29: Bolting Degradation or failure in Nuclear Power Plants." The applicant also stated that its bolting integrity procedures incorporate action items from NUREG-1339; EPRI NP-5769, "Degradation and Failure of Bolting in Nuclear Power Plants;" and EPRI TR-104213, "Bolted Joint Maintenance and Application Guide."

Staff Evaluation. During its audit, the staff reviewed the applicant's claim of consistency with the GALL Report. The staff also reviewed and confirmed that the plant's conditions are bounded by the conditions for which the GALL Report was evaluated.

The staff reviewed the applicant's design guide for bolted joints, as documented in the Audit Report, and noted explicit references to NUREG-1339, EPRI NP-5769, and EPRI NP-5067. The staff also noted that NUREG-1339, EPRI NP-5769, and EPRI TR-104213, but not EPRI NP-5067, are explicitly referenced in the GALL Report AMP, "Bolting Integrity." The staff further noted that the basis section of EPRI TR-104213 describes NP-5067 as a guidance document that nuclear power plant personnel have frequently used to develop and improve the plant's bolting program. In addition to the other guidance documents, the staff finds the applicant's use of NP-5067 acceptable because TR-104213 endorses NP-5067 as an appropriate guidance document for the development of plant bolting programs. Therefore, the staff does not consider use of NP-5067 to be an exception to the GALL Report AMP.

3-50

Aging Management Review Results

The staff compared elements one through six of the applicant's program to the corresponding elements of the GALL Report AMP, "Bolting Integrity." As discussed in the Audit Report, the staff confirmed that each element of the applicant's program is consistent with the corresponding elements of the GALL Report AMP, with the exception of the "scope of program," "preventive actions," "parameters monitored or inspected," "detection of aging effects," and "acceptance criteria" program elements. For these elements, the staff determined the need for additional clarification, which resulted in the issuance of the RAIs discussed below.

The staff noted that in the GALL Report AMP, "Bolting Integrity," the "scope of program," "parameters monitored or inspected," "detection of aging effects," and "acceptance criteria" program elements all include recommendations to manage the aging of structural bolting and indicate that structural bolting is within the scope of this program. The staff also noted that structural bolting is not within the scope of the applicant's Bolting Integrity Program, and the applicant used other AMPs to manage aging of structural bolting.

By letter dated December 29, 2009, the staff issued RAI B2.1.7-01. This RAI asked the applicant to explain why the applicant's use of AMPs different from the Bolting Integrity Program was not identified as an exception to the GALL Report, and to identify which AMPs manage aging of structural bolting. The staff also asked that the applicant provide justification that the credited AMPs are adequate to manage aging effects during the period of extended operation.

In its response, dated February 19, 2010, the applicant stated that the "detection of aging effects" program element of the GALL Report AMP XI.M18 indicates that structural bolts and fasteners are inspected under the structures monitoring or equivalent program. In addition, the applicant stated that GALL Report, Chapter III, items B2-7, B2-10, B3-7, B4-7, B4-10, and B5-7 all credit the GALL Report AMP XI.S6, "Structures Monitoring Program," for managing the aging effect of loss of material. The applicant, therefore, does not consider use of the Structures Monitoring Program as an exception to the GALL Report AMP, "Bolting Integrity." The applicant also stated that visual inspections performed by the Structures Monitoring Program are capable of detecting loss of material and are suitable for managing this aging effect in structural bolting. The applicant further stated that it avoids the aging effect of cracking in structural bolting by controlling lubricants that may cause contamination leading to cracking. The applicant manages the aging effect of loss of preload through the control of preload during installation and maintenance activities.

The staff finds the applicant's response to RAI B2.1.7-01 acceptable because the applicant's Structures Monitoring Program has established controls, inspections, and processes that are capable of managing loss of material and cracking in structural bolting. The staff's concern described in RAI B2.1.7-01 is resolved.

The program description and the "scope of program" and "preventive actions" program elements of the GALL Report AMP XI.M18 reference NUREG-1339, which recommends that thread lubricants containing molybdenum disulfide not be used because they increase the potential for SCC in bolting, especially high-strength steel bolting. During its review of the applicant's Reactor Head Closure Studs Program, the staff noted that the reactor head stud threads are lubricated with a lubricant containing molybdenum disulfide. Based on this information, the staff is unclear if the applicant uses thread lubricants containing molybdenum disulfide in other bolting applications.

By letter dated December 29, 2009, the staff issued RAI B2.1.7-03 requesting that the applicant clarify if it uses thread lubricants containing molybdenum disulfide for bolting that is included within the scope of the Bolting Integrity Program. If it does use such lubricants, the staff also requested that the applicant explain why it did not identify such use as an exception to the GALL Report and how it will manage the aging effects of concern in NUREG-1339 during the period of extended operation.

Aging Management Review Results

In its response, dated February 19, 2010, the applicant stated that review of its maintenance procedures and engineering observations confirmed that molybdenum disulfide lubricant is not used for bolting within the scope of the Bolting Integrity Program. The applicant stated that the common lubricants in its Bolting Integrity Program procedures have no molybdenum disulfide ingredients. The applicant further stated that field observations and discussions with maintenance and other engineering personnel confirm that molybdenum disulfide has not been used.

The staff finds the applicant's response acceptable because the applicant's does not use thread lubricants containing molybdenum disulfide for bolting within the scope of the Bolting Integrity Program. The staff's concern described in RAI B2.1.7-03 is resolved.

The staff also reviewed the portions of the "scope of program," "parameters monitored or inspected," and "monitoring and trending" program elements associated with the exceptions to determine if the program will be adequate to manage the aging effects for which it is credited. The staff's evaluation of these exceptions follows.

Exception 1. LRA Section B2.1.7 identifies an exception to the "scope of program" program element. The applicant stated that the GALL Report Bolting Integrity Program specifies use of ASME Code Section XI, 1995 edition with 1996 addenda, but its current ISI Program is based on ASME Code Section XI, 2001 edition with 2002 and 2003 addenda. In addition, during the period of extended operation, a different ASME Code edition will be used in accordance with the provisions of 10 CFR 50.55a.

The "scope of program" program element for the GALL Report AMP states, "The staff's recommendations and guidelines for comprehensive bolting integrity programs that encompass all safety-related bolting are delineated in NUREG-1339, which includes the criteria established in the 1995 edition through the 1996 addenda of ASME Code Section XI." The staff considers this reference to ASME Code Section XI, 1995 edition through 1996 addenda, to be a statement of historical fact related to issuance of NUREG-1339, not a recommendation that this specific ASME Code Section XI edition and addenda should be the basis for an applicant's Bolting Integrity Program. In addition, the GALL Report, in general, references ASME Code Section XI, 2001 edition with 2002 and 2003 addenda, which is the same edition and addenda on which the applicant bases its current ISI Program. On this basis, the staff finds the applicant's current use of ASME Code Section XI, 2001 edition with 2002 and 2003 addenda, and the future use of those ASME Code editions and addenda in accordance with the provisions of 10 CFR 50.55a, acceptable.

Exception 2. LRA Section B2.1.7 identifies an exception to the "parameters monitored or inspected" program element. The applicant stated that in EPRI NP-5769, the discussion of bolt preload indicates that job inspection torque is non-conservative because, for a given fastener tension, the torque required to restart the bolt is greater than the torque to which the bolt is initially tightened. The applicant further stated that it provides torque values in plant procedures, vendor instructions, and design documents or specifications. The torque values provided in procedures include consideration of the expected relaxation of the fastener over the life of the joint and gasket stress in application of pressure closure bolting.

The staff determined that the LRA does not provide sufficient information for the staff to clearly understand the details of this exception. By letter dated December 29, 2009, the staff issued RAI B2.1.7-02. This RAI asked that the applicant provide a clearer description of the exception and the differences between the applicant's Bolting Integrity Program and the recommendations in the GALL Report AMP. It also asked that the applicant justify how the actions taken under the Bolting Integrity Program are adequate to manage the aging effects, in lieu of the actions recommended in the GALL Report AMP.

Aging Management Review Results

In its response, dated February 19, 2010, the applicant amended its LRA by revising the description of this exception. The revised description states that loss of preload is not a parameter of inspection under the Bolting Integrity Program. The applicant stated that, in lieu of inspection for loss of preload, preload is managed by control of preload during installation or maintenance activities and that guidance for proper installation torque is provided in plant procedures, if not provided by vendor instructions or by design documents or specifications. The applicant also stated that EPRI NP-5769, Volume 2, Section 10, indicates that torque inspection is non-conservative because, for a given fastener tension, the amount of torque required to restart a bolt is not a good indication of the amount of torque applied at installation. The applicant further stated that techniques for measuring the amount of bolt tension in an assembled joint are both difficult and unreliable, and inspection of preload is not necessary if the installation method has been carefully followed. The applicant credits control of preload during installation or maintenance activities, in lieu of direct inspection, to determine whether reduction in preload may have occurred.

Based on its review of the LRA and the applicant's clarification of the exception in response to RAI B2.1.7-02, the staff finds this exception acceptable because direct inspection for reduction in preload is unreliable, and proper application of preload during installation and maintenance activities can adequately manage the aging effect of loss of preload. The staff's concern described in RAI B2.1.7-02 is resolved.

Exception 3. LRA Section B2.1.7 identifies an exception to the "monitoring and trending" program element. Under this program element, the GALL Report, "Bolting Integrity Program," states that if bolting connections for pressure retaining components (not covered by ASME Code Section XI) are reported to be leaking, then they may be inspected daily. If the leak rate does not increase, the applicant may decrease the inspection frequency to biweekly or weekly. However, the applicant stated that for pressure retaining components reported to be leaking, it will initiate the corrective action program and identify corrective actions, such as shortening the inspection frequency. The applicant bases these corrective actions on the analysis of trending data to ensure there is not a loss of intended function of the subject component. The applicant also stated that, when deemed necessary, it would implement preventive maintenance activities, such as gasket replacement or bolting tightness checks.

The staff finds this exception acceptable because the applicant's corrective action program is consistent with the requirements of 10 CFR Part 50, Appendix B. The applicant includes provisions for reporting, documenting, and evaluating safety significance as well as trending and implementing corrective actions for bolted pressure boundary components reported to be leaking. In addition, the corrective action program has provisions to determine an appropriate inspection frequency for a bolted pressure boundary component found to be leaking.

Based on its audit and review of the applicant's responses to RAIs B2.1.7-01, B2.1.7-02, and B2.1.7-03, the staff finds that elements one through six of the applicant's Bolting Integrity Program, with acceptable exceptions, are consistent with the corresponding program elements of the GALL Report AMP, "Bolting Integrity," and are, therefore, acceptable.

Operating Experience. LRA Section B2.1.7 summarizes operating experience related to the Bolting Integrity Program. The applicant stated that its Bolting Integrity Program incorporates applicable industry experience on bolting issues, including concerns with material control and certification, bolting installation and inspection practices, and use of lubricants and injection sealants. The applicant also stated that it has reviewed the NRC INs, bulletins, circulars, and GLs listed in NUREG-1339. The applicant also stated that a review of plant operating experience identified issues with corrosion, missing or loose bolts, inadequate thread engagement, and improper bolt application, but there has been no reported case of cracking of bolts due to SCC. The applicant further stated that, in all cases, it corrected the identified concern or evaluated and accepted as-is, it identified no significant safety events, and it

Aging Management Review Results

implemented additional actions, such as procedural enhancements, as needed to prevent recurrence.

The staff reviewed operating experience information, in the application and during the audit, to determine if the applicant reviewed the applicable aging effects and industry and plant-specific operating experience. As discussed in the Audit Report, the staff conducted an independent search of the plant operating experience information to determine if the applicant had adequately incorporated and evaluated operating experience related to this program. The staff found that the applicant had adequately identified and incorporated industry and applicable plant-specific operating experience.

Based on its audit and review of the application, the staff finds that operating experience related to the applicant's program demonstrates that it can adequately manage the detrimental effects of aging on SSCs within the scope of the program. The staff confirmed that the "operating experience" program element satisfies the criterion of SRP-LR Section A.1.2.3.10 and is, therefore, acceptable.

UFSAR Supplement. LRA Section A1.7 provides the UFSAR supplement for the Bolting Integrity Program. The staff reviewed this UFSAR supplement description of the program and notes that it conforms to the recommended description for this type of program, as described in SRP-LR Tables 3.1-2, 3.2-2, 3.3-2, and 3.4-2. The staff also notes that the applicant committed (Commitment No. 9) to the ongoing implementation of the existing Bolting Integrity Program to manage the aging of applicable components during the period of extended operation. The staff determines that the information in the UFSAR supplement is an adequate summary description of the program, as required by 10 CFR 54.21(d).

Conclusion. On the basis of its audit and review of the applicant's Bolting Integrity Program, the staff determines that those program elements for which the applicant claimed consistency with the GALL Report are consistent. In addition, the staff reviewed the exceptions and justifications and determines that the AMP, with the exceptions, is adequate to manage the aging effects for which the LRA credits it. The staff concludes that the applicant has demonstrated that the effects of aging will be adequately managed so that the intended function(s) will be maintained consistent with the CLB for the period of extended operation, as required by 10 CFR 54.21(a)(3). The staff also reviewed the UFSAR supplement for this AMP, and concludes that it provides an adequate summary description of the program, as required by 10 CFR 54.21(d).

3.0.3.2.4 Open-Cycle Cooling Water System Program

Summary of Technical Information in the Application. LRA Section B2.1.9 describes the existing Open-Cycle Cooling Water System Program as consistent, with an enhancement, with the GALL Report AMP XI.M20, "Open-Cycle Cooling Water System." The applicant stated that the program is used to manage the loss of material and reduction of heat transfer for those components exposed to raw water of the open-cycle cooling water system and that the program uses both surveillance and control techniques to manage the aging effects caused by biofouling, corrosion, erosion, and silting. The applicant also stated that the surveillance techniques include visual inspection and NDE of selected components in combination with thermal and hydraulic performance monitoring. The applicant further stated that the control techniques used in this program consist of water chemistry controls, flushes, and physical and chemical cleaning. Finally, the applicant considered the program consistent with commitments made in response to NRC GL 89-13.

Staff Evaluation. During its audit, the staff reviewed the applicant's claim of consistency with the GALL Report. The staff also reviewed and confirmed that the plant's conditions are bounded by the conditions for which the GALL Report was evaluated.

Aging Management Review Results

The staff compared elements one through six of the applicant's program to the corresponding elements of the GALL Report AMP, "Open-Cycle Cooling Water System." As discussed in the Audit Report, the staff confirmed that each element of the applicant's program is consistent with the corresponding element of the GALL Report AMP, with the exception of the "preventive actions" program element. For this element, the staff determined the need for additional clarification, which resulted in the issuance of an RAI, discussed below.

The GALL Report AMP, "Open-Cycle Cooling Water System," indicates that the system components are constructed of appropriate materials and lined or coated to protect the underlying metal surfaces from being exposed to aggressive cooling water environments. However, during the audit, the staff determined that the applicant did not consider the aging issues associated with the internal epoxy coating. By letter dated December 29, 2009, the staff issued RAI B2.1.9-1 requesting that the applicant provide information that accurately depicts the material arrangement of the open-cycle cooling water system, including linings or coatings. In addition, the staff asked the applicant to provide additional information on how the coating system is managed for aging effects so that its degradation will not affect the operability of the system.

In its response, dated February 19, 2010, the applicant stated that the coatings are not credited for aging management of the underlying material, and the basis document has been revised in order to clarify the coating material arrangement for the open-cycle cooling water system. The applicant also stated that it conducts visual inspections of the internal surfaces to detect coating failure or degradation, which would be evident by corrosion nodules, fresh rust stains, missing sections of coatings, or disbonded coatings. The applicant further stated that visual inspection procedures for heat exchanger channel heads and tubesheets require identification of any tube blockage or fouling. Finally, the applicant stated that the open-cycle cooling water system AMP basis document has also been revised to indicate the applicable plant procedures and procedure sections including the requirements for coating degradation and heat exchanger tube fouling with coating degradation products.

The staff finds the applicant's response acceptable because visual inspections will be used to identify coating degradation in the open-cycle cooling water system and heat exchanger plugging due to coating degradation products, which is consistent with guidance found in the GALL Report for ensuring protective coatings are inspected. The staff's concern described in RAI B2.1.9-1 is resolved.

The staff also reviewed the portions of the "detection of aging effects" and "acceptance criteria" program elements associated with the enhancement to determine if the program will be adequate to manage the aging effects for which it is credited. The staff's evaluation of this enhancement follows.

Enhancement. LRA Section B2.1.9, "Open-Cycle Cooling Water System," identifies an enhancement to the "detection of aging effects" and "acceptance criteria" program elements. The applicant stated that it will clarify the guidance for conducting heat exchanger and piping inspections using NDE techniques and the related acceptance criteria during onsite procedures before the period of extended operation.

The staff reviewed this enhancement against the corresponding program elements in the GALL Report, "Open-Cycle Cooling Water System" AMP, XI.M20. This enhancement is consistent with the GALL Report, "detection of aging effects" element, which states that nondestructive testing is an effective method to measure surface condition and the extent of wall thinning. Similarly, the enhancement is consistent with the GALL Report "acceptance criteria" element, which is based on the effective cleaning of biological fouling organisms and maintenance of protective coatings or linings. The degradation resulting from inadequately performing the cleaning of biological fouling organisms and maintaining the protective coatings or linings would

Aging Management Review Results

be detectable with NDE techniques. Based on its review, the staff finds this enhancement acceptable because, when implemented before the period of extended operation, it will make the program consistent with the recommendations in the GALL Report AMP.

Based on its audit and review of the applicant's response to RAI B2.1.9-1, the staff finds that elements one through six of the applicant's Open-Cycle Cooling Water System Program, with acceptable enhancement, are consistent with the corresponding program elements of the GALL Report AMP, "Open-Cycle Cooling Water System" and are, therefore, acceptable.

Operating Experience. LRA Section B2.1.9 summarizes operating experience related to the open-cycle cooling water system. The applicant stated that, in 1994, it implemented a new chemistry control program for the emergency spray pond systems and, from 1994–2006, it made additional changes to the program that led to spray pond chemistry that was more susceptible to fouling. The applicant also stated that, in 2006, it observed degraded performance in the open-cycle cooling water heat exchangers for all three units. The applicant took corrective action to return the systems to full operability and to ensure that root cause issues were corrected.

The staff reviewed operating experience information, in the application and during the audit, to determine if the applicant reviewed the applicable aging effects and industry and plant-specific operating experience. As discussed in the Audit Report, the staff conducted an independent search of the plant operating experience information to determine if the applicant had adequately incorporated and evaluated operating experience related to this program. The staff found that the applicant had adequately identified and incorporated industry and applicable plant-specific operating experience.

Based on its audit and review of the application, the staff finds that operating experience related to the applicant's program demonstrates that it can adequately manage the detrimental effects of aging on SSCs within the scope of the program, and that implementation of the program has resulted in the applicant taking corrective actions. The staff confirmed that the "operating experience" program element satisfies the criterion of SRP-LR Section A.1.2.3.10 and is, therefore, acceptable.

UFSAR Supplement. LRA Section A1.9 provides the UFSAR supplement for the Open-Cycle Cooling Water System Program. The staff reviewed this UFSAR supplement description of the program and notes that it conforms to the recommended description for this type of program, as described in SRP-LR Tables 3.2-2, 3.3-2, and 3.4-2. The staff also notes that the applicant committed (Commitment No. 11) to enhancing the Open-Cycle Cooling Water System Program before entering the period of extended operation. Specifically, the applicant committed to clarifying the guidance for inspections using NDE techniques and related acceptance criteria for heat exchangers. In addition, the applicant committed to clarifying the guidance for conducting piping inspections using NDE techniques and related acceptance criteria. The staff determines that the information in the UFSAR supplement is an adequate summary description of the program, as required by 10 CFR 54.21(d).

Conclusion. On the basis of its audit, answers to the RAI, and review of the applicant's Open-Cycle Cooling Water System Program, the staff determines that those program elements for which the applicant claimed consistency with the GALL Report are consistent. In addition, the staff reviewed the enhancement and confirmed that its implementation through Commitment No. 11, prior to the period of extended operation would make the existing AMP consistent with the GALL Report AMP to which it was compared. The staff concludes the applicant has demonstrated that the effects of aging will be adequately managed so that the intended function(s) will be maintained consistent with the CLB for the period of extended operation, as required by 10 CFR 54.21(a)(3). The staff also reviewed the UFSAR supplement for this AMP,

Aging Management Review Results

and concludes that it provides an adequate summary description of the program, as required by 10 CFR 54.21(d).

3.0.3.2.5 Closed-Cycle Cooling Water System

Summary of Technical Information in the Application. LRA Section B2.1.10 describes the existing Closed-Cycle Cooling Water System Program as consistent, with exceptions and enhancements, with the GALL Report AMP XI.M21, "Closed-Cycle Cooling Water System." The applicant stated that the program manages loss of material, cracking, and reduction in heat transfer for components in closed-cycle cooling water systems. In addition, the program includes both the maintenance of corrosion inhibitor concentration at the prescribed level and periodic testing and inspections to evaluate system and component performance. The applicant also stated that it maintains water chemistry, consistent with the EPRI Closed-Cycle Cooling Water Chemistry Guidelines, using nitrite as an iron corrosion inhibitor, tolyltriazole as a copper corrosion inhibitor, and glutaraldehyde as a biocide and as a means to control pH. The applicant further stated that the inspection processes include visual, eddy-current, and ultrasonic methods, and periodic testing includes functional demonstrations and monitoring of the thermal and hydraulic performance tests.

Staff Evaluation. During its audit, the staff reviewed the applicant's claim of consistency with the GALL Report. The staff also reviewed and confirmed that the plant's conditions are bounded by the conditions for which the GALL Report was evaluated.

The staff compared elements one through six of the applicant's program to the corresponding elements of the GALL Report, "Closed-Cycle Cooling Water System." As discussed in the Audit Report, the staff confirmed that each element of the applicant's program is consistent with the corresponding element of the GALL Report, "Closed-Cycle Cooling Water System," with the exception of the "scope of program" program element. For this element, the staff determined the need for additional clarification, which resulted in the issuance of an RAI, discussed below.

The GALL Report AMP XI.M21 recommends that the program include preventive measures to minimize corrosion and SCC as well as testing and inspection to monitor the effects of corrosion and SCC on the intended function of the components scoped into the license renewal process. However, during its audit, the staff noted that the applicant's Closed-Cycle Cooling Water System Program does not conduct internal inspections or performance testing for all components within the scope of license renewal under 10 CFR 54.4(a)(2).

By letter dated December 29, 2009, the staff issued RAI B2.1.10-1 requesting that the applicant provide justification for limiting the internal inspections and performance testing of components based upon the criteria that was used to scope these components into the license renewal process. In its response, dated February 19, 2010, the applicant stated that for each component, it selects a specific non-chemistry monitoring technique, from among those provided in EPRI TR-107396, to reflect the actual conditions of the component type, configuration, and license renewal intended functions to be managed. The applicant also stated that for those components with the function of material integrity, such as pressure boundary or leakage barrier, it verifies the effectiveness of the water chemistry conditions by the non-chemical monitoring technique of inspection, as provided in the EPRI document, Section 8, "Additional Monitoring Techniques." The applicant further stated that for components with heat transfer functions, it will verify the effectiveness of the chemical conditions through NDE techniques, heat transfer performance testing, or flow monitoring, as described in the EPRI document.

The staff finds the applicant's response acceptable because the applicant conducts non-chemistry monitoring techniques to ensure the effectiveness of the water chemistry conditions, which is consistent with the GALL Report guidance. The staff's concern described in RAI B2.1.10-1 is resolved.

3-57

Aging Management Review Results

The staff also reviewed the portions of the "preventive actions," "parameters monitored or inspected," "detection of aging effects," "monitoring and trending," and "acceptance criteria," program elements associated with the exceptions and enhancements to determine whether the program will be adequate to manage the aging effects for which it is credited. The staff's evaluation of these exceptions and enhancements follows.

Exception 1. LRA Section B2.1.10 identifies an exception to the "preventive actions" program element. In the GALL Report AMP, this program element recommends that the materials used in the closed-cycle cooling water system be appropriate for the type of service. The applicant stated that the essential cooling water system for each unit uses an aluminum "window" as the pressure boundary material between the closed-cycle cooling water and an ionization detector. The applicant also stated that its chemical treatment program does not include controls, described in EPRI TR–107396, that are appropriate for aluminum. The applicant further stated that it would maintain the integrity of the aluminum windows using the Inspection of Internal Surfaces in Miscellaneous Piping and Ducting Components Program.

The staff reviewed this exception to the GALL Report and noted that the applicant took the exception because it considered that its current water chemistry guidelines would not effectively manage the aging of aluminum components. However, further discussion with the applicant during the audit indicated that the material may not be aluminum. By letter dated December 29, 2009, the staff issued RAI B2.1.10-2 requesting that the applicant provide information on the actual material of the "window." If the "window" is not aluminum, the staff asked the applicant to provide additional information on the AMP to be used. If the window is aluminum, the staff asked the applicant to provide the technical basis for the adequacy of the Inspection of Internal Surfaces in Miscellaneous Piping and Ducting Components Program to inspect the aluminum component. In its response, dated February 19, 2010, the applicant stated that the window material is aluminum and that its strongly-bonded, surface oxide film protects it from general corrosion. The applicant also stated that corrosion of aluminum usually manifests in the form of pitting corrosion and that it controls the chloride concentration in the closed-cycle cooling water system so that rapid loss of material from pitting is not anticipated. The applicant further stated that it will use visual inspection, in accordance with the Internal Surfaces in Miscellaneous Piping and Ducting Components Program, to manage the age related degradation of the aluminum window component.

The staff finds the program exception acceptable because the applicant is using a visual inspection technique, which is appropriate for managing loss of material due to pitting for the aluminum window. The staff's concern described in RAI B2.1.10-2 is resolved.

Exception 2. LRA Section B2.1.10 identifies an exception to "parameters monitored or inspected" and "monitoring and trending" program elements. In the GALL Report Closed-Cycle Cooling Water System AMP, the "parameters monitored or inspected" program element recommends that testing and inspection be conducted as described in EPRI TR–107396 and that parameters monitored for pumps include flow, suction pressure, and discharge pressure. The "parameters monitored or inspected" program element of the GALL Report AMP XI.M21 also recommends that the parameters monitored for heat exchangers should include flow, inlet and outlet temperatures, and differential pressure. The "monitoring and trending" element of the GALL Report AMP further recommends that visual inspections and performance or functional tests should be performed to confirm the effectiveness of the closed-cycle cooling water program. The applicant stated the following exceptions to the GALL Report Closed-Cycle Cooling Water System AMP:

- Heat exchangers are not monitored for differential pressure for the essential cooling water, spent fuel cooling and cleanup, and shutdown cooling heat exchangers systems.

Aging Management Review Results

- The circulating pumps for the essential chilled water and essential cooling water system will not be subject to periodic internal visual inspections or casing NDE.

- The ventilation cooling coils in the essential chilled water system are not monitored for differential pressure or subject to visual inspection.

- The diesel generator jacket water heat exchanger, turbo air intercooler, turbocharger, and governor lube oil cooler are not individually monitored for flow, inlet and outlet temperatures, and differential pressure, and they are not visually inspected internally.

- The reactor coolant hot leg sample cooler heat exchanger is not periodically inspected or tested.

- Heat exchanger parameters are not monitored. Performance monitoring and inspections are not conducted to manage reduction in heat transfer for the letdown heat exchanger, auxiliary steam vent condenser, auxiliary steam radiation monitor cooler, cooling coils for the normal heating, ventilation, and air conditioning (HVAC) units, SG sampling coolers (hot leg, cold leg, downcomer blowdown), pressurizer steam space and surge line sample coolers, and safety injection sample coolers.

The staff reviewed this exception to the GALL Report and noted that the applicant took the exception because these systems were scoped into the LRA for reasons other than being safety-related. In some cases, the applicant indicated that it took other measures, such as conducting visual inspections of other components made out of the same material in the same environment. However, the staff noted that there were many instances that the applicant indicated it did not need to conduct visual inspection or other performance testing of components and did not provide alternatives to managing the aging of these components.

By letter dated December 29, 2009, the staff issued RAI B2.1.10-3 requesting that the applicant provide additional information for three items. The first item questioned why the applicant is not using an inspection technique to monitor loss of material for specific heat exchangers that do not have a license renewal heat transfer intended function, and how water chemistry control alone is adequate to manage this aging effect. In its response, dated February 19, 2010, the applicant stated that it verifies chemistry control measures for these heat exchangers (that do not have license renewal heat transfer functions but are evaluated as a pressure boundary or leakage barriers) through the visual inspection of internal surfaces of components made of similar materials and exposed to similar environments. The applicant also stated that the Closed-Cycle Cooling Water System Program basis document has been revised to identify the applicable plant procedures, procedure sections, and their inspection requirements for these internal visual inspections. The applicant further stated that it modified LRA Section B2.1.10 to reflect these additional changes.

The second item in RAI B2.1.10-3 requested that the applicant clarify how water chemistry control, in combination with preventative maintenance and performance testing, will adequately manage the aging affects associated with the ventilation cooling coils, especially for loss of material. In its response, dated February 19, 2010, the applicant stated that it will verify the effectiveness of the water chemistry control measures for the ventilation cooling coils by performance testing, which may include, but is not limited to, cooling coil performance tests to verify that the required flow rates are achieved. The applicant stated that the ventilation coils are not subject to visual inspection or NDE because of their internal diameter and geometry. The applicant further stated that the GALL Report indicates that performance testing is one of the specific non-chemistry monitoring techniques provided in EPRI TR-107396, and these

Aging Management Review Results

non-chemistry tests are considered preventative maintenance to ensure that the license renewal intended functions are maintained during the period of extended operation.

The third item in RAI B2.1.10-3 requested that the applicant provide additional information describing why it is not using an inspection technique to monitor the aging effect of the reactor coolant hot leg sample cooler and how water chemistry control alone is adequate to managing the aging degradation. In its response, dated February 19, 2010, the applicant stated that the reactor coolant hot leg sample cooler is a sealed unit, not subject to opening for routine inspection or maintenance. The applicant also stated that it will evaluate the effectiveness of the water chemistry control measures for this heat exchanger by conducting visual inspection of internal surfaces for selected components fabricated of similar materials and exposed to closed-cycle cooling water of the same corrosion inhibitor program. The applicant further stated that it revised the Closed-Cycle Cooling Water System Program basis document to indentify the plant procedures, procedure sections, and inspection requirements for the internal surfaces of the representative components to serve as the leading indicators of the effectiveness of water chemistry control.

Based on its review of the LRA and the information provided in the applicant's response to RAI B2.1.10-3, the staff finds the exception acceptable because the applicant will use non-chemistry monitoring techniques to verify the effectiveness of its water chemistry control. The applicant will conduct inspections of components of the same material in the same environment as the components not being inspected as well as performance testing of ventilation cooling coils in accordance with the guidance in EPRI TR-107396, which is consistent with the guidance provided in the GALL Report. The staff's concerns described in RAI B2.1.10-3 are resolved.

Exception 3. LRA Section B2.1.10 identifies an exception to "preventive actions," "parameters monitored or inspected," "detection of aging effects," "monitoring and trending," and "acceptance criteria" program elements. In the GALL Report AMP, "Closed-Cycle Cooling Water System," these program elements recommend the use of EPRI TR–107396, Revision 0, "Closed Cooling Water Chemistry Guidelines." However, the applicant indicated that it bases the Closed-Cycle Cooling Water System Program on Revision 1 of this document, published in 2004.

The staff reviewed this exception to the GALL Report and noted that the applicant took the exception because the EPRI Closed Cooling Water Chemistry Guidelines have been updated from the version cited in the GALL Report. The staff finds this exception acceptable because the newer version of the above EPRI guidelines contains all previous requirements and provides more recent operating experience information.

Enhancement. LRA Section B2.1.10 identifies an enhancement to the "preventive actions" and "acceptance criteria" program elements. The applicant indicated that this enhancement expands on the existing program elements by incorporating the guidance of EPRI TR–107396, "Closed-Cooling Water Chemistry Guidelines," with respect to water chemistry control for frequency of sampling and analysis, normal operating limits, action level concentrations, and time for implementing corrective actions upon attainment of action levels.

The staff reviewed this enhancement against the corresponding program elements in the GALL Report AMP. The staff finds the applicant's enhancement acceptable because it will make the program consistent with the GALL Report recommendation by setting the appropriate limits for water chemistry control with respect to frequency of sampling and analyses, normal operating limits, and action level concentrations.

Based on its audit and review of the applicant's responses to RAIs B2.1.10-1, B2.1.10-2, and B2.1.10-3, the staff finds that elements one through six of the applicant's Closed-Cycle Cooling Water System Program, with acceptable exceptions and enhancement, are consistent with the

Aging Management Review Results

corresponding program elements of the GALL Report Closed-Cycle Cooling Water System AMP and are, therefore, acceptable.

Operating Experience. LRA Section B2.1.10 summarizes operating experience related to the Closed-Cycle Cooling Water System Program. The applicant stated that there has been no evidence of significant fouling or loss of material that resulted in the loss of intended function for most of the closed-cycle cooling systems. However, the applicant also stated that in 2001, it identified elevated levels of chlorides and sulfates in the Unit 3 essential cooling water system due to a leak in a heat exchanger tube from the essential spray pond (ESP) system and that corrosion had occurred from the open-cycle cooling water side of the heat exchanger. The applicant plugged the tube to correct the leakage and expanded the inspection program to include100 percent of the other units' essential cooling water heat exchange tubes. The applicant stated that this inspection revealed no further corrosion.

The staff reviewed operating experience information, in the application and during the audit, to determine if the applicant reviewed the applicable aging effects and industry and plant-specific operating experience. As discussed in the Audit Report, the staff conducted an independent search of the plant operating experience information to determine if the applicant had adequately incorporated and evaluated operating experience related to this program. The staff found that the applicant had adequately identified and incorporated industry and applicable plant-specific operating experience.

Based on its audit and review of the application, the staff finds that operating experience related to the applicant's program demonstrates that it can adequately manage the detrimental effects of aging on SSCs within the scope of the program, and that implementation of the program has resulted in the applicant taking corrective actions. The staff confirmed that the "operating experience" program element satisfies the criterion in SRP-LR Section A.1.2.3.10 and is, therefore, acceptable.

UFSAR Supplement. LRA Section A1.10 provides the UFSAR supplement for the Closed-Cycle Cooling Water System Program. The staff reviewed this UFSAR supplement description of the program and notes that it conforms to the recommended description for this type of program, as described in SRP-LR Tables 3.1-2, 3.2-2, 3.3-2, and 3.4-2. The staff also notes that the applicant committed (Commitment No. 12) to the ongoing implementation of the existing Closed-Cooling Water System Program and to the enhancement of the Closed-Cycle Cooling Water System Program before entering the period of extended operation. Specifically, the applicant committed to incorporating the guidance in the EPRI TR-107396 with respect to water chemistry control for frequency of chemistry sampling, normal operating chemistry limits, action level concentrations, and time limits for implementing corrective actions if the action levels are reached. The staff determines that the information in the UFSAR supplement is an adequate summary description of the program, as required by 10 CFR 54.21(d).

Conclusion. On the basis of its audit and review of the applicant's Closed-Cycle Cooling Water System Program, the staff determines that those program elements for which the applicant claimed consistency with the GALL Report are consistent. In addition, the staff reviewed the exceptions and their justifications, and determines that the AMP, with the exceptions, is adequate to manage the aging effects for which the LRA credits it. The staff reviewed the enhancement and confirmed that its implementation through Commitment No. 12, prior to the period of extended operation, would make the existing AMP consistent with the GALL Report AMP to which it was compared. The staff concludes that the applicant has demonstrated that the effects of aging will be adequately managed so that the intended function(s) will be maintained consistent with the CLB for the period of extended operation, as required by 10 CFR 54.21(a)(3). The staff also reviewed the UFSAR supplement for this AMP, and concludes that it provides an adequate summary description of the program, as required by 10 CFR 54.21(d).

Aging Management Review Results

3.0.3.2.6 Inspection of Overhead Heavy Load and Light Load (Related to Refueling) Handling Systems Program

Summary of Technical Information in the Application. LRA Section B2.1.11 describes the existing Inspection of Overhead Heavy Load and Light Load (Related to Refueling) Handling Systems Program as consistent, with an enhancement, with the GALL Report AMP XI.M23, "Inspection of Overhead Heavy Load and Light Load (Related to Refueling) Handling Systems." The applicant stated that the Inspection of Overhead Heavy Load and Light Load (Related to Refueling) Handling Systems Program manages loss of material for all cranes, trolley and hoist structural components, fuel handling equipment, and applicable rails within the scope of license renewal. The applicant also stated that visual inspections will be used to assess conditions, such as loss of material due to corrosion and visible signs of rail wear.

Staff Evaluation. During its audit, the staff reviewed the applicant's claim of consistency with the GALL Report. The staff also reviewed and confirmed that the plant's conditions are bounded by the conditions for which the GALL Report was evaluated. The staff compared elements one through six of the applicant's program to the corresponding elements of the GALL Report AMP. As discussed in the Audit Report, the staff confirmed that these elements are consistent with the corresponding element of the GALL Report AMP. The staff also reviewed the portions of the "detection of aging effects" program element associated with the enhancement to determine whether the program will be adequate to manage the aging effects for which it is credited. The staff's evaluation of this enhancement follows.

Enhancement. LRA Section B2.1.11 identifies an enhancement to the "detection of aging effects" program element. Specifically, the applicant stated that it will enhance to inspect for loss of material due to corrosion or rail wear. The "detection of aging effects" program element of the GALL Report AMP states that crane rails and structural components are visually inspected on a routine basis for degradation. The staff finds this enhancement acceptable because, when implemented, it will make this existing program consistent with the recommendations in the GALL Report AMP.

Based on its audit, the staff finds that elements one through six of the applicant's Inspection of Overhead Heavy Load and Light Load (Related to Refueling) Handling Systems Program, with acceptable enhancement, are consistent with the corresponding program elements of the GALL Report AMP and are, therefore, acceptable.

Operating Experience. LRA Section B2.1.11 summarizes operating experience related to the Inspection of Overhead Heavy Load and Light Load (Related to Refueling) Handling Systems Program. The applicant stated that it has identified no occurrences of unacceptable corrosion of components within the scope of the program. Additionally, the applicant stated that since it has not operated any cranes, hoists, trolleys, or fuel-handling equipment outside of their design limits or beyond their design lifetime, no fatigue-related structural failures have occurred.

The staff reviewed operating experience information, in the application and during the audit, to determine if the applicant reviewed the applicable aging effects and industry and plant-specific operating experience. As discussed in the Audit Report, the staff conducted an independent search of the plant operating experience information to determine if the applicant had adequately incorporated and evaluated operating experience related to this program. The staff found that the applicant had adequately identified and incorporated industry and applicable plant-specific operating experience.

Based on its audit and review of the application, the staff finds that operating experience related to the applicant's program demonstrates that it can adequately manage the detrimental effects of aging on SSCs within the scope of the program. The staff confirmed that the "operating experience" program element satisfies the criterion of SRP-LR Section A.1.2.3.10 and is, therefore, acceptable.

Aging Management Review Results

UFSAR Supplement. LRA Section A1.11 provides the UFSAR supplement for the Overhead Heavy Load and Light Load (Related to Refueling) Handling Systems Program. The staff reviewed this UFSAR supplement description of the program and notes that it conforms to the recommended description for this type of program, as described in SRP-LR Table 3.3-2. The staff also notes that the applicant committed (Commitment No. 13) to enhancing the Overhead Heavy Load and Light Load (Related to Refueling) Handling Systems Program before entering the period of extended operation. Specifically, the applicant committed to using the existing program and, before the period of extended operation, enhancing procedures to inspect for loss of material due to corrosion or rail wear. The staff determines that the information in the UFSAR supplement is an adequate summary description of the program, as required by 10 CFR 54.21(d).

Conclusion. On the basis of its audit and review of the applicant's Inspection of Overhead Heavy Load and Light Load (Related to Refueling) Handling Systems Program, the staff determines that those program elements for which the applicant claimed consistency with the GALL Report are consistent. In addition, the staff reviewed the enhancement and confirmed that its implementation through Commitment No. 13, prior to the period of extended operation, would make the existing AMP consistent with the GALL Report AMP to which it was compared. The staff concludes that the applicant has demonstrated that the effects of aging will be adequately managed so that the intended function(s) will be maintained consistent with the CLB for the period of extended operation, as required by 10 CFR 54.21(a)(3). The staff also reviewed the UFSAR supplement for this AMP, and concludes that it provides an adequate summary description of the program, as required by 10 CFR 54.21(d).

3.0.3.2.7 Fire Protection Program

Summary of Technical Information in the Application. LRA Section B2.1.12 describes the Fire Protection Program as an existing program that is consistent, with exceptions and enhancements, with the GALL Report AMP XI.M26, "Fire Protection." The applicant stated that the program manages loss of material for fire rated doors, fire dampers, diesel-driven fire pumps, and the carbon dioxide and halon fire suppression systems, cracking, spalling, and loss of material for fire barrier walls, ceilings, and floors, and hardness and shrinkage due to weathering of fire barrier penetration seals.

Staff Evaluation. During its audit, the staff reviewed the applicant's claim of consistency with the GALL Report. The staff also reviewed and confirmed that the plant's conditions are bounded by the conditions for which the GALL Report was evaluated.

The staff compared elements one through six of the applicant's program to the corresponding elements of the GALL Report "Fire Protection" AMP. As discussed in the Audit Report, the staff confirmed that each element of the applicant's program is consistent with the corresponding element of the GALL Report AMP, with the exception of the "parameters monitored or inspected," and "acceptance criteria" program elements. For these elements, the staff determined the need for additional clarification, which resulted in the issuance of RAIs.

The "parameters monitored or inspected" program element of the GALL Report AMP, "Fire Protection," recommends that the diesel-driven fire pump be observed during performance tests such as flow and discharge tests, sequential starting capability tests, and controller function tests for detection of any degradation of the fuel supply line. The "acceptance criteria" program element of the GALL Report AMP, "Fire Protection," recommends that no corrosion is acceptable in the fuel supply line for the diesel-driven fire pump. The staff noted that the applicant's basis document for this program indicated that Fuel Oil Chemistry and External Surface Monitoring Programs manage the fuel oil supply line. The PVNGS basis document also states to visually inspect the diesel fuel oil supply line for signs of degradation and references the source of the inspection as the LRA. It is not clear to the staff which program or procedure

Aging Management Review Results

the applicant uses for performing this inspection and where the applicant specified the acceptance criterion for the inspection.

By letter dated December 29, 2009, the staff issued RAI B2.1.12-1 requesting that the applicant confirm which program is used to perform this inspection and identify where the acceptance criterion for the inspection is specified.

In its response, dated February 19, 2010, the applicant stated that the Fuel Oil Chemistry Program manages the aging of internal surfaces of components exposed to fuel oil, including the diesel-driven fire pump fuel oil supply piping line. Further, the applicant stated that the One-Time Inspection Program, consistent with the GALL Report, verifies the effectiveness of this program. The applicant also stated that the One-Time Inspection Program was revised to include the fuel oil supply line of one diesel-driven fire pump in the sample of components to be inspected. The applicant stated that the External Surfaces Monitoring Program manages the aging of external surfaces of components associated with the diesel-driven fire pump, and is consistent with the GALL Report.

The applicant stated that, consistent with the GALL Report AMP, "Fire Protection," plant procedures require demonstration of the diesel-driven fire pump operability by starting and running each pump on a monthly basis. This test verifies that the fuel oil day tank is above the minimum level and detects degradation of the fuel oil supply line by visual inspection while the diesel driven fire pump is in operation. The applicant also stated that it revised the Fire Protection Program basis document to clarify the AMPs applied to manage the aging of the diesel-driven fire pump.

The staff noted that the GALL Report, Chapter VII.G, "Auxiliary Systems - Fire Protection," recommends the Fuel Oil Chemistry and One-Time Inspection Programs to manage the aging effect of loss of material for piping exposed to fuel oil. The staff also noted that for external surfaces, the GALL Report recommends the External Surface Monitoring Program. The staff finds the applicant's response acceptable because it manages the aging effect of loss of material on the internal and external surfaces of the fuel oil supply line consistent with the GALL Report. In addition, the applicant includes the fuel oil supply line in the sample of components to be inspected in the One-Time Inspection Program, it performs the testing of the diesel-driven fire pump consistent with the GALL Report Fire Protection AMP, and it revised the Fire Protection Program basis document to clarify the AMPs applied to manage the aging of the diesel-driven fire pump. The staff's concern described in RAI B2.1.12-1 is resolved.

The staff also reviewed the portions of the "parameters monitored or inspected," "detection of aging effects," "monitoring and trending," and "acceptance criteria," program elements, associated with the exception and enhancements, to determine if the program will be adequate to manage the aging effects for which it is credited. The staff's evaluation of these exceptions and enhancements follows.

Exception. LRA Section B2.1.12 identifies an exception to the "parameters monitored or inspected," and "detection of aging effects," program elements. The exception states that halon and CO_2 fire suppression systems are visually inspected and functionally tested once every 18 months per the Technical Requirements Manual Surveillance Requirement (TSR). The "parameters monitored or inspected" and "detection of aging effects" program elements of the GALL Report AMP, "Fire Protection," recommend that periodic visual inspections and functional testing be performed at least once every six months to examine the halon and CO_2 fire suppression systems for signs of degradation.

The applicant stated that this functional test would identify any mechanical damage to the halon and CO_2 fire suppression systems that may prevent the system from performing its intended function. The applicant also stated that it considers the 18-month frequency sufficient to ensure

Aging Management Review Results

system availability and operability based on a review of the past 10 years of station operating history that has shown no degradation or loss of intended function between test intervals.

The staff reviewed the applicant's CLB and confirmed that it performs the visual inspections and functional testing of the halon and CO_2 fire suppression systems once every 18 months. The staff also reviewed the plant operating experience reports and did not find any evidence of age-related degradation in the halon or CO_2 systems. Based on its review of plant operating experience and the fact that the applicant is performing testing and inspection in accordance with its CLB, the staff finds that the inspection and testing frequency of once every 18 months is adequate to ensure the system maintains its function. The staff finds the exception acceptable.

However, as part of its annual update letter dated December 7, 2009, the applicant revised the exception such that the halon and CO_2 dampers are integrity-validated every 54 months, a change from the 18 month periodicity as defined in the exception in the LRA. By letter dated December 29, 2009, the staff issued RAI B2.1.12-2, requesting that the applicant provide technical justification for the 54-month testing and inspection interval.

In its response, dated February 19, 2010, the applicant stated the following:

> With respect to the 54 month destructive testing of the Electro-Thermal Links (ETLs), PVNGS has performed an engineering analysis, consistent with the methodology described within EPRI Technical Report 1006756 "Fire Protection Equipment Surveillance Optimization and Maintenance Guide 2003" to extend the frequency of the test so that the confidence of functionality obtained by successful completion of the test is aligned with reliability and logistical concerns of the test. The calculation indicates that a full functional test every six years of the dampers actuated by ETLs will maintain a 95% success rate assuming the same amount of failures as have occurred in the last 10 years and adjusting for uncertainty at the 99% level. The selection of a testing interval of 54 months, compared to the calculated value of 72 months for 95% success rate, provides an additional margin of protection.

> The 18 month halon and CO_2 fire suppression systems inspection intervals were previously in the initial PVNGS Units 1 and 2 Operating Licenses (OL), Appendix A, Technical Specifications 3.7.11.3 and 3.7.11.6. The NRC approved the relocation of the fire protection technical specifications to the UFSAR in OL Amendments 14 and 8 for Units 1 and 2, respectively, dated April 8, 1987 (prior to the issuance of the Unit 3 OL), using the guidance in Generic Letter 86-10. Subsequently, following the creation of the licensee-controlled Technical Requirements Manual (TRM) when the improved standard Technical Specifications were implemented in August 1998 (OL Amendment No. 117 for all three units), the fire protection technical specifications were added to the TRM. In accordance with PVNGS OL License Conditions, APS may make changes to the approved fire protection program without approval of the Commission only if those changes would not adversely affect the ability to achieve and maintain safe shutdown in the event of a fire. The 18 month halon and CO_2 fire suppression system circuit actuation testing and damper functional testing (destructive testing of the ETLs) intervals in the TRM were changed from 18 months to 54 months in November 2009 in accordance with the license condition as evaluated and documented in a PVNGS internal License Document Change Request.

The staff reviewed the applicant response and noted that the engineering evaluation was performed assuming the same number of failures as have occurred in the last 10 years accounting for uncertainties. However, the number of failures in the past 10 years was based on testing performed every 18 months. The staff noted that the number of failures should

Aging Management Review Results

increase when the applicant performs tests every 54 months. The staff finds that the change in testing frequency may adversely affect the ability of this program to adequately manage aging so that the intended function of this system will remain consistent with the CLB for the period of extended operation.

In a May 16, 2010 conference call, the staff discussed its concern with the applicant. The applicant agreed to supplement its response to RAI B2.1.12-2 and modify the commitment to address the staff's concern.

By letter dated July 21, 2010, the applicant submitted Amendment No. 20, which provided an updated commitment to the Fire Protection AMP. The applicant stated that the existing Fire Protection Program is credited for license renewal, and before the period of extended operation procedures, will be enhanced to perform the testing of the ETLs and functional testing of the halon and CO_2 dampers. This testing will occur every 18 months, or at the frequency specified in the CLB in effect, upon entry into the period of extended operation.

The staff finds that the applicant's commitment adequately addresses the issue because the functional testing frequency of the halon and CO_2 dampers will be once every 18 months or at the frequency specified in the CLB, and this frequency will assure that the intended function of this system will be maintained through the period of extended operation. The staff's concern in RAI B2.1.12-2 is resolved.

Enhancement 1. LRA Section B2.1.12 identifies an enhancement to the "parameters monitored or inspected," "detection of aging effects," "monitoring and trending," and "acceptance criteria" program elements to expand on the existing program elements by adding trending requirements for the diesel-driven fire pump and including visual inspections of the fuel supply line to detect degradation. The staff confirmed that the applicant included this enhancement as Commitment No. 14 in LRA Appendix A, Table A4-1.

This enhancement, when implemented, will make the Fire Protection Program consistent with the GALL Report AMP, "Fire Protection," which recommends that performance of the fire pump be monitored during the periodic test to detect any signs of degradation in the fuel supply lines. Based on its review, the staff finds the enhancement acceptable because it will make the program consistent with the GALL Report.

In its letter dated December 7, 2009, as part of its annual update, the applicant stated that it revised the procedures to incorporate the GALL Report AMP recommendation of visual inspections of the fuel supply line to detect degradation, and therefore, it deleted this part of enhancement 1. The applicant revised Commitment No. 14 to reflect that this enhancement of the procedure is complete.

Enhancement 2. LRA Section B2.1.12 identifies an enhancement to the "parameters monitored or inspected," "detection of aging effects," "monitoring and trending," and "acceptance criteria" program elements to expand on the existing program elements by adding criteria to inspect for mechanical damage, corrosion, and loss of material of the CO_2 system discharge nozzles. The staff confirmed that the applicant included this enhancement as Commitment No. 14 in LRA Appendix A, Table A4-1.

The staff notes that this enhancement, when implemented, will make the Fire Protection Program consistent with the GALL Report AMP, which recommends visual inspections to detect any sign of corrosion and mechanical damage of CO_2 systems. Based on its review, the staff finds the enhancement acceptable because it will make the program consistent with the GALL Report.

Enhancement 3. LRA Section B2.1.12 identifies an enhancement to the "parameters monitored or inspected," "detection of aging effects," "monitoring and trending," and "acceptance criteria" program elements to expand on the existing program elements by adding criteria to state the

Aging Management Review Results

qualification requirements for inspecting penetration seals, fire rated doors, fire barrier walls, ceilings, and floors. The staff confirmed that the applicant included this enhancement as Commitment No. 14 in LRA Appendix A, Table A4-1.

The staff notes that this enhancement, when implemented, will make the Fire Protection Program consistent with the GALL Report AMP, which recommends visual inspections by fire protection qualified inspectors of penetration seals, fire barrier wall ceilings and floors, and fire rated doors. Based on its review, the staff finds the enhancement acceptable because it will make the program consistent with the GALL Report.

Based on its audit, review of the LRA, and review of the applicant's responses to RAIs B2.1.12-1 and B2.1.12-2, the staff finds that elements one through six of the applicant's Fire Protection Program, with acceptable exceptions and enhancements, are consistent with the corresponding program elements of the GALL Report AMP, "Fire Protection," and are, therefore, acceptable.

Operating Experience. LRA Section B2.1.12 summarizes operating experience related to the Fire Protection Program. The applicant stated that there have been instances of Thermo-Lag degradation and cracking, and it has reworked these portions of affected Thermo-Lag envelopes according to its specifications. The applicant also stated that it has experienced door skin cracks, which have been weld repaired according to specifications. The applicant further stated that plant staff and other industry representatives performed a fire protection audit in 2005, which included multiple walkdowns of existing fire barriers and the halon and CO_2 systems. The applicant stated that the audit team found no degraded fire barriers.

The staff reviewed operating experience information, in the application and during the audit, to determine if the applicant reviewed the applicable aging effects and industry and plant-specific operating experience. As discussed in the Audit Report, the staff conducted an independent search of the plant operating experience information to determine if the applicant had adequately incorporated and evaluated operating experience related to this program. The staff found that the applicant had adequately identified and incorporated industry and applicable plant-specific operating experience.

Based on its audit and review of the application, the staff finds that operating experience related to the applicant's program demonstrates that it can adequately manage the detrimental effects of aging on SSCs within the scope of the program. The staff confirmed that the "operating experience" program element satisfies the criterion in SRP-LR Section A.1.2.3.10 and is, therefore, acceptable.

UFSAR Supplement. LRA Section A1.12 provides the UFSAR supplement for the Fire Protection Program. By letter dated December 7, 2009, the applicant submitted a revision to the UFSAR supplement to delete the enhancement to add trending requirements for the diesel-driven fire pump and include visual inspections of the fuel supply line to detect degradation because the procedure revisions are complete. The staff reviewed this UFSAR supplement description of the program and notes that it conforms to the recommended description for this type of program, as described in SRP-LR Table 3.3-2.

The staff also notes that the applicant committed (Commitment No.14) to enhancing the Fire Protection Program before entering the period of extended operation. Specifically, the applicant committed to enhancing procedures to include trending requirements for the diesel driven fire pump (enhancement complete), enhancing procedures to inspect for mechanical damage, corrosion and loss of material of the CO_2 system discharge nozzles, and enhancing procedures to state the qualification requirements for inspecting penetration seals, fire rated doors, fire barrier walls, ceilings, and floors. The staff determines that the information in the UFSAR supplement, as amended, is an adequate summary description of the program, as required by 10 CFR 54.21(d).

Aging Management Review Results

Conclusion. On the basis of its audit and review of the applicant's Fire Protection Program and the applicant's response to the staff's RAIs, the staff determines that those program elements for which the applicant claimed consistency with the GALL Report are consistent. The staff reviewed the exception and its justification, and determines that the AMP is adequate to manage the aging effects for which the LRA credits it. The staff also reviewed the enhancements and confirmed that their implementation through Commitment No. 14, prior to the period of extended operation, will make the existing AMP consistent with the GALL Report AMP. The staff concludes that the applicant has demonstrated that the effects of aging will be adequately managed so that the intended functions will be maintained consistent with the CLB for the period of extended operation, as required by 10 CFR 54.21(a)(3). The staff also reviewed the UFSAR supplement as amended, for this AMP and concludes that it provides an adequate summary description of the program, as required by 10 CFR 54.21(d).

3.0.3.2.8 Fire Water System Program

Summary of Technical Information in the Application. LRA Section B2.1.13 describes the existing Fire Water System Program as consistent, with exceptions and enhancements, with the GALL Report AMP XI.M27, "Fire Water System." The applicant stated that this program manages loss of material for water-based fire protection systems. The applicant also stated that the program performs periodic hydrant inspections, fire main flushing, sprinkler inspections, and flow tests, in accordance with National Fire Protection Association (NFPA) codes and standards, to ensure that the water-based fire protection systems are capable of performing their intended function.

Staff Evaluation. During its audit, the staff reviewed the applicant's claim of consistency with the GALL Report. The staff also reviewed and confirmed that the plant's conditions are bounded by the conditions for which the GALL Report was evaluated. The staff compared elements one through six of the applicant's program to the corresponding elements of the GALL Report AMP, "Fire Water System." As discussed in the audit report, the staff confirmed that these elements are consistent with the corresponding elements of the GALL Report AMP.

The staff also reviewed the portions of the "preventive actions," "parameters monitored or inspected," "detection of aging effects," "acceptance criteria," and "corrective actions" program elements, associated with the exceptions and enhancements, to determine whether the program will be adequate to manage the aging effects for which it is credited. The staff's evaluation of these exceptions and enhancements follows.

Exception 1. LRA Section B2.1.13 identifies an exception to the "detection of aging effects," program element to perform power block hose station gasket inspections once per 18 months. The applicant stated that the TSR defines the inspection frequency to be 18 months. The "detection of aging effects" program element of the GALL Report AMP, "Fire Water System," recommends that the gasket inspections be done annually.

The staff reviewed the applicant's CLB and confirmed that it performs gasket inspections once per 18 months. The staff noted that the applicant's basis document for this program stated that it replaces any gaskets showing signs of deterioration. On the basis that the applicant is performing the gasket inspections in accordance with its CLB and that it is replacing any gaskets showing signs of degradation, the staff finds the frequency of inspection of once per 18 months adequate to manage the aging effects and, therefore, finds the exception acceptable.

Exception 2. LRA Section B2.1.13 identifies an exception to the "detection of aging effects," program element to perform hydrostatic testing on fire hoses once every three years. The applicant stated that replacement fire hoses that have been hydrostatically tested are available, if needed, in lieu of performing a hydrostatic test. The applicant stated that its TSR defines the inspection frequency to be 36 months. The "detection of aging effects" program element of the

Aging Management Review Results

GALL Report AMP, "Fire Water System," recommends that the fire hydrant hose hydrostatic tests be performed annually.

The staff reviewed the applicant's CLB and confirmed that it performs hydrostatic testing of fire hoses once every 36 months. The staff also reviewed the applicant's surveillance procedure for hose hydrostatic testing and confirmed that the applicant replaces the hose with a hydrostatically tested hose in lieu of testing. The staff reviewed the plant operating experience and did not identify any age-related degradation for fire hoses. The staff finds the exception acceptable because the applicant is performing the hydrostatic test in accordance with the CLB and replaces the hoses with hydrostatically tested fire hoses, if needed, in lieu of the test. In addition, operating experience has not identified any age-related degradation associated with the fire hoses.

Enhancement 1. LRA Section B2.1.13 identifies an enhancement to the "preventive actions" and "acceptance criteria" program elements to expand on the existing program elements by adding review and approval requirements under the Nuclear Administrative Technical Manual (NATM). The GALL Report AMP, "Fire Water System," recommends that preventive actions be taken to ensure biofouling does not occur and that acceptance criteria ensure that no biofouling exists. The staff confirmed that the applicant included this enhancement as Commitment No. 15 in LRA Appendix A, Table A4-1.

The applicant stated that it will enhance its procedures to chemically treat fire water to mitigate biofouling to include review and approval requirements under the NATM. The staff finds that including this procedure under the scope of the NATM ensures that the applicant will review and approve any changes to this procedure in accordance with the requirements of its QA Program. The staff finds this enhancement acceptable because it will make the program consistent with the recommendations in the GALL Report AMP.

Enhancement 2. LRA Section B2.1.13 identifies an enhancement to the "parameters monitored or inspected" program element to expand on the existing program element by enhancing the procedures to be consistent with the current code of record, or NFPA 25, 2002 edition. The staff confirmed that the applicant included this enhancement as Commitment No. 15 in LRA Appendix A, Table A4-1. The GALL Report Fire Water System AMP recommends that periodic flow testing is performed using the guidelines of NFPA 25, 1998 and 2002 editions.

The applicant stated that it will enhance the procedures for its fire protection test program and fire suppression water system flow testing to be consistent with the current code of record or NFPA 25, 2002 edition. The staff finds this enhancement acceptable because it will make the program consistent with the recommendations in the GALL Report AMP.

Enhancement 3. LRA Section B2.1.13 identifies an enhancement to the "detection of aging effects" program element to expand on the existing program element by adding field service testing of a representative sample of sprinklers or replacing sprinklers prior to 50 years in service, and testing thereafter every 10 years to ensure that signs of degradation are detected in a timely manner. The staff confirmed that the applicant included this enhancement as Commitment No. 15 in LRA Appendix A, Table A4-1.

The GALL Report Fire Water System AMP recommends that sprinkler heads are inspected before the end of the 50-year sprinkler head service life and at 10-year intervals thereafter during the extended period of operation. The staff finds this enhancement acceptable because it will make the program consistent with the recommendations in the GALL Report AMP.

Enhancement 4. LRA Section B2.1.13 identifies an enhancement to the "detection of aging effects" program element to expand on the existing program elements by enhancing procedures to be consistent with NFPA 25, Sections 7.3.2.1, 7.3.2.2, 7.3.2.3, and 7.3.2.4. The staff

Aging Management Review Results

confirmed that the applicant included this enhancement as Commitment No. 15 in LRA Appendix A, Table A4-1.

The GALL Report AMP, "Fire Water System," recommends that fire hydrant testing is performed to assure that the system functions by maintaining required operating pressures. The staff reviewed NFPA 25, Sections 7.3.2.1, 7.3.2.2, 7.3.2.3, and 7.3.2.4, and determined that these sections provide guidelines on how the fire hydrant testing should be performed, and for how long the pressure should be maintained to assure that system is functioning properly. The staff finds this enhancement acceptable because it will make the program consistent with the recommendation in the GALL Report AMP.

Enhancement 5. LRA Section B2.1.13 identifies an enhancement to the "corrective action" program element to expand on the existing program elements by enhancing procedures so that the QA Programs will apply to fire protection SSCs that are within the scope of license renewal that are also part of the boundary of the water reclamation facility (WRF). The staff confirmed that the applicant included this enhancement as Commitment No. 15 in LRA Appendix A, Table A4-1.

The GALL Report Fire Water System AMP recommends that for fire water systems and components identified within scope that are subject to an AMR for license renewal, the applicant's 10 CFR Part 50, Appendix B, Program is used for corrective actions for aging management during the period of extended operation. The staff finds this enhancement acceptable because it will make the program consistent with the recommendation in the GALL Report AMP.

Based on its audit, the staff finds that elements one through six of the applicant's Fire Water System Program, with acceptable exceptions and enhancements, are consistent with the corresponding program elements of the GALL Report AMP, "Fire Water System," and are, therefore, acceptable.

Operating Experience. LRA Section B2.1.13 summarizes operating experience related to the Fire Water System Program. The applicant stated that with the addition of sodium hydroxide and sodium sulfite, the corrosion rate has been 0.3 mils per year, thus indicating successful corrosion control measures. The applicant also stated that it performed remote field eddy-current testing on about 7,721 feet of 12-inch pipe covering the fire water main loop, and test results indicated that there were several sections of pipe that had localized degradation in excess of the minimum wall thickness. Further, the applicant stated that 6,000 feet of pipe on the north loop and 4,500 feet of pipe on the south loop was replaced with epoxy-lined, reinforced fiberglass pipe. The flushes of the deluge system, fire hydrants, and underground pipe identified little or no debris in the lines. A review of the past 10 years of plant operating experience provided no reports of gasket or fire hose degradation during inspections conducted at intervals of 18 months and three years, respectively.

The staff reviewed operating experience information in the application and during the audit to determine if the applicant reviewed the applicable aging effects and industry and plant-specific operating experience. As discussed in the Audit Report, the staff conducted an independent search of the plant operating experience information to determine if the applicant had adequately incorporated and evaluated operating experience related to this program. The staff found that the applicant had adequately identified and incorporated industry and applicable plant-specific operating experience.

Based on its audit and review of the application, the staff finds that operating experience related to the applicant's program demonstrates that it can adequately manage the detrimental effects of aging on SSCs within the scope of the program and that implementation of the program has resulted in the applicant taking corrective actions. The staff confirmed that the "operating

Aging Management Review Results

experience" program element satisfies the criterion in SRP-LR Section A.1.2.3.10 and is, therefore, acceptable.

UFSAR Supplement. LRA Section A1.13 provides the UFSAR supplement for the Fire Water System Program. The staff reviewed this UFSAR supplement description of the program and notes that it conforms to the recommended description for this type of program, as described in SRP-LR Table 3.3-2.

The staff also notes that the applicant committed (Commitment No.15) to enhancing the Fire Water System Program before entering the period of extended operation. Specifically, the applicant committed that it will enhance specific procedures to include review and approval requirements under the NATM; and it will enhance procedures to be consistent with the current code of record or NFPA 25, 2002 edition. In addition, the applicant will enhance procedures to field service test a representative sample or replace sprinklers prior to 50 years in service and test thereafter every 10 years to ensure that signs of degradation are detected promptly. It will enhance procedures to be consistent with NFPA 25 Sections 7.3.2.1, 7.3.2.2, 7.3.2.3, and 7.3.2.4 and enhance procedures so that the QA Programs will apply to fire protection SSCs that are within the scope of license renewal that are also part of the boundary of the WRF.

The staff determines that the information in the UFSAR supplement is an adequate summary description of the program, as required by 10 CFR 54.21(d).

Conclusion. On the basis of its audit and the review of the applicant's Fire Water System Program, the staff determines that those program elements for which the applicant claimed consistency with the GALL Report are consistent. In addition, the staff reviewed the exceptions and justification and determines that the AMP, with the exceptions, is adequate to manage the aging effects for which the LRA credits it. In addition, the staff reviewed the enhancements and confirmed that their implementation through Commitment No.15, prior to the period of extended operation, will make the existing AMP consistent with the GALL Report AMP. The staff concludes that the applicant has demonstrated that the effects of aging will be adequately managed so that the intended functions will be maintained consistent with the CLB for the period of extended operation, as required by 10 CFR 54.21(a)(3). The staff also reviewed the UFSAR supplement for this AMP, and concludes that it provides an adequate summary description of the program, as required by 10 CFR 54.21(d).

3.0.3.2.9   Fuel Oil Chemistry Program

Summary of Technical Information in the Application. LRA Section B2.1.14 describes the existing Fuel Oil Chemistry Program as consistent with exceptions and enhancements with the GALL Report AMP XI.M30, "Fuel Oil Chemistry." The applicant stated that the Fuel Oil Chemistry Program manages loss of material on the internal surface of components in the emergency diesel generator (EDG) fuel oil storage and transfer system, diesel fire pump fuel oil system, and SBO generator system. The program includes procedures for testing and maintaining the quality of stored and new fuel oil, inspecting the fuel oil storage tanks, and performing a sample inspection of components in systems that contain fuel oil.

Staff Evaluation. During its audit, the staff reviewed the applicant's claim of consistency with the GALL Report. The staff also reviewed and confirmed that the plant's conditions are bounded by the conditions for which the GALL Report was evaluated.

The staff compared elements one through six of the applicant's program to the corresponding elements of the GALL Report Fuel Oil Chemistry AMP. The staff confirmed that these elements are consistent with the corresponding elements of the GALL Report AMP, with the exceptions of the "preventive actions," "parameters monitored or inspected," "monitoring and trending," and "acceptance criteria" program elements, and the enhancements of "scope of program,"

Aging Management Review Results

"preventive actions," "parameters monitored or inspected," "detection of aging effects," and "monitoring and trending" program elements.

Exception 1. LRA Section B2.1.14 identifies an exception to the "preventive actions" program element. The GALL Report Fuel Oil Chemistry AMP states that biocides, stabilizers, and corrosion inhibitors are added to fuel oil to maintain its quality. The applicant stated in the LRA that it does not add stabilizers and corrosion inhibitors to the diesel fuel oil based on negligible underground temperature swings, an arid outdoor environment, and operating experience showing no water in the diesel fuel oil. Based on the staff's audit of operational experience, which confirmed the absence of water in the diesel fuel oil, the staff finds this acceptable because the Surveillance and Monitoring Program will continue to verify the absence of water in the diesel fuel oil.

Exception 2. LRA Section B2.1.14 states exceptions to the "parameters monitored or inspected," and "acceptance criteria" program elements. The GALL Report Fuel Oil Chemistry AMP states that American Society for Testing and Materials (ASTM) Standards D1796 and D2709 are to be used for determination of water and sediment contamination in diesel fuel. The applicant stated in the LRA that TS 5.5.13 requires the use of ASTM Standard D1796-83. The staff finds this acceptable because both of these ASTM standards perform the same type of testing, and the use of ASTM D1796-83 means the applicant's program will still be consistent with the intent of monitoring for water and sediment contamination in diesel fuel in the "parameters monitored or inspected" and "acceptance criteria" program elements of the GALL Report Fuel Oil Chemistry AMP.

Exception 3. LRA Section B2.1.14 identifies an exception to the "monitoring and trending" program element. The GALL Report Fuel Oil Chemistry AMP states that water and biological activity or particulate contamination concentrations are monitored and trended in accordance with the plant's TS, or at least quarterly. The applicant stated in the LRA that it does not monitor biological activity. In discussions with the applicant, the applicant clarified that it proactively adds biocide to each delivery of fuel oil, and that it monitors particulate contamination in the fuel as part of the surveillance program using ASTM D2276. The staff finds this acceptable because the monitoring for particulate contamination in the diesel fuel oil makes the applicant's program consistent with the "monitoring and trending" program element of the GALL Report AMP, "Fuel Oil Chemistry."

Enhancement 1. LRA Section B2.1.14 states enhancements to the "scope of program," "parameters monitored or inspected," and "monitoring and trending" program elements. The applicant stated that, before the period of extended operation, it will enhance the procedures to extend the scope of the Fuel Oil Chemistry Program to include the SBO generator fuel oil storage tank and the SBO generator skid fuel tanks. The staff compared these enhancements to the appropriate program elements in the GALL Report Fuel Oil Chemistry AMP and, because the enhancements are consistent with the program elements in the GALL Report AMP, the staff finds them acceptable.

Enhancement 2. LRA Section B2.1.14 states enhancements to the "preventive actions," and "detection of aging effects" program elements. The applicant stated that, prior to the period of extended operation, it will enhance the procedures to include 10-year periodic draining, cleaning, and inspections of the diesel-driven fire pump day tanks, SBO generator fuel oil storage tank, and SBO generator skid fuel tanks. The staff compared these enhancements to the appropriate program elements in the GALL Report AMP, "Fuel Oil Chemistry," and, because the enhancements are consistent with the program elements in the GALL Report AMP, the staff finds them acceptable.

Enhancement 3. LRA Section B2.1.14 identifies an enhancement to the "detection of aging effects," program element. The applicant stated that once, prior to the period of extended

Aging Management Review Results

operation, it will perform a UT or pulsed eddy-current (PEC) thickness examination to detect corrosion-related thinning on the tank bottoms of the EDG fuel oil storage tanks, EDG fuel oil day tanks, diesel-driven fire pump day tanks, SBO generator fuel oil storage tank, and SBO generator skid fuel tanks. The applicant also stated that it will perform a UT or PEC thickness examination to detect corrosion-related wall thinning if degradation is found during the visual inspections of the tanks. The staff compared these enhancements to the appropriate program elements in the GALL Report Fuel Oil Chemistry AMP and, because the enhancements are consistent with the program elements in the GALL Report AMP, the staff finds them acceptable.

Based on its audit, the staff finds that elements one through six of the applicant's Fuel Oil Chemistry Program, with acceptable exceptions and enhancements, are consistent with the corresponding program elements of the GALL Report AMP, "Fuel Oil Chemistry," and are, therefore, acceptable.

Operating Experience. LRA Section B2.1.14 summarizes operating experience related to the Fuel Oil Chemistry Program. The applicant stated that their operating experience has shown no instances of microbes in the EDG fuel oil, negligible underground temperature swings, and an absence of water in the EDG fuel oil.

The staff reviewed operating experience information, in the application and during the audit, to determine if the applicant reviewed the applicable aging effects and industry and plant-specific operating experience. The applicant provided the following operational experience:

> In 2005, during the U2R12 refueling outage, strainers downstream of the EDG fuel oil day tank were found to be clogged. The cause was determined to be a buildup of sediment in the fuel oil day tank.

The applicant took corrective actions including re-filling the day tank from empty following inspection to remove the sediment on the tank bottom; writing one-time corrective maintenance work orders to clean, inspect, and flush the remaining fuel oil day tanks; removing a film of fuel oil sediment in one of the five tanks inspected; and establishing a 10-year periodic preventive maintenance task to inspect, clean, and flush the diesel fuel oil storage and day tanks.

The staff also conducted an independent search of the plant operating experience information to determine if the applicant had adequately incorporated and evaluated operating experience related to this program. The staff found that the applicant had adequately identified and incorporated industry and applicable plant-specific operating experience.

Based on its audit and review of the application, the staff finds that operating experience related to the applicant's program demonstrates that it can adequately manage the detrimental effects of aging on SSCs within the scope of the program, and that implementation of the program has resulted in the applicant taking corrective actions. The staff confirmed that the "operating experience" program element satisfies the criterion in SRP-LR Section A.1.2.3.10 and is, therefore, acceptable.

UFSAR Supplement. LRA Section A1.14 provides the UFSAR supplement for the Fuel Oil Chemistry Program. The staff reviewed this UFSAR supplement description of the program and notes that it conforms to the recommended description for this type of program, as described in SRP-LR Table 3.3-2. The staff also notes that the applicant committed (Commitment No. 16) to enhancing the Fuel Oil Chemistry Program before the period of extended operation. Specifically, the applicant committed to enhancing their procedures to extend the scope of the program to include the SBO generator fuel oil storage tank and SBO generator skid fuel tanks. In addition, the applicant will enhance its procedures to include 10-year periodic draining, cleaning, and inspections on the diesel-driven fire pump day tanks, the SBO generator fuel oil storage tanks, and SBO generator skid fuel tanks. In addition, the applicant will conduct UT or PEC thickness examinations once on the tank bottoms and also to detect corrosion-related wall

Aging Management Review Results

thinning if degradation is found during the visual inspections of the tanks. The one-time UT or PEC examination on the tank bottoms will be performed before the period of extended operation. The staff finds that the information in the UFSAR supplement is an adequate summary description of the program, as required by 10 CFR 54.21(d).

Conclusion. On the basis of its audit and review of the applicant's Fuel Oil Chemistry Program, the staff finds that those program elements for which the applicant claimed consistency with the GALL Report are consistent. In addition, the staff reviewed the exceptions and their justifications and finds that the AMP, with the exceptions, is adequate to manage the aging effects for which the LRA credits it. Also, the staff reviewed the enhancements and confirmed that their implementation through Commitment No. 16, prior to the period of extended operation, would make the existing AMP consistent with the GALL Report AMP. The staff concludes that the applicant has demonstrated that the effects of aging will be adequately managed so that the intended function(s) will be maintained consistent with the CLB for the period of extended operation, as required by 10 CFR 54.21(a)(3). The staff also reviewed the UFSAR supplement for this AMP, and concludes that it provides an adequate summary description of the program, as required by 10 CFR 54.21(d).

3.0.3.2.10 Reactor Vessel Surveillance Program

Summary of Technical Information in the Application. LRA Section B2.1.15, describes the RV Surveillance Program, stating that this existing program is consistent with enhancement with the GALL Report AMP XI.M31, "Reactor Vessel Surveillance." The applicant stated that the RV Surveillance Program manages loss of fracture toughness and consists of scheduled withdrawal and testing of RPV material surveillance coupons, consistent with 10 CFR Part 50, Appendix H, "Reactor Vessel Material Surveillance Program Requirements," and with ASTM E185-82, "Standard Practice for Conducting Surveillance Tests for Light-Water Cooled Nuclear Power Reactor Vessels." The applicant will revise the current schedule to withdraw the next capsule at the equivalent RPV clad-base metal exposure of approximately 54 effective full-power years (EFPYs) for the period of extended operation and to withdraw remaining standby capsules at equivalent RPV clad-base metal exposures not exceeding the 72 EFPYs expected for a possible second period of extended operation.

Staff Evaluation. The staff reviewed the applicant's RV Surveillance Program to confirm whether the applicant's claim of consistency with the GALL Report RV Surveillance AMP with enhancements is valid.

Appendix H of 10 CFR Part 50 specifies surveillance program criteria for 40 years of operation. The GALL Report AMP, "Reactor Vessel Surveillance," specifies additional criteria for 60 years of operation. The staff determined that compliance with 10 CFR Part 50, Appendix H criteria for capsule design, location, specimens, test procedures, and reporting remains appropriate for this AMP because these items, which satisfy 10 CFR Part 50, Appendix H, will stay the same throughout the period of extended operation. To ensure that all capsules in the RPV, removed and tested during the period of extended operation, still meet the test procedures and reporting requirements of ASTM E 185-82, the staff imposes the following license condition to address this specific concern:

> All capsules in the reactor vessel that are removed and tested must meet the test procedures and reporting requirements of American Society for Testing and Materials (ASTM) E 185-82 to the extent practicable for the configuration of the specimens in the capsule. Any changes to the capsule withdrawal schedule, including spare capsules, must be approved by the NRC prior to implementation. All capsules placed in storage must be maintained for future insertion. Any changes to storage requirements must be approved by the NRC.

Aging Management Review Results

The 10 CFR, Part 50, Appendix H capsule withdrawal schedule during the period of extended operation is addressed according to the GALL Report's consideration of eight criteria for an acceptable RPV surveillance program for 60 years of operation.

Enhancements. The staff reviewed the enhancements and the associated justifications to determine whether this AMP is adequate to manage the aging effects for which it is credited. The enhancements are to keep two surveillance capsules in each RPV beyond 32 EFPYs of plant operation and then withdraw one at the equivalent RPV clad-base metal exposure of approximately 54 EFPYs (40 EFPYs actual operation during the period of extended operation) and the other at 72 EFPYs (54 EFPYs actual operation). This meets Criterion 5 of the GALL Report RV Surveillance AMP, which is applicable to a surveillance program that consists of capsules with a projected fluence of less than the 60-year fluence at the end of 40 years. The staff's review of this AMP against the remaining seven criteria is discussed below.

Criteria 1, 2, and 3 of the GALL Report AMP, "Reactor Vessel Surveillance," regard evaluation of 60-year upper-shelf energy (USE) and pressure-temperature (P-T) limits, using RG 1.99, Revision 2, "Radiation Embrittlement of Reactor Vessel Materials." The LRA RV Surveillance Program states that the applicant determined neutron embrittlement effects consistent with RG 1.99, by using option 1b, "Neutron Embrittlement Using Surveillance Data." The staff found, from its review of LRA Section 4.2, "Reactor Vessel Neutron Embrittlement Analysis," that the applicant actually used option 1a, "Neutron Embrittlement Using Chemistry Tables," of the GALL Report RV Surveillance AMP to evaluate USE and P-T limits for 60 years. Hence, by letter dated November 3, 2009, the staff issued RAI B2.1.15-1 requesting the applicant clarify which option was used.

The applicant's response to RAI B2.1.15-1, dated December 18, 2009, clarified that "[LRA] Tables 4.2-3 through 4.2-8 only present projections of embrittlement effects before results of surveillance data are applied. They do not reflect results of the RPV surveillance program." The applicant further stated that its RPV program "will evaluate embrittlement parameters based on surveillance data that is, using GALL [AMP] XI.M31, option 1b, as described." Therefore, RAI B2.1.15-1 is resolved. Further, compliance with RG 1.99, Revision 2 ensures that the AMP meets criteria 1, 2, and 3 of the GALL Report RV Surveillance AMP.

For Criterion 4, regarding pulled and tested capsule specimens, the LRA RV Surveillance Program states that, "[f]ragments of surveillance specimens are retained for possible future use." Hence, Criterion 4 is satisfied.

The applicant's proposed withdrawal schedule during the period of extended operation for the two capsules, as discussed earlier, was confirmed to meet Criterion 5.

Criterion 6 does not apply to the AMP because it is for plants having capsules with a projected fluence exceeding the 60-year fluence at the end of 40 years. Criterion 7 also does not apply to the AMP because it is for plants having no surveillance capsules.

Criterion 8 asks for justification for not including nozzle specimens in the surveillance program. The applicant did not address this issue in the RV Surveillance Program. However, the applicant addressed this issue indirectly in LRA Section 4.2.1. As indicated in Section 4.2.1 of this SER, the staff concludes that the current 32 EFPY neutron fluence for RPVs bounds 54 EFPY neutron fluence, indicating that neutron embrittlement of RPV nozzle materials will remain low during the period of extended operation. Hence, similar to the CLB, nozzle specimens are not required to be included in the surveillance program during the period of extended operation.

Based on the above evaluation of the applicant's RV Surveillance Program, the staff concludes that the AMP has met the eight acceptance criteria of the GALL Report RV Surveillance AMP and is, therefore, acceptable.

Aging Management Review Results

Operating Experience. In LRA Section B2.1.15, the applicant stated what the recent examination results of the RPV surveillance data revealed regarding neutron fluence, USE, and the reference nil-ductility transition temperature (RT_{NDT}). Evaluation of operating experience of this AMP should not be limited to "the recent examination results." Hence, by letter dated November 3, 2009, the staff issued RAI B2.1.15-2 requesting the applicant to provide a discussion of all tested RPV surveillance data.

The applicant responded to RAI B2.1.15-2 in a letter dated December 18, 2009, stating that "[t]he results of the earlier coupon examinations are consistent with the most recent results, and are summarized in the Appendix D credibility evaluation of these three most-recent examination reports." Hence, the applicant has supplied information regarding evaluation of earlier surveillance data, and RAI B2.1.15-2 is resolved.

UFSAR Supplement. The applicant provided its UFSAR Supplement for the RV Surveillance Program in LRA Section A1.15. Appendix H of 10 CFR Part 50 requires licensees to submit proposed changes to their RV Surveillance Program withdrawal schedules to the NRC for review and approval. The staff reviewed this UFSAR supplement description of the program and notes that it conforms to the recommended description for this type of program, as described in SRP-LR Table 3.1-2. To ensure that this reporting requirement will carry forward through the period of extended operation, the staff has imposed a license condition to the applicant's RV Surveillance Program, as stated earlier in the staff's evaluation. The staff reviewed the UFSAR Supplement and determines that the information in the supplement, with the license condition, provides an adequate summary description of the program, as required by 10 CFR 54.21(d).

Conclusion. On the basis of its review of the applicant's RV Surveillance Program and RAI responses, the staff determines that those program elements for which the applicant claimed consistency with the GALL Report are consistent. The staff concludes that the applicant has demonstrated that the effects of aging will be adequately managed so that the intended functions will be maintained consistent with the CLB for the period of extended operation, as required by 10 CFR 54.21(a)(3). The staff also reviewed the UFSAR supplement for this AMP, and concludes that, with the license condition, it provides an adequate summary description of the program, as required by 10 CFR 54.21(d).

3.0.3.2.11 Selective Leaching of Materials Program

Summary of Technical Information in the Application. LRA Section B2.1.17 describes the new Selective Leaching of Materials Program as consistent, with an exception, with the GALL Report AMP XI.M33, "Selective Leaching of Materials." The applicant stated that this program manages the loss of material due to selective leaching for copper alloys with zinc content greater than 15 percent, copper alloys with aluminum content greater than eight percent, and gray cast iron components exposed to closed-cycle cooling water, demineralized water, secondary water, and raw water. The applicant also stated that this program includes a one-time visual inspection and mechanical methods to determine whether loss of material due to selective leaching is occurring. The applicant further stated that the program includes an engineering evaluation of components if it detects selective leaching.

Staff Evaluation. During its audit, the staff reviewed the applicant's claim of consistency with the GALL Report. The staff also reviewed and confirmed that the plant's conditions are bounded by the conditions for which the GALL Report was evaluated. The staff compared elements one through six of the applicant's program to the corresponding elements of the GALL Report AMP, "Selective Leaching of Materials." As discussed in the Audit Report, the staff confirmed that these elements are consistent with the corresponding elements of the GALL Report AMP.

Upon further review, however, the staff determined that additional information was needed regarding the "scope of program" element. The GALL Report AMP "Selective Leaching of

Materials" states in the "scope of program" element that the program should include a one-time visual inspection and hardness measurements of a selected set of sample components to determine whether loss of material due to selective leaching is occurring. However, the LRA did not specify sample size or selection criteria. The staff noted that due to the uncertainty in determining the most susceptible locations and the potential for aging to occur in other locations, large sample sizes may be required in order to adequately confirm that selective leaching is not occurring. By e-mail dated November 17, 2010, the staff requested that the applicant provide specific information regarding how the selected set of components to be sampled will be determined and the size of the sample of components that will be inspected.

In its response dated December 3, 2010, the applicant stated that the representative sample size and inspection locations will be determined based on the materials of fabrication. The applicant also stated that a sample size of 20 percent of the site's population (up to a maximum of 25 inspections) has been established for each group of components including different material and environment combinations. The applicant also stated that the specific inspection locations will be identified considering those components that are most susceptible to selective leaching and that sampling will not be repeated for equivalent system components between the three units. The staff finds the applicant's response acceptable because the applicant's selected set of components will be based on material and environment combinations, and the sample locations will focus on the leading indicator components. Additionally, the applicant's program includes a sample size adequate to confirm selective leaching is not occurring. The staff's concern described above is resolved.

The staff also reviewed the portions of the "scope of program," "preventive actions," "parameters monitored or inspected," and "detection of aging effects" program elements associated with the exception to determine if the program will be adequate to manage the aging effects for which it is credited. The staff's evaluation of this exception follows.

Exception. LRA Section B2.1.17 identifies an exception to the "scope of program," "preventive actions," "parameters monitored or inspected," and "detection of aging effects" program elements. The GALL Report AMP, "Selective Leaching of Materials," recommends hardness testing of selected components in addition to a one-time visual inspection. Alternatively, the applicant's program includes use of other mechanical means (i.e., scraping or chipping) as a qualitative determination of selective leaching. The applicant stated that hardness testing may not be feasible for most components due to form and configuration and that other mechanical means provide an equally valid means of identification. Based on its review, the staff finds this exception acceptable because mechanical methods, such as scraping and chipping, can identify selective leaching of copper alloy and gray cast iron components.

Based on its audit and review of the LRA, the staff finds that elements one through six of the applicant's Selective Leaching of Materials Program, with exception, are consistent with the corresponding program elements of the GALL Report Selective Leaching of Materials AMP, and are, therefore, acceptable.

Operating Experience. LRA Section B2.1.17 summarizes operating experience related to the Selective Leaching of Materials Program. The applicant stated that this is a new, one-time inspection program with no plant-specific operating experience. However, the applicant also stated that industry operating experience on selective leaching is documented in IN 94-59, "Accelerated Dealloying of Cast Aluminum-Bronze Valves Caused by Microbiologically Induced Corrosion." The applicant further stated that it chemically treats the open-cycle cooling water systems with biocides to prevent microbiologically induced corrosion (MIC), and that it periodically recirculates systems that are not in continuous use to ensure adequate chemical mixing.

Aging Management Review Results

The staff reviewed the operating experience information, in the application and during the audit, to determine if the applicant reviewed the applicable aging effects and industry and plant-specific operating experience. As discussed in the Audit Report, the staff conducted an independent search of the plant operating experience information to determine if the applicant had adequately incorporated and evaluated operating experience related to this program. The staff found that the applicant had adequately identified and incorporated industry and applicable plant-specific operating experience.

Based on its audit and review of the application, the staff finds that operating experience related to the applicant's program demonstrates that it can adequately manage the detrimental effects of aging on SSCs within the scope of the program. The staff confirmed that the "operating experience" program element satisfies the criterion in SRP-LR Section A1.2.3.10 and is, therefore, acceptable.

UFSAR Supplement. LRA Section A1.17 provides the UFSAR supplement for the Selective Leaching of Materials Program. The staff reviewed this UFSAR supplement description of the program and notes that it conforms to the recommended description for this type of program, as described in SRP-LR Tables 3.1-2, 3.2-2, and 3.3-2. The staff also notes that the applicant committed (Commitment No. 19) to implementing the new Selective Leaching of Materials Program prior to entering the period of extended operation to manage the aging of applicable components. The staff determines that the information in the UFSAR supplement is an adequate summary description of the program, as required by 10 CFR 54.21(d).

Conclusion. On the basis of its audit and review of the applicant's Selective Leaching of Materials Program, the staff determines that those program elements for which the applicant claimed consistency with the GALL Report are consistent. In addition, the staff reviewed the exception and justification, and determines that the AMP, with the exception, is adequate to manage the aging effects for which the LRA credits it. The staff concludes that the applicant has demonstrated that the effects of aging will be adequately managed so that the intended function(s) will be maintained consistent with the CLB for the period of extended operation, as required by 10 CFR 54.21(a)(3). The staff also reviewed the UFSAR supplement for this AMP, and concludes that it provides an adequate summary description of the program, as required by 10 CFR 54.21(d).

3.0.3.2.12 Buried Piping and Tanks Inspection Program

Summary of Technical Information in the Application. LRA Section B2.1.18 describes the new Buried Piping and Tanks Inspection Program as consistent, with exceptions, with the GALL Report AMP XI.M34, "Buried Piping and Tanks Inspection." The applicant stated that this program manages loss of material caused by corrosion of the external surfaces of buried components. The applicant also stated that the program includes opportunistic or planned visual inspections to monitor the condition of protective coatings and wrappings on carbon steel, gray cast iron, or ductile iron components and to assess the condition of stainless steel components with no coatings or wraps.

Staff Evaluation. During its audit, the staff reviewed the applicant's claim of consistency with the GALL Report. The staff also reviewed and confirmed that the plant's conditions are bounded by the conditions for which the GALL Report was evaluated. The staff compared elements one through six of the applicant's program to the corresponding elements of the GALL Report AMP, "Buried Piping and Tanks Inspection." As discussed in the Audit Report, the staff confirmed that these elements are consistent with the corresponding element of the GALL Report AMP. The staff also reviewed the portions of the "scope of program," "preventive actions," and "acceptance criteria" program elements associated with the exceptions to determine if the program will be adequate to manage the aging effects for which it is credited. The staff's evaluation of these exceptions follows.

Aging Management Review Results

Exception 1. LRA Section B2.1.18 identifies an exception to the "scope of program" and "acceptance criteria" program elements. The scope of the GALL Report Buried Piping and Tanks Inspection AMP only includes buried steel piping, whereas the applicant's program also includes buried stainless steel piping. The applicant stated that it will inspect buried stainless steel piping for loss of material due to general, pitting, and crevice corrosion as well as MIC. Based on its review, the staff finds this exception acceptable because the periodic inspections in the applicants program are adequate to identify the loss of material in buried stainless steel piping.

Exception 2. LRA Section B2.1.18 identifies an exception to the "scope of program," "preventive actions," and "acceptance criteria" program elements. The GALL Report AMP, "Buried Piping and Tanks Inspection" recommends preventive measures such as coatings and wrappings to mitigate corrosion. However, the applicant's program also includes buried stainless steel piping, portions of which may not be coated or wrapped. The applicant stated that it will inspect buried stainless steel piping for loss of material due to general, pitting, and crevice corrosion as well as MIC. During the AMP audit, the applicant stated that it had examined soil quality during the review of the fire protection system, and the results of soil chemistry analysis indicate that the soil is corrosive with chloride concentration greater than 250 parts per million and resistivity less than 2500 ohms per centimeter. The staff notes that given these soil conditions, the uncoated buried stainless steel piping is susceptible to pitting corrosion and, depending on the water table level where the pipe is buried, it may also be susceptible to MIC. The staff conducted a conference call with the applicant on May 20, 2010, during which the applicant agreed to revise their RAI response to answer staff questions.

In a supplemental response to RAI B2.1.18-1, dated June 21, 2010, the applicant stated that before the period of extended operation, and again within the first 10 years of extended operation, it will excavate and inspect at least two 10-foot sections of stainless steel piping, inclusive of chemical volume and control, condensate transfer and storage, or fire protection.

The staff finds the applicant's response acceptable because the applicant will be performing four inspections of stainless steel piping and, as discussed in the Operating Experience Section below, 10 other inspections of steel and fire protection piping in the period prior to and during extended operation. These inspections will provide a reasonable assurance of the condition of buried piping that exceeds the existing GALL Report Buried Piping and Tanks Inspection AMP recommendations.

Based on its audit and review of the applicant's response to RAI B2.1.18-1, the staff finds that elements one through six of the applicant's Buried Piping and Tanks Inspection Program, with acceptable exceptions, are consistent with the corresponding program elements of the GALL Report Buried Piping and Tanks Inspection Program and are, therefore, acceptable.

Operating Experience. LRA Section B2.1.18 summarizes operating experience related to the Buried Piping and Tanks Inspection Program. The applicant stated that the Buried Piping and Tanks Inspection Program is a new program and, as such, it will evaluate industry and plant-specific operating experience during the development and implementation of this program. The applicant also stated that it has found degraded buried piping in the fire protection system. Segments of this ductile iron piping had localized corrosion, and the applicant replaced it with epoxy-lined, reinforced fiberglass pipe.

The staff reviewed operating experience, in the application and during the audit, to determine if the applicant reviewed the applicable aging effects and industry and plant-specific operating experience. As discussed in the Audit Report, the staff conducted an independent search of the plant operating experience information to determine if the applicant had adequately incorporated and evaluated operating experience related to this program.

Aging Management Review Results

During the audit, the staff reviewed the applicant's condition reports. The staff noted that the applicant found coating damage on the external surfaces of underground fire protection piping. Abrasion from the bedding rock material caused this damage, which exposed the piping to corrosive attack. The staff also noted that the applicant conducted remote, field eddy-current inspections to assess the condition of the piping. The applicant also stated that it isolated and repaired the piping failures in the underground fire protection system without adversely affecting the functionality of the fire protection system.

During its review, the staff identified operating experience, which warranted additional information to ensure the program would be effective in adequately managing aging effects during the period of extended operation. Additionally, a number of recent industry events involve leakage from buried piping due to corrosion stemming from coating damage during backfill of piping, failure of fiberglass piping, and failure of buried piping in and around piping penetrations. Based on information in the program basis document, it is not clear to the staff how the applicant considered this relevant industry operating experience in its program. By letter dated December 29, 2009, the staff issued RAI B2.1.18-1, requesting that the applicant provide additional information as to how it considered relevant industry operating experience during the development of its Buried Piping and Tanks Inspection Program.

In its response, dated February 19, 2010, the applicant stated that it incorporated industry operating experience in the development of the Buried Piping and Tanks Inspection Program. The applicant also stated that it reviewed and included selected provisions of industry guideline documents and participated in industry forums on the subject. The applicant further stated that it used EPRI Report 1016456, "Recommendations for an Effective Program to Control the Degradation of Buried Pipe," dated December 2008, and NEI Technical Report (TR) 07-07, "Industry Ground Water Protection Initiative, Final Guidance Document," dated August 2007, in the development of this program. The applicant stated that it was engaged in the EPRI Buried Piping Integrity Group and the NEI initiative on buried piping to learn from the recent events, including the operating experience at Oyster Creek Nuclear Generating Station and Indian Point Nuclear Generating Units 2 and 3.

Based on its review, the staff found the applicant's response to RAI B2.1.18-1 unacceptable because the staff has not endorsed EPRI Report 1016456 and NEI TR 07-07. The staff conducted a follow-up conference call with the applicant on May 20, 2010, during which the applicant agreed to revise their RAI response. In the supplemental responses to RAI B2.1.18-1, dated June 21, 2010, and July 21, 2010, the applicant provided all of its plant-specific operating experience related to buried piping. Four of the five leaks or degradation occurred in the fire protection system. The applicant stated that in addition to risk ranking of piping included in their commitment to the NEI initiative on buried piping, it will excavate and visually inspect at least 10 feet of piping during the period prior to extended operation and again during the first 10-year period of extended operation at the following locations:

- two inspections of stainless steel piping in each unit
- two inspections of cathodically protected steel piping in each unit
- three inspections of fire protection piping with potentially degraded cathodic bonding straps

The applicant also stated that prior to the period of extended operation it will enact provisions to ensure that power is maintained to the cathodic protection system at least 90 percent of the time and that it will perform National Association of Corrosion Engineers cathodic protection surveys annually.

The staff found that the applicant's response did not specifically state that steel piping containing hazardous materials would be inspected, and there is no docketed information on the quality of the backfill used during installation of the buried chemical and volume control, diesel fuel storage and transfer, domestic water, and essential spray ponds piping. By letter dated September 27, 2010, the staff issued follow-up RAI B2.1.18-1 requesting that the applicant provide additional information on the adequacy of its aging management of piping containing hazardous materials and justify why backfill quality is considered adequate such that damage will not occur to coatings (and piping when no coating is installed) in the buried chemical and volume control, diesel fuel storage and transfer, domestic water, and essential spray ponds piping.

In its responses dated October 13, 2010, and November 10, 2010, the applicant stated that the essential spray pond system does not contain hazardous material. The applicant also stated that there are approximately 1697 feet of in-scope, buried diesel fuel oil piping of which only 82 feet is not under concrete or asphalt. The applicant also stated that as part of its participation in the NEI industry buried piping initiative, it will excavate and inspect 10 feet of buried diesel fuel oil piping between January 1, 2012, and December 31, 2015. Finally, the applicant committed to inspect at least one 10 foot segment of this piping in each 10 year interval during the period of extended operation. The applicant further stated that the specifications for backfill ensure that buried piping and its coatings would not be damaged; however, it did find evidence that coarse angular backfill was utilized during construction of the fire protection system. Portions of this system have been and continue to be replaced with the applicant using appropriate quality backfill. The applicant stated that excavations of essential spray pond piping have revealed proper installation of backfill and that the pipe coating was intact. The staff noted that the applicant's September 27, 2010, RAI response stated that it would augment the buried piping inspection plans to include three fire protection piping locations based on plant-specific operating experience of leaks.

The staff noted that as a result of the applicant aligning its first diesel fuel oil piping inspection with the industry buried piping initiative, the inspection, depending on which unit it is conducted, could precede the 10-year period prior to the period of extended operation by as little as 3½ years and as much as 6 years. Based on its review, the staff finds the applicant's response to follow-up RAI B.2.1.18-1 acceptable because, (a) even though the first inspection of the buried diesel fuel oil piping could occur at 24 years into plant life, there is enough operating time to determine if coating damage and subsequent pipe corrosion has occurred, (b) the applicant will conduct an inspection of buried diesel fuel oil piping in each of the two 10-year periods during the period of extended operation, (c) the essential spray pond piping system does not contain hazardous materials, (d) the specifications for backfill ensure that buried piping and its coatings were not damaged, and (e) it has augmented the fire protection inspections based on plant-specific operating experience. The staff's concern described in RAI B.2.1.18-1 is resolved.

The LRA states that the chemical and volume control, diesel fuel storage and transfer, domestic water, fire protection, and essential spray pond systems include buried piping. Based on a review of plant-specific operating experience, the staff notes that the only in-scope buried piping failures have been in fire protection piping. The staff also notes that the applicant has enhanced the buried pipe program with additional inspections of the fire protection piping to ensure that the extent of condition has been identified. The staff finds the applicant's response acceptable because (a) cathodic protection will be available 90 percent of the time and annual NACE qualified surveys will be conducted, (b) all carbon steel piping is coated, (c) specifications for backfill ensure that buried piping and its coatings were not damaged and has augmented the fire protection inspections based on plant-specific operating experience, (d) excavated visual inspections of at least 10 linear feet of buried pipe in each material group will be conducted, and (e) the applicant will conduct a 10-foot excavated direct visual inspection of buried diesel fuel oil piping as part of its participation in the NEI industry buried piping initiative.

Aging Management Review Results

The staff reviewed operating experience information in the application and during the audit to determine whether the applicable aging effects and industry and plant-specific operating experience were reviewed by the applicant. As discussed in the Audit Report, the staff conducted an independent search of the plant operating experience information to determine whether the applicant had adequately incorporated and evaluated operating experience related to this program.

Based on its audit and review of the application, and review of the applicant's response to RAI B2.1-18-1, the staff finds that operating experience related to the applicant's program demonstrates that it can adequately manage the detrimental effects of aging on SSCs within the scope of the program and that implementation of the program has resulted in the applicant taking corrective actions. The staff confirmed that the "operating experience" program element satisfies the criterion in SRP-LR Section A.1.2.3.10 and, therefore, the staff finds it acceptable.

UFSAR Supplement. LRA Section A1.18 provides the UFSAR supplement for the Buried Piping and Tanks Inspection Program. The staff reviewed this UFSAR supplement description of the program, including changes made as a result of RAIs, and notes that it conforms to the recommended description for this type of program, as described in SRP-LR Tables 3.2-2, 3.3-2, and 3.4-2. The staff also notes that the applicant committed (Commitment No. 20) to implement the new Buried Piping and Tanks Inspection Program, inspections, and operating provisions of the cathodic protection system beyond those recommended in GALL AMP XI.M34 prior to entering the period of extended operation to manage the aging of applicable components. The staff determines that the information in the UFSAR supplement is an adequate summary description of the program, as required by 10 CFR 54.21(d).

Conclusion. On the basis of its audit and review of the applicant's Buried Piping and Tanks Inspection Program, the staff determines that those program elements for which the applicant claimed consistency with the GALL Report are consistent. In addition, the staff reviewed the exceptions and justifications, and determines that the AMP, with the exceptions, is adequate to manage the aging effects for which the LRA credits it. The staff concludes that the applicant has demonstrated that the effects of aging will be adequately managed so that the intended functions will be maintained consistent with the CLB for the period of extended operation, as required by 10 CFR 54.21(a)(3). The staff also reviewed the UFSAR supplement for this AMP, and concludes that it provides an adequate summary description of the program, as required by 10 CFR 54.21(d).

3.0.3.2.13 One-Time Inspection of American Society of Mechanical Engineers Code Class 1 Small-Bore Piping

Summary of Technical Information in the Application. LRA Section B2.1.19 describes the existing One-Time Inspections of ASME Code Class 1 Small-Bore Piping Program as consistent, with an exception, with the GALL Report AMP XI.M35, "One-Time Inspections of ASME Code Class 1 Small-Bore Piping." The applicant stated that the One-Time Inspection of ASME Code Class 1 Small-Bore Piping Program manages cracking of ASME Code Class 1 piping, less than or equal to four inches.

Staff Evaluation. During its audit, the staff reviewed the applicant's claim of consistency with the GALL Report. The staff also reviewed and confirmed that the plant's conditions are bounded by the conditions for which the GALL Report was evaluated. The staff compared elements one through six of the applicant's program to the corresponding elements of the GALL Report AMP. As discussed in the Audit Report, the staff confirmed that each element of the applicant's program is consistent with the corresponding element of the GALL Report AMP, with the exception of "parameters monitored or inspected" program element. For this element, the staff determined the need for additional clarification, which resulted in the issuance of an RAI.

Aging Management Review Results

The "parameters monitored or inspected" program element of the GALL Report One-Time Inspections of ASME Code Class 1 Small-Bore Piping AMP recommends that inspections will detect cracking in ASME Code Class 1 small-bore piping. During its audit, the staff noted that socket welds that fall within the weld examination sample will be examined in accordance with ASME Section XI Code requirements. During its audit, the staff further noted that if a qualified volumetric examination procedure for socket welds, endorsed by the industry, is available and incorporated into ASME Section XI at the time of the small-bore socket weld inspections, then volumetric examinations will be conducted on small-bore socket welds. The staff noted that if a volumetric examination procedure for socket welds, endorsed by the industry, is not available and incorporated into the ASME Section XI at the time of the small-bore socket weld inspections, then present ASME Section XI Code requirements will be used for examination of socket welds. The staff also noted that present ASME Section XI Code only requires surface examination for small-bore piping, but surface examination will not detect cracking that initiates on the inside of the piping before leakage occurs.

By letter dated December 29, 2009, the staff issued RAI B2.1.19-1 requesting the applicant to clarify what examination method it will use to detect cracking that initiates from the inside of socket welds, if a volumetric examination procedure endorsed by the industry for socket welds is not available.

In its response, dated February 19, 2010, the applicant stated that if no volumetric examination procedure for ASME Code Class 1 small bore socket welds has been endorsed by the industry and incorporated into ASME Section XI at the time it performs inspections of socket welded ASME Class 1 small-bore piping, it will use a plant procedure for volumetric examination of ASME Code Class 1 small-bore piping with socket welds. The applicant also stated that it revised LRA Sections A1.19 and B2.1.19 to include the use of a plant procedure for volumetric examination of ASME Code Class 1 small-bore piping with socket welds in the event a volumetric examination procedure endorsed by the industry and incorporated into ASME Section XI is not available.

Based on its review, the staff finds the applicant's response to RAI B2.1.19-1 acceptable because the applicant has committed to volumetric examination of small-bore piping socket welds, which is capable of detecting cracking initiated from the inside wetted area of the weld and is consistent with the recommendations of the GALL Report AMP. The staff's concern described in RAI B2.1.19-1 is resolved.

The staff also reviewed the portions of the "scope of program" program element associated with exception to determine if the program will be adequate to manage the aging effects for which it is credited. The staff's evaluation of this exception follows.

Exception. LRA Section B2.1.19 identifies an exception to the "scope of program" program element. The applicant stated in the LRA that for the risk-informed process, it performs examination requirements consistent with EPRI TR-112657, "Revised Risk-Informed Inservice Inspection Evaluation Procedure," Rev. B-A, instead of EPRI Report 1000701, "Interim Thermal Fatigue Management Guideline." The staff noted that although the applicant no longer uses risk-informed methods to select locations for ISIs, the applicant uses the results of previous risk-informed evaluations to select small-bore piping locations to be subjected to this one-time inspection.

The applicant further stated that guidelines for identifying piping susceptible to potential effects of thermal stratification or turbulent penetration that are provided in EPRI Report 1000701 are also provided in EPRI TR-112657, and the recommended inspection volume for welds in EPRI Report 1000701 are identical to those for inspection of thermal fatigue in risk-informed ISI programs. Therefore, the risk-informed process examination requirements meet the recommendations of the GALL Report. The staff noted that, although the inspection volumes

Aging Management Review Results

are identical, it is not clear if the applicant will inspect welds with the highest likelihood of degradation (e.g. welds with the highest stress but not necessarily highest risk category). The staff also reviewed the applicant's selection of welds that would be subjected to volumetric one-time inspection based on the risk-informed method and found that only butt welds would be inspected. The staff noted that, although the butt welds to be inspected have the highest risk, the environment of butt welds is not the same as for socket welds due to the crevice inherent in socket welds; the crevice could lead to aging effects in socket welds, which could be missed if the applicant inspects only butt welds.

By letter dated December 29, 2009, the staff issued RAI B2.1.19-2 requesting the applicant to specify if locations that it will inspect, according to risk-informed methods, also bound the locations of the highest likelihood of degradation. The staff also requested the applicant to justify how it would identify the degradation of socket welds if the risk-informed selection of locations of small-bore piping for volumetric inspection does not include socket welds.

In its response, dated February 19, 2010, the applicant stated that Class 1 socket welds are in the category of low probability of failure and are not included in the risk-informed sample population for volumetric inspection. The applicant further stated that it will augment the inspection population to include at least one socket weld in each unit with a different socket weld location selected for each unit. The applicant also stated that it revised LRA Sections A1.19 and B2.1.19, as shown in Amendment 9, to include a volumetric inspection of at least one socket weld in each unit with a different socket weld location selected for each unit.

The staff finds that the selection of locations of ASME Class 1 small-bore piping subject to the one-time inspection is not adequate given that the applicant has experienced multiple failures in its Class 1 socket welds. The GALL Report AMP states that:

> This inspection should be performed at a sufficient number of locations to ensure an adequate sample. This number, or sample size, is based on susceptibility, inspectability, dose considerations, operating experience, and limiting locations of the total population of ASME Code Class 1 small-bore piping locations.

The applicant's statement in its response "at least one socket weld" did not provide a definitive number concerning the inspection sample size. This was previously identified as Confirmatory Item 3.0.3.2.13-1.

In a conference call with the applicant, dated July 21, 2010, the staff stated it did not have assurance that the applicant would select a sufficient number of samples, as recommended by the GALL Report AMP XI.M35, to ensure adequate aging management. The applicant stated that it would modify its One-Time Inspection Program (discussed in SER Section 3.0.3.1.6) to volumetrically inspect 10 percent of the socket weld population for each unit. Although its ISI program does not use risk-informed methodology, it plans to use its risk-informed methodology to select the most susceptible welds from its population. The applicant will provide an assessment to show that the number of samples it inspects will be statistically significant and that, with the screening methodology, the inspection will provide reasonable assurance that if aging (cracking) of socket welds exists, the program will detect it. The staff finds the sample selection methodology acceptable, as it is consistent with the GALL Report One-Time Inspection of ASME Code Class 1 Small Bore Piping, which states that the sample should be based on "susceptibility, inspectability, dose considerations, operating experience, and limiting locations of the total population of ASME Code Class 1 small-bore piping locations."

In its response dated July 30, 2010, and supplemented by letter dated December 3, 2010, the applicant revised its AMP to volumetrically inspect at least 10 percent of Class 1 socket welds per unit with a maximum of 25 welds. The applicant further stated that the weld sample selection will be based on risk insights and susceptibility to aging degradation. The applicant stated that the Class 1 socket weld population is approximately 300 to 400 per unit and

Aging Management Review Results

approximately 1000 for the entire station. The staff noted that the sampling of at least 10 percent with 25 welds maximum in each unit, results in examining of 75 welds total. In addition, the applicant stated that weld selection will be based on "risk insight" and "the potential for aging degradation."

The staff finds that the number of welds to be inspected and the weld selection methodology are consistent with GALL Report AMP XI.M35, "One-Time Inspections of ASME Code Class 1 Small-Bore Piping." This AMP recommends that, "[t]his number, or sample size, is based on susceptibility, inspectability, dose considerations, operating experience, and limiting locations of the total population of ASME Code Class 1 small-bore piping locations." Since the number of welds to be inspected and the selection methodology, which will include the most risk significant and most susceptible welds, is consistent with the GALL Report recommendations, the staff finds that aging management of Class 1 socket welds is adequately addressed, and Confirmatory Item 3.0.3.2.13-1 is closed.

By letter dated December 3, 2010, the applicant also provided information regarding its inspection schedule. The applicant stated that the One-Time Inspection Program will be implemented within the 6 years prior to the period of extended operation. The staff finds the applicant's proposed inspection schedule is consistent with the recommendations of the GALL Report regarding timely implementation of the small bore piping inspections and is, therefore, acceptable.

Based on its review of the applicant's responses to RAIs B2.1.19-1 and B2.1.19-2, the staff finds that elements one through six of the applicant's One-Time Inspections of ASME Code Class 1 Small Bore Piping Program, with an acceptable exception, are consistent with the corresponding program elements of the GALL Report AMP and are therefore, acceptable.

Operating Experience. LRA Section B2.1.19 summarizes operating experience related to the one-time inspections of the ASME Code Class 1 Small Bore Piping program. The LRA states that it has experienced cracking of stainless steel ASME Code Class 1 piping less than or equal to a nominal pipe size of 4 inches. Furthermore, a hairline weld failure was caused by cyclic fatigue due to vibration combined with inadequate support on a shutdown cooling suction line. The applicant performed piping modifications that have reduced the excessive vibration. The applicant's review of the second 10-year ISI Interval Summary Reports for Units 1, 2, and 3 indicate there were no code repairs or code replacements required for continued service of ASME IWB Code components during the second 10-year ISI interval.

The staff reviewed operating experience information, in the application and during the audit, to determine if the applicant reviewed the applicable aging effects and industry and plant-specific operating experience. As discussed in the Audit Report, the staff conducted an independent search of the plant operating experience information to determine if the applicant had adequately incorporated and evaluated operating experience related to this program.

The staff's independent search of plant-specific operating history found three instances of cracking of Class 1 small-bore piping. The staff noted that LER 528-1987-018, LER 528-1996-006, and LER 528-2004-001 document the relevant cases of cracking. The staff noted that the applicant needs to evaluate all of the cracking events and determine if it needs a plant-specific AMP for periodic inspections of ASME Code Class 1 small-bore piping, as recommended in the GALL Report AMP. By letter dated December 29, 2009, the staff issued RAI B2.1.19-3 asking the applicant to provide plans to manage the aging of small-bore piping.

In its response, dated February 19, 2010, the applicant stated that it has experienced three instances where failures have occurred in ASME Code Class 1 small-bore piping with socket welds; the failures were reported in Licensee Event Reports LER 528-1987-018, LER-528-1996-006, and LER 528-2004-001. The applicant further stated that it conducted evaluations to determine the cause of each of the failures. In each case, the applicant

Aging Management Review Results

determined that the failure was a design-related failure that resulted from improper support of the components. The applicant stated that since cracking due to stress corrosion or thermal and mechanical loading was determined not to be the cause, a plant-specific AMP is not necessary.

Based on its review, the staff finds the applicant's response to RAI B2.1.19-3 unacceptable because the information from the operating experience review indicated that all three failures were due to high-cycle fatigue. The staff noted that high-cycle fatigue is a form of mechanical loading. The applicant's response did not include an adequate basis to conclude that a plant-specific AMP is not necessary.

The GALL Report One-Time Inspection of ASME Code Class 1 Small-Bore Piping AMP recommends the use of the One-Time Inspection of ASME Code Class 1 Small-Bore Piping only for those plants that have not experienced cracking of ASME Code Class 1 small-bore piping resulting from stress corrosion or thermal and mechanical loading. It further states that for those plants that have experienced cracking, it recommends periodic inspection of the subject piping, managed by a plant-specific AMP.

In a conference call, dated July 21, 2010, the applicant clarified that one of the socket weld failures reported was out of the scope of license renewal since it was determined to be an ASME Code Class 2 component. The applicant stated that it had experienced two failures in ASME Code Class 1 small-bore welds due to the same design deficiency, therefore, it will revise its operating experience discussion and withdraw LER 528-1996-006. The applicant stated that it corrected the design deficiency and has not experienced failures for an extended period of time.

The staff agrees that the applicant does not have to count failures of non-class 1 components in the context of program applicability. The staff also agrees that the One-Time Inspection AMP still applies because the applicant implemented design changes to mitigate the causal factors. As part of the corrective action process, it also evaluated similar systems and components. The staff agrees that the applicant has not experienced failures for an extended period of time; however, the staff emphasized that the applicant must justify its proposed sample size of the One-Time Inspection Program as statistically significant. The applicant agreed to modify its commitment associated with this program to incorporate the expanded sample size. These actions are consistent with the GALL Report One-Time Inspection of ASME Code Class 1 Small-Bore Piping AMP.

The staff finds that the operating experience related to the applicant's program demonstrates that it can adequately manage the detrimental effects of aging on SSCs within the scope of the program and that implementation of the program has resulted in the applicant taking corrective actions. The staff confirmed that the "operating experience" program element satisfies the criterion in SRP-LR Section A.1.2.3.10 and is, therefore, acceptable.

UFSAR Supplement. LRA Section A1.19 provides the UFSAR supplement for the One-Time Inspections of ASME Code Class 1 Small-Bore Piping Program. The staff reviewed this UFSAR supplement description of the program against the recommended description for this type of program, as described in SRP-LR Table 3.1-2. The staff notes that the applicant committed (Commitment No. 21) to implement the One-Time Inspection of ASME Code Class 1 Small-Bore Piping Program prior to entering the period of extended operation to manage the aging of applicable components. The staff determines that the information in the UFSAR supplement is an adequate summary description of the program, as required by 10 CFR 54.21(d).

Conclusion. On the basis of its audit and review of the applicant's One-Time Inspection of ASME Code Class 1 Small-Bore Piping Program, the staff determines that those program elements for which the applicant claimed consistency with the GALL Report AMP are consistent. In addition, the staff reviewed the exception and justification, and determines that the AMP, with the exception, is adequate to manage the aging effects for which the LRA credits it. The staff concludes that the applicant has demonstrated that the effects of aging will be adequately managed so that the intended functions will be maintained consistent with the CLB for the period of extended operation, as required by 10 CFR 54.21(a)(3). The staff also reviewed the UFSAR supplement for this AMP, and concludes that it provides an adequate summary description of the program, as required by 10 CFR 54.21(d).

3.0.3.2.14 External Surfaces Monitoring Program

Summary of Technical Information in the Application. LRA Section B2.1.20 describes the new External Surfaces Monitoring Program as consistent, with an exception, with the GALL Report AMP XI.M36, "External Surfaces Monitoring." The applicant stated that the program manages loss of material on the external surfaces of steel, aluminum, and copper alloy components, and hardening and loss of strength of elastomeric materials. In addition, the applicant stated that the program includes the use of visual inspection and physical manipulation.

Staff Evaluation. During its audit, the staff reviewed the applicant's claim of consistency with the GALL Report. The staff also reviewed and confirmed that the plant's conditions are bounded by the conditions for which the GALL Report was evaluated. The staff compared elements one through six of the applicant's program to the corresponding elements of the GALL Report External Surfaces Monitoring AMP. As discussed in the Audit Report, the staff confirmed that each element of the applicant's program is consistent with the corresponding element of the GALL Report AMP, with the exception of the "parameters monitored or inspected" and "detection of aging effects" program elements. For these elements, the staff determined the need for additional clarification, which resulted in the issuance of RAIs. The staff's RAIs and the applicants responses are discussed under evaluation of the exception below.

The staff also reviewed the portions of the "scope of program," "preventive actions," "detection of aging effects," "monitoring and trending," and "acceptance criteria" program elements associated with the exception to determine whether the program will be adequate to manage the aging effects for which it is credited. The staff's evaluation of this exception follows.

Exception. LRA Section B2.1.20 identifies an exception to the "scope of program," "preventive actions," "detection of aging effects," "monitoring and trending," and "acceptance criteria" program elements. The staff also considers this exception to affect the "parameters monitored or inspected" program element. The exception concerns the inclusion of aluminum, copper alloy, and elastomeric materials in the applicant's program, whereas the GALL Report AMP only manages aging effects associated with steel components. The applicant's External Surfaces Monitoring Program includes a provision to conduct physical manipulations as part of the inspection of elastomeric materials to address the aging effect of loss of ductility. The applicant intends to apply the visual inspection methods recommended in the GALL Report AMP to inspect aluminum and copper alloys.

The GALL Report External Surfaces Monitoring AMP recommends visual inspection of component surfaces at least once per refueling cycle. The Gall Report AMP recommends that the applicant inspect surfaces that are inaccessible or not readily visible during plant operations and refueling outages at intervals that would ensure that the component intended function is maintained. In addition to visual inspections, the applicant's program includes physical manipulation to detect elastomeric material aging effects, such as hardening and surface texture changes. However, under the applicant's AMP, not all inaccessible elastomeric components are physically manipulated to detect aging effects.

By letter dated December 29, 2009, the staff issued RAI B2.1.20-1 requesting that the applicant provide details on its use of alternative methods to detect the effects of aging on those elastomeric components not accessible for physical manipulation. If the aging management method involves sampling or inspection of an equivalent or analogous component, material, and

Aging Management Review Results

environment combination, such that aging specific to these artifacts will be detected, the staff requested the applicant to provide the alternative methods.

In its response, dated February 19, 2010, the applicant stated that visual inspections are the primary method for detecting degradation of elastomeric components. The applicant specified that these visual inspections include observation of discoloration, checkering, and cracking, which are visually detectable age-related degradation effects for elastomers. The applicant also stated that physical manipulation can verify the absence of hardening or loss of strength for elastomers. The applicant further stated that it identified no specific elastomeric components expected to be inaccessible to visual inspection.

However, the staff notes that visual examination of the external surfaces of an elastomer may not always reveal the need for physical manipulation because hardening does not always display checkering or cracking. The staff finds the applicant's response to RAI B2.1.20-1 unacceptable because visual examination alone is not an appropriate method to determine whether an elastomer is experiencing hardening or loss of strength. In a letter dated June 21, 2010, the applicant supplemented its response to RAI B2.1.20-1, stating that it will conduct physical manipulation on all accessible elastomeric material. The staff finds this response acceptable because the combination of visual examination and physical manipulation of elastomers has proven to be effective in detecting degradation in elastomers. The staff's concern in RAI B2.1.20-1 is resolved.

The GALL Report External Surfaces Monitoring AMP consists of periodic visual inspections of steel components to manage loss of material on external surfaces. The applicant's program is also credited with managing aging of aluminum components. Compared to steel, aluminum is not as conducive to visual inspection to detect degradation because under the pertinent plant environments and conditions, the degradation of aluminum involves the formation of a thin aluminum oxide film, which is indistinguishable from its initial service condition. The extent of corrosion of aluminum is not detectable by the visual inspection methods used for steel until the degradation is so extensive that component functionality is compromised.

By letter dated December 29, 2009, the staff issued RAI B2.1.20-2 requesting that the applicant explain how visual inspections can assess loss of material on aluminum components. The staff also requested if the applicant intends to use other contact methods or optical instruments in the inspection process.

In its response, dated February 19, 2010, the applicant stated that it does not anticipate rapid and aggressive corrosion because the aluminum components are exposed to a mild environment. The applicant further stated that visual inspection of aluminum components includes the specific, visually observable corrosion effects of chipping, cracking, flaking, oxidizing, or missing paint and coatings and that presence of these corrosion effects would indicate degradation deficiencies before the loss of component intended function.

The staff finds the applicant's response acceptable because the program specifically includes the observation of aging defects on painted surfaces or protective coatings, and pitting corrosion is a visually detectable aging effect on non-painted or non-coated aluminum components. Inspection of paint or coatings on external surfaces provides an adequate method to manage aging effects of aluminum components because this method does not rely solely on detection of oxidation and general corrosion products on bare aluminum metal surfaces. The staff's concern described in RAI B2.1.20-2 is resolved.

Based on its audit, review of the application, and review of the applicant's responses to RAIs B2.1.20-1 and B2.1.20-2, the staff finds that elements one through six of the applicant's External Surfaces Monitoring Program, and an acceptable exception, are consistent with the corresponding program elements of the GALL Report External Surfaces Monitoring AMP and are, therefore, acceptable.

Aging Management Review Results

Operating Experience. LRA Section B2.1.20 summarizes operating experience related to the External Surfaces Monitoring Program. The applicant stated that, while this is a new program, existing system inspections via walkdowns have been effective in maintaining the material condition of plant systems. The applicant also stated that these inspections are consistent with industry practice. The applicant further stated that it initiates corrective actions for any deficiencies or adverse trends identified in the walkdowns.

The staff reviewed operating experience information, in the application and during the audit, to determine if the applicant reviewed the applicable aging effects and industry and plant-specific operating experience. As discussed in the Audit Report, the staff conducted an independent search of the plant operating experience information to determine if the applicant had adequately incorporated and evaluated operating experience relevant to this program. The staff found that the applicant had adequately identified and incorporated industry and applicable plant-specific operating experience.

Based on its audit and review of the application, the staff finds that operating experience related to the applicant's program demonstrates that it can adequately manage the detrimental effects of aging on SSCs within the scope of the program. The staff confirmed that the "operating experience" program element satisfies the criterion in SRP-LR Section A.1.2.3.10 and is, therefore, acceptable.

UFSAR Supplement. LRA Section A1.20 provides the UFSAR supplement for the External Surfaces Monitoring Program. The staff reviewed this UFSAR supplement description of the program and notes that it conforms to the recommended description for this type of program, as described in SRP-LR Tables 3.2-2, 3.3-2, and 3.4-2. The staff also notes that the applicant committed (Commitment No. 22) to implement the new External Surfaces Monitoring Program prior to entering the period of extended operation to manage the aging of applicable components. The staff determines that the information in the UFSAR supplement is an adequate summary description of the program, as required by 10 CFR 54.21(d).

Conclusion. On the basis of its audit and review of the applicant's External Surfaces Monitoring Program, the staff determines that those program elements for which the applicant claimed consistency with the GALL Report are, indeed, consistent. In addition, the staff reviewed the applicant's exception and its justification, and determines that the AMP, with the exception, is adequate to manage the aging effects for which the LRA credits it. The staff concludes that the applicant has demonstrated that the effects of aging will be adequately managed so that the intended functions will be maintained consistent with the CLB for the period of extended operation, as required by 10 CFR 54.21(a)(3). The staff also reviewed the UFSAR supplement for this AMP, and concludes that it provides an adequate summary description of the program, as required by 10 CFR 54.21(d).

3.0.3.2.15 Inspection of Internal Surfaces in Miscellaneous Piping and Ducting Components Program

Summary of Technical Information in the Application. LRA Section B2.1.22 describes the new Inspection of Internal Surfaces in Miscellaneous Piping and Ducting Components Program as consistent, with exceptions with the GALL Report AMP XI.M38, "Inspection of Internal Surfaces in Miscellaneous Piping and Ducting Components." The applicant stated that this program manages cracking and loss of material of internal surfaces of piping, piping components, and ducts made of steel, aluminum, and copper alloys. In addition, the program manages the hardening and loss of strength of elastomers. The applicant also stated that it will conduct visual inspections to identify conditions that could result in the loss of intended functions in SSCs. Inspections include surface or volumetric examinations for cracking of stainless steel exposed to diesel exhaust and external or internal physical manipulation of elastomers to assess their hardening and loss of strength. The applicant further stated that the program

Aging Management Review Results

pertains to the aging management of the above-listed hardware not supported by other programs, and this AMP will use a site-specific work control process to conduct and document the recommended inspections. The applicant also stated that within 10 years of the start of the period of extended operation, it will review the process to ensure that there are enough inspections of components covered by this AMP to establish the adequacy of the inspections. If it conducts an insufficient number of inspections, additional inspections will be carried out to assure completeness before entering the period of extended operation.

Staff Evaluation. During its audit, the staff reviewed the applicant's claim of consistency with the GALL Report. The staff also reviewed and confirmed that the plant's conditions are bounded by the conditions for which the GALL Report was evaluated.

The staff compared elements one through six of the applicant's program to the corresponding elements of the GALL Report Inspection of Internal Surfaces in Miscellaneous Piping and Ducting Components AMP. As discussed in the Audit Report, the staff confirmed that each element of the applicant's program is consistent with the corresponding element of the GALL Report AMP, with the exception of the "detection of aging effects" and the "acceptance criteria" program elements related to fire protection piping. For these elements, the staff determined the need for additional clarification, which resulted in the issuance of an RAI. By letter dated December 29, 2009, the staff issued RAI B2.1.22-1 requesting that the applicant provide clarify the "detection of aging effects" and "acceptance criteria" program elements for thinning and narrowing of pipe walls.

In its response by letter dated February 19, 2010, the applicant stated that for the "detection of aging effects" program element it will use visual inspections to assess the condition of piping. If it witnesses wall thinning, by evidence of pitting, erosion, or other corrosion mechanisms, then it will conduct ultrasonic inspections to verify the adequacy of pipes for continued operation. For the "acceptance criteria" program element, the applicant stated that it will follow the ASME or American National Standards Institute (ANSI) specifications or design calculations for minimum wall thickness. The applicant further stated that it clarified the "acceptance criteria" program element in the basis document to reflect the method used to quantify and track the thinning of pipes. The staff reviewed the applicant's response to RAI B2.1.22-1 and finds it acceptable because visual examinations of pipes, followed by UT, are suitable to characterize pipe wall thickness with its acceptance based on industry standards or engineering calculations.

The staff also reviewed the portions of the "scope of program," "parameters monitored or inspected," "detection of aging effects," and "monitoring and trending" program elements associated with exceptions to determine if the program will be adequate to manage the aging effects for which it is credited.

Exception 1. LRA Section B.2.1.22 identifies an exception to the "scope of program," "parameters monitored or inspected," "detection of aging effects," and "monitoring and trending" program elements. The applicant recognized that the inclusion of stainless steel, aluminum, copper alloys, and elastomers in the "scope of program," beyond steel, constitutes an exception to the GALL Report AMP. To detect aging effects and degradation of materials, the applicant stated that it will continue to use visual examination techniques to assure there is no loss of intended functions in piping, piping components, ducts, and other components. The staff noted that the GALL Report allows for visual examinations and observations for loss of material, provided that the inspected material is steel. The staff also noted that aluminum, stainless steel, and copper alloys typically do not have painted surfaces, which indicate the severity of aging on the material surface. The staff further noted that program element "detection of aging effects," of the GALL Report, calls for the applicant to identify and justify the inspection technique used for detecting the aging effects of concern.

Aging Management Review Results

Upon further review of the LRA, amendments, and RAI responses, the staff found that the applicant will use the Inspection of Internal Surfaces in Miscellaneous Piping and Ducting Components Program to manage aging of aluminum and copper alloy materials. In RAIs B2.1.10-2 and B2.1.20-2, dated December 29, 2009, the staff asked the applicant to provide input on the detection, monitoring, and trending of aging effects on internal surfaces of relevant hardware made of aluminum. By letter dated February 19, 2010, the applicant responded noting that the resistance to corrosion of aluminum and iron-based and copper-based alloys depends on their environment. In most environments, aluminum will exhibit excellent corrosion resistance due to its regenerating surface oxide film. For iron-based and copper-based alloys exposed to the closed-cycle cooling water chemistry, the applicant will use corrosion inhibitors to manage loss of material. For aluminum, in the closed-cycle cooling water chemistry, the applicant does not anticipate rapid and aggressive loss of material and pitting because the water chemistry moderates the chloride and total halide ion concentrations. When visually inspecting components made of iron-based and copper-based alloys, the applicant will look for pitting, discoloration, surface irregularities, cracking, and other signs of distress. If visual internal inspections indicate age-related degradation in excess of the established acceptance criteria, then the applicant will remedy any deficiencies via its corrective action program. The staff agrees with the applicant's response. The applicant addresses the staff's concern in the responses to RAIs B2.1.10-2 and B2.1.20-2, and it is resolved.

In programs where the Inspection of Internal Surfaces in Miscellaneous Piping and Ducting Components AMP serves in a supporting role, as in the Closed-Cycle Cooling Water System Program, the applicant stated that it manages corrosion via inhibitors.

With the information provided by the applicant, the staff finds this program exception acceptable because the applicant will accomplish detection, monitoring, and trending of aging effects in stainless steel, aluminum, and copper alloys through an established work control process. The applicant will remedy deficiencies per its corrective action program and, within 10 years before extended operation, it will review all systems within the scope of the program for aging effects to provide reasonable assurance that their intended functions will remain through the next inspection cycle. Exception 2 below provides the staff's evaluation of the exception specific to elastomers.

Exception 2. LRA Section B.2.1.22 states a second exception affecting the "scope of program," "parameters monitored or inspected," "detection of aging effects," and "monitoring and trending" program elements. The applicant stated this exception includes visual and physical inspections of elastomers to assess hardening and loss of strength. The staff acknowledged the inclusion of elastomers to be an exception to the "scope of program" element and the applicant's intent to track their hardening and loss of strength to be an exception to the "parameters monitored or inspected" element. The staff also noted that the applicant identified and justified the inspection technique as recommended by the GALL Report "detection of aging effects" program element, yet, the technique is in addition to the visual examination recommended by the GALL Report and affects the "monitoring and trending" program element. By letters dated December 29, 2009, and February 19, 2010, the staff issued RAIs B2.1.20-1 and 3.3.2.2.5-1, respectively, which requested the applicant to clarify any conditions limiting the testing and inspections of elastomers (i.e., inaccessible locations, thermal aging, etc.). By letters dated February 19, 2010, and March 24, 2010, the applicant responded to the RAIs, stating that there are no inaccessible elastomers for inspection and that it will inspect all elastomeric flexible components based on plant procedures irrespective of the prevailing temperatures.

Upon review of the LRA and the applicant's responses to the RAIs, the staff finds this program exception acceptable because the applicant has defined appropriate inspection techniques (physical manipulations and supplemental tests if necessary) to assure the elastomers are properly monitored and inspected and that their pliability and ductility remains through the next inspection cycle. In addition, the applicant will remedy deficiencies per its corrective action program and, within 10 years of entering the period of extended operation, it will check elastomers for aging effects to provide reasonable assurance that their intended functions will remain through the next inspection cycle.

Exception 3. LRA Section B.2.1.22 states a third exception to the "scope of program," "parameters monitored or inspected," "detection of aging effects," and "monitoring and trending" program elements. In this exception the applicant stated that for stainless steel components exposed to diesel exhaust, it will use a volumetric evaluation to detect SCC of applicable internal surfaces. The staff noted that NUREG 1833, "Technical Bases for Revision to the License Renewal Guidance Documents," recommends a plant-specific AMP to evaluate whether the hot diesel exhaust gases will negatively affect the internal surfaces of the diesel engine exhaust piping and piping components. The staff evaluated and accepted this exception because the applicant uses Subsection IWA-2000 of the ASME Code Section XI for volumetric and visual examination techniques as valid inspection methods for the detection of cracking in metallic components. In addition, the applicant will remedy deficiencies per its corrective action program and will check the internal surfaces of the diesel exhaust piping and its components for aging effects within 10 years before entering the period of extended operation.

Based on its audit and review of the applicant's response to RAIs, the staff finds that elements one through six of the applicant's Inspection of Internal Surfaces in Miscellaneous Piping and Ducting Components Program, with acceptable exceptions, are consistent with the corresponding program elements of the GALL Report Inspection of Internal Surfaces in Miscellaneous Piping and Ducting Components AMP and are, therefore, acceptable.

Operating Experience. LRA Section B2.1.22 summarizes operating experience related to the Internal Surfaces in Miscellaneous Piping and Ducting Components Program. The applicant stated that this is a new program; therefore, there is no programmatic operating experience. In discussions between the applicant and the staff, the staff concluded that the operating experience identified by the staff's independent database search, supplemented by the applicant during the audit, is bounded by industry operating experience (i.e. no previously unknown aging effects were identified by the applicant or the staff).

The staff, however, noted that any operating experience resulting from maintenance and inspection activities of systems and components that is supportive to this AMP, should be included. By letter dated December 29, 2009, the staff issued RAI B2.1.16-1 requesting the applicant to provide a summary of operating experience resulting from observations for loss of material, cracking, and loss of heat transfer related maintenance, inspections, and associated corrective action activities. In its response, dated February 19, 2010, the applicant stated that the plant-specific operating experience compiled from 1996–October 2009 resulted in 16,000 aging effects documents. These instances, however, were more relevant to other programs such as the Lubricating Oil Analysis Programs (i.e., the applicant exhibited the capacity to effectively identify the existence of water in the lubricating oil which could result in a loss of material of internal surfaces of such environments).

Based on its audit, review of the application, and response to the RAI, the staff finds that operating experience related to the applicant's program demonstrates that it can adequately inspect and assess detrimental conditions affecting the aging of SSCs within the scope of the program. The staff confirmed that the "operating experience" program element satisfies the criterion in SRP-LR Section A.1.2.3.10 and is, therefore, acceptable.

UFSAR Supplement. LRA Section A1.22 provides the UFSAR supplement for the Internal Surfaces in Miscellaneous Piping and Ducting Components Program. The staff reviewed this UFSAR supplement description of the program against the recommended description for this type of program, as described in SRP-LR Tables 3.2-2, 3.3-2, and 3.4-2. The staff also notes

Aging Management Review Results

that the applicant committed (Commitment No. 24) to implement the new Inspection of Internal Surfaces in Miscellaneous Piping and Ducting Components Program prior to entering the period of extended operation to manage the aging of applicable components. The staff determines that the information in the UFSAR supplement is an adequate summary description of the program, as required by 10 CFR 54.21(d).

Conclusion. On the basis of its audit and review of the applicant's Inspection of Internal Surfaces in Miscellaneous Piping and Ducting Components Program, the staff determines that those program elements for which the applicant claimed consistency with the GALL Report are consistent. In addition, the staff reviewed the exceptions and justifications, and determines that the AMP, with the exceptions, is adequate to manage the aging effects for which the LRA credits it. The staff concludes that the applicant has demonstrated that the effects of aging will be adequately managed so that the intended functions will be maintained consistent with the CLB for the period of extended operation, as required by 10 CFR 54.21(a)(3). The staff also reviewed the UFSAR supplement for this AMP, and concludes that it provides an adequate summary description of the program, as required by 10 CFR 54.21(d).

3.0.3.2.16 Lubricating Oil Analysis Program

Summary of Technical Information in the Application. LRA Section B2.1.23 describes the existing Lubricating Oil Analysis Program as consistent, with exceptions, with the GALL Report AMP XI.M39, "Lubricating Oil Analysis." The applicant stated that the Lubricating Oil Analysis Program manages loss of material and reduction of heat transfer for components with surfaces exposed to lubricating and hydraulic oils. The applicant also stated that the program ensures that it maintains lubricating and hydraulic oil environments in mechanical systems to the required quality based on vendor and industry guidelines. The program monitors and controls oil contaminants, primarily water and particulates within acceptable limits, thereby preserving an environment that is not conducive to aging effects and that selected components' aging will be tracked so that corrective actions can be taken before a loss of intended function occurs. The program is implemented through plant procedures that address sampling, testing, and management of samples and schedules. Its effectiveness is verified through the One-Time Inspection Program (see SER Section 3.0.3.2.13).

Staff Evaluation. During its audit, the staff reviewed the applicant's claim of consistency with the GALL Report. The staff also reviewed and confirmed that the plant's conditions are bounded by the conditions for which the GALL Report was evaluated. The staff compared elements one through six of the applicant's program to the corresponding elements of the GALL Report Lubricating Oil Analysis AMP. As discussed in the Audit Report, the staff confirmed that each element of the applicant's program is consistent with the corresponding element of the GALL Report AMP, with the exception of the "parameters monitored or inspected" and "acceptance criteria" program elements. For these elements, the staff determined the need for additional clarifications, which resulted in the issuance of RAIs.

The "acceptance criteria" program element of the GALL Report Lubricating Oil Analysis AMP recommends particle concentration to be determined in accordance with industry standards. For each component, water and particle concentration should be based on the manufacturer's recommendations or industry standards. Viscosity band tolerances and metal limits, determined by spectral analysis and ferrography, are also based on manufacturer's recommendations, industry standards, or other justified basis. The applicant, however, stated that testing criteria, as described in its basis document, may be exceeded, but this would not necessarily mean the lubricating oil is "non-conforming." The staff noted that the applicant did not substantiate the basis for using lubrication oil with parameters outside the limits of the acceptance criteria. By letter dated December 29, 2009, the staff issued RAI B2.1.23-1 requesting the applicant to provide its sources of the acceptance criteria and its justification for the continued use of lubricating oil outside manufacturer's recommendations or industry standards.

Aging Management Review Results

In its response, dated February 19, 2010, the applicant revised the Lubricating Oil Analysis AMP basis document to reflect the background for program element "acceptance criteria" and clarify the method by which the continued use of lubricating oil outside manufacturer's recommendations or industry standards is allowed, subject to the approval of the lubrication engineer. Justification for the continued use is based on industry guidelines, national standards, suppliers' and manufacturers' specifications, and academic literature.

The staff reviewed the applicant's input and compared the references used by the applicant to those listed in the GALL Report Lubricating Oil Analysis AMP. For example, the industry standard ASTM D 4378—also acceptable by the International Standardization Organization—provides background and guidance for monitoring, sampling, testing, validating, and ensuring lubricants remain within acceptable parameters throughout their life cycle in power generating turbines. ASTM D 6224 provides background and guidance for monitoring mineral oils based on testing, rather than their in-service or calendar time, to save operating and maintenance costs while minimizing machine problems. The staff finds these references applicable and the applicant's response to RAI B2.1.23-1 acceptable because the applicant revised the Lubricating Oil Analysis AMP basis document element "acceptance criteria" to reflect current industry practices while maintaining a firm foundation on fundamentals. The staff finds the applicant's Lubricating Oil Analysis Program element "acceptance criteria" to conform to the GALL Report Lubricating Oil Analysis AMP. The staff's concern described in RAI B2.1.23-1 is resolved.

The staff also reviewed the portions of the "parameters monitored or inspected" and "acceptance criteria" program elements associated with exceptions to determine whether the program will be adequate to manage the aging effects for which it is credited. The staff's evaluation of these exceptions follows.

Exception 1. LRA Section B2.1.23 identifies an exception to the "parameters monitored or inspected" and "acceptance criteria" program elements of the GALL Report. For components with periodic oil changes, the GALL Report recommends oil to be sampled for particle count and water contamination. These can indicate wear and corrosion in systems and components, which limit their intended functions. The LRA Lubricating Oil Analysis Program and its basis document state that diesel engine oils are evaluated based on elemental analysis techniques as described in ASTM D 6595, "Determination of Wear Metals and Contaminants in Used Lubricating Oils or Used Hydraulic Fluids by Rotating Disc Electrode Atomic Emissions Spectroscopy," instead of particle count and water checks. These examination techniques are more thorough than those found in the GALL Report. The staff reviewed the applicant's claim and noted that, even though emissions spectroscopy could determine the presence of wear metals and contaminants accurately, neither ASTM D 6595 nor the applicant's basis document lists any acceptance criteria for the elemental analysis. By a letter dated December 29, 2009, the staff issued RAI B2.1.23-3 requesting the applicant to provide the acceptance criteria for elemental wear metals in lubricating oil and further information indicating that the elemental analysis gives a greater degree of insight into a lubricant's condition over the particle counting technique.

In its response, dated February 19, 2010, the applicant stated that ASTM D6595 does not have rigid acceptance criteria. Instead, the applicant uses it to define precision and accuracy in the collected data, determining trends in machine acceptability. The applicant conducts additional testing for data outliers and deals with operability issues through its corrective action program. The applicant also stated that National Academy of Sciences 1638, "Cleanliness Requirements of Parts Used in Hydraulic Systems," listed in the GALL Report, advocates particle counting but does not discriminate between contaminants; whereas, the spectrometer testing and elemental analysis distinguishes between oil additives for performance enhancement versus wear metals. The staff finds that acceptance of lubricating oil based on trend analysis, with data collected through elemental analysis and then further trended, acceptable because of the applicant's

Aging Management Review Results

close monitoring of oil and machine conditions. These techniques enhance its ability to take immediate corrective actions, thus ensuring SSCs will maintain their intended functions. The staff's concern described in RAI B2.1.23-3 is resolved. Therefore, this exception to the GALL Report Lubricating Oil Analysis AMP is acceptable.

Exception 2. LRA Section B2.1.23 identifies an exception to the "parameters monitored or inspected" and "acceptance criteria" program elements of the GALL Report Lubricating Oil Analysis Program. In the GALL Report AMP, program element "parameters monitored or inspected," recommends that for components that do not have regular oil changes, the flash point is used to verify that the oil is suitable for continued use. The applicant stated that its Lubricating Oil Analysis Program considers flash point as an indicator of fuel oil contamination of the lubrication oil; therefore, only lubricating oil in components with a potential of contamination with fuel oil are subject to flash point testing. The staff noted that it is not necessary to monitor flash point for non-diesel applications because the potential for fuel oil contamination is minimal. Therefore, this exception to the GALL Report Lubricating Oil Analysis AMP is acceptable.

Exception 3. LRA Section B2.1.23 identifies an exception to the "parameters monitored or inspected" and "acceptance criteria" program elements. In the GALL Report AMP, program element 3, "parameters monitored or inspected," recommends that for components that do not have regular oil changes, the neutralization number is determined to verify the oil suitability for continued use. The applicant stated that its Lubricating Oil Analysis Program tests diesel engine lubrication oils using the "total base number" parameter. The applicant does not use the "total acid number" parameter for evaluations of lubricants because of its limited utility in engine applications. It was not clear to the staff why the applicant uses only the "total base number" to monitor the lubricating oil of diesel engines and what lubricating oils of other components will be monitored with the "total acid number." By letter dated December 29, 2009, the staff issued RAI B2.1.23-2 requesting that the applicant provide a justification for monitoring only the "total base number" parameter for lubricating oil in diesel engines and also to provide information as to where the "neutralization number," the "total acid number" and "total base number" is used for monitoring lubricating oil in other components.

In its response, dated February 19, 2010, the applicant stated that it uses "neutralization" to identify the amount of titration injected to neutralize an oil sample. For example, the "total base number" indicates the amount of an acid added to the sample to neutralize its base, while the "total acid number" indicates the amount of base materials added to do the same for acidic oils. The applicant also stated that diesel oils by design are basic, but turn acidic during the combustion process. Base number testing is, therefore, admissible on diesel oils. The applicant also stated that its other lubricating oils do not have the base additive as part of their formulation. Testing of these oils for "total acid number" is, therefore, appropriate because they become increasingly acidic as they degrade. The staff finds this exception to the GALL Report Lubricating Oil Analysis AMP, element "parameters monitored or inspected," acceptable because the applicant applies the appropriate neutralization number depending on the lubricating oil used in specific oil systems. The staff's concern described in RAI B2.1.23-2 is resolved. Therefore, this exception to the GALL Report Lubricating Oil Analysis AMP is acceptable.

Operating Experience. LRA Section B2.1.23 summarizes operating experience related to the Lubricating Oil Analysis Program. The applicant stated that the site specific operating experience revealed no patterns or events involving loss of the intended function of any systems or components as a result of aging effects related to lubricating oil contamination or degradation.

The staff reviewed operating experience information, in the application and during the audit, to determine if the applicant reviewed the applicable aging effects and industry and plant-specific

Aging Management Review Results

operating experience. As discussed in the Audit Report, the staff conducted an independent search of the plant operating experience information to determine if the applicant had adequately incorporated and evaluated operating experience related to this program. The staff found that the applicant had adequately identified and incorporated industry and applicable plant-specific operating experience.

Based on its audit and review of the application, the staff finds that operating experience related to the applicant's program demonstrates that it can adequately manage the detrimental effects of aging on SSCs within the scope of the program. The staff confirmed that the "operating experience" program element satisfies the criterion in SRP-LR Section A.1.2.3.10 and is, therefore, acceptable.

UFSAR Supplement. LRA Section A1.23 provides the UFSAR supplement for the Lubricating Oil Analysis Program. The staff reviewed this UFSAR supplement description of the program against the recommended description for this type of program, as described in SRP-LR Tables 3.2-2, 3.3-2, and 3.4-2. The staff also notes that the applicant committed (Commitment No. 25) to the ongoing implementation of the existing Lubricating Oil Analysis Program to manage the aging of applicable components during the period of extended operation. The staff determines that the information in the UFSAR supplement is an adequate summary description of the program, as required by 10 CFR 54.21(d).

Conclusion. On the basis of its audit and review of the applicant's Lubricating Oil Analysis Program, including responses to RAIs, the staff determines that those program elements for which the applicant claimed consistency with the GALL Report are consistent. In addition, the staff reviewed the exceptions and justifications, and determines that the AMP, with the exceptions, is adequate to manage the aging effects for which the LRA credits it. The staff concludes that the applicant has demonstrated that the effects of aging will be adequately managed so that the intended functions will be maintained consistent with the CLB for the period of extended operation, as required by 10 CFR 54.21(a)(3). The staff also reviewed the UFSAR supplement for this AMP, and concludes that it provides an adequate summary description of the program, as required by 10 CFR 54.21(d).

3.0.3.2.17 Electrical Cables and Connections Not Subject to 10 CFR 50.49 Environmental Qualification Requirements Used in Instrumentation Circuits Program

Summary of Technical Information in the Application. LRA Section B2.1.25 describes the existing Electrical Cables and Connections Not Subject to 10 CFR 50.49 EQ Requirements Used in Instrumentation Circuits Program as consistent, with enhancements, to the GALL Report AMP XI.E2, "Electrical Cables and Connections Not Subject to 10 CFR 50.49 EQ Requirements Used in Instrumentation Circuits." The applicant stated that the scope of the LRA program includes the cables and connections used in sensitive instrumentation circuits with sensitive, high voltage, low-level signals within the ex-core neutron monitoring system. The applicant also stated that the program manages embrittlement, cracking, melting, discoloration, swelling, or loss of dielectric strength leading to reduced insulation resistance.

Staff Evaluation. During its audit, the staff reviewed the applicant's claim of consistency with the GALL Report AMP. The staff also reviewed the enhancements to determine whether the enhancements will make the program consistent with the one described in the GALL Report. The staff also reviewed and confirmed that the plant's conditions are bounded by the conditions for which the GALL Report was evaluated. The staff compared elements one through six of the applicant's program to the corresponding elements of the GALL Report AMP. As discussed in the Audit Report, the staff confirmed that each element of the applicant's program is consistent with the corresponding element of the GALL Report AMP, with the exception of the radiation monitoring question discussed below. The staff determined the need for additional clarification, which resulted in the issuance of an RAI.

Aging Management Review Results

The GALL Report AMP under the "scope of program" element states that this program applies to electrical cables and connections (cable system) used in circuits with sensitive, high-voltage, low-level signals such as radiation monitoring and nuclear instrumentation that are subject to an AMR. The LRA AMP, under the same program attribute only includes the ex-core neutron monitoring system cable system (nuclear instrumentation), and the staff noted that this does not include radiation monitoring. In a letter dated December 29, 2009, the staff issued RAI B2.1.25-1. This RAI requested the applicant to explain how the scope of the Electrical Cables and Connections Not Subject to 10 CFR 50.49 EQ Requirements Used in Instrumentation Circuits Program is consistent with the corresponding GALL Report AMP, considering the fact that the LRA AMP does not include radiation monitoring cables. In response to the staff's request, in a letter dated February 19, 2010, the applicant stated that all radiation monitors within the scope of license renewal, except for two, are either EQ or are active components with no external high-voltage, low-signal cable. The two non-EQ area radiation monitors, RU-37 and RU-38, had previously been evaluated as low-voltage instrument circuits and should have been evaluated as instrument circuits with sensitive, high-voltage, low-level signals. The applicant revised the scope of its program to include the non-EQ area radiation monitors, RU-37 and RU-38. The applicant also stated that it revised LRA Appendix A, Section A1.25 and Commitment No. 27 in Table A4-1, and Appendix B, Section B2.1.25 to include the non-EQ area radiation monitors within the scope of the program. The applicant further stated that it uses calibration surveillance tests to manage the aging of the cable insulation and connections for non-EQ area radiation monitors within the scope of license renewal.

The staff finds the applicant response acceptable because the applicant included non-EQ area radiation monitors (RU-37 and RU-38) in the scope of the Electrical Cables and Connections Not Subject to 10 CFR 50.49 EQ Requirements Used in Instrumentation Circuits Program. Radiation monitoring cables that are environmentally qualified are subject to 10 CFR 50.49 requirements and are not required to be in scope of the Electrical Cables and Connectors Not Subject to 10 CFR 50.49 EQ Requirements Used in Instrumentation Circuits Program. The cables that are inside active components are parts of an active assembly and are not subject to AMR. The staff's concern in RAI B2.1.25-1 is resolved.

The GALL Report Electrical Cables and Connections Not Subject to 10 CFR 50.49 EQ Requirements Used in Instrumentation Circuits Program states, under "detection of aging effects," that in cases where a calibration or surveillance program does not include the cabling system in the surveillance, the applicant will perform cable system testing. In the LRA AMP, under the same program attribute, the applicant states the ex-core neutron monitoring system is calibrated every 18 months, in accordance with scheduled surveillance and maintenance testing procedures. The GALL Report AMP recommends that cables disconnected during scheduled surveillance are to be tested separately.

In a letter dated December 29, 2009, the staff issued RAI B2.1.25-2 requesting the applicant to explain whether the ex-core neutron monitoring cables are disconnected during the 18-month scheduled surveillance. If they are, the staff requested the applicant to explain why it does not perform cable testing in accordance with the GALL Report AMP. If they are not, the staff requested the applicant to identify plant surveillance procedures that show that these cables are not disconnected.

In a letter dated February 19, 2010, the applicant responded to the staff's request stating that the ex-core neutron monitoring cables are disconnected during the 18-month scheduled surveillance and that it revised the LRA AMP to require testing of the ex-core neutron monitoring cables. The applicant also stated that it revised LRA Appendix A, Section A1.25 and Commitment No. 27 in Table A4-1 and Section B2.1.25 to require testing of the ex-core neutron monitoring cables. The applicant further stated that it conducts cable tests, such as insulation

Aging Management Review Results

resistance testing or other tests, to detect deterioration of the cable insulation system. The applicant will test the cable before the period of extended operation and every 10 years thereafter. Before testing, the applicant will determine acceptance criteria based on the type of cable and type of test performed. The staff finds the applicant's response acceptable because the ex-core neutron monitoring cables are disconnected during the 18-month surveillance and then tested separately. The testing frequency and methods are consistent with those in the GALL Report AMP. The staff's concern in RAI B2.1.25-2 is resolved.

The staff also reviewed the portions of the "scope of program," "detection of aging effects," and "corrective actions" program elements associated with enhancement to determine if the program will be adequate to manage the aging effects for which it is credited. The staff's evaluation of this enhancement follows.

Enhancements. In the LRA Electrical Cables and Connections Not Subject to 10 CFR 50.49 EQ Requirements Used in Instrumentation Circuits Program, the applicant stated that, before the period of extended operation, the following enhancement will be implemented in the following program elements:

> Scope of Program – Element 1, Parameters Monitored or Inspected - Element 3, Detection of Aging Effects - Element 4, Acceptance Criteria - Element 6, and Corrective Actions – Element 7
>
> Procedures will be enhanced to identify license renewal scope, require cable testing of ex-core neutron monitoring cables, require an evaluation of the calibration results for non-EQ area radiation monitors and require that acceptance criteria for cable testing be established based on the type of cable and type of test performed.

The staff finds the enhancements acceptable because the action will be taken before the period of extended operation and will make the applicant's existing program consistent with the recommendations in the GALL Report AMP. Further, the applicant has committed (Commitment No. 27) to implement these actions prior to the period of extended operation.

Based on its audit and review, the staff finds that, with enhancements, elements one through six of the applicant's Electrical Cables and Connections Not Subject to 10 CFR 50.49 EQ Requirements Used in Instrumentation Circuits Program are consistent with the corresponding program elements of the GALL Report AMP and are, therefore, acceptable.

Operating Experience. LRA Section B2.1.25 summarizes operating experience related to the program. The applicant stated that industry operating experience has identified occurrences of cable and connection insulation degradation in high-voltage, low-level instrumentation circuits performing radiation monitoring and nuclear instrumentation functions. The majority of occurrences are related to cable and connection insulation degradation inside of containment near the RV or to a change in an instrument readout associated with a proximate change in temperature inside the containment.

The applicant stated that a review of plant operating experience identified issues with ex-core noise and spiking. The applicant performed a root cause analysis and carried out corrective actions, including system walk-downs and testing, which identified cable and connection characterization. The applicant also stated that it continued coaxial connector replacements, utilization of ferrite beads, and improved grounding—all of which have been effective in improving overall performance. The staff noted that industry operating experience has identified a case where a change in temperature across a high-range radiation monitor cable in containment resulted in a substantial change in the reading of the monitor. The staff further noted that degradation of the circuit cable can cause changes in instrument calibration.

Aging Management Review Results

The staff reviewed the operating experience, in the application and during the audit, to determine if the applicant reviewed the applicable aging effects and industry and plant-specific operating experience. As discussed in the Audit Report, the staff conducted an independent search of the plant operating experience information to determine if the applicant had adequately incorporated and evaluated operating experience related to this program. The staff found that the applicant had adequately identified and incorporated industry and applicable plant-specific operating experience.

Based on its audit and review of the application, the staff finds that operating experience related to the applicant's program demonstrates that it can adequately manage the detrimental effects of aging on SSCs within the scope of the program. The staff confirmed that the "operating experience" program element satisfies the criterion in SRP-LR Section A.1.2.3.10 and is, therefore, acceptable.

UFSAR Supplement. LRA Section A1.25 provides the UFSAR supplement for the Electrical Cables and Connections Not Subject to 10 CFR 50.49 EQ Requirements Used in Instrumentation Circuits Program. The staff reviewed this UFSAR supplement description of the program and notes that it conforms to the recommended description for this type of program, as described in SRP-LR Table 3.6-2. The staff also noted that the applicant committed (Commitment No. 27) to enhance the program prior to entering the period of extended operation to manage the aging of applicable components. The staff determines that the information in the UFSAR supplement, as amended, is an adequate summary description of the program, as required by 10 CFR 54.21(d).

Conclusion. On the basis of its audit, review of the LRA including responses to RAIs, and review of the applicant's Electrical Cables and Connections Not Subject to 10 CFR 50.49 EQ Requirements Used in Instrumentation Circuits Program, the staff finds that, with enhancements, all program elements are consistent with the GALL Report. The staff concludes that the applicant has demonstrated that effects of aging will be adequately managed so that the intended functions will be maintained consistent with the CLB for the period of extended operation as required by 10 CFR 54.21(a)(3). The staff also reviewed the UFSAR supplement for this AMP, and concludes that it provides an adequate summary description of the program, as required by 10 CFR 54.21(d).

3.0.3.2.18 American Society of Mechanical Engineers Section XI, Subsection IWE Program

Summary of Technical Information in the Application. LRA Section B2.1.27 describes the existing ASME Section XI, Subsection IWE Program with exceptions that are consistent with the program elements in GALL Report AMP XI.SI, "ASME Section XI, Subsection IWE." The ASME Section XI, Subsection IWE Program manages loss of material and loss of sealing of the steel liner of the concrete containment building. For the inspection interval from July 18, 2008 to July 17, 2018, for Unit 1; from March 18, 2007 to March 17, 2017, for Unit 2; and from January 11, 2008 to January 10, 2018, for Unit 3; PVNGS performs containment ISIs in accordance with the 2001 Edition of ASME Section XI, Subsection IWE, with the 2002 and 2003 addenda, supplemented with the applicable requirements of 10 CFR 50.55a(b)(2)(ix).

The applicant conducts inspections to identify and manage any containment liner degradation due to loss of material that could result in loss of intended function. Included in this inspection program are the containment liner plate and its integral attachments, such as piping and electrical penetrations, access hatches, the fuel transfer tube, and pressure-retaining bolting. The applicant uses a general visual examination to identify indications of degradation. All areas requiring augmented examination, per criteria IWE-1240 and IWE-2420, receive a detailed visual inspection. Article IWE-3000 specifies acceptance criteria for components subject to IWE exam requirements.

Aging Management Review Results

Staff Evaluation. During its audit, the staff reviewed the applicant's claim of consistency with the GALL Report. The staff also reviewed and confirmed that the plant's conditions are bounded by the conditions for which the GALL Report was evaluated. The staff compared elements one through six of the applicant's program to the corresponding elements of the GALL Report ASME Section XI, Subsection IWE program. As discussed in the Audit Report, the staff confirmed that the program elements "preventive actions," "parameters monitored or inspected," "detection of aging effects," and "monitoring and trending" of the LRA AMP were consistent with the corresponding elements of the GALL Report AMP. Sufficient information was not available to determine if elements "scope of program," and "acceptance criteria" of the LRA AMP were consistent with the corresponding elements of the GALL Report AMP.

In order to obtain the information necessary to verify if the LRA program element "scope of program" is consistent with the corresponding element of the GALL Report AMP, the staff issued RAI B2.1.27-2 in a letter dated December 29, 2009. This RAI requested the applicant to explain why it did not include pressure retaining bolts in the scope of the LRA ASME Section XI, Subsection IWE AMP. In addition, the applicant was requested to explain how it implements the recommendations contained in EPRI NP-5769, EPRI TR-104213 and NUREG-1339 to prevent or mitigate degradation or failure of structural bolts with actual yield strength of 150 kilo-pounds per square inch (ksi) for containment high-strength, pressure-retaining bolts.

In its response to RAI B2.1.27-2, dated February 19, 2010, the applicant stated that the LRA program inspects pressure-retaining, high-strength bolts for aging management as part of the ASME Section XI, Subsection IWE Program. In addition, it has revised the AMP to add "pressure retaining bolting" to the list of in-scope components in the program element "scope of program."

The applicant also stated that there are no containment pressure retaining bolts used that are subject to the recommendations contained in EPRI NP-5769, EPRI TR-104213, and NUREG-1339. The operating experience addressed in these documents concerns bolting material with yield strength above 150 ksi. PVNGS specifications require that bolting subjected to the internal containment design pressure shall conform to ASME SA-320, Grade L43, or ASME SA-325. ASME SA-320, Grade L43 bolting material has a specified minimum yield strength of 105 ksi. ASME SA-325 bolting material has a specified minimum yield strength of 92 ksi or 81 ksi, depending on the bolt size.

The staff finds the applicant's response to B2.1.27-2 acceptable because the applicant has revised the AMP to add pressure retaining bolts to the ASME Section XI, Subsection IWE AMP list of in-scope components. The pressure retaining bolts used have a minimum yield strength of between 81 ksi and 105 ksi, and there is a reasonable assurance that the actual yield strength of these bolts will not exceed 150 ksi. Therefore, additional evaluations for SCC, as recommended in NUREG 1339, EPRI NP-5769, and EPRI TR-104213, are not necessary for the pressure retaining bolts used by the applicant. Based on the preceding discussion, the staff has determined that the "scope of program" element of the LRA ASME Section XI, ISI IWE AMP is consistent with corresponding element of the GALL Report AMP.

The program element "acceptance criteria" of the GALL Report AMP requires that containment steel material loss exceeding 10 percent of the nominal containment wall thickness, or material loss that is projected to exceed 10 percent wall thickness before the next examination, is documented. Such areas are to be accepted by engineering evaluation or corrected by repair or replacement in accordance with ASME Code, Subsection IWE-3122. During the audit, it was not clear to the staff how the applicant addresses this requirement in the AMP, since an applicant inspection report documented local degradation of the containment liner plate with loss of thickness of 0.04 inch. This local loss of thickness of 0.04 inch is more than 10 percent of the measured containment liner plate thickness of 0.263 to 0.27 inch minus the coating.

Aging Management Review Results

Therefore, in RAI B2.1.27-4, the staff requested the applicant to explain the basis for acceptance of local loss of thickness of greater than 10 percent of the nominal wall thickness.

In response to the RAI B2.1.27-4, dated February 19, 2010, the applicant stated that the containment liner plate was examined to evaluate a minor loss of material, apparently caused by the removal of a temporary lug. This slight damage to the containment liner plate resulted from original construction activities. The depth of the gouge was found to exceed 10 percent of the nominal thickness. In support of the engineering evaluation, the surrounding area was examined by UT to determine the actual thickness of the liner, and the responsible engineer evaluated the condition in accordance with the design specification. The condition was deemed acceptable with no corrective action required, and the damage was determined not to be aging-related and would not affect the ability of the liner plate to perform its intended function during the period of extended operation.

The staff finds the response to RAI B2.1.27-4 acceptable because the applicant has performed an engineering evaluation in accordance with ASME Code, Section IWE-3122 to accept a material loss exceeding 10 percent of the nominal containment wall thickness. In addition, since the loss was determined not to be age-related, the applicant did not have to project material loss until the next IWE examination. Therefore, the staff has determined that element "acceptance criteria" of the LRA ASME Section XI, Subsection IWE AMP is consistent with the corresponding element of the GALL Report AMP.

The staff also reviewed the portions of the program elements "scope of program," "parameter monitored/inspected," "monitoring and trending," and "acceptance criteria," associated with exceptions, to determine whether the program will be adequate to manage the aging effects for which it is credited. The staff's evaluation of these exceptions follows.

Exception 1. LRA Section B2.1.27 takes an exception to the "scope of the program" element. The applicant stated it will implement the 2001 edition of ASME Section XI, Subsection IWE. Pressure retaining containment seals and gaskets are not addressed by the 2001 edition of ASME Section XI, Subsection IWE (with the 2002 and 2003 addenda). These components are evaluated under 10 CFR Part 50, Appendix J.

The staff finds the exception to the "scope of the program" program element acceptable because GALL AMP XI.S1, "ASME Section XI, Subsection IWE," states that IWE evaluation shall be in accordance with the ASME Section XI, Subsection IWE 2001 edition including the 2002 and 2003 Addenda. Pressure retaining seals and gaskets inspections were originally included in the 1995 edition of the ASME Section XI, Subsection IWE code but were eliminated from the 2001 edition (with the 2002 and 2003 addenda) of the code.

Exception 2. LRA Section B2.1.27, takes an exception to program element 3, "parameters monitored or inspected." In the GALL Report AMP, XI.S1, "ASME Section XI, Subsection IWE," the program element provides seven categories of examination per Table IWE-2500-1. However, the program element in the LRA states that the program is in accordance with the 2001 Edition of the ASME Section XI code, which does not specify seven categories of examination in Table IWE-2500-1.

The staff finds this exception acceptable because the seven categories of examination listed in the GALL Report AMP were included in the 1995 Edition of the ASME Section XI code; however, they were deleted from the 2001 Edition of the code. The GALL Report AMP states that IWE evaluation shall be in accordance with the ASME Section XI, Subsection IWE 2001 edition including the 2002 and 2003 addenda.

Exception 3. LRA Section B2.1.27 takes an exception to program element 5 "monitoring and trending." In the GALL Report AMP XI.S1, "ASME Section XI, Subsection IWE," the program element recommends reexamining flaws accepted by engineering evaluation for three

Aging Management Review Results

consecutive inspection periods. The GALL Report AMP program element also discusses additional examinations per IWE-2430. Alternatively, the program element in the LRA states that the program is in accordance with the 2001 Edition of the ASME Section XI code, which recommends reexamining flaws during the next inspection period; it deletes IWE-2430.

The staff finds this exception acceptable because it complies with the intent of the GALL Report AMP XI.S1, "ASME Section XI, Subsection IWE," to comply with IWE-2430 inspection periodicity. The 2001 Edition of the ASME Section XI, Subsection IWE, article IWE-2420 requires reexamination of the flaws during the next inspection period and deletes IWE-2430 which was included in 1995 edition of the code.

Exception 4. LRA Section B2.1.27 takes an exception to the program element 6 (acceptance criteria). In the GALL Report AMP, this program element refers to acceptance criteria discussed in Table IWE-3410-1. Alternatively, these program elements in the LRA state, that Table IWE-3410-1 was deleted before the issuance of the 2001 Edition of ASME Section XI. The LRA further states that the acceptance standards previously specified in Table IWE-3410-1 are now given in Section IWE-3500.

The staff finds this exception acceptable because it complies with the intent of the GALL Report AMP XI.S1, "ASME Section XI, Subsection IWE." The applicant's use of IWE-3500 is appropriate since IWE-3410-1 was deleted from the ASME Section XI, Subsection IWE, 2001 Edition of the code, but the intent is for applicants to continue to use this acceptance criteria.

Operating Experience. LRA Section B2.1.27 summarizes operating experience related to the ASME Section XI ISI, IWE program. The applicant stated that it conducted the latest ASME Section XI, Subsection IWE program inspections at Units 1, 2, and 3 between April 2005 and May 2006. An evaluation of the results from the ISI examination indicated that the integrity of the containment system has been maintained. All discrepancies were corrected or determined "use-as-is," in accordance with work control practices and ASME Section XI.

The staff reviewed operating experience information, in the application and during the audit, to determine if the applicant reviewed the applicable aging effects and industry and plant-specific operating experience. As discussed in the Audit Report, the staff conducted an independent search of the plant operating experience information to determine if the applicant had adequately incorporated and evaluated operating experience related to this program.

The LRA ASME Section XI, Subsection IWE AMP states that the applicant conducted general visual examinations for period two of the first interval of the program for Unit 3. It gave special attention to the areas at the floor level mentioned in IN 2004-09, "Corrosion of Steel Containment and Containment Liner." It did not observe abnormal conditions or signs of degradation. However, it was not clear from the review of supporting documentation if the applicant considered liner plate corrosion concerns, identified in IN 2004-09, and recent industry operating experience related to Beaver Valley Power Station for liner plate corrosion. Therefore, in RAI B2.1.27-3 the staff requested the applicant to explain if it considered the applicability of IN 2004-09 and Beaver Valley Power Station containment liner plate corrosion for PVNGS containments to avoid similar problems. PVNGS containment liner plate is of identical design and construction to the Beaver Valley Power Station.

In response to the RAI B2.1.27-3, in a letter dated February 19, 2010, the applicant repeated the information contained in LRA Section B2.1.27-3 concerning IN 2004-09, including that the there were no areas that have evidence of water or moisture contacting the liner plate or penetrating the joint between the liner plate and the floor. The applicant also stated that EPRI performed a self-assessment of the IWE and IWL Programs for all three units and had no concerns. The EPRI self-assessment concluded that the IWE and IWL Programs address all issues listed in IN 2004-09. The applicant further stated that no liner plate corrosion similar to the Beaver Valley Power Station operating experience has been identified. This conclusion is

Aging Management Review Results

based on the applicant's inspection of the Unit 3 liner plate at different elevations after the coating was removed. These areas had scratches or blisters in the coating that required replacement of the coating. No signs of liner plate corrosion were detected in the base metal of the liner prior to re-coating. The staff finds the response to RAI B2.1.27-3 acceptable because the applicant has not found any liner plate degradation or evidence of water or moisture at the joint between the containment floor and liner plate as described in IN 2004-09. The applicant has not found any evidence of through-wall corrosion of the liner plate.

In LRA, Appendix B, Section B2.1.32, "Structures Monitoring Program," it states that no credit is taken for coatings in the determination of aging effects for the underlying materials. Although the coatings are not credited for aging management, the staff believed their failure could affect the functioning of safety systems. SER Section 3.0.3.3.2 documents the staff's evaluation of the applicant's coatings assessment program, which the applicant has in lieu of an AMP.

Based on its audit and review of the application, the staff finds that operating experience related to the applicant's program demonstrates that it can adequately manage the detrimental effects of aging on SSCs within the scope of the program and that implementation of the program has resulted in the applicant taking corrective actions. The staff confirmed that the "operating experience" program element satisfies the criterion in SRP-LR Section A.1.2.3.10 and is, therefore, acceptable.

UFSAR Supplement. LRA Section A1.27 provides the UFSAR supplement for the ASME Section XI, Subsection IWE Program. The staff reviewed this UFSAR supplement description of the program and notes that it conforms to the recommended description for this type of program, as described in SRP-LR Table 3.5-2. The staff also notes that the applicant committed (Commitment No. 29) to the ongoing implementation of the existing ASME Section XI, Subsection IWE Program to manage the aging of the steel liner of the concrete containment building during the period of extended operation. The staff determines that the information in the UFSAR supplement is an adequate summary description of the program, as required by 10 CFR 54.21(d).

Conclusion. On the basis of its review of the applicant's ASME Section XI, Subsection IWE Program, the staff finds all program elements consistent with the GALL Report AMP. In addition, the staff reviewed the exceptions and justifications, and determines that the AMP, with the exceptions, is adequate to manage the aging effects for which the LRA credits it. The staff concludes that the applicant has demonstrated that the effects of aging will be adequately managed so that the intended functions will be maintained consistent with the CLB for the period of extended operation, as required by 10 CFR 54.21(a)(3). The staff also reviewed the UFSAR supplement for this AMP, and concludes that it provides an adequate summary description of the program, as required by 10 CFR 54.21(d).

3.0.3.2.19 Masonry Wall Program

Summary of Technical Information in the Application. LRA Section B2.1.31 describes the existing Masonry Wall Program as being consistent, with enhancement, with the GALL Report AMP XI.S5, "Masonry Wall Program." In the LRA, the applicant states that the Masonry Wall Program is part of the Structures Monitoring Program that implements monitoring requirements for structures, as specified in 10 CFR 50.65. For Seismic Category I structures, the Masonry Wall Program manages cracking of masonry walls and structural steel restraint systems of the masonry walls within the scope of license renewal based on guidance provided in the Office of Inspection and Enforcement (IE) Bulletin 80-11, "Masonry Wall Design," and IN 87-67, "Lessons Learned from Regional Inspections of Licensee Actions in Response to NRC IE Bulletin 80-11." In the LRA, the applicant explains that some masonry walls are located in non-Category I structures and are within scope of license renewal based on UFSAR commitments to satisfy fire protection. The applicant stated that the guidance of IE Bulletin 80-11 does not apply; however,

Aging Management Review Results

aging management of these walls is evaluated under the Fire Protection Program. The applicant explained that the two non-seismic Category I structures that contain masonry walls within the scope of license renewal are the turbine building and fire pump house.

Staff Evaluation. During its audit, the staff reviewed the applicant's claim of consistency with the GALL Report. The staff also reviewed and confirmed that the plant's conditions are bounded by the conditions for which the GALL Report was evaluated. The staff compared elements one through six of the applicant's program to the corresponding elements of the GALL Report Masonry Wall Program. As discussed in the Audit Report, the staff confirmed that each element of the applicant's program is consistent with the corresponding element of the GALL Report AMP, with the exception of "detection of aging effects" and "acceptance criteria." For these elements, the staff determined the need for additional clarification, which resulted in the issuance of RAIs.

While reviewing the "detection of aging effects" program element, the staff noted that inspections for in-scope SSCs, including masonry walls, were scheduled to result in the complete observation of all systems of one 'equivalent unit' on a frequency of approximately 10 years. Observations would be conducted in different areas of different units to include a cross section of all three units. Using this method would ensure that, within a 30-year cycle, all units and all areas of each unit would be monitored. In the GALL Report Masonry Wall Program, it states that the primary parameter monitored or inspected is wall cracking that could potentially invalidate the evaluation basis and that masonry walls may be inspected as part of the Structures Monitoring Program (GALL Report, Chapter XI.S6) conducted under the Maintenance Rule. Industry standards (e.g., American Concrete Institute (ACI) 349.3R-96), identified in the GALL Report Structures Monitoring Program, suggest a five-year inspection frequency for structures exposed to natural environment, structures inside primary containment, continuous fluid-exposed structures, and structures retaining fluid or pressure. Industry standards recommend a 10-year inspection frequency for below-grade structures and structures in a controlled interior environment. It is not clear to the staff if all SSCs at each unit inspected under this AMP are in accordance with the industry standard inspection frequency or if only representative SSCs at each plant will be inspected within a 30-year period, implying that SSCs at each unit are completely inspected only once during the 30-year period.

By letter dated December 29, 2009, the staff issued RAI B2.1.32-1, asking the applicant to explain the inspection frequency for each unit and the plant in general. If the inspection interval exceeds the industry standard, the applicant must clearly explain the basis for extending the interval and explain how the chosen interval will adequately manage aging during the period of extended operation.

By letter dated February 19, 2010, the applicant responded to the RAI. SER Section 3.0.3.2.20 for the Structures Monitoring Program includes a summary of the RAI response and a detailed discussion of the staff's review of this issue.

While reviewing the "acceptance criteria" program element, the staff noted that the applicant points to the Structures Monitoring Program, which includes the Masonry Wall Program. The Structures Monitoring Program provides guidance for the determination of performance criteria of SSCs included within the scope of the Maintenance Rule. A component observation report is to be prepared after each inspection that considers all the individual observations in relation to the ability of the structure to provide the necessary support and protection for the SSCs included within the structure. If any areas are found to have significant aging effects, engineering notifications are made to determine appropriate corrective action. SSC deficiencies are categorized as minor, adverse, or critical. It is unclear to the staff what criteria the applicant uses to classify a deficiency as minor, adverse, or critical. By letter dated December 29, 2010, the staff issued RAI B2.1.32-2, asking the applicant to provide the criteria used to categorize a

SSC deficiency as minor, adverse, or critical and include references to site documents or procedures that contain the categorization criteria.

By letter dated February 19, 2010, the applicant responded to the RAI. SER Section 3.0.3.2.20 for the Structures Monitoring Program includes a summary of the RAI response and a detailed discussion of the staff's review of this issue. The staff also reviewed the portions of the "detection of aging effects" program element associated with the enhancement to determine whether the program will be adequate to manage the aging effects for which it is credited. The staff's evaluation of this enhancement follows.

Enhancement. LRA Section B2.1.31 identifies an enhancement to "detection of aging effects" to specify ACI 349.3R-96 as the reference for qualification of personnel to inspect structures under the Structures Monitoring Program. The staff finds this enhancement acceptable because when implemented, the AMP B2.1.31, "Masonry Wall Program," will be consistent with the GALL Report Masonry Wall Program. Inspector qualifications will be commensurate with industry codes, standards and guidelines, which will help provide assurance that the applicant will adequately manage the effects of aging.

Based on its audit and review of the applicant's responses to RAIs B2.1.32-1 and B2.1.32-2, the staff finds that elements one through six of the applicant's Masonry Wall Program, with acceptable enhancement, are consistent with the corresponding program elements of the GALL Report Masonry Wall Program and are, therefore, acceptable.

Operating Experience. LRA Section B2.1.31 summarizes operating experience related to the Masonry Wall Program. The LRA discusses cracking in the Unit 1 control building, which was found to be acceptable by engineering evaluation but required additional monitoring to verify the cracks were not progressing. The applicant inspected the same areas in Units 2 and 3 and found no cracks. During the audit, the staff reviewed that status of the cracks and found that they had been monitored appropriately, and that none of the cracks had grown. The LRA also explains that it has found no cracks larger than one-eighth of an inch and that engineering has evaluated all identified cracks. The LRA further explains that the applicant has found no degraded steel bracing.

The staff also reviewed operating experience information, in the application and during the audit, to determine if the applicant reviewed the applicable aging effects and industry and plant-specific operating experience. As discussed in the Audit Report, the staff conducted an independent search of the plant operating experience information to determine if the applicant had adequately incorporated and evaluated operating experience related to this program. The staff found that the applicant had adequately identified and incorporated industry and applicable plant-specific operating experience.

Based on its audit and review of the application, the staff finds that operating experience related to the applicant's program demonstrates that it can adequately manage the detrimental effects of aging on SSCs within the scope of the program and that implementation of the program has resulted in the applicant taking corrective actions. The staff confirmed that the "operating experience" program element satisfies the criterion in SRP-LR Section A.1.2.3.10 and, therefore, the staff finds it acceptable.

UFSAR Supplement. In LRA Section A1.31, the applicant provided the UFSAR supplement for the Masonry Wall Program. The staff reviewed this UFSAR supplement section and notes that it conforms to the recommended description for this type of program as described in SRP-LR Table 3.5-2. The staff also notes that the applicant committed (Commitment No. 33) to enhance the Masonry Wall Program prior to entering the period of extended operation. Specifically, the applicant committed to specifying ACI 349.3R-96 as the reference for qualification of personnel to inspect structures under the Structures Monitoring Program. The staff determines that the

Aging Management Review Results

information in the UFSAR supplement is an adequate summary description of the program, as required by 10 CFR 54.21(d).

Conclusion. On the basis of its audit and review of the applicant's Masonry Wall Program, including responses to RAIs, the staff determines that those program elements for which the applicant claimed consistency with the GALL Report are consistent. In addition, the staff reviewed the enhancement and confirmed that its implementation through Commitment No. 33, prior to the period of extended operation, would make the existing AMP consistent with the GALL Report AMP to which it was compared. The staff concludes that the applicant has demonstrated that the effects of aging will be adequately managed so that the intended function(s) will be maintained consistent with the CLB for the period of extended operation, as required by 10 CFR 54.21(a)(3). The staff also reviewed the UFSAR supplement for this AMP, and concludes that it provides an adequate summary description of the program, as required by 10 CFR 54.21(d).

3.0.3.2.20 Structures Monitoring Program

Summary of Technical Information in the Application. LRA Section B2.1.32 describes the existing Structures Monitoring Program as being consistent, with enhancement, with the GALL Report AMP XI.S6, "Structures Monitoring Program." The program implements the requirements of 10 CFR 50.65 (Maintenance Rule) and is consistent with the guidance of NUMARC 93-01, Rev. 2 and RG 1.160, Rev. 2. The Structures Monitoring Program provides inspection guidelines for concrete elements, structural steel, masonry walls, structural features (e.g., caulking, sealants, and roofs), structural supports, and miscellaneous components such as doors. The Structural Monitoring Program manages cracking, loss of material, and change in material properties by monitoring the condition of structures and structural supports that are within the scope of license renewal. The program includes all masonry walls and water-control structures within the scope of license renewal. The program monitors settlement for each major structure and inspects equipment, piping, conduits, cable trays, HVAC equipment, and instrument components. The applicant inspects each of the spray ponds every five years and performs settlement monitoring for each major structure every five years. For other inspections, it monitors representative SSCs at each of the three units, such that it inspects the equivalent of one complete unit every 10 years. All three units will be 100-percent inspected (with the possible exception of inaccessible areas) within a 30-year period.

Staff Evaluation. During its audit, the staff reviewed the applicant's claim of consistency with the GALL Report. The staff also reviewed and confirmed that the plant's conditions are bounded by the conditions for which the GALL Report was evaluated. The staff compared elements one through six of the applicant's program to the corresponding elements of the GALL Report Structures Monitoring Program. As discussed in the Audit Report, the staff confirmed that each element of the applicant's program is consistent with the corresponding element of the GALL Report, with the exception of "detection of aging effects," and "acceptance criteria." For these elements, the staff determined the need for additional clarification, which resulted in the issuance of RAIs.

While reviewing the "detection of aging effects" program element, the staff noted that inspections for in-scope SSCs are scheduled to result in total observation of all systems of one "equivalent Unit" on a frequency of approximately 10 years. To include a cross section of all three units, the applicant conducts observations in different areas of different units. Within a 30-year cycle, it monitors all units and all areas of each unit. The staff further noted that industry standards identified in the GALL Report Structures Monitoring Program (e.g., ACI 349.3R-96) suggest a five-year inspection frequency for structures exposed to a natural environment, structures inside primary containment, continuous fluid-exposed structures, and structures retaining fluid or pressure; and a 10-year inspection frequency for below-grade structures and structures in a controlled interior environment. It is not clear to the staff that all

SSCs at each unit, inspected under this AMP, are consistent with the industry standards inspection frequency (e.g., as noted in ACI 349.3R-96) or if only representative SSCs at each plant will be inspected within a 30-year period, implying that SSCs at each unit are completely inspected only once during the 30-year period.

By letter dated December 29, 2009, the staff issued RAI B2.1.32-1, asking the applicant to explain in more detail the inspection frequency for each unit and the plant in general. If the inspection interval exceeds the industry standard, the applicant must clearly explain the basis for extending the interval and explain how the chosen interval will adequately manage aging during the period of extended operation.

By letter dated February 19, 2010, the applicant responded to the RAI and explained that of the three units at PVNGS, a complete "representative unit" is inspected over every 10-year period. The applicant further explained that to include a cross section of all three units, it conducts observations in different areas of different units. In addition, other site programs look at civil SSCs, over and above those required by the Maintenance Rule. In addition, the applicant stated that structures are not subject to sustained aggressive adverse environmental conditions.

The staff reviewed the applicant's response and found it unacceptable because the staff does not believe that an inspection of a "representative unit" on the suggested 10-year interval will capture degradation throughout the three units. The staff believes that the applicant must inspect each unit on an interval not to exceed 10 years. The staff discussed this issue with the applicant during the license renewal inspection held the week of February 22, 2010. By letter dated April 1, 2010, the applicant supplemented their response to RAI B2.1.32-1, committing to enhancing the Structures Monitoring Program to inspect structures within the scope of license renewal on a 10-year frequency.

Although this response addressed the staff's concern with the "representative unit" approach and with structures being inspected at intervals greater than 10 years, it did not completely align the inspection interval with the guidance in ACI 349.3R. The guidance recommends a five-year inspection interval for structures within primary containment and structures exposed to a natural environment. The staff discussed these issues with the applicant during a conference call on April 12, 2010.

By letter dated May 21, 2010, the applicant supplemented its response to RAI B2.1.32-1 to align the inspection frequencies of safety-related structures (structures within primary containment) with the guidance in ACI 349.3R (Commitment No. 34). For non-safety-related structures, the applicant maintained a 10-year inspection interval for exterior surfaces. The applicant explained that it based this frequency on the associated risk, the non-aggressive exterior environment, and site operating experience. Specifically, the applicant explained that the site experiences very low rainfall, the water table is well below the lowest structures, and freeze cycles are infrequent and usually occur during dry conditions. The applicant also explained that there is minimal site experience with external concrete degradation and that any degradation detected during the period of extended operation will lead to a reassessment of the inspection frequency.

The staff reviewed the applicant's supplemental responses and found them acceptable because for safety-related structures the applicant has aligned the inspection frequency with the recommended guidance in ACI 349.3R. Further, the staff finds the 10-year inspection frequency acceptable for the non-safety-related structures exposed to an external environment because the site has a relatively benign environment. The low rainfall, minimal exposure to groundwater, and negligible weathering region classification (per ASTM C33), are all factors that reduce the degradation potential of the environment on exposed concrete. In addition, the site has minimal experience with concrete degradation due to exterior exposure and, if degradation is detected, the adequacy of the inspection frequency will be reassessed. The staff's concerns

Aging Management Review Results

in RAI B2.1.32-1 are resolved. Based on the review of the applicant's responses, the staff finds the "detection of aging effects" program element acceptable.

While reviewing the "acceptance criteria" program element, the staff noted the LRA Structures Monitoring Program states that acceptance criteria are to be commensurate with industry codes, standards, and guidelines, and are to consider industry and plant-specific operating experience. The Structures Monitoring Program provides guidance for the determination of performance criteria for SSCs included within the scope of the Maintenance Rule. These guidelines are used to establish inspection attributes for SSCs monitored by the Structures Monitoring Program with deficiencies categorized as minor, adverse, or critical, and depending on the deficiency categorization, the SSCs are considered acceptable or unacceptable. In the GALL Report Structures Monitoring Program, ACI 349.3R-96 provides an acceptable basis for developing acceptance criteria for concrete structural elements, steel liners, joints, coatings, and waterproofing membranes. It is unclear to the staff if the applicant has used ACI 349.3R-96 to provide the basis to establish the deficiency categorizations or if it uses some other basis and what criteria it uses to categorize a SSC deficiency as minor, adverse, or critical.

By letter dated December 29, 2010, the staff issued RAI B2.1.32-2, asking the applicant to provide the criteria used to categorize a SSC deficiency as minor, adverse, or critical and include references to site documents or procedures that contain the categorization criteria.

By letter dated February 19, 2010, the applicant defined the deficiency categories as follows:

Critical – A deficiency that requires corrective action (repair or more frequent inspection) to provide confidence that the associated structure will continue to perform its design function until the next regular inspection. For the purpose of this definition, the regular inspection interval is 10 years, (i.e., until the same structural element is reinspected in any unit), except for the spray ponds where the observation frequency is 5 years. A critical deficiency in one Unit area will precipitate an inspection of the corresponding structural elements in the other two Units.

Adverse – A deficiency that should be repaired (or inspected more frequently), when repair or more frequent inspection is not required to maintain the structure's functional capability. Such repairs may be done to prevent further degradation, maintain general material conditions, or to promote overall plant appearance. An Adverse deficiency in one Unit area will precipitate an inspection of the corresponding structural elements in the other two Units.

Minor – A deficiency that is acceptable as is, with no action required.

The applicant's response further specified the deficiency categories for concrete as follows:

Critical Deficiency – Conditions of degradation that must be repaired to continue to maintain the concrete component's functional capability or to restore the concrete component to its applicable functional capability. Transportability concerns shall be evaluated for all critical deficiencies.

Adverse Deficiency – Conditions of degradation that do not expose the embedded steel re-enforcement, do not impact the design function of the concrete component, or is passive (non-active condition) where it is not required to maintain the concrete component's functional capability. However, such repairs should be performed to prevent further degradation or to promote overall plant appearance.

Minor Deficiency – Conditions of degradation that do not expose the embedded steel re-enforcement, do not impact the design function of the concrete component, or is passive (non-active condition) shall be acceptable without further evaluation.

The staff reviewed the applicant's response and found it vague and qualitative, compared to the detailed evaluation criteria provided in ACI 349.3R for concrete structural elements, steel liners, joints, coatings, and waterproofing membranes. Therefore, the staff finds the applicant's deficiency criteria unacceptable for the period of extended operation.

By letter dated May 21, 2010, the applicant supplemented the original response and committed to enhance the Structures Monitoring Program, before the period of extended operation, to quantify the acceptance criteria and critical parameters for monitoring degradation (Commitment No. 34). The applicant further stated that the program will be enhanced to incorporate applicable industry codes, standards, and guidelines (e.g. ACI 349.3R-96, ANSI/ASCE 11-90) for acceptance criteria.

The staff reviewed the applicant's supplemental response and found it acceptable because the applicant committed to enhance the Structures Monitoring Program to include the acceptance criteria discussed in the GALL Report Structures Monitoring Program. The staff's concern in RAI B2.1.32-2 is resolved. Based on the review of the applicant's responses, the staff finds the "acceptance criteria" program element acceptable.

The staff also reviewed the portions of the "detection of aging effects" program element associated with an enhancement to determine whether the program will be adequate to manage the aging effects for which it is credited. The staff's evaluation of this enhancement follows.

Enhancement. LRA Section B2.1.32 identifies an enhancement to "detection of aging effects" to specify ACI 349.3R-96 as the reference for qualification of personnel to inspect structures under the Structures Monitoring Program. The staff found this enhancement acceptable because when implemented the LRA Structures Monitoring Program will be consistent with the GALL Report Structures Monitoring Program relative to inspector qualifications being commensurate with industry codes, standards, and guidelines to help provide assurance that the applicant will adequately manage the effects of aging.

Based on its audit, review of the LRA Structures Monitoring Program, and review of the applicant's responses to RAIs B2.1.32-1 and B2.1.32-2, the staff finds that elements one through six of the applicant's Structures Monitoring Program, with acceptable enhancement, are consistent with the corresponding program elements of the GALL Report Structures Monitoring Program and are, therefore, acceptable.

Operating Experience. LRA Section B2.1.32 summarizes operating experience related to the Structures Monitoring Program. The staff interviewed the applicant's technical personnel during the audit to confirm that plant-specific operating experience revealed no degradation not bounded by industry experience. The staff reviewed operating experience information, in the application and during the audit, to determine if the applicant reviewed the applicable aging effects and industry and plant-specific operating experience. As discussed in the Audit Report, the staff conducted an independent search of the plant operating experience information to determine if the applicant had adequately incorporated and evaluated operating experience related to this program.

During its review, the staff identified two condition report disposition requests related to leakage of SFP water. The staff determined the need for additional information, which resulted in the issuance of an RAI.

By letter dated December 29, 2010, the staff issued RAI B2.1.32-3, asking the applicant to discuss any apparent cause analysis performed to identify the source of leakage as well as corrective actions taken to stop leakage. The staff asked the applicant to explain how the leakage has affected the condition of the concrete and what steps it has taken or will take to

Aging Management Review Results

ensure adequacy of the concrete during the period of extended operation. The staff also asked the applicant to discuss any actions taken to ensure the drain system remains free and clear allowing the system to properly prevent water from accumulating behind the liner.

By letter dated February 19, 2010, the applicant responded and explained that the SFP is equipped with a telltale drain system, designed to capture leakage through the SFP liner. In July of 2005, in Unit 1, borated water leakage was discovered at two locations on the exterior face of the SFP concrete walls. The applicant explained that this leakage was a result of backed-up water within the telltale drain system, due to the telltale drain valves being in the closed position for an extended period. With no place to drain, the water in the telltale drain system eventually migrated through the SFP concrete walls via extremely small cracks. Once the drain valves were re-opened, each drain line released a large amount of borated water.

The applicant further explained that currently, and continuing through the period of extended operation, plant personnel open the valves and record drained water on a daily basis. The applicant explained that this procedure should keep water from backing up and leaking through the concrete. The applicant will measure and trend drainage from the telltale system and investigate any abnormalities through the corrective action program. The applicant also explained that in order to ensure the drain lines were clear, it inspected all telltale drain lines in all units via boroscope between 2008 and 2009, and it will re-inspect the lines on a two-year frequency. The applicant further explained that Construction Technology Laboratories (CTL) performed an NDE of the SFP concrete walls; the CTL inspection report concluded that the borated water leakage did not have an adverse impact on the concrete.

The staff reviewed the applicant's response and found that the actions taken to unclog the drain system and monitor it in the future would provide some assurance of reduced leakage through the concrete. However, the staff requested the applicant to provide more information on the extent of the leakage as well as a more detailed discussion of their confidence that the leakage is no longer passing through the concrete walls.

By letter dated May 21, 2010, the applicant supplemented their original response and provided the CTL inspection report, which documented the NDE conducted after the leakage incident in 2005. In their response, the applicant explained that leakage through the concrete in areas that were initially identified as showing leakage stopped completely, based on the fact that there are no longer wetted areas visible. In addition, the applicant explained that in 2006 and 2007, shallow aquifer wells were installed down-gradient of each unit. These wells are sampled periodically and no radioactivity has been detected. The applicant further explained that it determined that approximately eight ounces of fluid leaked from both leakage locations in 2005. The applicant also stated that no indications of leakage have been identified in Units 2 or 3.

The staff reviewed the applicant's response and found it acceptable because the NDE investigation showed no significant degradation to the concrete as a result of the 2005 leakage event. The amount of leakage through the concrete was minimal, and inspections of accessible concrete since the event show no indications of continued leakage. In addition, the applicant has plans in place to properly maintain the telltale drain system. This reduces the likelihood of future leakage migrating through the concrete and causing degradation. The staff's concern in RAI B2.1.32-3 is resolved.

Based on its audit, review of the application, and review of the applicant's response to RAI B2.1.32-3, the staff finds that operating experience related to the applicant's program demonstrate that it can adequately manage the detrimental effects of aging on SSCs within the scope of the program and that implementation of the program has resulted in the applicant taking corrective actions. The staff confirmed that the "operating experience" program element satisfies the criterion in SRP-LR Section A.1.2.3.10 and, therefore, the staff finds it acceptable.

Aging Management Review Results

In LRA, Appendix B, Section B2.1.32, "Structures Monitoring Program," it states that no credit is taken for coatings in the determination of aging effects for the underlying materials. Although the coatings are not credited for aging management, the staff believed their failure could affect the functioning of safety systems. SER Section 3.0.3.3.2 documents the staff's evaluation of the applicant's coatings assessment program, which the applicant has in lieu of an AMP.

UFSAR Supplement. In LRA Section A1.32, the applicant provided the UFSAR supplement for the Structures Monitoring Program. The staff reviewed this UFSAR supplement section and notes that it conforms to the recommended description for this type of program as described in SRP-LR Table 3.5-2. The staff also notes that the applicant committed (Commitment No. 34) to enhance the Structures Monitoring Program prior to entering the period of extended operation. Specifically, the applicant committed to specifying ACI 349.3R-96 as the reference for qualification of personnel to inspect structures under the Structures Monitoring Program.

The applicant also committed to align inspection frequencies with the guidance in ACI 349.3R, except for the exterior surfaces of non-safety-related structures, for which the applicant maintained a 10-year inspection interval. The Structures Monitoring Program will be enhanced to quantify the acceptance criteria and critical parameters for monitoring degradation, and to provide guidance for identifying unacceptable conditions requiring further technical evaluation or corrective action. Procedures will also be enhanced to incorporate applicable industry codes, standards and guidelines for acceptance criteria.

Conclusion. On the basis of its audit and review of the applicant's Structures Monitoring Program, as well as the RAIs discussed above, the staff determines that those program elements for which the applicant claimed consistency with the GALL Report are consistent. Also, the staff reviewed the enhancement and confirmed that its implementation through Commitment No. 34, prior to the period of extended operation, would make the existing AMP consistent with the GALL Report AMP to which it was compared. The staff concludes that the applicant has demonstrated that the effects of aging will be adequately managed so that the intended function(s) will be maintained consistent with the CLB for the period of extended operation, as required by 10 CFR 54.21(a)(3). The staff also reviewed the UFSAR supplement for this AMP, and concludes that it provides an adequate summary description of the program, as required by 10 CFR 54.21(d).

3.0.3.2.21 Regulatory Guide 1.127, "Inspection of Water-Control Structures Associated with Nuclear Power Plants"

Summary of Technical Information in the Application. LRA Section B2.1.33 describes the existing RG 1.127, "Inspection of Water-Control Structures Associated with Nuclear Power Plants" program as consistent, with an enhancement, with the GALL Report AMP XI.S7, "RG 1.127, Inspection of Water-Control Structures Associated with Nuclear Power Plants." In the LRA, the applicant explains that PVNGS is not committed to RG 1.127; however, a Structures Monitoring Program is in place that includes all water-control structural components within the scope of RG 1.127. The applicant further explains that the program manages aging for the water-control structures associated with emergency cooling water systems, and that the inspections are performed on a frequency of at least once every five years, based on acceptable inspection results from previous inspections.

Staff Evaluation. During its audit, the staff reviewed the applicant's claim of consistency with the GALL Report. The staff also reviewed and confirmed that the plant's conditions are bounded by the conditions for which the GALL Report was evaluated. The staff compared elements one through six of the applicant's program to the corresponding elements of the RG 1.127, Inspection of Water-Control Structures Associated with Nuclear Power Plants AMP in the GALL Report. As discussed in the Audit Report, the staff confirmed that each element of the applicant's program is consistent with the corresponding element of the GALL Report AMP, with

Aging Management Review Results

the exception of the "detection of aging effects" and "acceptance criteria" program elements. For these elements, the staff determined the need for additional clarification, which resulted in the issuance of RAIs.

The "detection of aging effects" element of the GALL Report AMP discusses special inspections immediately following the occurrence of significant natural phenomena, such as large floods, earthquakes, hurricanes, tornadoes, and intense local rainfalls. In the corresponding element of the AMP basis document, the applicant states that PVNGS has no earthen dams or other water control structures in-scope for license renewal that would require special inspections after the occurrence of significant natural phenomena. By letter dated March 2, 2010, the staff issued RAI B2.1.33-1 requesting the applicant to explain why it is unnecessary to inspect the spray pond structures after unusual natural events.

In its response, dated April 1, 2010, the applicant explained that the spray pond structures are Seismic Category I and designed to remain functional after a safe shutdown earthquake, extreme wind phenomena, and other design basis events. The PVNGS Technical Requirements Manual, TLCO 3.3.103, requires corrective action to be initiated to evaluate the effect upon facility features important to safety following a seismic event greater than or equal to 0.02 gravity. The applicant further stated that it would enter other unusual natural events into the corrective action plan to assess the need for inspection of the spray ponds. The staff reviewed the applicant's response and finds it acceptable because it explains that the applicant will take actions to assess the need for spray pond inspections after significant natural phenomena. The staff's concern in RAI B2.1.33-1 is resolved.

The "acceptance criteria" element of the GALL Report AMP states that the "evaluation criteria" provided in Chapter 5 of ACI 349.3R provides acceptance criteria for determining the adequacy of observed aging effects and specifies criteria for further evaluation. In the corresponding element of the AMP basis document, it states that that SSC deficiencies are categorized as minor, adverse, or critical and, depending on the deficiency categorization, the SSCs are considered to be acceptable or unacceptable. It is not clear to the staff that these statements are consistent because reviewed basis documents do not include criteria to categorize a SSC deficiency as minor, adverse, or critical. Therefore, by letter dated December 29, 2009, the staff issued RAI B2.1.32-2 requesting the applicant to provide the criteria used to categorize a SCC deficiency as minor, adverse or critical. This issue applies to all the AMPs that follow the plant procedures related to the Structures Monitoring Program (i.e. Structures Monitoring, Masonry Wall, and Water-Control Structures).

By letter dated February 19, 2010, the applicant responded to the RAI. SER Section 3.0.3.2.20 includes a summary of the RAI response and a detailed discussion of the staff's review of this issue.

The staff also reviewed the portions of the "detection of aging effects" program element associated with the enhancement to determine if the program will be adequate to manage the aging effects for which it is credited. The staff's evaluation of this enhancement follows.

Enhancement. LRA Section B2.1.33 identifies an enhancement to the "detection of aging effects" program element. The LRA explains that the applicant will enhance procedures to specify that the essential spray pond inspections include concrete below the water level. The GALL Report recommends inspections of concrete below the water level to detect aging effects. As such, implementation of the enhancement through Commitment No. 35 will make the applicant's program consistent with the GALL Report and, therefore, acceptable.

Based on its audit, review of the LRA AMP, and review of the applicant's responses to RAIs B2.1.33-1 and B2.1.32-2, the staff finds that elements one through six of the applicant's RG 1.127, Inspection of Water-Control Structures Associated with Nuclear Power Plants

Program, with acceptable enhancement, are consistent with the corresponding program elements of the GALL Report AMP, and are, therefore, acceptable.

Operating Experience. LRA Section B2.1.33 summarizes operating experience related to the RG 1.127, Inspection of Water-Control Structures Associated with Nuclear Power Plants Program. The LRA discusses cracking on the outer surfaces of the sidewalls of the spray ponds and explains that there are no visual indications associated with the cracks that would indicate the structural integrity of the walls is compromised. The LRA also discusses a leak that was discovered on the south end of the west wall of the Unit 1 Spray Pond A. The applicant determined that the source of the leak was a damaged expansion joint, which was repaired. The applicant examined similar locations in the other ponds and found no evidence of leakage.

The staff reviewed operating experience information, in the application and during the audit, to determine if the applicant reviewed the applicable aging effects and industry and plant-specific operating experience. As discussed in the Audit Report, the staff conducted an independent search of the plant operating experience information to determine if the applicant had adequately incorporated and evaluated operating experience related to this program.

During its review, the staff identified operating experience that required clarification and resulted in the issuance of an RAI as discussed below.

During the on-site operating experience review, the applicant provided several condition reports that discussed degradation of the spray pond concrete walls. During walkdowns, the staff observed cracking and spalling near the top of the spray pond walls. Therefore, by letter dated March 2, 2010, the staff issued RAI B2.1.33-2 asking the applicant to explain how the AMP is addressing degradation of the spray pond walls and how the structural stability of the walls will be maintained during the period of extended operation.

In its response, dated April 1, 2010, the applicant stated that it would continue to inspect the spray pond walls on a five-year frequency during the period of extended operation. The applicant also explained that the existing condition of the spray pond structures has been assessed and will be reworked prior to the period of extended operation.

The staff reviewed the applicant's response and found it unacceptable because it did not clearly explain what 'rework' meant, and it did not discuss the criteria used to identify the degraded locations for rework. The staff discussed these with the applicant in a conference call on May 14, 2010, during which the applicant stated it would submit a supplemental response that would also include a repair plan.

By letter dated June 21, 2010, the applicant submitted a supplemental response to RAI B2.1.33-2. The response explained that the spray pond concrete restoration will be completed using existing approved engineering specifications for concrete work. The response further explained that current inspections have identified degradation on the surface of the spray pond walls and delamination in the top 6–8 inches of the walls. Currently, the applicant is developing corrective actions, including the possibility of removing and re-pouring the top of the concrete walls, and repairing reinforcement as needed. The corrective actions will be completed for each spray pond, with priority given to the spray pond with the most degradation, before the period of extended operation. The applicant's response further explained that it plans to begin the corrective actions in 2011 with Unit 1 and should be completed on all units by 2015. The response also stated that the wall degradation was evaluated; the evaluation concluded that the structural integrity of the spray ponds is not challenged and the spray ponds remain operable. The applicant also added Commitment No. 56 to address the spray pond repair schedule.

The staff reviewed the applicant's response and found it acceptable because it explains the degradation is in the top 6–8 inches of the spray pond walls and does not affect the structural

Aging Management Review Results

integrity of the walls. The response also outlines a plan to fix the degradation prior to the period of extended operation (Commitment No. 56). The staff's concern in RAI B2.1.33-2 is resolved.

Based on its audit, review of the application, and review of RAI B2.1.33-2 response, the staff finds that operating experience related to the applicant's program demonstrates that the program can adequately manage the detrimental effects of aging on SSCs within the scope of the program and that implementation of the existing program has resulted in the applicant taking corrective actions. The staff confirmed that the "operating experience" program element satisfies the criterion in SRP-LR Section A.1.2.3.10 and, therefore, the staff finds it acceptable.

UFSAR Supplement. LRA Section A1.33 provides the UFSAR supplement for the RG 1.127, Inspection of Water-Control Structures Associated with Nuclear Power Plants Program. The staff reviewed this UFSAR supplement description of the program and notes that it conforms to the recommended description for this type of program, as described in SRP-LR Table 3.5-2. The staff also notes that the applicant has committed (Commitment No. 35) to the ongoing implementation of the existing RG 1.127, Inspection of Water-Control Structures Associated with Nuclear Power Plants Program to manage the aging of applicable components during the period of extended operation. The applicant also committed (Commitment No. 35) to enhancing the procedures, prior to the period of extended operation, to specify inspections of concrete below the water level. The applicant further committed (Commitment No. 56) to develop a repair plan for the spray pond wall degradation and to implement the repairs beginning in 2011 and expecting to complete repairs by 2015. The staff determines that the information in the UFSAR supplement is an adequate summary description of the program, as required by 10 CFR 54.21(d).

Conclusion. On the basis of its audit and review of the applicant's RG 1.127, Inspection of Water-Control Structures Associated with Nuclear Power Plants Program, and the applicant's responses to the RAIs, the staff determines that those program elements for which the applicant claimed consistency with the GALL Report are consistent. The staff reviewed the enhancement and confirmed that its implementation through Commitment No. 35, prior to the period of extended operation, would make the existing AMP consistent with the GALL Report AMP to which it was compared. Also, the staff reviewed Commitment No. 56, which was added to address the spray pond wall degradation prior to the period of extended operation, and found it acceptable. The staff concludes that the applicant has demonstrated that the effects of aging will be adequately managed so that the intended function(s) will be maintained consistent with the CLB for the period of extended operation, as required by 10 CFR 54.21(a)(3). Finally, the staff reviewed the UFSAR supplement for this AMP and concludes that it provides an adequate summary description of the program, as required by 10 CFR 54.21(d).

3.0.3.2.22 Metal Fatigue of Reactor Coolant Pressure Boundary Program

Summary of Technical Information in the Application. LRA Section B3.1 describes the existing Metal Fatigue of RCPB program as consistent, with enhancements, with the GALL Report AMP X.M1, "Metal Fatigue of Reactor Coolant Pressure Boundary." In the LRA, the applicant identifies the Metal Fatigue of RCPB program (Metal Fatigue AMP) as an existing program that, when enhanced, will be consistent with the program elements in the GALL Report AMP.

In a letter dated May 27, 2010, the applicant submitted LRA Amendment No. 16. In this amendment, the applicant modified the program description, enhancements, and "operating experience" program element description for the LRA AMP, in order to reconcile inconsistencies between the AMP and corresponding discussions in LRA Section 4.3.1.

Staff Evaluation. During its audit, the staff reviewed the applicant's claim of consistency with the GALL Report. The staff also reviewed and confirmed that the plant's conditions are bounded by the conditions for which the GALL Report was evaluated.

Aging Management Review Results

The staff compared elements one through six of the applicant's program to the corresponding elements of the GALL Report Metal Fatigue AMP. As discussed in the Audit Report, the staff confirmed that each element of the applicant's program is consistent with the corresponding element of the GALL Report AMP, with the exception of the "operating experience" program element, the enhancements, and the corresponding discussions in LRA Section 4.3.1 "Fatigue Aging Management Program." For these areas, the staff determined the need for additional clarification, which resulted in the issuance of RAIs. SER Section 4.3.1 discusses these RAIs in detail.

The staff noted that the program did not account for applicable TS tracking requirements for the design basis transients that are discussed and evaluated in UFSAR Section 3.9.1.1 and in UFSAR Tables 3.9.1-1 and 3.9-1. These TS tracking requirements are in accordance with TS 5.5.5, which requires the applicant to implement the following administrative controls:

> 5.5.5 Component Cyclic or Transient Limit
>
> This program provides controls to track the UFSAR Section 3.9.1.1 cyclic and transient occurrences to ensure that components are maintained within the design limits.

Thus, the staff questioned why the applicant did not account for applicable TS tracking requirements in the program description and "scope of program" element of the AMP.

The staff also noted that the program's monitoring methods included both cycle-count (CC) monitoring, cycle-based fatigue (CBF) monitoring, and stress-based fatigue (SBF) monitoring methods. The staff noted that of these methods, only the CC and CBF monitoring methods involve tracking and counting the number of design basis transients occurring at each of the units. The staff noted there were apparent inconsistencies between the transients monitored by the AMP and those tracked under TS 5.5.5. By letter dated April 28, 2010, the applicant submitted Amendment 14 to the LRA to address and correct the inconsistencies. The staff noted that, of these methods, only the CBF method would perform periodic updates of the cumulative usage factor (CUF) values in accordance with the "detection of aging effects" program element recommendations in the GALL Report. The applicant indicated that the CC monitoring method would not perform periodic updates of CUF values. In order for the CC monitoring method to be consistent with the "detection of aging effects" program element in the GALL Report Metal Fatigue AMP, the applicant needs to clarify whether this method performs a CUF update calculation if the acceptance criterion on cycle limits is reached.

By letter dated April 28, 2010, the applicant submitted Amendment 14 to the LRA to address and amend the CC monitoring basis. The staff verified that the amended CC monitoring basis performs periodic CUF updates when a CC action limit is reached. Based on this review, the staff finds that the applicant's CC monitoring basis is acceptable because the amended basis is consistent with the staff recommendation in the "detection of aging effects" program element in the GALL Report Metal Fatigue AMP.

The applicant's SBF monitoring method credits performing periodic updates of the CUF values for specific components. The staff noted that the SBF monitoring method tracks the stress, pressure, and temperature parameters for the components and uses the changes in these parameters to perform stress based updates of the CUF calculations. The staff finds this consistent with the recommendation in the "detection of aging effects" program element in the GALL Report Metal Fatigue AMP for performance of periodic CUF updates.

The staff noted that in LRA Section 4.3.2, the applicant had indicated that it would use SBF monitoring of one component with a high CUF value as a bounding SBF monitoring basis for other component locations that have high existing CUF values. The applicant did not explain in LRA Section 4.3.1.2 or Appendix B, Section B3.1 why it was valid to propose SBF monitoring on

Aging Management Review Results

a bounding basis. The staff noted that if a bounding basis was valid, the applicant did not explain how it would apply corrective actions to the non-monitored locations if the monitored location reaches the action limit. The staff determines that the applicant needs to address why it would be permissible to apply SBF on a bounding basis.

By letter dated April 28, 2010, the applicant submitted Amendment 14, which stated that it no longer credits its SBF monitoring on a bounding basis. The applicant modified Commitment No. 39 to reflect the use of a fatigue monitoring software program and a methodology for SBF monitoring that will implement a three-dimensional, six-element tensor stress analysis method, and conform to the requirements of ASME Section III, Article NB-3200. The staff noted that this commitment has been placed on both the UFSAR supplement for this TLAA and the UFSAR supplement for the applicant's enhanced Metal Fatigue AMP for purposes of addressing the technical issues raised and discussed in RIS 2008-30, "Fatigue Analysis of Nuclear Power Plant Components." The staff finds that this revision resolves the staff's concern because it is no longer using the monitoring methodology of concern to the staff as identified in RIS-2008-30.

As stated in SER Section 4.3.1.2, the staff was concerned with the acceptance criteria for the applicant's program and the corrective action options proposed by the applicant for cycle counting activities and CUF monitoring activities.

The staff discussed these apparent inconsistencies with the applicant in a public meeting that was held at NRC Headquarters, Rockville, MD, on May 6, 2010 (meeting summary memorandum to the applicant dated June 25, 2010).

The staff noted that in LRA Amendment 16, the applicant amended the LRA Appendix B, Section B3.1 such that the program description, the "operating experience" program element, and the enhancements are consistent with the conforming changes that the applicant had made to LRA Section 4.3.1 in LRA Amendment 14, dated April 28, 2010. The staff noted the applicant made the following changes in LRA Amendment 16:

- amended the program description and enhancement 1 on the "scope of program" element to reflect the applicable TS tracking and counting requirements in TS 5.5.5

- amended the program description to provide clear indication of the differences between the current program implemented at the facility and the enhanced version of the program that will be implemented during the period of extended operation

- amended the program description, enhancement 1 for the "scope of program" element; enhancement 3 for the "parameters monitored or inspected" and "monitoring and trending" program elements; and enhancement 4 for the "detection of aging effects" program element to reflect use of appropriate monitoring software packages

- amended enhancement 2 for the "preventative actions," "acceptance criteria," and "corrective actions," program elements to reflect updated action limits and corrective actions for both CC and CUF monitoring activities

- amended the "operating experience" program element to incorporate an administrative change to the previous operating experience discussion

The staff finds the changes to the program descriptions for the program acceptable because they accounted for applicable TS tracking requirements. Also, the enhanced version of the program will implement appropriate monitoring software methods for both the CC and SBF methods.

LRA Section B3.1 states that the calculated design lifetime CUF is defined by Subparagraph NB 3222.4 of the Section III of the ASME Boiler and Pressure Vessel (B&PV) Code, and an equivalent term is defined for valves in Paragraph NB-3552. The staff noted that, ASME B&PV

Code, Section III, Subsection NB-3552 defines "excluded cycles," whereas Subsection NB-3553 defines "fatigue usage." By letter dated December 29, 2009, the staff issued RAI B3.1-1 asking the applicant to clarify which ASME Code section will be used in the calculation of the valve fatigue usage term.

In its response dated February 19, 2010, the applicant stated it calculated the valve fatigue usage using ASME B&PV Code, Section III, 1974 edition, up to and including winter 1975, summer 1976, Subsection NB-3550, as shown on LRA Table 4.3-9. The applicant further stated the use of NB-3552 in Appendix B was a typographical error and has been corrected to NB-3550.

Based on its review, the staff finds the applicant's response to RAI B3.1-1 acceptable because the applicant clarified that this was a typographical error. In addition, the applicant calculated valve fatigue usage I(t) in accordance with ASME B&PV Code, Section III, Subsection NB-3550, which includes Subsection NB-3553 defines "Fatigue Usage," and the use of ASME B&PV Code, Section III, 1974 edition is in accordance with 10 CFR 50.55a. The staff's concern described in RAI B3.1-1 is resolved.

During its audit, the staff reviewed the applicant's program basis document and noted that the "acceptance criteria" program element will be enhanced with action limits that further ensure that fatigue usage factors for RCPB components are maintained below the CUF of 1.0, as established by Section III Subsection NB of the ASME B&PV Code, and that other limits assumed as the basis for safety determinations are maintained. The staff noted that the applicant did not define the term "other limits." By letter dated December 29, 2009, the staff issued RAI B3.1-2 requesting the applicant to clarify what are the "other limits" that are assumed as the basis for safety determinations to be maintained, as described in the enhancement of the "acceptance criteria" program element.

In its response dated February 19, 2010, the applicant stated the "other limits" are those considerations not addressed as ASME Section III fatigue analyses that depend on an assumed number of load cycles. The applicant further stated that these "other limits" are still within the bounds of the ASME Code, but are not directly related to fatigue (i.e. the monitoring of crack propagation of an embedded flaw and the determination of high energy line break locations). The applicant stated that it outlined the basis for the action limits required in each case in the "disposition" of each of these subsections of the LRA. The applicant further stated that the LRA identified the safety determinations listed below that are addressed as other limits in the Metal Fatigue Monitoring Program:

- high energy line break locations in Class 1 RCP boundary piping (LRA 4.3.2.14)
- linear elastic fracture mechanics fatigue crack growth analysis of indications in a Unit 2 pressurizer support skirt forging weld (LRA 4.3.2.4)
- fatigue crack growth and fracture mechanics stability analyses of half-nozzle repairs to alloy 600 material in reactor coolant hot legs (LRA 4.7.4)

SER Section 4.3.2.14 documents the staff's evaluation for high-energy line break (HELB) locations in Class 1 reactor coolant pump (RCP) boundary piping. SER Section 4.3.2.4 documents the staff's evaluation for the linear elastic fracture mechanics fatigue crack growth analysis of indications in a Unit 2 pressurizer support skirt-forging weld. SER Section 4.7.4 documents the staff's evaluation for the fatigue crack growth and fracture mechanics stability analyses of half-nozzle repairs to Alloy 600 material in reactor coolant hot legs. The staff noted that the applicant's "other limits" are associated with an assumed number of load cycles, which is tracked by the applicant's enhanced Metal Fatigue AMP. Furthermore, the staff noted these "other limits" are associated with ASME Section III fatigue analyses, but not directly related to fatigue; the use of the design limit of 1.0 for cumulative fatigue usage is not applicable.

Aging Management Review Results

Based on its review, the staff finds the applicant's response to RAI B3.1-2 acceptable because, the applicant clarified the "other limits" as part of its "acceptance criteria" program element. In addition, the applicant will establish limits to ensure the analytical bases of ASME Section III fatigue analyses that are not directly related to cumulative fatigue usage and the applicant's establishment of these limits is consistent with the recommendations of SRP-LR Section A.1.2.3.6. Further, SER Section 4.3 documents the staff's evaluations of these ASME Section III fatigue analyses that are not directly related to cumulative fatigue usage. The staff's concern described in RAI B3.1-2 is resolved.

During its audit, the staff reviewed the applicant's program basis document and noted that the FatiguePro® software, utilizes a one dimensional stress-intensity transfer function to calculate the fatigue effects of transient cycles used by the Metal Fatigue of RCPB Program. The staff noted that the applicant did not indicate consideration of RIS 2008-30 in its development of this program. By letter dated December 29, 2009, the staff issued RAI B3.1-3 requesting the applicant describe how it considered RIS 2008-30 in the development of the Metal Fatigue of RCPB Program and how it incorporated the results of this review into the enhanced Metal Fatigue AMP.

In its response, dated February 19, 2010, the applicant stated that before the issuance of RIS 2008-30 on December 16, 2008, it was aware of staff concerns regarding the use of single element stress models to evaluate metal fatigue CUF using single element stress models similar to those in FatiguePro® to perform an evaluation of NUREG-6260 locations. The applicant stated that it performed an initial screening of the plant specific NUREG-6260 locations and selected three locations for further analysis. The applicant stated that an ASME III NB3200 three-dimensional, six-element analysis evaluated these locations (charging nozzle safe end, shutdown cooling elbow, and pressurizer surge line elbow). The staff noted that the applicant reviewed RIS 2008-30 in August 2009 and determined that no additional actions were required. The applicant stated it determined that the staff concerns also applied to the FatiguePro® single element model being used in the SBF monitoring module of FatiguePro®. The staff noted that the applicant amended its LRA by letter dated February 19, 2010, to remove the text that states FatiguePro® will be used to monitor SBF locations. The staff also noted that LRA Sections A2.1 and B3.1 and Commitment No. 39 have also been amended to reflect the use of a software monitoring program that incorporates a three-dimensional, six-element stress model. The staff noted that Commitment No. 39, as amended by letter dated May 27, 2010, specifically states, in part, the following:

> The SBF method will use a fatigue monitoring software program that incorporates a three-dimensional, six-component stress tensor method meeting ASME III NB-3200 requirements.

Based on its review, the staff finds the applicant's response to RAI B3.1-3 acceptable because, the applicant committed (Commitment No. 39) to using a software that incorporates a three-dimensional, six-component stress tensor method meeting ASME III NB-3200 requirements. In addition, the applicant's use of such software addresses the concerns in RIS 2008-30 of using a single element stress models. The staff's concern described in RAI B3.1-3 is resolved.

The staff noted that LRA Section 4.3 states that the Metal Fatigue AMP monitors and tracks the number of critical thermal and pressure transients for selected RCS components. The staff further noted that LRA Section 4.3.1.4, "Present and Projected Status of Monitored Locations," states that a composite worst-case (composite-unit) envelope of operating transients was created, which included only the highest accumulation of each transient experienced among the three units from 1985–2005. The staff noted that the applicant did not provide individual plant data used for each unit to develop the composite-unit envelope. By letter dated December 29, 2009, the staff issued RAI B3.1-4 requesting the applicant provide the

Aging Management Review Results

accumulation of transients, for each of the three units, used to develop the composite-unit envelope for the period from 1985–2005.

The applicant provided its response by letter dated February 19, 2010; however, the applicant amended its response to RAI B3.1-4 in its entirety and revised LRA Section 4.3.1 in Amendment 14, dated April 28, 2010. The applicant stated that it made the following changes:

- revised Table 4.3-2 to clearly show correspondence between the LRA and UFSAR

- revised Table 4.3-2 to identify transients that are tracked, provided justification for those transients that are not tracked, and clarified UFSAR limits

- changed "global" monitoring to "cycle counting"

- simplified the transient projection process and clarified that it is not intended to be used for action

- revised the discussion on how the transient count data was recovered

- revised Table 4.3-3 to be consistent with the UFSAR transients, replaced the worst case unit with actual totals for all three units, and provided the new simplified projections

- revised the location-specific monitoring points (i.e. Table 4.3-4 now identifies the NUREG/CR-6260 locations and the pressurizer spray nozzle location)

- Incorporated miscellaneous clarifications and editorial changes

The staff noted that the applicant amended LRA Section 4.3.1.4 Subsection "Recount Method" to state the following, in part:

> Several APS employees and contractor personnel were designated based on their long-term familiarity with PVNGS to perform document reviews. The reviewers examined the microfilmed control room logs, NRC Monthly Operating Reports and LERs for the period prior to January 1996 for all three PVNGS units. The personal recollections and records of unit personnel were used to supplement the record review, and a best-source total was determined for each monitored transient. The best-source total was added to the actual count of events following 1995 to obtain a best-source total as of the end of 2005.

The staff noted from LRA Section 4.3.1.4 that, before January 1996, the applicant's CC procedure did not contain all transients listed in UFSAR Section 3.9.1.1. Therefore, the applicant reconstituted the number of cycles that occurred at all three units before January 1996 in order to obtain the number of cycles for each transient that has occurred to date. The staff further noted that based on the applicant's reconstitution of transient events, the applicant is no longer using the method of "composite worst-case unit accumulation of cycles."

Based on its review, the staff finds the applicant's response to RAI B3.1-4, as amended, acceptable because the applicant revised LRA Section 4.3.1.4 to remove the use of its "composite worst-case unit accumulation of cycles" methodology and reconstituted the number of transient occurrences for each unit by reviewing control room logs, NRC Monthly Operating Reports, and LERs to obtain an accurate count of transients for each unit. The staff's concern described in RAI B3.1-4 is resolved.

The staff noted LRA Section B3.1 states that the locations in which fatigue effects are controlled by "a simple comparison" counting method are those with relatively low design fatigue usage

Aging Management Review Results

values. The staff required further information regarding the use of "a simple comparison" counting method. By letter dated December 29, 2009, the staff issued RAI B3.1-5 requesting the applicant to identify the locations selected for "a simple comparison" counting method, explain how it selected these locations, and define the criteria used to classify fatigue usage values as relatively low fatigue usage values.

In its response, dated February 19, 2010, the applicant stated the locations listed in LRA Table 4.3-4 with "Fatigue Management Method" labeled "Global" are those locations monitored by the "simple comparison" counting method. The staff noted that by letter dated April 28, 2010, the applicant amended LRA Table 4.3-4 and revised the use of the term "Global" with "cycle counting." The staff noted that the revision to the terminology was made to remove confusion from the LRA, but the applicant methodology did not change.

The staff noted that the applicant reviewed all locations with a current design fatigue analysis and selected any locations with either low fatigue usage values or CUF bounded by a higher-usage location in the same plant system for the CC monitoring method. The applicant further described its criteria used to select these locations for the cycle counting method. The staff noted that Commitment No. 39 as amended by letter dated May 27, 2010, states the following, in part:

> The enhanced Metal Fatigue of Reactor Coolant Pressure Boundary program will provide action limits on cycles and on CUF that will initiate corrective actions before the licensing basis limits on fatigue effects at any location are exceeded.
>
> - In order to ensure sufficient cycle count margin to accommodate occurrence of a low-probability transient, corrective actions must be taken before the remaining number of allowable occurrences for any specified transient becomes less than 1.
>
> - CUF action limits will be established to require corrective action when the calculated CUF (from cycle-based or stress-based monitoring) for any monitored location is projected to reach 1.0 within the next 2 or 3 operating cycles. In order to ensure sufficient margin to accommodate occurrence of a low probability transient, corrective actions will be taken while there is still sufficient margin to accommodate at least one occurrence of the worst-case design transient event (i.e., with the highest fatigue usage per event cycle).

Based on its review, the staff finds the applicant's response to RAI B3.1-5 acceptable because the applicant clarified its method and criteria for selecting locations that will be monitored by CC, it committed (Commitment No. 39) to establishing actions limits such that corrective actions will be taken prior to exceeding allowable cycles, and it will continue to monitor these transients to ensure allowable cycles are not exceeded. The staff's concern described in RAI B3.1-5 is resolved.

During its audit, the staff reviewed the applicant's program basis document and Commitment No. 39 and noted that the program will be enhanced to include additional locations with high CUFs. However, the staff noted that the applicant did not identify the locations or provide justification for their use. By letter dated December 29, 2009 the staff issued RAI B3.1-6 requesting the applicant provide additional information to explain which locations it included in the Metal Fatigue AMP as enhancements, explain how it selected these locations, and define the criteria it used to classify fatigue usage values as high fatigue usage values.

In its response, dated February 19, 2010, the applicant stated that all locations in LRA Table 4.3-4, except the pressurizer spray nozzle, are being added as enhancements to the program. The staff noted that LRA Table 4.3-4, as amended, lists the pressurizer spray nozzle.

Aging Management Review Results

The applicant stated that the existing program includes a simplified cycle-based CUF calculation for the pressurizer spray nozzle in each unit, while fatigue in all other locations is currently managed by manual CC only, with current action levels at 90 percent of the number of cycles assumed by the design basis.

The applicant stated that it reviewed all locations with a current design fatigue analysis and selected any locations that did not have low fatigue usage values or were identified in NUREG/CR-6260 for explicit monitoring, using either CBF or SBF methodologies. LRA Table 4.3-4, as amended, lists seven locations, associated with NUREG/CR-6260 and the pressurizer spray nozzle, as monitored by either CBF or SBF methodologies. The applicant further described its criteria used to select these locations for the CBF or SBF methodologies. The staff noted that in Commitment No. 39, as amended, the applicant will establish CUF action limits such that there will be two or three operating cycles to initiate and complete corrective actions.

Based on its review, the staff finds the applicant's response to RAI B3.1-6 acceptable because the applicant clarified the locations that will be added to the program as part of the enhancement. In addition, the applicant committed (Commitment No. 39) to establishing actions limits such that it will take corrective actions before exceeding the CUF design limit. Further, the applicant will continue to monitor these locations to ensure the design limit of 1.0 is not exceeded. The staff's concern described in RAI B3.1-6 is resolved.

During its review, the staff was unclear whether the applicant verified that the plant-specific locations listed in LRA Table 4.3-11 per NUREG/CR-6260 were bounding for the generic NUREG/CR-6260 components as well as for the plant. The staff requested the applicant to address this issue; the staff's evaluation of the applicant's response is found in SER Section 4.3.4.2.

During its audit, the staff reviewed the applicant's program basis document and noted that Commitment No. 39 stated that the Metal Fatigue AMP will be enhanced with additional CC and fatigue usage action limits. However, the staff noted the applicant did not provide information on what additional CC and fatigue usage action limits will be included in the Metal Fatigue AMP. By letter dated December 29, 2009, the staff issued RAI B3.1-7 asking the applicant to provide the additional CC and fatigue usage action limit that will be included in the Metal Fatigue AMP as enhancements.

In its response, dated February 19, 2010, the applicant stated that the current surveillance test procedure requires action when 90 percent of the allowable cycles are achieved for any monitored transient. The applicant further stated that during extended operation, projections indicate that certain allowable cycles and fatigue limits may be approached. Therefore, specific and targeted action limits are necessary to ensure actual fatigue limits are not exceeded. The applicant stated that it has not yet developed those action limits and, as it implements the transition to FatiguePro®, there are certain embedded administrative tools in FatiguePro® that will allow for specification of action limits based on projected fatigue usage at specific locations that account for actual cumulative fatigue. The applicant stated that the action limits can be based on the time required to implement expected or projected mitigating actions (such as component replacements or revisions to ASME Code Fatigue Analysis of Record) prior to exceeding actual fatigue limits.

The staff noted that the applicant is required to maintain actual fatigue limits below the design limits. The staff further noted that the applicant may select actions limits to implement mitigating actions, such as component replacements or revisions to ASME Code Fatigue Analysis of Record, as the applicant deems appropriate. The staff noted that the applicant's Commitment No. 39, as amended, will require the applicant to perform corrective actions before the remaining number of allowed occurrences for any specified transient becomes less than 1.0.

Aging Management Review Results

The staff also noted that the CUF action limits will be set so that there will be two or three operating cycles to initiate and complete corrective actions to ensure the design limit is not exceeded.

Based on its review, the staff finds the applicant's response to RAI B3.1-7 acceptable because, the applicant's current action limits of 90 percent of the allowable cycles for any monitored transient provides time for corrective actions to be taken before exceeding design limits. In addition, the CC limits for the enhanced program provides for corrective actions to be taken prior exceeding the design allowable limits, and the CUF action limits for the enhanced program will provide two or three operating cycles so corrective actions are taken prior to exceeding the design limit. The staff's concern described in RAI B3.1-7 is resolved.

During its audit, the staff reviewed the applicant's program basis document and noted that the scope of the Metal Fatigue AMP will be enhanced with a revised list of monitored plant transients that contribute to high usage factor as described for the "parameters monitored or inspected" and "monitoring and trending" program elements. The staff noted that the applicant did not describe this enhancement in Commitment No. 39 for the Metal Fatigue AMP. By letter dated December 29, 2009, the staff issued RAI B3.1-8 requesting the applicant provide additional information on how Commitment No. 39 will be revised to incorporate the enhancement to the Metal Fatigue AMP related to the revised list of monitored plant transients that contribute to high usage factor. The staff also asked that the applicant clarify the implementation schedule for the fatigue usage calculations described in Commitment No. 39.

In its response, dated February 19, 2010, the applicant amended Commitment No. 39. Subsequently, by letter dated May 27, 2010, the applicant further modified LRA Section B3.1 in Amendment 16, removing the following enhancement:

> Parameters Monitored or Inspected – Element 3/Monitoring and Trending – Element 5
>
> The scope of the Metal Fatigue of Reactor Coolant Pressure Boundary program will be enhanced with a revised list of monitored plant transients that contribute to high usage factor, and with a revised list of monitored locations in Class 1 piping and vessels and in parts of the Class 2 steam generators that have a Class 1 analysis.

The staff noted that in Amendment 16, for the enhancement of the "scope of program" element, the applicant stated, in part:

> Metal Fatigue of Reactor Coolant Pressure Boundary program will monitor plant transients as required by PVNGS Technical Specification 5.5.5.

The staff noted that the applicant will monitor those transients that are specified in TS 5.5.5, which refers to the transients listed in UFSAR Section 3.9.1.1. The staff finds this acceptable because the applicant will monitor its design transients consistent with its TS and CLB. The staff noted that the applicant's implementation schedule of "no later than two years prior to the period of extended operation" is consistent with the SRP-LR Section 3.0.1.

Based on its review, the staff finds the applicant's response to RAI B3.1-8 acceptable because, the applicant clarified that its enhanced program will monitor plant transients as required by TS 5.5.5 and CLB, and the applicant's implementation of the enhancement will be before the period of extended operation, consistent with the SRP-LR. The staff's concern described in RAI B3.1-8 is resolved.

The staff also reviewed the portions of the program elements associated with the enhancements to determine if the program will be adequate to manage the aging effects for which it is credited. The staff's evaluation of these enhancements follows.

Aging Management Review Results

Enhancement 1. In LRA Section B3.1, as amended, the applicant identifies an enhancement to the "scope of program" element for implementation of the following activities:

- CUF tracking for environmental component locations (NUREG/CR-6260 locations) that are not monitored using the applicant's CC monitoring method

- use of FatiguePro® software as the CC and CBF monitoring method basis and use of a six-component stress tensor, three-dimensional software program for implementation of the applicant's SBF monitoring methodology

- tracking of plant transients required by TS 5.5.5

The staff's review determined that this enhancement has been appropriately accounted for in the changes to Commitment No. 39, as submitted to the staff in a letter dated May 27, 2010. The staff also determined that the enhancement ensures that the "scope of program" element will be consistent with the GALL Report Metal Fatigue AMP when the enhanced program is implemented before the period of extended operation.

The staff's review included applicable resolution of RAI 4.3-2, and RAIs 4.3-7 through RAI 4.3-10, associated with the scope of the applicant's program (see SER Section 4.3.2.1 for a detailed discussion of these RAIs). On the basis of its review, the staff finds this enhancement acceptable because, when implemented prior to the period of extended operation, it will make the program consistent with the recommendations in the GALL Report Metal Fatigue AMP.

Enhancement 2. LRA Section B3.1, as amended, identifies an enhancement to the "preventative actions," "acceptance criteria," and "corrective actions" program elements of the Metal Fatigue AMP. This enhancement updates the summary descriptions to reflect the applicable changes made to the action limits and corrective actions for CC and CUF activities.

The staff's review determined that this enhancement has been appropriately accounted for in the changes to Commitment No. 39. Further, the staff verified that the enhancement program elements for this AMP will be consistent with the corresponding program elements in the GALL Report Metal Fatigue AMP when the enhanced program is implemented during the period of extended operation.

The staff noted that GALL Report Metal Fatigue AMP acceptance criteria maintains the fatigue usage below the design code limit of 1.0, considering environmental fatigue effects and corrective actions that prevent the usage factor from exceeding the design code limit during the period of extended operation. The staff noted that the applicant's program when enhanced also will contain action limits that ensure timely corrective actions such that the CUF of a particular component does not exceed the design limit of 1.0.

The review included applicable resolution of RAIs 4.3-11 and RAI 4.3-12, which are associated with the "acceptance criteria" and "corrective actions" program elements for the applicant's program as documented in SER Section 4.3.1. The staff's evaluation of environmentally-assisted fatigue is documented in SER Section 4.3.4.2. On the basis of its review and responses to RAIs 4.3-11 and RAI 4.3-12, the staff finds this enhancement acceptable because when it is implemented prior to the period of extended operation and it will make the program consistent with the recommendations in the GALL Report AMP.

Enhancement 3 and 4. LRA Section B3.1, as amended, identifies an enhancement to the "parameters monitored or inspected," "detection of aging effects," and "monitoring and trending" program elements. The staff noted that these enhancements reflect the applicant's proposed use of FatiguePro® software and use of a three-dimensional, six-element stress tensor software program.

The staff noted that the applicant modified Commitment No. 39 to reflect the use of a fatigue monitoring software program and a methodology for SBF monitoring that will implement a

Aging Management Review Results

three-dimensional, six-element stress tensor analysis method and will conform to the requirements of ASME Section III Article NB-3200. This resolves the staff's concern because the applicant is no longer using the monitoring methodology that was of concern in RIS-2008-30. The staff also noted that the use of this software allows the applicant to monitor cycles and the CUF to ensure that the design limit of 1.0 is not exceeded consistent with the recommendations in the GALL Report AMP.

The staff's review determines this enhancement is appropriately accounted for in the changes to Commitment No. 39, and the enhancement in the commitment ensures that the "parameters monitored or inspected," "detection of aging effects," and "monitoring and trending" program elements will be consistent with the corresponding program elements in the GALL Report Metal Fatigue AMP when the enhanced program is implemented during the period of extended operation.

Based on its audit and review of the applicant's responses to RAIs, the staff finds that elements one through six of the applicant's Metal Fatigue of Reactor Coolant Pressure Boundary Program, with enhancements 1 through 4, are consistent with the corresponding program elements of the GALL Report Metal Fatigue AMP and are, therefore, acceptable.

Operating Experience. LRA Section B3.1 summarizes operating experience related to the Metal Fatigue AMP. The staff noted that this section discusses the operating experience that led to the industry's development of the FatiguePro® software. The applicant provided the following examples where it had implemented corrective actions based on the existing program's CC and CUF monitoring bases:

- implementation of weld overlays on pressurizer surge line and spray line nozzles and hot leg surge nozzles in order to address the potential for thermal stratification

- replacements of the auxiliary spray line and main spray line components to address concerns raised in Bulletin 88-08

- implementation of linear elastic fracture mechanics evaluations for fatigue flaws detected in the Unit 2 pressurizer support skirt forging weld during an ISI

- actions to address CE Owners Group recommendations and initiatives on fatigue induced surge line microcracking

The staff reviewed operating experience information, in the application and during the audit, to determine if the applicant reviewed the applicable aging effects and industry and plant-specific operating experience. As discussed in the Audit Report, the staff conducted an independent search of the plant operating experience information to determine if the applicant had adequately incorporated and evaluated operating experience related to this program. The staff found that the applicant had adequately identified and incorporated industry and applicable plant-specific operating experience.

Based on its audit and review of the application, including LRA amendments, the staff finds that operating experience related to the applicant's program demonstrates that it can adequately manage the detrimental effects of aging on SSCs within the scope of the program and that it satisfies the criterion in SRP-LR Section A.1.2.3.10.

UFSAR Supplement. LRA Section A2.1 provides the UFSAR supplement for the Metal Fatigue AMP, as amended by LRA Amendment 16 (May 27, 2010). The staff reviewed this UFSAR supplement description of the program against the recommended description for this type of program, as described in SRP-LR Table 4.3-2.

Amendment 16 modifies LRA Appendix A, Section A2.1 in order to make it consistent with the changes made in LRA Appendix B, Section B3.1. The staff noted the applicant's changes included updates of the summary description to reflect:

Aging Management Review Results

- applicable TS 5.5.5 cycle tracking requirements

- applicable changes to the action limits and corrective actions on CC monitoring activities

- applicable changes to the action limits and corrective actions on CUF monitoring activities

- proposed use of FatiguePro® for the program's CC and CBF monitoring bases and of a three-dimensional, six-component stress tensor software methodology for the program's SBF monitoring basis

The staff also noted that the amendment included changes to Commitment No. 39 in order to reflect the changes to the four enhancements for the program. The staff determines that the information in the UFSAR supplement is an adequate summary description of the program, as required by 10 CFR 54.21(d).

Conclusion. On the basis of its audit, review of the applicant's Metal Fatigue of Reactor Coolant Pressure Boundary Program, and responses to the staff's RAIs, the staff determines that those program elements for which the applicant claimed consistency with the GALL Report are consistent. Also, the staff reviewed the enhancements and confirmed that their implementation through Commitment No. 39 prior to the period of extended operation would make the existing AMP consistent with the GALL Report AMP to which it was compared. The staff concludes that the applicant has demonstrated that the effects of aging will be adequately managed so that the intended functions will be maintained consistent with the CLB for the period of extended operation, as required by 10 CFR 54.21(a)(3). The staff also reviewed the UFSAR supplement for this AMP and concludes that it provides an adequate summary description of the program, as required by 10 CFR 54.21(d).

3.0.3.2.23 Concrete Containment Tendon Prestress Program

Summary of Technical Information in the Application. LRA Section B3.3 describes the existing Concrete Containment Tendon Prestress Program with enhancements that are consistent with the program elements in GALL Report AMP X.S1, "Concrete Containment Tendon Prestress." The applicant states that the Concrete Containment Tendon Prestress Program and is within the ASME Section XI Subsection IWL Program. The program manages the loss of tendon prestress aging effect in the post-tensioning system and is consistent with supplemental requirements of 10 CFR 50.55a.

The applicant stated that prior to September 1996, the tendon examinations were controlled by RG 1.35, "Inservice Inspection of Ungrouted Tendons in Prestressed Concrete Containments." Since 2001, ASME XI Subsection IWL, 1992 edition with 1992 addenda controls this program. The beginning of the second 10-year inspection interval will be August 1, 2011, for all three units, and the program will be updated for subsequent intervals as required by 10 CFR 50.55a(b)(2)(vi) and (viii), and 10 CFR 50.55a(g)(4)(ii).

The applicant further stated that tendon lift-off surveillances were performed for Units 1 and 3 at one, three and five year intervals following the post-structural-integrity test, then at 5-year intervals. Unit 2 tendons were examined using visual and other methods, but that the lift-off test surveillances were encompassed within the Unit 1 tests under rules applicable at that time to two unit plants with virtually-identical containments. A licensing change under RR-4 imposed similar lift-off testing on Unit 2, beginning with its 20th year, but it extended the surveillance interval to 10 years.

The applicant went on to state the following:

> The PVNGS post-tensioning system consists of inverted-U-shaped vertical tendons, extending up through the basemat, through the full height of the cylindrical walls and over the dome; and horizontal circumferential (hoop)

3-125

tendons, at intervals from the basemat to about the 45-degree elevation of the dome. The tendons are ungrouted, in grease-filled ducts.

The design basis of the containment system requires that the average prestresses of the tendons in the horizontal dome and cylinder hoop tendon subgroups, and in the vertical tendon group, remain above their respective minimum required values (MRVs). The MRVs are from the original design bases and assumptions.

In order to ensure that the design basis continues to be met, the acceptance criteria require that the prestress in each tendon remain above, or within a stated tolerance below, the predicted force. The predicted lower limit (PLL) described in NUREG-1801 is functionally equivalent to the first action level of the PVNGS and IWL 3221.1 acceptance criteria, at 95 percent of the predicted force line. The surveillance program predicted mean prestress force lines and their tolerance-band upper and lower limit lines, and the predicted forces for each surveillance tendon, were developed from the loss of prestress model used for the original design, and are consistent with Regulatory Guide 1.35.1, Proposed Revision 0.

The LRA states that the first set of regression analyses of tendon lift-off data were performed in support of this LRA. The applicant asserts that the regression analyses of surveillance data are consistent with IN 99-10, "*Degradation of Prestressing Tendon Systems in Prestressed Concrete Containments*, Attachment 3" and that the program will be enhanced to continue to compare regression analysis trend lines. The comparisons will include the individual lift-off values of tendons surveyed to date in each of the vertical and hoop tendon groups, with the MRV and PLL for each tendon group to the end of the licensed operating period. The applicant will take appropriate corrective action if future values indicated by the regression analysis trend line drop below the PLL or MRV. The applicant will updated regression analyses for tendons of the affected unit and for a combined data set of all three units following each inspection.

Staff Evaluation. During its audit, the staff reviewed the applicant's claim of consistency with the GALL Report. The staff also reviewed and confirmed that the plant's conditions are bounded by the conditions for which the GALL Report was evaluated. The staff compared elements one through six of the applicant's program to the corresponding elements of the GALL Report AMP. As discussed in the Audit Report, the staff confirmed that elements "scope of program," "preventive actions," "parameters monitored or inspected," and "detection of aging effects" of the LRA AMP were consistent with the corresponding elements of the GALL Report AMP.

Enhancement. The applicant has committed (Commitment No. 41) to enhancing the AMP program elements "monitoring and trending" and "acceptance criteria." These enhancements expand on the existing program elements, and require an update of the plant procedures for the regression analysis methods, including the use of individual tendon data in accordance with IN 99-10, "Degradation of Prestressing Tendon Systems in Prestressed Concrete Containments." The staff has determined that these enhancements are acceptable because the program description section of the GALL Report Concrete Containment Tendon Prestress AMP recommends the use of IN 99-10 for constructing the trend lines (regression analysis).

Operating Experience. LRA Section B3.3 summarizes operating experience related to the Concrete Containment Tendon Prestress Program. The applicant stated that qualified personnel, familiar with the tendon performance history and with current issues and practices, conduct the tendon surveillance and lift-off tests. The program employs examination procedures which invoke, and are developed from, the design criteria (MRVs), inspection acceptance criteria (predicted force lines and their tolerance band upper and lower limit lines

Aging Management Review Results

and their application to action levels and corrective actions), inspection schedule, sample selection, and effective stress calculation methods.

The applicant further stated that the tendon inspections to date have shown the following:

- no evidence of significant corrosion or other effects that might damage wires
- minimum wire breakage (after initial installation)
- no accelerated loss of prestress due to high temperatures or other causes

The only significant findings were pinhole grease leaks in Units 2 and 3, which were evaluated as having no detrimental effects on the containment structure.

The applicant also stated that the most recent results of the Concrete Containment Tendon Prestress Program demonstrated that average prestress in both the vertical tendon group and the horizontal cylinder and horizontal dome tendon subgroups should remain above the applicable MRVs for at least 60 years of operation. Therefore, all tendons should maintain their design basis function for the extended period of operation. The material condition of other components (e.g., concrete, bearing surfaces, grease, buttonheads, etc.) showed only minor degradation in a few areas; none indicating a need for significant corrective action.

The staff reviewed operating experience information, in the application and during the audit, to determine if the applicant reviewed the applicable aging effects and industry and plant-specific operating experience. As discussed in the Audit Report, the staff conducted an independent search of the plant operating experience information to determine if the applicant had adequately incorporated and evaluated operating experience related to this program.

The GALL Report Concrete Containment Tendon Prestress AMP states that IN 99-10 provides guidance for constructing the trend line. However, the LRA Concrete Containment Tendon Prestress Program, "monitoring and trending" element states that the program will be enhanced to require a regression analysis for each tendon group after every surveillance. The applicant performed tendon surveillance for Units 1, 2, and 3 during 2008, 2006, and 2002, respectively. However, according to the LRA AMP, the applicant has not revised the Concrete Containment Tendon Prestress Program document until now. Therefore, the staff requested, in RAI B3.3-1 issued in a letter dated December 27, 2009, that the applicant to provide the status and conclusions of the regression analysis performed in accordance with IN 99-10.

In response to the RAI B3.3-1, in a letter dated February 19, 2010, the applicant stated that it performed regression analyses of tendon lift-off data in support of the LRA. The regression analyses of surveillance data are consistent with IN 99-10. The applicant will enhance the program to continue to compare regression analysis trend lines of the individual lift-off values of tendons surveyed to date. LRA Table A4-1, Commitment No. 41, documents this enhancement.

The applicant further stated that it extended a regression analysis of the lift-off data of the most recent results to date to 60 years. This analysis demonstrated that the average prestress in tendons should remain above the applicable minimum required values for at least 60 years of operation and that all tendons should, therefore, maintain their design basis function for the extended period of operation.

The staff finds the response to RAI B3.3-1 acceptable because the applicant has performed regression analysis using the individual tendon data to construct the lift-off trend lines in accordance with IN 99-10 for both the vertical tendon group and the horizontal cylinder and horizontal dome tendon subgroups. In addition, the applicant has committed to enhancing the tendon surveillance procedures prior to the period of extended operation. These procedures will require an update of the regression analysis for each tendon group of each unit.

Based on its audit and review of the application, the staff finds that operating experience related to the applicant's program demonstrates that it can adequately manage the detrimental effects

Aging Management Review Results

of aging on SSCs within the scope of the program and that implementation of the program has resulted in the applicant taking corrective actions. The staff confirmed that the "operating experience" program element satisfies the criterion in SRP-LR Section A.1.2.3.10 and is, therefore, acceptable.

UFSAR Supplement. LRA Section A2.3 provides the UFSAR supplement for the Concrete Containment Tendon Prestress Program. The staff reviewed this UFSAR supplement description of the program and notes that it conforms to the recommended description for this type of program, as described in SRP-LR Table 3.5-2. The staff also notes that the applicant committed (Commitment No. 41) to the ongoing implementation of the existing Concrete Containment Tendon Prestress Program with enhancements to manage the aging of the concrete containment and post-tensioning system during the period of extended operation. The staff determines that the information in the UFSAR supplement is an adequate summary description of the program, as required by 10 CFR 54.21(d).

Conclusion. On the basis of its review of the applicant's Concrete Containment Tendon Prestress Program, the staff finds all program elements consistent with the GALL Report. Also, the staff reviewed the enhancement and confirmed that its implementation through Commitment No. 41 before the period of extended operation would make the existing AMP consistent with the GALL Report AMP to which it was compared. The staff concludes that the applicant has demonstrated that the effects of aging will be adequately managed so that the intended functions will be maintained consistent with the CLB for the period of extended operation, as required by 10 CFR 54.21(a)(3). The staff also reviewed the UFSAR supplement for this AMP, and concludes that it provides an adequate summary description of the program, as required by 10 CFR 54.21(d).

3.0.3.3 Aging Management Programs Not Consistent with or Not Addressed in the Generic Aging Lessons Learned Report

In LRA Appendix B, the applicant identified the following AMPs as plant-specific:

- Nickel-Alloy AMP
- Protective Coatings Monitoring and Maintenance Program

For AMPs not consistent with or not addressed in the GALL Report, the staff performed a complete review to determine their adequacy to monitor or manage aging. The following section documents the staff's review of these plant-specific AMPs.

3.0.3.3.1 Nickel-Alloy Aging Management Program

Summary of Technical Information in the Application. LRA Section B2.1.34 describes the existing Nickel-Alloy AMP as plant-specific. The applicant stated that the Nickel-Alloy AMP manages cracking due to SCC in most components fabricated from alloy 600 (including alloy 82/182 weld metal) in the RCS and engineering safety features (ESF) systems. The applicant also stated that the Nickel-Alloy AMP uses inspections, mitigation techniques, repair or replace activities and monitoring of operating experience to manage the aging of alloy 600.

Staff Evaluation. The staff reviewed program elements 1–6 and 10 of the applicant's program against the acceptance criteria for the corresponding elements, as stated in SRP-LR Section A.1.2.3. The staff's review focused on how the applicant's program manages aging effects through the effective incorporation of these program elements. The staff's evaluation of each of these elements follows.

Prior to that evaluation, it may prove helpful to consider the manner in which this AMP is used to manage aging. This AMP appears only in LRA Table 2 items, which are subordinate to LRA Table 1, items 3.1.1.31 and 3.1.1.34. These table 1 items are subject to further evaluation in LRA paragraphs 3.1.2.2.13 and 3.1.2.2.16.1.

Aging Management Review Results

For LRA Table 1, items 3.1.1.31 and 3.1.1.34 (see GALL Report, Table 3.1-1, items 31 and 34), the GALL Report recommends that aging management be accomplished using the ASME Section XI ISIs IWB, IWC, and IWD AMP and the Water Chemistry AMP. For Nickel Alloys, the GALL Report further recommends that applicants comply with all NRC orders, and provide a commitment in the UFSAR supplement to implement applicable bulletins, GLs, and staff-accepted industry guidelines.

For each LRA Table 2 item that cites the Nickel-Alloy AMP, it also includes the applicant's ASME Section XI ISI IWB, IWC, and IWD AMP; Water Chemistry AMP; and commitment to comply with NRC orders, bulletins, GLs, and staff-accepted industry guidelines. That is, the approach to aging management for these components offered by the applicant, contains all aspects recommended by the GALL Report, in addition to the Nickel-Alloy AMP.

The staff's review of the ASME Section XI ISI IWB, IWC, and IWD AMP is in SER Section 3.0.3.1.1 and the Water Chemistry AMP is in SER Section 3.0.3.2.1. Given that other sections of this SER evaluate these LRA AMPs, and have found them consistent with the GALL Report, the staff finds that the use of this AMP is beneficial, but not essential, for the applicant to achieve consistency with the GALL Report.

Based on this finding, the remainder of this evaluation is based on the consistency of the AMP with the SRP-LR and the sufficiency of the AMP to accomplish its stated objectives.

Scope of the Program. LRA Section B2.1.34 states that the Nickel-Alloy AMP is an existing program that manages the effects of aging in alloy 600 components (including alloy 82/182 weld metal) found in the RCS and the ESF system. The scope of the program element contains a specific list of components, which are included within the scope, as well as a general specification of components that are excluded.

The staff reviewed the applicant's "scope of the program" program element against the criteria in SRP-LR Section A.1.2.3.1, which states that the program should include the specific SCs for which the program manages aging.

Based on the list provided, which addresses materials and components included within the scope of the AMP, the staff confirmed that the "scope of the program" program element satisfies the criterion defined in SRP-LR Section A.1.2.3.1 and is, therefore, acceptable.

Preventive Actions. LRA Section B2.1.34 states that several techniques are available to mitigate cracking due to PWSCC. These techniques remove one or more of the conditions necessary to cause cracking (i.e., susceptible material, tensile stress, specific environment). The section provides a specific list of components and potential preventive actions.

LRA Section B2.1.34 also states that the "Water Chemistry Program (B2.1.2) provides preventive actions for monitoring and control of the supporting environment for PWSCC."

The staff reviewed the applicant's "preventive actions" program element against the criteria in SRP-LR Section A.1.2.3.2, which states that the applicant should describe activities for prevention and mitigation programs.

Based on the LRA description of the available mitigative techniques and the description of where these techniques may be employed, the staff confirmed that the "preventive actions" program element satisfies the criterion defined in SRP-LR Section A.1.2.3.2 and is, therefore, acceptable.

Parameters Monitored or Inspected. LRA Section B2.1.34 states that the program monitors for cracking due to PWSCC through a combination of visual, surface, and volumetric exams. These exams directly detect cracking or detect the presence of boric acid, which may be deposited on visible surfaces as a result of a through-wall crack.

Aging Management Review Results

The staff reviewed the applicant's "parameters monitored or inspected" program element against the criteria in SRP-LR Section A.1.2.3.3. This section states that the applicant should identify the parameters to be monitored or inspected and link them to the degradation of the particular SC intended function. The parameter monitored or inspected, in a condition monitoring program, should detect the presence and extent of aging effects.

The staff finds that, for the components under consideration, cracking is the degradation mechanism that will affect the intended function and that a combination of visual, surface, and volumetric exams will directly detect cracks or secondary evidence of cracks (e.g., boric acid). Based on this finding, the staff confirmed that the "parameters monitored or inspected" program element satisfies the criterion defined in SRP-LR Section A.1.2.3.3 and is, therefore, acceptable.

Detection of Aging Effects. LRA Section B2.1.34 states that the applicant uses visual, surface, and volumetric exams to detect cracking due to PWSCC in alloy 600 components. In this element, the applicant also provides a list of components to be inspected, the inspection methods to be used, and the reference documents containing the inspection requirement (e.g., CFR, ASME Code, ASME Code Case). These documents contain procedures for conducting the inspection as well as allowable inspection intervals.

The staff reviewed the applicant's "detection of aging effects" program element against the criteria in SRP-LR Section A.1.2.3.4, which states that detection of aging effects should occur before there is a loss of the SC intended function(s). The criteria also states that parameters to be monitored or inspected should be appropriate to ensure that the SC intended function will be adequately maintained for license renewal under all CLB design conditions. The criteria further state that a program based solely on detecting SC failure should not be considered as an effective AMP for license renewal. The criteria states that this program element describes "when," "where," and "how" program data are collected (i.e., all aspects of activities to collect data as part of the program). The criteria continue by stating that the method or technique and frequency may be linked to plant-specific or industry-wide operating experience.

In its review, the staff determined that cracking is an appropriate parameter to monitor to ensure the maintenance of intended function of the components under consideration. The staff also determined that a combination of visual, surface, and volumetric test methods were capable of detecting aging prior to loss of intended function. The staff further determined that, while observation of boric acid or other secondary evidence of cracking was not fully consistent with detection of aging prior to loss of function (pressure boundary in this case), it was an effective method of identifying aging prior to loss of actual function of the component (failure of a piping system to supply adequate water to cool the reactor core). The staff additionally determined that this element of the AMP refers to the CFR, ASME Code, and various code cases and that these documents contain the specifications (how, where, when) for these inspections. The staff finally determined that there is no industry or plant-specific operating experience, which necessitates deviating from the inspections proposed in this program element.

Based on the above evaluation, the staff confirmed that the "detection of aging effects" program element satisfies the criterion defined in SRP-LR Section A.1.2.3.4 and is, therefore, acceptable.

Monitoring and Trending. LRA Section B2.1.34 provides a detailed list of inspections to be conducted and, in conjunction with program element 4, frequencies of those inspections. The list of inspections refers to governing documents (CFR, ASME, ASME Code Cases). These documents provide guidance regarding the evaluation of inspection data against acceptance criteria and the timing of subsequent inspections to ensure that component intended function is not lost before the next inspection.

Aging Management Review Results

The staff reviewed the applicant's "monitoring and trending" program element against the criteria in SRP-LR Section A.1.2.3.5, which state that the applicant should describe monitoring and trending activities, which should provide predictability of the extent of degradation and, thus, ensure timely corrective or mitigative actions. The criteria also state that plant-specific or industry-wide operating experience may be considered in evaluating the appropriateness of the technique and frequency. The criteria further state that this program element describes "how" the applicant evaluates data collected. It may also include trending for a forward look, including an evaluation of the results against the acceptance criteria and a prediction of the rate of degradation, in order to confirm that timing of the next scheduled inspection will occur before a loss of SC intended function.

In this review, the staff determined that this program element adequately describes the monitoring and trending proposed. The staff also determined that the governing documents for the inspections to be monitored and trended provide sufficient guidance concerning inspection frequency and the modification of that frequency based on past inspections or other plant-specific or industry operating experience to provide timely corrective action or mitigation or additional inspections prior to loss of intended function. The staff further determined that the program element and the governing documents provided sufficient guidance to allow the applicant to compare collected data to applicable acceptance standards.

Based on the above evaluation, the staff confirmed that the "monitoring and trending" program element satisfies the criterion defined in SRP-LR Section A.1.2.3.5 and is, therefore, acceptable.

Acceptance Criteria. LRA Section B2.1.34 states that acceptance criteria for this program are contained in governing documents (Materials Reliability Program (MRP) 139, CFR, ASME Code, and ASME Code Cases).

The staff reviewed the applicant's "acceptance criteria" program element against the criteria in SRP-LR Section A.1.2.3.6, which states the acceptance criteria of the program and its basis should be described, including ensuring that the SC intended function(s) are maintained under all CLB design conditions during the period of extended operation. Acceptance criteria could be specific numerical values or could consist of a discussion of the process for calculating specific numerical values of conditional acceptance criteria to ensure that the SC intended function(s) will be maintained under all CLB design conditions. The applicant may cite information from available references. The acceptance criteria, which permit some degradation, are based on maintaining the intended function under all CLB design loads. The criteria further state that the applicant should conduct qualitative inspections to the same predetermined criteria as quantitative inspections by personnel, in accordance with ASME Code and through approved site-specific programs.

In its review, the staff determined that the acceptance criteria for these inspections are clearly defined in the program element or in the governing documents. The staff also has no reason to believe that these values, many of which carry the force of regulation, would not allow for the intended function of the components under consideration to be maintained during the period of extended operation under all CLB design loads.

Based on the above review, the staff confirmed that the "acceptance criteria" program element satisfies the criterion defined in SRP-LR Section A.1.2.3.6 and is, therefore, acceptable.

Operating Experience. LRA Section B2.1.34 summarizes operating experience related to the Nickel-Alloy AMP. In this program element, the applicant provided a detailed list of components that had been inspected and had shown no failures, components that had been inspected and had shown failures, and components that had shown no failures but had been proactively replaced or mitigated based solely on the potential for cracking.

Aging Management Review Results

The staff reviewed this information against the acceptance criteria in SRP-LR Section A.1.2.3.10, which states that the operating experience information provided should present objective evidence that the applicant will adequately manage the effects of aging so that the intended function(s) of the in-scope components and structures are maintained during the period of extended operation.

In this review, the staff found very few instances in which the inspections performed showed significant indications. In each of these cases, the applicant took actions indicating that this AMP was effective in addressing adverse inspection findings. Also during this review, the staff identified numerous instances where the applicant made proactive repairs or replacements. The staff views these activities as indications that the AMP is effective in indicating the potential for degradation and the need for preventive action. The staff considers the replacement of the RV heads to be significant evidence of the applicant's commitment to adequate aging management of nickel-alloy components.

Based on this review, the staff finds that operating experience related to the applicant's program demonstrates that it can adequately manage the detrimental effects of aging on SSCs within the scope of the program and that implementation of the program has resulted in the applicant taking corrective actions. The staff confirmed that the "operating experience" program element satisfies the criterion in SRP-LR Section A.1.2.3.10 and is, therefore, acceptable.

UFSAR Supplement. LRA Section A1.34 provides the UFSAR supplement for the Nickel-Alloy AMP. The staff reviewed this UFSAR supplement description of the program and notes that it conforms to the recommended description for this type of program, as described in SRP-LR Table 3.1-2. The staff also notes that the applicant committed (Commitment Nos. 7 and 36) to the ongoing implementation of the existing Nickel-Alloy Penetration Nozzles Welded to the Upper RV Closure Heads of PWRs Program and Nickel-Alloy AMP, to manage the aging of the RV head and penetrations nozzles during the period of extended operation. The staff determines that the information in the UFSAR supplement is an adequate summary description of the program, as required by 10 CFR 54.21(d).

Conclusion. The staff concludes that, based on the AMR items present in the application and the scope of other AMPs, the use of this AMP is not necessary to obtain consistency with the GALL Report. The staff also concludes that this AMP contains valuable information, which will help in the management of aging of nickel alloys.

On the basis of its technical review of the applicant's Nickel-Alloy AMP, the staff concludes that the applicant has demonstrated that, through the use of this AMP, the effects of aging of nickel alloys may be adequately managed so that the intended functions of the components under consideration will be maintained consistent with the CLB for the period of extended operation, as required by 10 CFR 54.21(a)(3). The staff also reviewed the UFSAR supplement for this AMP, and concludes that it provides an adequate summary description of the program, as required by 10 CFR 54.21(d).

3.0.3.3.2 Protective Coatings Monitoring and Maintenance Program

Summary of Technical Information in the Application. LRA Section 2.1.6.4 describes the generic safety issue (GSI)-191, "Assessment of Debris Accumulation on PWR Sump Performance," and states the following:

> By letter no. 102-05336, dated September 1, 2005, APS submitted to the NRC a response to Generic Letter (GL) 2004-02, "Potential Impact of Debris Blockage on Emergency Recirculation During Design Basis Accidents at Pressurized Water Reactors". The issues identified in GSI-191 and Generic Letter 2004-02 are not aging-related issues. Also, the issues are not related to the 40-year term of the current operating license, and, therefore, are not time-limited aging

Aging Management Review Results

analyses. The containment sumps are evaluated in Section 2.[4].1, Containment Building.

In LRA, Appendix B, Section B2.1.32, "Structures Monitoring Program," the applicant states that it takes no credit for coatings in the determination of aging effects for the underlying materials. Although the coatings are not credited for aging management, the staff believes their failure could impact the functioning of safety systems.

Staff Evaluation. The staff reviewed LRA Section 2.1.6.4 and Table B2, line item XI.S.8, "Protective Coating Monitoring and Maintenance Program," using GALL XI.S8 as guidance. As the applicant did not credit the Protective Coatings Monitoring and Maintenance Program for aging management, the staff was unable to review the program elements against the acceptance criteria, as stated in SRP-LR Section A.1.2.3. As a result, the staff determined the need for additional clarification, which resulted in issuance of RAI B2.1.32-1.

In RAI B2.1.32-1, dated August 11, 2009, the staff requested that the applicant provide details of the Coatings Assessment Program, referenced in the supplemental response to GL 2004-02, dated February 29, 2008. These details were needed to provide adequate assurance that protective coatings in containment would be properly maintained and not become a debris source that might challenge the ECCS.

In its response, by letter dated September 10, 2009, the applicant provided the following information:

> As noted in the RAI, the coatings program is not credited for Palo Verde Nuclear Generating Station (PVNGS) license renewal aging management and, as such, this program has not been included within the scope of license renewal. However, Arizona Public Service Company (APS) agrees that the coatings program is important for the Emergency Core Cooling System (ECCS) performance.
>
> The Palo Verde Containment Coatings Condition Assessment procedure defines the inservice monitoring program for containment coatings to include service level I coatings and unqualified coatings. Inspection results are used to maintain coating system integrity and as an input into the mechanical design of the ECCS sump and input into the Mechanical Design/Nuclear Fuel Management Design Basis Accident (DBA) Loss-of-Coolant Accident (LOCA) Analysis. The procedure directs visual inspections of accessible coated surfaces of each area or room of containment to look for defects such as:
>
> - Blistering
> - Cracking
> - Flaking/Peeling delamination
> - Rusting
> - Mechanical or physical damage
>
> Inspections are performed every operating cycle. Coatings identified as being degraded are documented in the corrective action process. The coating specialist reviews newly identified defects or damage to determine if further testing is required to identify the extent of the defect or damage and to answer any operability questions. Possible corrective actions include:
>
> - Recoating
> - Coating removal

Aging Management Review Results

- Repair at a future date (bare substrate addressed)
- Monitor and trend (coatings that are not added to unqualified coatings log)

Coatings inspectors are qualified to a certified coatings inspection program or are ANSI N45.2.6 certified.

The staff reviewed the response to RAI B2.1.32-1 and, during a follow-up discussion with the applicant, confirmed that the applicant was going to submit a letter to the NRC, with a commitment to incorporate the coating inspections described in the GL 2004-02 response into the UFSAR. This commitment would require coating inspections to continue into the period of extended operation. By letter dated May 7, 2010, the licensee submitted the commitment described in their response to RAI B2.1.32-1.

The staff finds the inspection frequency of the containment coatings to be acceptable since, at the frequencies stated above, it will provide adequate assurance that there is proper maintenance of the protective coatings so that they will not degrade and become a debris source that may challenge the ECCS. The staff also finds the scope of the program acceptable since it includes all accessible coated areas inside containment. The acceptance criterion is found to be acceptable since the staff has accepted and confirmed the acceptability of the ASTM standards proposed above using RG 1.54, Revision 1. The method of performing the coatings inspection is acceptable since the staff has confirmed that the ASTM standard proposed is acceptable using RG 1.54, Revision 1. The qualification of personnel who perform the inspection is acceptable since the staff has confirmed that the ANSI standard proposed is found to be acceptable using RG 1.54, Revision 1. Therefore, the staff's concern in RAI B2.1.32-1 is resolved.

The commitment, discussed under GSI-191 above, is a regulatory commitment under PVNGS CLB, and it is not captured in license renewal; however, since this commitment incorporates actions into the UFSAR, this will ensure GSI-191 related inspections continue into the period of extended operation.

Conclusion. On the basis of its technical review of the applicant's Protective Coating Monitoring and Maintenance Program, the staff concludes that the applicant has demonstrated that effects of aging will be adequately managed so that the intended function(s) will be maintained consistent with the CLB for the period of extended operation, as required by 10 CFR 54.21(a)(1).

3.0.4 Quality Assurance Program Attributes Integral to Aging Management Programs

3.0.4.1 Summary of Technical Information in the Application

In Appendix A, "Updated Final Safety Analysis Report Supplement," Section A1, "Summary Descriptions of Aging Management Programs," and Appendix B, "Aging Management Programs," Section B1.3, "Quality Assurance Program and Administrative Controls," of the LRA, the applicant described the elements of "corrective action," "confirmation process," and "administrative controls" that are applied to the AMPs for both safety-related and nonsafety-related components. Appendix A, Section A1 and Appendix B, Section B1.3 of the LRA state that the QA Program implements the requirements of 10 CFR Part 50, Appendix B, "Quality Assurance Criteria for Nuclear Power Plants and Fuel Reprocessing Plants," and is consistent with the SRP-LR.

3.0.4.2 Staff Evaluation

Under 10 CFR 54.21(a)(3), an applicant is required to demonstrate that the effects of aging on SCs subject to an AMR will be adequately managed so that their intended functions will be

Aging Management Review Results

maintained consistent with the CLB for the period of extended operation. The SRP-LR, branch technical position (BTP) RLSB-1, "Aging Management Review - Generic," describes ten attributes of an acceptable AMP. Three of these ten attributes are associated with the QA activities of corrective action (element 7), confirmation process (element 8), and administrative controls (element 9). Table A.1-1, "Elements of an Aging Management Program for License Renewal," of BTP RLSB-1 provides the following description of these quality attributes:

> **Corrective Actions** – Corrective actions, including root cause determination and prevention of recurrence, should be timely.
>
> **Confirmation Process** – Confirmation process should ensure that preventive actions are adequate and that appropriate corrective actions have been completed and are effective.
>
> **Administrative Controls** – Administrative controls should provide a formal review and approval process.

The SRP-LR, BTP IQMB-1, "Quality Assurance for Aging Management Programs," states that those aspects of the AMP that affect quality of safety-related SSCs are subject to the QA requirements of 10 CFR Part 50, Appendix B. Additionally, for nonsafety-related SCs subject to an AMR, the applicant's existing 10 CFR Part 50, Appendix B, QA Program may address the elements of corrective action, confirmation process, and administrative control. BTP IQMB-1 provides the following guidance with regard to the QA attributes of AMPs:

> Safety-related structures and components are subject to 10 CFR Part 50 Appendix B requirements, which are adequate to address all quality-related aspects of an aging management program consistent with the CLB of the facility for the period of extended operation.

The SRP-LR, Appendix A.2, BTP IQMB-1 states that for nonsafety-related SCs that are subject to an AMR for license renewal, an applicant has an option to expand the scope of its Appendix B to 10 CFR Part 50 Program to include these SCs to address corrective action, confirmation process, and administrative control for aging management during the period of extended operation. In this case, the applicant should document such a commitment in the UFSAR supplement, in accordance with 10 CFR 54.21(d).

The staff reviewed the applicant's AMPs described in Appendix A and Appendix B of the LRA as well as the associated implementing procedures. The purpose of this review was to ensure that the QA attributes (e.g., corrective action, confirmation process, and administrative controls) were consistent with the staff's guidance described in BTP IQMB-1. The staff also audited these attributes and reported the results in an Audit Report dated April 7, 2010. Based on the staff's evaluation and audit, the descriptions of the AMPs and their associated quality attributes provided in Appendix A, Section A1, and Appendix B, Section B1.3, of the LRA are consistent with the staff's position regarding QA for aging management.

3.0.4.3 Conclusion

On the basis of the NRC staff's evaluation, the descriptions and applicability of the PVNGS AMPs and their associated quality attributes provided in Appendix A, Section A1, and Appendix B, Section B1.3, of the LRA, were determined to be consistent with the staff's position regarding QA for aging management. The staff concludes that the QA attributes (corrective action, confirmation process, and administrative control) of the applicant's AMPs are consistent with 10 CFR 54.21(a)(3).

Aging Management Review Results

3.1 Aging Management of Reactor Vessel, Internals and Reactor Coolant System

This section of the SER documents the staff's review of the applicant's AMR results for the RV, internals, and RCS components and component groups for the following:

- RV and internals
- Reactor coolant system pressurizer
- Steam generators

3.1.1 Summary of Technical Information in the Application

LRA Section 3.1 provides AMR results for the RV, RV internals, and RCS components and component groups. LRA Table 3.1.1, "Summary of Aging Management Evaluations in Chapter IV of NUREG-1801 [GALL Report] for the RV, Internals, and Reactor Coolant System," is a summary comparison of the applicant's AMRs with those evaluated in the GALL Report for the RV, RV internals, and RCS components and component groups.

The applicant's AMRs evaluated and incorporated applicable plant-specific and industry operating experience in the determination of AERMs. The plant-specific evaluation included condition reports and discussions with appropriate site personnel to identify AERMs. The applicant's review of industry operating experience included a review of the GALL Report and operating experience issues identified since the issuance of the GALL Report.

3.1.2 Staff Evaluation

The staff reviewed LRA Section 3.1 to determine if the applicant provided sufficient information to demonstrate that it will adequately manage the effects of aging for the RV, RV internals, and RCS components, within the scope of license renewal and subject to an AMR, so that the intended function(s) will be maintained consistent with the CLB for the period of extended operation, as required by 10 CFR 54.21(a)(3).

The staff reviewed AMRs to ensure the applicant's claim that certain AMRs were consistent with the GALL Report. The staff did not repeat its review of the matters described in the GALL Report; however, the staff did verify that the material presented in the LRA was applicable and that the applicant identified the appropriate GALL Report AMRs. SER Section 3.0.3 documents the staff's evaluations of the AMPs, and SER Section 3.1.2.1 documents the staff's evaluations.

The staff also reviewed AMRs consistent with the GALL Report and for which further evaluation is recommended. The staff confirmed that the applicant's further evaluations were consistent with the SRP-LR Section 3.1.2.2 acceptance criteria. SER Section 3.1.2.2 documents the staff's evaluations.

The staff also conducted a technical review of the remaining AMRs not consistent with, or not addressed in, the GALL Report. The technical review evaluated whether the applicant noted all plausible aging effects and whether the aging effects listed were appropriate for the material-environment combinations specified. SER Section 3.1.2.3 documents the staff's evaluations.

For SSCs that the applicant claimed were not applicable or required no aging management, the staff reviewed the AMR items and the plant's operating experience to verify the applicant's claims.

Table 3.1-1 summarizes the staff's evaluation of components, aging effects or mechanisms, and AMPs listed in LRA Section 3.1 and addressed in the GALL Report.

Table 3.1-1. Staff Evaluation for Reactor Vessel, Reactor Vessel Internals and Reactor Coolant System Components in the GALL Report

Component Group (GALL Report Item No.)	Aging Effect/ Mechanism	AMP in GALL Report	Further Evaluation in GALL Report	AMP in LRA, Supplements, or Amendments	Staff Evaluation
Steel pressure vessel support skirt and attachment welds (3.1.1-1)	Cumulative fatigue damage	TLAA, evaluated in accordance with 10 CFR 54.21(c)	Yes	Not applicable	Not applicable to PVNGS. See SER Section 3.1.2.1.1.
Steel; stainless steel; steel with nickel-alloy or stainless steel cladding; nickel-alloy RV components: flanges; nozzles; penetrations; safe ends; thermal sleeves; vessel shells, heads and welds (3.1.1-2)	Cumulative fatigue damage	TLAA, evaluated in accordance with 10 CFR 54.21(c) and environmental effects are to be addressed for Class 1 components	Yes	Not applicable	Not applicable to PWRs. See SER Section 3.1.2.1.1.
Steel; stainless steel; steel with nickel-alloy or stainless steel cladding; nickel-alloy RCPB piping, piping components, and piping elements exposed to reactor coolant (3.1.1-3)	Cumulative fatigue damage	TLAA, evaluated in accordance with 10 CFR 54.21(c) and environmental effects are to be addressed for Class 1 components	Yes	Not applicable	Not applicable to PWRs. See SER Section 3.1.2.1.1.
Steel pump and valve closure bolting (3.1.1-4)	Cumulative fatigue damage	TLAA, evaluated in accordance with 10 CFR 54.21(c) check Code limits for allowable cycles (less than 7000 cycles) of thermal stress range	Yes	Not applicable	Not applicable to PWRs. See SER Section 3.1.2.1.1.
Stainless steel and nickel-alloy RV internals components (3.1.1-5)	Cumulative fatigue damage	TLAA, evaluated in accordance with 10 CFR 54.21(c)	Yes	TLAA	Consistent with GALL Report. See SER Section 3.1.2.2.1.
Nickel-alloy tubes and sleeves in a reactor coolant and secondary feedwater/steam environment (3.1.1-6)	Cumulative fatigue damage	TLAA, evaluated in accordance with 10 CFR 54.21(c)	Yes	TLAA	Consistent with GALL Report. See SER Section 3.1.2.2.1.

Aging Management Review Results

Component Group (GALL Report Item No.)	Aging Effect/ Mechanism	AMP in GALL Report	Further Evaluation in GALL Report	AMP in LRA, Supplements, or Amendments	Staff Evaluation
Steel and stainless steel RCPB closure bolting, head closure studs, support skirts and attachment welds, pressurizer relief tank components, SG components, piping and components external surfaces and bolting (3.1.1-7)	Cumulative fatigue damage	TLAA, evaluated in accordance with 10 CFR 54.21(c)	Yes	TLAA	Consistent with GALL Report. See SER Section 3.1.2.2.1.
Steel; stainless steel; and nickel-alloy RCPB piping, piping components, piping elements; flanges; nozzles and safe ends; pressurizer vessel shell heads and welds; heater sheaths and sleeves; penetrations; and thermal sleeves (3.1.1-8)	Cumulative fatigue damage	TLAA, evaluated in accordance with 10 CFR 54.21(c) and environmental effects are to be addressed for Class 1 components	Yes	TLAA	Consistent with GALL Report. See SER Section 3.1.2.2.1.
Steel; stainless steel; steel with nickel-alloy or stainless steel cladding; nickel-alloy RV components: flanges; nozzles; penetrations; pressure housings; safe ends; thermal sleeves; vessel shells, heads and welds (3.1.1-9)	Cumulative fatigue damage	TLAA, evaluated in accordance with 10 CFR 54.21(c) and environmental effects are to be addressed for Class 1 components	Yes	TLAA	Consistent with GALL Report. See SER Section 3.1.2.2.1.
Steel; stainless steel; steel with nickel-alloy or stainless steel cladding; nickel-alloy SG components (flanges; penetrations; nozzles; safe ends, lower heads and welds) (3.1.1-10)	Cumulative fatigue damage	TLAA, evaluated in accordance with 10 CFR 54.21(c) and environmental effects are to be addressed for Class 1 components	Yes	TLAA	Consistent with GALL Report. See SER Section 3.1.2.2.1.
Steel top head enclosure (without cladding) top head nozzles (vent, top head spray or reactor core isolation cooling, and spare) exposed to reactor coolant (3.1.1-11)	Loss of material due to general, pitting and crevice corrosion	Water Chemistry and One-Time Inspection	Yes	Not applicable	Not applicable to PWRs. See SER Section 3.1.2.2.2(1).

Aging Management Review Results

Component Group (GALL Report Item No.)	Aging Effect/ Mechanism	AMP in GALL Report	Further Evaluation in GALL Report	AMP in LRA, Supplements, or Amendments	Staff Evaluation
Steel SG shell assembly exposed to secondary feedwater and steam (3.1.1-12)	Loss of material due to general, pitting and crevice corrosion	Water Chemistry and One-Time Inspection	Yes	Not applicable	Not applicable to PVNGS. See SER Section 3.1.2.2.2(1).
Steel and stainless steel isolation condenser components exposed to reactor coolant (3.1.1-13)	Loss of material due to general (steel only), pitting and crevice corrosion	Water Chemistry and One-Time Inspection	Yes	Not applicable	Not applicable to PWRs. See SER Section 3.1.2.2.2(2).
Stainless steel, nickel alloy, and steel with nickel-alloy or stainless steel cladding RV flanges, nozzles, penetrations, safe ends, vessel shells, heads and welds (3.1.1-14)	Loss of material due to pitting and crevice corrosion	Water Chemistry and One-Time Inspection	Yes	Not applicable	Not applicable to PWRs. See SER Section 3.1.2.2.2(3).
Stainless steel; steel with nickel-alloy or stainless steel cladding; and nickel-alloy RCPB components exposed to reactor coolant (3.1.1-15)	Loss of material due to pitting and crevice corrosion	Water Chemistry and One-Time Inspection	Yes	Not applicable	Not applicable to PWRs. See SER Section 3.1.2.2.2(3).
Steel SG upper and lower shell and transition cone exposed to secondary feedwater and steam (3.1.1-16)	Loss of material due to general, pitting and crevice corrosion	ISI (IWB, IWC, and IWD), and Water Chemistry and, for Westinghouse Model 44 and 51 S/G, if general and pitting corrosion of the shell is known to exist, additional inspection procedures are to be developed.	Yes	ISI and Water Chemistry	Consistent with GALL Report. See SER Section 3.1.2.2.2(4).
Steel (with or without stainless steel cladding) RV beltline shell, nozzles, and welds (3.1.1-17)	Loss of fracture toughness due to neutron irradiation embrittlement	TLAA, evaluated in accordance with 10 CFR 50, Appendix G, and RG 1.99. The applicant may choose to demonstrate that the materials of the nozzles are not controlling for the TLAA evaluations.	Yes	TLAA	Consistent with GALL Report. See SER Section 3.1.2.2.3(1).

Aging Management Review Results

Component Group (GALL Report Item No.)	Aging Effect/ Mechanism	AMP in GALL Report	Further Evaluation in GALL Report	AMP in LRA, Supplements, or Amendments	Staff Evaluation
Steel (with or without stainless steel cladding) RV beltline shell, nozzles, and welds; safety injection nozzles (3.1.1-18)	Loss of fracture toughness due to neutron irradiation embrittlement	RV Surveillance	Yes	RV Surveillance	Consistent with GALL Report. See SER Section 3.1.2.2.3(2).
Stainless steel and nickel-alloy top head enclosure vessel flange leak detection line (3.1.1-19)	Cracking due to SCC and IGSS	A plant-specific AMP is to be evaluated.	Yes	Not applicable	Not applicable to PWRs. See SER Section 3.1.2.2.4(1).
Stainless steel isolation condenser components exposed to reactor coolant (3.1.1-20)	Cracking due to SCC and IGSCC	ISI (IWB, IWC, and IWD), Water Chemistry, and plant-specific verification program	Yes	Not applicable	Not applicable to PWRs. See SER Section 3.1.2.2.4(2).
RV shell fabricated of SA508-Cl 2 forgings clad with stainless steel using a high-heat-input welding process (3.1.1-21)	Crack growth due to cyclic loading	TLAA	Yes	Not applicable	Not applicable to PVNGS. See SER Section 3.1.2.2.5.
Stainless steel and nickel-alloy RV internals components exposed to reactor coolant and neutron flux (3.1.1-22)	Loss of fracture toughness due to neutron irradiation embrittlement, void swelling	FSAR supplement commitment to (1) participate in industry RV internals aging programs (2) implement applicable results (3) submit for NRC approval > 24 months before the extended period an RV internals inspection plan based on industry recommendation.	No, but licensee commitment needs to be confirmed	Commitment and RCS	Consistent with GALL Report. See SER Section 3.1.2.2.6.
Stainless steel RV closure head flange leak detection line and bottom-mounted instrument guide tubes (3.1.1-23)	Cracking due to SCC	A plant-specific AMP is to be evaluated.	Yes	ISI and Water Chemistry	Consistent with GALL Report. See SER Section 3.1.2.2.7(1).
Class 1 CASS piping, piping components, and piping elements exposed to reactor coolant (3.1.1-24)	Cracking due to SCC	Water Chemistry and, for CASS components that do not meet the NUREG-0313 guidelines, a plant-specific AMP	Yes	Not applicable	Not applicable to PVNGS. See SER Section 3.1.2.2.7(2).

Aging Management Review Results

Component Group (GALL Report Item No.)	Aging Effect/ Mechanism	AMP in GALL Report	Further Evaluation in GALL Report	AMP in LRA, Supplements, or Amendments	Staff Evaluation
Stainless steel jet pump sensing line (3.1.1-25)	Cracking due to cyclic loading	A plant-specific AMP is to be evaluated.	Yes	Not applicable	Not applicable to PWRs. See SER Section 3.1.2.2.8(1).
Steel and stainless steel isolation condenser components exposed to reactor coolant (3.1.1-26)	Cracking due to cyclic loading	ISI (IWB, IWC, and IWD) and plant-specific verification program	Yes	Not applicable	Not applicable to PWRs. See SER Section 3.1.2.2.8(2).
Stainless steel and nickel-alloy RV internals screws, bolts, tie rods, and hold-down springs (3.1.1-27)	Loss of preload due to stress relaxation	FSAR supplement commitment to (1) participate in industry RV internals aging programs (2) implement applicable results (3) submit for NRC approval > 24 months before the extended period an RV internals inspection plan based on industry recommendation.	No, but licensee commitment needs to be confirmed	Commitment and Reactor Coolant Supplement	Consistent with GALL Report. See SER Section 3.1.2.2.9.
Steel SG feedwater impingement plate and support exposed to secondary feedwater (3.1.1-28)	Loss of material due to erosion	A plant-specific AMP is to be evaluated.	Yes	Not applicable	Not applicable to PVNGS. See SER Section 3.1.2.2.10.
Stainless steel steam dryers exposed to reactor coolant (3.1.1-29)	Cracking due to flow-induced vibration	A plant-specific AMP is to be evaluated.	Yes	Not applicable	Not applicable to PWRs. See SER Section 3.1.2.2.11.

Aging Management Review Results

Component Group (GALL Report Item No.)	Aging Effect/ Mechanism	AMP in GALL Report	Further Evaluation in GALL Report	AMP in LRA, Supplements, or Amendments	Staff Evaluation
Stainless steel RV internals components (e.g., upper internals assembly, rod cluster control assembly, guide tube assemblies, baffle/former assembly, lower internal assembly, shroud assemblies, plenum cover and plenum cylinder, upper grid assembly, control rod guide tube assembly, core support shield assembly, core barrel assembly, lower grid assembly, flow distributor assembly, thermal shield, instrumentation support structures) (3.1.1-30)	Cracking due to SCC, IASCC	Water Chemistry and FSAR supplement commitment to (1) participate in industry RV internals aging programs (2) implement applicable results (3) submit for NRC approval > 24 months before the extended period an RV internals inspection plan based on industry recommendation.	No, but licensee commitment needs to be confirmed	Water Chemistry, Commitment, and Reactor Coolant Supplement	Consistent with GALL Report. See SER Section 3.1.2.2.12.
Nickel alloy and steel with nickel-alloy cladding piping, piping component, piping elements, penetrations, nozzles, safe ends, and welds (other than RV head); pressurizer heater sheaths, sleeves, diaphragm plate, manways and flanges; core support pads/core guide lugs (3.1.1-31)	Cracking due to PWSCC	ISI (IWB, IWC, and IWD) and Water Chemistry and FSAR supplement commitment to implement applicable plant commitments to (1) NRC Orders, Bulletins, and Generic Letters associated with nickel- alloys and (2) staff-accepted industry guidelines.	No, but licensee commitment needs to be confirmed	ISI, Water Chemistry, and Nickel- Alloy Aging Management Programs, Commitment, and Reactor Coolant Supplement	Consistent with GALL Report. See SER Section 3.1.2.2.13.
Steel SG feedwater inlet ring and supports (3.1.1-32)	Wall thinning due to FAC	A plant-specific AMP is to be evaluated.	Yes	SG Tubing Integrity and Water Chemistry	Consistent with GALL Report. See SER Section 3.1.2.2.14.
Stainless steel and nickel- alloy RV internals components (3.1.1-33)	Changes in dimensions due to void swelling	FSAR supplement commitment to (1) participate in industry RV internals aging programs (2) implement applicable results (3) submit for NRC approval > 24 months before the extended period an RV internals inspection plan based on industry recommendation.	No, but licensee commitment needs to be confirmed	Commitment and Reactor Coolant Supplement	Consistent with GALL Report. See SER Section 3.1.2.2.15.

Aging Management Review Results

Component Group (GALL Report Item No.)	Aging Effect/ Mechanism	AMP in GALL Report	Further Evaluation in GALL Report	AMP in LRA, Supplements, or Amendments	Staff Evaluation
Stainless steel and nickel-alloy reactor control rod drive head penetration pressure housings (3.1.1-34)	Cracking due to SCC and PWSCC	ISI (IWB, IWC, and IWD) and Water Chemistry and for nickel alloy, comply with applicable NRC Orders and provide a commitment in the FSAR supplement to implement applicable (1) Bulletins and Generic Letters and (2) staff-accepted industry guidelines.	No, but licensee commitment needs to be confirmed	ISI, Water Chemistry, and Nickel-Alloy Management Programs, Commitment, and Reactor Coolant Supplement	Consistent with GALL Report. See SER Section 3.1.2.2.16(1).
Steel with stainless steel or nickel-alloy cladding primary side components; SG upper and lower heads, tubesheets and tube-to-tube sheet welds (3.1.1-35)	Cracking due to SCC and PWSCC	ISI (IWB, IWC, and IWD) and Water Chemistry and for nickel alloy, comply with applicable NRC Orders and provide a commitment in the FSAR supplement to implement applicable (1) Bulletins and Generic Letters and (2) staff-accepted industry guidelines.	No, but licensee commitment needs to be confirmed	Plant-specific Commitment	. See SER Section 3.1.2.2.16(1).
Nickel-alloy, stainless steel pressurizer spray head (3.1.1-36)	Cracking due to SCC and PWSCC	Water Chemistry and One-Time Inspection and, for nickel-alloy welded spray heads, comply with applicable NRC Orders and provide a commitment in the FSAR supplement to implement applicable (1) Bulletins and Generic Letters and (2) staff-accepted industry guidelines.	No, but licensee commitment needs to be confirmed	Water Chemistry, One-Time Inspection, and Reactor Coolant Supplement	Consistent with GALL Report. See SER Section 3.1.2.2.16(2).
Stainless steel and nickel-alloy RV internals components (e.g., upper internals assembly, rod cluster control assembly, guide tube assemblies, lower internal assembly, CEA shroud assemblies, core shroud assembly, core support shield assembly, core barrel assembly, lower grid assembly, flow distributor assembly) (3.1.1-37)	Cracking due to SCC, PWSCC, IASCC	Water Chemistry and FSAR supplement commitment to (1) participate in industry RV internals aging programs (2) implement applicable results (3) submit for NRC approval > 24 months before the extended period an RV internals inspection plan based on industry recommendation.	No, but licensee commitment needs to be confirmed	Water Chemistry, Commitment, and Reactor Coolant Supplement	Consistent with GALL Report. See SER Section 3.1.2.2.17.

Aging Management Review Results

Component Group (GALL Report Item No.)	Aging Effect/ Mechanism	AMP in GALL Report	Further Evaluation in GALL Report	AMP in LRA, Supplements, or Amendments	Staff Evaluation
Steel (with or without stainless steel cladding) control rod drive return line nozzles exposed to reactor coolant (3.1.1-38)	Cracking due to cyclic loading	BWR Control Rod Drive Return Line Nozzle	No	Not applicable	Not applicable to PWRs. See SER Section 3.1.2.1.1.
Steel (with or without stainless steel cladding) feedwater nozzles exposed to reactor coolant (3.1.1-39)	Cracking due to cyclic loading	BWR Feedwater Nozzle	No	Not applicable	Not applicable to PWRs. See SER Section 3.1.2.1.1.
Stainless steel and nickel-alloy penetrations for control rod drive stub tubes instrumentation, jet pump instrumentation, standby liquid control, flux monitor, and drain line exposed to reactor coolant (3.1.1-40)	Cracking due to SCC, IGSCC, cyclic loading	BWR Penetrations and Water Chemistry	No	Not applicable	Not applicable to PWRs. See SER Section 3.1.2.1.1.
Stainless steel and nickel-alloy piping, piping components, and piping elements greater than or equal to 4 NPS; nozzle safe ends and associated welds (3.1.1-41)	Cracking due to SCC and IGSCC	BWR SCC and Water Chemistry	No	Not applicable	Not applicable to PWRs. See SER Section 3.1.2.1.1.
Stainless steel and nickel-alloy vessel shell attachment welds exposed to reactor coolant (3.1.1-42)	Cracking due to SCC and IGSCC	BWR Vessel ID Attachment Welds and Water Chemistry	No	Not applicable	Not applicable to PWRs. See SER Section 3.1.2.1.1.
Stainless steel fuel supports and control rod drive assemblies control rod drive housing exposed to reactor coolant (3.1.1-43)	Cracking due to SCC and IGSCC	BWR Vessel Internals and Water Chemistry	No	Not applicable	Not applicable to PWRs. See SER Section 3.1.2.1.1.
Stainless steel and nickel-alloy core shroud, core plate, core plate bolts, support structure, top guide, core spray lines, spargers, jet pump assemblies, control rod drive housing, nuclear instrumentation guide tubes (3.1.1-44)	Cracking due to SCC, IGSCC, IASCC	BWR Vessel Internals and Water Chemistry	No	Not applicable	Not applicable to PWRs. See SER Section 3.1.2.1.1.

Aging Management Review Results

Component Group (GALL Report Item No.)	Aging Effect/ Mechanism	AMP in GALL Report	Further Evaluation in GALL Report	AMP in LRA, Supplements, or Amendments	Staff Evaluation
Steel piping, piping components, and piping elements exposed to reactor coolant (3.1.1-45)	Wall thinning due to FAC	FAC	No	Not applicable	Not applicable to PWRs. See SER Section 3.1.2.1.1.
Nickel-alloy core shroud and core plate access hole cover (mechanical covers) (3.1.1-46)	Cracking due to SCC, IGSCC, IASCC	ISI (IWB, IWC, and IWD), and Water Chemistry	No	Not applicable	Not applicable to PWRs. See SER Section 3.1.2.1.1.
Stainless steel and nickel-alloy RV internals exposed to reactor coolant (3.1.1-47)	Loss of material due to pitting and crevice corrosion	ISI (IWB, IWC, and IWD), and Water Chemistry	No	Not applicable	Not applicable to PWRs. See SER Section 3.1.2.1.1.
Steel and stainless steel Class 1 piping, fittings and branch connections < NPS 4 exposed to reactor coolant (3.1.1-48)	Cracking due to SCC, IGSCC (for stainless steel only), and thermal and mechanical loading	ISI (IWB, IWC, and IWD), Water chemistry, and One-Time Inspection of ASME Code Class 1 Small-bore Piping	No	Not applicable	Not applicable to PWRs. See SER Section 3.1.2.1.1.
Nickel-alloy core shroud and core plate access hole cover (welded covers) (3.1.1-49)	Cracking due to SCC, IGSCC, IASCC	ISI (IWB, IWC, and IWD), Water Chemistry, and, for BWRs with a crevice in the access hole covers, augmented inspection using UT or other demonstrated acceptable inspection of the access hole cover welds	No	Not applicable	Not applicable to PWRs. See SER Section 3.1.2.1.1.
High-strength low alloy steel top head closure studs and nuts exposed to air with reactor coolant leakage (3.1.1-50)	Cracking due to SCC and IGSCC	Reactor Head Closure Studs	No	Not applicable	Not applicable to PWRs. See SER Section 3.1.2.1.1.
CASS jet pump assembly castings; orificed fuel support (3.1.1-51)	Loss of fracture toughness due to thermal aging and neutron irradiation embrittlement	Thermal Aging and Neutron Irradiation Embrittlement of CASS	No	Not applicable	Not applicable to PWRs. See SER Section 3.1.2.1.1.

Aging Management Review Results

Component Group (GALL Report Item No.)	Aging Effect/ Mechanism	AMP in GALL Report	Further Evaluation in GALL Report	AMP in LRA, Supplements, or Amendments	Staff Evaluation
Steel and stainless steel reactor coolant pressure boundary (RCPB) pump and valve closure bolting, manway and holding bolting, flange bolting, and closure bolting in high-pressure and high-temperature systems (3.1.1-52)	Cracking due to SCC, loss of material due to wear, loss of preload due to thermal effects, gasket creep, and self-loosening	Bolting Integrity	No	Bolting Integrity	Consistent with GALL Report.
Steel piping, piping components, and piping elements exposed to closed cycle cooling water (3.1.1-53)	Loss of material due to general, pitting and crevice corrosion	Closed-Cycle Cooling Water System	No	Closed-Cycle Cooling Water System	Consistent with GALL Report.
Copper alloy piping, piping components, and piping elements exposed to closed cycle cooling water (3.1.1-54)	Loss of material due to pitting, crevice, and galvanic corrosion	Closed-Cycle Cooling Water System	No	Not applicable	Not applicable to PVNGS. See SER Section 3.1.2.1.1.
CASS Class 1 pump casings, and valve bodies and bonnets exposed to reactor coolant > 250°C (> 482°F) (3.1.1-55)	Loss of fracture toughness due to thermal aging embrittlement	ISI (IWB, IWC, and IWD). Thermal aging susceptibility screening is not necessary, ISI requirements are sufficient for managing these aging effects. ASME Code Case N-481 also provides an alternative for pump casings.	No	ISI	Consistent with GALL Report.
Copper alloy > 15% Zn piping, piping components, and piping elements exposed to closed cycle cooling water (3.1.1-56)	Loss of material due to selective leaching	Selective Leaching of Materials	No	Not applicable	Not applicable to PVNGS. See SER Section 3.1.2.1.1.
Cast austenitic stainless steel Class 1 piping, piping component, and piping elements and control rod drive pressure housings exposed to reactor coolant > 250°C (> 482°F) (3.1.1-57)	Loss of fracture toughness due to thermal aging embrittlement	Thermal Aging Embrittlement of CASS	No	Not applicable	Not applicable to PVNGS. See SER Section 3.1.2.1.1.
Steel RCPB external surfaces exposed to air with borated water leakage (3.1.1-58)	Loss of material due to boric acid corrosion	Boric Acid Corrosion	No	Boric Acid Corrosion	Consistent with GALL Report.

Aging Management Review Results

Component Group (GALL Report Item No.)	Aging Effect/ Mechanism	AMP in GALL Report	Further Evaluation in GALL Report	AMP in LRA, Supplements, or Amendments	Staff Evaluation
Steel SG steam nozzle and safe end, feedwater nozzle and safe end, AFW nozzles and safe ends exposed to secondary feedwater/steam (3.1.1-59)	Wall thinning due to FAC	FAC	No	FAC	Consistent with GALL Report.
Stainless steel flux thimble tubes (with or without chrome plating) (3.1.1-60)	Loss of material due to wear	Flux Thimble Tube Inspection	No	Not applicable	Not applicable to PVNGS. See SER Section 3.1.2.1.1.
Stainless steel, steel pressurizer integral support exposed to air with metal temperature up to 288°C (550°F) (3.1.1-61)	Cracking due to cyclic loading	ISI (IWB, IWC, and IWD)	No	ISI	Consistent with GALL Report.
Stainless steel, steel with stainless steel cladding RCS cold leg, hot leg, surge line, and spray line piping and fittings exposed to reactor coolant (3.1.1-62)	Cracking due to cyclic loading	ISI (IWB, IWC, and IWD)	No	ISI	Consistent with GALL Report. See SER Section 3.1.2.1.3.
Steel RV flange, stainless steel and nickel-alloy RV internals exposed to reactor coolant (e.g., upper and lower internals assembly, CEA shroud assembly, core support barrel, upper grid assembly, core support shield assembly, lower grid assembly) (3.1.1-63)	Loss of material due to wear	ISI (IWB, IWC, and IWD)	No	ISI	Consistent with GALL Report.
Stainless steel and steel with stainless steel or nickel-alloy cladding pressurizer components (3.1.1-64)	Cracking due to SCC, PWSCC	ISI (IWB, IWC, and IWD) and Water Chemistry	No	ISI and Water Chemistry	Consistent with GALL Report.

Aging Management Review Results

Component Group (GALL Report Item No.)	Aging Effect/ Mechanism	AMP in GALL Report	Further Evaluation in GALL Report	AMP in LRA, Supplements, or Amendments	Staff Evaluation
Nickel-alloy RV upper head and control rod drive penetration nozzles, instrument tubes, head vent pipe (top head), and welds (3.1.1-65)	Cracking due to PWSCC	ISI (IWB, IWC, and IWD) and Water Chemistry and Nickel-Alloy Penetration Nozzles Welded to the Upper RV Closure Heads of PWRs	No	ISI, Water Chemistry, and Nickel-Alloy Penetration Nozzles Welded to the Upper RV Closure Heads of Pressurized Water	Consistent with GALL Report.
Steel SG secondary manways and handholds (cover only) exposed to air with leaking secondary-side water or steam (3.1.1-66)	Loss of material due to erosion	ISI (IWB, IWC, and IWD) for Class 2 components	No	Not applicable	Not applicable to PVNGS. See SER Section 3.1.2.1.1.
Steel with stainless steel or nickel-alloy cladding; or stainless steel pressurizer components exposed to reactor coolant (3.1.1-67)	Cracking due to cyclic loading	ISI (IWB, IWC, and IWD), and Water Chemistry	No	ISI and Water Chemistry	Consistent with GALL Report.
Stainless steel, steel with stainless steel cladding Class 1 piping, fittings, pump casings, valve bodies, nozzles, safe ends, manways, flanges, CRD housing; pressurizer heater sheaths, sleeves, diaphragm plate; pressurizer relief tank components, RCS cold leg, hot leg, surge line, and spray line piping and fittings (3.1.1-68)	Cracking due to SCC	ISI (IWB, IWC, and IWD), and Water Chemistry	No	ISI and Water Chemistry	Consistent with GALL Report.
Stainless steel, nickel-alloy safety injection nozzles, safe ends, and associated welds and buttering exposed to reactor coolant (3.1.1-69)	Cracking due to SCC, PWSCC	ISI (IWB, IWC, and IWD), and Water Chemistry	No	ISI and Water Chemistry	Consistent with GALL Report.
Stainless steel; steel with stainless steel cladding Class 1 piping, fittings and branch connections < NPS 4 exposed to reactor coolant (3.1.1-70)	Cracking due to SCC, thermal and mechanical loading	ISI (IWB, IWC, and IWD), Water chemistry, and One-Time Inspection of ASME Code Class 1 Small-bore Piping	No	ISI, Water Chemistry and One Time Inspection of ASME Code Class 1 Small-Bore Piping	Consistent with GALL Report.

Aging Management Review Results

Component Group (GALL Report Item No.)	Aging Effect/ Mechanism	AMP in GALL Report	Further Evaluation in GALL Report	AMP in LRA, Supplements, or Amendments	Staff Evaluation
High-strength low alloy steel closure head stud assembly exposed to air with reactor coolant leakage (3.1.1-71)	Cracking due to SCC; loss of material due to wear	Reactor Head Closure Studs	No	Reactor Head Closure Studs	Consistent with GALL Report.
Nickel-alloy SG tubes and sleeves exposed to secondary feedwater/steam (3.1.1-72)	Cracking due to outer-diameter SCC and intergranular attack, loss of material due to fretting and wear	SG Tube Integrity and Water Chemistry	No	SG Integrity and Water Chemistry	Consistent with GALL Report.
Nickel-alloy SG tubes, repair sleeves, and tube plugs exposed to reactor coolant (3.1.1-73)	Cracking due to PWSCC	SG Tube Integrity and Water Chemistry	No	SG Integrity and Water Chemistry	Consistent with GALL Report.
Chrome plated steel, stainless steel, nickel-alloy SG anti-vibration bars exposed to secondary feedwater/steam (3.1.1-74)	Cracking due to SCC, loss of material due to crevice corrosion and fretting	SG Tube Integrity and Water Chemistry	No	SG Integrity and Water Chemistry	Consistent with GALL Report.
Nickel-alloy once-through SG tubes exposed to secondary feedwater/steam (3.1.1-75)	Denting due to corrosion of carbon steel tube support plate	SG Tube Integrity and Water Chemistry	No	Not applicable	Not applicable to PVNGS. See SER Section 3.1.2.1.1.
Steel SG tube support plate, tube bundle wrapper exposed to secondary feedwater/steam (3.1.1-76)	Loss of material due to erosion, general, pitting, and crevice corrosion, ligament cracking due to corrosion	SG Tube Integrity and Water Chemistry	No	SG Integrity and Water Chemistry	Consistent with GALL Report.
Nickel-alloy SG tubes and sleeves exposed to phosphate chemistry in secondary feedwater/steam (3.1.1-77)	Loss of material due to wastage and pitting corrosion	SG Tube Integrity and Water Chemistry	No	Not applicable	Not applicable to PVNGS. See SER Section 3.1.2.1.1.
Steel SG tube support lattice bars exposed to secondary feedwater/steam (3.1.1-78)	Wall thinning due to FAC	SG Tube Integrity and Water Chemistry	No	Not applicable	Not applicable to PVNGS. See SER Section 3.1.2.1.1.

Aging Management Review Results

Component Group (GALL Report Item No.)	Aging Effect/ Mechanism	AMP in GALL Report	Further Evaluation in GALL Report	AMP in LRA, Supplements, or Amendments	Staff Evaluation
Nickel-alloy SG tubes exposed to secondary feedwater/steam (3.1.1-79)	Denting due to corrosion of steel tube support plate	SG Tube Integrity; Water Chemistry and, for plants that could experience denting at the upper support plates, evaluate potential for rapidly propagating cracks and then develop and take corrective actions consistent with NRC Bulletin 88-02.	No	SG Tube Integrity and Water Chemistry	Not applicable to PVNGS. See SER Section 3.1.2.1.1.
CASS RV internals (e.g., upper internals assembly, lower internal assembly, CEA shroud assemblies, control rod guide tube assembly, core support shield assembly, lower grid assembly) (3.1.1-80)	Loss of fracture toughness due to thermal aging and neutron irradiation embrittlement	Thermal Aging and Neutron Irradiation Embrittlement of CASS	No	Not applicable	Not applicable to PVNGS. See SER Section 3.1.2.1.1.
Nickel-alloy or nickel-alloy clad SG divider plate exposed to reactor coolant (3.1.1-81)	Cracking due to PWSCC	Water Chemistry	No	Water Chemistry and Plant-specific Commitment	Consistent with GALL Report. See SER Section 3.1.2.1.2.
Stainless steel SG primary side divider plate exposed to reactor coolant (3.1.1-82)	Cracking due to SCC	Water Chemistry	No	Not applicable	Not applicable to PVNGS. See SER Section 3.1.2.1.1.
Stainless steel; steel with nickel-alloy or stainless steel cladding; and nickel-alloy RV internals and RCPB components exposed to reactor coolant (3.1.1-83)	Loss of material due to pitting and crevice corrosion	Water Chemistry	No	Water Chemistry	Consistent with GALL Report.
Nickel-alloy SG components such as, secondary side nozzles (vent, drain, and instrumentation) exposed to secondary feedwater/steam (3.1.1-84)	Cracking due to SCC	Water Chemistry and One-Time Inspection or ISI (IWB, IWC, and IWD).	No	Water Chemistry, One-Time Inspection, or ISI	Consistent with GALL Report.
Nickel-alloy piping, piping components, and piping elements exposed to air-indoor uncontrolled (external) (3.1.1-85)	None	None	NA	None	Consistent with GALL Report.

Aging Management Review Results

Component Group (GALL Report Item No.)	Aging Effect/ Mechanism	AMP in GALL Report	Further Evaluation in GALL Report	AMP in LRA, Supplements, or Amendments	Staff Evaluation
Stainless steel piping, piping components, and piping elements exposed to air-indoor uncontrolled (External); air with borated water leakage; concrete; gas (3.1.1-86)	None	None	NA	None	Consistent with GALL Report.
Steel piping, piping components, and piping elements in concrete (3.1.1-87)	None	None	NA	Not applicable	Not applicable to PVNGS. See SER Section 3.1.2.1.1.

The staff's review of the RV, RV internals, and RCS component groups followed one of three categories. One category, documented in SER Section 3.1.2.1, reviewed AMR results for components that the applicant indicated are consistent with the GALL Report and require no further evaluation. Another category, documented in SER Section 3.1.2.2, reviewed AMR results for components that the applicant indicated are consistent with the GALL Report and for which further evaluation is recommended. The third category, documented in SER Section 3.1.2.3, reviewed AMR results for components that the applicant indicated are not consistent with, or not addressed in, the GALL Report. SER Section 3.0.3 documents the staff's review of AMPs credited to manage or monitor aging effects of the RV, RV internals, and RCS.

3.1.2.1 Aging Management Review Results Consistent with the Generic Aging Lessons Learned Report

LRA Section 3.1.2.1 identifies the materials, environments, AERMs, and the following programs that manage aging effects for the RV, RV internals, and RCS components:

- ASME Section XI ISI, Subsections IWB, IWC, and IWD
- Bolting Integrity
- Boric Acid Corrosion
- Closed-Cycle Cooling Water System
- External Surfaces Monitoring Program
- Flow-Accelerated Corrosion
- Lubricating Oil Analysis
- Nickel-Alloy Aging Management Program
- Nickel-Alloy Penetration Nozzles Welded to the Upper RV Closure Heads of PWRs
- One-Time Inspection
- One-Time Inspection of ASME Code Class 1 Small-Bore Piping
- Reactor Coolant System Supplement
- Reactor Head Closure Studs

Aging Management Review Results

- RV Surveillance
- Steam Generator Tube Integrity
- Water Chemistry

LRA Tables 3.1.2-1– 3.1.2-4 summarize AMRs for the RV, RV internals, and RCS components and indicate AMRs claimed to be consistent with the GALL Report.

The staff reviewed the LRA to confirm that the applicant: (a) provided a brief description of the system, components, materials, and environments; (b) stated that the applicable aging effects were reviewed and evaluated in the GALL Report; and (c) identified those aging effects for the RV, RV internals, and RCS components that are subject to an AMR. Based on its audit and review, the staff determines that, for AMRs not requiring further evaluation, as identified in LRA Table 3.1.1, the applicant's references to the GALL Report are acceptable and no further staff review is required.

3.1.2.1.1 Aging Management Review Results Identified as Not Applicable

LRA Table 3.1.1, item 3.1.1.01 states that this is a CE vessel with no support skirt, so the applicable GALL Report line was not used. The staff noted that according to the SRP-LR and the GALL Report, this item is applicable to boiling water reactors (BWRs) only. Because the PVNGS is a PWR design, this item is not applicable.

LRA Table 3.1.1, items 3.1.1.02–3.1.1.04 and 3.1.1.38–3.1.1.51 state that these items are applicable only to BWRs. The staff verified that these items do not apply because the units are a PWR design. Based on this determination, the staff finds that the applicant has provided an acceptable basis for concluding AMR items 3.1.1.02–3.1.1.04 and 3.1.1.38–3.1.1.51 are not applicable.

LRA Table 3.1.1, item 3.1.1.54 addresses copper alloy piping, piping components, and piping elements exposed to closed-cycle cooling water subject to loss of material due to pitting, crevice, and galvanic corrosion for this component group. The applicant stated that this item is not applicable because it has no in-scope copper alloy piping, piping components, or piping elements exposed to closed-cycle cooling water in the RCS, so the applicable GALL Report line was not used. The staff reviewed LRA Sections 2.3.1 and 3.1 and confirmed that the applicant's LRA does not have any AMR results for the RCS that include copper alloy piping, piping components, and piping elements exposed to closed-cycle cooling water. The staff reviewed the applicant's UFSAR and confirmed that no in-scope copper alloy piping, piping components, and piping elements exposed to closed-cycle cooling water are present in these systems and, therefore, finds the applicant's determination acceptable.

LRA Table 3.1.1, item 3.1.1.56 addresses copper alloy greater than 15-percent zinc piping, piping components, and piping elements exposed to closed-cycle cooling water subject to loss of material due to selective leaching for this component group. The applicant stated that this item is not applicable because it has no in-scope copper alloy greater than 15-percent zinc components exposed to closed-cycle cooling water in the RCS, so the applicable GALL Report item was not used. The staff reviewed LRA Sections 2.3.1 and 3.1 and confirmed that the applicant's LRA does not have any AMR results for the RCS that include copper alloy greater than 15-percent zinc piping, piping components, and piping elements exposed to closed-cycle cooling water. The staff reviewed the applicant's UFSAR and confirmed that no in-scope copper alloy greater than 15-percent zinc piping, piping components, and piping elements exposed to closed-cycle cooling water are present in these systems and, therefore, finds the applicant's determination acceptable.

LRA Table 3.1.1, item 3.1.1.57 addresses cast austenitic stainless steel (CASS) Class 1 piping, piping component, and piping elements and control rod drive pressure housings exposed to

Aging Management Review Results

reactor coolant greater than 250 degrees C (greater than 482 degrees F) subject to loss of fracture toughness due to thermal aging embrittlement for this component group. The applicant stated that this item is not applicable because the RCS does not have CASS piping, piping components, or piping elements exposed to reactor coolant, and the control rod drive pressure housings are made of stainless steel and nickel alloy, so that the applicable GALL Report items were not used. The staff reviewed LRA Sections 2.3.1 and 3.1 and confirmed that the applicant's LRA does not have any AMR results for the RCS that include CASS Class 1 piping, piping component, and piping elements and control rod drive pressure housings exposed to reactor coolant greater than 250 degrees C (greater than 482 degrees F). The staff reviewed the applicant's UFSAR and confirmed that no in-scope CASS Class 1 piping, piping component, and piping elements and control rod drive pressure housings exposed to reactor coolant greater than 250 degrees C (greater than 482 degrees F) are present in these systems; therefore, it finds the applicant's determination acceptable. The staff also confirmed that the applicant addresses the control rod drive pressure housings, which are fabricated of stainless steel and nickel alloy, in LRA Table 3.1.1, item 3.1.1.34, for cracking due to SCC and PWSCC and stainless steel; steel with nickel-alloy or stainless steel cladding; and nickel-alloy RV internals and RCPB components exposed to reactor coolant in LRA Table 3.1.1, item 3.1.1.83, for loss of material due to pitting and crevice corrosion. SER Section 3.1.2.2.16(1) documents the staff's review and its evaluation of LRA Table 3.1.1, item 3.1.1.34. The staff noted that for the control rod drive pressure housings that referenced LRA Table 3.1.1, item 3.1.1.83, the applicant proposes to manage loss of material due to pitting and crevice corrosion with its Water Chemistry Program, which is consistent with the recommendations of the GALL Report and is, therefore, acceptable.

LRA Table 3.1.1, item 3.1.1.60 addresses stainless steel flux thimble tubes (with or without chrome plating) exposed to reactor coolant subject to loss of material due to wear for this component group. The applicant stated that this item is not applicable because it has a CE-design RV and internals, and the subject GALL Report line is applicable to Westinghouse-design RV and internals only. The staff reviewed the applicant's UFSAR and confirmed that its RV is a CE design. The staff reviewed the GALL Report and confirmed that LRA Table 3.1.1, item 3.1.1.60 and GALL Report, AMR item IV.B2-13 is specifically applicable to Westinghouse-design RV and internals and, therefore, finds the applicant's determination acceptable.

LRA Table 3.1.1, item 3.1.1.66 addresses steel SG secondary manways and handholds (cover only) exposed to air with leaking secondary-side water or steam subject to loss of material due to erosion for this component group. The applicant stated that this item is not applicable because it has recirculating SGs, so the applicable GALL Report line was not used. The staff noted that LRA Table 3.1.1, item 3.1.1.66 references GALL Report, item IV.D2-5, which is applicable to once-through SGs. The staff reviewed the applicant's UFSAR Section 5.1, Figures 5.4-8A and 5D-1B and confirmed that the applicant's SGs are recirculating-type SGs, therefore, finds the applicant's determination acceptable.

LRA Table 3.1.1, item 3.1.1.75 addresses nickel-alloy once-through SG tubes exposed to secondary feedwater and steam subject to denting due to corrosion of carbon steel tube support plate for this component group. The applicant stated that this item is not applicable because it has recirculating SGs, so the applicable GALL Report line was not used. The staff noted that LRA Table 3.1.1, item 3.1.1.75 references GALL Report, AMR item IV.D2-13, which is applicable to once-through SGs. The staff reviewed the applicant's UFSAR Section 5.1, Figures 5.4-8A and 5D-1B and confirmed that the applicant's SGs are recirculating-type SGs, therefore, finds the applicant's determination acceptable.

LRA Table 3.1.1, item 3.1.1.77 addresses nickel-alloy SG tubes and sleeves exposed to phosphate chemistry in secondary feedwater and steam subject to loss of material due to

Aging Management Review Results

wastage and pitting corrosion for this component group. The applicant stated that this item is not applicable because it does not operate on phosphate chemistry in secondary feedwater or steam with the replacement SGs, so the applicable GALL Report line was not used. The staff noted that the applicant's Water Chemistry Program is consistent with the guidelines provided in EPRI TR-1008224, "PWR Secondary Water Chemistry Guidelines," Revision 6. The staff noted that this is a later revision to EPRI TR-102134 and that its use is acceptable because it is consistent with GALL AMP XI.M2. The staff reviewed EPRI TR-1008224 and UFSAR Section 10.3.5 and confirmed that the applicant does not operate on phosphate chemistry in the secondary side, therefore, finds the applicant's determination acceptable.

LRA Table 3.1.1, item 3.1.1.78 addresses steel SG tube support lattice bars exposed to secondary feedwater and steam subject to wall thinning due to FAC for this component group. The applicant stated that this item is not applicable because its SGs do not contain steel tube support lattice bars, so the applicable GALL Report line was not used. The staff noted that in LRA Section B2.1.8, the applicant stated its design is a two-loop CE plant with two identical replacement SGs designed by Asea Brown Boveri-CE, considered a modified CE System 80 design. The applicant further stated that it replaced the original SGs in Units 1, 2, and 3 during the fall of 2005, 2003, and 2007, respectively. The applicant stated that the tube support system is fabricated of Type 409 ferritic stainless steel. The staff's safety evaluation of the applicant's amendment for replacement SGs and uprated power operations for Unit 2 is documented in a letter dated September 23, 2003 (ADAMS No. ML032720538) and for Units 1 and 3 is documented in a letter dated November 16, 2005 (ADAMS No. ML053130275). The staff confirmed that the applicant's tube support system is fabricated of Type 409 ferritic stainless steel in the above mentioned safety evaluations; therefore, it finds the applicant's determination acceptable.

LRA Table 3.1.1, item 3.1.1.79 addresses denting due to corrosion of carbon steel tube support plate in nickel-alloy SG tubes exposed to secondary feedwater and steam. The applicant stated that this item is not applicable.

However, the staff noted that LRA Table 3.1.2-4 did not include the item addressing the GALL Report, item IV.D1-17, corresponding to ligament cracking due to corrosion in steel tube support plates, whereas in this same table, the applicant addressed the GALL Report, item IV.D1-19, corresponding to SG tube denting due to corrosion of carbon steel tube support plates. The staff also noted that in LRA Section B2.1.8, the applicant stated that the tube support system is similar to the original design and, like the original design, is fabricated from 409 ferritic stainless steel. The staff noted that this type of stainless steel is not susceptible to the general corrosion that affects carbon steel tube support plates, which induces SG tube denting because of the buildup and expansion of corrosion products in the annulus between the SG tubes external surface and the tube support plates. The staff requested the applicant to explain this apparent inconsistency in a conference call on October 28, 2010.

In a letter dated November 10, 2010, the applicant revised LRA Section 3.1.2.1.4, Tables 3.1.1 and 3.1.2-4, and Sections A1.8 and B2.1.8, to correct the aging effect for SG tubes to be consistent with the GALL Report line IV.D1-19. The applicant stated that LRA Table 3.1.1, item 3.1.1.79 is not applicable because its SGs do not have a carbon steel tube support system. The applicant further stated that its steam generator tube support system is fabricated from 409 ferritic stainless steel, so denting is not an applicable aging effect. Accordingly, the applicant deleted the denting aging effect from LRA Section 3.1.2.1.4, and the LRA AMR item corresponding to the GALL Report line IV.D1-19 from LRA Table 3.1.2.1-4, and revised the description of the Steam Generator Tube Integrity Program in LRA Section B2.1.8 and UFSAR Supplement A1.8.

The staff reviewed the applicant's clarification about SG tube support plates and the subsequent LRA modifications, and finds it acceptable that the applicant concluded that LRA Table 3.1.1,

Aging Management Review Results

item 3.1.1.79 is not applicable, consistent with GALL Report item IV.D1-19, because the applicant's SGs contain 409 ferritic stainless steel SG tube support plates, which does not induce SG tube denting, as described above.

LRA Table 3.1.1, item 3.3.1.80 addresses loss of fracture toughness due to thermal aging and neutron irradiation embrittlement in CASS RV internals (e.g., upper internals assembly, lower internal assembly, control element assembly (CEA) shroud assemblies, control rod guide tube assembly, core support shield assembly, and lower grid assembly). The applicant stated that LRA Table 3.1.1, item 3.3.1.80 is not applicable and that its RV internals do not contain CASS, so it did not use the applicable GALL Report items. The staff reviewed the applicant's UFSAR and noted that UFSAR Section 4.5.2.1, "Reactor Internals Materials," Subsection B, states that the upper guide structure (UGS) assembly contains "ASTM A-351, Grade CF8," which is a CASS material. The staff noted that there is a discrepancy between the applicant's UFSAR and LRA as to whether the RV internals contain CASS. By letter dated March 2, 2010, the staff issued RAI 3.1.1-1 requesting the applicant to resolve the discrepancy between UFSAR Section 4.5.2.1, which states the UGS assembly contains CASS components and LRA Table 3.1.1, which states the RV internals do not contain CASS components. The staff also asked that the applicant provide the necessary revisions to the LRA if the RV internals contain CASS components.

In its response, dated April 1, 2010, the applicant stated that UFSAR Section 4.5.2.1 incorporated sections from the CE Standard Safety Analysis Report and reflected initial design information that was not incorporated into the as-built design. The applicant stated that it performed a review of the as-built Reactor Internals Bill of Materials and confirmed that no CASS material was used in the UGS assembly. The applicant clarified that it used the Reactor Internals Bill of Materials during the license renewal aging evaluations. The applicant further stated that it is tracking the discrepancy within its corrective actions program.

Based on its review, the staff finds the applicant's response to RAI 3.1.1-1, and the applicant's claim that the RV internals does not contain CASS, acceptable because the applicant confirmed in the as-built Reactor Internals Bill of Materials that CASS was not used in the UGS assembly, and the applicant entered this discrepancy in the UFSAR into its corrective actions program. The staff's concern described in RAI 3.1.1-1 is resolved.

LRA Table 3.1.1, item 3.1.1.82 addresses stainless steel SG primary side divider plate exposed to reactor coolant subject to cracking due to SCC for this component group. The applicant stated that this item is not applicable because its SG primary channel dividers are made of nickel alloy, so it did not use the applicable GALL Report line. The staff reviewed UFSAR Section 5.2 and confirmed the applicant's divider plate is fabricated of nickel alloy. The staff also noted that, in LRA Table 3.1.2-4, the applicant stated that its SG primary head divider plate is fabricated of nickel alloy and is managed for cracking with its Water Chemistry Program, consistent with the recommendations of the GALL Report AMR, item IV.D1-6. Based on its review as described above, the staff finds the applicant's determination acceptable.

LRA Table 3.1.1, item 3.1.1.87 addresses steel piping, piping components, and piping elements in concrete. The GALL Report states that there is no AERM. The applicant stated that this item is not applicable because its RV, internals, and RCS have no in-scope steel piping, piping components, or piping elements embedded in concrete, so it did not use the applicable GALL Report line. The staff reviewed LRA Sections 2.3.1 and 3.1 and confirmed that the applicant's LRA does not have any AMR results for the RCS that include steel piping, piping components, and piping elements in concrete. The staff reviewed the applicant's UFSAR and confirmed that no in-scope steel piping, piping components, and piping elements in concrete are present in the systems; therefore, it finds the applicant's determination acceptable.

Aging Management Review Results

The staff evaluated the applicant's claim of consistency with the GALL Report. The staff also reviewed information pertaining to the applicant's consideration of recent operating experience and proposals for managing aging effects. Based on its review, the staff concludes that the AMR results, which the applicant claimed to be consistent with the GALL Report, are indeed consistent.

3.1.2.1.2   Cracking Due to Primary Water Stress Corrosion Cracking

LRA Table 3.1.1, item 3.1.1-81 addresses cracking due to PWSCC for nickel-alloy or nickel-alloy clad SG divider plates exposed to reactor coolant. The LRA states that the SG primary channel dividers are made of nickel alloy. The applicant credited its Water Chemistry Program to manage the cracking due to PWSCC, consistent with the GALL Report.

The staff noted that, from international operating experience in SGs, extensive cracking due to PWSCC has been identified in SG divider plates fabricated from Alloy 600, even with proper primary water chemistry. The staff noted that cracks have been detected very close to the tubesheet and with depths of almost a quarter of the divider plate thickness. Therefore, the staff noted that the Primary Water Chemistry Program alone may not be effective in managing the aging effect of cracking due to PWSCC in SG divider plate assembly components fabricated from Alloy 600 and its associated weld metals.

The staff noted that these SG divider plate cracks could impact adjacent items such as the tubesheet and the channel head if they propagate to the boundary with these items. The staff further noted that for the tubesheet, PWSCC cracks in the divider plate assembly components fabricated from Alloy 600 and its associated weld metals could propagate to the tubesheet cladding with possible consequences to the integrity of the tube-to-tubesheet welds. Furthermore, for the channel head, the PWSCC cracks in the divider plate could propagate to the SG triple point and potentially affect the pressure boundary of the SG channel head.

UFSAR, Section 1.2.3.3, states that a vertical divider plate separates the inlet and outlet plenums in the lower head of the SGs, but the staff did not find information about the materials of the divider plate assembly nor its junction to the lower head and to the tubesheet in the UFSAR or the LRA.

The staff held conference calls on October 22, November 3 and 19, 2010, with the applicant to discuss and clarify the staff's concerns. The staff asked the applicant to clarify how the SG divider plate is assembled to the lower head and to the tubesheet and to identify the materials of the divider plate and associated welds. During the discussion, the staff also asked the applicant to provide information on how it will manage the possible effects of PWSCC on these welds if the compositions of the SG divider plate divider bar welds (all areas) are susceptible to PWSCC, thereby potentially compromising the RCS pressure boundary. The staff also requested information concerning the inspection method since it should be capable of detecting PWSCC. The applicant agreed to provide information on its management of this aging effect in these components.

By letter dated November 23, 2010, the applicant described how the SG primary side divider plates are attached to the channel head, stay cylinder, and tubesheet via a tongue-in-groove connection. The applicant stated that all components are manufactured from Alloy 690 material, and the SG specifications show the divider plate bars welded to the channel head, stay cylinder, and tubesheet cladding using Alloy 52, 82, 152, and 182 filler materials, but not all detailed information of SG specifications, especially about filler materials, was included in the UFSAR. The applicant further stated that there is no routine inspection requirement for the divider bar welds because (a) these welds do not provide a reactor coolant system pressure boundary; (b) these welds do not provide structural support to the SGs; (c) the divider plate "floats" in the tongue and groove, and the force on the divider plate transferred to the divider plate bar welds is the relatively low differential pressure between the SG inlet and outlet (compared with RCS pressure); and (d) a crack in the divider bar weld due to PWSCC would need to propagate from the divider bar weld through the channel head cladding to get to the base metal.

However, in response to the staff's concern regarding potential failure of the RCS pressure boundary due to possible PWSCC of SG divider plate bar welds, the applicant committed (Commitment No. 61) to one of the following:

> 1. Perform an inspection of each PVNGS SG to assess the condition of the divider plate bar welds. The examination technique(s) will be capable of detecting PWSCC in the divider plate bar welds.
>
> 2. Perform an analytical evaluation of the SG divider plate bar welds in order to establish a technical basis which concludes that the SG RCS pressure boundary is adequately maintained with the presence of SG divider plate bar weld cracking.
>
> 3. If results of industry and NRC studies and operating experience document that potential failure of the SG RCS pressure boundary due to PWSCC cracking of SG divider plate bar welds is not a credible concern, the commitment will be revised to reflect that conclusion.

Moreover, the applicant stated that if the first option were selected, it would be completed for each SG in each unit during a SG tube eddy-current inspection outage. This inspection would be conducted between 20 and 25 calendar years of SG operation, according to the dates of SGs replacement for Units 1, 2 and 3 (fall of 2005, 2003, and 2007 respectively). The applicant clarified that for Units 1 and 3, this would approximately correspond to the first 5 years after entering the period of extended operation (i.e., for Unit 1, between September 1, 2025, and December 1, 2030; and for Unit 3, between September 1, 2027, and December 1, 2032). For Unit 2, this would correspond to a time period between 3 years prior to and 2 years after entering the period of extended operation (i.e., September 1, 2023, and December 1, 2028). The applicant further stated that if the second or third option were selected, it would be completed prior to September 1, 2023, when the first replaced SGs (Unit 2) would reach 20 years of operation.

By letter dated February 25, 2011, the applicant corrected information in the November 23, 2010, letter by stating that it determined, from reviewing each unit's SG as-built documentation, that the divider plate bars in Unit 2 were made of Alloy 600 as a result of a change report issued during fabrication. The applicant further stated that it had reviewed the as-built documentation to determine if there were other differences in SG primary-side materials between the units, and no other differences were found. However, the applicant also identified that the divider bar set screws and the divider patch plate cap screws in the SGs are made of materials other than Alloy 690.

In order to address potential PWSCC of the Unit 2 Alloy 600 SG divider plate bars, in its letter dated February 25, 2011, the applicant expanded Commitment No. 61 to include the Unit 2 SG divider plate bars within the scope of the committed analyses. The applicant also committed to include the exposed portions of the Unit 2 SG divider plate bars within the scope of the committed inspections. The applicant stated that inspection or analysis of the screws is not being included in this commitment because any possible PWSCC that may occur in the screws would not be expected to propagate to the reactor coolant pressure boundary material.

By letter dated March 17, 2011, the applicant modified Commitment No. 61 from inspecting the "exposed portions" of the divider plate bars to inspecting "accessible surfaces" of the divider plate bars in order to clarify the inspection of the divider plate bars in the Unit 2 SGs. This change was intended to use standard industry terminology to refer to surfaces that can be accessed for examination. The applicant also clarified its letter dated February 25, 2011, stating

Aging Management Review Results

that the installed divider patch plate cap screws in all SGs were made of Alloy 690. The applicant also clarified that the divider bar set screws were made of stainless steel and, since they are under a compressive stress, are not susceptible to PWSCC. Further, the set screws are welded in place.

In the final version of Commitment No. 61, the applicant commits to perform one of the following options:

1. Perform an inspection of each Palo Verde Unit 1, 2, and 3 steam generator to assess the condition of the divider plate bar welds in all units, and the accessible surfaces of the divider plate bars in Unit 2. The examination technique(s) will be capable of detecting PWSCC in the divider plate bar welds in all units, and in the accessible surfaces of the divider plate bars in Unit 2.

2. Perform an analytical evaluation of the steam generator divider plate bar welds in all units, and the divider plate bars in Unit 2, in order to establish a technical basis which concludes that the SG reactor coolant system pressure boundary is adequately maintained with the presence of steam generator divider plate bar weld cracking.

3. If results of industry and NRC studies and operating experience document that potential failure of the SG reactor coolant system pressure boundary due to PWSCC cracking of SG divider plate bar welds and the divider plate bars in Unit 2 is not a credible concern, this commitment will be revised to reflect that conclusion.

Based on its review, the staff finds the applicant's options and associated revised Commitment No. 61 acceptable because the applicant identified which parts of the divider plates were made of Alloy 600 or associated weld materials. Further, the applicant will assess the condition of the divider plate bar welds in all units and the accessible surfaces of the divider plate bars in Unit 2 using an appropriate option. If the applicant inspects each SG divider plate bar weld, it will do so with appropriate examination technique and in a time period consistent with the detection of potential PWSCC. The staff finds that the timing of this inspection for each unit is acceptable because the proposed implementation schedule allows operation of the SGs for between 20 and 25 years, and it is unlikely that significant detrimental PWSCC cracking will have initiated before this time. The staff also noted that the applicant could alternatively perform an evaluation of the welds or use the results of NRC and industry operating experience to rule out this aging effect.

The staff concludes that the applicant has demonstrated that the effects of aging for these components will be adequately managed so that their intended functions will be maintained consistent with the CLB during the period of extended operation, as required by 10 CFR 54.21(a)(3).

3.1.2.2 Aging Management Review Results Consistent with the Generic Aging Lessons Learned Report for Which Further Evaluation Is Recommended

In LRA Section 3.1.2.2, the applicant further evaluates aging management, as recommended by the GALL Report, for the RV, internals, and RCS components and provides information concerning how it will manage the following aging effects:

- cumulative fatigue damage
- loss of material due to general, pitting, and crevice corrosion
- loss of fracture toughness due to neutron irradiation embrittlement
- cracking due to SCC and IGSCC
- crack growth due to cyclic loading

Aging Management Review Results

- loss of fracture toughness due to neutron irradiation embrittlement and void swelling
- cracking due to SCC
- cracking due to cyclic loading
- loss of preload due to stress relaxation
- loss of material due to erosion
- cracking due to flow-induced vibration
- cracking due to SCC and IASCC
- cracking due to PWSCC
- wall thinning due to FAC
- changes in dimensions due to void swelling
- cracking due to SCC and PWSCC
- cracking due to SCC, PWSCC, and IASCC
- QA for aging management of nonsafety-related components

For component groups evaluated in the GALL Report, for which the applicant claimed consistency with the report and for which the report recommends further evaluation, the staff reviewed the applicant's evaluation to determine if it adequately addressed the issues further evaluated. In addition, the staff reviewed the applicant's further evaluations against the criteria contained in SRP-LR Section 3.1.2.2. The staff's review of the applicant's further evaluation follows.

3.1.2.2.1   Cumulative Fatigue Damage

In LRA Section 3.1.2.2.1, the applicant stated that the analysis of cumulative fatigue damage in the RPV pressure boundary piping, valves, and other components; and of those SG secondary side components with a fatigue analysis are TLAAs as defined in 10 CFR 54.3 and are evaluated in accordance with 10 CFR 54.21(c)(1).

The applicant identified that the following AMRs in LRA Table 3.1.1 are applicable and stated the following for each applicable item:

> **Item 3.1.1.5**, PVNGS RV internals are designed to ASME III Subsection NG, some with a fatigue analysis, LRA Section 4.3.3 describes the evaluation of these TLAAs
>
> **Item 3.1.1.06**, Cumulative fatigue damage of SG tubes is not a TLAA as defined in 10 CFR 54.3, see LRA Section 4.3.2.5.
>
> **Item 3.1.1.07**, Reactor coolant pressure boundary closure bolting (reactor pressure vessel [RPV] head studs, pump, valve, and pressurizer and SG manway and port bolting) and pressurizer vessel support skirts and attachment welds are designed to ASME III Class 1, with a fatigue analysis. Both the SG primary and secondary shells, integral supports, nozzles, and bolting have a Class 1 fatigue analysis; the pressurizer relief tank is not an ASME III Class 1 component, nor is it designed to other fatigue or cyclic design rules, and therefore has no fatigue TLAA
>
> - LRA Section 4.3.2.1 describes the evaluation of these TLAAs for RV closure bolting and welded attachments
> - LRA Section 4.3.2.3 describes the evaluation of these TLAAs for the reactor coolant pump, its closure bolting, and its integral supports
> - LRA Section 4.3.2.4 describes the evaluation of these TLAAs for pressurizer closure bolting, its support skirt, and welded attachments

3-159

- LRA Section 4.3.2.5 describes the evaluation of these TLAAs for SG primary and secondary-side pressure boundaries, feedwater nozzles, closure bolting and welded attachments

- LRA Section 4.3.2.6 describes the evaluation of these TLAAs for Class 1 valves, including their bolting

- LRA Section 4.3.2.7 describes the evaluation of these TLAAs for piping and piping components

Item 3.1.1.08, Reactor coolant pressure boundary piping and the pressurizer are designed to ASME III Class 1, with fatigue analyses

- LRA Section 4.3.2.7 describes the evaluation of these TLAAs for piping and other piping components.

Item 3.1.1.09, The RV pressure boundary is designed to ASME III Class 1, with fatigue analyses

- LRA Section 4.3.2.1 describes the evaluation of these TLAAs for the RV, including the shell, heads, flanges, penetrations, welds, nozzles, and safe end butters

- LRA Section 4.3.2.2 describes the evaluation of these TLAAs for the control element assembly (CEA) housings

Item 3.1.1.10, The SG primary and secondary pressure boundaries are designed respectively to ASME III Class 1 and 2, but both the SG primary and secondary shells and nozzles have a Class 1 fatigue analysis.

- LRA Section 4.3.2.5 describes the evaluation of these TLAAs for SG primary and secondary-side pressure boundaries including the heads, feedwater nozzles, other nozzles and safe end butters, and closures

The staff reviewed LRA Section 3.1.2.2.1 against the criteria in SRP-LR Section 3.1.2.2.1, which states that fatigue is a TLAA, as defined in 10 CFR 54.3. Under 10 CFR 54.21(c)(1), TLAAs must be evaluated. Section 4.3, "Metal Fatigue Analysis," of the SRP-LR addresses this TLAA separately. The staff finds that the applicant's AMR results are consistent with the recommendations of the GALL Report and SRP-LR except for those areas identified below.

The staff noted that for LRA Table 3.1.1, item 3.1.1.6, for the recirculating SG tubes, the GALL Report identifies cumulative fatigue damage as an applicable aging effect for Class 1 tubes and sleeves and recommends that an applicant's metal fatigue analysis be used to manage this aging effect during the period of extended operation. The staff verified that in LRA Table 3.1.2-4, the applicant credited its SG Tubing Integrity Program, as the condition monitoring program to manage cracking in these tubes.

However, the staff noted that the applicant performed a CUF calculation of the replacement SG tubes because the tubes are ASME Code Class 1 components, designed to ASME Section III. The staff noted that the various degradation mechanisms reference SG tube cracking, induced either by SCC or by any other mechanisms. Cracking induced by these mechanisms has no relationship to cracking induced by high-cycle or low-cycle fatigue mechanisms. The staff noted that cracking of SG tubes has been induced by IGSCC, PWSCC, outer-diameter SCC, or intergranular attack mechanisms and that the ISI of the tubes required by plant TS have largely been implemented to detect cracking induced by these mechanisms. The staff also noted that these mechanisms do not have a relationship to the use of CUF calculations to qualify the tubes for cracking by fatigue and do not constitute a valid basis for concluding the CUF values do not qualify the tubes for fatigue-induced cracking during their design life. It is not clear to the staff

Aging Management Review Results

why the CUF value for the SG tubes is zero. By letter dated July 21, 2010, the staff issued RAI 4.3-13, requesting that the applicant justify its basis for concluding that the CUF calculation for the SG tubes does not need to be identified as a TLAA. Additionally, the staff asked the applicant to provide its basis for the CUF value of zero for the SG tubes. The staff previously identified this as part of Open Item 4.3-1.

The staff noted that in LRA Section 3.1.2.2.1, the applicant stated that the pressurizer support skirts and attachment welds were designed to ASME Section III requirements and had received an applicable ASME Section III CUF analysis. The staff determined that neither LRA Table 3.1.2-2 nor LRA Table 3.1.2-3 include any applicable items for management of cumulative fatigue damage in the pressurizer support skirts and attachment welds. By letter July 21, 2010, the staff issued RAI 4.3-13, asking that the applicant justify its basis for omitting applicable AMR items for cumulative fatigue damage of the pressurizer support skirts and pressurizer attachment weld components. The staff previously identified this as part of Open Item 4.3-1.

In its response dated August 12, 2010, the applicant stated that the SG tube CUF value was taken from the applicable design report for each unit. The applicant further clarified that the zero value for the SG tube CUF, included in the design reports, is based on the cyclic stress range being below the endurance limit. The staff noted that the applicant amended LRA Section 4.3.2.5 to identify the SG tube fatigue analysis as a TLAA and to disposition the TLAA for the SG tubes in accordance with 10 CFR 54.21(c)(1)(i). The staff also noted that the applicant amended LRA Table 3.1.2-4 to include the associated AMR item consistent with GALL Report AMR, item IV.D1-21. Furthermore, the staff noted that LRA Table 3.1.2-3 was amended to include the associated AMR line items for the pressurizer support skirt and attachment weld consistent with GALL Report AMR, item IV.C2-10. The staff confirmed that these additional AMR items are consistent with the associated GALL Report AMR items. The staff's evaluation of the pressurizer support skirt and SG tubes is documented in SER Sections 4.3.2.4.2 and 4.3.2.5.2, respectively.

Based on its review of the amended LRA Tables 3.1.2-3 and 3.1.2-4, the staff finds the applicant's response to RAI 4.3-13, parts 1 and 2, and the additions of the AMR line items, acceptable because they are consistent with the associated GALL Report AMR items for the pressurizer support skirt and attachment weld and the SG tubes. The staff's concern described in RAI 4.3-13 is resolved, and this portion of Open Item 4.3-1 is closed.

Based on its review, the staff concludes that the applicant's proposal to manage cumulative fatigue damage in ASME Code Class 1 components meets the SRP-LR Section 3.1.2.2.1 criteria. For those items that apply to LRA Section 3.1.2.2.1, the staff determines that the LRA is consistent with the GALL Report. The staff also finds that the applicant has demonstrated that it will adequately manage the effects of aging so that the intended function(s) will be maintained consistent with the CLB during the period of extended operation, as required by 10 CFR 54.21(a)(3). SER Section 4.3 documents the staff's review of the applicant's evaluation of the TLAA for these components.

3.1.2.2.2 Loss of Material Due to General, Pitting, and Crevice Corrosion

The staff reviewed LRA Section 3.1.2.2.2 against the following criteria in SRP-LR Section 3.1.2.2.2:

- LRA Section 3.1.2.2.2 states that PVNGS has a recirculating SG, not a once-through SG, so the applicant did not use the applicable GALL Report row.

    SRP-LR Section 3.1.2.2.2 states that loss of material due to general, pitting, and crevice corrosion may occur in the steel PWR SG shell assembly exposed to secondary feedwater and steam. Loss of material due to general, pitting, and crevice corrosion also may occur in the steel top head enclosure (without cladding) top head nozzles

3-161

Aging Management Review Results

(vent, top head spray or reactor core isolation cooling, and spare) exposed to reactor coolant. The existing program controls reactor water chemistry to mitigate corrosion. However, control of water chemistry does not preclude loss of material due to pitting and crevice corrosion at locations with stagnant flow conditions; therefore, the effectiveness of water chemistry control programs should be verified to ensure that corrosion does not occur. The GALL Report recommends further evaluation of programs to verify the effectiveness of water chemistry control programs. A one-time inspection of selected components at susceptible locations is an acceptable method to determine whether an aging effect is occurring or is slowly progressing such that the component's intended functions will be maintained during the period of extended operation.

SRP-LR Section 3.1.2.2.2.1 does not define the scope of applicability for this aging effect. GALL Report Table IV.D1, "Steam Generator (Recirculating)," which is applicable to the PVNGS units shows that only GALL Report AMR, item IV.D1-12 for SG upper and lower shell and transition cone is applicable for this aging effect. SER Section 3.1.2.2.2(4) discusses this separately. GALL Report Table IV.D2, "Steam Generator (Once-Through)," which is not applicable to PVNGS, Units 1, 2, and 3, however, shows the identical component name—SG shell assembly—as GALL Report AMR, item IV.D2-8 for this aging effect. Since SER Section 3.1.2.2.2(4) separately discusses the only item for this aging effect in GALL Report Table IV.D1, and GALL Report Table IV.D2 is for a different type of SG, the staff agrees that this issue is not applicable to PVNGS units.

LRA Table 3.1.1, item 3.1.1.11, which also addresses loss of material due to general, pitting, and crevice corrosion in the steel top head enclosure (without cladding) top head nozzles (vent, top head spray or reactor core isolation cooling, and spare) exposed to reactor coolant is identified as not applicable because it applies to BWRs only. Because the PVNGS units are PWRs, the staff finds that this component and aging effect combination does not apply to PVNGS.

- LRA Section 3.1.2.2.2.2 states that the aging effect is not applicable to PVNGS; it is applicable to BWRs only. SRP-LR Section 3.1.2.2.2, item 2 states that loss of material due to pitting and crevice corrosion may occur in stainless steel BWR isolation condenser components exposed to reactor coolant. Loss of material due to general, pitting, and crevice corrosion may occur in steel BWR isolation condenser components. The staff finds that SRP-LR Section 3.1.2.2.2, item 2 is not applicable to PVNGS because the PVNGS units are PWRs, and the staff guidance in this SRP-LR section is only applicable to BWRs with an isolation condenser.

- LRA Section 3.1.2.2.2.3 states that the aging effect is not applicable to PVNGS; it is applicable to BWRs only. SRP-LR Section 3.1.2.2.2, item 3 states that loss of material due to pitting and crevice corrosion may occur in stainless steel, nickel alloy, and steel with stainless steel or nickel-alloy cladding flanges, nozzles, penetrations, pressure housings, safe ends, and vessel shells, heads, and welds exposed to reactor coolant. This section of the SRP-LR is cross-referenced to the GALL Report, Table IV.C1, which is for BWRs. The staff finds that SRP-LR Section 3.1.2.2.2, item 3 is not applicable to PVNGS because the PVNGS units are PWRs, and the staff guidance in this SRP-LR section is only applicable to BWRs.

- LRA Section 3.1.2.2.2.4 addresses loss of material due to general, pitting, and crevice corrosion in the steel SG shell and transition cone exposed to secondary feedwater and steam, stating that augmented inspection is recommended for Westinghouse Model 44 and 51 SGs, where a high stress region exists at the shell to transition cone weld, if general and pitting corrosion of the shell is known to exist. The SGs at PVNGS are CE-modified System 80, so the augmented inspection is not applicable.

Aging Management Review Results

SRP-LR Section 3.1.2.2.2.4 states that loss of material due to general, pitting, and crevice corrosion may occur in the steel PWR SG upper and lower shell and transition cone exposed to secondary feedwater and steam. The existing program controls chemistry to mitigate corrosion and ISI to detect loss of material. The extent and schedule of the existing SG inspections are designed to ensure that flaws cannot attain a depth sufficient to threaten the integrity of the welds; however, according to IN 90-04, the program may not be sufficient to detect pitting and crevice corrosion, if general and pitting corrosion of the shell is known to occur. The GALL Report recommends augmented inspection to manage this aging effect. Furthermore, the GALL Report clarifies that this issue is limited to Westinghouse Model 44 and 51 SGs with a high-stress region at the shell to transition cone weld.

Based on the clarification provided by the SRP-LR regarding the type of SGs that are affected by the subject degradation mechanisms, the staff agrees with the applicant that the augmented inspection is not applicable because they do not have Westinghouse Model 44 and 51 SGs.

Based on the programs identified above, the staff concludes that the applicant's programs meet SRP-LR Section 3.1.2.2.2 criteria. For those items that apply to LRA Section 3.1.2.2.2, the staff determines that the LRA is consistent with the GALL Report and that the applicant has demonstrated that it will adequately manage the effects of aging so that the intended function(s) will be maintained consistent with the CLB during the period of extended operation, as required by 10 CFR 54.21(a)(3).

3.1.2.2.3 Loss of Fracture Toughness Due to Neutron Irradiation Embrittlement

The staff reviewed LRA Section 3.1.2.2.3 against the following criteria in SRP-LR Section 3.1.2.2.3:

- LRA Section 3.1.2.2.3.1 addresses loss of fracture toughness due to certain aspects of neutron irradiation embrittlement as an aging effect that the applicant will manage through conducting TLAAs, consistent with the SRP-LR. The applicant states that LRA Section 4.2 describes the evaluation of these neutron embrittlement TLAAs.

    SRP-LR Section 3.1.2.2.3.1 states that certain aspects of neutron irradiation embrittlement are TLAAs, as defined in 10 CFR 54.3. TLAAs are required to be evaluated in accordance with 10 CFR 54.21(c)(1).

    Loss of fracture toughness due to neutron irradiation embrittlement is limited to RPV materials having a neutron fluence greater than 1×10^{17} n/cm^2 (for energy values greater than 1.0 mega electron-volt (MeV)) at the end of the period of extended operation. SER Section 4.2 accepted the applicant's evaluation of RPV neutron embrittlement in terms of USE, pressurized thermal shock, and P-T limits, which represent a complete set of analytical means for predicting and managing loss of fracture toughness due to neutron irradiation embrittlement. Therefore, the staff concludes that the applicant's program meets the SRP-LR Section 3.1.2.2.3.1 criterion. The staff also confirmed that LRA Table 3.1.2-1 correctly identified the GALL Report Table IV.A2 item under this aging mechanism (IV.A2-23 for RPV shell). LRA Table 3.1.2-1 did not, however, list GALL Report AMR, item IV.A2-16 for RPV nozzles under this aging mechanism. This is acceptable because the estimated neutron fluence at the end of the period of extended operation for PVNGS, Units 1, 2, and 3 RPV nozzles is less than 1×10^{17} n/cm^2 (for energy values greater than 1.0 MeV).

    SER Section 4.2 documents the staff's review of the applicant's evaluation of this TLAA.

- LRA Section 3.1.2.2.3.2 addresses loss of fracture toughness due to neutron irradiation embrittlement as an aging effect that the applicant will manage, consistent with the

Aging Management Review Results

SRP-LR, by the RV Surveillance Program. This LRA section stated that due primarily to low-leakage cores, the revised 54 EFPY fluence projections are less than the original 32 EFPY projections. Further, it stated that PVNGS retains sufficient unexposed archived material to provide two additional sets of test specimens for each material, sufficient to support the program for the period of extended operation.

SRP-LR Section 3.1.2.2.3 states that loss of fracture toughness due to neutron irradiation embrittlement may occur in BWR and PWR RV beltline shell, nozzle, and welds exposed to reactor coolant and neutron flux. A RV Materials Surveillance Program monitors neutron irradiation embrittlement of the RV. RV surveillance programs are plant-specific, depending on matters such as the composition of limiting materials, availability of surveillance capsules, and projected fluence levels. In accordance with 10 CFR Part 50, Appendix H, an applicant is required to submit its proposed withdrawal schedule for approval before implementation. Untested capsules placed in storage must be maintained for future insertion. Thus, further staff evaluation is required for license renewal. GALL Report Chapter XI, Section M31 provides specific recommendations for an acceptable AMP.

The staff noted that the LRA Table 3.1.2-1 subcomponent that credits the RV Surveillance Program for managing its loss of fracture toughness aging effect is "RV Shell," which is not consistent with the corresponding GALL Report, AMR item IV.A2-24, "Vessel shell...(including beltline welds)." The staff determined that the applicant's subcomponent "RV Shell" meant to include beltline welds because its RV Surveillance Program meets the ASTM E185-82 requirements and contains weld specimens for monitoring their neutron irradiation embrittlement. The staff accepted the applicant's RV Surveillance Program, as indicated in SER Section 3.0.3.2.10. Hence, the staff concludes that the applicant's program meets SRP-LR Section 3.1.2.2.3.2 criteria. The staff also confirmed that LRA Table 3.1.2-1 identified all GALL Report Table IV.A2 AMR items under this aging mechanism (IV.A2-17 and IV.A2-24).

Based on the TLAA and the program identified above, the staff concludes that the applicant's programs meet SRP-LR Section 3.1.2.2.3.1 and Section 3.1.2.2.3.2 criteria. For those AMR items that apply to LRA Section 3.1.2.2.3, the staff concludes that the LRA is consistent with the GALL Report. The staff also finds that the applicant has demonstrated that it will adequately manage the effects of aging so that the intended function(s) will be maintained consistent with the CLB during the period of extended operation, as required by 10 CFR 54.21(a)(3).

3.1.2.2.4   Cracking Due to Stress Corrosion Cracking and Intergranular Stress Corrosion Cracking

The staff reviewed LRA Section 3.1.2.2.4 against the following criteria in SRP-LR Section 3.1.2.2.4:

- LRA Section 3.1.2.2.4.1 states that this aging effect is not applicable PVNGS; it is applicable to BWRs only. SRP-LR Section 3.1.2.2.4, item 1 states that cracking due to SCC and IGSCC may occur in the stainless steel and nickel-alloy BWR top head enclosure vessel flange leak detection lines. The staff finds that SRP-LR Section 3.1.2.2.4, item 1 is not applicable to PVNGS because the PVNGS units are PWRs, and the staff guidance in this SRP-LR section is only applicable to BWRs.

- LRA Section 3.1.2.2.4.2 states that this aging effect is not applicable PVNGS, that it is applicable to BWRs only. SRP-LR Section 3.1.2.2.4, item 2 states that cracking due to SCC and IGSCC may occur in stainless steel BWR isolation condenser components exposed to reactor coolant. The staff finds that SRP-LR Section 3.1.2.2.4, item 2 is not

Aging Management Review Results

applicable to PVNGS because the PVNGS units are PWRs, and the staff guidance in this SRP-LR section is only applicable to BWRs with an isolation condenser.

Based on the above, the staff concludes that SRP-LR Section 3.1.2.2.4 criteria do not apply.

3.1.2.2.5 Crack Growth Due to Cyclic Loading

LRA Section 3.1.2.2.5 addresses crack growth of underclad flaws in RPV forgings due to cyclic loading as a potential aging effect that may be managed through a TLAA, consistent with the SRP-LR. However, the applicant's evaluation concludes that underclad cracking is not a TLAA for PVNGS, Units 1, 2, and 3.

SRP-LR Section 3.1.2.2.5 states that crack growth due to cyclic loading could occur in RV shell forgings clad with stainless steel using a high-heat-input welding process. Growth of intergranular separations (underclad cracks) in the heat-affected zone under austenitic stainless steel cladding is a TLAA to be evaluated for the period of extended operation for all the SA 508-Cl 2 forgings where the cladding was deposited with a high heat input welding process. The methodology for evaluating the underclad flaw should be consistent with the current well-established flaw evaluation procedure and criterion in the ASME Section XI Code.

As evaluated in SER Section 4.7.6, the staff agrees with the applicant's conclusion that RPV underclad cracking is not a TLAA for PVNGS, Units 1, 2, and 3 because high-heat-input, submerged-arc-welding processes, which caused the underclad cracking, were not used for the fabrication of the PVNGS, Units 1, 2, and 3 cladding over the RPV nozzles and flange. This aging effect is not applicable to the PVNGS units.

3.1.2.2.6 Loss of Fracture Toughness Due to Neutron Irradiation Embrittlement and Void Swelling

The staff reviewed LRA Section 3.1.2.2.6 against the criteria in SRP-LR Section 3.1.2.2.6.

LRA Section 3.1.2.2.6 addresses loss of fracture toughness due to neutron irradiation, embrittlement, and void swelling as an aging effect that the applicant will manage, consistent with the SRP-LR. It will manage this effect by participating in the industry programs for investigating and managing aging effects on reactor internals and evaluating and implementing the results of the industry programs as applicable to the reactor internals. Upon completion of these programs, but not less than 24 months before entering the period of extended operation, the applicant will submit an inspection plan for reactor internals to the NRC for review and approval.

SRP-LR Section 3.1.2.2.6 states that loss of fracture toughness due to neutron irradiation embrittlement and void swelling may occur in stainless steel and nickel-alloy RV internals components exposed to reactor coolant and neutron flux. The GALL Report recommends no further AMR if the applicant commits in the Final Safety Analysis Report (FSAR) supplement to participate in industry programs for investigating and managing aging effects on reactor internals and to evaluate and implement the results of the industry programs as applicable to the reactor internals. In addition, upon completion of these programs, but not less than 24 months before entering the period of extended operation, the applicant must commit to submitting an inspection plan for reactor internals for the staff's review and approval.

As described in LRA Section 3.1.2.2.6, the applicant made a commitment to incorporate all three GALL Report recommendations, stated above, to manage this aging mechanism. The RCS Supplement (LRA, Appendix B, Section B2.1.21) contains this commitment (Commitment No. 23). Commitment No. 23 is also identified in the UFSAR Supplement A1.21. Therefore, the staff concludes that the applicant's program meets the SRP-LR Section 3.1.2.2.6 criteria for managing the aging effects due to neutron irradiation embrittlement and void swelling. The staff also examined LRA Table 3.1.2-1 to determine if the RPV internals subjected to these aging

Aging Management Review Results

effects are consistent with those listed in GALL Report Table IV.B3. The staff confirmed that LRA Table 3.1.2-1 identified GALL Report AMR, items IV.B3-12, IV.B3-16, and IV.B3-20 under this aging mechanism. However, this LRA table did not specifically list core shroud assembly bolts (GALL Report AMR, item IV.B3-10) and tie rods (GALL Report AMR, item IV.B3-12) under this aging mechanism. In addition, this LRA table did not list core support plate, fuel alignment pins, and core support column bolts as part of the GALL Report AMR, item IV.B3-20 under this aging mechanism. Therefore, by letter dated January 28, 2010, the staff issued RAI 3.1.2.2.6-1 asking that the applicant confirm that the unit core shroud assemblies are welded structures that do not have bolts and tie rods. The staff also asked the applicant to clarify why the LRA table did not specifically list core support plate, fuel alignment pins, and core support column bolts under this aging effect as part of the GALL Report, AMR item IV.B3-20.

The staff reviewed the applicant's response dated March 1, 2010, to RAI 3.1.2.2.6-1. The applicant confirmed in the response that the PVNGS unit core shroud assemblies are welded structures that do not have bolts and tie rods; therefore, the GALL Report, AMR item IV.B3-10 is not applicable to the PVNGS units. As to the core support plate and core support column bolts, the response states that the PVNGS units used, instead, the core shroud assembly and the core shroud end plate to position and support the reactor core and provide control of the reactor coolant flow into each fuel assembly. Further, fuel alignment pins are included in the RV internals core support structure lower support structure assembly. Hence, RAI 3.1.2.2.6-1 is resolved. Based on the RAI resolution and the staff's evaluation presented earlier, the staff concludes that the applicant's program meets the SRP-LR Section 3.1.2.2.6 criteria. The applicant has demonstrated that it will adequately manage the effects of aging so that the intended function(s) will be maintained consistent with the CLB during the period of extended operation, as required by 10 CFR 54.21(a)(3).

3.1.2.2.7 Cracking Due to Stress Corrosion Cracking

The staff reviewed LRA Section 3.1.2.2.7 against the following criteria in SRP-LR Section 3.1.2.2.7:

- LRA Section 3.1.2.2.7.1 refers to LRA Table 3.1.1, item 3.1.1.23 and addresses stainless steel bottom-mounted instrument guide tubes exposed to reactor coolant, which are being managed for cracking due to SCC by the Water Chemistry Program augmented by the ASME Section XI ISI, Subsections IWB, IWC, and IWD Program.

    The staff reviewed LRA Section 3.1.2.2.7.1 against the criteria in SRP-LR Section 3.1.2.2.7, item 1, which states that cracking due to SCC could occur in PWR stainless steel bottom-mounted instrument guide tubes exposed to reactor coolant. The SRP-LR also states that the GALL Report recommends that a plant-specific AMP be evaluated to ensure that this aging effect is adequately managed. BTP RLSB-1 (Appendix A.1 of the SRP-LR) describes the acceptance criteria.

    SER Sections 3.0.3.2.1 and 3.0.3.1.1 document the staff's evaluation of the applicant's Water Chemistry and ASME Section XI ISI, Subsections IWB, IWC, and IWD Programs, respectively. In its review of components associated with LRA Table 3.1.1, item 3.1.1.23, the staff finds the applicant's proposal to manage aging using the Water Chemistry and ASME Section XI ISI Programs acceptable because the Water Chemistry Program will mitigate the potential development and progress of the aging effect, while the ASME Section XI ISI, Subsections IWB, IWC and IWD Program will verify the effectiveness of the Water Chemistry Program.

    Based on the programs identified, the staff concludes that the applicant's programs meet SRP-LR Section 3.1.2.2.7, item 1 criterion. For those items that apply to LRA Section 3.1.2.2.7.1, the staff determines that the LRA is consistent with the GALL Report. The staff also finds that the applicant has demonstrated that the effects of aging

Aging Management Review Results

will be adequately managed so that the intended function will be maintained consistent with the CLB during the period of extended operation, as required by 10 CFR 54.21(a)(3).

- LRA Section 3.1.2.2.7.2 refers to LRA Table 3.1.1, item 3.1.1.24 and addresses the aging management of cracking due to SCC of CASS reactor coolant piping and components exposed to reactor coolant. The applicant stated, in LRA Section 3.1.2.2.7.2, that the RCS does not have CASS piping, piping components, and piping elements exposed to reactor coolant; therefore, this item is not applicable.

  The staff reviewed LRA Section 3.1.2.2.7.2 against the criteria in SRP-LR Section 3.1.2.2.7, item 2, which states that cracking due to SCC could occur Class 1 PWR CASS RCS piping, piping components, and piping elements exposed to reactor coolant. The existing program relies on control of water chemistry to mitigate SCC; however, SCC could occur for CASS components that do not meet the NUREG-0313 guidelines with regard to ferrite and carbon content. The GALL Report recommends further evaluation of a plant-specific program for these components to ensure that this aging effect is adequately managed.

  The staff reviewed LRA Section 3.1.2.2.7.2 and compared it to GALL AMR, item IV.C2-3 (R-05) for CASS in the reactor vessel, internals, and RCS. The GALL Report recommends use of monitoring and control of primary water chemistry and material selection to manage the aging effect. SRP-LR Section 3.1.2.2.7.2 is not applicable, as the applicant does not have CASS material exposed to the reactor coolant in this system. The staff verified that the RCS does not contain CASS piping, piping components, and piping elements by reviewing LRA Table 3.1.1, items 3.1.1.24 and 3.1.1.57 and LRA Table 3.1.2-2, which refer to the AMPs, materials, and components for the RCS.

  The applicant stated that cracking due to SCC in CASS filters is addressed by LRA Table 3.1.1, item 3.1.1.68. The staff noted that LRA Table 3.1.1, item 3.1.1.24 relies on the control of water chemistry and a plant-specific program, and LRA Table 3.1.1, item 3.1.1.68 specifies the ASME Section XI ISI, Subsections IWB, IWC, and IWD Program and Water Chemistry Program. The staff reviewed the applicant's ASME Section XI ISI, Subsections IWB, IWC, and IWD Program, and SER Section 3.0.3.1.1 documents its evaluation. The staff determined that the applicant's program performs periodic visual, volumetric, or surface examinations of Class 1, 2, and 3 pressure-retaining components. This program is capable of detecting cracking due to SCC and is an acceptable plant-specific program for managing this aging effect. The staff finds it acceptable that the applicant addressed cracking due to SCC of these CASS filters under LRA Table 3.1.1, item 3.1.1.68 because the applicant proposed its ASME Section XI ISI, Subsections IWB, IWC, and IWD Program that is capable of detecting this aging effect and its Water Chemistry Program, that is consistent with the recommendations of the GALL Report.

  Based on its review, the staff concludes that the SRP-LR Section 3.1.2.2.7, item 2 criteria is not applicable to the PVNGS units because the RCS does not contain CASS piping, piping components, and piping elements exposed to reactor coolant, other than filters addressed by item 3.1.1.68.

3.1.2.2.8 Cracking Due to Cyclic Loading

The staff reviewed LRA Section 3.1.2.2.8 against the following criteria in SRP-LR Section 3.1.2.2.8:

3-167

Aging Management Review Results

- LRA Section 3.1.2.2.8.1 states that the aging effect is not applicable to PVNGS; it is applicable to BWRs only. SRP-LR Section 3.1.2.2.8, item 1 states that cracking due to cyclic loading may occur in the stainless steel BWR jet pump sensing lines. The staff finds that SRP-LR Section 3.1.2.2.8, item 1 is not applicable to PVNGS because the PVNGS units are PWRs, and the staff guidance in this SRP-LR section is only applicable to BWRs.

- LRA Section 3.1.2.2.8.2 states that the aging effect is not applicable to PVNGS; it is applicable to BWRs only. SRP-LR Section 3.1.2.2.8, item 2 states that cracking due to cyclic loading may occur in steel and stainless steel BWR isolation condenser components exposed to reactor coolant. The staff finds that SRP-LR Section 3.1.2.2.8, item 1 is not applicable to PVNGS because the PVNGS units are PWRs, and the staff guidance in this SRP-LR section is only applicable to BWRs.

Based on the above, the staff concludes that SRP-LR Section 3.1.2.2.8 criteria do not apply.

3.1.2.2.9 Loss of Preload Due to Stress Relaxation

The staff reviewed LRA Section 3.1.2.2.9 against the criteria in SRP-LR Section 3.1.2.2.9.

LRA Section 3.1.2.2.9 addresses loss of preload due to stress relaxation for PVNGS stainless steel screws, bolts, and tie rods of the CEA shroud assembly components exposed to reactor coolant as an aging effect that the applicant will manage, consistent with the SRP-LR, by the commitment of PVNGS AMP B2.1.21.

SRP-LR Section 3.1.2.2.9 states that loss of preload due to stress relaxation may occur in stainless steel and nickel-alloy PWR RV internals screws, bolts, tie rods, and hold-down springs exposed to reactor coolant. The GALL Report recommends no further AMR if the applicant commits in the FSAR supplement to participate in the industry programs for investigating and managing aging effects on reactor internals and to evaluate and implement the results of the industry programs as applicable to the reactor internals. In addition, upon completion of these programs, but not less than 24 months before entering the period of extended operation, the applicant must submit an inspection plan for reactor internals to the staff for review and approval.

As described in LRA Section 3.1.2.2.9, the applicant made a commitment to incorporate all three GALL Report recommendations, stated above, to manage this aging mechanism. The RCS Supplement contains this commitment (Commitment No. 23) and it is also identified in the UFSAR Supplement A1.21. Therefore, the staff concludes that the applicant's program meets the SRP-LR Section 3.1.2.2.9 criteria for managing the aging effects due to loss of preload due to stress relaxation. The staff also examined LRA Table 3.1.2-1 to determine if the RPV internals subjected to these aging effects are consistent with those listed in GALL Report Table IV.B3. The staff confirmed that LRA Table 3.1.2-1 identified GALL Report Table IV.B3, item IV.B3-6 under this aging mechanism. This LRA Table did not list core shroud assembly bolts and tie rods (GALL Report AMR, item IV.B3-7) under this aging mechanism because the PVNGS unit core shroud assemblies are welded structures, as confirmed by the applicant in its response to RAI 3.1.2.2.6-1. Therefore, the staff concludes that the LRA is consistent with the GALL Report, and that the applicant has demonstrated that it will adequately manage the effects of aging so that the intended function(s) will be maintained consistent with the CLB during the period of extended operation, as required by 10 CFR 54.21(a)(3).

3.1.2.2.10 Loss of Material Due to Erosion

The staff reviewed LRA Section 3.1.2.2.10 against the criteria in SRP-LR Section 3.1.2.2.10.

LRA Section 3.1.2.2.10 addresses loss of material due to erosion for PVNGS steel SG feedwater impingement plates and supports exposed to secondary feedwater. The applicant

Aging Management Review Results

stated that the PVNGS SGs do not have feedwater impingement plates. Hence, this aging mechanism is not applicable to PVNGS, Units 1, 2, and 3.

SRP-LR Section 3.1.2.2.10 states that loss of material due to erosion could occur in steel SG feedwater impingement plates and supports exposed to secondary feedwater.

Since the PVNGS SGs were designed with no feedwater impingement plates, the staff agrees with the applicant's conclusion that this SRP-LR criterion does not apply to the PVNGS units.

3.1.2.2.11 Cracking Due to Flow-Induced Vibration

The staff reviewed LRA Section 3.1.2.2.11 against the criteria in SRP-LR Section 3.1.2.2.11.

LRA Section 3.1.2.2.11 states that the aging effect is not applicable to PVNGS; it is applicable to BWRs only. SRP-LR Section 3.1.2.2.11 states that cracking due to flow-induced vibration could occur for the BWR stainless steel steam dryers exposed to reactor coolant. The staff finds that SRP-LR Section 3.1.2.2.11 is not applicable to PVNGS because the PVNGS units are PWRs, and the staff guidance in this SRP-LR section is only applicable to BWRs.

3.1.2.2.12 Cracking Due to Stress Corrosion Cracking and Irradiation-Assisted Stress Corrosion Cracking

The staff reviewed LRA Section 3.1.2.2.12 against the criteria in SRP-LR Section 3.1.2.2.12.

LRA Section 3.1.2.2.12 addresses cracking due to SCC and IASCC of stainless steel RPV internals exposed to reactor coolant as an aging effect that the applicant will manage, consistent with the SRP-LR, through its Water Chemistry Program and the commitment in the RCS Supplement.

SRP-LR Section 3.1.2.2.12 states that cracking due to SCC and IASCC may occur in PWR stainless steel reactor internals exposed to reactor coolant. The existing program controls water chemistry to mitigate these aging effects. The GALL Report recommends no further AMR if the applicant commits in the UFSAR supplement to participate in the industry programs for investigating and managing aging effects on reactor internals and to evaluate and implement the results of the industry programs as applicable to the reactor internals. In addition, upon completion of these programs, but not less than 24 months before entering the period of extended operation, the applicant must submit an inspection plan for reactor internals to the staff for review and approval.

As indicated in SER Section 3.0.3.2.1, the staff accepts the Water Chemistry Program for mitigating the aging effects due to SCC and IASCC, meeting one of the requirements mentioned in SRP-LR Section 3.1.2.2.12. Further, as described in LRA Section 3.1.2.2.12, the applicant made a commitment to incorporate all three GALL Report recommendations, stated above, to manage this aging mechanism. The PVNGS RCS Supplement contains this commitment (Commitment No. 23). UFSAR Supplement A1.21 also identifies Commitment No. 23. Therefore, the staff concludes that the applicant's program meets the SRP-LR Section 3.1.2.2.12 criteria. The staff also confirmed that LRA Table 3.1.2-1 identified the following GALL Report Table IV.B3 AMR items under this aging mechanism: IV.B3-2, IV.B3-11, IV.B3-15, IV.B3-21, and IV.B3-28. However, this LRA table does not cover all RPV internals in GALL Report Table IV.B3 under this aging mechanism. Therefore, by letter dated January 28, 2010, the staff issued RAI 3.1.2.2.12-1 and asked the applicant to clarify the disposition of the core support plate and core support column of the lower internal assembly (IV.B3-21) and the fuel alignment plate, the fuel alignment plate guide lugs, and guide lug inserts of the upper internals assembly (IV.B3-28).

The staff reviewed the applicant's response to RAI 3.1.2.2.12-1, dated March 1, 2010. Consistent with the response to RAI 3.1.2.2.6-1, the applicant clarified that, instead of using the core support plate and core support column, the PVNGS units used the core shroud assembly

Aging Management Review Results

and the core shroud end plate to position and support the reactor core and provide control of the reactor coolant flow into each fuel assembly. For GALL Report, AMR item IV.B3-28 components, this response provides a detailed component list, including two of the three components specified in the GALL Report (the UGS support plate and the fuel alignment plate). The applicant evaluates the listed components to ensure consistency with GALL Report, AMR item IV.B3-28 for the aging effect of cracking due to SCC and IASCC. The third GALL Report-specified component, fuel alignment plate guide lugs and guide lug inserts, is not in the list. However, the applicant identified them to be integral parts of the PVNGS unit core shroud assembly and added the fuel alignment plate guide lugs and guide lug inserts to this component under GALL Report, AMR item IV.B3-11. This is acceptable because the aging mechanism and recommended AMP for both GALL Report items are identical. Hence, RAI 3.1.2.2.12-1 is resolved. Based on this RAI response and the staff's evaluation presented earlier, the staff concludes that the applicant's program meets the SRP-LR Section 3.1.2.2.12 criteria. The applicant has demonstrated that it will adequately manage the effects of aging so that the intended function(s) will be maintained consistent with the CLB during the period of extended operation, as required by 10 CFR 54.21(a)(3).

3.1.2.2.13 Cracking Due to Primary Water Stress Corrosion Cracking

The staff reviewed LRA Section 3.1.2.2.13 against the criteria in SRP-LR Section 3.1.2.2.13.

LRA Section 3.1.2.2.13 refers to LRA Table 3.1.1, item 3.1.1.31 and addresses nickel-alloy components, including RCPB components and penetrations inside the RCS. These components include pressurizer heater sheaths and sleeves, nozzles, and other internal components exposed to reactor coolant, which are being managed for cracking due to PWSCC by the Water Chemistry Program, ASME Section XI ISI, Subsections IWB, IWC, and IWD Program, and the Nickel-Alloy AMP. In addition, the applicant provides in the UFSAR Supplement a commitment (Commitment No. 23) to implement applicable NRC orders, bulletins, and GLs associated with nickel alloys as well as staff-accepted industry guidelines. Further, the applicant participates in industry initiatives, such as owners group programs and the EPRI Materials Reliability Program, to manage the aging effects associated with nickel alloys. Upon completion of these programs, but not less than 24 months before entering the period of extended operation, the applicant will submit an inspection plan for RCS nickel-alloy pressure boundary components for the staff's review and approval. The applicant addressed the further evaluation requirements by stating that the management of cracking, due to PWSCC of nickel-alloy components exposed to reactor coolant, will be performed by water chemistry and ISI, augmented by a plant-specific Nickel-Alloy AMP. In addition to these programs, the applicant will implement applicable NRC orders, bulletins, and GLs associated with nickel alloys as well as staff-accepted industry guidelines.

The staff reviewed LRA Section 3.1.2.2.13 against the criteria described in SRP–LR Section 3.1.2.2.13, which states that cracking due to PWSCC could occur in nickel alloy and steel with nickel-alloy cladding PWR components. These components include RCPB components and penetrations inside the RCS such as pressurizer heater sheathes and sleeves, nozzles, and other internal components exposed to reactor coolant. The SRP-LR also states that, with the exception of RV upper head nozzles and penetrations, the GALL Report recommends ASME Section XI ISI (for Class 1 components) and control of water chemistry. For nickel-alloy components, no further AMR is necessary if the applicant complies with applicable NRC orders and provides a commitment in the UFSAR supplement to implement applicable bulletins, GLs, and staff-accepted industry guidelines.

SER Sections 3.0.3.2.1, 3.0.3.1.1, and 3.0.3.3.1 document the staff's evaluations of the applicant's Water Chemistry Program, ASME Section XI ISI, Subsections IWB, IWC, and IWD Program, and Nickel-Alloy AMP, respectively. The staff noted that the Water Chemistry Program controls the chemical environment to ensure that the aging effects due to

Aging Management Review Results

contaminants are limited by managing the primary and secondary water. The staff noted that this is accomplished by limiting the concentration of chemical species known to cause corrosion and adding chemical species known to inhibit degradation by their influence on pH and dissolved oxygen levels. The staff also noted that this program is effective in creating an environment that is not conducive for cracking to occur. The staff noted that the ASME Section XI ISI, Subsections IWB, IWC, and IWD Program includes requirements for scheduling of examinations and tests for Class 1, 2, and 3 components. The staff further noted that this program requires periodic visual, surface, volumetric examinations and leakage tests of Class 1, 2, and 3 pressure-retaining components, provides measures for monitoring to detect aging effects before the loss of intended function, and provides measures for the repair and replacement of components with aging effects. The staff noted that the Nickel-Alloy AMP will augment the ASME Section XI ISI, Subsections IWB, IWC, and IWD Program for nickel-alloy components, which consists of inspections, mitigation techniques, repair or replace activities, and monitoring of operating experience to manage the aging of these components.

The staff finds the applicant's use of its Water Chemistry Program, ASME Section XI ISI, Subsections IWB, IWC, and IWD Program, and Nickel-Alloy AMP acceptable to manage this aging effect because by controlling water chemistry, the applicant will create an environment that is not conducive for cracking to occur. In addition, the applicant will conduct a combination of periodic visual, surface, and volumetric examinations that are proven capable of detecting cracking and are consistent with ASME Section XI, ASME Code Case N-729-1, subject to the conditions specified in 10 CFR 50.55a(g)(6)(ii)(D)(2)–(6), and ASME Code Case N-722, subject to the conditions listed in 10 CFR 50.55a(g)(6)(ii)(E)(2)–(4). Further, the applicant has credited programs consistent with the GALL Report in addition to crediting its plant-specific Nickel-Alloy AMP and has provided Commitment No. 23, consistent with the SRP-LR.

Based on the program identified, the staff concludes that the applicant's program meets SRP-LR Section 3.1.2.2.13 criteria. For those items that apply to LRA Section 3.1.2.2.13, the staff determines that the LRA is consistent with the GALL Report. The applicant has demonstrated that it will adequately manage the effects of aging for these components so that their intended function will be maintained consistent with the CLB during the period of extended operation, as required by 10 CFR 54.21(a)(3).

3.1.2.2.14 Wall Thinning Due to Flow-Accelerated Corrosion

LRA Table 3.1.1, item 3.1.1.32, addresses the steel SG feedwater inlet ring (feedring) and supports exposed to secondary feedwater or steam, which are being managed for wall thinning due to FAC by the SG Tube Integrity Program and Water Chemistry Program. The applicant addressed the further evaluation criteria of the SRP-LR by stating that feedring wall thinning, as addressed in IN 91-19, is not applicable to PVNGS due to the model of SGs in use, and that no action is required. However, the applicant stated that the Water Chemistry Program and the SG Tubing Integrity Program are conservatively credited to manage wall thinning due to FAC for the feedring. In LRA Table 3.1.2-4, plant-specific Note 1, the applicant states that feedring wall thinning, as described in IN 9-19, has been detected only in certain CE pre-System 80 SGs and that its SGs are CE modified System 80. The applicant further stated that because no operating experience at its plant or other units with CE modified System 80 SGs suggests that degradation of the feedrings is occurring, it has determined this condition is not applicable to its plant, and no action is required.

The staff reviewed LRA Section 3.1.2.2.14 against criteria in SRP-LR Section 3.1.2.2.14, which states that wall thinning due to FAC could occur in steel feedrings and supports. The GALL Report references IN 91-19, "Steam Generator Feedwater Distribution Piping Damage," for evidence of FAC in SGs and recommends that a plant-specific AMP be evaluated because existing programs may not be capable of mitigating or detecting wall thinning due to FAC.

3-171

Aging Management Review Results

In its review of components associated with LRA item 3.1.1.32, the staff noted that the applicant assigned generic note E to the AMR item, indicating that the GALL Report recommends a plant-specific AMP be evaluated for this combination of component, material, environment, and aging effect combination. Because the GALL Report recommends that a plant-specific AMP be evaluated, and the applicant credits the Water Chemistry Program and the SG Tube Integrity Program to manage wall thinning in these components, the staff finds the applicant's use of generic note E to be acceptable.

The staff does not consider IN 91-19 to be limited to CE SGs, however, and the staff was unclear why the applicant stated no action was required for addressing FAC of the feedring. The applicant's description of the SG design in LRA Sections 2.3.1.4 and B2.1.8 does not provide sufficient detail of the feedwater inlet ring and supports for the staff to determine if FAC could potentially occur in the replacement SG design.

Moreover, the applicant stated that it conservatively credits its Water Chemistry Program and SG Tubing Integrity Program to manage wall thinning due to FAC. In LRA Section B2.1.8, the applicant stated that tube support degradation is monitored by the presence of normal support signals at expected tube locations and by visual inspection of the secondary side. It further stated that its SG management procedure specifies that it will visually inspect SGs, as required, on the secondary side at the accessible portions of the following locations: tubesheet region, both hot and cold leg, tube supports, flow distribution plate, and upper steam drum internals. The staff was not clear if the SG feedwater ring was included in the scope of the SG Tube Integrity Program. This was previously identified as Confirmatory Item 3.1.2.2.14-1.

During a conference call on July 9, 2010, and in a letter dated July 30, 2010, the applicant clarified that the material of the SG feedring is fabricated from P11 steel and, therefore, is FAC resistant. The applicant also explained that the Steam Generator Tube Integrity Program considers wall thinning of the SG feedring and applicable operating experience as part of the secondary side SG Degradation Assessment, performed before every outage. The staff finds this information acceptable and Confirmatory Item 3.1.2.2.14-1 is closed.

SER Sections 3.0.3.2.1 and 3.0.3.1.5 document the staff's evaluation of the applicant's Water Chemistry Program and SG Tube Integrity Program, respectively. The staff noted that the Water Chemistry Program manages wall thinning in the secondary water system, including maintaining appropriate chemical concentrations in the SG secondary side and the secondary systems, to limit aging effects associated with corrosion mechanisms. The staff further noted that the SG Tube Integrity Program addresses wall thinning due to FAC of the SG feedring through the SG degradation assessment. Since the SG feedring is fabricated from FAC-resistant material, the staff finds this aging effect unlikely to occur in this component. Therefore, the staff finds the applicant's proposal to manage the aging effect of wall thinning due to FAC for the steel SG feedring and supports by using the Water Chemistry Program and the SG Tube Integrity Program acceptable because, 1) the Water Chemistry Program provides mitigation for this aging effect and its use is consistent with the recommendations of the GALL Report, and 2) the SG Tube Integrity Program is adequate for verifying the effectiveness of the Water Chemistry Program for the secondary side SG internals and for managing wall thinning due to FAC before tube integrity is compromised.

The staff concludes that the applicant's programs meet SRP-LR Section 3.1.2.2.14 criteria. For those items that apply to LRA Section 3.1.2.2.14, the staff determines the LRA is consistent with the GALL Report. The applicant has demonstrated that it will adequately manage the effects of aging so that the intended functions will be maintained consistent with the CLB during the period of extended operation, as required by 10 CFR 54.21(a)(3).

3.1.2.2.15 Changes in Dimensions Due to Void Swelling

The staff reviewed LRA Section 3.1.2.2.15 against the criteria in SRP-LR Section 3.1.2.2.15.

Aging Management Review Results

LRA Section 3.1.2.2.15 addresses changes in dimension due to void swelling for stainless steel and nickel-alloy reactor internal components exposed to reactor coolant as an aging effect that the applicant will manage, consistent with the SRP-LR, by the commitment of PVNGS RCS Supplement.

SRP-LR Section 3.1.2.2.15 states that changes in dimensions due to void swelling may occur in stainless steel and nickel-alloy PWR internal components exposed to reactor coolant. The GALL Report recommends no further AMR if the applicant commits in the FSAR supplement to participate in the industry programs for investigating and managing aging effects on reactor internals and to evaluate and implement the results of the industry programs as applicable to the reactor internals. In addition, upon completion of these programs, but not less than 24 months before entering the period of extended operation, the applicant must submit an inspection plan for reactor internals for the staff's review and approval.

As described in LRA Section 3.1.2.2.15, the applicant made a commitment to incorporate all three GALL Report recommendations, stated above, to manage this aging mechanism. PVNGS AMP B2.1.21 contains this commitment (Commitment No. 23). UFSAR Supplement A1.21 also identifies Commitment No. 23. Therefore, the staff concludes that the applicant's program meets the SRP-LR Section 3.1.2.2.15 criteria. The staff also confirmed that LRA Table 3.1.2-1 identified the following GALL Report Table IV.B3 AMR items under this aging mechanism: IV.B3-4, IV.B3-13, IV.B3-14, IV.B3-19, and IV.B3-27. However, this LRA table does not cover all RPV internals in GALL Report Table IV.B3 under this aging mechanism. Therefore, by letter dated January 28, 2010, the staff issued RAI 3.1.2.2.15-1 and asked the applicant to clarify the disposition of the core support plate, fuel alignment pins, and core support column bolts of the lower internal assembly (IV.B3-19) and the fuel alignment plate, the fuel alignment plate guide lugs, and guide lug inserts of the upper internals assembly (IV.B3-27). Additionally, the staff asked the applicant to discuss the relationship between RV internals in-core instrumentation support structures (identified in LRA Table 3.1.2-1) and the core support plate, fuel alignment pins, and core support column bolts of the lower internal assembly (listed in the GALL Report Table IV.B3).

The staff reviewed the applicant's response to RAI 3.1.2.2.15-1, dated March 1, 2010. For GALL Report-specified items IV.B3-19 and IV.B3-27 components, some do not exist and some take different names in the PVNGS units. The only PVNGS unit component that seems inconsistent with the GALL Report is the fuel alignment plate guide lugs and guide lug inserts. Instead of classifying it under AMR item IV.B3-27 as in the GALL Report, this component is placed under GALL Report, AMR item IV.B3-13 as part of the PVNGS unit core shroud assembly. This is acceptable because the aging mechanism and recommended AMP for both GALL Report AMR items are identical. The applicant further clarified that the RV internals in-core instrumentation support structures are evaluated as part of the lower support structure assembly. Hence, RAI 3.1.2.2.15-1 is resolved. Based on the applicant's response to this RAI and the staff's evaluation presented earlier, the staff concludes that the applicant's program meets the SRP-LR Section 3.1.2.2.15 criteria. The applicant has demonstrated that it will adequately manage the effects of aging so that the intended function(s) will be maintained consistent with the CLB during the period of extended operation, as required by 10 CFR 54.21(a)(3).

3.1.2.2.16 Cracking Due to Stress Corrosion Cracking and Primary Water Stress Corrosion Cracking

The staff reviewed LRA Section 3.1.2.2.16 against the following criteria in SRP-LR Section 3.1.2.2.16:

- LRA Section 3.1.2.2.16, item 1 refers to LRA Table 3.1.1, item 3.1.1.34 and addresses stainless steel and nickel-alloy reactor control rod drive head penetration pressure housings exposed to reactor coolant (internal), which are being managed for cracking due to SCC and PWSCC. The applicant addressed the further evaluation criteria of the SRP-LR by stating that the nickel alloy portion of the RV control element drive mechanism housing (lower) credits the Water Chemistry Program and the ASME Section XI ISI, Subsections IWB, IWC, and IWD Program, which will be augmented by the Nickel-Alloy AMP. The applicant further stated that it will comply with applicable NRC orders and the UFSAR Commitment. The applicant also stated for the stainless steel RV control element drive mechanism housing (upper and lower) it credits the Water Chemistry Program and the ASME Section XI ISI, Subsections IWB, IWC, and IWD Program. LRA Section 3.1.2.2.16.1 also refers to LRA Table 3.1.1, item 3.1.1-35 and addresses steel with stainless steel or nickel-alloy cladding primary side components. These components include SG upper and lower heads, tubesheets, and tube-to-tube sheet welds exposed to reactor coolant (internal) subject to cracking due to SCC and PWSCC. The applicant stated that LRA Table 3.1.1, item 3.1.1-35 is not applicable because the SGs are the recirculating type.

The staff reviewed LRA Section 3.1.2.2.16.1 against the criteria in SRP-LR Section 3.1.2.2.16, item 1, which states that cracking due to PWSCC could occur on the nickel-alloy control rod drive head penetration pressure housings. The GALL Report recommends ASME Section XI ISI and control of water chemistry to manage this aging and recommends no further AMR for PWSCC of nickel alloy if the applicant complies with applicable NRC orders and provides a commitment in the FSAR supplement to implement applicable bulletins, GLs, and staff-accepted industry guidelines.

SER Sections 3.0.3.2.1, 3.0.3.1.1 and 3.0.3.3.1 document the staff's evaluations of the applicant's Water Chemistry Program, ASME Section XI ISI, Subsections IWB, IWC, and IWD Program and Nickel-Alloy AMP, respectively. The staff noted that the Water Chemistry Program controls the chemical environment to ensure that the aging effects due to contaminants are limited by managing the primary and secondary water. The staff noted that this is accomplished by limiting the concentration of chemical species known to cause corrosion and adding chemical species known to inhibit degradation by their influence on pH and dissolved oxygen levels. The staff also noted that this program is effective in creating an environment that is not conducive for cracking to occur. The staff noted that the ASME Section XI ISI, Subsections IWB, IWC, and IWD Program includes requirements for the scheduling of examinations and tests for Class 1, 2, and 3 components. The staff further noted that this program requires periodic visual, surface, volumetric examinations, and leakage tests of Class 1, 2, and 3 pressure-retaining components. This program also provides measures for monitoring to detect aging effects before the loss of intended function and provides measures for the repair and replacement of components with aging effects. The staff noted that the Nickel-Alloy AMP will augment the ASME Section XI ISI, Subsections IWB, IWC, and IWD Program for nickel-alloy components. Furthermore, the Nickel-Alloy AMP consists of inspections, mitigation techniques, repair or replace activities, and monitoring of operating experience to manage the aging.

The SRP-LR states that no further AMR for PWSCC of nickel alloy is necessary if the applicant complies with applicable NRC orders and provides a commitment in the UFSAR supplement to implement applicable bulletins, GLs, and staff-accepted industry guidelines. In addition, the applicant must credit its Water Chemistry Program and ASME Section XI ISI, Subsections IWB, IWC, and IWD Program for aging management. The staff noted that the applicant's commitment (Commitment No. 23) in LRA Appendix A, Section A1.21 states that it will implement applicable NRC orders, bulletins, and GLs associated with nickel alloys as well as staff-accepted industry guidelines. In addition, the applicant will participate in the industry initiatives, such as owners group programs

Aging Management Review Results

and the EPRI Materials Reliability Program, to manage the aging effects associated with nickel alloys. Upon completion of these programs, but not less than 24 months before entering the period of extended operation, the applicant will submit an inspection plan for RCS nickel-alloy pressure boundary components to the NRC for review and approval. The staff noted that the applicant's commitment includes the aspects from the SRP-LR recommendations and finds that it is consistent with the commitment described in SRP-LR 3.1.2.2.16, item 1. The staff also notes that all of the nickel-alloy AMR results lines that refer to LRA Table 3.1.1, item 3.1.1.34 are aligned with the applicant's commitment, as described in LRA Appendix A, Section A1.21. The staff finds the applicant's proposal acceptable because the applicant credits its Water Chemistry Program and ASME Section XI ISI, Subsections IWB, IWC, and IWD Program, augmented by its Nickel-Alloy AMP for nickel-alloy components. The applicant has provided the appropriate commitment in the UFSAR Supplement, and the AMR results lines refer to the commitment, consistent with the recommendations in the GALL Report and SRP-LR.

The staff reviewed GALL Report, AMR item IV.D2-4, which is associated with LRA Table 3.1.1, item 3.1.1.35. The staff noted that LRA Table 3.1.1, item 3.1.1.35 and GALL Report, AMR item IV.D2-4 recommendations for aging management are specific to the primary side components—upper and lower heads and tube sheets and tube-to-tube sheet welds for once-through SGs. The LRA also states that this item is not applicable because the SGs are recirculating-type. UFSAR Table 5.1-2 states that the SG tubes are fabricated from Alloy 690TT and that the tubesheet in contact with the reactor coolant is clad with weld deposited NiCrFe alloy, which is described as Alloy 600 cladding in LRA Section B2.1.34.

The staff noted that the components associated with SRP-LR Table 3.1.1, item 3.1.1-35, are applicable to the once-through type SGs that are found in Babcock & Wilcox PWRs as discussed in the following paragraphs.

SRP-LR Section 3.1.2.2.16.1 identifies that cracking due to PWSCC could occur on the primary coolant side of PWR steel SG tube-to-tube sheet welds made or clad with nickel alloy. The GALL Report recommends ASME Code, Section XI, ISI, Subsections IWB, IWC, and IWD, and Water Chemistry Programs to manage this aging effect. The SRP-LR also recommends no further AMR for PWSCC of nickel alloy if the applicant complies with applicable NRC Orders and provides a commitment in its UFSAR supplement to implement applicable NRC bulletins, generic letters, and staff-accepted industry guidelines. The GALL Report, revision 1 addresses this aging effect in item IV.D2-4, which is only applicable to once-through SGs and not applicable to recirculating SGs.

The staff noted that ASME Code, Section XI does not require inspection of the tube-to-tubesheet welds. In addition, no specific NRC orders or bulletins address inspection requirements for these welds. The staff is concerned that the region of the autogenous tube-to-tubesheet welds may have insufficient chromium content to prevent initiation of PWSCC if the tubesheet cladding or associated weld materials are Alloy 600. This may be the case even when the SG tubes are made from Alloy 690TT, which has been shown to have sufficient chromium content to prevent this aging effect. Consequently, a PWSCC crack initiated in the cladding region, close to a tube, may propagate into or through the weld, causing a failure of the weld and of the RCP boundary, even for recirculating SGs. For some plants, the RCP boundary in this area has been redefined by a license amendment such that the autogenous tube-to-tubesheet weld is no longer included in the RCP boundary. Since the staff has not approved such a redefinition of the RCP boundary for the PVNGS SGs, the staff

Aging Management Review Results

considers that the effectiveness of the Primary Water Chemistry Program should be verified to ensure PWSCC is not occurring and the RCP boundary is not breached.

The staff held conference calls with the applicant on October 22, November 3, and November 19, 2010, to discuss and clarify the staff's concerns. The staff asked the applicant how it managed PWSCC in SG tube-to-tubesheet welds if the tubesheet cladding is Alloy 600. The applicant agreed to provide information on its management of this aging mechanism.

By letter dated November 23, 2010, the applicant explained that the SGs tubes are manufactured from Alloy 690TT with a chromium content of 30 percent. The tubesheet cladding is composed of Alloy 82 with a chromium content of 18–20 percent and that the tube-to-tubesheet weld is an autogenous weld, which is created by melting the corner of the tubesheet clad to the tube end without adding filler metal. The applicant described statements from an industry review (MRP-115) that identified a threshold for PWSCC resistance for Alloys 600/82/182 with a chromium content of 22–30 percent. In comparison, the applicant stated it expected the chromium content of the tube-to-tubesheet welds to be 20–30 percent. The staff does not find this information to be a sufficient basis for precluding its concern about potential failure of the SG primary-to-secondary pressure boundary due to PWSCC of tube-to-tubesheet welds.

The applicant stated that the visual inspection performed every refueling outage on Alloy 82 repairs of several Alloy 600 high temperature components (half nozzle replacements using Alloy 690 nozzles welded with Alloy 82) have detected no leakage. However, the staff noted that the applicant did not provide information that would confirm the absence of cracking in these repaired areas. Further, the staff noted that differences in geometric configuration and fabrication do not allow for comparison of these repairs with the SG tube-to-tubesheet welds.

In response to the staff's concern, the applicant committed (Commitment No. 62) to the following:

In response to the NRC staff concern regarding potential failure of the steam generator primary-to-secondary pressure boundary due to PWSCC cracking of tube-to-tubesheet welds, APS commits to perform one of the following two resolution options:

    1. Perform a one-time inspection of a representative number of tube-to-tubesheet welds in each steam generator to determine if PWSCC cracking is present. If weld cracking is identified:

    a. The condition will be resolved through repair or engineering evaluation to justify continued service, as appropriate.

    b. An ongoing monitoring program will be established to perform routine tube-to-tubesheet weld inspections for the remaining life of the steam generators.

    2. Perform an analytical evaluation of the steam generator tube-to-tubesheet welds in order to:

    a. Establish a technical basis which concludes that the structural integrity of the steam generator tube-to-tubesheet interface is adequately maintained with the presence of tube-to-tubesheet weld cracking.

    b. Establish a technical basis which concludes that the steam generator tube-to-tubesheet welds are not required to perform a reactor coolant pressure boundary function.

Aging Management Review Results

Moreover, the applicant stated that if the first option is selected, it would be completed for each SG in each unit during an eddy-current inspection outage This outage would be chosen such that it is between 20 and 25 calendar years of SG operation, according to the dates of SG replacement for Units 1, 2 and 3 (fall of 2005, 2003, and 2007, respectively). For Units 1 and 3, the applicant stated the inspection would approximately correspond to the first 5 years after entering the period of extended operation (i.e., September 1, 2025, to December 1, 2030, and September 1, 2027, to December 1, 2032, respectively). For Unit 2, this would approximately correspond to 3 years prior to and 2 years after entering the period of extended operation (i.e., September 1, 2023, to December 1, 2028). The applicant further stated that if the second option is selected, it would be completed prior to September 1, 2023, the date when the first replaced SGs (Unit 2) will reach 20 years of operation.

Based on its review, the staff finds the applicant's commitment (Commitment No. 62) acceptable because it will manage the aging effect of cracking due to PWSCC in the SG tube-to-tubesheet welds either by demonstrating that those welds do not have a structural integrity or pressure boundary function or by implementing a one-time inspection. This one-time inspection will be capable of detecting PWSCC cracking on a representative number of tube-to-tubesheet welds for each SG in a time period consistent with the detection of potential PWSCC. The staff finds the timing of these inspections to be acceptable because the proposed implementation schedule allows operation of the SGs for between 20 and 25 years, and it is unlikely that significant detrimental PWSCC cracking will have initiated before this time. The staff also noted that, if the aging effect is revealed, this one-time inspection is accompanied by corrective actions, including an evaluation of the degradation and the implementation of routine inspections of the tube-to-tubesheet welds for the remaining life of the SGs.

- LRA Section 3.1.2.2.16.2 addresses nickel alloy and stainless steel pressurizer spray heads exposed to reactor coolant. The GALL Report recommends use of GALL AMP XI.M2 "Water Chemistry," and GALL AMP XI.M32 "One-Time Inspection." In addition, for nickel-alloy welded spray heads, the applicant must comply with applicable NRC orders and provide a commitment in the UFSAR supplement to implement applicable bulletins, GLs, and staff-accepted industry guidelines to manage cracking due to SCC and PWSCC for this component group. The applicant stated that this item is not applicable because it has determined that the pressurizer spray heads are not included in scope of license renewal; therefore, it did not use the applicable GALL Report line.

The staff reviewed LRA Section 3.1.2.2.16.2 against the criteria in SRP-LR Section 3.1.2.2.16, item 2, which states that cracking due to SCC could occur on stainless steel pressurizer spray heads, and cracking due to PWSCC could occur on nickel-alloy pressurizer spray heads when exposed to reactor coolant. The SRP-LR also states the existing program relies on control of water chemistry to mitigate this aging effect. The GALL Report recommends one-time inspection to confirm that cracking is not occurring. For nickel-alloy welded spray heads, the GALL Report recommends no further AMR if the applicant complies with applicable NRC orders and provides a commitment in the UFSAR supplement to implement applicable bulletins, GLs, and staff-accepted industry guidelines.

The staff reviewed the LRA scoping and screening results for the pressurizer, which indicate that the spray heads are not included in the scope of the license renewal. In addition, the staff reviewed the LRA aging management evaluation tables and did not identify the inclusion of the pressurizer spray heads. In its review, the staff further noted that LRA Section 3.1.2.2.16.2 indicates that the pressurizer spray heads are not included in the scope of the license renewal. However, the LRA section does not provide a

Aging Management Review Results

technical basis for why the pressurizer spray heads are not in the scope of the license renewal process and why this component is not managed by an AMP.

By letter dated April 1, 2010, the applicant stated that LRA Sections 3.1.2.1.3 and 3.1.2.2.16.2 and Tables 2.3.1-3, 3.1.1, and 3.1.2-3 have been revised to add the pressurizer spray heads to the scope of license renewal. The applicant stated that the Water Chemistry Program and One-Time Inspection Program are credited to manage the aging effects of cracking due to SCC and PWSCC of the nickel-alloy components. The applicant also stated that since the pressurizer spray head is not a pressure-retaining component and is not part of the RCPB, it is not included in the Alloy 600 Management Program Plan. The applicant further stated that it complies with applicable NRC orders and provides a commitment in the UFSAR supplement to implement applicable bulletins, GLs, and staff-accepted industry guidelines.

SER Sections 3.0.3.2.1 and 3.0.3.1.6 document the staff's evaluations of the applicant's Water Chemistry Program and One-Time Inspection Program, respectively. Based on its review, the staff finds the LRA revision and the applicant's proposal to manage the aging effect of the pressurizer spray head acceptable because (1) the Water Chemistry Program monitors the water chemistry control parameters against the established parameter limits and, if a parameter exceeds the limit, the program performs adequate actions such that the water chemistry control continues to mitigate the aging effect, (2) the One-Time Inspection Program includes a one-time inspection of selected components to verify the effectiveness of the Water Chemistry Program, (3) the use of the Water Chemistry Program and One-Time Inspection Program to manage the aging effect is consistent with the GALL Report and SRP-LR, (4) the applicant also committed to comply with applicable NRC orders and provided a commitment in the UFSAR supplement to implement applicable bulletins, GLs, and staff-accepted industry guidelines in accordance with the SRP-LR and GALL Report. Based on its review, the staff's concern, described above, is resolved.

Based on the programs identified, the staff concludes that the applicant's programs and Commitment No. 23 meet SRP-LR Section 3.1.2.2.16 criteria. For those items that apply to LRA Section 3.1.2.2.16, the staff determines that the LRA is consistent with the GALL Report. In addition, the applicant has demonstrated that it will adequately manage the effects of aging so that the intended function(s) will be maintained consistent with the CLB during the period of extended operation as required by 10 CFR 54.21(a)(3).

3.1.2.2.17 Cracking Due to Stress Corrosion Cracking, Primary Water Stress Corrosion Cracking, and Irradiation-Assisted Stress Corrosion Cracking

The staff reviewed LRA Section 3.1.2.2.17 against criteria in SRP-LR 3.1.2.2.17 which states cracking due to SCC, PWSCC, and IASCC could occur in PWR stainless steel and nickel-alloy RV internals components. The SRP-LR also states the existing program relies on control of water chemistry to mitigate these effects. It further states that no further AMR is necessary if the applicant provides a commitment in the UFSAR Supplement to participate in the industry programs for investigating and managing aging effects on reactor internals as well as to evaluate and implement the results of the industry programs as applicable to the reactor internals. In addition, upon completion of these programs, but not less than 24 months before entering the period of extended operation, the applicant must submit an inspection plan for reactor internals for the staff's review and approval. The staff noted that the applicant's commitment (Commitment No. 23) in LRA Appendix A, Section A1.21 is consistent with the commitment described in SRP-LR 3.1.2.2.17. The staff also notes that all of the AMR results lines that refer to Table 3.1.1, item 3.1.1-37 are aligned with the applicant's commitment as described in LRA Appendix A, Section A1.21. The staff finds the applicant's proposal acceptable because the applicant credits its Water Chemistry Program and has provided the

Aging Management Review Results

appropriate commitment in the UFSAR Supplement. In addition, the AMR results lines refer to the commitment, consistent with the recommendations in the GALL Report and SRP-LR.

Based on the programs identified above, the staff concludes that the applicant's programs meet SRP-LR Section 3.1.2.2.17 criteria. For those items that apply to LRA Section 3.1.2.2.17, the staff determines that the LRA is consistent with the GALL Report. In addition, the applicant has demonstrated that it will adequately manage the effects of aging so that the intended function(s) will be maintained consistent with the CLB during the period of extended operation, as required by 10 CFR 54.21(a)(3).

3.1.2.2.18 Quality Assurance for Aging Management of Nonsafety-Related Components

SER Section 3.0.4 documents the staff's evaluation of the applicant's QA program.

3.1.2.3 Aging Management Review Results Not Consistent with or Not Addressed in the Generic Aging Lessons Learned Report

In LRA Tables 3.1.2-1 through 3.1.2-4, the staff reviewed additional details of the AMR results for material, environment, AERM, and AMP combinations not consistent with, or not addressed in, the GALL Report.

In LRA Tables 3.1.2-1 through 3.1.2-4, via notes F–J, the applicant indicated which combinations of component type, material, environment, and AERM do not correspond to an item in the GALL Report. The applicant provided further information about how it will manage the aging effects. Specifically, Note F states that the GALL Report does not evaluate the material for the AMR item component. Note G states that the GALL Report does not evaluate the environment for the AMR item component. Note H states that the GALL Report does not evaluate the aging effect for the AMR item component, material, and environment combination. Note I indicates that the aging effect identified in the GALL Report for the item component, material, and environment combination is not applicable. Note J indicates that the GALL Report does not evaluate either the component or the material and environment combination for the item.

For component type, material, and environment combinations not evaluated in the GALL Report, the staff reviewed the applicant's evaluation to determine if the applicant has demonstrated that it will adequately manage the effects of aging so that the intended function(s) will be maintained consistent with the CLB for the period of extended operation. The following sections document the staff's evaluation.

3.1.2.3.1 Reactor Vessel and Internals—Summary of Aging Management Review—License Renewal Application Table 3.1.2-1

The staff reviewed LRA Table 3.1.2-1, which summarizes the results of AMR evaluations for the RV and internals component groups.

In LRA Table 3.1.2-1, the applicant stated that for nickel-alloy RV control element drive mechanism housing (lower) and nozzles, RV flange leak monitoring tube, RV head vent penetration, and RV in-core instrumentation nozzle exposed to borated water leakage there is no aging effect and no AMP is proposed. In LRA Table 3.1.2-2, the applicant stated that for nickel-alloy piping and thermowells exposed to borated water leakage, there is no aging effect and no AMP is proposed. In LRA Table 3.1.2-3, the applicant stated that for nickel-alloy pressurizer heater sheaths and sleeves and pressurizer instrument penetrations exposed to borated water leakage, there is no aging effect and no AMP is proposed. In LRA Table 3.1.2-4, the applicant stated that for nickel-alloy SG primary nozzles and safe ends exposed to borated water leakage, there is no aging effect and no AMP is proposed. The AMR item cites generic note G.

Aging Management Review Results

The staff reviewed the associated items in the LRA and confirmed that no aging effect is applicable for this component, material, and environmental because austenitic materials such as nickel alloys are not subject to loss of material or cracking when exposed to this environment. In addition, these materials are used as corrosion-resistant replacement materials where other materials have degraded. The staff noted that according to EPRI NP-5769, "Degradation and Failure of Bolting in Nuclear Power Plants," Volumes 1 and 2, April 1988, corrosion-resistant materials, such as austenitic and martensitic stainless steels and high strength nickel base alloys, offer good protection against loss of material due to boric acid corrosion. The staff also noted that the conditions required for cracking due to a variety of mechanisms (SCC, PWSCC, IASCC, and IGSCC) to occur, such as being exposed to an aqueous solution (reactor coolant or other corrosive solutions) and high temperatures, do not exist on the surfaces of these components when exposed to borated water leakage. Therefore, the staff finds no AMP is necessary for nickel alloys in a borated water leakage environment.

Based on its review, the staff finds that the applicant has appropriately evaluated the AMR results of material, environment, AERM, and AMP combinations not evaluated in the GALL Report. The staff finds that the applicant has demonstrated that it will adequately manage the effects so that the intended function(s) will be maintained consistent with the CLB for the period of extended operation, as required by 10 CFR 54.21(a)(3).

3.1.2.3.2 Reactor Coolant System—Summary of Aging Management Review—License Renewal Application Table 3.1.2-2

The staff reviewed LRA Table 3.1.2-2, which summarizes the results of AMR evaluations for the RCS component groups.

SER Section 3.1.2.3.1 documents the staff's evaluation for nickel-alloy components, exposed to borated water leakage, that are not subject to an AERM, with generic note G.

Based on its review, the staff finds that the applicant has appropriately evaluated the AMR results of material, environment, AERM, and AMP combinations not evaluated in the GALL Report. The staff finds that the applicant has demonstrated that it will adequately manage the effects of aging so that the intended function(s) will be maintained consistent with the CLB for the period of extended operation, as required by 10 CFR 54.21(a)(3).

3.1.2.3.3 Pressurizer—Summary of Aging Management Review—License Renewal Application Table 3.1.2-3

The staff reviewed LRA Table 3.1.2-3, which summarizes the results of AMR evaluations for the pressurizer component groups.

SER Section 3.1.2.3.1 documents the staff's evaluation for nickel-alloy components, exposed to borated water leakage, that are not subject to an AERM, with generic note G.

Based on its review, the staff finds that the applicant has appropriately evaluated the AMR results of material, environment, AERM, and AMP combinations not evaluated in the GALL Report. The staff finds that the applicant has demonstrated that it will adequately manage the effects of aging so that the intended function(s) will be maintained consistent with the CLB for the period of extended operation, as required by 10 CFR 54.21(a)(3).

3.1.2.3.4 Steam Generators—Summary of Aging Management Review—License Renewal Application Table 3.1.2-3

The staff reviewed LRA Table 3.1.2-4, which summarizes the results of AMR evaluations for the SG component groups.

SER Section 3.1.2.3.1 documents the staff's evaluation for nickel-alloy components, exposed to borated water leakage, that are not subject to an AERM, with generic note G.

Based on its review, the staff finds that the applicant has appropriately evaluated the AMR results of material, environment, AERM, and AMP combinations not evaluated in the GALL Report. The staff finds that the applicant has demonstrated that it will adequately manage the effects of aging so that the intended function(s) will be maintained consistent with the CLB for the period of extended operation, as required by 10 CFR 54.21(a)(3).

3.1.3 Conclusion

The staff concludes that the applicant has provided sufficient information to demonstrate that the effects of aging for the RV, internals, and RCS components within the scope of license renewal and subject to an AMR, will be adequately managed so that the intended function(s) will be maintained consistent with the CLB for the period of extended operation, as required by 10 CFR 54.21(a)(3).

3.2 Aging Management of Engineered Safety Features Systems

This section of the SER documents the staff's review of the applicant's AMR results for the ESF systems components and the following component groups:

- containment leak test system
- containment purge system
- containment hydrogen control system
- safety injection and shutdown cooling system

3.2.1 Summary of Technical Information in the Application

LRA Section 3.2 provides AMR results for the ESF systems components and component groups. LRA Table 3.2.1, "Summary of Aging Management Evaluations in Chapter V of NUREG-1801 [the GALL Report] for Engineered Safety Features," is a summary comparison of the applicant's AMRs with those evaluated in the GALL Report for the ESF systems components and component groups.

The applicant's AMRs evaluated and incorporated applicable plant-specific and industry operating experience in the determination of AERMs. The plant-specific evaluation included condition reports and discussions with appropriate site personnel to identify AERMs. The applicant's review of industry operating experience included a review of the GALL Report and operating experience issues identified since the issuance of the GALL Report.

3.2.2 Staff Evaluation

The staff reviewed LRA Section 3.2 to determine if the applicant provided sufficient information to demonstrate that it would adequately manage the effects of aging for the ESF systems components within the scope of license renewal and subject to an AMR, so that the intended function(s) will be maintained consistent with the CLB for the period of extended operation, as required by 10 CFR 54.21(a)(3).

The staff reviewed AMRs to ensure the applicant's claim that certain AMRs were consistent with the GALL Report. The staff did not repeat its review of the matters described in the GALL Report; however, the staff did verify that the material presented in the LRA was applicable and that the applicant identified the appropriate GALL Report AMRs. SER Section 3.0.3 documents the staff's evaluations of the AMPs. Details of the staff's evaluations are documented in SER Section 3.2.2.1.

The staff also reviewed AMRs consistent with the GALL Report and for which further evaluation is recommended. The staff confirmed that the applicant's further evaluations were consistent

Aging Management Review Results

with the SRP-LR Section 3.2.2.2 acceptance criteria. SER Section 3.2.2.2 documents the staff's evaluations.

The staff also conducted a technical review of the remaining AMRs not consistent with, or not addressed in, the GALL Report. The technical review evaluated if the applicant identified all plausible aging effects and if the aging effects listed were appropriate for the material-environment combinations specified. SER Section 3.2.2.3 documents the staff's evaluations.

For SSCs that the applicant claimed were not applicable or required no aging management, the staff reviewed the AMR items and the plant's operating experience to verify the applicant's claims.

Table 3.2-1 summarizes the staff's evaluation of components, aging effects or mechanisms, and AMPs, listed in LRA Section 3.2 and addressed in the GALL Report.

Table 3.2-1. Staff Evaluation for Engineered Safety Features Systems Components in the GALL Report

Component Group (GALL Report Item No.)	Aging Effect/ Mechanism	AMP in GALL Report	Further Evaluation in GALL Report	AMP in LRA, Supplements, or Amendments	Staff Evaluation
Steel and stainless steel piping, piping components, and piping elements in ECCS (3.2.1-1)	Cumulative fatigue damage	TLAA, evaluated in accordance with 10 CFR 54.21(c)	Yes	TLAA	Consistent with GALL Report. See SER Section 3.2.2.2.1.
Steel with stainless steel cladding pump casing exposed to treated borated water (3.2.1-2)	Loss of material due to cladding breach	A plant-specific AMP is to be evaluated. Reference NRC IN 94-63, "Boric Acid Corrosion of Charging Pump Casings Caused by Cladding Cracks"	Yes	Not applicable	Not applicable to PVNGS. See SER Section 3.2.2.2.2.
Stainless steel containment isolation piping and components internal surfaces exposed to treated water (3.2.1-3)	Loss of material due to pitting and crevice corrosion	Water Chemistry and One-Time Inspection	Yes	Water Chemistry and One-Time Inspection	Consistent with GALL Report. See SER Section 3.2.2.2.3(1).
Stainless steel piping, piping components, and piping elements exposed to soil (3.2.1-4)	Loss of material due to pitting and crevice corrosion	A plant-specific AMP is to be evaluated.	Yes	Not applicable	Not applicable to PVNGS. See SER Section 3.2.2.2.3(2).
Stainless steel and aluminum piping, piping components, and piping elements exposed to treated water (3.2.1-5)	Loss of material due to pitting and crevice corrosion	Water Chemistry and One-Time Inspection	Yes	Not applicable	Not applicable to PWRs. See SER Section 3.2.2.2.3(3).

Aging Management Review Results

Component Group (GALL Report Item No.)	Aging Effect/ Mechanism	AMP in GALL Report	Further Evaluation in GALL Report	AMP in LRA, Supplements, or Amendments	Staff Evaluation
Stainless steel and copper alloy piping, piping components, and piping elements exposed to lubricating oil (3.2.1-6)	Loss of material due to pitting and crevice corrosion	Lubricating Oil Analysis and One-Time Inspection	Yes	Not applicable	Not applicable to PVNGS. See SER Section 3.2.2.2.3(4).
Partially encased stainless steel tanks with breached moisture barrier exposed to raw water (3.2.1-7)	Loss of material due to pitting and crevice corrosion	A plant-specific AMP is to be evaluated for pitting and crevice corrosion of tank bottoms because moisture and water can egress under the tank due to cracking of the perimeter seal from weathering.	Yes	Not applicable	Not applicable to PVNGS. See SER Section 3.2.2.2.3(5).
Stainless steel piping, piping components, piping elements, and tank internal surfaces exposed to condensation (internal) (3.2.1-8)	Loss of material due to pitting and crevice corrosion	A plant-specific AMP is to be evaluated.	Yes	Inspection of Internal Surfaces in Miscellaneous Piping and Ducting Components	Consistent with GALL Report. See SER Section 3.2.2.2.3(6).
Steel, stainless steel, and copper alloy heat exchanger tubes exposed to lubricating oil (3.2.1-9)	Reduction of heat transfer due to fouling	Lubricating Oil Analysis and One-Time Inspection	Yes	Not applicable	Not applicable to PVNGS. See SER Section 3.2.2.2.4(1).
Stainless steel heat exchanger tubes exposed to treated water (3.2.1-10)	Reduction of heat transfer due to fouling	Water Chemistry and One-Time Inspection	Yes	Not applicable	Not applicable to PVNGS. See SER Section 3.2.2.2.4(2).
Elastomer seals and components in standby gas treatment system exposed to air-indoor uncontrolled (3.2.1-11)	Hardening and loss of strength due to elastomer degradation	A plant-specific AMP is to be evaluated.	Yes	Not applicable	Not applicable to PWRs. See SER Section 3.2.2.2.5.
Stainless steel HPSI (charging) pump miniflow orifice exposed to treated borated water (3.2.1-12)	Loss of material due to erosion	A plant-specific AMP is to be evaluated for erosion of the orifice due to extended use of the centrifugal HPSI pump for normal charging.	Yes	Not applicable	Not applicable to PVNGS. See SER Section 3.2.2.2.6.

Aging Management Review Results

Component Group (GALL Report Item No.)	Aging Effect/ Mechanism	AMP in GALL Report	Further Evaluation in GALL Report	AMP in LRA, Supplements, or Amendments	Staff Evaluation
Steel drywell and suppression chamber spray system nozzle and flow orifice internal surfaces exposed to air-indoor uncontrolled (internal) (3.2.1-13)	Loss of material due to general corrosion and fouling	A plant-specific AMP is to be evaluated.	Yes	Not applicable	Not applicable to PWRs. See SER Section 3.2.2.2.7.
Steel piping, piping components, and piping elements exposed to treated water (3.2.1-14)	Loss of material due to general, pitting, and crevice corrosion	Water Chemistry and One-Time Inspection	Yes	Not applicable	Not applicable to PWRs. See SER Section 3.2.2.2.8(1).
Steel containment isolation piping, piping components, and piping elements internal surfaces exposed to treated water (3.2.1-15)	Loss of material due to general, pitting, and crevice corrosion	Water Chemistry and One-Time Inspection	Yes	Not applicable	Not applicable to PVNGS. See SER Section 3.2.2.2.8(2).
Steel piping, piping components, and piping elements exposed to lubricating oil (3.2.1-16)	Loss of material due to general, pitting, and crevice corrosion	Lubricating Oil Analysis and One-Time Inspection	Yes	Not applicable	Not applicable to PVNGS. See SER Section 3.2.2.2.8(3).
Steel (with or without coating or wrapping) piping, piping components, and piping elements buried in soil (3.2.1-17)	Loss of material due to general, pitting, crevice, and microbiologically-influenced corrosion	Buried Piping and Tanks Surveillance or Buried Piping and Tanks Inspection	No Yes	Not applicable	Not applicable to PWRs. See SER Section 3.2.2.2.9.
Stainless steel piping, piping components, and piping elements exposed to treated water > 60°C (> 140°F) (3.2.1-18)	Cracking due to SCC and IGSCC	BWR SCC and Water Chemistry	No	Not applicable	Not applicable to PWRs. See SER Section 3.2.2.1.1.
Steel piping, piping components, and piping elements exposed to steam or treated water (3.2.1-19)	Wall thinning due to FAC	FAC	No	Not applicable	Not applicable to PWRs. See SER Section 3.2.2.1.1
CASS piping, piping components, and piping elements exposed to treated water (borated or unborated) > 250°C (> 482°F) (3.2.1-20)	Loss of fracture toughness due to thermal aging embrittlement	Thermal Aging Embrittlement of CASS	No	Not applicable	Not applicable to PWRs. See SER Section 3.2.2.1.1.
High-strength steel closure bolting exposed to air with steam or water leakage (3.2.1-21)	Cracking due to cyclic loading, SCC	Bolting Integrity	No	Not applicable	Not applicable to PVNGS. See SER Section 3.2.2.1.1.

Aging Management Review Results

Component Group (GALL Report Item No.)	Aging Effect/ Mechanism	AMP in GALL Report	Further Evaluation in GALL Report	AMP in LRA, Supplements, or Amendments	Staff Evaluation
Steel closure bolting exposed to air with steam or water leakage (3.2.1-22)	Loss of material due to general corrosion	Bolting Integrity	No	Not applicable	Not applicable to PVNGS. See SER Section 3.2.2.1.1.
Steel bolting and closure bolting exposed to air - outdoor (external), or air - indoor uncontrolled (external) (3.2.1-23)	Loss of material due to general, pitting, and crevice corrosion	Bolting Integrity	No	Bolting Integrity	Consistent with GALL Report.
Steel closure bolting exposed to air - indoor uncontrolled (external) (3.2.1-24)	Loss of preload due to thermal effects, gasket creep, and self-loosening	Bolting Integrity	No	Bolting Integrity	Consistent with GALL Report.
Stainless steel piping, piping components, and piping elements exposed to closed-cycle cooling water > 60°C (> 140°F) (3.2.1-25)	Cracking due to stress corrosion SCC	Closed-Cycle Cooling Water System	No	Closed-Cycle Cooling Water System	Consistent with GALL Report.
Steel piping, piping components, and piping elements exposed to closed-cycle cooling water (3.2.1-26)	Loss of material due to general, pitting, and crevice corrosion	Closed-Cycle Cooling Water System	No	Not applicable	Not applicable to PVNGS. See SER Section 3.2.2.1.1.
Steel heat exchanger components exposed to closed-cycle cooling water (3.2.1-27)	Loss of material due to general, pitting, crevice, and galvanic corrosion	Closed-Cycle Cooling Water System	No	Closed-Cycle Cooling Water System	Consistent with GALL Report.
Stainless steel piping, piping components, piping elements, and heat exchanger components exposed to closed-cycle cooling water (3.2.1-28)	Loss of material due to pitting and crevice corrosion	Closed-Cycle Cooling Water System	No	Closed-Cycle Cooling Water System	Consistent with GALL Report.
Copper alloy piping, piping components, piping elements, and heat exchanger components exposed to closed-cycle cooling water (3.2.1-29)	Loss of material due to pitting, crevice, and galvanic corrosion	Closed-Cycle Cooling Water System	No	Not applicable	Not applicable to PVNGS. See SER Section 3.2.2.1.1.
Stainless steel and copper alloy heat exchanger tubes exposed to closed-cycle cooling water (3.2.1-30)	Reduction of heat transfer due to fouling	Closed-Cycle Cooling Water System	No	Closed-Cycle Cooling Water System	Consistent with GALL Report.

Aging Management Review Results

Component Group (GALL Report Item No.)	Aging Effect/ Mechanism	AMP in GALL Report	Further Evaluation in GALL Report	AMP in LRA, Supplements, or Amendments	Staff Evaluation
External surfaces of steel components including ducting, piping, ducting closure bolting, and containment isolation piping external surfaces exposed to air-indoor uncontrolled (external); condensation (external) and air-outdoor (external) (3.2.1-31)	Loss of material due to general corrosion	External Surfaces Monitoring	No	External Surfaces Monitoring	Consistent with GALL Report.
Steel piping and ducting components and internal surfaces exposed to air - indoor uncontrolled (Internal) (3.2.1-32)	Loss of material due to general corrosion	Inspection of Internal Surfaces in Miscellaneous Piping and Ducting Components	No	Inspection of Internal Surfaces in Miscellaneous Piping and Ducting Components	Consistent with GALL Report.
Steel encapsulation components exposed to air-indoor uncontrolled (internal) (3.2.1-33)	Loss of material due to general, pitting, and crevice corrosion	Inspection of Internal Surfaces in Miscellaneous Piping and Ducting Components	No	Not applicable	Not applicable to PVNGS. See SER Section 3.2.2.1.1.
Steel piping, piping components, and piping elements exposed to condensation (internal) (3.2.1-34)	Loss of material due to general, pitting, and crevice corrosion	Inspection of Internal Surfaces in Miscellaneous Piping and Ducting Components	No	Not applicable	Not applicable to PWRs. See SER Section 3.2.2.1.1.
Steel containment isolation piping and components internal surfaces exposed to raw water (3.2.1-35)	Loss of material due to general, pitting, crevice, and MIC and fouling	Open-Cycle Cooling Water System	No	Not applicable	Not applicable to PVNGS. See SER Section 3.2.2.1.1.
Steel heat exchanger components exposed to raw water (3.2.1-36)	Loss of material due to general, pitting, crevice, galvanic, and MIC and fouling	Open-Cycle Cooling Water System	No	Not applicable	Not applicable to PVNGS. See SER Section 3.2.2.1.1.
Stainless steel piping, piping components, and piping elements exposed to raw water (3.2.1-37)	Loss of material due to pitting, crevice, and MIC	Open-Cycle Cooling Water System	No	Not applicable	Not applicable to PVNGS. See SER Section 3.2.2.1.1.
Stainless steel containment isolation piping and components internal surfaces exposed to raw water (3.2.1-38)	Loss of material due to pitting, crevice, and MIC and fouling	Open-Cycle Cooling Water System	No	Not applicable	Not applicable to PVNGS. See SER Section 3.2.2.1.1.

Aging Management Review Results

Component Group (GALL Report Item No.)	Aging Effect/ Mechanism	AMP in GALL Report	Further Evaluation in GALL Report	AMP in LRA, Supplements, or Amendments	Staff Evaluation
Stainless steel heat exchanger components exposed to raw water (3.2.1-39)	Loss of material due to pitting, crevice, and MIC and fouling	Open-Cycle Cooling Water System	No	Not applicable	Not applicable to PVNGS. See SER Section 3.2.2.1.1.
Steel and stainless steel heat exchanger tubes (serviced by open-cycle cooling water) exposed to raw water (3.2.1-40)	Reduction of heat transfer due to fouling	Open-Cycle Cooling Water System	No	Not applicable	Not applicable to PVNGS. See SER Section 3.2.2.1.1.
Copper alloy > 15% Zn piping, piping components, piping elements, and heat exchanger components exposed to closed-cycle cooling water (3.2.1-41)	Loss of material due to selective leaching	Selective Leaching of Materials	No	Not applicable	Not applicable to PVNGS. See SER Section 3.2.2.1.1.
Gray cast iron piping, piping components, piping elements exposed to closed-cycle cooling water (3.2.1-42)	Loss of material due to selective leaching	Selective Leaching of Materials	No	Not applicable	Not applicable to PVNGS. See SER Section 3.2.2.1.1.
Gray cast iron piping, piping components, and piping elements exposed to soil (3.2.1-43)	Loss of material due to selective leaching	Selective Leaching of Materials	No	Not applicable	Not applicable to PVNGS. See SER Section 3.2.2.1.1.
Gray cast iron motor cooler exposed to treated water (3.2.1-44)	Loss of material due to selective leaching	Selective Leaching of Materials	No	Not applicable	Not applicable to PVNGS. See SER Section 3.2.2.1.1.
Aluminum, copper alloy > 15% Zn, and steel external surfaces, bolting, and piping, piping components, and piping elements exposed to air with borated water leakage (3.2.1-45)	Loss of material due to Boric acid corrosion	Boric Acid Corrosion	No	Boric Acid Corrosion	Consistent with GALL Report.
Steel encapsulation components exposed to air with borated water leakage (internal) (3.2.1-46)	Loss of material due to general, pitting, crevice and boric acid corrosion	Inspection of Internal Surfaces in Miscellaneous Piping and Ducting Components	No	Not applicable	Not applicable to PVNGS. See SER Section 3.2.2.1.1.
CASS piping, piping components, and piping elements exposed to treated borated water > 250°C (> 482°F) (3.2.1-47)	Loss of fracture toughness due to thermal aging embrittlement	Thermal Aging Embrittlement of CASS	No	Not applicable	Not applicable to PVNGS. See SER Section 3.2.2.1.1.

Aging Management Review Results

Component Group (GALL Report Item No.)	Aging Effect/ Mechanism	AMP in GALL Report	Further Evaluation in GALL Report	AMP in LRA, Supplements, or Amendments	Staff Evaluation
Stainless steel or stainless-steel-clad steel piping, piping components, piping elements, and tanks (including safety injection tanks/accumulators) exposed to treated borated water > 60°C (> 140°F) (3.2.1-48)	Cracking due to SCC	Water Chemistry	No	Water Chemistry	Consistent with GALL Report.
Stainless steel piping, piping components, piping elements, and tanks exposed to treated borated water (3.2.1-49)	Loss of material due to pitting and crevice corrosion	Water Chemistry	No	Water Chemistry	Consistent with GALL Report. Except for shutdown cooling heat exchanger; see SER Section 3.2.2.1.2
Aluminum piping, piping components, and piping elements exposed to air - indoor uncontrolled (internal/external) (3.2.1-50)	None	None	NA	None	Consistent with GALL Report.
Galvanized steel ducting exposed to air - indoor controlled (external) (3.2.1-51)	None	None	NA	None	Consistent with GALL Report.
Glass piping elements exposed to air - indoor uncontrolled (external), lubricating oil, raw water, treated water, or treated borated water (3.2.1-52)	None	None	NA	None	Consistent with GALL Report.
Stainless steel, copper alloy, and nickel-alloy piping, piping components, and piping elements exposed to air - indoor uncontrolled (external) (3.2.1-53)	None	None	NA	None	Consistent with GALL Report.
Steel piping, piping components, and piping elements exposed to air - indoor controlled (external) (3.2.1-54)	None	None	NA	Not applicable	Not applicable to PVNGS. See SER Section 3.2.2.1.1.

Aging Management Review Results

Component Group (GALL Report Item No.)	Aging Effect/ Mechanism	AMP in GALL Report	Further Evaluation in GALL Report	AMP in LRA, Supplements, or Amendments	Staff Evaluation
Steel and stainless steel piping, piping components, and piping elements in concrete (3.2.1-55)	None	None	NA	None	Consistent with GALL Report.
Steel, stainless steel, and copper alloy piping, piping components, and piping elements exposed to gas (3.2.1-56)	None	None	NA	None	Consistent with GALL Report.
Stainless steel and copper alloy < 15% Zn piping, piping components, and piping elements exposed to air with borated water leakage (3.2.1-57)	None	None	NA	None	Consistent with GALL Report.

The staff's review of the ESF systems component groups fell into three categories. One category, documented in SER Section 3.2.2.1, reviewed AMR results for components that the applicant indicated are consistent with the GALL Report and require no further evaluation. Another category, documented in SER Section 3.2.2.2, reviewed AMR results for components that the applicant indicated are consistent with the GALL Report and for which further evaluation is recommended. A third category, documented in SER Section 3.2.2.3, reviewed AMR results for components that the applicant indicated are not consistent with, or not addressed in, the GALL Report. SER Section 3.0.3 documents the staff's review of AMPs credited to manage or monitor aging effects of the ESF systems components.

3.2.2.1 Aging Management Review Results Consistent with the Generic Aging Lessons Learned Report

LRA Section 3.2.2.1 identifies the materials, environments, AERMs, and the following programs that manage aging effects for the ESF systems components:

- ASME Section XI ISI, Subsections IWB, IWC, and IWD
- Bolting Integrity
- Boric Acid Corrosion
- Closed-Cycle Cooling Water System
- External Surfaces Monitoring
- Inspection of Internal Surfaces in Miscellaneous Piping and Ducting Components
- Nickel-Alloy Aging Management
- One-Time Inspection
- One-Time Inspection of ASME Code Class 1 Small-Bore Piping
- RCS Supplement
- Water Chemistry

LRA Tables 3.2.2-1 through 3.2.2-4 summarize AMRs for the ESF systems components and indicate AMRs claimed to be consistent with the GALL Report.

Aging Management Review Results

The staff audited and reviewed the information in the LRA. The staff did not repeat its review of the matters described in the GALL Report; however, the staff did verify that the material presented in the LRA was applicable and that the applicant identified the appropriate GALL Report AMRs. The staff's evaluation follows.

3.2.2.1.1 Aging Management Review Results Identified as Not Applicable

LRA Table 3.2.1, items 3.2.1.18 through 3.2.1.20 and 3.2.1.34 discuss the applicant's determination that these items are applicable only to BWRs. The staff verified that these items do not apply because the units are a PWR design. Based on this determination, the staff finds that the applicant has provided an acceptable basis for concluding AMR items 3.2.1.18 through 3.2.1.20 and 3.2.1.34 are not applicable.

LRA Table 3.2.1, item 3.2.1.21 addresses high-strength steel closure bolting, exposed to air with steam or water leakage, in the ESF. The GALL Report recommends use of GALL Report AMP XI.M18, "Bolting Integrity," to manage cracking due to cyclic loading or SCC for this component group. The applicant stated that this item is not applicable because there is no in-scope, high-strength steel closure bolting exposed to air with steam or water leakage in its ESF systems. By letter dated February 19, 2010, the staff issued RAI 3.3-1, part (a), asking that the applicant clarify if its statement means that it does not use high-strength steel closure bolting or that the bolting is used, but not exposed to the stated environment. In its response, dated March 24, 2010, the applicant stated that it does not use high-strength (greater than 150 kilo pounds per square inch) steel closure bolting in the containment leak test, containment purge, or containment hydrogen control systems, per plant specification. The applicant also stated that its safety injection and shutdown cooling system uses stainless steel bolting that is evaluated in a borated water leakage environment. The staff noted that the applicant's ESF systems is comprised of four plant systems:

(1) the containment leak test system

(2) the containment purge system

(3) the containment hydrogen control system

(4) the safety injection and shutdown cooling system

The staff also noted that the applicant's response addresses all four systems, stating that for three of the systems, high-strength steel closure bolting is not used and for the fourth system, stainless steel, not steel, closure bolting is used. The staff reviewed LRA Sections 2.3.2 and 3.2 and confirmed that the applicant's LRA does not have any AMR results that include high-strength steel closure bolting exposed to air with steam or water leakage. Because the applicant does not use high-strength steel closure bolting in any of the four plant systems comprising its ESF systems, the staff finds that the applicant's response to RAI 3.3-1, part (a) is acceptable. The staff's concern in RAI 3.3-1, part (a) is resolved. Further, the staff accepts the applicant's determination that LRA Table 3.2.1, item 3.2.1.21, is not applicable.

LRA Table 3.2.1, item 3.2.1.22, addresses steel closure bolting, exposed to air with steam or water leakage, in the ESF systems. The GALL Report recommends use of GALL Report AMP XI.M18, "Bolting Integrity," to manage loss of material due to general corrosion for this component group. The applicant stated that this item is not applicable because there is no closure bolting in its ESF exposed to an environment of "water [sic] with steam or water leakage." By letter dated February 19, 2010, the staff issued RAI 3.3-1, part (b), requesting that the applicant correct an apparent wording error where the word "water" was incorrectly used in place of the word "air" The staff also asked that the applicant justify its claim that steel closure bolting in its ESF is not exposed to an environment of air with steam or water leakage. In its response, dated March 24, 2010, the applicant corrected the wording error in the LRA and stated that air with water leakage would be an event-driven environment, not considered normal

Aging Management Review Results

for the ESF systems. The applicant also stated that for steel closure bolting, the aging effect and mechanism in LRA Table 3.2.1, item 3.2.1.22, is included in the aging effect and mechanism for LRA Table 3.2.1, item 3.2.1.23; the Bolting Integrity Program is credited in both lines to manage loss of material due to general corrosion for steel closure bolting. The applicant further stated that it evaluated all steel closure bolting in its ESF systems as part of LRA Table 3.2.1, item 3.2.1.23, rather than in item 3.2.1.22. The staff confirmed that, for steel closure bolting, the aging effect stated in LRA Table 3.2.1, item 3.2.1.22, is included in item 3.2.1.23, and that the Bolting Integrity Program is credited to manage loss of material for both items. The staff finds the applicant's response to RAI 3.3-1, part (b) acceptable because it corrects the error and clarifies that components that could have been evaluated under item 3.2.1.22 are included in the evaluations under item 3.2.1.23. The staff's concern in RAI 3.3-1, part (b) is resolved. The staff finds the applicant's determination that LRA Table 3.2.1, item 3.2.1.22 is not applicable, to be acceptable.

LRA Table 3.2.1, item 3.2.1.26 addresses steel piping, piping components, and piping elements exposed to closed-cycle cooling water subject to loss of material due to general, pitting, and crevice corrosion. The applicant stated that this item is not applicable because it has no in-scope steel piping, piping components, and piping elements exposed to closed-cycle cooling water in the ESF systems, so the applicable GALL Report line was not used. The staff reviewed LRA Sections 2.3.2 and 3.2 and confirmed that the applicant's LRA does not have any AMR results for the ESF systems that include steel piping, piping components, and piping elements exposed to closed-cycle cooling water. The staff reviewed the applicant's UFSAR and confirmed that no in-scope steel piping, piping components, and piping elements exposed to closed-cycle cooling water are present in these systems and, therefore, finds the applicant's determination acceptable.

LRA Table 3.2.1, item 3.2.1.29 addresses copper alloy piping, piping components, piping elements, and heat exchanger components exposed to closed-cycle cooling water subject to loss of material due to pitting, crevice, and galvanic corrosion for this component group. The applicant stated that this item is not applicable because it has no in-scope copper alloy piping, piping components, piping elements, and heat exchanger components exposed to closed-cycle cooling water in the ESF systems, so the applicable GALL Report lines were not used. The staff reviewed LRA Sections 2.3.2 and 3.2 and confirmed that the applicant's LRA does not have any AMR results for the ESF systems that include copper alloy piping, piping components, piping elements, and heat exchanger components exposed to closed-cycle cooling water. The staff reviewed the applicant's UFSAR and confirmed that no in-scope copper alloy piping, piping components, piping elements, and heat exchanger components exposed to closed-cycle cooling water are present in these systems and, therefore, finds the applicant's determination acceptable.

LRA Table 3.2.1, item 3.2.1.33 addresses steel encapsulation components exposed to air-indoor uncontrolled (internal) subject to loss of material due to general, pitting, and crevice corrosion. The applicant stated that this item is not applicable because it has no in-scope steel encapsulation components exposed to air-indoor uncontrolled (internal) in the ESF systems, so the applicable GALL Report line was not used. The staff reviewed LRA Sections 2.3.2 and 3.2 and confirmed that the applicant's LRA does not have any AMR results for the ESF that include steel encapsulation components exposed to air-indoor uncontrolled (internal). The staff reviewed the applicant's UFSAR and confirmed that no in-scope steel encapsulation components exposed to air-indoor uncontrolled (internal) are present in these systems and, therefore, finds the applicant's determination acceptable.

LRA Table 3.2.1, item 3.2.1.35 addresses steel containment isolation piping and components internal surfaces exposed to raw water that are subject to loss of material due to general, pitting, crevice, and microbiologically-influenced corrosion and fouling. The applicant stated that this

Aging Management Review Results

item is not applicable because the containment isolation components were evaluated in the systems in which the components were found to have the function of containment integrity, so the applicable GALL Report line was not used. By letter dated January 28, 2010, the staff issued RAI 3.2.2-1 requesting the applicant to clarify the statement that the containment isolation components were evaluated in the system in which the components were found to have the function of containment integrity. In addition, the staff asked that the applicant provide additional information on the adequacy of the AMP used to manage steel containment isolation piping.

In its response, dated March 1, 2010, the applicant stated that its plant equipment list does not contain a separate system for mechanical containment isolation components. As such, the applicant stated that the containment isolation components are evaluated as part of the plant system to which they are assigned and are consistent with the expectations in the GALL Report. For the ESF systems, the applicant stated that GALL Report, Table 3.2-1, item 3 summarizes the mechanical containment isolation components. Gall Report, Table 3.3-1, item 24 for auxiliary systems and Table 3.4-1, item 16, for steam and power conversion systems also recommend further evaluations for this item. The staff finds the applicant's response acceptable because it clarified where and how it evaluated the associated items in the LRA. The staff's concern described in RAI 3.2.2-1 is resolved.

The staff reviewed LRA Sections 2.3.2 and 3.2 and confirmed that the LRA does not have any AMR results for the ESF systems that include steel containment isolation piping and components internal surfaces exposed to raw water. The staff reviewed the applicant's UFSAR and confirmed no in-scope steel containment isolation piping and components internal surfaces exposed to raw water are present in these systems and, therefore, finds the applicant's determination acceptable.

LRA Table 3.2.1, item 3.2.1.36 addresses steel heat exchanger components exposed to raw water subject to loss of material due to general, pitting, crevice, galvanic, and microbiologically-influenced corrosion and fouling. The applicant stated that this item is not applicable because it has no in-scope steel heat exchanger components exposed to raw water in the ESF systems, so the applicable GALL Report lines were not used. The staff reviewed LRA Sections 2.3.2 and 3.2 and confirmed that the applicant's LRA does not have any AMR results for the ESF systems that include steel heat exchanger components exposed to raw water. The staff reviewed the applicant's UFSAR and confirmed that no in-scope steel heat exchanger components exposed to raw water are present in these systems and, therefore, finds the applicant's determination acceptable.

LRA Table 3.2.1, item 3.2.1.37 addresses stainless steel piping, piping components, and piping elements exposed to raw water subject to loss of material due to pitting, crevice, and microbiologically-influenced corrosion. The applicant stated that this item is not applicable because it has no in-scope stainless steel piping, piping components, and piping elements exposed to raw water in the ECCS, so the applicable GALL Report line was not used. The staff reviewed LRA Sections 2.3.2 and 3.2 and confirmed that the applicant's LRA does not have any AMR results for the ECCS that include stainless steel piping, piping components, and piping elements exposed to raw water. The staff reviewed the applicant's UFSAR and confirmed that no in-scope stainless steel piping, piping components, and piping elements exposed to raw water are present in these systems and, therefore, finds the applicant's determination acceptable.

LRA Table 3.2.1, item 3.2.1.38 addresses stainless steel containment isolation piping and component internal surfaces exposed to raw water, which are subject to loss of material due to pitting, crevice, and microbiologically-influenced corrosion and fouling. The applicant stated that this item is not applicable because it has no in-scope stainless steel components exposed to raw water in the ESF systems, so the applicable GALL Report line was not used. The staff

Aging Management Review Results

reviewed LRA Sections 2.3.2 and 3.2 and confirmed that the applicant's LRA does not have any AMR results for the ESF systems that include stainless steel containment isolation piping and components internal surfaces exposed to raw water. The staff reviewed the applicant's UFSAR and confirmed that no in-scope stainless steel containment isolation piping and components internal surfaces exposed to raw water are present in these systems and, therefore, finds the applicant's determination acceptable.

LRA Table 3.2.1, item 3.2.1.39 addresses stainless steel heat exchanger components exposed to raw water subject to loss of material due to pitting, crevice, and microbiologically-influenced corrosion and fouling for this component group. The applicant stated that this item is not applicable because it has no in-scope stainless steel heat exchanger components exposed to raw water in the ESF systems, so the applicable GALL Report lines were not used. The staff reviewed LRA Sections 2.3.2 and 3.2 and confirmed that the applicant's LRA does not have any AMR results for the ESF systems that include stainless steel heat exchanger components exposed to raw water. The staff reviewed the applicant's UFSAR and confirmed that no in-scope stainless steel heat exchanger components exposed to raw water are present in these systems and, therefore, finds the applicant's determination acceptable.

LRA Table 3.2.1, item 3.2.1.40 addresses steel and stainless steel heat exchanger tubes (serviced by open-cycle cooling water) exposed to raw water subject to reduction of heat transfer due to fouling for this component group. The applicant stated that this item is not applicable because it has no in-scope steel and stainless steel heat exchanger tubes (serviced by open-cycle cooling water) exposed to raw water in the ESF systems, so the applicable GALL Report lines were not used. The staff reviewed LRA Sections 2.3.2 and 3.2 and confirmed that the applicant's LRA does not have any AMR results for the ESF systems that include steel and stainless steel heat exchanger tubes (serviced by open-cycle cooling water) exposed to raw water. The staff reviewed the applicant's UFSAR and confirmed that no in-scope steel and stainless steel heat exchanger tubes (serviced by open-cycle cooling water) exposed to raw water are present in these systems and, therefore, finds the applicant's determination acceptable.

LRA Table 3.2.1, item 3.2.1.41 addresses copper alloy with greater than 15-percent zinc piping, piping components, piping elements, and heat exchanger components exposed to closed-cycle cooling water subject to loss of material due to selective leaching. The applicant stated that this item is not applicable because it has no in-scope copper alloy with greater than 15-percent zinc piping, piping components, piping elements, and heat exchanger components exposed to closed-cycle cooling water in the ESF systems, so the applicable GALL Report lines were not used. The staff reviewed LRA Sections 2.3.2 and 3.2 and confirmed that the applicant's LRA does not have any AMR results for the ESF system that include copper alloy with greater than 15-percent zinc piping, piping components, piping elements, and heat exchanger components exposed to closed-cycle cooling water. The staff reviewed the applicant's UFSAR and confirmed that no in-scope copper alloy with greater than 15-percent zinc piping, piping components, piping elements, and heat exchanger components exposed to closed-cycle cooling water are present in these systems and, therefore, finds the applicant's determination acceptable.

LRA Table 3.2.1, item 3.2.1.42 addresses gray cast iron piping, piping components, and piping elements exposed to closed-cycle cooling water subject to loss of material due to selective leaching. The applicant stated that this item is not applicable because it has no in-scope gray cast iron piping, piping components, and piping elements exposed to closed-cycle cooling water in the ESF systems, so the applicable GALL Report line was not used. The staff reviewed LRA Sections 2.3.2 and 3.2 and confirmed that the applicant's LRA does not have any AMR results for the ESF system that include gray cast iron piping, piping components, and piping elements exposed to closed-cycle cooling water. The staff reviewed the applicant's UFSAR and

Aging Management Review Results

confirmed that no in-scope gray cast iron piping, piping components, and piping elements exposed to closed-cycle cooling water are present in these systems and, therefore, finds the applicant's determination acceptable.

LRA Table 3.2.1, item 3.2.1.43 addresses gray cast iron piping, piping components, and piping elements exposed to soil subject to loss of material due to selective leaching. The applicant stated that this item is not applicable because it has no in-scope gray cast iron piping, piping components, and piping elements exposed to soil in the ESF systems, so it did not use the applicable GALL Report line. The staff reviewed LRA Sections 2.3.2 and 3.2 and confirmed that the applicant's LRA does not have any AMR results for the ESF system that include gray cast iron piping, piping components, and piping elements exposed to soil. The staff reviewed the applicant's UFSAR and confirmed that no in-scope gray cast iron piping, piping components, and piping elements exposed to soil are present in these systems and, therefore, it finds the applicant's determination acceptable.

LRA Table 3.2.1, item 3.2.1.44 addresses gray cast iron motor cooler exposed to treated water subject to loss of material due to selective leaching. The applicant stated that this item is not applicable because it has no in-scope gray cast iron motor cooler exposed to treated water in the ESF systems, so the applicable GALL Report lines were not used. The staff reviewed LRA Sections 2.3.2 and 3.2 and confirmed that the applicant's LRA does not have any AMR results for the ESF system that include gray cast iron motor cooler exposed to treated water. The staff reviewed the applicant's UFSAR and confirmed that no in-scope gray cast iron motor cooler exposed to treated water are present in these systems and, therefore, it finds the applicant's determination acceptable.

LRA Table 3.2.1, item 3.2.1.46 addresses steel encapsulation components exposed to air with borated water leakage (internal) subject to loss of material due to general, pitting, crevice, and boric acid corrosion. The applicant stated that this item is not applicable because it has no in-scope steel encapsulation components exposed to air with borated water leakage (internal) in the ESF systems, so the applicable GALL Report line was not used. The staff reviewed LRA Sections 2.3.2 and 3.2 and confirmed that the applicant's LRA does not have any AMR results for the ESF systems that include steel encapsulation components exposed to air with borated water leakage (internal). The staff reviewed the applicant's UFSAR and confirmed that no in-scope steel encapsulation components exposed to air with borated water leakage (internal) are present in these systems and, therefore, finds the applicant's determination acceptable.

LRA Table 3.2.1, item 3.2.1.47 addresses CASS piping, piping components, and piping elements exposed to treated borated water greater than 250 degrees C (greater than 482 degrees F) subject to loss of fracture toughness due to thermal-aging embrittlement. The applicant stated that this item is not applicable because it has no in-scope CASS piping, piping components, and piping elements exposed to treated borated water greater than 250 degrees C in the ECCS, so the applicable GALL Report line was not used. The staff reviewed LRA Sections 2.3.2 and 3.2 and confirmed that the applicant's LRA does not have any AMR results for the ESF systems that include CASS piping, piping components, and piping elements exposed to treated borated water greater than 250 degrees C (greater than 482 degrees F). The staff reviewed the applicant's UFSAR and confirmed that no in-scope CASS piping, piping components, and piping elements exposed to treated borated greater than 250 degrees C (greater than 482 degrees F) are present in these systems and, therefore, finds the applicant's determination acceptable.

LRA Table 3.2.1, item 3.2.1.54 addresses steel piping, piping components, and piping elements exposed to air-indoor controlled (external). The GALL Report recommends that there is no aging effect requiring management. The applicant stated that this item is not applicable because it has no in-scope steel piping, piping components, and piping elements exposed to air-indoor controlled (external) in the ESF systems, so the applicable GALL Report line was not

Aging Management Review Results

used. The staff reviewed LRA Sections 2.3.2 and 3.2 and confirmed that the applicant's LRA does not have any AMR results for the ESF systems that include steel piping, piping components, and piping elements exposed to air-indoor controlled (external). The staff reviewed the applicant's UFSAR and confirmed that no in-scope steel piping, piping components, and piping elements exposed to air-indoor controlled (external) are present in the systems and, therefore, finds the applicant's determination acceptable.

3.2.2.1.2 Loss of Material Due to Boric Acid Corrosion

LRA Table 3.2.1, item 3.2.1.49 addresses stainless steel piping, piping components, piping elements, and tanks exposed to treated borated water, which are managed for loss of material due to pitting and crevice corrosion. The LRA credits the Water Chemistry Program to manage this aging effect. The GALL Report recommends AMP XI.M2, "Water Chemistry Program," to ensure that this aging effect is adequately managed. The AMR item cites generic note A, indicating that the item is consistent with the GALL Report item for component, material, environment, and aging effect, and the LRA AMP is consistent with the GALL Report AMP.

In its review of components subordinate to item 3.2.1.49, for which the applicant assigned generic note A for the heat exchanger (shutdown cooling), the staff noted that the GALL Report, under items V.A-27 and V.D1-30, is for stainless steel material. The shutdown cooling heat exchanger is a carbon steel heat exchanger clad with stainless steel, which is not the same material as indicated in the GALL Report. By letter dated December 29, 2009, the staff issued RAI B2.1.4-1 requesting that the applicant justify why the Water Chemistry Program is more appropriate than the Boric Acid Corrosion Control Program to manage the clad carbon steel.

In its response, dated February 19, 2010, the applicant described the environments and potential degradation effects for the material types composing the heat exchanger and explained the selection of AMPs for each of those material types and environmental combinations. The applicant also stated that the exterior carbon steel surfaces of the heat exchanger are potentially susceptible to corrosion in the event of boric acid leakage and that the Boric Acid Corrosion Program will manage component materials. In addition, the applicant provided justification for the use of other AMPs to cover the remaining construction materials of the heat exchanger.

The applicant's response reported that the LRA was revised in Amendment No. 9 to reflect AMP selections for the individual shutdown cooling heat exchanger components, as described above. Specifically, LRA Table 3.2.2-4 was revised to add GALL Report, Table 2, item 3.2.1.45 for loss of material on the carbon steel exterior surfaces for the shutdown cooling heat exchanger exposed to an environment of borated water leakage, managed by the Boric Acid Corrosion Program. In addition, LRA Table 3.2.2-4 was revised to change the standard note from "A" to "C" for the AMR results associated with loss of material and cracking of the stainless steel cladding of the shutdown cooling heat exchanger channel cylinder and tubesheet, exposed to treated borated water, managed by the Water Chemistry Program.

The staff verified that the noted changes to the LRA were provided in Amendment No. 9 of Enclosure 2 of its above response. Based on its review, the staff finds the applicant's response to RAI B2.1.4-1 acceptable because the amended LRA added Table 3.2.1, item 3.2.1.45 to account for the potential loss of material for the shutdown cooling heat exchanger's exterior carbon steel surfaces due to borated water leakage. The staff's concern in RAI B2.1.4-1 is resolved.

The staff concludes that the applicant has demonstrated that it will adequately manage the effects of aging for the subject components so that the intended function(s) will be maintained consistent with the CLB during the period of extended operation as required by 10 CFR 54.21(a)(3).

Aging Management Review Results

3.2.2.2 Aging Management Review Results Consistent with the Generic Aging Lessons Learned Report for Which Further Evaluation Is Recommended

In LRA Section 3.2.2.2, the applicant further evaluates aging management, as recommended by the GALL Report, for the ESF systems components and provides information concerning how it will manage the following aging effects:

- cumulative fatigue damage
- loss of material due to cladding
- loss of material due to pitting and crevice corrosion
- reduction of heat transfer due to fouling
- hardening and loss of strength due to elastomer degradation
- loss of material due to erosion
- loss of material due to general corrosion and fouling
- loss of material due to general, pitting, and crevice corrosion
- loss of material due to general, pitting, crevice, and microbiologically-influenced corrosion
- QA for aging management of nonsafety-related components

For component groups evaluated in the GALL Report, for which the applicant claimed consistency with the report and for which the report recommends further evaluation, the staff reviewed the applicant's evaluation to determine if it adequately addressed the issues further evaluated. In addition, the staff reviewed the applicant's further evaluations against the criteria contained in SRP-LR Section 3.2.2.2. The staff's review of the applicant's further evaluation follows.

3.2.2.2.1 Cumulative Fatigue Damage

In LRA Section 3.2.2.2.1, the applicant stated that the evaluation of fatigue is a TLAA, as defined in 10 CFR 54.3, and TLAAs are evaluated in accordance with 10 CFR 54.21(c)(1). The applicant further stated that its piping, designed to ASME III Class 2, Class 3, and ANSI B31.1 standards, assumes a reduction in the allowable secondary stress range if more than 7,000 full-range thermal cycles are expected in a design lifetime. LRA Section 4.3.5 describes the evaluation of these cyclic-design TLAAs. The applicant further stated that HPSI and low pressure safety injection pumps are ASME III Class 2 components, designed with a specified number of thermal transient cycles. LRA Section 4.3.2.11 describes the evaluation of these cyclic-design TLAAs.

The staff reviewed LRA Section 3.2.2.2.1 against the criteria in SRP-LR Section 3.2.2.2.1. Fatigue is a TLAA, as defined in 10 CFR 54.3. TLAAs are required to be evaluated in accordance with 10 CFR 54.21(c)(1). Section 4.3, "Metal Fatigue Analysis," of the SRP-LR addresses this TLAA separately. The staff finds the applicant's AMR results are consistent with the recommendations of the GALL Report and SRP-LR, except for the following identified areas.

The staff noted that the summary description in LRA Section 4.3.5 states that the implicit fatigue analyses, discussed in the section, are applicable to all ASME Code Class 2 and 3 and ANSI B31.1 piping, piping components, and piping elements. The staff noted that it is not clear if the LRA includes all corresponding AMR items for applicable ASME Code Class 2 and 3 or ANSI B31.1 piping, piping components, and piping elements in scope for license renewal. The staff also noted that this includes those components in the ESF systems (LRA Section 3.2), auxiliary systems (LRA Section 3.3), and steam and power conversion systems (LRA

Section 3.4). By letter dated July 21, 2010 the staff issued RAI 4.3-13 asked the applicant to clarify if the LRA includes all applicable AMR items with an aging effect of cumulative fatigue damage for those components in scope for license renewal. If not, the staff asked the applicant to explain why the LRA does not include all corresponding AMR items on cumulative fatigue damage for applicable ASME Code Class 2 and 3 or ANSI B31.1 piping, piping components, and piping elements in scope for license renewal. The staff also asked the applicant to identify all component types that are within the scope of the implicit fatigue analyses for ASME Code Class 2 and 3 components and B31.1 components in LRA Section 4.3.5. The staff previously identified this as Open Item 4.3-1.

In its response dated August 12, 2010, the applicant stated that, for the ESF systems, no additions are required because LRA Table 3.2.2-4 includes AMR items with an aging effect of cumulative fatigue damage. The staff reviewed LRA Table 3.2.2-4 and confirmed that it contains applicable AMR items associated with the aging effect of cumulative fatigue damage for piping and piping components.

Based on its review, the staff finds the applicant's response to RAI 4.3-13, part 3, related to LRA Section 3.2.2.2.1, acceptable because the applicant confirmed that no additional AMR items associated with ASME Code Class 2 and 3 or ANSI B31.1 piping, piping components, and piping elements were subject to aging management review in accordance with 10 CFR Part 54.21(a)(1). The staff confirmed that the applicant included AMR items associated with ASME Code Class 2 and 3 or ANSI B31.1 piping, piping components, and piping elements with an aging effect of cumulative fatigue damage in LRA Table 3.2.2-4. The staff's concern described in RAI 4.3-13 is resolved, and this portion of the Open Item is closed.

Based on its review, the staff concludes that the applicant's proposal to manage cumulative fatigue damage in ASME III Class 2, Class 3, and ANSI B31.1 components meets the SRP-LR Section 3.2.2.2.1 criteria. For those line items that apply to LRA Section 3.2.2.2.1, the staff determines that the LRA is consistent with the GALL Report and that the applicant has demonstrated that the effects of aging will be adequately managed so that the intended function(s) will be maintained consistent with the CLB during the period of extended operation, as required by 10 CFR 54.21(a)(3). SER Section 4.3 documents the staff's review of the applicant's evaluation of the TLAA for these components.

3.2.2.2.2 Loss of Material Due to Cladding

LRA Section 3.2.2.2.2 addresses stainless steel clad pump casings exposed to treated borated water in the ECCS. The GALL Report recommends use of a plant-specific AMP to manage the loss of material due to cladding breach for this component group. The applicant stated that this item is not applicable because it has no in-scope steel pump casings clad with stainless steel exposed to treated borated water in the ECCS. The staff reviewed LRA Sections 2.3.2 and 3.2 and confirmed that the applicant's LRA does not have any AMR results for the ECCS that include steel pump casings with stainless steel cladding exposed to treated borated water. The staff reviewed the applicant's UFSAR and confirmed that there are no in-scope steel pump casings with stainless cladding exposed to treated borated water in the ECCS systems and, therefore, finds the applicant's determination acceptable.

3.2.2.2.3 Loss of Material Due to Pitting and Crevice Corrosion

The staff reviewed LRA Section 3.2.2.2.3 against the following criteria in the SRP-LR Section 3.2.2.2.3:

- LRA Section 3.2.2.2.3.1 addresses the loss of material due to pitting and crevice corrosion of internal surfaces for stainless steel containment isolation components exposed to treated water in the ESF systems. The applicant indicated that this item is

Aging Management Review Results

not applicable because there are no stainless steel components within the scope of license renewal exposed to treated water in the ESF systems.

The staff reviewed the LRA Table 3.2.2-4 for the safety injection and shutdown cooling system and, contrary to the applicant's statement, noted that the safety injection and shutdown cooling system has stainless steel piping in a treated (demineralized) water environment. The staff further noted the applicant's intent to evaluate the related items with the Water Chemistry and One-Time Inspection Programs, is appropriate; however, the applicant's statement that this item is "not applicable" appeared incorrect. In order to resolve this issue, the staff held a teleconference with the applicant on March 12, 2010, wherein the applicant indicated it was revising this section of the LRA.

In a letter dated April 1, 2010, the applicant modified LRA Section 3.2.2.2.3.1 by deleting the "not applicable" aspect. It added that the Water Chemistry and One-Time Inspection Programs will manage the loss of material due to pitting and crevice corrosion for stainless steel containment isolation piping and components exposed to treated water, including demineralized water. The applicant further stated that the One-Time Inspection Program will include selected components at susceptible locations where contaminants could accumulate.

The staff finds the applicant's response acceptable because the applicant corrected its LRA by recognizing that this item was applicable and by describing how it will manage the loss of material due to pitting and crevice corrosion for stainless steel containment isolation piping and components exposed to treated water. The staff's concern described in the above discussion is resolved.

The staff reviewed LRA Section 3.2.2.2.3.1 against the criteria in SRP-LR Section 3.2.2.2.3.1, which states that loss of material due to pitting and crevice corrosion could occur on the internal surfaces of stainless steel components for containment isolation piping, piping components, and piping elements exposed to treated water. The SRP-LR notes that monitoring and control of water chemistry will mitigate degradation; however, water chemistry control does not preclude this aging effect in locations with stagnant flow conditions. It continues by stating that the applicant should verify the effectiveness of the Water Chemistry Control Program and that a one-time inspection at susceptible locations is an acceptable method.

SER Sections 3.0.3.2.1 and 3.0.3.1.6 document the staff's evaluation of the applicant's Water Chemistry and One-Time Inspection Programs, respectively. The staff finds the applicant's proposal to manage aging using the Water Chemistry Control Program, augmented by the One-Time Inspection Program, acceptable because the applicant's Water Chemistry Program limits the concentrations of chemical species known to cause corrosion and adds chemical species known to inhibit degradation. In addition, the staff notes that the One-Time Inspection Program verifies the effectiveness of the Water Chemistry Program and evaluates aging effects, including loss of material.

Based on the programs identified, the staff concludes that the applicant meets SRP-LR Section 3.2.2.2.3.1 criteria. For those items that apply to LRA Section 3.2.2.2.3.1, the staff determines that the LRA is consistent with the GALL Report. The applicant has demonstrated that it will adequately manage the effects of aging so that the intended functions will be maintained consistent with the CLB during the period of extended operation, as required by 10 CFR 54.21(a)(3).

- LRA Section 3.2.2.2.3.2 addresses loss of material due to pitting and crevice corrosion in stainless steel piping, piping components, and piping elements exposed to soil. The applicant stated that this item is not applicable because the ECCS does not contain stainless steel piping, piping components, and piping elements in a soil environment.

Aging Management Review Results

The staff reviewed LRA Sections 2.3.2 and 3.2 and confirmed that the applicant's LRA does not have any AMR results for the ECCS that include stainless steel piping, piping components, and piping elements exposed to soil. The staff reviewed the applicant's UFSAR and confirmed that no in-scope stainless steel piping, piping components, and piping elements exposed to soil are present in the ECCS and, therefore, finds the applicant's determination acceptable.

- LRA Section 3.2.2.2.3.3 addresses loss of material due to pitting and crevice corrosion in BWR stainless steel and aluminum piping, stating that this aging effect is not applicable since it is applicable to BWRs only. SRP-LR Section 3.2.2.2.3 states that loss of material due to pitting and crevice corrosion may occur in BWR stainless steel and aluminum piping, piping components, and piping elements exposed to treated water. The staff finds that SRP-LR Section 3.2.2.2.3, item 3 is not applicable because the PVNGS units are PWRs, and the staff guidance in this SRP-LR section is only applicable to BWRs.

- LRA Section 3.2.2.2.3.4 addresses stainless steel and copper alloy piping, piping components, and elements exposed to lubricating oil. The GALL Report recommends use of GALL AMPs XI.M39, "Lubricating Oil Analysis," and XI.M32, "One-Time Inspection" to manage loss of material due to pitting and crevice corrosion for this component group. The applicant stated that this item is not applicable because "PVNGS has no in-scope stainless steel and copper alloy piping, piping components, and piping elements exposed to lubricating oil." The staff reviewed LRA Sections 2.3.2 and 3.2 and confirmed that the applicant's LRA does not have any AMR results for the ESF systems that include stainless steel and copper alloy components exposed to lube oil in the ESF systems. The staff also reviewed the UFSAR to verify the same. Based on its review of the LRA and UFSAR, the staff confirmed that the applicant's plant does not have any in-scope stainless steel and copper alloy piping, piping components, and piping elements exposed to lubricating oil ESF systems and, therefore, finds the applicant's determination acceptable.

- LRA Section 3.2.2.2.3.5 addresses loss of material due to pitting and crevice corrosion in partially encased stainless steel tanks exposed to raw water. The applicant stated that this item is not applicable because the ECCS does not have any in-scope stainless steel tanks with a moisture barrier configuration in a raw water environment. The staff reviewed LRA Sections 2.3.2 and 3.2 and confirmed that the applicant's LRA does not have any AMR results for the ECCS that include stainless steel tanks exposed to raw water. The staff reviewed the applicant's UFSAR and confirmed that no in-scope stainless steel tanks exposed to raw water are present in the ECCS and, therefore, finds the applicant's determination acceptable.

- LRA Section 3.2.2.2.3.6 addresses the loss of material due to pitting and crevice corrosion of the stainless steel piping, components, and tanks exposed to internal condensation. The applicant indicated that the Inspection of Internal Surfaces in Miscellaneous Piping and Ducting Components Program will manage this aging issue.

  The staff reviewed LRA Section 3.2.2.2.3.6 against the criteria in SRP-LR Section 3.2.2.2.3.6, which states that loss of material due to pitting and crevice corrosion could occur for stainless steel piping, piping components, piping elements, and tanks exposed to internal condensation. The SRP-LR further indicates that a plant-specific program is to manage this aging issue.

  SER Section 3.0.3.2.15 documents the staff's review of the applicant's Inspection of Internal Surfaces in Miscellaneous Piping and Ducting Components Program. The staff noted that the applicant's program includes stainless steel materials and performs visual

Aging Management Review Results

inspections to detect loss of material. The staff finds the applicant's use of the Inspection of Internal Surfaces in Miscellaneous Piping and Ducting Components Program acceptable to manage the loss of material due to pitting and corrosion, because the program includes the relevant material and performs appropriate inspections capable of detecting this aging effect.

Based on the program identified, the staff concludes that the applicant's program meets SRP-LR Section 3.2.2.2.3.6 criteria. For those items that apply to LRA Section 3.2.2.2.3.6, the staff determines that the LRA is consistent with the GALL Report. The applicant has demonstrated that it will adequately manage the effects of aging so that the intended functions will be maintained consistent with the CLB during the period of extended operation, as required by 10 CFR 54.21(a)(3).

Based on the programs identified above, the staff concludes that the applicant's programs meet SRP-LR Section 3.2.2.2.3 criteria. For those items that apply to LRA Section 3.2.2.2.3, the staff determines that the LRA is consistent with the GALL Report. The staff also finds that the applicant has demonstrated that it will adequately manage the effects of aging so that the intended function(s) will be maintained consistent with the CLB during the period of extended operation, as required by 10 CFR 54.21(a)(3).

3.2.2.2.4 Reduction of Heat Transfer Due to Fouling

The staff reviewed LRA Section 3.2.2.2.4 against the following criteria in the SRP-LR Section 3.2.2.2.4:

- LRA Section 3.2.2.2.4.1 addresses the reduction of heat transfer due to fouling of steel, stainless steel, and copper alloy heat exchanger tubes exposed to lubricating oil for ESF systems. The applicant stated that this item is not applicable because PVNGS has no in-scope steel, stainless steel, and copper alloy heat exchanger tubes exposed to lubricating oil in the ESF systems. The staff reviewed LRA Sections 2.3.2 and 3.2 and confirmed that the applicant's LRA does not have any AMR results for the ESF systems that include steel, stainless steel, and copper alloy heat exchanger tubes exposed to lubricating oil. The staff reviewed the UFSAR and confirmed that there are no in-scope heat exchangers constructed of steel, stainless steel, or copper alloy exposed to lubricating oil in the ESF systems and, therefore, finds the applicant's determination acceptable.

- LRA Section 3.2.2.2.4.2 addresses the reduction of heat transfer due to fouling of stainless steel heat exchanger tubes exposed to treated water for ESF systems. The applicant stated that this item is not applicable because PVNGS has no in-scope stainless steel heat exchanger tubes exposed to treated water with the aging effect of reduction of heat transfer in the ESF systems. The staff reviewed LRA Sections 2.3.2 and 3.2 and confirmed that the applicant's LRA does not have any AMR results for the ESF systems that include stainless steel heat exchanger tubes exposed to treated water. The staff reviewed the UFSAR and confirmed that no in-scope heat exchangers constructed of stainless steel exposed to treated water are present in the ESF systems, therefore, finds the applicant's determination acceptable.

Based on the programs identified above, the staff concludes that the applicant meets SRP-LR Section 3.2.2.2.4 criteria. For those items that apply to LRA Section 3.2.2.2.4, the staff determines that the LRA is consistent with the GALL Report. In addition, the applicant has demonstrated that it will adequately manage the effects of aging so that the intended function(s) will be maintained consistent with the CLB during the period of extended operation, as required by 10 CFR 54.21(a)(3).

Aging Management Review Results

3.2.2.2.5 Hardening and Loss of Strength Due to Elastomer Degradation

The staff reviewed LRA Section 3.2.2.2.5 against the criteria in the SRP-LR Section 3.2.2.2.5.

LRA Section 3.2.2.2.5 addresses hardening and loss of strength due to elastomer degradation, stating that this aging effect is not applicable since it is only applicable to BWRs. The SRP-LR Section 3.2.2.2.5 states that hardening and loss of strength due to elastomer degradation may occur in elastomer seals and components of the BWR standby gas treatment system ductwork and filters exposed to air-indoor uncontrolled. The staff finds that SRP-LR Section 3.2.2.2.5 is not applicable because the PVNGS units are PWRs, and the staff guidance in this SRP-LR section is applicable to components within the standby gas treatment system in BWRs.

3.2.2.2.6 Loss of Material Due to Erosion

LRA Section 3.2.2.2.6 addresses the stainless steel minimum flow recirculation line orifice for the HPSI pump exposed to treated borated water. The GALL Report recommends a plant-specific AMP to evaluate for erosion of the orifice due to extended use of the centrifugal HPSI pump for normal charging. The applicant stated that this item is not applicable because PVNGS does not use the HPSI pumps for normal charging. The staff reviewed the applicant's UFSAR and confirmed that the HPSI pumps are isolated from the RCS during normal operation. Because the applicant does not use HPSI pumps for normal charging, erosion of the HPSI minimum flow recirculation line orifice is not likely to occur and, therefore, the applicant's determination is acceptable.

3.2.2.2.7 Loss of Material Due to General Corrosion and Fouling

LRA Section 3.2.2.2.7 addresses loss of material due to general corrosion and fouling on steel drywell and suppression chamber spray system nozzle and flow orifice internal surfaces exposed to air-indoor uncontrolled. This section states that this aging effect is not applicable; it is applicable to BWRs only. SRP-LR Section 3.2.2.2.7 states that loss of material due to general corrosion and fouling may occur on steel drywell and suppression chamber spray system nozzle and flow orifice internal surfaces exposed to air-indoor uncontrolled and may cause plugging of the spray nozzles and flow orifices. The staff finds that SRP-LR Section 3.2.2.2.7 is not applicable because the PVNGS units are PWRs, and the staff guidance in this SRP-LR section is only applicable to drywell and suppression chamber spray systems in BWRs.

3.2.2.2.8 Loss of Material Due to General, Pitting, and Crevice Corrosion

The staff reviewed LRA Section 3.2.2.2.8 against the following criteria in the SRP-LR Section 3.2.2.2.8:

- LRA Section 3.2.2.2.8.1 addresses loss of material due to general, pitting, and crevice corrosion in BWR steel piping, piping components, and piping elements exposed to treated water. This section states that this aging effect is not applicable; it is applicable to BWRs only. SRP-LR Section 3.2.2.2.8 states that loss of material due to general, pitting, and crevice corrosion may occur in BWR steel piping, piping components, and piping elements exposed to treated water. The staff finds that SRP-LR Section 3.2.2.2.8 is not applicable to PVNGS because the PVNGS units are PWRs, and the staff guidance in this SRP-LR section is only applicable to steel piping, piping components, and piping elements exposed to treated water in BWRs.

- LRA Section 3.2.2.2.8.2 addresses the loss of material due to general, pitting and crevice corrosion from the internal surfaces of steel containment isolation piping and components exposed to treated water. The applicant stated that this item is not applicable, because it evaluated the containment isolation components in the systems in which the components were found to have the function of containment integrity. By

Aging Management Review Results

letter dated January 28, 2010, the staff issued RAI 3.2.2-1 requesting the applicant to clarify what it meant by the statement that it evaluated the containment isolation components n the system in which the components were found to have the function of containment integrity. In addition, the staff requested that the applicant provide additional information on the adequacy of the AMPs used to manage steel containment isolation piping.

In its response, dated March 1, 2010, the applicant stated that the plant equipment list does not contain a separate system for mechanical containment isolation components. As such, the applicant indicated that it evaluates the containment isolation components as part of the plant equipment list system to which they are assigned and are consistent with the expectations in the GALL Report. For the ESF systems, the applicant stated that GALL Report Table 3.2-1, item 3 summarizes the mechanical containment isolation components and that Table 3.3-1, item 24 for auxiliary systems and Table 3.4-1, item 16 for steam and power conversion systems recommend further evaluations for this item. The staff finds the applicant's response acceptable because it clarified where and how it evaluated the associated items in the LRA. The staff's concern described in RAI 3.2.2-1 is resolved.

The staff reviewed LRA 2.3.2 and 3.2 and confirmed that the applicant's LRA does not have any AMR results for the ESF systems that include steel containment isolation piping, piping components, and piping elements exposed to treated water. The staff reviewed the applicant's UFSAR and confirmed no in-scope steel containment isolation piping, piping components, and piping elements exposed to treated water are present in the ESF systems and, therefore, the staff finds the applicant's determination acceptable.

- LRA Section 3.2.2.2.8.3 addresses carbon steel piping, piping components and elements exposed to lubricating oil. The GALL Report recommends use of GALL AMPs XI.M39, "Lubricating Oil Analysis," and XI.M32, "One-Time Inspection" to manage loss of material due to general, pitting, and crevice corrosion for this component group. The applicant stated that this item is not applicable because "PVNGS has no in-scope carbon steel components exposed to lubricating oil in the ESF systems, so the applicable GALL Report lines were not used." The staff reviewed LRA Sections 2.3.2 and 3.2 and confirmed that the applicant's LRA does not have any AMR results for the ESF systems that include steel components exposed to lube oil in the ESF systems. The staff also reviewed the UFSAR to verify the same. Based on its review of the LRA and UFSAR, the staff finds the applicant's determination acceptable.

Based on the programs identified above, the staff concludes that the applicant meets SRP-LR Section 3.2.2.2.8 criteria. For those items that apply to LRA Section 3.2.2.2.8, the staff determines that the LRA is consistent with the GALL Report. Further, the applicant has demonstrated that it will adequately manage the effects of aging so that the intended function(s) will be maintained consistent with the CLB during the period of extended operation, as required by 10 CFR 54.21(a)(3).

3.2.2.2.9   Loss of Material Due to General, Pitting, Crevice, and Microbiologically-Influenced Corrosion

LRA Section 3.2.2.2.9 addresses loss of material due to general, pitting, crevice, and microbiologically-influenced corrosion in steel piping, piping components, and piping elements, with or without coating or wrapping, buried in soil. The applicant stated that this item is only applicable to BWRs. The staff reviewed SRP-LR Section 3.2.2.2.9 and Table 3.2-1, which refers to the GALL Report item E-42. The staff reviewed the GALL Report Table V.B. and confirmed that it is only applicable to the standby gas treatment system for BWR plants. The staff finds the applicant's determination acceptable.

3-202

Aging Management Review Results

3.2.2.2.10 Quality Assurance for Aging Management of Nonsafety-Related Components

SER Section 3.0.4 documents the staff's evaluation of the applicant's QA program.

3.2.2.3 Aging Management Review Results Not Consistent with or Not Addressed in the Generic Aging Lessons Learned Report

In LRA Tables 3.2.2-1 through 3.2.2-4, the staff reviewed additional details of the AMR results for material, environment, AERM, and AMP combinations not consistent with, or not addressed in, the GALL Report.

In LRA Tables 3.2.2-1 through 3.2.2-4, the applicant indicated, using Notes F–J, which combinations of component type, material, environment, and AERM do not correspond to an item in the GALL Report. The applicant provided information about how it will manage the aging effects. Specifically, Note F indicates that the material for the AMR item component is not evaluated in the GALL Report. Note G indicates that the environment for the AMR item component and material is not evaluated in the GALL Report. Note H indicates that the aging effect for the AMR item component, material, and environment combination is not evaluated in the GALL Report. Note I indicates that the aging effect identified in the GALL Report for the item component, material, and environment combination is not applicable. Note J indicates that neither the component nor the material and environment combination for the item is evaluated in the GALL Report.

For component type, material, and environment combinations not evaluated in the GALL Report, the staff reviewed the applicant's evaluation to determine if the applicant has demonstrated that it will adequately manage the effects of aging so that the intended function(s) will be maintained consistent with the CLB for the period of extended operation. The staff's evaluation is documented in the following sections.

3.2.2.3.1 Containment Leak Test System—Summary of Aging Management Review—License Renewal Application Table 3.2.2-1

The staff reviewed LRA Table 3.2.2-1, which summarize the results of AMR evaluations for the containment leak test system component groups. The staff's review did not identify any items with notes F–J, indicating that the combinations of component type, material, environment, and AERM for this system are consistent with the GALL Report.

SER Section 3.0.2.2 documents the staff's evaluation of the items with Notes A–E.

3.2.2.3.2 Containment Purge System—Summary of Aging Management Review—License Renewal Application Table 3.2.2-2

The staff reviewed LRA Table 3.2.2-2, which summarizes the results of AMR evaluations for the containment purge system component groups.

In LRA Table 3.2.2-2, the applicant stated that the Bolting Integrity Program manages carbon steel closure bolting exposed to atmosphere or weather, for loss of preload. In LRA Table 3.2.2-3, the applicant also stated that the Bolting Integrity Program manages stainless steel closure bolting, exposed to plant indoor air, for loss of preload. The AMR items cite generic note G, indicating that the environment is not in the GALL Report for this component and material combination. The applicant also cited plant-specific note 1, stating that loss of preload is considered applicable for all closure bolting. The staff reviewed all AMR result lines in the GALL Report where the aging effect is loss of preload and confirmed there are no AMR results for carbon steel bolting in an environment of atmosphere or weather or for stainless steel bolting in an environment of plant indoor air.

SER Section 3.0.3.2.3 documents the staff's evaluation of the applicant's Bolting Integrity Program. The staff noted that the mechanisms listed in the GALL Report that cause loss of

Aging Management Review Results

preload are thermal effects, gasket creep, and self-loosening, and that these aging effects are not dependent on the specific bolting material or environment. The staff also noted that activities in the Bolting Integrity Program that manage loss of preload are equally effective for carbon steel and stainless steel bolts. The staff further noted that the GALL Report recommends using the Bolting Integrity Program to manage loss of preload in carbon steel bolts exposed to air-indoor uncontrolled. The staff finds the applicant's use of the Bolting Integrity Program to manage loss of preload in carbon steel and stainless steel closure bolting acceptable because it is consistent with the GALL Report recommendations for managing loss of preload in carbon steel closure bolting exposed to air. In addition, the Bolting Integrity Program's activities for managing loss of preload are applicable for both carbon steel and stainless steel bolts.

Based on its review, the staff finds that the applicant has appropriately evaluated the AMR results of material, environment, AERM, and AMP combinations not evaluated in the GALL Report. The staff finds that the applicant has demonstrated that it will adequately manage the effects of aging so that the intended function(s) will be maintained consistent with the CLB for the period of extended operation, as required by 10 CFR 54.21(a)(3).

3.2.2.3.3 Containment Hydrogen Control System–Summary of Aging Management Review–License Renewal Application Table 3.2.2-3

The staff reviewed LRA Table 3.2.2-3, which summarizes the results of AMR evaluations for the containment hydrogen control system component groups.

SER Section 3.2.2.3.2 documents the staff's evaluation for stainless steel closure bolting exposed to plant indoor air being managed for loss of preload by the Bolting Integrity Program, citing generic note G.

In LRA Table 3.2.2-3, the applicant stated that for nickel-alloy flexible hoses exposed internally to dry gas, there is no aging effect and no AMP is proposed. The AMR item cites generic Note G, indicating that for the item, the environment is not in the GALL Report for this component and material.

The staff reviewed the associated items in the LRA and confirmed that no aging effect is applicable for this component, material, and environment because austenitic materials such as nickel alloys are not subject to loss of material or cracking when exposed to this environment, and these materials are used as corrosion-resistant replacement materials where other materials have degraded. The staff noted that corrosion-resistant materials, such as austenitic and martensitic stainless steels and high strength nickel base alloys, offer good protection against loss of material. The staff also noted that the conditions required for cracking, due to a variety of mechanisms (SCC, PWSCC, IASCC and IGSCC) to occur, such as being exposed to an aqueous solution (reactor coolant or other corrosive solutions) and high temperatures, do not exist on the surfaces of these components when exposed to dry gas. The staff noted that GALL Report AMR item IV.E-1 states that nickel alloy exposed to air-indoor uncontrolled is not subject to an AERM. The staff noted that an air-indoor uncontrolled environment is more aggressive than a dry gas environment because it is possible for condensation to occur in an air-indoor uncontrolled environment. The staff finds the applicant's determination acceptable because nickel alloy is a highly corrosion resistant material and is not subject to conditions where cracking is possible. In addition, this component is exposed to a less aggressive environment when compared to GALL, AMR item IV.E-1, for which there is no AERM.

Based on its review, the staff finds that the applicant has appropriately evaluated the AMR results of material, environment, AERM, and AMP combinations not evaluated in the GALL Report. The staff finds that the applicant has demonstrated that it will adequately manage the effects of aging so that the intended function(s) will be maintained consistent with the CLB for the period of extended operation, as required by 10 CFR 54.21(a)(3).

Aging Management Review Results

3.2.2.3.4 Safety Injection and Shutdown Cooling System—Summary of Aging Management Review—License Renewal Application Table 3.2.2-4

The staff reviewed LRA Table 3.2.2-4, which summarizes the results of AMR evaluations for the safety injection and shutdown cooling system component groups.

In LRA Table 3.2.2-4, the applicant stated that for flow indicators made of glass exposed to borated water leakage (external) there is no aging effect and no AMP is proposed. The AMR item cites generic Note G, indicating that for the item, the environment is not in the GALL Report for this component and material.

The staff reviewed all AMR items in the GALL Report where the material is glass, and confirmed that, for this environment, there are no entries in the GALL Report for this component and material. The staff notes that for glass components, the GALL Report does list a treated water environment in contrast to borated water leakage, but this is not significant to the evaluation because the GALL Report AMR items state that there is no aging effect and no recommended AMP. The staff finds the applicant's proposal acceptable because glass in an environment of borated water leakage is expected to have the same aging effects as glass in an environment of treated borated water, and the GALL Report states that there are no AERMs for glass piping components in a treated borated water environment.

In LRA Table 3.2.2-4, the applicant stated that for nickel-alloy piping exposed to borated water leakage there is no aging effect and no AMP is proposed. The AMR item cites generic Note G, indicating that for the item, the environment is not in the GALL Report for this component and material.

The staff reviewed the associated items in the LRA and confirmed that no aging effect is applicable for this component, material, and environment because austenitic materials such as nickel alloys are not subject to loss of material or cracking when exposed to this environment, and these materials are used as corrosion-resistant replacement materials where other materials have degraded. The staff noted that according to EPRI NP-5769, "Degradation and Failure of Bolting in Nuclear Power Plants," Volumes 1 and 2, April 1988, corrosion-resistant materials, such as austenitic and martensitic stainless steels and high-strength, nickel-base alloys, offer good protection against loss of material due to boric acid corrosion. The staff also noted that the conditions required for cracking to occur (e.g., SCC, PWSCC, IASCC and IGSCC), such as being exposed to high temperatures and reactor coolant or other corrosive solutions, do not exist on the surfaces of these components when exposed to borated water leakage. Therefore, the staff finds no AMP is necessary for nickel alloys in a borated water leakage environment.

In LRA Table 3.2.2-4, the applicant stated that the Water Chemistry and One-Time Inspection Programs manage the stainless steel shutdown cooling heat exchangers exposed to treated borated water, for reduction of heat transfer. The AMR item cites generic Note H, indicating that the aging effect is not in the GALL Report for this component, material, and environment combination. The item associated with the stainless steel shutdown cooling heat exchangers exposed to treated borated water cites plant-specific Note 1, which states that reduction in heat transfer due to fouling is a potential aging effect for stainless steel heat exchanger components in treated borated water. This statement is based upon the component, material, aging effects, and AMP combination of GALL Report, AMR item VII.E1-4.

The staff reviewed all AMR result items in the GALL Report where the component and material is stainless steel heat exchangers exposed to treated borated water and noted that the applicant's plant-specific Note 1 reference does not directly relate to the aging effect, whereas GALL Report items EP-34, AP-62, and SP-40 do address heat transfer fouling. The staff also noted that the GALL Report recommends that the aging be managed by the Water Chemistry

Aging Management Review Results

Program and verified by the One Time Inspection Program, the same programs that the applicant has selected to manage aging.

SER Sections 3.0.3.2.1 and 3.0.3.1.6 document the staff's evaluations of the applicant's Water Chemistry and One-Time Inspection Programs, respectively. The staff finds the applicant's proposal to manage aging using the Water Chemistry and One-Time Inspection Programs acceptable because the Water Chemistry Program minimizes the potential development and progress of heat exchanger fouling, and the One-Time Inspection Program verifies the effectiveness of the Water Chemistry Program using inspections with specific attributes related to reduction of heat transfer.

Based on its review, the staff finds that the applicant has appropriately evaluated the AMR results of material, environment, and AMP combinations not addressed in the GALL Report. The staff finds that the applicant has demonstrated that it will adequately manage the effects of aging so that the intended function(s) will be maintained consistent with the CLB for the period of extended operation, as required by 10 CFR.21(a)(3).

In LRA Tables 3.3.2-4 and 3.3.2-10, the applicant stated that calcium silicate and mineral wool insulation exposed to borated water leakage have no AERM, and no AMP is proposed. The AMR items cite generic Note J.

The staff reviewed the associated items in the LRA and noted that the applicant only included piping insulation in-scope for license renewal for those systems where the insulation has an intended function. The staff also noted that both calcium silicate and mineral wool insulation materials can experience loss of insulating properties when exposed to moisture, due to effects such as compression of the material or change in material properties, but that proper jacketing of the insulation can be effective at preventing moisture intrusion. The staff further noted that both materials can retain moisture well after exposure, prolonging the contact time of the moisture with the piping being insulated. By letter dated July 30, 2010, the applicant submitted confirmatory information that the affected piping is jacketed with overlapping seams such that moisture intrusion is not a concern. The staff finds the applicant's determination, that calcium silicate and mineral wool insulation in this environment have no AERM, acceptable because these materials perform well when exposed to air and are properly jacketed to prevent moisture intrusion.

Based on its review, the staff finds that the applicant has appropriately evaluated the AMR results of material, environment, AERM, and AMP combination not addressed in the GALL Report. The staff finds that the applicant demonstrated that it will adequately manage the effects of aging for these components so that their intended functions will be maintained consistent with the CLB during the period of extended operation, as required by 10 CFR 54.21(a)(3).

3.2.3 Conclusion

The staff concludes that the applicant has provided sufficient information to demonstrate that the effects of aging for the ESF systems components, within the scope of license renewal and subject to an AMR, will be adequately managed so that the intended function(s) will be maintained consistent with the CLB for the period of extended operation, as required by 10 CFR 54.21(a)(3).

3.3 Aging Management of Auxiliary Systems

This section of the SER documents the staff's review of the applicant's AMR results for the auxiliary systems components and component groups of:

- fuel handling and storage system

- spent fuel pool and cleanup system
- essential cooling water system
- essential chilled water system
- normal chilled water system
- nuclear cooling water system
- essential spray pond system
- nuclear sampling system
- compressed air system
- chemical volume and control system (CVCS)
- control building HVAC system
- auxiliary building HVAC system
- fuel building HVAC system
- containment building HVAC system
- diesel generator building HVAC system
- radwaste building HVAC system
- turbine building HVAC system
- miscellaneous site structures/spray pond pump house HVAC system
- fire protection system
- diesel generator fuel oil storage and transfer system
- diesel generator
- domestic water system
- demineralized water system
- water reclamation facility (WRF) fuel system
- service gases (N^2 and H^2) system
- gaseous radwaste system
- radioactive waste drains system
- station blackout generator
- cranes, hoists, and elevators
- miscellaneous auxiliary systems in-scope only for criterion 10 CFR 54.4(a)(2)

3.3.1 Summary of Technical Information in the Application

LRA Section 3.3 provides AMR results for the auxiliary systems components and component groups. LRA Table 3.3.1, "Summary of Aging Management Evaluations in Chapter VII of NUREG-1801 [the GALL Report] for Auxiliary Systems," is a summary comparison of the applicant's AMRs with those evaluated in the GALL Report for the auxiliary systems components and component groups.

The applicant's AMRs evaluated and incorporated applicable plant-specific and industry operating experience in the determination of AERMs. The plant-specific evaluation included condition reports and discussions with appropriate site personnel to identify AERMs. The applicant's review of industry operating experience included a review of the GALL Report and operating experience issues identified since the issuance of the GALL Report.

3.3.2 Staff Evaluation

The staff reviewed LRA Section 3.3 to determine if the applicant provided sufficient information to demonstrate that it will adequately manage the effects of aging for the auxiliary systems components, within the scope of license renewal and subject to an AMR, so that the intended function(s) will be maintained consistent with the CLB for the period of extended operation, as required by 10 CFR 54.21(a)(3).

Aging Management Review Results

The staff reviewed AMRs to ensure the applicant's claim that certain AMRs were consistent with the GALL Report. The staff did not repeat its review of the matters described in the GALL Report; however, the staff did verify that the material presented in the LRA was applicable and that the applicant identified the appropriate GALL Report AMRs. SER Section 3.0.3 documents the staff's evaluations of the AMPs, and SER Section 3.3.2.1 documents details of the staff's evaluation.

The staff also reviewed AMRs consistent with the GALL Report and for which further evaluation is recommended. The staff confirmed that the applicant's further evaluations were consistent with the SRP-LR Section 3.3.2.2 acceptance criteria. SER Section 3.3.2.2 documents the staff's evaluations.

The staff also conducted a technical review of the remaining AMRs not consistent with, or not addressed in, the GALL Report. The technical review evaluated if the applicant identified all plausible aging effects and if the aging effects listed were appropriate for the material-environment combinations specified. SER Section 3.3.2.3 documents the staff's evaluations.

For SSCs that the applicant claimed were not applicable or required no aging management, the staff reviewed the AMR items and the plant's operating experience to verify the applicant's claims.

Table 3.3-1 summarizes the staff's evaluation of components, aging effects or mechanisms, and AMPs listed in LRA Section 3.3 and addressed in the GALL Report.

Table 3.3-1. Staff Evaluation for Auxiliary System Components in the GALL Report

Component Group (GALL Report Item No.)	Aging Effect/ Mechanism	AMP in GALL Report	Further Evaluation in GALL Report	AMP in LRA, Supplements, or Amendments	Staff Evaluation
Steel cranes-structural girders exposed to air-indoor uncontrolled (external) (3.3.1-1)	Cumulative fatigue damage	TLAA to be evaluated for structural girders of cranes. See the SRP-LR, Section 4.7 for generic guidance for meeting the requirements of 10 CFR 54.21(c)(1).	Yes	TLAA	Consistent with GALL Report. See SER Section 3.3.2.2.1.
Steel and stainless steel piping, piping components, piping elements, and heat exchanger components exposed to air - indoor uncontrolled, treated borated water or treated water (3.3.1-2)	Cumulative fatigue damage	TLAA, evaluated in accordance with 10 CFR 54.21(c)	Yes	TLAA	Consistent with GALL Report. See SER Section 3.3.2.2.1.

Aging Management Review Results

Component Group (GALL Report Item No.)	Aging Effect/ Mechanism	AMP in GALL Report	Further Evaluation in GALL Report	AMP in LRA, Supplements, or Amendments	Staff Evaluation
Stainless steel heat exchanger tubes exposed to treated water (3.3.1-3)	Reduction of heat transfer due to fouling	Water Chemistry and One-Time Inspection	Yes	Not applicable	The staff did not agree with the applicant's determination of 'not applicable.' See SER Sections 3.3.2.2.2 and 3.3.2.3.2.
Stainless steel piping, piping components, and piping elements exposed to sodium pentaborate solution > 60°C (> 140°F) (3.3.1-4)	Cracking due to SCC	Water Chemistry and One-Time Inspection	Yes	Not applicable	Not applicable to PWRs. See SER Section 3.3.2.2.3(1).
Stainless steel and stainless clad steel heat exchanger components exposed to treated water > 60°C (> 140°F) (3.3.1-5)	Cracking due to SCC	A plant specific AMP is to be evaluated.	Yes	Not applicable	Not applicable. See SER Section 3.3.2.2.3(2).
Stainless steel diesel engine exhaust piping, piping components, and piping elements exposed to diesel exhaust (3.3.1-6)	Cracking due to SCC	A plant specific AMP is to be evaluated.	Yes	Inspection Of Internal Surfaces In Miscellaneous Piping And Ducting Components	Consistent with GALL Report. See SER Section 3.3.2.2.3(3).
Stainless steel non-regenerative heat exchanger components exposed to treated borated water > 60°C (> 140°F) (3.3.1-7)	Cracking due to SCC and cyclic loading	Water Chemistry and a plant-specific verification program. An acceptable verification program is to include temperature and radioactivity monitoring of the shell side water, and eddy current testing of tubes.	Yes	Water Chemistry and One-Time Inspection	Consistent with GALL Report. See SER Section 3.3.2.2.4(1).
Stainless steel regenerative heat exchanger components exposed to treated borated water > 60°C (> 140°F) (3.3.1-8)	Cracking due to SCC and cyclic loading	Water Chemistry and a plant-specific verification program. The AMP is to be augmented by verifying the absence of cracking due to SCC and cyclic loading. A plant specific AMP is to be evaluated.	Yes	Water Chemistry and One-Time Inspection	Consistent with GALL Report. See SER Section 3.3.2.2.4(2).

Aging Management Review Results

Component Group (GALL Report Item No.)	Aging Effect/ Mechanism	AMP in GALL Report	Further Evaluation in GALL Report	AMP in LRA, Supplements, or Amendments	Staff Evaluation
Stainless steel high-pressure pump casing in PWR CVCS (3.3.1-9)	Cracking due to SCC and cyclic loading	Water Chemistry and a plant-specific verification program. The AMP is to be augmented by verifying the absence of cracking due to SCC and cyclic loading. A plant specific AMP is to be evaluated.	Yes	Water Chemistry and One-Time Inspection	Consistent with GALL Report. See SER Section 3.3.2.2.4(3).
High-strength steel closure bolting exposed to air with steam or water leakage. (3.3.1-10)	Cracking due to SCC, cyclic loading	Bolting Integrity. The AMP is to be augmented by appropriate inspection to detect cracking if the bolts are not otherwise replaced during maintenance.	Yes	Not applicable	Not applicable to PVNGS. See SER Section 3.3.2.2.4(4).
Elastomer seals and components exposed to air - indoor uncontrolled (internal/external) (3.3.1-11)	Hardening and loss of strength due to elastomer degradation	A plant specific AMP is to be evaluated.	Yes	Inspection of Internal Surfaces in Miscellaneous Piping and Ducting Components and External Surfaces Monitoring	Consistent with GALL Report. See SER Section 3.3.2.2.5(1).
Elastomer lining exposed to treated water or treated borated water (3.3.1-12)	Hardening and loss of strength due to elastomer degradation	A plant-specific AMP is to be evaluated.	Yes	Inspection of Internal Surfaces in Miscellaneous Piping and Ducting Components	Consistent with GALL Report. See SER Section 3.3.2.2.5(2).
Boral, boron steel spent fuel storage racks neutron-absorbing sheets exposed to treated water or treated borated water (3.3.1-13)	Reduction of neutron-absorbing capacity and loss of material due to general corrosion	A plant specific AMP is to be evaluated.	Yes	Not applicable	Not applicable to PVNGS. See SER Section 3.3.2.2.6.
Steel piping, piping component, and piping elements exposed to lubricating oil (3.3.1-14)	Loss of material due to general, pitting, and crevice corrosion	Lubricating Oil Analysis and One-Time Inspection	Yes	Lubricating Oil Analysis and One-Time Inspection	Consistent with GALL Report. See SER Section 3.3.2.2.7(1).
Steel RCP oil collection system piping, tubing, and valve bodies exposed to lubricating oil (3.3.1-15)	Loss of material due to general, pitting, and crevice corrosion	Lubricating Oil Analysis and One-Time Inspection	Yes	Lubricating Oil Analysis and One-Time Inspection	Consistent with GALL Report. See SER Section 3.3.2.2.7(1).

Aging Management Review Results

Component Group (GALL Report Item No.)	Aging Effect/ Mechanism	AMP in GALL Report	Further Evaluation in GALL Report	AMP in LRA, Supplements, or Amendments	Staff Evaluation
Steel RCP oil collection system tank exposed to lubricating oil (3.3.1-16)	Loss of material due to general, pitting, and crevice corrosion	Lubricating Oil Analysis and One-Time Inspection to evaluate the thickness of the lower portion of the tank	Yes	Lubricating Oil Analysis and One-Time Inspection	Consistent with GALL Report. See SER Section 3.3.2.2.7(1).
Steel piping, piping components, and piping elements exposed to treated water (3.3.1-17)	Loss of material due to general, pitting, and crevice corrosion	Water Chemistry and One-Time Inspection	Yes	Not applicable	Not applicable to PWRs. See SER Section 3.3.2.2.7(2).
Stainless steel and steel diesel engine exhaust piping, piping components, and piping elements exposed to diesel exhaust (3.3.1-18)	Loss of material/general (steel only), pitting and crevice corrosion	A plant specific AMP is to be evaluated.	Yes	Inspection of Internal Surfaces in Miscellaneous Piping and Ducting Components	Consistent with GALL Report. See SER Section 3.3.2.2.7(3).
Steel (with or without coating or wrapping) piping, piping components, and piping elements exposed to soil (3.3.1-19)	Loss of material due to general, pitting, crevice, and MIC	Buried Piping and Tanks Surveillance or Buried Piping and Tanks Inspection	No / Yes	Buried Piping and Tanks Inspection	Consistent with GALL Report. See SER Section 3.3.2.2.8.
Steel piping, piping components, piping elements, and tanks exposed to fuel oil (3.3.1-20)	Loss of material due to general, pitting, crevice, and MIC and fouling	Fuel Oil Chemistry and One-Time Inspection	Yes	Fuel Oil Chemistry and One-Time Inspection	Consistent with GALL Report. See SER Section 3.3.2.2.9(1).
Steel heat exchanger components exposed to lubricating oil (3.3.1-21)	Loss of material due to general, pitting, crevice, and MIC and fouling	Lubricating Oil Analysis and One-Time Inspection	Yes	Lubricating Oil Analysis and One-Time Inspection	Consistent with GALL Report. See SER Section 3.3.2.2.9(2).
Steel with elastomer lining or stainless steel cladding piping, piping components, and piping elements exposed to treated water and treated borated water (3.3.1-22)	Loss of material due to pitting and crevice corrosion (only for steel after lining/cladding degradation)	Water Chemistry and One-Time Inspection	Yes	Water Chemistry and One-Time Inspection	Not applicable to PVNGS. See SER Section 3.3.2.2.10(1).
Stainless steel and steel with stainless steel cladding heat exchanger components exposed to treated water (3.3.1-23)	Loss of material due to pitting and crevice corrosion	Water Chemistry and One-Time Inspection	Yes	Not applicable	Not applicable to PWRs. See SER Section 3.3.2.2.10(2).

Aging Management Review Results

Component Group (GALL Report Item No.)	Aging Effect/ Mechanism	AMP in GALL Report	Further Evaluation in GALL Report	AMP in LRA, Supplements, or Amendments	Staff Evaluation
Stainless steel and aluminum piping, piping components, and piping elements exposed to treated water (3.3.1-24)	Loss of material due to pitting and crevice corrosion	Water Chemistry and One-Time Inspection	Yes	Not applicable	See SER Section 3.3.2.2.10(2).
Copper alloy HVAC piping, piping components, piping elements exposed to condensation (external) (3.3.1-25)	Loss of material due to pitting and crevice corrosion	A plant-specific AMP is to be evaluated.	Yes	Inspection of Internal Surfaces in Miscellaneous Piping and Ducting Components and External Surfaces Monitoring	Consistent with GALL Report. See SER Section 3.3.2.2.10(3).
Copper alloy piping, piping components, and piping elements exposed to lubricating oil (3.3.1-26)	Loss of material due to pitting and crevice corrosion	Lubricating Oil Analysis and One-Time Inspection	Yes	Lubricating Oil Analysis and One-Time Inspection	Consistent with GALL Report. See SER Section 3.3.2.2.10(4).
Stainless steel HVAC ducting and aluminum HVAC piping, piping components and piping elements exposed to condensation (3.3.1-27)	Loss of material due to pitting and crevice corrosion	A plant-specific AMP is to be evaluated.	Yes	Inspection Of Internal Surfaces In Miscellaneous Piping And Ducting Components	Consistent with GALL Report. See SER Section 3.3.2.2.10(5).
Copper alloy fire protection piping, piping components, and piping elements exposed to condensation (internal) (3.3.1-28)	Loss of material due to pitting and crevice corrosion	A plant-specific AMP is to be evaluated.	Yes	Inspection Of Internal Surfaces In Miscellaneous Piping And Ducting Components	Consistent with GALL Report. See SER Section 3.3.2.2.10(6).
Stainless steel piping, piping components, and piping elements exposed to soil (3.3.1-29)	Loss of material due to pitting and crevice corrosion	A plant-specific AMP is to be evaluated.	Yes	Buried Piping and Tanks Inspection	Consistent with GALL Report. See SER Section 3.3.2.2.10(7).
Stainless steel piping, piping components, and piping elements exposed to sodium pentaborate solution (3.3.1-30)	Loss of material due to pitting and crevice corrosion	Water Chemistry and One-Time Inspection	Yes	Not applicable	Not applicable to PWRs. See SER Section 3.3.2.2.10(8).
Copper alloy piping, piping components, and piping elements exposed to treated water (3.3.1-31)	Loss of material due to pitting, crevice, and galvanic corrosion	Water Chemistry and One-Time Inspection	Yes	Not applicable	Not applicable to PWRs. See SER Section 3.3.2.2.11.

Aging Management Review Results

Component Group (GALL Report Item No.)	Aging Effect/ Mechanism	AMP in GALL Report	Further Evaluation in GALL Report	AMP in LRA, Supplements, or Amendments	Staff Evaluation
Stainless steel, aluminum and copper alloy piping, piping components, and piping elements exposed to fuel oil (3.3.1-32)	Loss of material due to pitting, crevice, and MIC	Fuel Oil Chemistry and One-Time Inspection	Yes	Fuel Oil Chemistry and One-Time Inspection	Consistent with GALL Report. See SER Section 3.3.2.2.12(1).
Stainless steel piping, piping components, and piping elements exposed to lubricating oil (3.3.1-33)	Loss of material due to pitting, crevice, and MIC	Lubricating Oil Analysis and One-Time Inspection	Yes	Lubricating Oil Analysis and One-Time Inspection	Consistent with GALL Report. See SER Section 3.3.2.2.12(2).
Elastomer seals and components exposed to air-indoor uncontrolled (internal or external) (3.3.1-34)	Loss of material due to wear	A plant specific AMP is to be evaluated.	Yes	Not applicable	Not applicable to PVNGS. See SER Section 3.3.2.2.13.
Steel with stainless steel cladding pump casing exposed to treated borated water (3.3.1-35)	Loss of material due to cladding breach	A plant-specific AMP is to be evaluated. Reference NRC IN 94-63, "Boric Acid Corrosion of Charging Pump Casings Caused by Cladding Cracks."	Yes	Not applicable	Not applicable to PVNGS. See SER Section 3.3.2.2.14.
Boraflex spent fuel storage racks neutron-absorbing sheets exposed to treated water (3.3.1-36)	Reduction of neutron-absorbing capacity due to boraflex degradation	Boraflex Monitoring	No	Not applicable	Not applicable to PVNGS. See SER Section 3.3.2.1.1.
Stainless steel piping, piping components, and piping elements exposed to treated water > 60°C (> 140°F) (3.3.1-37)	Cracking due to SCC, IGSCC	BWR Reactor Water Cleanup System	No	Not applicable	Not applicable to PWRs. See SER Section 3.3.2.1.1.
Stainless steel piping, piping components, and piping elements exposed to treated water > 60°C (> 140°F) (3.3.1-38)	Cracking due to SCC	BWR SCC and Water Chemistry	No	Not applicable	Not applicable to PWRs. See SER Section 3.3.2.1.1.
Stainless steel BWR spent fuel storage racks exposed to treated water > 60°C (> 140°F) (3.3.1-39)	Cracking due to SCC	Water Chemistry	No	Not applicable	Not applicable to PWRs. See SER Section 3.3.2.1.1.
Steel tanks in diesel fuel oil system exposed to air - outdoor (external) (3.3.1-40)	Loss of material due to general, pitting, and crevice corrosion	Aboveground Steel Tanks	No	Not applicable	Not applicable to PVNGS. See SER Section 3.3.2.1.1.

Aging Management Review Results

Component Group (GALL Report Item No.)	Aging Effect/ Mechanism	AMP in GALL Report	Further Evaluation in GALL Report	AMP in LRA, Supplements, or Amendments	Staff Evaluation
High-strength steel closure bolting exposed to air with steam or water leakage (3.3.1-41)	Cracking due to cyclic loading, SCC	Bolting Integrity	No	Not applicable	Not applicable to PVNGS. See SER Section 3.3.2.1.1.
Steel closure bolting exposed to air with steam or water leakage (3.3.1-42)	Loss of material due to general corrosion	Bolting Integrity	No	Not applicable	Not applicable to PVNGS. See SER Section 3.3.2.1.1.
Steel bolting and closure bolting exposed to air-indoor uncontrolled (external) or air-outdoor (external) (3.3.1-43)	Loss of material due to general, pitting, and crevice corrosion	Bolting Integrity	No	Bolting Integrity	Consistent with GALL Report.
Steel compressed air system closure bolting exposed to condensation (3.3.1-44)	Loss of material due to general, pitting, and crevice corrosion	Bolting Integrity	No	Not applicable	Not applicable to PVNGS. See SER Section 3.3.2.1.1.
Steel closure bolting exposed to air-indoor uncontrolled (external) (3.3.1-45)	Loss of preload due to thermal effects, gasket creep, and self-loosening	Bolting Integrity	No	Bolting Integrity	Consistent with GALL Report.
Stainless steel and stainless clad steel piping, piping components, piping elements, and heat exchanger components exposed to closed cycle cooling water > 60°C (> 140°F) (3.3.1-46)	Cracking due to SCC	Closed-Cycle Cooling Water System	No	Closed-Cycle Cooling Water System	Consistent with GALL Report.
Steel piping, piping components, piping elements, tanks, and heat exchanger components exposed to closed cycle cooling water (3.3.1-47)	Loss of material due to general, pitting, and crevice corrosion	Closed-Cycle Cooling Water System	No	Closed-Cycle Cooling Water System	Consistent with GALL Report.
Steel piping, piping components, piping elements, tanks, and heat exchanger components exposed to closed cycle cooling water (3.3.1-48)	Loss of material due to general, pitting, crevice, and galvanic corrosion	Closed-Cycle Cooling Water System	No	Closed-Cycle Cooling Water System	Consistent with GALL Report.
Stainless steel; steel with stainless steel cladding heat exchanger components exposed to closed cycle cooling water (3.3.1-49)	Loss of material due to MIC	Closed-Cycle Cooling Water System	No	Not applicable	Not applicable to PWRs. See SER Section 3.3.2.1.1.

Aging Management Review Results

Component Group (GALL Report Item No.)	Aging Effect/ Mechanism	AMP in GALL Report	Further Evaluation in GALL Report	AMP in LRA, Supplements, or Amendments	Staff Evaluation
Stainless steel piping, piping components, and piping elements exposed to closed cycle cooling water (3.3.1-50)	Loss of material due to pitting and crevice corrosion	Closed-Cycle Cooling Water System	No	Closed-Cycle Cooling Water System	Consistent with GALL Report.
Copper alloy piping, piping components, piping elements, and heat exchanger components exposed to closed cycle cooling water (3.3.1-51)	Loss of material due to pitting, crevice, and galvanic corrosion	Closed-Cycle Cooling Water System	No	Closed-Cycle Cooling Water System	Consistent with GALL Report.
Steel, stainless steel, and copper alloy heat exchanger tubes exposed to closed cycle cooling water (3.3.1-52)	Reduction of heat transfer due to fouling	Closed-Cycle Cooling Water System	No	Closed-Cycle Cooling Water System	Consistent with GALL Report.
Steel compressed air system piping, piping components, and piping elements exposed to condensation (internal) (3.3.1-53)	Loss of material due to general and pitting corrosion	Compressed Air Monitoring	No	Inspection of Internal Surfaces in Miscellaneous Piping and Ducting Components	Alternative program used. See SER Section 3.3.2.1.2
Stainless steel compressed air system piping, piping components, and piping elements exposed to internal condensation (3.3.1-54)	Loss of material due to pitting and crevice corrosion	Compressed Air Monitoring	No	Inspection of Internal Surfaces in Miscellaneous Piping and Ducting Components	Alternative program used. See SER Section 3.3.2.1.2.
Steel ducting closure bolting exposed to air - indoor uncontrolled (external) (3.3.1-55)	Loss of material due to general corrosion	External Surfaces Monitoring	No	External Surfaces Monitoring	Consistent with GALL Report.
Steel HVAC ducting and components external surfaces exposed to air-indoor uncontrolled (external) (3.3.1-56)	Loss of material due to general corrosion	External Surfaces Monitoring	No	External Surfaces Monitoring	Consistent with GALL Report.
Steel piping and components external surfaces exposed to air-indoor uncontrolled (External) (3.3.1-57)	Loss of material due to general corrosion	External Surfaces Monitoring	No	External Surfaces Monitoring	Consistent with GALL Report.

Aging Management Review Results

Component Group (GALL Report Item No.)	Aging Effect/ Mechanism	AMP in GALL Report	Further Evaluation in GALL Report	AMP in LRA, Supplements, or Amendments	Staff Evaluation
Steel external surfaces exposed to air-indoor uncontrolled (external), air-outdoor (external), and condensation (external) (3.3.1-58)	Loss of material due to general corrosion	External Surfaces Monitoring	No	External Surfaces Monitoring and Inspection of Internal Surfaces in Miscellaneous Piping And Ducting Components	Consistent with GALL Report or alternative program used. See SER Section 3.3.2.1.3.
Steel heat exchanger components exposed to air-indoor uncontrolled (external) or air-outdoor (external) (3.3.1-59)	Loss of material due to general, pitting, and crevice corrosion	External Surfaces Monitoring	No	External Surfaces Monitoring	Consistent with GALL Report.
Steel piping, piping components, and piping elements exposed to air-outdoor (external) (3.3.1-60)	Loss of material due to general, pitting, and crevice corrosion	External Surfaces Monitoring	No	External Surfaces Monitoring	Consistent with GALL Report.
Elastomer fire barrier penetration seals exposed to air-outdoor or air-indoor uncontrolled (3.3.1-61)	Increased hardness, shrinkage and loss of strength due to weathering	Fire Protection	No	Fire Protection	Consistent with GALL Report.
Aluminum piping, piping components, and piping elements exposed to raw water (3.3.1-62)	Loss of material due to pitting and crevice corrosion	Fire Protection	No	Not applicable	Not applicable to PVNGS. See SER Section 3.3.2.1.1.
Steel fire rated doors exposed to air-outdoor or air-indoor uncontrolled (3.3.1-63)	Loss of material due to wear	Fire Protection	No	Fire Protection	Consistent with GALL Report.
Steel piping, piping components, and piping elements exposed to fuel oil (3.3.1-64)	Loss of material due to general, pitting, and crevice corrosion	Fire Protection and Fuel Oil Chemistry	No	Not applicable	Not applicable. See SER Section 3.3.2.1.1.
Reinforced concrete structural fire barriers-walls, ceilings and floors exposed to air-indoor uncontrolled (3.3.1-65)	Concrete cracking and spalling due to aggressive chemical attack, and reaction with aggregates	Fire Protection and Structures Monitoring Program	No	Fire Protection and Structures Monitoring Program	Consistent with GALL Report.

Aging Management Review Results

Component Group (GALL Report Item No.)	Aging Effect/ Mechanism	AMP in GALL Report	Further Evaluation in GALL Report	AMP in LRA, Supplements, or Amendments	Staff Evaluation
Reinforced concrete structural fire barriers-walls, ceilings and floors exposed to air-outdoor (3.3.1-66)	Concrete cracking and spalling due to freeze thaw, aggressive chemical attack, and reaction with aggregates	Fire Protection and Structures Monitoring Program	No	Fire Protection and Structures Monitoring Program	Consistent with GALL Report.
Reinforced concrete structural fire barriers-walls, ceilings and floors exposed to air-outdoor or air-indoor uncontrolled (3.3.1-67)	Loss of material due to corrosion of embedded steel	Fire Protection and Structures Monitoring Program	No	Fire Protection and Structures Monitoring Program	Consistent with GALL Report.
Steel piping, piping components, and piping elements exposed to raw water (3.3.1-68)	Loss of material due to general, pitting, crevice, and MIC and fouling	Fire Water System	No	Fire Water System	Consistent with GALL Report.
Stainless steel piping, piping components, and piping elements exposed to raw water (3.3.1-69)	Loss of material due to pitting and crevice corrosion, and fouling	Fire Water System	No	Fire Water System	Consistent with GALL Report.
Copper alloy piping, piping components, and piping elements exposed to raw water (3.3.1-70)	Loss of material due to pitting, crevice, and MIC and fouling	Fire Water System	No	Fire Water System	Consistent with GALL Report.
Steel piping, piping components, and piping elements exposed to moist air or condensation (internal) (3.3.1-71)	Loss of material due to general, pitting, and crevice corrosion	Inspection of Internal Surfaces in Miscellaneous Piping and Ducting Components	No	Inspection of Internal Surfaces in Miscellaneous Piping and Ducting Components	Consistent with GALL Report.
Steel HVAC ducting and components internal surfaces exposed to condensation (internal) (3.3.1-72)	Loss of material due to general, pitting, crevice, and (for drip pans and drain lines) MIC	Inspection of Internal Surfaces in Miscellaneous Piping and Ducting Components	No	Inspection of Internal Surfaces in Miscellaneous Piping and Ducting Components	Consistent with GALL Report.
Steel crane structural girders in load handling system exposed to air-indoor uncontrolled (external) (3.3.1-73)	Loss of material due to general corrosion	Inspection of Overhead Heavy Load and Light Load (Related to Refueling) Handling Systems	No	Inspection of Overhead Heavy Load and Light Load (Related to Refueling) Handling Systems	Consistent with GALL Report.

Aging Management Review Results

Component Group (GALL Report Item No.)	Aging Effect/ Mechanism	AMP in GALL Report	Further Evaluation in GALL Report	AMP in LRA, Supplements, or Amendments	Staff Evaluation
Steel cranes-rails exposed to air-indoor uncontrolled (external) (3.3.1-74)	Loss of material due to Wear	Inspection of Overhead Heavy Load and Light Load (Related to Refueling) Handling Systems	No	Inspection of Overhead Heavy Load and Light Load (Related to Refueling) Handling Systems	Consistent with GALL Report.
Elastomer seals and components exposed to raw water (3.3.1-75)	Hardening and loss of strength due to elastomer degradation; loss of material due to erosion	Open-Cycle Cooling Water System	No	Not applicable	Not applicable to PVNGS. See SER Section 3.3.2.1.1.
Steel piping, piping components, and piping elements (without lining/coating or with degraded lining/coating) exposed to raw water (3.3.1-76)	Loss of material due to general, pitting, crevice, and MIC, fouling, and lining/coating degradation	Open-Cycle Cooling Water System	No	Open-Cycle Cooling Water System and Inspection Of Internal Surfaces in Miscellaneous Piping And Ducting Components	Consistent with GALL Report or alternative program used. See SER Section 3.3.2.1.4.
Steel heat exchanger components exposed to raw water (3.3.1-77)	Loss of material due to general, pitting, crevice, galvanic, and MIC and fouling	Open-Cycle Cooling Water System	No	Open-Cycle Cooling Water System	Consistent with GALL Report.
Stainless steel, nickel alloy, and copper alloy piping, piping components, and piping elements exposed to raw water (3.3.1-78)	Loss of material due to pitting and crevice corrosion	Open-Cycle Cooling Water System	No	Open-Cycle Cooling Water System	Consistent with GALL Report.
Stainless steel piping, piping components, and piping elements exposed to raw water (3.3.1-79)	Loss of material due to pitting and crevice corrosion, and fouling	Open-Cycle Cooling Water System	No	Inspection Of Internal Surfaces In Miscellaneous Piping And Ducting Components	Alternative program used. See SER Section 3.3.2.1.5.
Stainless steel and copper alloy piping, piping components, and piping elements exposed to raw water (3.3.1-80)	Loss of material due to pitting, crevice, and microbiologically influenced corrosion	Open-Cycle Cooling Water System	No	Open-Cycle Cooling Water System	Consistent with GALL Report.

Aging Management Review Results

Component Group (GALL Report Item No.)	Aging Effect/ Mechanism	AMP in GALL Report	Further Evaluation in GALL Report	AMP in LRA, Supplements, or Amendments	Staff Evaluation
Copper alloy piping, piping components, and piping elements, exposed to raw water (3.3.1-81)	Loss of material due to pitting, crevice, and MIC and fouling	Open-Cycle Cooling Water System	No	Inspection Of Internal Surfaces In Miscellaneous Piping And Ducting Components	Alternative program used. See SER Section 3.3.2.1.6.
Copper alloy heat exchanger components exposed to raw water (3.3.1-82)	Loss of material due to pitting, crevice, galvanic, and MIC and fouling	Open-Cycle Cooling Water System	No	Open-Cycle Cooling Water System	Consistent with GALL Report.
Stainless steel and copper alloy heat exchanger tubes exposed to raw water (3.3.1-83)	Reduction of heat transfer due to fouling	Open-Cycle Cooling Water System	No	Open-Cycle Cooling Water System	Consistent with GALL Report.
Copper alloy > 15% Zn piping, piping components, piping elements, and heat exchanger components exposed to raw water, treated water, or closed cycle cooling water (3.3.1-84)	Loss of material due to selective leaching	Selective Leaching of Materials	No	Selective Leaching of Materials	Consistent with GALL Report.
Gray cast iron piping, piping components, and piping elements exposed to soil, raw water, treated water, or closed-cycle cooling water (3.3.1-85)	Loss of material due to selective leaching	Selective Leaching of Materials	No	Selective Leaching of Materials	Consistent with GALL Report.
Structural steel (new fuel storage rack assembly) exposed to air - indoor uncontrolled (external) (3.3.1-86)	Loss of material due to general, pitting, and crevice corrosion	Structures Monitoring Program	No	Not applicable	Not applicable to PVNGS. See SER Section 3.3.2.1.1.
Boraflex spent fuel storage racks neutron-absorbing sheets exposed to treated borated water (3.3.1-87)	Reduction of neutron-absorbing capacity due to boraflex degradation	Boraflex Monitoring	No	Not applicable	Not applicable to PVNGS. See SER Section 3.3.2.1.1.
Aluminum and copper alloy > 15% Zn piping, piping components, and piping elements exposed to air with borated water leakage (3.3.1-88)	Loss of material due to boric acid corrosion	Boric Acid Corrosion	No	Not applicable	Not applicable to PVNGS. See SER Section 3.3.2.1.1.

Aging Management Review Results

Component Group (GALL Report Item No.)	Aging Effect/ Mechanism	AMP in GALL Report	Further Evaluation in GALL Report	AMP in LRA, Supplements, or Amendments	Staff Evaluation
Steel bolting and external surfaces exposed to air with borated water leakage (3.3.1-89)	Loss of material due to boric acid corrosion	Boric Acid Corrosion	No	Boric Acid Corrosion	Consistent with GALL Report.
Stainless steel and steel with stainless steel cladding piping, piping components, piping elements, tanks, and fuel storage racks exposed to treated borated water > 60°C (> 140°F) (3.3.1-90)	Cracking due to SCC	Water Chemistry	No	Water Chemistry	Consistent with GALL Report.
Stainless steel and steel with stainless steel cladding piping, piping components, and piping elements exposed to treated borated water (3.3.1-91)	Loss of material due to pitting and crevice corrosion	Water Chemistry	No	Water Chemistry	Consistent with GALL Report.
Galvanized steel piping, piping components, and piping elements exposed to air-indoor uncontrolled (3.3.1-92)	None	None	NA	None	Consistent with GALL Report.
Glass piping elements exposed to air, air-indoor uncontrolled (external), fuel oil, lubricating oil, raw water, treated water, and treated borated water (3.3.1-93)	None	None	NA	None	Consistent with GALL Report.
Stainless steel and nickel-alloy piping, piping components, and piping elements exposed to air-indoor uncontrolled (external) (3.3.1-94)	None	None	NA	None	Consistent with GALL Report.
Steel and aluminum piping, piping components, and piping elements exposed to air-indoor controlled (external) (3.3.1-95)	None	None	NA	None	Consistent with GALL Report.
Steel and stainless steel piping, piping components, and piping elements in concrete (3.3.1-96)	None	None	NA	None	Consistent with GALL Report.

Aging Management Review Results

Component Group (GALL Report Item No.)	Aging Effect/ Mechanism	AMP in GALL Report	Further Evaluation in GALL Report	AMP in LRA, Supplements, or Amendments	Staff Evaluation
Steel, stainless steel, aluminum, and copper alloy piping, piping components, and piping elements exposed to gas (3.3.1-97)	None	None	NA	None	Consistent with GALL Report.
Steel, stainless steel, and copper alloy piping, piping components, and piping elements exposed to dried air (3.3.1-98)	None	None	NA	None	Consistent with GALL Report.
Stainless steel and copper alloy < 15% Zn piping, piping components, and piping elements exposed to air with borated water leakage (3.3.1-99)	None	None	NA	None	Consistent with GALL Report.

The staff's review of the auxiliary systems component groups followed one of three categories. One category, documented in SER Section 3.3.2.1, reviewed AMR results for components that the applicant indicated are consistent with the GALL Report and require no further evaluation. Another category, documented in SER Section 3.3.2.2, reviewed AMR results for components that the applicant indicated are consistent with the GALL Report and for which further evaluation is recommended. A third category, documented in SER Section 3.3.2.3, reviewed AMR results for components that the applicant indicated are not consistent with, or not addressed in, the GALL Report. SER Section 3.0.3 documents the staff's review of AMPs credited to manage or monitor aging effects of the auxiliary systems components.

3.3.2.1 Aging Management Review Results Consistent with the Generic Aging Lessons Learned Report

LRA Section 3.3.2.1 identifies the materials, environments, AERMs, and the following programs that manage aging effects for the auxiliary systems components:

- ASME Section XI ISI, Subsections IWB, IWC, and IWD
- Bolting Integrity
- Boric Acid Corrosion
- Buried Piping and Tanks Inspection
- Closed-Cycle Cooling Water System
- External Surfaces Monitoring
- Fire Water System
- FAC
- Fuel Oil Chemistry
- Inspection of Internal Surfaces in Miscellaneous Piping and Ducting Components

3-221

Aging Management Review Results

- Inspection of Overhead Heavy Load and Light Load (Related to Refueling) Handling System
- Lubricating Oil Analysis
- Nickel-Alloy Aging Management
- One-Time Inspection
- One-Time Inspection of ASME Code Class I Small-Bore Piping
- Open-Cycle Cooling Water System
- RCS Supplement
- Selective Leaching Of Materials
- Water Chemistry

LRA Tables 3.3.2-1 through 3.3.2-30 summarize the AMRs for the auxiliary system components and indicate AMRs claimed to be consistent with the GALL Report.

The staff audited and reviewed the information in the LRA. The staff did not repeat its review of the matters described in the GALL Report; however, the staff did verify that the material presented in the LRA was applicable and that the applicant identified the appropriate GALL Report AMRs. The staff's evaluation follows.

3.3.2.1.1   Aging Management Review Results Identified as Not Applicable

LRA Table 3.3.1, items 3.3.1.36 through 3.3.1.39 and 3.3.1.49 discuss the applicant's determination that these items are applicable only to BWRs. The staff verified that these items do not apply because the units are a PWR design. Based on this determination, the staff finds that the applicant has provided an acceptable basis for concluding the AMR items 3.3.1.36 through 3.3.1.39 and 3.3.1.49 are not applicable.

LRA Table 3.3.1, item 3.3.1.40, addresses steel tanks in the diesel fuel oil system exposed to outdoor air (external). The applicant stated that this item is not applicable because there are no steel tanks in the EDG fuel oil storage and transfer system that are exposed to the outdoor air (external) environment. The staff reviewed LRA Sections 2.3.3 and 3.3 and confirmed that the applicant's LRA does not have any AMR results for the diesel generator fuel oil storage and transfer system that includes steel tanks exposed to outdoor air (external). The staff also reviewed the applicant's information in the UFSAR and confirmed that no in-scope steel tanks exposed to outdoor air (external) are present in the diesel generator fuel oil storage and transfer system and, therefore, finds the applicant's determination acceptable.

LRA Table 3.3.1, item 3.3.1.41, addresses high-strength steel closure bolting exposed to air with steam or water leakage in the auxiliary systems. The GALL Report recommends use of AMP XI.M18, "Bolting Integrity," to manage cracking due to cyclic loading or SCC for this component group. The applicant stated that this item is not applicable because high strength steel closure bolting is not used in the auxiliary systems. The staff reviewed LRA Sections 2.3.3 and 3.3 and confirmed that the applicant's LRA does not have any AMR results for the auxiliary systems that include high strength steel closure bolting exposed to air with steam or water leakage. During its review of the UFSAR, operating experience, and applicant interviews associated with the Bolting Integrity Program, the staff did not identify the use of high strength steel closure bolting in the auxiliary systems within the scope of license renewal. Based on its review of the LRA and the applicant's Bolting Integrity Program, the staff confirmed that there is no high-strength steel closure bolting exposed to air with steam or water leakage in the auxiliary systems and, therefore, finds the applicant's determination applicable.

Aging Management Review Results

LRA Table 3.3.1, items 3.3.1.42 and 3.3.1.44, address steel closure bolting exposed to air with steam or water leakage (item 42) or condensation (item 44) in the auxiliary systems. The GALL Report recommends use of AMP XI.M18, "Bolting Integrity," to manage loss of material due to general corrosion for this component group. The applicant stated that the items are not applicable because it has no in-scope steel closure bolting exposed to air with steam or water leakage or condensation in the auxiliary systems. By letter dated February 19, 2010, the staff issued RAI 3.3-1, part (b), requesting that the applicant justify its claim that steel closure bolting in its auxiliary systems are not exposed to an environment of air with steam or water leakage or condensation. In its response, dated March 24, 2010, the applicant stated that air with water leakage would be an event-driven environment, not considered normal for the auxiliary systems. The applicant also stated that for steel closure bolting, the aging effect and mechanism in LRA Table 3.3.1, items 3.3.1.42 and 3.3.1.44 are both included in the aging effect and mechanisms for LRA Table 3.3.1, item 3.3.1.43. The Bolting Integrity Program is credited in all three lines to manage the aging effect of loss of material due to general corrosion. The applicant further stated that it evaluated steel closure bolting in its auxiliary systems as part of LRA Table 3.3.1, item 3.3.1.43, rather than in item 3.3.1.42. The staff confirmed that for steel closure bolting the aging effect stated in LRA Table 3.3.1, item 3.3.1.42, is included in item 3.3.1.43 and that the Bolting Integrity Program is credited to manage the aging effect.

The staff finds the applicant's response to RAI 3.3.1, part (b) acceptable because it clarifies that components that could have been evaluated under items 3.3.1.42 and 3.3.1.44 are appropriately included in the evaluations under item 3.3.1.43. The staff also finds it acceptable for the applicant to evaluate steel closure bolting subject to general corrosion under item 3.3.1.43 because the aging effect and mechanism identified in items 3.3.1.42 and 3.3.1.44 are included in the aging effect and mechanisms identified in item 3.3.1.43. Further, the AMP recommended by both items is the same; therefore, the staff further finds the applicant's determination, that items 3.3.1.42 and 3.3.1.44 are not applicable, to be acceptable.

LRA Table 3.3.1, item 3.3.1.62 addresses aluminum piping, piping components, and piping elements exposed to raw water. The GALL Report recommends the use of AMP XI.M27, "Fire Protection Program," to manage loss of material due to pitting and crevice corrosion for this component group. The applicant stated that this item is not applicable because it has no in-scope aluminum components exposed to raw water in the fire protection system. The staff reviewed LRA Sections 2.3.3 and 3.3 and confirmed that the applicant's LRA does not have any AMR results for the fire protection system that include aluminum piping, piping components, and piping elements exposed to raw water. The staff also reviewed the applicant's UFSAR and confirmed that there are no in-scope aluminum piping, piping components, and piping elements exposed to raw water in the fire protection system, and therefore, finds the applicant's determination acceptable.

LRA Table 3.3.1, item 3.3.1.64 addresses steel piping, piping components, and piping elements exposed to fuel oil. The GALL Report recommends use of AMPs XI.M27, "Fire Protection Program," and AMP XI.M30, "Fuel Oil Chemistry Program," to manage loss of material due to general, pitting, and crevice corrosion for this component group. The applicant stated that this item is not applicable because it used other available applicable GALL Report items. The staff reviewed LRA Table 3.3.2-19, "Fire Protection System," and confirmed that the applicant addressed steel piping, piping components, and piping elements exposed to fuel oil and credited the Fuel Oil Chemistry and One-Time Inspection Programs to manage loss of material due to general, pitting, and crevice corrosion. This is consistent with GALL Report item VII.H2-24 and is addressed by LRA Table 3.3.1, item 3.3.1.20 and LRA Subsection 3.3.2.2.9.1. Based on its review of the LRA, the staff concludes that the applicant has properly addressed steel piping, piping components, and piping elements exposed to fuel oil using other available GALL Report items, and therefore, finds the applicant's determination acceptable.

Aging Management Review Results

LRA Table 3.3.1, item 3.3.1.75 addresses elastomer seals and components exposed to raw water. The GALL Report recommends use of AMP XI.M20, "Open-Cycle Cooling Water System," to manage hardening and loss of strength due to elastomer degradation and loss of material due to erosion for this component group. The applicant stated that this item is not applicable because it has no in-scope elastomer components exposed to raw water in the open-cycle cooling water systems. The staff reviewed LRA Sections 2.3.3 and 3.3 and identified several items that could be classified as elastomer seals and components exposed to raw water: in Table 3.3.2-7, it identified polyvinyl chloride (PVC) components; in Table 3.3.2-22, it identified polyethylene items; and in Tables 3.3.2–3.3.22 and 3.3.2-30, it identified carbon steel with elastomer linings.

The staff held a teleconference call with the applicant on July 8, 2010, to discuss its concerns regarding how the components discussed above, or any other elastomer components exposed to raw water in auxiliary systems, are managed for aging during the period of extended operation. The applicant indicated it would provide information demonstrating that the PVC and polyethylene components are not susceptible to loss of material due to erosion, and that the AMPs proposed to manage aging of the elastomer lined carbon steel piping are appropriate. The applicant stated it would submit a formal response to RAI 3.3.1-1. This was previously identified as Confirmatory Item 3.3.2.1.1-1.

In its response dated July 30, 2010, the applicant stated that PVC and polyethylene are thermoplastics that are rigid and have good resistance to abrasion and erosion, and loss of material due to erosion only occurs if the fluid contains particulates and fluid velocities are high. The applicant also stated that the PVC components are in the essential spray pond and are not subject to high fluid velocities. Further, the applicant stated that the polyethylene components are in the well water portion of the domestic water system, that this system contains minimal particulates, and the components are not subject to high flow velocities. The staff finds this acceptable because the applicant provided sufficient information to show that the subject components will not have a loss of material due to erosion requiring management during the period of extended operation.

For the carbon steel pipe with elastomer linings exposed to raw water, the applicant stated that loss of material and potential consequences are managed by the Fire Water System Program, which uses internal visual inspections performed in accordance with the Inspection of Internal Surfaces in Miscellaneous Piping and Ducting Components Program. The applicant further stated that the potential consequences are also addressed by the Fire Water System Program through periodic flow testing of the fire water loops and the fire suppression water system. The staff finds the applicant's response acceptable because the inspections and flow tests are capable of identifying the loss of material.

For the carbon steel pipe with elastomer linings exposed to oily and non-radioactive waste, the applicant stated that the loss of material is managed by internal visual inspections in accordance with the Inspection of Internal Surfaces in Miscellaneous Piping and Ducting Components Program. The staff finds the applicant's proposal to manage aging using the cited AMP acceptable because the surveillance techniques can identify and adequately manage this aging effect. Based on the above information, the concern described in RAI 3.3.1-1 is resolved, and Confirmatory Item 3.3.2.1.1-1 is closed.

LRA Table 3.3.1, item 3.3.1.86 addresses structural steel new fuel storage rack assemblies exposed to indoor uncontrolled air. The applicant stated that this item is not applicable because its new fuel storage assemblies are made of stainless steel. The staff reviewed LRA Sections 2.3.3 and 3.3 and confirmed that the applicant's LRA does not have any AMR results for the auxiliary systems that include steel fuel storage racks exposed to indoor uncontrolled air. The staff also reviewed the applicant's UFSAR and confirmed that no in-scope steel fuel storage assemblies are present in the auxiliary systems and therefore, finds the applicant's determination acceptable.

LRA Table 3.3.1, item 3.3.1.87 addresses boraflex spent fuel storage racks neutron absorbing sheets exposed to treated borated water. The applicant stated that this item is not applicable because it does not have any boraflex spent fuel storage racks exposed to treated borated water. The staff reviewed LRA Sections 2.3.3 and 3.3 and confirmed that the applicant's LRA does not have any AMR results for the auxiliary systems that include boraflex spent fuel storage racks neutron absorbing sheets exposed to treated borated water. The staff also reviewed the applicant's UFSAR and confirmed that no boraflex spent fuel storage racks neutron absorbing sheets are present in the auxiliary systems and therefore, finds the applicant's determination acceptable.

LRA Table 3.3.1, item 3.3.1.88 addresses aluminum and copper alloy with greater than 15-percent zinc piping, piping components, and piping elements exposed to air with borated water leakage. The GALL Report, items VII.A3-4, VII.E1-10, VII.I-12, and VIII.E-39, recommend use of AMP XI.M10, "Boric Acid Corrosion," to manage loss of material due to boric acid corrosion for this component group. The applicant stated that this item is not applicable because there are no in-scope aluminum and copper alloy greater than 15-percent zinc piping, piping components, and piping elements in the auxiliary systems exposed to air with borated water leakage. The staff reviewed LRA Sections 2.3.3 and 3.3 and confirmed that the applicant's LRA does not have any AMR results in-scope aluminum and copper alloy greater than 15-percent zinc piping, piping components, and piping elements in the auxiliary systems exposed to air with borated water leakage. The staff also reviewed the applicant's information in the UFSAR associated with Table 3.3.1, item 3.3.1.88 and confirmed that no in-scope aluminum and copper alloy greater than 15-percent zinc piping, piping components, and piping elements are present in the auxiliary systems. Therefore, the staff finds the applicant's determination acceptable.

3.3.2.1.2 Loss of Material Due to General and Pitting Corrosion of Steel and Stainless Steel Internal Surfaces Exposed to Condensation

LRA Table 3.3.1, items 3.3.1.53 and 3.3.1.54, address stainless steel and steel piping, piping components, and piping elements exposed to internal condensation. The GALL Report recommends use of AMP XI.M24, "Compressed Air Monitoring," to manage loss of material for this component group. The applicant stated that these items are not applicable because no applicable components were in the scope of license renewal. The staff conducted audit interviews and observed the plant compressed air system as part of a walkdown of plant environments and materials during the AMP audit. Based on this information, the staff determined that there were stainless steel and steel piping components exposed to internal condensation.

By letter dated December 29, 2009, the NRC issued RAI B2.1.20-3 asking that the applicant clarify how it will manage the aging effects on piping and valves within the compressed air system that are exposed to condensation for loss of material and other potential aging effects.

In its response, dated February 19, 2010, the applicant stated that several components within the compressed air system were incorrectly identified as having an environment of dry gas, further noting that the components that supply nitrogen to the spent fuel pool gate seals were correctly identified as having a dry gas environment. The applicant also stated that it revised LRA Tables 3.3.1 and 3.3.2-9 to reflect the addition of wetted air to the environments to which carbon steel and stainless steel piping and components in the compressed air system are exposed. The applicant also added plant specific footnote 1, as follows:

> AMP XI.M24, "Compressed Air Monitoring" applies to monitoring the piping and components associated with the air compressors and dryers. The air

Aging Management Review Results

compressor, dryer piping, and components are not in-scope for Palo Verde. In-scope piping and components for Palo Verde are associated with containment penetrations and nitrogen gas piping/components for backup to the spent fuel pool gate seals. Therefore, XI.M24 is not considered appropriate to Palo Verde and alternate AMPs are specified for the in-scope piping and components.

The applicant further stated that it credits the Inspection of Internal Surfaces in Miscellaneous Piping and Ducting Components Program for managing loss of material aging effects.

Based on its review, the staff finds the applicant's response acceptable for the following reasons:

- GALL Report AMP XI.M24, "Compressed Air Monitoring," is based on the applicant's response to NRC GL 88-14. Based on a review of the applicant's response to this GL and LRA Section 2.3.3.9, there are no in scope stainless steel and steel piping, piping components, and piping elements exposed to internal condensation that are subject to the requirements of the GL and this AMR item.

- The staff notes that only the portions of the compressed air system which provide containment isolation for the instrument air, service, and breathing air containment penetration piping are in the scope of license renewal based on the criteria of 10 CFR 54.4(a)(1). The staff also notes that only the nonsafety-related portions of the instrument air subsystem in the auxiliary and containment buildings that attach to safety-related containment building penetration piping, and the safety-related backup nitrogen supply tubing to the spent fuel pool gate seals are within the scope of license renewal as nonsafety-related components affecting safety-related components, based on the criterion of 10 CFR 54.4(a)(2). With the exception of the atmospheric dump valves, all of the air-operated valves that support fire protection, EQ, and SBO requirements (10 CFR 54.4(a)(3) criterion) fail to a safe position upon loss of instrument air. The motive force supply for the atmospheric dump valves is dry nitrogen gas; thus, these valves are not in the scope of this AMR item.

- In a conference call between the staff and the applicant on June 9, 2010, the applicant stated that the solenoid valves that vent air from all of the air-operated valves that must fail to the safe position are full port valves. Therefore, the staff does not have a concern related to potential blockage if corrosion products should travel through the system to the solenoid valves.

- Given the safety function of the in-scope components (i.e., pressure retaining boundary), the Inspection of Internal Surfaces in Miscellaneous Piping and Ducting Components Program is adequate to manage loss of material due to general and pitting corrosion. This program will perform visual inspections to detect aging effects that could result in a loss of component intended function during periodic maintenance.

The staff's concern described in RAI B2.1.20-3 is resolved.

3.3.2.1.3 Loss of Material Due to General Corrosion of Steel External Surfaces Exposed to Air

LRA Table 3.3.1, item 3.3.1.58 addresses carbon steel and gray cast iron valves and piping exposed to weather and plant indoor air which are being managed for loss of material. The LRA credits the Internal Surfaces in Miscellaneous Piping and Ducting Components Program to manage the loss of material aging effect. The GALL Report recommends AMP XI.M36, "External Surfaces Monitoring Program," to ensure that the applicant adequately manages these aging effects. The associated AMR item cites generic Note E, indicating that the LRA AMR is

Aging Management Review Results

consistent with the GALL Report item for material, environment and aging effect, but a different AMP is credited.

For those items associated with generic Note E, the GALL Report recommends the External Surfaces Monitoring Program, which recommends using visual inspections to assess the condition of SSCs for loss of material when managing the aging effects of these items. In its review of components associated with item 3.3.1.58, for which the applicant cited generic Note E, the staff noted that the Internal Surfaces in Miscellaneous Piping and Ducting Components Program proposes to manage the aging of piping and valves made of carbon steel and gray cast iron through the use of visual inspection techniques. The staff also noted, under item 3.3.1.58, that the applicant specifically plans to use the Internal Surfaces in Miscellaneous Piping and Ducting Components Program to manage the aging effects of the internal surfaces of the fire protection components that are normally vented to the air. The applicant further stated that there is no difference between the environmental conditions to which the surfaces for these components are exposed, thus eliminating the distinction between internal and external environments.

SER Section 3.0.3.2.15 documents the staff's evaluation of the applicant's Internal Surfaces in Miscellaneous Piping and Ducting Components Program. The staff evaluated this program to determine if it is adequate to inspect the internal surfaces of piping, piping components, and elements (including the internal surfaces of fire protection components) for loss of material. The staff also reviewed the recommendations set in the program elements of both GALL Report programs, External Surfaces Monitoring and Inspection of Internal Surfaces in Miscellaneous Piping and Ducting Components. The staff noted that, while both programs focus on visual inspection of SSC for loss of material, in the "detection of aging effects" program element, the GALL Report External Surfaces Monitoring AMP is time-dependant with inspections performed at least once per refueling cycle. GALL Report Internal Surfaces in Miscellaneous Piping and Ducting Components AMP, on the other hand, is an opportunistic AMP where inspections are performed when systems and components are available for inspection.

The staff noted that the applicant stated that within 10 years of entering the period of extended operation, it will review all systems within the scope of the Internal Surfaces in Miscellaneous Piping and Ducting Components Program to determine the number of inspection opportunities afforded for systems or components within the scope of the program. When necessary, it will follow with additional inspections to provide reasonable assurance that the intended functions are maintained. For the fire protection system piping, the applicant stated that it will augment the visual inspection, when necessary, with volumetric inspections to monitor loss of material and pipe thinning. The staff also noted that in NUREG-1833, "Technical Bases for Revision to the License Renewal Guidance Documents," external surfaces of steel components exposed to air, moisture, and humidity are vulnerable to general corrosion, and the selected AMP is the External Surfaces Monitoring Program. NUREG-1833 also notes that for steel piping, components and elements exposed to outdoor air, the Inspection of Internal Surfaces in Miscellaneous Piping and Ducting Components AMP provides an acceptable means to manage aging for these components. In its review of components associated with item 3.3.1.58, the staff finds the applicant's proposal to manage aging using the Internal Surfaces in Miscellaneous Piping and Ducting Components Program acceptable because: (1) it will use similar visual inspections as those recommended by the External Surfaces Monitoring AMP and where necessary will augment these with volumetric inspections; (2) the environment applicable to the two programs are similar per the GALL Report; and (3) periodic monitoring of SSCs will be performed to assure that their intended functions are maintained.

The staff concludes that the applicant has demonstrated that it will adequately manage the effects of aging for this component type, in the specified environment, so that its intended

Aging Management Review Results

function(s) will be maintained consistent with the CLB during the period of extended operation, as required by 10 CFR 54.21(a)(3).

3.3.2.1.4 Loss of Material Due to General, Pitting, Crevice, and Microbiologically-Influenced Corrosion, Fouling, and Lining and Coating Degradation

LRA Table 3.3.1, item 76 addresses steel piping, piping components, and piping elements (without lining or coating or with degraded lining or coating) exposed to raw water, which are managed for loss of material due to general, pitting, crevice, and microbiologically-influenced corrosion, fouling, and lining or coating degradation. The LRA credits the Inspection of Internal Surfaces in Miscellaneous Piping and Ducting Components Program to manage the aging effect for a subsection of these components. The GALL Report recommends AMP XI.M20, "Open-Cycle Cooling Water System," to ensure that these aging effects are adequately managed. The associated AMR item cites generic Note E, indicating that the LRA AMR is consistent with GALL Report item for material, environment, and aging effect, but it credits a different AMP.

For those items associated with generic Note E, the GALL Report recommends the Open-Cycle Cooling Water System AMP, which recommends using surveillance and control techniques to manage aging of these items. In its review of components associated with item 76, for which the applicant cited generic Note E, the staff noted that the Inspection of Internal Surfaces in Miscellaneous Piping and Ducting Components Program proposes to manage the aging of steel piping, piping components, and piping elements (without lining or coating or with degraded lining or coating) through the use of visual inspections of plant components for evidence of degradation (see staff evaluation of this AMP in SER Section 3.0.3.2.15). The staff also noted that the applicant identified components such as floor drains and building sumps that may be exposed to a variety of types of treated and untreated water as well as to raw water for the determination of aging effects. The staff finally noted that this water is not monitored by a chemistry program and, therefore, the Open-Cycle Cooling Water Program could not be used to properly manage the aging effect in this environment. The staff finds the applicant's use of the Inspection of Internal Surfaces in Miscellaneous Piping and Ducting Components Program is acceptable because its surveillance techniques can identify and adequately manage this aging effect.

The staff concludes that the applicant has demonstrated that it will adequately manage the effects of aging of these components so that their intended functions will be maintained consistent with the CLB during the period of extended operation as required by 10 CFR 54.21(a)(3).

3.3.2.1.5 Loss of Material Due to Pitting and Crevice Corrosion and Fouling

LRA Table 3.3.1, item 3.3.1.79 addresses stainless steel piping, piping components, and piping elements exposed to raw water which are managed for loss of material due to pitting and crevice corrosion and fouling. The LRA credits the Inspection of Internal Surfaces in Miscellaneous Piping and Ducting Components Program to manage the aging effect. The GALL Report recommends AMP XI.M20, "Open-Cycle Cooling Water System," to ensure that the applicant adequately manages these aging effects. The associated AMR items cite generic Note E, indicating that the LRA AMR is consistent with the GALL Report item for material, environment, and aging effect, but a different AMP is credited.

For those items associated with generic Note E, the GALL Report recommends the Open-Cycle Cooling Water System AMP, which recommends using surveillance and control techniques to manage the aging of these components. In its review of components associated with item 79, for which the applicant cited generic Note E, the staff noted that the Inspection of Internal Surfaces in Miscellaneous Piping and Ducting Components Program proposes to manage the aging of stainless steel piping, piping components, and piping elements through the use of

Aging Management Review Results

visual inspections of plant components for evidence of degradation (see staff evaluation of this program in SER Section 3.0.3.2.15). The staff also noted that the applicant identified components such as floor drains and building sumps that may be exposed to a variety of types of treated and untreated water as well as to raw water for the determination of aging effects. The applicant stated that these environments may contain contaminants, including oil and boric acid, as well as originally treated water that is not monitored by a chemistry program. The staff further noted there are no water chemistry controls in these environments (i.e., drains and sumps) and, therefore, the Open-Cycle Cooling Water System Program could not properly manage this system. The staff finds that the Inspection of Internal Surfaces in Miscellaneous Piping and Ducting Components AMP is appropriate because its surveillance techniques can identify and will adequately manage this aging effect.

The staff concludes that the applicant has demonstrated that it will adequately manage the effects of aging of these components so that their intended functions will be maintained consistent with the CLB during the period of extended operation, as required by 10 CFR 54.21(a)(3).

3.3.2.1.6 Loss of Material Due to Pitting, Crevice, and Microbiologically-Influenced Corrosion and Fouling

LRA Table 3.3.1, item 81 addresses copper alloy piping, piping components, and piping elements exposed to raw water, which are managed for loss of material due to pitting, crevice, and microbiologically-influenced corrosion and fouling. The LRA credits the Inspection of Internal Surfaces in Miscellaneous Piping and Ducting Components Program to manage the aging effect. The GALL Report recommends AMP XI.M20, "Open-Cycle Cooling Water System," to ensure that the applicant adequately manages these aging effects. The associated AMR item cites generic Note E, indicating that the LRA AMR is consistent with the GALL Report item for material, environment, and aging effect, but a different AMP is credited.

For those items associated with generic Note E, the GALL Report recommends the Open-Cycle Cooling Water System AMP, which recommends using surveillance and control techniques to manage the aging of these items. In its review of components associated with item 3.3.1.81, for which the applicant cited generic Note E, the staff noted that the Inspection of Internal Surfaces in Miscellaneous Piping and Ducting Components Program proposes to manage the aging of copper alloy piping, piping components, and piping elements through the use of visual inspections of plant components for evidence of degradation (see staff evaluation of this program in SER Section 3.0.3.2.15). The staff also noted that the applicant identified components such as floor drains and building sumps that may be exposed to a variety of types of treated and untreated water as well as to raw water for the determination of aging effects. The staff further noted that there are no water chemistry controls in these environments (i.e., drains and sumps) and therefore, the Open-Cycle Cooling Water System Program could not properly manage this aging effect. The staff finds the applicant's use of the Inspection of Internal Surfaces in Miscellaneous Piping and Ducting Components Program acceptable because its surveillance techniques can identify and adequately manage the effects of aging.

The staff concludes that the applicant has demonstrated that it will adequately manage the effects of aging of these components so that their intended functions will be maintained consistent with the CLB during the period of extended operation, as required by 10 CFR 54.21(a)(3).

The staff evaluated the applicant's claim of consistency with the GALL Report. The staff also reviewed information pertaining to the applicant's consideration of recent operating experience and proposals for managing aging effects. On the basis of its review, the staff concludes that the AMR results, which the applicant claimed to be consistent with the GALL Report, are consistent. Therefore, the staff concludes that the applicant has demonstrated that the effects

Aging Management Review Results

of aging for these components will be adequately managed so that their intended function(s) will be maintained consistent with the CLB during the period of extended operation, as required by 10 CFR 54.21(a)(3).

3.3.2.2 Aging Management Review Results Consistent with the Generic Aging Lessons Learned Report for Which Further Evaluation Is Recommended

In LRA Section 3.3.2.2, the applicant further evaluates aging management, as recommended by the GALL Report, for the auxiliary system components and provides information concerning how it will manage the following aging effects:

- cumulative fatigue damage
- reduction of heat transfer due to fouling
- cracking due to SCC
- cracking due to SCC and cyclic loading
- hardening and loss of strength due to elastomer degradation
- reduction of neutron-absorbing capacity and loss of material due to general corrosion
- loss of material due to general, pitting, and crevice corrosion
- loss of material due to general, pitting, crevice, and microbiologically-influenced corrosion
- loss of material due to general, pitting, crevice, microbiologically-influenced corrosion and fouling
- loss of material due to pitting and crevice corrosion
- loss of material due to pitting, crevice, and galvanic corrosion
- loss of material due to pitting, crevice, and microbiologically-influenced corrosion
- loss of material due to wear
- loss of material due to cladding breach
- QA for aging management of nonsafety-related components

For component groups evaluated in the GALL Report, for which the applicant claimed consistency with the report and for which the report recommends further evaluation, the staff reviewed the applicant's evaluation to determine if it adequately addressed the issues further evaluated. In addition, the staff reviewed the applicant's further evaluations against the criteria contained in SRP-LR Section 3.3.2.2. The staff's review of the applicant's further evaluation follows.

3.3.2.2.1 Cumulative Fatigue Damage

In LRA Section 3.3.2.2.1, the applicant stated that evaluation of cumulative fatigue damage of auxiliary system piping and heat exchangers, and the number of significant lifts assumed for design of fuel handling equipment is a TLAA, as defined in 10 CFR 54.3. TLAAs are evaluated in accordance with 10 CFR 54.21(c)(1). The applicant further stated that LRA Section 4.7.1 describes the evaluation of fuel-handling equipment TLAAs. The applicant stated that its piping outside the RCPB is designed to ASME III Class 2, Class 3, and ANSI B31.1, all of which require a reduction in the allowable secondary stress range if more than 7,000 full-range thermal cycles are expected in a design lifetime. LRA Section 4.3.5 describes the evaluation of these cyclic piping design TLAAs. The applicant further stated that a survey of other than

Aging Management Review Results

ASME III Class 1 pressure-retaining components (vessels, heat exchangers, pumps, and valves) discovered two Class 2 heat exchangers in each unit, the CVCS letdown and regenerative heat exchangers. LRA Section 4.3.2.9 describes the evaluation of this TLAA.

The staff reviewed LRA Section 3.3.2.2.1 against the criteria in SRP-LR Section 3.3.2.2.1, which states that fatigue is TLAA, as defined in 10 CFR 54.3. TLAAs are required to be evaluated in accordance with 10 CFR 54.21(c)(1). Section 4.3, "Metal Fatigue Analysis," of the SRP-LR addresses this TLAA separately. The staff finds the applicant's AMR results are consistent with the recommendations of the GALL Report and SRP-LR, except for the following noted areas.

The staff noted that the Summary Description in LRA Section 4.3.5 states that the implicit fatigue analyses discussed in the section are applicable to all ASME Code Class 2 and 3 and ANSI B31.1 piping, piping components, and piping elements. The staff noted that it is not clear if the LRA includes all corresponding AMR items for applicable ASME Code Class 2 and 3 or ANSI B31.1 piping, piping components, and piping elements within the scope of license renewal. The staff also noted that this includes those components in the ESF Systems (LRA Section 3.2), Auxiliary Systems (LRA Section 3.3), and Steam and Power Conversion Systems (LRA Section 3.4). By letter dated July 21, 2010, the staff issued RAI 4.3-13 requesting the applicant to clarify if the LRA includes all applicable AMR items with an aging effect of cumulative fatigue damage for those components within the scope of license renewal. If not, the staff asked the applicant to justify why the LRA does not include all corresponding AMR items on cumulative fatigue damage for applicable ASME Code Class 2 and 3 or ANSI B31.1 piping, piping components, and piping elements within the scope of license renewal. Further, the staff asked the applicant to identify all component types that are within the scope of the implicit fatigue analyses for ASME Code Class 2 and 3 components and B31.1 components in LRA Section 4.3.5 and, therefore, should be within the scope of applicable component-specific AMR items on cumulative fatigue damage. The staff previously identified this as Open Item 4.3-1.

In its response dated August 12, 2010, the applicant stated that, for auxiliary systems, additional AMR line items were added to the LRA. The applicant further stated that LRA Tables 3.3.2-8, 3.3.2-21, and 3.3.2-30 required the addition of AMR line items. The staff also noted that the applicant amended LRA Table 3.3.2-8 to include the associated AMR item consistent with the GALL Report AMR item VII.E-16, LRA Table 3.3.2-21 to include the associated AMR item consistent with the GALL Report AMR item VII.E1-8, and LRA Table 3.3.2-30 to include the associated AMR item consistent with the GALL Report AMR item VIII B1-10. The staff confirmed that these additional AMR items are consistent with the associated GALL Report AMR items. The staff's evaluation for ANSI B31.1 and ASME III Class 2 and 3 piping is documented in SER Section 4.3.5.2.

Based on its review of the amended LRA Tables 3.3.2-8, 3.3.2-21 and 3.3.2-30, the staff finds the applicant's response to RAI 4.3-13, part 3, and the additions of these AMR items acceptable because they are consistent with the GALL Report AMR items VII.E-16, VII.E1-8 and VIII B1-10. The staff's concern described in RAI 4.3-13 is resolved, and this portion of Open Item 4.3-1 is closed.

Based on its review, the staff concludes that the applicant's proposal to manage cumulative fatigue damage in auxiliary system piping and heat exchangers, and the number of significant lifts assumed for design of fuel handling equipment meets the SRP-LR Section 3.3.2.2.1 criteria. For those items that apply to LRA Section 3.3.2.2.1, the staff determines that the LRA is consistent with the GALL Report. The applicant has demonstrated that it will adequately manage the effects of aging so that the intended function(s) will be maintained consistent with the CLB during the period of extended operation, as required by 10 CFR 54.21(a)(3). SER Section 4.3 documents the staff's review of the applicant's evaluation of the TLAA for these components.

Aging Management Review Results

3.3.2.2.2 Reduction of Heat Transfer Due to Fouling

LRA Section 3.3.2.2.2 refers to LRA Table 3.3.1, item 3.3.1.3, and addresses reduction of heat transfer due to fouling of stainless steel heat exchanger tubes exposed to treated water for auxiliary systems. The applicant stated that this item is not applicable because it only applies to BWRs.

The staff reviewed SRP-LR Table 3.3-1, item 3, and noted that, contrary to the applicant's statement, this item applies to both BWRs and PWRs. Additionally, although not listed in the GALL Report Table VII.A3, associated with the PWR spent fuel pool cooling system, the staff noted that the documented basis from NUREG-1833 for adding the related item, AP-62, was a precedent established in the R.E. Ginna SER, NUREG-1786. The environment specifically noted in that SER was "treated water-borated." As such, the omission from the GALL Report Table VII.A3 appears to have been inadvertent and is being addressed in the update to the GALL Report currently in progress.

The staff reviewed LRA Sections 2.3.3 and Section 3.3 and identified two items with a reduction of heat transfer due to fouling for stainless steel heat exchanger tubes exposed to treated borated water in Table 3.3.2-2, "Spent Fuel Pool Cooling and Cleanup System," and Table 3.3.2-8, "Nuclear Sampling System." In both instances, the applicant cited generic Note H, indicating that for the items, the aging effect is not in the GALL Report for this component, material, and environment combination. The LRA states that the Water Chemistry and One-Time Inspection Programs would manage the reduction of heat transfer for these components. SER Section 3.3.2.3.2 discusses the staff's review for these items.

3.3.2.2.3 Cracking Due to Stress Corrosion Cracking

The staff reviewed LRA Section 3.3.2.2.3 against the following criteria in SRP-LR Section 3.3.2.2.3:

- LRA Section 3.3.2.2.3.1 addresses cracking due to SCC in the stainless steel components of a BWR standby liquid control system, stating that this aging effect is not applicable to PVNGS; it is applicable to BWRs only. SRP-LR Section 3.3.2.2.3.1 states that cracking due to SCC could occur in the stainless steel piping, piping components, and piping elements of the BWR standby liquid control system that are exposed to sodium pentaborate solution greater than 60 degrees C (140 degrees F). The staff finds that SRP-LR Section 3.3.2.2.3, item 1 is not applicable because the PVNGS units are PWRs, and the staff guidance in this SRP-LR section is only applicable to BWRs.

- LRA Section 3.3.2.2.3.2 refers to LRA Table 3.3.1, item 3.3.1.5, and addresses cracking due to SCC in stainless steel and stainless steel clad steel heat exchangers exposed to treated water greater than 60 degrees C (140 degrees F). The applicant stated that this item is not applicable to PVNGS, because it only applies to BWRs.

    The staff reviewed LRA Section 3.3.2.2.3.2 against the criteria in SRP-LR Section 3.3.2.2.3, item 2, which states that cracking due to SCC could occur in stainless steel and stainless clad steel heat exchanger components exposed to treated water greater than 60 degrees C (140 degrees F). It continued by recommending further evaluation of a plant-specific AMP to ensure this aging effect was adequately managed and noted that acceptance criteria are described in Appendix A.1 of the SRP-LR. In addition, the staff noted that, contrary to the applicant's statement, the related item applies to both BWRs and PWRs.

    Furthermore, in reviewing LRA Table 3.3.2-30, the staff noted that the applicant listed the sample cooler heat exchanger with an aging mechanism of cracking for stainless steel exposed to secondary water. As such, the basis for the applicant's determination, that this item was not applicable to PVNGS, was not clear to the staff. By letter dated

Aging Management Review Results

April 28, 2010, the staff issued RAI 3.3.2.2.3-1 requesting the applicant to provide its basis for why SRP-LR Section 3.3.2.2.3.2 was not applicable or, if it was applicable, to explain how aging will be managed.

In its response, dated May 21, 2010, the applicant stated that SRP-LR Table 3.3.1, item 5 referenced SRP-LR Section 3.3.2.2.3.2, which is related to the GALL Report, items VII.E3-3 and VIIE3-19. These related items only appear in the GALL Report tables associated with BWR reactor water cleanup systems; however, the applicant also noted that the stainless steel sample cooler exposed to secondary water will experience cracking, which is managed by the Water Chemistry and One-Time Inspections Programs, consistent with the GALL Report, item VIII.F-3. The staff noted that this item is related to SRP-LR further evaluation Section 3.4.2.2.6.

The staff reviewed the applicant's response to RAI 3.3.2.2.3-1 and finds the response acceptable because the applicant addressed the in-scope heat exchangers components constructed of stainless steel and stainless clad steel exposed to treated water greater than 60 degrees C (140 degrees F) using item VIII.F-3, and it will manage these components as indicated in SRP-LR Section 3.4.2.2.6. The staff finds the applicant's evaluation of SRP-LR Section 3.3.2.2.3, item 2 through SRP-LR further evaluation in Section 3.4.2.2.6 an acceptable approach.

- LRA Section 3.3.2.2.3 refers to LRA Table 3.3.1, item 3.3.1.6, and addresses stainless steel diesel engine exhaust piping, piping components, and piping elements exposed to diesel exhaust, which the Inspection of Internal Surfaces in Miscellaneous Piping and Ducting Components Program manages cracking due to SCC. The applicant addressed the further evaluation criteria of the SRP-LR by stating in the AMP that this program will perform visual inspections to detect aging effects that could result in loss of component intended function.

   The staff reviewed LRA Section 3.3.2.2.3.3 against the criteria in SRP-LR Section 3.3.2.2.3, item 3, which states that cracking due to SCC could occur in stainless steel diesel engine exhaust piping, piping components, and piping elements exposed to diesel exhaust. The SRP-LR also states that the GALL Report recommends further evaluation of a plant-specific AMP to ensure that the applicant adequately manages these aging effects. SER Section 3.0.3.2.15 documents the staff's evaluation of the applicant's Inspection of Internal Surfaces in Miscellaneous Piping and Ducting Components Program. In its review of components associated with item 3.3.1.6, the staff finds the applicant's proposal to manage aging using the Internal Surfaces in Miscellaneous Piping and Ducting Components Program acceptable. This program will include volumetric evaluations of the internal surfaces of stainless steel components exposed to diesel exhaust and will perform visual inspections during periodic maintenance, predictive maintenance, surveillance testing, and corrective maintenance. Both volumetric evaluations and visual inspections are capable of detecting cracking due to SCC that could result in a loss of component intended function.

   Based on the program identified, the staff concludes that the applicant's programs meet the SRP-LR Section 3.3.2.2.3, item 3, criteria. For those items that apply to LRA Section 3.3.2.2.3, item 3, the staff determines that the LRA is consistent with the GALL Report. The staff also finds that the applicant has demonstrated that the effects of aging will be adequately managed so that the intended functions will be maintained consistent with the CLB during the period of extended operation, as required by 10 CFR 54.21(a)(3).

The staff determines that the LRA is consistent with the GALL Report and that the applicant has demonstrated that the effects of aging will be adequately managed so that the intended

Aging Management Review Results

function(s) will be maintained consistent with the CLB during the period of extended operation, as required by 10 CFR 54.21(a)(3).

3.3.2.2.4 Cracking Due to Stress Corrosion Cracking and Cyclic Loading

The staff reviewed LRA Section 3.3.2.2.4 against the following criteria in SRP-LR Section 3.3.2.2.4:

- LRA Section 3.3.2.2.4.1 refers to Table 3.3.1, item 7, and addresses cracking due to SCC and cyclic loading in stainless steel non-regenerative heat exchanger components exposed to borated water greater than 60 degrees C (140 degrees F). The LRA states that the Water Chemistry and One-Time Inspection Programs will manage cracking due to SCC and cyclic loading for the stainless steel CVCS letdown (non-regenerative) heat exchanger components exposed to treated borated water, where temperature and radioactivity of the shell-side water are monitored by installed instrumentation, and the One-Time Inspection Program is selected in lieu of eddy current testing of tubes. The applicant also noted that the staff accepted this position in NUREG-1785, "Safety Evaluation Report Related to the License Renewal of H.B. Robinson Steam Electric Plant, Unit 2."

The staff reviewed LRA Section 3.3.2.2.4, item 1, against the criteria in SRP-LR Section 3.3.2.2.4, item 1, which states that the existing program relies on monitoring and control of primary water chemistry in PWRs to manage cracking due to SCC, and that the effectiveness of the Water Chemistry Control Program should be verified to ensure that cracking does not occur. SRP-LR Section 3.3.2.2.4.1 further states that the GALL Report recommends a plant-specific AMP to verify the absence of cracking due to SCC and cyclic loading and that an acceptable verification program includes temperature and radioactivity monitoring of the shell side water and eddy current testing of tubes.

The staff reviewed the applicant's Water Chemistry and One-Time Inspection Programs in SER Sections 3.0.3.2.1 and 3.0.3.1.6, respectively. The staff noted that the various non-regenerative heat exchanger components can be subjected to enhanced visual inspection or volumetric inspection to detect cracking. The applicant's One-time Inspection Program uses representative sampling to detect aging of components with similar environments since all these components are subjected to a treated borated water environment. If the applicant discovers any cracking during execution of its One-Time Inspection Program, it will evaluate the cracking through its Corrective Action Program where the need for expanded inspection sites, periodic inspections, and data trending. However, the applicant did not specify the NDE methodology that would be used as an alternative to eddy current testing of the heat exchanger tubes.

On July 8, 2010, the staff held a teleconference call with the applicant to discuss its concerns associated with the NDE method that will be used during the one-time inspection and the justification of the proposed methodology based on plant-specific and industry operating experience. In LRA Amendment 20, dated July 21, 2010, the applicant clarified its use of the One-Time Inspection Program by stating that it will select heat exchanger tubes in a similar environment and made of similar material to the non-regenerative heat exchanger tubes. The applicant also stated that this program will conduct eddy current testing of the stainless steel heat exchanger tubes in a borated water environment that is above the threshold temperature for cracking in stainless steel. The staff finds this clarification acceptable because eddy current testing, the NDE method specified by the applicant, is capable of identifying cracking in stainless steel heat exchanger tubes and, therefore, is able to verify the absence of cracking due to SCC and cyclic loading.

Aging Management Review Results

The staff concludes that the applicant's programs meet SRP-LR Section 3.3.2.2.4, item 1 criteria. For those line items that apply to LRA Section 3.3.2.2.4, item 1, the staff determines that the LRA is consistent with the GALL Report and that the applicant has demonstrated that the effects of aging will be adequately managed so that the intended functions will be maintained consistent with the CLB during the period of extended operation, as required by 10 CFR 54.21(a)(3).

- LRA Section 3.3.2.2.4.2 refers to Table 3.3.1, item 8, and addresses cracking due to SCC and cyclic loading in stainless steel PWR regenerative heat exchanger components exposed to borated water greater than 60 degrees C (140 degrees F). The LRA states that the Water Chemistry and the One-Time Inspection Programs will manage this aging effect for the stainless steel CVCS and nuclear sampling systems heat exchanger components exposed to treated borated water. The LRA also noted that the one-time inspection will include selected components at susceptible locations.

  The staff reviewed LRA Section 3.3.2.2.4, item 2 against the criteria in SRP-LR Section 3.3.2.2.4, item 2, which states that management of this aging effect relies on monitoring and control of primary water chemistry in PWRs, but control of water chemistry does not preclude cracking due to SCC and cyclic loading. The SRP-LR further states that the effectiveness of the Water Chemistry Control Program should be verified and recommends that the applicant evaluate a plant-specific AMP to ensure that it adequately manages these aging effects. Appendix A.1 of the SRP-LR describes the acceptance criteria.

  The staff reviewed the applicant's Water Chemistry and One-Time Inspection Programs in SER Sections 3.0.3.2.1 and 3.0.3 1.6, respectively. The staff also noted that the various regenerative heat exchanger components can be subjected to enhanced visual inspection or volumetric inspection to detect cracking. The applicant's One-time Inspection Program uses representative sampling to detect aging of components with similar environments since all these components are subjected to a treated borated water environment. If the applicant discovers any cracking during execution of its One-Time Inspection Program, it will evaluate the cracking through its Corrective Action Program, which includes the need for expanded inspection sites, periodic inspections, and data trending. The staff concludes that cracking in stainless steel PWR regenerative heat exchanger components exposed to borated water, will be adequately managed by the applicant's One-time Inspection Program in lieu of a plant-specific program, through the period of extended operation because detection methods will detect cracking and corrective action will consider expansion of the number of inspection sites, periodic inspection, and trending of data.

  Based on the programs identified, the staff concludes that the applicant's programs meet SRP-LR Section 3.3.2.2.4.2 criteria. For those items that apply to LRA Section 3.4.2.2.4.2, the staff determines that the LRA is consistent with the GALL Report. The applicant has demonstrated that the effects of aging will be adequately managed so that the intended function(s) will be maintained consistent with the CLB during the period of extended operation, as required by 10 CFR 54.21(a)(3).

- LRA Section 3.3.2.2.4.3 refers to LRA Table 3.3.1, item 3.3.1.9, and addresses stainless steel high-pressure pump casing in PWR CVCS exposed to treated borated water (internal), which are being managed for cracking due to SCC and cyclic loading by the Water Chemistry Program and One-Time Inspection Program. The applicant addressed the further evaluation criteria of the SRP-LR by stating that the Water Chemistry Program and the One-Time Inspection Program will manage cracking due to SSC and cyclic loading for stainless steel pump casings exposed to treated borated water. The

3-235

Aging Management Review Results

applicant stated the one-time inspection will include selected components at susceptible locations.

The staff reviewed LRA Section 3.3.2.2.4.3 against the criteria in SRP-LR Section 3.3.2.2.4, item 3, which states that the existing AMP relies on monitoring and control of primary water chemistry in PWRs to manage the aging effects of cracking due to SCC. The SRP-LR further states that that control of water chemistry does not preclude cracking due to SCC and cyclic loading, and the effectiveness of the Water Chemistry Control Program should be verified to ensure that cracking does not occur. The SRP-LR also states that the GALL Report recommends that the applicant evaluate a plant-specific AMP to verify the absence of cracking due to SCC and cyclic loading to ensure that it adequately manages these aging effects.

SER Sections 3.0.3.2.1 and 3.0.3.1.6 document the staff's evaluation of the applicant's Water Chemistry Program and One-Time Inspection Program, respectively. The staff noted that the Water Chemistry Program controls the chemical environment to ensure that the aging effects due to contaminants are limited by managing the primary and secondary water. The staff noted that this is accomplished by limiting the concentration of chemical species known to cause corrosion and adding chemical species known to inhibit degradation by their influence on pH and dissolved oxygen levels. The staff also noted that this program is effective in creating an environment that is not conducive for cracking to occur in areas of intermediate and high flow, where thorough mixing takes place and the monitoring samples are representative of actual conditions. The staff noted that the applicant's One-Time Inspection Program will conduct NDE inspections of a representative group of components in order to verify the effectiveness of the Water Chemistry Program in low flow and stagnant areas. The applicant's proposal to manage the SCC and cyclic loading of stainless steel high-pressure pump casing is consistent with GALL Report item VII.E1-7. In its review of components associated with item 9, the staff finds the applicant's proposal to manage aging using the Water Chemistry Program and One-Time Inspection Program acceptable. because the Water Chemistry Program will create an environment that is not conducive for cracking to occur, the One-Time Inspection Program will verify the effectiveness of the water chemistry, and the applicant's use of these programs is consistent with the GALL Report.

Based on the programs identified, the staff concludes that the applicant's programs meet SRP-LR Section 3.3.2.2.4, item 3 criteria. For those items that apply to LRA Section 3.3.2.2.4.3, the staff determines that the LRA is consistent with the GALL Report. The applicant has demonstrated that the effects of aging will be adequately managed so that the intended function(s) will be maintained consistent with the CLB during the period of extended operation as required by 10 CFR 54.21(a)(3).

- LRA section 3.3.2.2.4 refers to Table 3.3.1, item 3.3.1.10, and addresses high-strength steel closure bolting exposed to air with steam or water leakage in the auxiliary systems. The GALL Report recommends use of AMP XI.M18, "Bolting Integrity," to manage cracking due to SCC and cyclic loading for this component group. The applicant stated that this item is not applicable because it has no in-scope, high-strength steel closure bolting exposed to air with steam or water leakage in the auxiliary systems. The staff reviewed LRA Sections 2.3.3 and 3.3 and confirmed that the applicant's LRA does not have any AMR results for the auxiliary systems that include high-strength steel closure bolting in the auxiliary systems exposed to air with steam or water leakage. During its on-site audit of the applicant's Bolting Integrity Program, the staff also confirmed that the applicant does not have any in-scope, high-strength closure bolting in its auxiliary systems during audit interviews. Based on its review of the LRA and applicant audit interviews, the staff confirmed that there is no in-scope steel closure bolting exposed to

Aging Management Review Results

air with steam or water leakage in the auxiliary systems, and, therefore, finds the applicant's determination acceptable.

Based on the programs identified above, the staff concludes that the applicant's programs meet SRP-LR Section 3.3.2.2.4 criteria. For those items that apply to LRA Section 3.3.2.2.4, the staff determines that the LRA is consistent with the GALL Report. The applicant has demonstrated that the effects of aging will be adequately managed so that the intended function(s) will be maintained consistent with the CLB during the period of extended operation, as required by 10 CFR 54.21(a)(3).

3.3.2.2.5 Hardening and Loss of Strength Due to Elastomer Degradation

The staff reviewed LRA Section 3.3.2.2.5 against the following criteria in SRP-LR Section 3.3.2.2.5:

- LRA Section 3.3.2.2.5.1 refers to LRA Table 3.3.1, item 3.3.1.11, and addresses elastomeric seals of heating and ventilation systems exposed to air indoor uncontrolled, which are managed for hardening and loss of strength due to degradation by the External Surfaces Monitoring and Inspection of Internal Surfaces in Miscellaneous Piping and Ducting Components Programs. The applicant addressed the further evaluation acceptance criteria of the SRP-LR by stating that the programs will manage hardening and loss of strength from degradation for elastomeric internal and external surfaces exposed to ventilation atmosphere in locations where the ambient temperature cannot be shown to be less than 95 degrees F.

The staff reviewed LRA Section 3.3.2.2.5.1 against the criteria in SRP-LR Section 3.3.2.2.5, item1, which states that hardening and loss of strength due to elastomer degradation could occur in elastomeric seals and components of heating and ventilation systems exposed to air-indoor uncontrolled (internal or external). The SRP-LR recommends further evaluation of a plant-specific AMP to ensure that the aging effect is adequately managed. Acceptance criteria are described in Branch Technical Position RLSB-1 (Appendix A.1 of the SRP-LR).

In its review of components associated with LRA Table 3.3.1, item 11, the staff noted that the GALL Report, in Table XI.D, "Environments," provides a basis for the applicant to use 95 degrees F as a threshold temperature, below which thermal aging of organic elastomers can be considered insignificant during the period of extended operation. However, the GALL Report does not provide a basis for using this temperature to preclude other potential elastomeric degradation due to exposure to ozone, oxidation, or radiation. By letter dated February 19, 2010, the staff issued RAI 3.3.2.2.5.1-1, asking the applicant to identify the systems containing in-scope elastomeric components that will be inspected and to determine if use of the 95 degree F criterion results in excluding any in-scope elastomeric components from aging management.

In its response, dated March 24, 2010, the applicant identified seven mechanical systems included in LRA Section 3.3.2.2.5.1:

(1) the containment purge system

(2) the fuel building HVAC system

(3) the auxiliary building HVAC system

(4) the containment HVAC system

(5) the diesel building HVAC system

(6) the control building HVAC system

3-237

Aging Management Review Results

(7) the miscellaneous buildings HVAC system.

The applicant stated that none of the elastomeric flexible connectors in these systems are excluded from aging management based on the 95 degree F criterion. The staff noted that the applicant's list of HVAC systems, where aging management of elastomeric components is not excluded, is similar to the list of systems for which the GALL Report recommends aging management of elastomeric components. The staff finds the applicant's response acceptable because the systems listed by the applicant are similar to the systems recommended in the GALL Report, and the use of the 95 degree F criterion does not result in unacceptable exclusion of components from aging management. The staff's concern described in RAI 3.3.2.2.5.1-1 is resolved.

SER Sections 3.0.3.2.14 and 3.0.3.2.15 document the staff's evaluation of the applicant's External Surfaces Monitoring Program and Inspection of Internal Surfaces in Miscellaneous Piping and Ducting Components Program, respectively. The staff noted that both GALL Report AMPs manage loss of material for steel components; however, for each of these programs, the applicant has taken exceptions to the GALL Report AMPs. These exceptions increase the scope of the materials managed to include elastomers and augment the visual inspections specified in the GALL Report with physical manipulations to verify absence of hardening or loss of strength for elastomers. The staff finds the applicant's proposal to manage aging using the External Surfaces Monitoring Program and Inspection of Internal Surfaces in Miscellaneous Piping and Ducting Components Program acceptable because the applicant will perform, 1) visual inspections of external surfaces during engineering walkdowns; 2) visual inspections of internal surfaces during periodic maintenance, predictive maintenance, surveillance testing, and corrective maintenance; and 3) physical manipulation may be used during the visual inspections to verify absence of hardening or loss of strength for elastomers.

Based on the programs identified, the staff concludes that the applicant's programs meet SRP-LR Section 3.3.2.2.5, item 1 criteria. For those items that apply to LRA Section 3.3.2.2.5.1, the staff determines that the LRA is consistent with the GALL Report and that the applicant has demonstrated that the effects of aging will be adequately managed so that the intended function(s) will be maintained consistent with the CLB during the period of extended operation, as required by 10 CFR 54.21(a)(3).

- LRA Section 3.3.2.2.5.2 refers to Table 3.3-1, item 12, and addresses the elastomeric boot seal for the safety injection pump suction strainer in the refueling water tank exposed to treated borated water, which is being managed for hardening and loss of strength due to degradation by the Internal Surfaces in Miscellaneous Piping and Ducting Components Program. The applicant addressed the further evaluation criteria by stating that the program will manage hardening and loss of strength from degradation for the boot seal.

The staff reviewed LRA Section 3.3.2.2.5.2 against the criteria in SRP-LR Section 3.3.2.2.5, item 2. This states that hardening and loss of strength due to elastomeric degradation could occur in elastomeric linings of filters, valves, and ion exchangers in spent fuel pool cooling and cleanup systems exposed to treated water or to treated borated water. It also recommends further evaluation of a plant-specific AMP to ensure that the aging effect is adequately managed. BTP RLSB-1 describes acceptance (Appendix A.1 of the SRP-LR).

LRA Section 3.3.2.2.5.2 states that the Inspection of Internal Surfaces in Miscellaneous Piping and Ducting Components AMP will manage hardening and loss of strength for the elastomer boot seal for the safety injection pump suction strainer in the refueling water tank that is exposed to treated borated water. The staff notes that the applicant did not address elastomeric linings of filters, valves, and ion exchangers in spent fuel pool cooling and cleanup systems. The staff also notes that a search of the applicant's UFSAR confirmed that no in-scope elastomeric linings of filters, valves, and ion exchangers exposed to treated water or to treated borated water are present in the spent fuel pool cooling and cleanup systems except for the boot seal.

SER Section 3.0.3.2.15 documents the staff's evaluation of the applicant's Internal Surfaces in Miscellaneous Piping and Ducting Components Program. The staff noted that the GALL Report Inspection of Internal Surfaces in Miscellaneous Piping and Ducting Components AMP manages loss of material for steel components. However, the applicant has taken exceptions to the GALL Report AMP that increase the scope of the materials managed to include elastomers and augment the visual inspections specified in the GALL Report with physical manipulations to verify absence of hardening or loss of strength for elastomers. In its review of components associated with item 12, the staff finds the applicant's proposal to manage aging using the Internal Surfaces in Miscellaneous Piping and Ducting Components Program acceptable because the applicant will perform visual inspections during periodic maintenance, predictive maintenance, surveillance testing, and corrective maintenance, and the visual inspections will be augmented by physical manipulation to verify absence of hardening or loss of strength for elastomers.

Based on the programs identified, the staff concludes that the applicant's program meets SRP-LR Section 3.3.2.2.5, item 2 criteria. For those items that apply to LRA Section 3.3.2.2.5.2, the staff determines that the LRA is consistent with the GALL Report. The applicant has demonstrated that the effects of aging will be adequately managed so that the intended functions will be maintained consistent with the CLB during the period of extended operation, as required by 10 CFR 54.21(a)(3).

Based on the programs identified above, the staff concludes that the applicant's programs meet SRP-LR Section 3.3.2.2.5 criteria. For those items that apply to LRA Section 3.3.2.2.5, the staff determines that the LRA is consistent with the GALL Report. The staff finds that the applicant has demonstrated that the effects of aging will be adequately managed so that the intended function(s) will be maintained consistent with the CLB during the period of extended operation, as required by 10 CFR 54.21(a)(3).

3.3.2.2.6 Reduction of Neutron-Absorbing Capacity and Loss of Material Due to General Corrosion

LRA Section 3.3.2.2.6 refers to Table 3.3.1, item 3.3.1.13, which addresses boral, boron steel spent fuel storage racks, and neutron-absorbing sheets exposed to treated water or treated borated water, which are being managed for a reduction of neutron-absorbing capacity and loss of material due to general corrosion. The GALL Report recommends use of a plant-specific AMP to manage aging. The applicant stated that this item is not applicable because it does not use boral or boron steel in its spent fuel storage racks to maintain subcriticality. The staff reviewed LRA Sections 2.3.3 and 3.3 and confirmed that the applicant's LRA does not have any AMR results for the auxiliary systems that include boral or boron steel or other neutron-absorbing sheets in the auxiliary systems exposed to treated water or treated borated water. The staff also reviewed the spent fuel pool criticality analysis in the UFSAR and confirmed that the analysis does not rely on boral or boron steel neutron-absorbing materials. Based on its review of the LRA and UFSAR, the staff confirmed that there is no boral, boron steel spent fuel storage racks, neutron-absorbing sheets exposed to treated water or treated borated water in the auxiliary systems, and, therefore, finds the applicant's determination acceptable.

Aging Management Review Results

3.3.2.2.7   Loss of Material Due to General, Pitting, and Crevice Corrosion

The staff reviewed LRA Section 3.3.2.2.7 against the following criteria in SRP-LR Section 3.3.2.2.7:

- LRA Section 3.3.2.2.7 refers to LRA Table 3.3.1, items 3.3.1.14, 3.3.1.15, and 3.3.1.16, and addresses cast iron and carbon steel piping and their components and elements including tubing, valves, and tanks in the RCP oil collection system, exposed to lubricating oil. The Lubricating Oil Analysis and One-Time Inspection Programs manage these components for loss of material due to general, pitting, and crevice corrosion and SCC. The applicant addressed the further evaluation criteria of the SRP-LR by stating that it will include a one-time inspection of selected components at susceptible locations where contaminants such as water could accumulate. The applicant further stated that the one-time inspection will assess the thickness of the lower portion of a representative sample of RCP lubricating oil collection tanks.

    The staff reviewed LRA Section 3.3.2.2.7.1 against the criteria in SRP-LR Section 3.3.2.2.7, item 1, which states that loss of material due to general, pitting, and crevice corrosion could occur in steel piping, their components, and elements, including the tubing, valves, and tanks in the RCP oil collection system, exposed to lubricating oil (as part of the fire protection system). It also states that the existing AMP relies on the periodic sampling and analysis of lubricating oil to maintain contaminants within acceptable limits, thereby preserving an environment that is not conducive to corrosion. It further states that control of lube oil contaminants may not always have been adequate to preclude corrosion, and corrosion may occur at locations in the RCP oil collection tank where water from wash downs may accumulate. The effectiveness of the program, therefore, should be verified with a one-time inspection to ensure that corrosion is not occurring.

    SER Sections 3.0.3.2.16 and 3.0.3.1.6 document the staff's review of the LRA Lubricating Oil Analysis Program and One-Time Inspection Program, respectively. In its review of the cast iron and carbon steel components associated with the LRA items listed above, the staff finds the applicant's proposal to manage aging using the Lubricating Oil Analysis and One-Time Inspection Programs acceptable because the Lubricating Oil Analysis Program provides for periodic sampling of lubricating oil to maintain contaminants at acceptable limits to preclude loss of material due to general, pitting, and crevice corrosion. In addition, the applicant will perform one-time inspections of select steel piping, piping components, and piping elements, including the tubing, valves, and tanks in the RCP oil collection system, exposed to lubricating oil, for loss of material due to general, pitting, and crevice corrosion. This one-time inspection will verify the effectiveness of the Lubricating Oil Analysis Program in applicable auxiliary systems, following the GALL Report recommendation that the "One Time Inspection" is an acceptable AMP to verify the effectiveness of the applicant's Lubricating Oil Analysis Program.

    Based on the programs identified, the staff finds that the applicant's programs meet SRP-LR Section 3.3.2.2.7, item 1 criteria. For those items that apply to LRA Section 3.3.2.2.7.1, the staff determines that the LRA is consistent with the GALL Report and that the applicant has demonstrated that the effects of aging will be adequately managed so that the intended function(s) will be maintained consistent with the CLB during the period of extended operation, as required by 10 CFR 54.21(a)(3).

- LRA Section 3.3.2.2.7 addresses loss of material due to general, pitting, and crevice corrosion in steel components in the BWR reactor water cleanup and shutdown cooling systems exposed to treated water, stating that this aging effect is not applicable; it is

Aging Management Review Results

applicable to BWRs only. The staff finds that SRP-LR Section 3.3.2.2.7 is not applicable because the PVNGS units are PWRs, and the staff guidance in this SRP-LR section is only applicable to BWRs.

- LRA Section 3.3.2.2.7 refers to LRA Table 3.3.1, item 18, and addresses stainless steel and steel diesel engine exhaust piping, piping components, and piping elements exposed to diesel exhaust, which the Inspection of Internal Surfaces in Miscellaneous Piping and Ducting Components Program manages for loss of material due to general (steel only), pitting, and crevice corrosion. The applicant addressed the further evaluation criteria of the SRP-LR in the AMP by stating that this program will perform visual inspections to detect aging effects that could result in loss of component intended function.

   The staff reviewed LRA Section 3.3.2.2.7.3 against the criteria in SRP-LR Section 3.3.2.2.7, item 3, which states that loss of material due to general (steel only), pitting, and crevice corrosion could occur for steel and stainless steel diesel engine exhaust piping, piping components, and piping elements exposed to diesel exhaust. The SRP-LR also states that the GALL Report recommends further evaluation of a plant-specific AMP to ensure that the applicant adequately manages these aging effects.

   SER Section 3.0.3.1.15 documents the staff's evaluation of the applicant's Inspection of Internal Surfaces in Miscellaneous Piping and Ducting Components Program. In its review of components associated with item 3.3.1-18, the staff finds the applicant's proposal to manage aging using the Inspection of Internal Surfaces in Miscellaneous Piping and Ducting Components Program acceptable because the program will perform visual inspections during periodic maintenance, predictive maintenance, surveillance testing, and corrective maintenance. These inspections are capable of detecting pitting and crevice corrosion that could result in a loss of component intended function.

   Based on the programs identified, the staff concludes that the applicant's programs meet the SRP-LR Section 3.3.2.2.7, item 3 criteria. For those items that apply to LRA Section 3.3.2.2.7.3, the staff determines that the LRA is consistent with the GALL Report. The applicant has demonstrated that the effects of aging will be adequately managed so that the intended functions will be maintained consistent with the CLB during the period of extended operation, as required by 10 CFR 54.21(a)(3).

Based on the programs identified above, the staff concludes that the applicant's programs meet SRP-LR Section 3.3.2.2.7 criteria. For those items that apply to LRA Section 3.3.2.2.7, the staff determines that the LRA is consistent with the GALL Report. The staff finds that the applicant has demonstrated that the effects of aging will be adequately managed so that the intended function(s) will be maintained consistent with the CLB during the period of extended operation, as required by 10 CFR 54.21(a)(3).

3.3.2.2.8   Loss of Material Due to General, Pitting, Crevice, and Microbiologically-Influenced Corrosion

LRA Section 3.3.2.2.8 refers to Table 3.3.1, item 3.3.1.19, and addresses steel piping, piping components, and piping elements buried in soil which are being managed for loss of material due to general, pitting, crevice, and microbiologically-influenced corrosion by the Buried Piping and Tanks Inspection Program. The applicant addressed the further evaluation criteria of the SRP-LR by stating that the Buried Piping and Tanks Inspection Program will manage the loss of material due to general, pitting, crevice, and microbiologically-influenced corrosion for carbon steel external surfaces of buried components.

The staff reviewed LRA Section 3.3.2.2.8 against the criteria in SRP-LR Section 3.3.2.2.8, which states that loss of material due to general, pitting, crevice, and MIC could occur for steel piping,

piping components, and piping elements, with or without coating or wrapping, in a soil environment. The SRP-LR also states that the Buried Piping and Tanks Inspection Program relies on industry practice, frequency of pipe excavation, and operating experience to manage the effects of loss of material from general, pitting, and crevice corrosion and MIC and the effectiveness of the Buried Piping and Tanks Inspection Program should be verified to evaluate the inspection frequency and operating experience with buried components, ensuring that loss of material is not occurring.

SER Section 3.0.3.2.12 documents the staff's evaluation of the applicant's Buried Piping and Tanks Inspection Program. The staff finds the applicant's proposal to manage aging using the Buried Piping and Tanks Inspection Program acceptable because the program requires periodic visual inspections of the external surface of buried steel piping, prior to and within the period of extended operation, to ensure that the applicant will adequately manage corrosion of external surfaces.

Based on the program identified, the staff concludes that the applicant's program meets SRP-LR Section 3.3.2.2.8 criteria. For those items that apply to LRA Section 3.3.2.2.8, the staff determines that the LRA is consistent with the GALL Report and that the applicant has demonstrated that the effects of aging will be adequately managed so that their intended function(s) will be maintained consistent with the CLB during the period of extended operation, as required by 10 CFR 54.21(a)(3).

3.3.2.2.9 Loss of Material Due to General, Pitting, Crevice, Microbiologically-Influenced Corrosion and Fouling

The staff reviewed LRA Section 3.3.2.2.9 against the following criteria in SRP-LR Section 3.3.2.2.9.

- LRA Section 3.3.2.2.9.1 refers to Table 3.3.1, item 20, and addresses the loss of material due to general, pitting, and crevice corrosion; MIC; and fouling for carbon steel components in the fuel oil system. The applicant stated that the Fuel Oil Chemistry Program (reviewed in SER Section 3.0.3.2.9) and the One-Time Inspection Program (reviewed in SER Section 3.0.3.1.6) manage this aging effect. The applicant also stated that the one-time inspection will include selected components at susceptible locations where contaminants could accumulate (e.g. stagnant flow locations and tank bottoms).

    The staff reviewed LRA Section 3.3.2.2.9.1 against the criteria in SRP-LR Section 3.3.2.2.9, item 1, which states that loss of material due to general, pitting, and crevice corrosion, MIC, and fouling could occur for steel piping, piping components, piping elements, and tanks exposed to fuel oil. The GALL Report recommends that these aging effects be managed through the use of the Fuel Oil Chemistry Program and that the effectiveness of this program be verified through the use of the One-Time Inspection Program. The GALL Report also recommends further evaluation because corrosion or fouling may occur at locations where contaminants accumulate.

    The staff noted that LRA Section 3.3.2.2.9.1 identifies appropriate AMPs and lists the critical conditions requiring further review (i.e., the potential for fuel oil chemistry control to be ineffective in locations where contaminants accumulate). The staff finds that LRA Section 3.3.2.2.9.1 is consistent with SRP-LR Section 3.3.2.2.9, item 1.

    In its review of LRA Section 3.3.2.2.9.1, the staff also reviewed AMR items, which refer to LRA Table 3.3.1, item 3.3.1.20 and are associated with LRA Section 3.3.2.2.9.1. In this review, the staff noted that the applicant proposes that the AMR items associated with item 3.3.1.20 are consistent with the GALL Report in all respects, except the applicant has taken some exceptions to the GALL Report AMP (generic Note B) or are

Aging Management Review Results

consistent with the GALL Report, except that the component is different and that the applicant has taken some exceptions to the GALL Report AMP (generic Note D).

Based on the programs identified, the staff concludes that the applicant's programs meet SRP-LR Section 3.3.2.2.9.1 criteria. For those items that apply to LRA Section 3.3.2.2.9.1, the staff determines that the LRA is consistent with the GALL Report and that the applicant has demonstrated that the effects of aging will be adequately managed so that the intended function(s) will be maintained consistent with the CLB during the period of extended operation, as required by 10 CFR 54.21(a)(3).

- LRA Section 3.3.2.2.9.2 refers to LRA Table 3.3.1, item 21, and addresses steel heat exchanger components exposed to lubricating oil, which the Lubricating Oil Analysis Program and the One-Time Inspection Program manage for loss of material due to general, pitting, or crevice corrosion, MIC, and fouling. The applicant addressed the further evaluation criteria of the SRP-LR by stating that it will include a one-time inspection of selected components at susceptible locations where contaminants such as water could accumulate.

The staff reviewed LRA Section 3.3.2.2.9.2 against the criteria in SRP-LR Section 3.3.2.2.9, item 2, which states that loss of material due to general, pitting, or crevice corrosion, MIC, and fouling could occur for steel heat exchanger components exposed to lubricating oil. SRP-LR Section 3.3.2.2.9 item 2 further states that: (1) the existing AMP relies on the periodic sampling and analysis of lubricating oil to maintain contaminants within acceptable limits, thereby preserving an environment that is not conducive to corrosion; (2) the effectiveness of lubricating oil control should be verified to ensure that corrosion is not occurring; and (3) the GALL Report recommends further evaluation of programs to manage corrosion by verifying their effectiveness with a one-time inspection of selected components at susceptible locations.

SER Sections 3.0.3.2.16 and 3.0.3.1.6 document the staff's review of the applicant's Lubricating Oil Analysis Program and One-Time Inspection Program, respectively. In its review of the steel components associated with the LRA item listed above, the staff finds the applicant's proposal to manage aging effects using the Lubricating Oil Analysis and One-Time Inspection Programs acceptable because: (1) the Lubricating Oil Analysis Program provides for periodic sampling of lubricating oil to maintain contaminants at acceptable limits to preclude loss of material due to general, pitting, or crevice corrosion, MIC, and fouling; and (2) the applicant will perform one-time inspections of steel heat exchanger components exposed to lubricating oil for loss of material due to general, pitting, crevice, MIC, and fouling to verify the effectiveness of the Lubricating Oil Analysis Program. This follows the GALL Report recommendation that the "One Time Inspection" is an acceptable AMP to verify the effectiveness of the applicant's Lubricating Oil Analysis Program.

Based on the programs identified, the staff concludes that the applicant's programs meet SRP-LR Section 3.3.2.2.9.2 criteria. For those items that apply to LRA Section 3.3.2.2.9, item 2, the staff determines that the LRA is consistent with the GALL Report and that the applicant has demonstrated that the effects of aging will be adequately managed so that the intended function(s) will be maintained consistent with the CLB during the period of extended operation, as required by 10 CFR 54.21(a)(3).

Based on the programs identified above, the staff concludes that the applicant's programs meet SRP-LR Section 3.3.2.2.9 criteria. For those items that apply to LRA Section 3.3.2.2.9, the staff determines that the LRA is consistent with the GALL Report. The staff also finds that the applicant has demonstrated that the effects of aging will be adequately managed so that the

Aging Management Review Results

intended function(s) will be maintained consistent with the CLB during the period of extended operation, as required by 10 CFR 54.21(a)(3).

3.3.2.2.10 Loss of Material Due to Pitting and Crevice Corrosion

The staff reviewed LRA Section 3.3.2.2.10 against the following criteria in SRP-LR Section 3.3.2.2.10:

- LRA Section 3.3.2.2.10.1 refers to Table 3.3.1, item 3.3.1.22 and addresses steel piping with either elastomeric liners or stainless steel cladding exposed to treated water and treated borated water if the cladding or lining is degraded and affected by loss of material due to pitting and crevice corrosion. The applicant stated that this item is not applicable because it has no in-scope components constructed of steel with elastomeric lining or steel with stainless steel cladding exposed to treated or treated borated water in the spent fuel pool cooling system. The staff reviewed LRA Sections 2.3.3 and 3.3 and confirmed that the applicant's LRA does not have any AMR results for the spent fuel pool cooling system that include steel piping with either elastomeric liners or stainless steel cladding exposed to treated water and treated borated water if the cladding or lining is degraded. The staff reviewed the applicant's UFSAR, and confirmed that no in-scope steel piping with either elastomeric liners or stainless steel cladding exposed to treated water and treated borated water are present in the spent fuel pool cooling system and, therefore, it finds the applicant's determination acceptable.

- LRA Section 3.3.2.2.10.2 refers to Table 3.3.1, items 3.3.1.23 and 3.3.1.24 and addresses the loss of material of the stainless steel, aluminum, and stainless steel clad heat exchanger components exposed to treated water in the auxiliary systems. The applicant indicated that this item is only applicable to boiling water reactors and noted this in LRA Table 3.3.1, for items 3.3.1.23 and 3.3.1.24.

    For item 3.3.1.23, the staff verified that this item does not apply because the units are a PWR design. Based on this determination, the staff finds that the applicant has provided an acceptable basis for concluding the AMR, item 3.3.1.23 is not applicable.

    The staff reviewed LRA Section 3.3.2.2.10.2 against the criteria in SRP-LR Section 3.3.2.2.10.2, which states that loss of material due to pitting and crevice corrosion could occur for stainless steel and aluminum piping, piping components, and piping elements exposed to treated water. In addition, the SRP-LR states that monitoring and controlling water chemistry manages this aging effect, but high concentrations of impurities at crevices and stagnant flow locations could cause pitting or crevice corrosion. The SRP-LR states that the effectiveness of the chemistry control program should be verified and notes that the GALL Report recommends a one-time inspection of select components as susceptible locations to ensure that corrosion is not occurring. The staff also noted that, contrary to the applicant's statement, SRP-LR Table 3.3-1, item 24, applies to both PWRs and BWRs.

    The staff reviewed the tables in LRA Section 3.3 for the auxiliary systems and found multiple stainless steel components in this commodity group for multiple systems that were exposed to treated (demineralized) water with an aging effect given as loss of material. In every case, the applicant cited Table 3.4.1, item 16, which is the subject of a further evaluation in SRP-LR Section 3.4.2.2.7, item 1. Also in every case, the applicant indicated that the Water Chemistry Program, augmented by the One-Time Inspection Program, was managing this aging effect. SER Sections 3.0.3.2.1 and 3.0.3.1.6 document the staff's evaluation of the applicant's Water Chemistry and One-Time Inspection Programs, respectively. In its review of components associated with item 3.3.1.24, which the applicant evaluated using Table 3.4.1, item 3.4.1.16, the staff finds the applicant's proposal to manage aging using the Water Chemistry Program

Aging Management Review Results

augmented by the One-Time Inspection Program acceptable because the applicant's Water Chemistry Program limits the concentrations of chemical species known to cause corrosion and adds chemical species known to inhibit degradation. In addition, the staff notes that the One-Time Inspection Program verifies the effectiveness of the Water Chemistry Program and evaluates aging effects, including loss of material.

Based on the programs identified, the staff concludes that the applicant's programs meet SRP-LR Section 3.3.2.2.10, item 2 criteria. For those items that apply to LRA Section 3.3.2.2.10.2, the staff determines that the LRA is consistent with the GALL Report. The applicant has demonstrated that it will adequately manage the effects of aging so that the intended function(s) will be maintained consistent with the CLB during the period of extended operation, as required by 10 CFR 54.21(a)(3).

- LRA Section 3.3.2.2.10.3 refers to LRA Table 3.3.1, item 25 and addresses copper alloy HVAC piping, piping components, and piping elements exposed to condensation, which are being managed for loss of material due to pitting and crevice corrosion by the External Surfaces Monitoring and Inspection of Internal Surfaces in Miscellaneous Piping and Ducting Components Programs. The applicant addressed the further evaluation criteria of the SRP-LR by stating that it will use the External Surfaces Monitoring Program to perform inspections on the external surfaces of the components exposed to plant indoor air, and it will use the Inspection of Internal Surfaces in Miscellaneous Piping and Ducting Components Program to perform inspections on the internal surfaces of the components exposed to the ventilation atmosphere.

  The staff reviewed LRA Section 3.3.2.2.10.3 against the criteria in SRP-LR Section 3.3.2.2.10, item 3, which states that loss of material due to pitting and crevice corrosion could occur for copper alloy HVAC piping, piping components, and piping elements exposed to condensation. The SRP-LR also states that the GALL Report recommends further evaluation of a plant-specific AMP.

  SER Sections 3.0.3.2.14 and 3.3.3.2.15 document the staff's evaluation of the applicant's External Surfaces Monitoring and Inspection of Internal Surfaces in Miscellaneous Piping and Ducting Components Programs, respectively. The staff finds the applicant's use of the External Surfaces Monitoring and Inspection of Internal Surfaces in Miscellaneous Piping and Ducting Components Programs acceptable because both these programs perform visual inspections that are capable of detecting loss of material for the components being managed.

  Based on the programs identified, the staff concludes that the applicant's programs meet SRP-LR Section 3.3.2.2.10, item 3 criteria. For those items that apply to LRA Section 3.3.2.2.10.3, the staff determines that the LRA is consistent with the GALL Report. The applicant has demonstrated that it will adequately manage the effects of aging so that the intended function(s) will be maintained consistent with the CLB during the period of extended operation, as required by 10 CFR 54.21(a)(3).

- LRA Section 3.3.2.2.10.4 refers to LRA Table 3.3.1, item 26 and addresses copper alloy piping and components and elements exposed to lubricating oil, which the Lubricating Oil Analysis Program and the One-Time Inspection Program are managing for loss of material due to pitting and crevice corrosion. The applicant addressed the further evaluation criteria of the SRP-LR by stating that it will include a one-time inspection of selected components at susceptible locations where contaminants, such as water, could accumulate.

  The staff reviewed LRA Section 3.3.2.2.10.4 against the criteria in SRP-LR Section 3.3.2.2.10, item 4, which states that loss of material due to pitting and crevice corrosion could occur for copper alloy piping, piping components, and piping elements

Aging Management Review Results

exposed to lubricating oil. The SRP-LR Section 3.3.2.2.10 item 4, further states that: 1) the existing AMP relies on the periodic sampling and analysis of lubricating oil to maintain contaminants within acceptable limits, thereby preserving an environment that is not conducive to corrosion; (2) a one-time inspection of selected components at susceptible locations is an acceptable method to ensure that corrosion is not occurring; and (3) the GALL Report recommends further evaluation of programs to manage corrosion by verifying their effectiveness with a one-time inspection of selected components at susceptible locations.

SER Sections 3.0.3.2.16 and 3.0.3.1.6 document the staff's review of the applicant's Lubricating Oil Analysis and One-Time Inspection Programs, respectively. In its review of the components associated with the LRA item listed above, the staff finds the applicant's proposal to manage aging effects using the Lubricating Oil Analysis and One-Time Inspection Programs acceptable because: (1) the Lubricating Oil Analysis Program provides for periodic sampling of lubricating oil to maintain contaminants at acceptable limits to preclude loss of material; (2) the applicant will perform one-time inspections of select copper alloy piping, piping components, and piping elements exposed to lubricating oil for loss of material due to pitting and crevice corrosion to verify the effectiveness of the Lubricating Oil Analysis Program following The GALL Report recommendation that the "One Time Inspection" is an acceptable AMP to verify the effectiveness of the applicant's Lubricating Oil Analysis Program.

Based on the programs identified, the staff concludes that the applicant's programs meet SRP-LR Section 3.3.2.2.10, item 4 criteria. For those items that apply to LRA Section 3.3.2.2.10.4, the staff determines that the LRA is consistent with the GALL Report and that the applicant has demonstrated that the effects of aging will be adequately managed so that the intended function(s) will be maintained consistent with the CLB during the period of extended operation, as required by 10 CFR 54.21(a)(3).

- LRA Section 3.3.2.2.10.5 refers to LRA Table 3.3.1, item 27 and addresses loss of material due to pitting and crevice corrosion in HVAC stainless steel ducting and components and aluminum piping and components exposed to condensation. The applicant stated that the Inspection of Internal Surfaces in Miscellaneous Piping and Ducting Components Program will manage the loss of material due to pitting and crevice corrosion for stainless steel and aluminum internal surfaces exposed to ventilation atmosphere and wetted gas.

  The staff reviewed LRA Section 3.3.2.2.10.5 against the criteria in SRP-LR Section 3.3.2.2.10, item 5, which states that loss of material due to pitting and crevice corrosion could occur in HVAC aluminum piping, piping components, and piping elements and stainless steel ducting and components exposed to condensation. It also recommends further evaluation of a plant-specific AMP to ensure this aging effect is adequately managed.

  The applicant stated that the Inspection of Internal Surfaces in Miscellaneous Piping and Ducting Components Program will conduct visual inspections during periodic maintenance, predictive maintenance, surveillance testing, and corrective maintenance to detect aging effects that could result in a loss of component intended function. The staff noted that the GALL Report AMP manages aging effect of loss of material for steel components. However, the applicant has taken an exception to increase the scope of the materials to include aluminum and stainless steel alloy. The staff reviewed the exception as part of its review of the Inspection of Internal Surfaces in Miscellaneous Piping and Ducting Components Program, and its evaluation is documented in SER Section 3.0.3.2.15.

Aging Management Review Results

Based on the programs identified, the staff concludes that the applicant's programs meet SRP-LR Section 3.3.2.2.10 item 5 criteria. For those items that apply to LRA Section 3.3.2.2.10.5, the staff determines that the LRA is consistent with the GALL Report and that the applicant has demonstrated that the effects of aging will be adequately managed so that the intended functions will be maintained consistent with the CLB during the period of extended operation, as required by 10 CFR 54.21(a)(3).

- LRA Section 3.3.2.2.10.6 refers to LRA Table 3.3.1, item 28 and addresses copper alloy fire protection system piping, piping components, and piping elements exposed to condensation, which are being managed for loss of material due to pitting and crevice corrosion by the applicant's Inspection of Internal Surfaces in Miscellaneous Piping and Ducting Components Program. The applicant addressed the further evaluation criteria of the SRP-LR by stating that it will use the Inspection of Internal Surfaces in Miscellaneous Piping and Ducting Components Program to perform inspections on the internal surfaces of the components exposed to wetted gas.

  The staff reviewed LRA Section 3.3.2.2.10.6 against the criteria in SRP-LR Section 3.3.2.2.10, item 6, which states that loss of material, due to pitting and crevice corrosion, could occur for copper alloy fire protection system piping, piping components, and piping elements exposed to internal condensation. The SRP-LR also states that the GALL Report recommends further evaluation of a plant specific AMP to ensure that the aging effects are adequately managed.

  SER Section 3.0.3.2.15 documents the staff's evaluation of the applicant's Inspection of Internal Surfaces in Miscellaneous Piping and Ducting Components Program. The staff finds the applicant's use of the Inspection of Internal Surfaces in Miscellaneous Piping and Ducting Components Program acceptable because it performs visual inspections that can detect loss of material for copper alloy components.

  Based on the programs identified, the staff concludes that the applicant's programs meet SRP-LR Section 3.3.2.2.10, item 6 criteria. For those items that apply to LRA Section 3.3.2.2.10.6, the staff determines that the LRA is consistent with the GALL Report and that the applicant has demonstrated that the effects of aging will be adequately managed so that the intended function(s) will be maintained consistent with the CLB during the period of extended operation as required by 10 CFR 54.21(a)(3).

- LRA Section 3.3.2.2.10.7 refers to Table 3.3.1, item 29 and addresses stainless steel piping, piping components, and piping elements exposed to soil which are being managed for loss of material due to pitting and crevice corrosion by the Buried Piping and Tanks Inspection Program. The applicant addressed the further evaluation criteria of the SRP-LR by stating that the Buried Piping and Tanks Inspection Program will manage the loss of material due to pitting and crevice corrosion in external surfaces of stainless steel piping, piping components, and piping elements in a soil environment.

  The staff reviewed LRA Section 3.3.2.2.10.7 against the criteria described in SRP-LR Section 3.3.2.2.10, item 7, which states that loss of material due to pitting and crevice corrosion could occur for stainless steel piping, piping components, and piping elements exposed to soil. The SRP-LR also recommends further evaluation of a plant-specific AMP to ensure that these aging effects are adequately managed. BTP RLSB-1 describes acceptance criteria are (Appendix A.1 of the SRP-LR).

  SER Section 3.0.3.2.12 documents the staff's evaluation of the applicant's Buried Piping and Tanks Inspection Program. The staff finds the applicant's proposal to manage aging using the Buried Piping and Tanks Inspection Program acceptable because it requires periodic visual inspections of the external surface of buried steel piping prior to and within the period of extended operation.

Aging Management Review Results

Based on the programs identified, the staff concludes that the applicant's programs meet SRP-LR Section 3.3.2.2.10, item 7 criteria. For those items that apply to LRA Section 3.3.2.2.10.7, the staff determines that the LRA is consistent with the GALL Report and that the applicant has demonstrated that the effects of aging will be adequately managed so that the intended function(s) will be maintained consistent with the CLB during the period of extended operation as required by 10 CFR 54.21(a)(3).

- LRA Section 3.3.2.2.10.8 references Table 3.3.1, item 30 and addresses loss of material due to pitting and crevice corrosion in stainless steel piping, piping components, and piping elements of the BWR Standby Liquid Control System that are exposed to sodium pentaborate solution, stating that this aging effect is not applicable; it is applicable to BWRs only. The staff finds that SRP-LR Section 3.3.2.2.10 item 8 is not applicable because the PVNGS units are PWRs, and the staff guidance in this SRP-LR section is only applicable to BWRs.

Based on the programs identified above, the staff concludes that the applicant's programs meet SRP-LR Section 3.3.2.2.10 criteria. For those items that apply to LRA Section 3.3.2.2.10, the staff determines that the LRA is consistent with the GALL Report and that the applicant has demonstrated that the effects of aging will be adequately managed so that the intended function(s) will be maintained consistent with the CLB during the period of extended operation, as required by 10 CFR 54.21(a)(3).

3.3.2.2.11 Loss of Material Due to Pitting, Crevice, and Galvanic Corrosion

LRA Section 3.3.2.2.11 refers to Table 3.3.1, item 3.3.1.31 and addresses the loss of material due to pitting, crevice, and galvanic corrosion. The applicant indicated that this is only applicable to BWRs. The staff reviewed LRA Section 3.3.2.2.11 against the criteria in SRP-LR Section 3.3.2.2.11 and agrees that this item is not applicable to a PWR unit, and therefore, finds the applicant's determination acceptable.

3.3.2.2.12 Loss of Material Due to Pitting, Crevice, and Microbiologically-Influenced Corrosion

The staff reviewed LRA Section 3.3.2.2.12 against the following criteria in SRP-LR Section 3.3.2.2.12:

- LRA Section 3.3.2.2.12.1 refers to Table 3.3.1, item 3.3.1.32 and addresses the loss of material due to pitting and crevice corrosion and MIC for stainless steel, aluminum, and copper components exposed to fuel oil. The applicant stated that the Fuel Oil Chemistry Program (LRA B2.1.14), reviewed in SER Section 3.0.3.2.9, and the One-Time Inspection Program (LRA B2.1.16), reviewed in SER Section 3.0.3.1.6, manage this aging effect. The applicant also stated that the one-time inspection will include selected components at susceptible locations where contaminants could accumulate (e.g. stagnant flow locations and tank bottoms).

    The staff reviewed LRA Section 3.3.2.2.12.1 against the criteria in SRP-LR Section 3.3.2.2.12.1, which states that loss of material due to pitting and crevice corrosion and MIC could occur in stainless steel, aluminum, and copper alloy piping, piping components, and piping elements exposed to fuel oil. The GALL Report recommends that these aging effects be managed by the Fuel Oil Chemistry Program (GALL Report AMP XI.M30) and that the effectiveness of this program be verified through the use of the One-Time Inspection Program (GALL Report AMP XI.M32). The GALL Report also recommends further evaluation because corrosion may occur at locations were contaminants accumulate.

    The staff notes that in LRA Section 3.3.2.2.12.1, the applicant has identified appropriate AMPs and has identified the critical conditions requiring further review, i.e., the potential for fuel oil chemistry control to be ineffective in locations where contaminants

3-248

Aging Management Review Results

accumulate. The staff finds that LRA Section 3.3.2.2.12.1 is consistent with SRP-LR Section 3.3.2.2.12, item 1.

Based on the programs identified, the staff concludes that the applicant's programs meet SRP-LR Section 3.3.2.2.12.1 criteria. For those items that apply to LRA Section 3.3.2.2.12.1, the staff determines that the LRA is consistent with the GALL Report and that the applicant has demonstrated that the effects of aging will be adequately managed so that the intended function(s) will be maintained consistent with the CLB during the period of extended operation, as required by 10 CFR 54.21(a)(3)

- LRA Section 3.3.2.2.12 refers to LRA Table 3.3.1, item 3.3.1.33 and addresses stainless steel piping, piping components, and elements exposed to lubricating oil, which are being managed for loss of material due to pitting and crevice corrosion and MIC by the Lubricating Oil Analysis Program and the One-Time Inspection Program. The applicant addressed the further evaluation criteria of the SRP-LR by stating that it will include a one-time inspection of selected components at susceptible locations where contaminants such as water could accumulate.

  The staff reviewed LRA Section 3.3.2.2.12 item 2 against the criteria in SRP-LR Section 3.3.2.2.12, item 2, which states that loss of material due to pitting, crevice, and MIC could occur in stainless steel piping, piping components, and piping elements exposed to lubricating oil. SRP-LR Section 3.3.2.2.12.2 further states that: (1) the existing program relies on the periodic sampling and analysis of lubricating oil to maintain contaminants within acceptable limits, thereby preserving an environment that is not conducive to corrosion; (2) the effectiveness of the lubricating oil program is verified through one-time inspection of selected components at susceptible locations to ensure that corrosion is not occurring and that the component's intended function will be maintained during the period of extended operation; and (3) the GALL Report recommends further evaluation of programs to manage corrosion by verifying their effectiveness with a one-time inspection of selected components at susceptible locations.

  The staff reviewed the applicant's Lubricating Oil Analysis Program and One-Time Inspection Program in SER Sections 3.0.3.2.16 and 3.0.3.1.6, respectively. In its review of the components associated with the LRA items listed above, the staff finds the applicant's proposal to manage aging using the Lubricating Oil Analysis and One-Time Inspection Programs acceptable because: (1) the Lubricating Oil Analysis Program provides for periodic sampling of lubricating oil to maintain contaminants at acceptable limits to preclude loss of material; (2) the applicant will perform one-time inspections of select stainless steel piping, piping components, and piping elements exposed to lubricating oil for loss of material due to pitting and crevice corrosion and MIC. This one-time inspection verifies the effectiveness of the Lubricating Oil Analysis Program in applicable auxiliary systems, following The GALL Report recommendation that the "One Time Inspection" is an acceptable AMP to verify the effectiveness of the applicant's Lubricating Oil Analysis Program.

  Based on the programs identified, the staff concludes that the applicant's programs meet SRP-LR Section 3.3.2.2.12.2 criteria. For those items that apply to LRA Section 3.3.2.2.12.2, the staff determines that the LRA is consistent with the GALL Report and that the applicant has demonstrated that the effects of aging will be adequately managed so that the intended function(s) will be maintained consistent with the CLB during the period of extended operation, as required by 10 CFR 54.21(a)(3).

Based on the programs identified above, the staff concludes that the applicant's programs meet SRP-LR Section 3.3.2.2.12 criteria. For those items that apply to LRA Section 3.3.2.2.12, the

Aging Management Review Results

staff determines that the LRA is consistent with the GALL Report. The staff also finds that the applicant has demonstrated that the effects of aging will be adequately managed so that the intended function(s) will be maintained consistent with the CLB during the period of extended operation, as required by 10 CFR 54.21(a)(3).

3.3.2.2.13 Loss of Material Due to Wear

LRA Section 3.3.2.2.13 refers to Table 3.3.1, item 34 and addresses elastomeric components exposed to air-indoor uncontrolled (internal or external) affected by loss of material due to wear. The applicant stated that this item is not applicable because it has no in scope elastomeric components exposed to relative motion with other components to produce an aging effect of loss of material due to wear. The staff reviewed LRA Sections 2.3.3 and 3.3 and confirmed that the applicant's LRA does not have any AMR results for the auxiliary systems that include elastomeric components exposed to air-indoor uncontrolled (internal or external) that are subject to relative motion. The staff notes that the SRP-LR references GALL Report tables that limit the scope of this item to HVAC systems. The staff reviewed the applicant's UFSAR and confirmed that no in-scope elastomeric components exposed to air-indoor uncontrolled (internal or external) are present in the auxiliary systems that are exposed to relative motion and, therefore, it finds the applicant's determination acceptable.

3.3.2.2.14 Loss of Material Due to Cladding Breach

LRA Section 3.3.2.2.14 refers to Table 3.3.1, item 3.3.1.35 and addresses the loss of material due to cladding breach for steel pumps with stainless steel cladding exposed to treated borated water in the auxiliary systems. The applicant stated that PVNGS has no in-scope stainless steel clad pump casings exposed to treated borated water in the CVCS. The staff reviewed LRA Sections 2.3.3 and 3.3 and confirmed that the applicant's LRA does not have any AMR results for the auxiliary systems that include steel pumps with stainless steel cladding exposed to treated borated water. The staff reviewed the applicant's UFSAR to verify the design of the CVCS pumps. The staff confirmed, using the UFSAR, that these charging pump casings were fabricated from solid pieces of stainless steel; therefore, the staff finds the applicant's determination acceptable.

3.3.2.2.15 Quality Assurance for Aging Management of Nonsafety-Related Components

SER Section 3.0.4 documents the staff's evaluation of the applicant's QA program.

3.3.2.3   Aging Management Review Results Not Consistent with or Not Addressed in the Generic Aging Lessons Learned Report

In LRA Tables 3.3.2-1 through 3.3.2-30, the staff reviewed additional details of the AMR results for material, environment, AERM, and AMP combinations not consistent with or not addressed in the GALL Report.

In LRA Tables 3.3.2-1 through 3.3.2-30, via Notes F–J, the applicant indicated which combinations of component type, material, environment, and AERM do not correspond to an item in the GALL Report. The applicant provided further information about how it will manage the aging effects. Specifically, Note F indicates that the material for the AMR item component is not evaluated in the GALL Report. Note G indicates that the environment for the AMR item component and material is not evaluated in the GALL Report. Note H indicates that the aging effect for the AMR item component, material, and environment combination is not evaluated in the GALL Report. Note I indicates that the aging effect identified in the GALL Report for the item component, material, and environment combination is not applicable. Note J indicates that neither the component nor the material and environment combination for the item is evaluated in the GALL Report.

Aging Management Review Results

For component type, material, and environment combinations not evaluated in the GALL Report, the staff reviewed the applicant's evaluation to determine if the applicant has demonstrated that it will adequately manage the effects of aging so that the intended function(s) will be maintained consistent with the CLB for the period of extended operation. The following sections document the staff's evaluation.

3.3.2.3.1 Fuel Handling and Storage System—Summary of Aging Management Review—License Renewal Application Table 3.3.2-1

The staff reviewed LRA Table 3.3.2-1, which summarizes the results of AMR evaluations for the fuel-handling and storage system component groups.

The staff's review did not find any items indicating plant-specific Notes F–J, where the combination of component type, material, environment, and AERM does not correspond to an item in the GALL Report.

Ser Section 3.0.2.1 documents the staff's evaluation of the items with Notes A–E.

3.3.2.3.2 Spent Fuel Pool Cooling and Cleanup System—Summary of Aging Management Review—License Renewal Application Table 3.3.2-2

The staff reviewed LRA Table 3.3.2-2, which summarizes the results of AMR evaluations for the spent fuel pool cooling and cleanup system component groups.

In LRA Tables 3.3.2-2, 3.3.2-7, and 3.3.2-20 the applicant stated that the Bolting Integrity Program manages stainless steel closure bolting exposed to treated borated water and raw water, and carbon steel closure bolting exposed to diesel fuel oil, for loss of preload. The AMR items cite generic Note G, indicating that the environment is not in the GALL Report for the component, material, and environment.

The staff noted that for each of the AMR results, the closure bolting is in a liquid environment, and the bolted fittings have the same environment both internally and externally. The staff also noted that in such an environment visual examination during system walkdowns would not readily detect indications of leakage around the bolted joints. By letter dated April 28, 2010, the staff issued RAI 3.3.2-2 asking that the applicant explain what activities of the Bolting Integrity Program it will use to detect loss of preload for closure bolting in a liquid environment. The staff also asked the applicant to clarify if there are any indirect indicators that may identify reduction of preload for closure bolting in the liquid environments.

In its response, dated May 21, 2010, the applicant stated that for the submerged bolting in the spent fuel pool cooling and cleanup system and the diesel fuel oil storage and transfer system, the AMSE Section XI, Subsections IWB, IWC, and IWD Program manages the inspection of safety related bolting and supplements the Bolting Integrity Program. The applicant also stated that for these bolting components, inspections detect loss of material due to corrosion and evidence of leakage. The applicant further stated that the extent and schedule of ASME Section XI, Subsections IWB, IWC, and IWD Program inspections, combined with periodic system walkdowns, assure detection of leakage before the leakage becomes excessive. In addition, any unusual indications of system performance, such as reduced ability to hold pressure or pump and piping vibration or noise, observed during periodic system walkdowns and pressure testing are identified and entered into the Corrective Action Program. The applicant stated that inability to successfully complete an ASME Code Section XI pressure test or observation of unusual indications of system performance would be used as indicators to reveal a potential loss of preload or other aging effects for closure bolting. The applicant's response also indicated that additional direct inspections are not performed for the stainless steel closure bolting exposed to raw water in the essential spray system and that indirect indicators such as unusual changes in system performance would be used indirectly to detect reduction of preload for these components.

Aging Management Review Results

The staff notes that the applicant has adequately addressed loss of preload in bolting components exposed to a liquid environment by crediting additional ASME Code Section XI inspections, where they are applicable, and by using indirect performance indicators to aid in detecting loss of bolting preload, if ASME inspections are not required. The staff finds these inspection methods, together with control of bolting preload during design and maintenance activities, acceptable to manage loss of preload in these bolting components exposed to a liquid environment.

In LRA Tables 3.3.2-2 and 3.3.2-8, the applicant stated that the Water Chemistry and One-Time Inspection Programs manage the stainless steel fuel pool cooling and reactor hot leg sample cooler heat exchangers exposed to treated borated water for reduction of heat transfer. The AMR item cites generic Note H, indicating that for the items, the aging effect is not in the GALL Report for this component, material, and environment combination. The items associated with the stainless steel fuel pool cooling and reactor hot leg sample cooler heat exchangers exposed to treated borated water, Tables 3.2.2-2 and 3.3.2-8 cite plant specific notes 2 and 1, respectively, which state, "Reduction in heat transfer due to fouling is a potential aging effect for stainless steel heat exchanger components in treated borated water. This non-GALL Report item is based upon the component, material, aging effects, and aging management program combination of GALL Report item VII.E1-4."

The staff reviewed all AMR result items in the GALL Report where the component and material is stainless steel heat exchangers exposed to treated borated water and noted that these items relate to SPR Section 3.3.2.2.2. This is inconsequential since the associated GALL Report items recommend that the Water Chemistry Program, verified by the One-Time Inspection Program, manages aging, and the applicant selected the same programs to manage aging.

SER Sections 3.0.3.2.1 and 3.0.3.1.6 document the staff's evaluation of the applicant's Water Chemistry and a One-Time Inspection Programs, respectively.

For those items that apply to LRA Section 3.3.2.2.2, the staff determines that the LRA contains appropriate AMR line items to ensure consistency with the GALL Report and that the applicant has demonstrated that the effects of aging will be adequately managed so that the intended function(s) will be maintained consistent with the CLB during the period of extended operation, as required by 10 CFR 54.21(a)(3).

3.3.2.3.3 Essential Cooling Water System—Summary of Aging Management Review— License Renewal Application Table 3.3.2-3

The staff reviewed LRA Table 3.3.2-3, which summarizes the results of AMR evaluations for the essential cooling water system component groups.

In LRA Tables 3.3.2-3, 3.3.2-4, 3.3.2-5, 3.3.2-7, 3.3.2-21, 3.3.2-22, 3.3.2-23, 3.3.2-27, 3.3.2-28, and 3.3.2-30, the applicant stated that the Boiling Integrity Program manages stainless steel closure bolting exposed to plant indoor air for loss of preload. The AMR items cite generic Note G, indicating that the environment is not in the GALL Report for this component and material. Some of the AMR items also cite plant-specific note 1, indicating that loss of preload is considered to be applicable for all closure bolting.

SER Section 3.0.3.2.3 documents the staff's evaluation of the applicant's Bolting Integrity Program. The staff noted that the mechanisms identified in the GALL Report as causing loss of preload in carbon steel bolts are thermal effects, gasket creep, and self-loosening, which are not all dependent on the material or the environment. The staff also noted that activities in the Bolting Integrity Program that control and manage loss of preload are effective for various bolting materials and environments. The staff further noted that the GALL Report, item VII.I-5 (AP-26) recommends using the Bolting Integrity Program to manage the aging effect of loss of preload in carbon steel bolts exposed to air-indoor uncontrolled.

Aging Management Review Results

On the basis that the GALL Report recommends the Bolting Integrity Program for managing loss of preload in carbon steel bolting exposed to air-indoor uncontrolled, and the Bolting Integrity Program's activities for managing loss of preload are applicable for other bolting materials and similar environments, the staff finds the applicant's use of the Bolting Integrity Program to manage loss of preload in stainless steel bolting exposed to plant indoor air to be acceptable.

In LRA Tables 3.3.2-3, 3.3.2-10, 3.3.2-11, and 3.3.2-22, the applicant stated that aluminum chambers, carbon steel heat exchanger (seal injection), piping, strainer, and tank, cast iron pump and flow indicator, and stainless steel valve components exposed to closed-cycle cooling water, borated water leakage, and potable water are managed for loss of material by the Inspection of Internal Surfaces in Miscellaneous Piping and Ducting Components Program. The AMR items cite generic Note G, indicating that for the item(s) the environment is not in the GALL Report for this component and material

The staff reviewed the applicant's Inspection of Internal Surfaces in Miscellaneous Piping and Ducting Components Program, and SER Section 3.0.3.2.15 documents its evaluation. The staff finds the applicant's proposal to manage aging using the Internal Surfaces in Miscellaneous Piping and Ducting Components Program acceptable because the applicant will: (1) conduct visual inspections during periodic maintenance, predictive maintenance, surveillance testing and corrective maintenance to detect loss of material for the steel, copper alloy, cast iron, aluminum and stainless steel components exposed to borated water leakage, potable water, or closed-cycle cooling water: (2) remedy deficiencies per the applicant's Corrective Action Program; and (3) within 10 years prior to extended operation will review all systems within the scope of the program for aging effects to provide reasonable assurance that their intended functions will remain through the next inspection cycle.

On the basis of its review, the staff finds that the applicant has appropriately evaluated the AMR results of material, environment, AERM, and AMP combinations not evaluated in the GALL Report. The staff finds that the applicant has demonstrated that the effects of aging will be adequately managed so that the intended function(s) will be maintained consistent with the CLB for the period of extended operation, as required by 10 CFR 54.21(a)(3).

3.3.2.3.4 Essential Chilled Water System—Summary of Aging Management Review—License Renewal Application Table 3.3.2-4

The staff reviewed LRA Table 3.3.2-4, which summarizes the results of AMR evaluations for the essential chilled water system component groups.

In LRA Tables 3.3.2-4, 3.3.2-10, and 3.3.2-21, the applicant stated that calcium silicate and mineral wool insulation exposed to borated water leakage have no AERM and no proposed AMP. The AMR items cite generic Note J.

The staff reviewed the associated items in the LRA and noted that the applicant only included piping insulation in scope for license renewal for those systems where the insulation has an intended function. The staff also noted that both calcium silicate and mineral wool insulation materials can experience loss of insulating properties when exposed to moisture, due to effects such as compression of the material or change in material properties, but that proper jacketing of the insulation can be effective at preventing moisture intrusion. The staff further noted that both materials can retain moisture well after exposure, prolonging the contact time of the moisture with the piping being insulated. The staff confirmed that the affected piping is jacketed with overlapping seams, such that moisture intrusion is not a concern. The staff finds the applicant's determination that calcium silicate and mineral wool insulation in this environment have no AERM acceptable because these materials perform well when exposed to air and are properly jacketed to prevent moisture intrusion.

Aging Management Review Results

On the basis of its review, the staff finds that the applicant has appropriately evaluated the AMR results of material, environment, AERM, and AMP combination not addressed in the GALL Report. The staff finds that the applicant demonstrated that the effects of aging for these components will be adequately manage so that their intended functions will be maintained consistent with the CLB during the period of extended operation, as required by 10 CFR 54.21(a)(3).

In LRA Tables 3.3.2-4, 3.3.2-5, and 3.3.2-6, the applicant stated that the Closed-Cycle Cooling Water System Program manages the nickel-alloy flexible hoses, exposed internally to closed-cycle cooling water, for loss of material. The AMR items cite generic Note G.

The staff reviewed the associated items in the LRA and confirmed that the applicant has identified the correct aging effects for this component, material, and environmental combination because, similar to GALL Report AMR item VII.C2-10, these components are also subject to loss of material when exposed to closed-cycle cooling water. The staff noted that the conditions required for cracking, due to a variety of mechanisms (SCC, PWSCC, IASCC and IGSCC), such as high fluid temperatures, do not exist for these components when exposed to the closed-cycle cooling water in this system.

SER Section 3.0.3.2.5 documents the staff's evaluation of the applicant's Closed-Cycle Cooling Water System Program. The staff noted that this program includes maintenance of system corrosion inhibitor concentrations to minimize aging effects and periodic testing and inspections to evaluate system and component performance. The staff noted that controlling the chemistry of the closed-cycle cooling water will create an environment that is not conducive to corrosion. Furthermore, the applicant's program includes periodic inspection processes such as visual, eddy current, and ultrasonic examinations. The program also includes periodic testing methods, such as functional demonstrations, monitoring, and thermal and hydraulic performance testing, to confirm the effectiveness of the chemistry control and ensure that degradation is not occurring. The staff finds the applicant's proposal to manage aging using the Closed-Cycle Cooling Water System Program acceptable because the applicant is controlling the chemistry of the water to create an environment that is not conducive for degradation, and it will perform periodic inspections or testing methods, as described above, to confirm the effectiveness of the chemistry control.

In LRA Table 3.3.2-4, the applicant stated that, for sight gauges made of glass exposed to dry gas, there is no aging effect and no proposed AMP. The AMR item cites generic Note G, indicating that for the item, the environment is not in the GALL Report for this component and material.

The staff noted that in LRA Table 3.0-1 the applicant described dry gas as dry air and inert or non-reactive gases, including compressed instrument air, nitrogen, oxygen, hydrogen, helium, halon, or freon. The staff finds the applicant's proposal acceptable because the GALL Report does not identify an aging effect for any glass material exposed to any environment (e.g., lubricating oil, air, treated borated water). The dry gas environment is less aggressive than the examples in the GALL Report; therefore, no AMP is required.

SER Section 3.3.2.3.3 documents the staff's evaluation for the Bolting Integrity Program, managing the stainless steel closure bolting, exposed to plant indoor air, for loss of preload, citing generic Note G.

On the basis of its review, the staff finds that the applicant has appropriately evaluated the AMR results of material, environment, AERM, and AMP combinations not evaluated in the GALL Report. The staff finds that the applicant has demonstrated that the effects of aging will be adequately managed so that the intended function(s) will be maintained consistent with the CLB for the period of extended operation, as required by 10 CFR 54.21(a)(3).

Aging Management Review Results

3.3.2.3.5 Normal Chilled Water System—Summary of Aging Management Review—License Renewal Application Table 3.3.2-5

The staff reviewed LRA Table 3.3.2-5, which summarizes the results of AMR evaluations for the normal chilled water system component groups.

SER Section 3.3.2.3.3 documents the staff's evaluation for the Bolting Integrity Program managing stainless steel closure bolting, exposed to plant indoor air, for loss of preload, citing generic Note G.

SER Section 3.3.2.3.4 documents the staff's evaluation for nickel-alloy components, exposed to closed-cycle cooling water, being managed for loss of material by the Closed-Cycle Cooling Water System Program, with generic Note G.

3.3.2.3.6 Nuclear Cooling Water System—Summary of Aging Management Review—License Renewal Application Table 3.3.2-6

The staff reviewed LRA Table 3.3.2-6, which summarizes the results of AMR evaluations for the nuclear cooling water system component groups.

SER Section 3.3.2.3.4 documents the staff's evaluation for the Closed-Cycle Cooling Water System Program managing nickel-alloy components, exposed to closed-cycle cooling water, for loss of material, with generic Note G.

3.3.2.3.7 Essential Spray Pond System—Summary of Aging Management Review—License Renewal Application Table 3.3.2-7

The staff reviewed LRA Table 3.3.2-7, which summarizes the results of AMR evaluations for the ESP system component groups.

In LRA Tables 3.3.2-7, 3.3.2-30, and 3.3.2-31, the applicant stated that the piping and piping components (valve, strainer, flow indicator, tubing) and a corrosion test rack made of PVC exposed to plant indoor air (external) and raw or secondary water (internal) do not have an AERM and do not require an AMP. In LRA Table 3.3.2-19, the applicant stated that piping made of fiberglass reinforced plastic exposed to a buried environment (external) and raw water (internal) does not have an AERM and does not require an AMP. In LRA Table 3.3.2-22, the applicant stated that piping made of polyethylene exposed to plant indoor air (external) or a buried environment (external) and to raw water (internal) does not have an AERM and does not require an AMP. In LRA Table 3.3.2-23, the applicant stated that thermoplastic demineralizers exposed to plant indoor air (external) and demineralized water (internal) do not have an AERM and do not require an AMP. The AMR items cite Note F, indicating that the GALL Report does not evaluate the component, material, and environment combinations.

The staff noted that for the evaluated components in the ESP system (Table 3.3.2-7), fire protection system (Table 3.3.2-19), domestic water system (Table 3.3.2-22), demineralized water system (Table 3.3.2-23), and oily waste and non radioactive waste system (Table 3.3.2-31)—none of these systems are high-temperature and high-pressure systems. The staff confirmed that the components are used in applications where sustained exposure to ultraviolet light, high radiation, and ozone concentrations is not expected. The staff also notes that, based on its review of technical literature (e.g., Roff, 1956), current industry research, and operating experience related to PVC, polyethylene, thermo plastics, and fiberglass-reinforced plastic piping and piping components, in the absence of specific environmental stressors such as ultraviolet light, high radiation, or ozone concentrations, piping components made of these materials do not exhibit aging effects of concern during the period of extended operation. For the PVC components from Table 3.3.2 30, "Miscellaneous Auxiliary Systems In-Scope ONLY based on Criterion 10 CFR 54.4(a)(2)," that are exposed to secondary water, the staff notes that these items are located in the secondary chemical control system and, based on plant drawings,

Aging Management Review Results

located in a cold lab in the auxiliary building. The staff also notes that these items would not be exposed to direct ultraviolet lighting, high radiation, or ozone. Therefore, the staff finds the applicant's proposal acceptable because the subject components have no aging effects that cause degradation during the period of extended operation.

On the basis of its review, the staff finds that the applicant has appropriately evaluated the AMR results of material, environment, AERM, and AMP combinations not addressed in the GALL Report. The staff finds that the applicant has demonstrated that the effects of aging for these components will be adequately managed so that their intended function(s) will be maintained consistent with the CLB during the period of extended operation, as required by 10 CFR 54.21(a)(3).

In LRA Tables 3.3.2-7, 3.3.2-9, 3.3.2-22, and 3.3.2-30, the applicant stated that the Bolting Integrity Program manages copper alloy and copper alloy with greater than 8-percent aluminum closure bolting exposed to plant indoor for loss of preload. The AMR items cite generic Notes F and G, indicating that the material is not in the GALL Report for this component.

SER Section 3.0.3.2.3 documents the staff's evaluation of the applicant's Bolting Integrity Program. The staff noted that the mechanisms identified in the GALL Report as causing loss of preload in carbon steel bolts are thermal effects, gasket creep, and self-loosening and that these mechanisms can cause loss of preload in copper alloy bolts. The staff also noted that activities in the Bolting Integrity Program that control and manage loss of preload are effective for other metal bolting materials. The staff further noted that the GALL Report, item VII.I-5 (AP-26) recommends using the Bolting Integrity Program to manage loss of preload in carbon steel bolts exposed to air-indoor uncontrolled. The staff finds the applicant's use of the Bolting Integrity Program to manage loss of preload in copper alloy and copper alloy with greater than 8-percent aluminum bolting acceptable because it is consistent with the GALL Report recommendations for managing loss of preload in carbon steel bolting. In addition, the Bolting Integrity Program's activities for managing loss of preload are applicable for other metal bolting materials.

In LRA Tables 3.3.2-7, 3.3.2-10, 3.3.2-13, 3.3.2-19, 3.3.2-23, 3.3.2-24, 3.3.2-27, 3.4.2-2, and 3.5.2-10, the applicant stated that for stainless steel and CASS components exposed to atmosphere or weather conditions, there are no AERM and that no AMP will be implemented. The applicant referenced generic Note G for these items, indicating that the environment is not listed in the GALL Report for this material and component combination.

The staff reviewed all AMR results in the GALL Report where the material is stainless steel and the aging effect is loss of material and confirmed that there are no entries for this environment in the GALL Report for the component and material. However, the staff noted that the GALL Report does include AMR results for stainless steel components exposed to indoor uncontrolled air and that the GALL Report recommends no AMP be used because there is no aging effect for this material and environment combination. The staff also noted that the climate at this location is dry and arid with high average temperatures (68–108 degrees F) throughout the year and low average rainfall (8–10 inches per year), making the aging effects of the atmosphere or weather at this location similar to those of indoor uncontrolled air. The staff finds the applicant's determination that no AMP is required acceptable because stainless steel components exposed to the atmosphere or weather would not be expected to experience an aging effect at this location.

In LRA Table 3.3.2-7, the applicant stated that for nickel-alloy spray nozzles exposed to atmosphere or weather, there is no aging effect and no proposed AMP. The AMR item cites generic Note G. The applicant also stated, in its plant-specific notes 4 and 5, that these nickel-alloy components are located outside with an uncontrolled external air environment and are not exposed to aggressive chemical species. Furthermore, the plant outdoor environment is

Aging Management Review Results

not subject to industry air pollution or saline environment and alternate wetting and drying has shown a tendency to "wash" the surface material rather than concentrate contaminants.

The staff reviewed the associated items in the LRA and confirmed that no aging effect is applicable for this component, material, and environment because austenitic materials such as nickel alloys are not subject to loss of material or cracking when exposed to this environment, and these materials are used as corrosion-resistant replacement materials where other materials have degraded. The staff noted that corrosion-resistant materials, such as austenitic and martensitic stainless steels and high strength nickel base alloys, offer good protection against loss of material. The staff also noted that the conditions required for cracking, due to a variety of mechanisms (SCC, PWSCC, IASCC and IGSCC), to occur such as being exposed to an aqueous solution (reactor coolant or other corrosive solutions) and high temperatures, do not exist on the surfaces of these components when exposed to atmosphere or weather. The staff noted the applicant's definition of atmosphere or weather in LRA Table 3.0-1, "Mechanical Environments," states, in part, that there is no exposure to salt spray or other aggressive contaminants. Based on its review, the staff finds the applicant's determination acceptable because nickel alloys are highly corrosion-resistant, and these components are not exposed to an environment that contains salt spray or other contaminants that create an environment conducive to loss of material or corrosion. In addition, these components are not subject to the conditions required to induce cracking.

SER Section 3.3.2.3.2 documents the staff's evaluation of the Bolting Integrity Program, citing generic Note G, managing stainless steel closure bolting exposed to raw water for loss of preload.

SER Section 3.3.2.3.3 documents the staff's evaluation of the Bolting Integrity Program, citing generic Note G, managing stainless steel closure bolting exposed to plant indoor air for loss of preload.

On the basis of its review, the staff finds that the applicant has appropriately evaluated the AMR results of material, environment, AERM, and AMP combinations not evaluated in the GALL Report. The staff finds that the applicant has demonstrated that the effects of aging will be adequately managed so that the intended function(s) will be maintained consistent with the CLB for the period of extended operation, as required by 10 CFR 54.21(a)(3).

3.3.2.3.8 Nuclear Sampling System—Summary of Aging Management Review—License Renewal Application Table 3.3.2-8

The staff reviewed LRA Table 3.3.2-8, which summarizes the results of AMR evaluations for the nuclear sampling system component groups.

Section 3.3.2.3.2 documents the staff's evaluation for stainless steel heat exchanger components exposed to treated borated water managed by the Water Chemistry and One-Time Inspection Programs for reduction of heat transfer, with generic Note H.

On the basis of its review, the staff finds that the applicant has appropriately evaluated the AMR results of material, environment, AERM, and AMP combinations not evaluated in the GALL Report. The staff finds that the applicant has demonstrated that the effects of aging will be adequately managed so that the intended function(s) will be maintained consistent with the CLB for the period of extended operation, as required by 10 CFR 54.21(a)(3).

3.3.2.3.9 Compressed Air System—Summary of Aging Management Review—License Renewal Application Table 3.3.2-9

The staff reviewed LRA Table 3.3.2-9, which summarizes the results of AMR evaluations for the compressed air system component groups.

Aging Management Review Results

SER Section 3.3.2.3.7 documents the staff's evaluation for copper alloy and copper alloy with greater than 8 percent aluminum closure bolting exposed to plant indoor air being managed for loss of preload by the Bolting Integrity Program, citing generic Notes F and G.

On the basis of its review, the staff finds that the applicant has appropriately evaluated the AMR results of material, environment, AERM, and AMP combinations not evaluated in the GALL Report. The staff finds that the applicant has demonstrated that the effects of aging will be adequately managed so that the intended function(s) will be maintained consistent with the CLB for the period of extended operation, as required by 10 CFR 54.21(a)(3).

3.3.2.3.10 Chemical and Volume Control System—Summary of Aging Management Review—License Renewal Application Table 3.3.2-10

The staff reviewed LRA Table 3.3.2-10, which summarizes the results of AMR evaluations for the CVCS component groups.

In LRA Tables 3.3.2-4, 3.3.2-10, and 3.3.2-21, the applicant stated that calcium silicate and mineral wool insulation exposed to borated water leakage have no AERM and no proposed AMP. The AMR items cite generic Note J. Section 3.3.2.3.4 documents the staff's review.

In LRA Table 3.3.2-10, the applicant stated that for flow indicators and sight gauges made of glass exposed to borated water leakage (external) there is no aging effect and no proposed AMP. The AMR items cite generic Note G, indicating that for the items, the environment is not in the GALL Report for this component and material.

The staff reviewed all AMR result items in the GALL Report where the material is glass, and confirmed that for this environment, there are no entries in the GALL Report for this component and material. The staff notes that for glass components, the GALL Report does list a treated water environment in contrast to borated water leakage, but this is non consequential to the evaluation because the GALL Report items state that there is no aging effect and no recommended AMP.

The staff finds the applicant's proposal acceptable because glass, in an environment of borated water leakage, is expected to have the same aging effects as glass in an environment of treated borated water. The GALL Report states that there are no AERMs for glass piping components in a treated borated water environment.

In LRA Table 3.3.2-10, the applicant stated that for nickel-alloy piping exposed to borated water leakage there is no aging effect and no proposed AMP. The AMR items cite generic Note G.

The staff reviewed the associated items in the LRA and confirmed that no aging effect is applicable for this component, material, and environment because austenitic materials such as nickel alloys are not subject to loss of material or cracking when exposed to this environment, and these materials are used as corrosion-resistant replacement materials where other materials have degraded. The staff noted that according to EPRI NP-5769, "Degradation and Failure of Bolting in Nuclear Power Plants, Volumes 1 and 2," April 1988, corrosion-resistant materials, such as austenitic and martensitic stainless steels and high strength nickel base alloys, offer good protection against loss of material due to boric acid corrosion. The staff also noted that the conditions required for cracking due to a variety of mechanisms (SCC, PWSCC, IASCC and IGSCC) to occur, such as being exposed to an aqueous solution (reactor coolant or other corrosive solutions) and high temperatures, do not exist on the surfaces of these components when exposed to borated water leakage. Therefore, the staff finds no AMP is necessary for nickel alloys in a borated water leakage environment.

SER Section 3.3.2.3.3 documents the staff's evaluation for aluminum, carbon steel, copper alloy, cast iron, and stainless steel components exposed to closed-cycle cooling water, borated water leakage, and potable water, with an aging effect of loss of material managed by the

Aging Management Review Results

Inspection of Internal Surfaces in Miscellaneous Piping and Ducting Components Program, that cite generic Note G.

SER Section 3.3.2.3.7 documents the staff's evaluation for stainless steel and CASS components exposed to atmosphere or weather conditions, where there are no AERM and no AMP, citing generic Note G.

On the basis of its review, the staff finds that the applicant has appropriately evaluated the AMR results of material, environment, AERM, and AMP combinations not evaluated in the GALL Report. The staff finds that the applicant has demonstrated that the effects of aging will be adequately managed so that the intended function(s) will be maintained consistent with the CLB for the period of extended operation, as required by 10 CFR 54.21(a)(3).

3.3.2.3.11 Control Building Heating, Ventilation, and Air Conditioning System—Summary of Aging Management Review—License Renewal Application Table 3.3.2-11

The staff reviewed LRA Table 3.3.2-11, which summarizes the results of AMR evaluations for the control building HVAC system component groups.

SER Section 3.3.2.3.3 documents the staff's evaluation for aluminum, carbon steel, copper alloy, cast iron, and stainless steel components exposed to closed-cycle cooling water, borated water leakage, and potable water, with an aging effect of loss of material managed by the Inspection of Internal Surfaces in Miscellaneous Piping and Ducting Components Program, that cite generic Note G.

On the basis of its review, the staff finds that the applicant has appropriately evaluated the AMR results of material, environment, AERM, and AMP combinations not evaluated in the GALL Report. The staff finds that the applicant has demonstrated that the effects of aging will be adequately managed so that the intended function(s) will be maintained consistent with the CLB for the period of extended operation, as required by 10 CFR 54.21(a)(3).

3.3.2.3.12 Auxiliary Building Heating, Ventilation, and Air Conditioning System—Summary of Aging Management Review—License Renewal Application Table 3.3.2-12

The staff reviewed LRA Table 3.3.2-12, which summarizes the results of AMR evaluations for the auxiliary building HVAC system component groups. The staff's review did not find any items indicating plant-specific Notes F–J, where the combination of component type, material, environment, and AERM does not correspond to an item in the GALL Report.

SER Section 3.0.2.1 documents the staff's evaluation of the items with Notes A–E.

3.3.2.3.13 Fuel Building Heating, Ventilation, and Air Conditioning System—Summary of Aging Management Review—License Renewal Application Table 3.3.2-13

The staff reviewed LRA Table 3.3.2-13, which summarizes the results of AMR evaluations for the fuel building HVAC system component groups.

In LRA Tables 3.3.2-13, 3.3.2-17, 3.3.2-19, 3.3.2-20, 3.3.2-23, and 3.3.2-24, the applicant stated that carbon steel closure bolting exposed to atmosphere or weather are managed for loss of preload by the Bolting Integrity Program. The AMR items cite generic Note G, indicating that the environment is not in the GALL Report for this component and material. The AMR items also cite plant-specific note 1, indicating that loss of preload is considered to be applicable for all closure bolting.

SER Section 3.0.3.2.3 documents the staff's evaluation of the applicant's Bolting Integrity Program. The staff noted that the mechanisms identified in the GALL Report as causing loss of preload in carbon steel bolts are thermal effects, gasket creep, and self-loosening, which are not all dependent on the environment. The staff also noted that activities in the Bolting Integrity Program that control and manage loss of preload are effective for various bolting environments.

Aging Management Review Results

The staff further noted that the GALL Report, item VII.I-5 (AP-26) recommends using the Bolting Integrity Program to manage the aging effect of loss of preload in carbon steel bolts exposed to air-indoor uncontrolled.

On the basis that the GALL Report recommends the Bolting Integrity Program for managing loss of preload in carbon steel bolting exposed to atmosphere or weather, and the Bolting Integrity Program's activities for managing loss of preload are applicable for other similar bolting environments, the staff finds the applicant's use of the Bolting Integrity Program to manage loss of preload in carbon steel bolting exposed to atmosphere or weather to be acceptable.

SER Section 3.3.2.3.7 documents the staff's evaluation for stainless steel and CASS components exposed to atmosphere or weather conditions, where there are no AERM and no AMP, that cite generic Note G.

On the basis of its review, the staff finds that the applicant has appropriately evaluated the AMR results of material, environment, AERM, and AMP combinations not evaluated in the GALL Report. The staff finds that the applicant has demonstrated that the effects of aging will be adequately managed so that the intended function(s) will be maintained consistent with the CLB for the period of extended operation, as required by 10 CFR 54.21(a)(3).

3.3.2.3.14 Containment Building Heating, Ventilation, and Air Conditioning System—Summary of Aging Management Review—License Renewal Application Table 3.3.2-14

The staff reviewed LRA Table 3.3.2-14, which summarizes the results of AMR evaluations for the containment building HVAC system component groups.

The staff's review did not find any items, indicating plant-specific Notes F–J, where the combination of component type, material, environment, and AERM does not correspond to an item in the GALL Report.

SER Section 3.0.2.1 documents the staff's evaluation of the items with Notes A–E.

3.3.2.3.15 Diesel Generator Building Heating, Ventilation, and Air Conditioning System—Summary of Aging Management Review—License Renewal Application Table 3.3.2-15

The staff reviewed LRA Table 3.3.2-15, which summarizes the results of AMR evaluations for the diesel generator building HVAC system component groups.

The staff's review did not find any items, indicating plant-specific Notes F–J, where the combination of component type, material, environment, and AERM does not correspond to an item in the GALL Report.

SER Section 3.0.2.1 documents the staff's evaluation of the items with Notes A–E.

3.3.2.3.16 Radwaste Building Heating, Ventilation, and Air Conditioning System—Summary of Aging Management Review—License Renewal Application Table 3.3.2-16

The staff reviewed LRA Table 3.3.2-16, which summarizes the results of AMR evaluations for the radwaste building HVAC system component groups.

The staff's review did not find any items, indicating plant-specific Notes F–J, where the combination of component type, material, environment, and AERM does not correspond to an item in the GALL Report.

SER Section 3.0.2.1 documents the staff's evaluation of the items with Notes A–E.

Aging Management Review Results

3.3.2.3.17 Turbine Building Heating, Ventilation, and Air Conditioning System—Summary of Aging Management Review—License Renewal Application Table 3.3.2-17

The staff reviewed LRA Table 3.3.2-17, which summarizes the results of AMR evaluations for the turbine building HVAC system component groups.

In LRA Tables 3.3.2-17, 3.3.2-19, 3.3.2-20, 3.3.2-24, and 3.3.2-28, the applicant stated that the aluminum dampers, flame arrestors, valves, vents, and heat exchangers exposed to atmosphere or weather (external and internal) are managed for loss of material by the External Surfaces Monitoring Program. The AMR items cite generic Note G, indicating that for the item the environment is not in the GALL Report for this component and material. SER Section 3.0.3.2.14 documents the staff's evaluation of the applicant's External Surfaces Monitoring Program. The staff noted that all items but one (flame arrestor) are exposed to external air. For the excepted item, the applicant has cited plant-specific note 4, which states that these items are vented or open to the outside atmosphere so the distinction between internal and external is not relevant for aging purposes.

By letter dated December 29, 2009, the staff issued RAI B2.1.20-2 asking the applicant provide additional information on the detection, monitoring, and trending aging effects on hardware made of aluminum. By letter dated February 19, 2010, the applicant responded stating that aluminum due to its oxide formed on its surface exhibits a good resistance to corrosion except when exposed to halide or chloride-aerated solutions. In the mild environment to which aluminum components are exposed, the applicant also stated that "...rapid and aggressive corrosion of aluminum is not anticipated and visual inspection [performed] for loss of material and general corrosion, degraded material or physical conditions, and chipping, cracking, flaking, oxidizing, or missing paint and coatings as defined in plant procedures will identify degradation deficiencies prior to the loss of intended function." In addition, the applicant further stated that plant procedures require that degradation deficiencies be documented and evaluated and corrective actions taken in accordance with standards or site-specific methods. The staff finds the applicant's proposal to manage aging using the External Surfaces Monitoring Program acceptable because the applicant assures that aluminum is not exposed to detrimental environments and, if material degradation becomes evident, it will document and address the degradation through existing plant procedures before the loss of intended functions.

SER Section 3.3.2.3.13 documents the staff's evaluation for carbon steel closure bolting exposed to atmosphere or weather and fuel oil managed for loss of preload by the Bolting Integrity Program, citing generic Note G.

On the basis of its review, the staff finds that the applicant has appropriately evaluated the AMR results of material, environment, AERM, and AMP combinations not evaluated in the GALL Report. The staff finds that the applicant has demonstrated that the effects of aging will be adequately managed so that the intended function(s) will be maintained consistent with the CLB for the period of extended operation, as required by 10 CFR 54.21(a)(3).

3.3.2.3.18 Miscellaneous Site Structure and Spray Pond Pump House Heating, Ventilation, and Air Conditioning System—Summary of Aging Management Review—License Renewal Application Table 3.3.2-18

The staff reviewed LRA Table 3.3.2-18, which summarizes the results of AMR evaluations for the miscellaneous site structures and spray pond pump house HVAC system component groups.

The staff's review did not find any items, indicating plant-specific Notes F–J, where the combination of component type, material, environment, and AERM does not correspond to an item in the GALL Report.

SER Section 3.0.2.1 documents the staff's evaluation of the items with Notes A–E.

Aging Management Review Results

3.3.2.3.19 Fire Protection System—Summary of Aging Management Review—License Renewal Application Table 3.3.2-19

The staff reviewed LRA Table 3.3.2-19, which summarizes the results of AMR evaluations for the fire protection system component groups.

In LRA Tables 3.3.2-19 and 3.3.2-21, the applicant stated that copper alloy with greater than 15-percent zinc piping and valve components exposed to wetted gas are managed for loss of material by the Selective Leaching of Materials Program. The applicant cited generic Note G, indicating that the environment is not in the GALL Report for this component and material combination.

SER Section 3.0.3.2.11 documents the staff's evaluation of the applicant's Selective Leaching of Materials Program. The staff noted that the GALL Report, in the definitions section, states that copper alloys with greater than 15-percent zinc are susceptible to selective leaching. The staff finds the applicant's use of Selective Leaching of Materials Program acceptable because it uses visual inspection and mechanical methods to determine if loss of material due to selective leaching is occurring and, if signs of selective leaching are present, it performs metallurgical examinations and additional inspections.

In LRA Tables 3.3.2-19, 3.3.2-22, and 3.3.2-24, the applicant stated that copper alloys and copper alloys with Zinc greater than 15 percent for piping, piping components, sight gauges, strainers, and valves exposed to atmosphere or weather (external and internal) are managed for loss of material by the External Surfaces Monitoring Program. The AMR items cite generic Note G in all but two of the valve items, indicating that for these items the environment is not in the GALL Report for this component and material. The two valve items listed as Note B, should be more appropriately listed as G, because for these items the environment is not in the GALL Report for this component and material. SER Section 3.0.3.2.14 documents the staff's evaluation of the applicant's External Surfaces Monitoring Program. The staff noted in NUREG-1833, "Technical Bases for Revision to the License Renewal Guidance Documents" that copper alloys in dried air environment or in an indoor uncontrolled air environment exhibit no aging effects and no AMPS are designated. The staff also noted that NUREG-1833 states that comprehensive tests, conducted over a 20-year period under the supervision of ASTM, have confirmed the suitability of copper and copper alloys for atmospheric exposure as cited in Metals Handbook, Volume 13, "Corrosion," American Society for Metals, 1987. The staff finds the applicant's proposal to manage aging using the External Surfaces Monitoring Program acceptable because PVNGS is located in AZ, where the climate is extremely dry. Hence, according to NUREG-1833, no aging effects should be anticipated, and the applicant will still manage aging effects in the hardware through visual inspections crediting the External Surfaces Monitoring Program.

In LRA Table 3.3.2-19, the applicant stated that for copper alloy valves internally exposed to plant indoor air there is no aging effect and no proposed AMP. The AMR item cites generic Note G, indicating that the environment is not in the GALL Report for this component and material. The staff reviewed all AMR result items in the GALL Report where the material is copper alloy internally exposed to plant indoor air and confirmed that there are no aging effect entries in the GALL Report for this component, material, and environment combination.

The staff notes that the applicant's definition for the environment of plant indoor air as "indoor air on systems with temperatures higher than the dew point" is the same as the GALL Report definition for "air-indoor uncontrolled" in Table XI.D. The staff also notes that copper alloy valves are subject to the same aging effects (or lack of aging effects) on both the internal and external surface of the component exposed to a plant indoor air environment. The staff further notes that GALL Report, item VIII.I-2 identifies no aging effect or AMP for copper alloy components exposed to uncontrolled indoor air (external). Since the LRA components are

Aging Management Review Results

similar to other GALL Report items for the material and environment (i.e., GALL Report, item VIII.I-2, where the AERM is listed as "none," the AMP is listed as "none," and no further evaluation is required), the staff concurs that the effect of plant indoor air on the internal surface of copper alloy valves will not result in aging that will be of concern during the period of extended operation.

In LRA Table 3.3.2-19, the applicant stated that the Fire Protection Program manages thermo-lag fire barrier seals, externally exposed to plant indoor air, for loss of material and cracking. The AMR items cite generic Note J, indicating that neither the component nor the material and environment combination is evaluated in the GALL Report.

The staff reviewed the applicant's Fire Protection Program, and SER Section 3.0.3.2.7 documents its evaluation. The staff noted that the applicant's Fire Protection Program provides for visual inspection of fire barriers once every 18 months for detection of cracking and loss of material. The staff also noted that thermo-lag is primarily used to provide fire barriers for cable trays and conduits. The staff finds the applicant's proposal to manage aging using the Fire Protection Program acceptable because the program performs visual inspections of fire barriers that are capable of detecting loss of material and cracking for thermo-lag fire barriers.

SER Section 3.3.2.3.13 documents the staff's evaluation for carbon steel closure bolting exposed to atmosphere or weather and fuel oil managed for loss of preload by the Bolting Integrity Program, citing generic Note G.

SER Section 3.3.2.3.7 documents the staff's evaluation for PVC, fiberglass, and polyethylene piping and piping components exposed to plant indoor air, buried external environments, and raw water internal environment with no aging effects and no AMP, citing generic Note F.

SER Section 3.3.2.3.7 documents the staff's evaluation for stainless steel and CASS components exposed to atmosphere or weather conditions, where there are no AERM and no AMP, citing generic Note G.

SER Section 3.3.2.3.17 documents the staff's evaluation for aluminum dampers, flame arrestors, valves, vents, and heat exchangers exposed to atmosphere or weather (external and internal) and managed for loss of material corrosion by the External Surfaces Monitoring Program with generic Note G.

3.3.2.3.20 Diesel Generator Fuel Oil Storage and Transfer System—Summary of Aging Management Review—License Renewal Application Table 3.3.2-20

The staff reviewed LRA Table 3.3.2-20, which summarizes the results of AMR evaluations for the diesel generator fuel oil storage and transfer system component groups.

SER Section 3.3.2.3.2 documents the staff's evaluation for stainless steel closure bolting exposed to treated borated water managed for loss of preload by the Bolting Integrity Program, citing generic Note G.

SER Section 3.3.2.3.13 documents the staff's evaluation for carbon steel closure bolting exposed to atmosphere or weather and fuel oil managed for loss of preload by the Bolting Integrity Program, citing generic Note G.

SER Section 3.3.2.3.17 documents the staff's evaluation for aluminum dampers, flame arrestors, valves, vents, and heat exchangers exposed to atmosphere or weather (external and internal), managed for loss of material corrosion by the External Surfaces Monitoring Program, citing generic Note G.

On the basis of its review, the staff finds that the applicant has appropriately evaluated the AMR results of material, environment, AERM, and AMP combinations not evaluated in the GALL Report. The staff finds that the applicant has demonstrated that the effects of aging will be adequately managed so that the intended function(s) will be maintained consistent with the CLB for the period of extended operation, as required by 10 CFR 54.21(a)(3).

3.3.2.3.21 Diesel Generator—Summary of Aging Management Review—License Renewal Application Table 3.3.2-21

The staff reviewed LRA Table 3.3.2-21, which summarizes the results of AMR evaluations for the diesel generator component groups.

In LRA Tables 3.3.2-4, 3.3.2-10, and 3.3.2-21, the applicant stated that calcium silicate and mineral wool insulation exposed to borated water leakage have no AERM and no proposed AMP. The AMR items cite generic Note J. SER Section 3.3.2.3.4 documents the staff's review.

In LRA Tables 3.3.2-21, 3.3.2-28, and 3.4.2-3, the applicant stated that the aluminum heat exchanger (governor oil cooler), valve, and filter exposed to lubricating oil are managed for loss of material by the Lubricating Oil Analysis and One-Time Inspection Programs. The AMR items cite generic Note G, indicating that for the items, the environment is not in the GALL Report for this component and material.

The staff also reviewed the applicant's Lubricating Oil Analysis and One-Time Inspection Programs, and SER Sections 3.0.3.2.16 and 3.0.3.1.6 document their evaluations, respectively. The staff finds the applicant's proposed AMPs acceptable because the Lubricating Oil Analysis Program will control contaminates within limits to preclude loss of material, and the One-Time Inspection Program will verify the effectiveness of the Lubricating Oil Analysis Program by verifying that loss of material is not occurring.

In LRA Table 3.3.2-21, the applicant stated that for nickel-alloy valves exposed internally to dry gas there is no aging effect and no proposed AMP. The AMR item cites generic Note G.

The staff reviewed the associated items in the LRA and confirmed that no aging effect is applicable for this component, material, and environment because austenitic materials such as nickel alloys are not subject to loss of material or cracking when exposed to this environment, and these materials are used as corrosion-resistant replacement materials where other materials have degraded. The staff noted that corrosion-resistant materials, such as austenitic and martensitic stainless steels and high strength nickel base alloys, offer good protection against loss of material. The staff also noted that the conditions required for cracking due to a variety of mechanisms (SCC, PWSCC, IASCC and IGSCC) to occur, such as being exposed to an aqueous solution (reactor coolant or other corrosive solutions) and high temperatures, do not exist on the surfaces of these components when exposed to dry gas. The staff noted that GALL Report, AMR item IV.E-1 states that nickel alloys exposed to air-indoor uncontrolled is not subject to an aging effect requiring management. The staff noted that an air-indoor uncontrolled environment is more aggressive than a dry gas environment because it is possible for condensation to occur in an air-indoor uncontrolled environment. The staff finds the applicant's determination acceptable because nickel alloy is a highly corrosion-resistant material and it is not subject to conditions where cracking is possible. In addition, this component is exposed to a less aggressive environment when compared to GALL Report, AMR item IV.E-1, for which there is no AERM.

SER Section 3.3.2.3.3 documents the staff's evaluation for stainless steel closure bolting exposed to plant indoor air managed for loss of preload by the Bolting Integrity Program, citing generic note G.

SER Section 3.3.2.3.19 documents the staff's evaluation for copper alloy with greater than 15-percent zinc piping and valve components exposed to wetted gas managed for loss of material for by the Selective Leaching of Materials Program, citing generic Note G.

Aging Management Review Results

On the basis of its review, the staff finds that the applicant has appropriately evaluated the AMR results of material, environment, AERM, and AMP combinations not evaluated in the GALL Report. The staff finds that the applicant has demonstrated that the effects of aging will be adequately managed so that the intended function(s) will be maintained consistent with the CLB for the period of extended operation, as required by 10 CFR 54.21(a)(3).

3.3.2.3.22 Domestic Water System—Summary of Aging Management Review—License Renewal Application Table 3.3.2-22

The staff reviewed LRA Table 3.3.2-22, which summarizes the results of AMR evaluations for the domestic water system component groups.

In LRA Tables 3.3.2-22 and 3.3.2-24, the applicant stated that copper alloy sight gauges, strainers, and valves exposed to atmosphere or weather are managed for loss of material by the External Surfaces Monitoring Program. The applicant cited generic Note G, indicating that the environment is not in the GALL Report for this component and material combination.

SER Section 3.0.3.2.14 documents the staff's evaluation of the applicant's External Surfaces Monitoring Program. The applicant stated that the External Surfaces Monitoring Program conducts visual inspections on external surfaces to detect aging effects that could result in a loss of component intended function. The staff finds the applicant's use of the External Surfaces Monitoring Program acceptable because this program conducts visual inspections that are capable of detecting loss of material on the external surfaces of the components being managed.

In LRA Tables 3.3.2-22 and 3.3.2-30, the applicant stated that copper alloy piping, pumps, strainers, and valves exposed to potable water are managed for loss of material by the Inspection of Internal Surfaces in Miscellaneous Piping and Ducting Components Program. The applicant cited generic Note G, indicating that the environment is not in the GALL Report for this component and material combination.

SER Section 3.0.3.1.15 documents the staff's evaluation of the applicant's Inspection of Internal Surfaces in Miscellaneous Piping and Ducting Components Program. The applicant stated that the Inspection of Internal Surfaces in Miscellaneous Piping and Ducting Components Program conducts visual inspections during periodic maintenance, predictive maintenance, surveillance testing, and corrective maintenance and is capable of detecting loss of material that could result in a loss of component intended function. The staff finds the applicant's use of Internal Surfaces in Miscellaneous Piping and Ducting Components Program to manage loss of material for these components acceptable because it uses visual inspection to detect loss of material.

SER Section 3.3.2.3.7 documents the staff's evaluation for copper alloy and copper alloy with greater than 8-percent aluminum closure bolting exposed to plant indoor air managed for loss of preload by the Bolting Integrity Program, citing generic Notes F and G.

SER Section 3.3.2.3.3 documents the staff's evaluation for stainless steel closure bolting exposed to plant indoor air managed for loss of preload by the Bolting Integrity Program, citing generic Note G.

SER Section 3.3.2.3.19 documents the staff's evaluation for copper alloys and copper alloys with zinc greater than 15-percent piping, piping components, sight gauges, strainers, and valves exposed to atmosphere or weather (external and internal) managed for loss of material by the External Surfaces Monitoring Program, citing generic Note G.

SER Section 3.3.2.3.7 documents the staff's evaluation for PVC, fiberglass, and polyethylene piping and piping components exposed to plant indoor air, buried external environments, and raw water internal environment with no aging effects and no AMP, citing generic Note F.

Aging Management Review Results

SER Section 3.3.2.3.3 documents the staff's evaluation for aluminum, carbon steel, copper alloy, cast iron, and stainless steel components exposed to closed-cycle cooling water, borated water leakage, and potable water, with an aging effect of loss of material managed by the Inspection of Internal Surfaces in Miscellaneous Piping and Ducting Components Program, citing generic Note G.

On the basis of its review, the staff finds that the applicant has appropriately evaluated the AMR results of material, environment, AERM, and AMP combinations not evaluated in the GALL Report. The staff finds that the applicant has demonstrated that the effects of aging will be adequately managed so that the intended function(s) will be maintained consistent with the CLB for the period of extended operation, as required by 10 CFR 54.21(a)(3).

3.3.2.3.23 Demineralized Water System—Summary of Aging Management Review—License Renewal Application Table 3.3.2-23

The staff reviewed LRA Table 3.3.2-23, which summarizes the results of AMR evaluations for the demineralized water system component groups.

SER Section 3.3.2.3.3 documents the staff's evaluation for stainless steel closure bolting exposed to plant indoor air managed for loss of preload by the Bolting Integrity Program, citing generic Note G.

SER Section 3.3.2.3.13 documents the staff's evaluation for carbon steel closure bolting exposed to atmosphere or weather and fuel oil managed for loss of preload by the Bolting Integrity Program, citing generic Note G.

SER Section 3.3.2.3.7 documents the staff's evaluation for stainless steel and CASS components exposed to atmosphere or weather conditions, where there are no AERM and no AMP, that cite generic Note G.

On the basis of its review, the staff finds that the applicant has appropriately evaluated the AMR results of material, environment, AERM, and AMP combinations not evaluated in the GALL Report. The staff finds that the applicant has demonstrated that the effects of aging will be adequately managed so that the intended function(s) will be maintained consistent with the CLB for the period of extended operation, as required by 10 CFR 54.21(a)(3).

3.3.2.3.24 Water Reclamation Facility Fuel System—Summary of Aging Management Review—License Renewal Application Table 3.3.2-24

The staff reviewed LRA Table 3.3.2-24, which summarizes the results of AMR evaluations for the WRF fuel system component groups.

In LRA Table 3.3.2-24, the applicant stated that for sight gauges made of glass exposed to atmosphere or weather (external) there is no aging effect and no proposed AMP. The AMR item cites generic Note G, indicating that for the item, the environment is not in the GALL Report for this component and material.

The staff noted that in LRA Table 3.0-1, the applicant described atmosphere weather as moist, ambient temperatures, humidity, and exposure to weather, including precipitation and wind, with temperature extremes of 11 degrees F to 121 degrees F. The staff finds the applicant's proposal acceptable because the GALL Report does not identify an aging effect for any glass material exposed to any environment (e.g., lubricating oil, air, treated borated water) and the atmosphere weather environment is less aggressive than the examples in the GALL Report; therefore, no AMP is required.

SER Section 3.3.2.3.13 documents the staff's evaluation for carbon steel closure bolting exposed to atmosphere or weather and fuel oil managed for loss of preload by the Bolting Integrity Program, citing generic Note G.

SER Section 3.3.2.3.19 documents the staff's evaluation for copper alloys and copper alloys with Zinc greater than 15-percent piping, piping components, sight gauges, and valves exposed to atmosphere or weather (external and internal) managed for loss of material by the External Surfaces Monitoring Program with generic Note G.

SER Section 3.3.2.3.7 documents the staff's evaluation for stainless steel and CASS components exposed to atmosphere or weather conditions, where there are no AERM and no AMP, citing generic Note G.

SER Section 3.3.2.3.17 documents the staff's evaluation for aluminum dampers, flame arrestors, valves, vents, and heat exchangers exposed to atmosphere or weather (external and internal) managed for loss of material corrosion by the External Surfaces Monitoring Program with generic Note G.

On the basis of its review, the staff finds that the applicant has appropriately evaluated the AMR results of material, environment, AERM, and AMP combinations not evaluated in the GALL Report. The staff finds that the applicant has demonstrated that the effects of aging will be adequately managed so that the intended function(s) will be maintained consistent with the CLB for the period of extended operation, as required by 10 CFR 54.21(a)(3).

3.3.2.3.25 Service Gases (Nitrogen and Hydrogen) System—Summary of Aging Management Review—License Renewal Application Table 3.3.2-25

The staff reviewed LRA Table 3.3.2-25, which summarizes the results of AMR evaluations for the service gases (N^2 and H^2) system component groups.

The staff's review did not find any items indicating plant-specific Notes F–J, where the combination of component type, material, environment, and AERM does not correspond to an item in the GALL Report.

SER Section 3.0.2.1 documents the staff's evaluation of the items with Notes A–E.

3.3.2.3.26 Gaseous Radwaste System—Summary of Aging Management Review—License Renewal Application Table 3.3.2-26

The staff reviewed LRA Table 3.3.2-26, which summarizes the results of AMR evaluations for the gaseous radwaste system component groups.

The staff's review did not find any items indicating plant-specific Notes F–J, where the combination of component type, material, environment, and AERM does not correspond to an item in the GALL Report.

SER Section 3.0.2.1 documents the staff's evaluation of the items with Notes A–E.

3.3.2.3.27 Radioactive Waste Drains System—Summary of Aging Management Review— License Renewal Application Table 3.3.2-27

The staff reviewed LRA Table 3.3.2-27, which summarizes the results of AMR evaluations for the radioactive waste drains system component groups.

SER Section 3.3.2.3.3 documents the staff's evaluation for stainless steel closure bolting exposed to plant indoor air managed for loss of preload by the Bolting Integrity Program, citing generic Note G.

SER Section 3.3.2.3.7 documents the staff's evaluation for stainless steel and CASS components exposed to atmosphere or weather conditions, where there are no AERM and no AMP, that cite generic Note G.

On the basis of its review, the staff finds that the applicant has appropriately evaluated the AMR results of material, environment, AERM, and AMP combinations not evaluated in the GALL

Aging Management Review Results

Report. The staff finds that the applicant has demonstrated that the effects of aging will be adequately managed so that the intended function(s) will be maintained consistent with the CLB for the period of extended operation, as required by 10 CFR 54.21(a)(3).

3.3.2.3.28 Station Blackout Generator—Summary of Aging Management Review—License Renewal Application Table 3.3.2-28

The staff reviewed LRA Table 3.3.2-28, which summarizes the results of AMR evaluations for the SBO generator component groups.

SER Section 3.3.2.3.3 documents the staff's evaluation for stainless steel closure bolting exposed to plant indoor air managed for loss of preload by the Bolting Integrity Program, citing generic Note G.

SER Section 3.3.2.3.17 documents the staff's evaluation for aluminum dampers, flame arrestors, valves, vents, and heat exchangers exposed to atmosphere or weather (external and internal) managed for loss of material corrosion by the External Surfaces Monitoring Program with generic Note G.

SER Section 3.3.2.3.21 documents the staff's evaluation for aluminum heat exchanger (governor oil cooler), valve, and filter exposed to lubricating oil and managed for loss of material by the Lubricating Oil Analysis and One-Time Inspection Programs with generic Note G.

On the basis of its review, the staff finds that the applicant has appropriately evaluated the AMR results of material, environment, AERM, and AMP combinations not evaluated in the GALL Report. The staff finds that the applicant has demonstrated that the effects of aging will be adequately managed so that the intended function(s) will be maintained consistent with the CLB for the period of extended operation, as required by 10 CFR 54.21(a)(3).

3.3.2.3.29 Cranes, Hoists, and Elevators—Summary of Aging Management Review—License Renewal Application Table 3.3.2-29

The staff reviewed LRA Table 3.3.2-29, which summarizes the results of AMR evaluations for the cranes, hoists, and elevators component groups.

The staff's review did not find any items indicating plant-specific Notes F–J, where the combination of component type, material, environment, and AERM does not correspond to an item in the GALL Report.

SER Section 3.0.2.1 documents the staff's evaluation of the items with Notes A–E.

3.3.2.3.30 Miscellaneous Auxiliary Systems In-Scope Only for Criterion Under Title 10, Part 54.4(a)(2) of the Code of Federal Regulations—Summary of Aging Management Review—License Renewal Application Table 3.3.2-30

The staff reviewed LRA Table 3.3.2-30, which summarizes the results of AMR evaluations for the miscellaneous auxiliary systems in-scope only for criterion 10 CFR 54.4(a)(2) component groups.

SER Section 3.3.2.3.7 documents the staff's evaluation for copper alloy and copper alloy with greater than 8-percent aluminum closure bolting exposed to plant indoor air managed for loss of preload by the Bolting Integrity Program, citing generic Notes F and G.

SER Section 3.3.2.3.3 documents the staff's evaluation for stainless steel closure bolting exposed to plant indoor air managed for loss of preload by the Bolting Integrity Program, citing generic Note G.

SER Section 3.3.2.3.22 documents the staff's evaluation for copper alloy piping, pumps, strainers, and valves exposed to potable water managed for loss of material by the Inspection of

Aging Management Review Results

Internal Surfaces in Miscellaneous Piping and Ducting Components Program, citing generic Note G.

On the basis of its review, the staff finds that the applicant has appropriately evaluated the AMR results of material, environment, AERM, and AMP combinations not evaluated in the GALL Report. The staff finds that the applicant has demonstrated that the effects of aging will be adequately managed so that the intended function(s) will be maintained consistent with the CLB for the period of extended operation, as required by 10 CFR 54.21(a)(3).

3.3.2.3.31 Auxiliary Systems—Summary of Aging Management Evaluation—Oily Waste and Non-Radioactive Waste System—License Renewal Application Table 3.3.2-31

SER Section 3.3.2.3.7 documents the staff's evaluation for copper alloy and copper alloy with greater than 8-percent aluminum closure bolting exposed to plant indoor air managed for loss of preload by the Bolting Integrity Program, citing generic Notes F and G.

SER Section 3.3.2.3.3 documents the staff's evaluation for stainless steel closure bolting exposed to plant indoor air managed for loss of preload by the Bolting Integrity Program, citing generic Note G.

SER Section 3.3.2.3.3 documents the staff's evaluation for aluminum, carbon steel, copper alloy, cast iron, and stainless steel components exposed to closed-cycle cooling water, borated water leakage and potable water, with an aging effect of loss of material managed by the Inspection of Internal Surfaces in Miscellaneous Piping and Ducting Components Program, citing generic Note G.

On the basis of its review, the staff finds that the applicant has appropriately evaluated the AMR results of material, environment, AERM, and AMP combinations not evaluated in the GALL Report. The staff finds that the applicant has demonstrated that the effects of aging will be adequately managed so that the intended function(s) will be maintained consistent with the CLB for the period of extended operation, as required by 10 CFR 54.21(a)(3).

3.3.3 Conclusion

The staff concludes that the applicant has provided sufficient information to demonstrate that the effects of aging for the auxiliary systems components within the scope of license renewal and subject to an AMR will be adequately managed so that the intended function(s) will be maintained consistent with the CLB for the period of extended operation, as required by 10 CFR 54.21(a)(3).

3.4 Aging Management of Steam and Power Conversion Systems

This section of the SER documents the staff's review of the applicant's AMR results for the steam and power conversion systems components and component groups of the following systems:

- main steam system
- condensate storage and transfer system
- auxiliary feedwater (AFW) system

3.4.1 Summary of Technical Information in the Application

LRA Section 3.4 provides AMR results for the steam and power conversion systems components and component groups. LRA Table 3.4.1, "Summary of Aging Management Evaluations in Chapter VIII of NUREG-1801 [GALL Report] for Steam and Power Conversion Systems," is a summary comparison of the applicant's AMRs with those evaluated in the GALL Report for the steam and power conversion systems components and component groups.

Aging Management Review Results

The applicant's AMRs evaluated and incorporated applicable plant-specific and industry operating experience in the determination of AERMs. The plant-specific evaluation included condition reports and discussions with appropriate site personnel to identify AERMs. The applicant's review of industry operating experience included a review of the GALL Report and operating experience issues identified since the issuance of the GALL Report.

3.4.2 Staff Evaluation

The staff reviewed LRA Section 3.4 to determine if the applicant provided sufficient information to demonstrate that it will adequately manage the effects of aging for the steam and power conversion systems components within the scope of license renewal and subject to an AMR, so that the intended function(s) will be maintained consistent with the CLB for the period of extended operation, as required by 10 CFR 54.21(a)(3).

The staff reviewed AMRs to ensure the applicant's claim that certain AMRs were consistent with the GALL Report. The staff did not repeat its review of the matters described in the GALL Report; however, the staff did verify that the material presented in the LRA was applicable and that the applicant identified the appropriate GALL Report AMRs. SER Section 3.0.3 documents the staff's evaluations of the AMPs, and SER Section 3.4.2.1 details the staff's evaluation.

The staff also reviewed AMRs consistent with the GALL Report and for which further evaluation is recommended. The staff confirmed that the applicant's further evaluations were consistent with the SRP-LR Section 3.4.2.2 acceptance criteria. SER Section 3.4.2.2 documents the staff's evaluations.

The staff also conducted a technical review of the remaining AMRs not consistent with or not addressed in the GALL Report. The technical review evaluated if the applicant noted all plausible aging effects and if the aging effects listed were appropriate for the material-environment combinations specified. SER Section 3.4.2.3 documents the staff's evaluations.

For SSCs that the applicant claimed were not applicable or required no aging management, the staff reviewed the AMR items and the plant's operating experience to verify the applicant's claims.

Table 3.4-1 summarizes the staff's evaluation of components, aging effects or mechanisms, and AMPs listed in LRA Section 3.4 and addressed in the GALL Report.

Table 3.4-1. Staff Evaluation for Steam and Power Conversion Systems Components in the GALL Report

Component Group (GALL Report Item No.)	Aging Effect/ Mechanism	AMP in GALL Report	Further Evaluation in GALL Report	AMP in LRA, Supplements, or Amendments	Staff Evaluation
Steel piping, piping components, and piping elements exposed to steam or treated water (3.4.1-1)	Cumulative fatigue damage	TLAA, evaluated in accordance with 10 CFR 54.21(c)	Yes	TLAA	Consistent with GALL Report. See SER Section 3.4.2.2.1.
Steel piping, piping components, and piping elements exposed to steam (3.4.1-2)	Loss of material due to general, pitting and crevice corrosion	Water Chemistry and One-Time Inspection	Yes	Not applicable	Not applicable to PVNGS. See SER Section 3.4.2.2.5(3).

Aging Management Review Results

Component Group (GALL Report Item No.)	Aging Effect/ Mechanism	AMP in GALL Report	Further Evaluation in GALL Report	AMP in LRA, Supplements, or Amendments	Staff Evaluation
Steel heat exchanger components exposed to treated water (3.4.1-3)	Loss of material due to general, pitting and crevice corrosion	Water Chemistry and One-Time Inspection	Yes	Water Chemistry and One-Time Inspection	Consistent with GALL Report. See SER Section 3.4.2.2.2(1).
Steel piping, piping components, and piping elements exposed to treated water (3.4.1-4)	Loss of material due to general, pitting and crevice corrosion	Water Chemistry and One-Time Inspection	Yes	Water Chemistry and One-Time Inspection	Consistent with GALL Report. See SER Section 3.4.2.2.2(1).
Steel heat exchanger components exposed to treated water (3.4.1-5)	Loss of material due to general, pitting, crevice, and galvanic corrosion	Water Chemistry and One-Time Inspection	Yes	Not applicable	Not applicable to PWRs. See SER Section 3.4.2.1.1.
Steel and stainless steel tanks exposed to treated water (3.4.1-6)	Loss of material due to general (steel only) pitting and crevice corrosion	Water Chemistry and One-Time Inspection	Yes	Water Chemistry and One-Time Inspection	Consistent with GALL Report. See SER Section 3.4.2.2.7(1).
Steel piping, piping components, and piping elements exposed to lubricating oil (3.4.1-7)	Loss of material due to general, pitting and crevice corrosion	Lubricating Oil Analysis and One-Time Inspection	Yes	Lubricating Oil Analysis and One-Time Inspection	Consistent with GALL Report. See SER Section 3.4.2.2.2(2).
Steel piping, piping components, and piping elements exposed to raw water (3.4.1-8)	Loss of material due to general, pitting, crevice, and MIC and fouling	Plant specific	Yes	Not Applicable	Not applicable to PVNGS. See SER Section 3.4.2.2.3.
Stainless steel and copper alloy heat exchanger tubes exposed to treated water (3.4.1-9)	Reduction of heat transfer due to fouling	Water Chemistry and One-Time Inspection	Yes	Not applicable	Not applicable to PVNGS. See SER Section 3.4.2.2.4(1).
Steel, stainless steel, and copper alloy heat exchanger tubes exposed to lubricating oil (3.4.1-10)	Reduction of heat transfer due to fouling	Lubricating Oil Analysis and One-Time Inspection	Yes	Lubricating Oil Analysis and One-Time Inspection	Consistent with GALL Report. See SER Section 3.4.2.2.4(2).
Buried steel piping, piping components, piping elements, and tanks (with or without coating or wrapping) exposed to soil (3.4.1-11)	Loss of material due to general, pitting, crevice, and MIC	Buried Piping and Tanks Surveillance or Buried Piping and Tanks Inspection	No Yes	Not applicable	Not applicable to PVNGS. See SER Section 3.4.2.2.5(1).

Aging Management Review Results

Component Group (GALL Report Item No.)	Aging Effect/ Mechanism	AMP in GALL Report	Further Evaluation in GALL Report	AMP in LRA, Supplements, or Amendments	Staff Evaluation
Steel heat exchanger components exposed to lubricating oil (3.4.1-12)	Loss of material due to general, pitting, crevice, and MIC	Lubricating Oil Analysis and One-Time Inspection	Yes	Lubricating Oil Analysis and One-Time Inspection	Consistent with GALL Report. See SER Section 3.4.2.2.5(2).
Stainless steel piping, piping components, piping elements exposed to steam (3.4.1-13)	Cracking due to SCC	Water Chemistry and One-Time Inspection	Yes	Not applicable	Not applicable to PWRs. See SER Section 3.4.2.1.1.
Stainless steel piping, piping components, piping elements, tanks, and heat exchanger components exposed to treated water > 60°C (> 140°F) (3.4.1-14)	Cracking due to SCC	Water Chemistry and One-Time Inspection	Yes	Water Chemistry and One-Time Inspection	Consistent with GALL Report. See SER Section 3.4.2.2.6.
Aluminum and copper alloy piping, piping components, and piping elements exposed to treated water (3.4.1-15)	Loss of material due to pitting and crevice corrosion	Water Chemistry and One-Time Inspection	Yes	Water Chemistry and One-Time Inspection	Consistent with GALL Report. See SER Section 3.4.2.2.7(1).
Stainless steel piping, piping components, and piping elements; tanks, and heat exchanger components exposed to treated water (3.4.1-16)	Loss of material due to pitting and crevice corrosion	Water Chemistry and One-Time Inspection	Yes	Water Chemistry and One-Time Inspection	Consistent with GALL Report. See SER Section 3.4.2.2.7(1).
Stainless steel piping, piping components, and piping elements exposed to soil (3.4.1-17)	Loss of material due to pitting and crevice corrosion	Plant specific	Yes	Not applicable	Not applicable to PVNGS. See SER Section 3.4.2.2.7(2).
Copper alloy piping, piping components, and piping elements exposed to lubricating oil (3.4.1-18)	Loss of material due to pitting and crevice corrosion	Lubricating Oil Analysis and One-Time Inspection	Yes	Not applicable	Not applicable to PVNGS. See SER Section 3.4.2.2.7(3).
Stainless steel piping, piping components, piping elements, and heat exchanger components exposed to lubricating oil (3.4.1-19)	Loss of material due to pitting, crevice, and MIC	Lubricating Oil Analysis and One-Time Inspection	Yes	Not applicable	Not applicable to PVNGS. See SER Section 3.4.2.2.8.
Steel tanks exposed to air - outdoor (external) (3.4.1-20)	Loss of material, general, pitting, and crevice corrosion	Aboveground Steel Tanks	No	Not applicable	Not applicable to PVNGS. See SER Section 3.4.2.1.1.

Aging Management Review Results

Component Group (GALL Report Item No.)	Aging Effect/ Mechanism	AMP in GALL Report	Further Evaluation in GALL Report	AMP in LRA, Supplements, or Amendments	Staff Evaluation
High-strength steel closure bolting exposed to air with steam or water leakage (3.4.1-21)	Cracking due to cyclic loading, SCC	Bolting Integrity	No	Not applicable	Not applicable to PVNGS. See SER Section 3.4.2.1.1.
Steel bolting and closure bolting exposed to air with steam or water leakage, air - outdoor (external), or air - indoor uncontrolled (external); (3.4.1-22)	Loss of material due to general, pitting and crevice corrosion; loss of preload due to thermal effects, gasket creep, and self-loosening	Bolting Integrity	No	Bolting Integrity	Consistent with GALL Report.
Stainless steel piping, piping components, and piping elements exposed to closed-cycle cooling water > 60°C (> 140°F) (3.4.1-23)	Cracking due to SCC	Closed-Cycle Cooling Water System	No	Not applicable	Not applicable to PVNGS. See SER Section 3.4.2.1.1.
Steel heat exchanger components exposed to closed cycle cooling water (3.4.1-24)	Loss of material due to general, pitting, crevice, and galvanic corrosion	Closed-Cycle Cooling Water System	No	Closed-Cycle Cooling Water System	Consistent with GALL Report.
Stainless steel piping, piping components, piping elements, and heat exchanger components exposed to closed cycle cooling water (3.4.1-25)	Loss of material due to pitting and crevice corrosion	Closed-Cycle Cooling Water System	No	Closed-Cycle Cooling Water System	Consistent with GALL Report.
Copper alloy piping, piping components, and piping elements exposed to closed cycle cooling water (3.4.1-26)	Loss of material due to pitting, crevice, and galvanic corrosion	Closed-Cycle Cooling Water System	No	Not applicable	Not applicable to PVNGS. See SER Section 3.4.2.1.1.
Steel, stainless steel, and copper alloy heat exchanger tubes exposed to closed cycle cooling water (3.4.1-27)	Reduction of heat transfer due to fouling	Closed-Cycle Cooling Water System	No	Not applicable	Not applicable to PVNGS. See SER Section 3.4.2.1.1.
Steel external surfaces exposed to air - indoor uncontrolled (external), condensation (external), or air outdoor (external) (3.4.1-28)	Loss of material due to general corrosion	External Surfaces Monitoring	No	External Surfaces Monitoring	Consistent with GALL Report.
Steel piping, piping components, and piping elements exposed to steam or treated water (3.4.1-29)	Wall thinning due to FAC	FAC	No	FAC	Consistent with GALL Report.

Aging Management Review Results

Component Group (GALL Report Item No.)	Aging Effect/ Mechanism	AMP in GALL Report	Further Evaluation in GALL Report	AMP in LRA, Supplements, or Amendments	Staff Evaluation
Steel piping, piping components, and piping elements exposed to air outdoor (internal) or condensation (internal) (3.4.1-30)	Loss of material due to general, pitting, and crevice corrosion	Inspection of Internal Surfaces in Miscellaneous Piping and Ducting Components	No	Inspection of Internal Surfaces in Miscellaneous Piping and Ducting Components	Consistent with GALL Report.
Steel heat exchanger components exposed to raw water (3.4.1-31)	Loss of material due to general, pitting, crevice, galvanic, and MIC and fouling	Open-Cycle Cooling Water System	No	Not applicable	Not applicable to PVNGS. See SER Section 3.4.2.1.1.
Stainless steel and copper alloy piping, piping components, and piping elements exposed to raw water (3.4.1-32)	Loss of material due to pitting, crevice, and MIC	Open-Cycle Cooling Water System	No	Not applicable	Not applicable to PVNGS. See SER Section 3.4.2.1.1.
Stainless steel heat exchanger components exposed to raw water (3.4.1-33)	Loss of material due to pitting, crevice, and MIC and fouling	Open-Cycle Cooling Water System	No	Not applicable	Not applicable to PVNGS. See SER Section 3.4.2.1.1.
Steel, stainless steel, and copper alloy heat exchanger tubes exposed to raw water (3.4.1-34)	Reduction of heat transfer due to fouling	Open-Cycle Cooling Water System	No	Not applicable	Not applicable to PVNGS. See SER Section 3.4.2.1.1.
Copper alloy > 15% Zn piping, piping components, and piping elements exposed to closed cycle cooling water, raw water, or treated water (3.4.1-35)	Loss of material due to selective leaching	Selective Leaching of Materials	No	Selective Leaching of Materials	Consistent with GALL Report.
Gray cast iron piping, piping components, and piping elements exposed to soil, treated water, or raw water (3.4.1-36)	Loss of material due to selective leaching	Selective Leaching of Materials	No	Selective Leaching of Materials	Consistent with GALL Report.
Steel, stainless steel, and nickel-based alloy piping, piping components, and piping elements exposed to steam (3.4.1-37)	Loss of material due to pitting and crevice corrosion	Water Chemistry	No	Water Chemistry	Consistent with GALL Report.
Steel bolting and external surfaces exposed to air with borated water leakage (3.4.1-38)	Loss of material due to boric acid corrosion	Boric Acid Corrosion	No	Not applicable	Not applicable to PVNGS. See SER Section 3.4.2.1.1.

Aging Management Review Results

Component Group (GALL Report Item No.)	Aging Effect/ Mechanism	AMP in GALL Report	Further Evaluation in GALL Report	AMP in LRA, Supplements, or Amendments	Staff Evaluation
Stainless steel piping, piping components, and piping elements exposed to steam (3.4.1-39)	Cracking due to SCC	Water Chemistry	No	Water Chemistry	Consistent with GALL Report.
Glass piping elements exposed to air, lubricating oil, raw water, and treated water (3.4.1-40)	None	None	NA	None	Consistent With GALL Report.
Stainless steel, copper alloy, and nickel-alloy piping, piping components, and piping elements exposed to air - indoor uncontrolled (external) (3.4.1-41)	None	None	NA	None	Consistent with GALL Report.
Steel piping, piping components, and piping elements exposed to air - indoor controlled (external) (3.4.1-42)	None	None	NA	Not applicable	Not applicable to PVNGS. See SER Section 3.4.2.1.1.
Steel and stainless steel piping, piping components, and piping elements in concrete (3.4.1-43)	None	None	NA	None	Consistent with GALL Report.
Steel, stainless steel, aluminum, and copper alloy piping, piping components, and piping elements exposed to gas (3.4.1-44)	None	None	NA	None	Consistent with GALL Report.

The staff's review of the steam and power conversion systems component groups followed any one of several approaches. One approach, documented in SER Section 3.4.2.1, reviewed AMR results for components that the applicant indicated are consistent with the GALL Report and require no further evaluation. Another approach, documented in SER Section 3.4.2.2, reviewed AMR results for components that the applicant indicated are consistent with the GALL Report and for which further evaluation is recommended. A third approach, documented in SER Section 3.4.2.3, reviewed AMR results for components that the applicant indicated are not consistent with, or not addressed in, the GALL Report. SER Section 3.0.3 documents the staff's review of AMPs credited to manage or monitor aging effects of the steam and power conversion systems components.

3.4.2.1 Aging Management Review Results Consistent with the Generic Aging Lessons Learned Report

LRA Section 3.4.2.1 identifies the materials, environments, AERMs, and the following programs that manage aging effects for the steam and power conversion systems components:

- Bolting Integrity Program

Aging Management Review Results

- External Surfaces Monitoring Program
- FAC Program
- One-Time Inspection Program
- Water Chemistry Program

LRA Tables 3.4.2-1 through 3.4.2-3 summarize AMRs for the steam and power conversion systems components and indicate AMRs claimed to be consistent with the GALL Report.

The staff audited and reviewed the information in the LRA. The staff did not repeat its review of the matters described in the GALL Report; however, the staff did verify that the material presented in the LRA was applicable and that the applicant identified the appropriate GALL Report AMRs.

The staff reviewed the LRA to confirm that the applicant provided a brief description of the system, components, materials, and environments; stated that the applicable aging effects were reviewed and evaluated in the GALL Report; and identified those aging effects for the steam and power conversion systems components that are subject to an AMR. On the basis of its review, the staff determines that, for AMRs not requiring further evaluation, as identified in LRA Table 3.4.1, the applicant's references to the GALL Report are acceptable and no further staff review is required.

3.4.2.1.1 Aging Management Review Results Identified as Not Applicable

LRA Table 3.4.1, items 3.4.1.5 and 3.4.1.13 discuss the applicant's determination that these items are applicable only to BWRs. The staff verified that these items do not apply because the units are a PWR design. Based on this determination, the staff finds that the applicant has provided an acceptable basis for concluding AMR items 3.4.1.5 and 3.4.1.13 are not applicable.

LRA Table 3.4.1, item 3.4.1.20 addresses steel tanks exposed to outdoor air (external). The applicant stated that this item is not applicable because there are no steel tanks in the condensate or AFW systems that are exposed to the outdoor air (external) environment. The staff reviewed LRA Sections 2.3.4 and 3.4 and confirmed that the applicant's LRA does not have any AMR results for the condensate storage and transfer and AFW systems that include steel tanks exposed to outdoor air (external). The staff also reviewed the applicant's information in the UFSAR and confirmed that no in-scope steel tanks exposed to outdoor air (external) are present in the condensate storage and transfer and AFW systems. Therefore, the staff finds the applicant's determination acceptable.

LRA Table 3.4.1, item 3.4.1.21 addresses high-strength steel closure bolting exposed to air with steam or water leakage in the steam and power conversion systems. The GALL Report recommends use of AMP XI.M18, "Bolting Integrity," to manage cracking due to cyclic loading or SCC for this component group. The applicant stated that this item is not applicable because high-strength closure bolting is not used in the steam and power conversion systems. The staff reviewed LRA Sections 2.3.4 and 3.4 and the UFSAR and confirmed that the applicant's LRA does not have any AMR results for the steam and power conversion systems that include high-strength steel closure bolting exposed to air with steam or water leakage. During its review of operating experience and applicant interviews associated with the Bolting Integrity Program, the staff did not find any evidence of high-strength steel closure bolting in the steam and power conversion systems. Based on its review of the LRA, UFSAR, and the applicant's Bolting Integrity Program, the staff confirmed that there is no high-strength steel closure bolting exposed to air with steam or water leakage in the steam and power conversion systems and, therefore, it finds the applicant's determination acceptable.

LRA Table 3.4.1, item 3.4.1.23 addresses stainless steel piping, piping components, and piping elements exposed to closed-cycle cooling water greater than 60 degrees C (140 degrees F). The GALL Report recommends use of AMP XI.M21, "Closed-Cycle Cooling Water," to manage

Aging Management Review Results

cracking due to SCC for this component group. The applicant stated that this item is not applicable because it has no in-scope stainless steel piping, piping components, and piping elements exposed to closed-cycle cooling water greater than 60 degrees C (140 degrees F) in the condensate, blowdown, or AFW systems. The staff reviewed LRA Sections 2.3.4 and 3.4 and confirmed that the applicant's LRA does not have any AMR results for the steam and power conversion system that include stainless steel piping, piping components, and piping elements exposed to closed-cycle cooling water greater than 60 degrees C (140 degrees F). Further, the staff identifies no closed-cycle cooling water environment within the steam and power conversion system. The staff reviewed the applicant's UFSAR and confirmed that no in-scope stainless steel piping, piping components, and piping elements exposed to closed-cycle cooling water greater than 60 degrees C (140 degrees F) are present in the steam and power conversion system. Therefore, the staff finds the applicant's determination acceptable.

LRA Table 3.4.1, item 3.4.1.26 addresses copper alloy piping, piping components, and piping elements exposed to closed-cycle cooling water. The GALL Report recommends use of AMP XI.M21, "Closed-Cycle Cooling Water," to manage loss of material due to pitting, crevice, and galvanic corrosion for this component group. The applicant stated that this item is not applicable because it has no in-scope copper alloy components exposed to closed-cycle cooling water in the condensate, blowdown, or AFW systems. The staff reviewed LRA Sections 2.3.4 and 3.4 and confirmed that the applicant's LRA does not have any AMR results for the steam and power conversion system that include copper alloy piping, piping components, and piping elements exposed to closed-cycle cooling water. The staff reviewed the applicant's UFSAR and confirmed that no in-scope copper alloy piping, piping components, and piping elements exposed to closed-cycle cooling water are present in the steam and power conversion system and, therefore, it finds the applicant's determination acceptable.

LRA Table 3.4.1, item 3.4.1.27 addresses steel, stainless steel, and copper alloy heat exchanger tubes exposed to closed-cycle cooling water. The GALL Report recommends use of AMP XI.M21, "Closed-Cycle Cooling Water," to manage reduction of heat transfer due to fouling for this component group. The applicant stated that this item is not applicable because it has no in-scope copper alloy components exposed to closed-cycle cooling water in the condensate, blowdown, or AFW systems. The staff reviewed LRA Sections 2.3.4 and 3.4 and confirmed that the applicant's LRA does not have any AMR results for the steam and power conversion system that include steel, stainless steel, and copper alloy heat exchanger tubes exposed to closed-cycle cooling water. The staff reviewed the applicant's UFSAR and confirmed that no in-scope steel, stainless steel, and copper alloy heat exchanger tubes exposed to closed-cycle cooling water are present in the steam and power conversion system; therefore, it finds the applicant's determination acceptable.

LRA Table 3.4.1, item 3.4.1.31 addresses steel heat exchanger components exposed to raw water. The GALL Report recommends use of AMP XI.M20, "Open-Cycle Cooling Water," to manage loss of material due to pitting, crevice, and MIC and fouling for this component group. The applicant stated that this item is not applicable because it has no in-scope steel heat exchanger components exposed to raw water in the condensate, blowdown, or AFW systems. The staff reviewed LRA Sections 2.3.4 and 3.4 and confirmed that the applicant's LRA does not have any AMR results for the steam and power conversion systems that include steel heat exchanger components exposed to raw water. The staff reviewed the applicant's UFSAR and confirmed that no in-scope steel heat exchanger components exposed to raw water are present in the steam and power conversion systems and, therefore, it finds the applicant's determination acceptable.

LRA Table 3.4.1, item 3.4.1.32 addresses stainless steel and copper alloy piping, piping components, and piping elements exposed to raw water. The GALL Report recommends use of AMP XI.M20, "Open-Cycle Cooling Water," to manage loss of material due to pitting, crevice

Aging Management Review Results

and MIC, for this component group. The applicant stated that this item is not applicable because it has no in-scope stainless steel or copper alloy components exposed to raw water in the steam turbine, condensate, blowdown, or AFW systems. The staff reviewed LRA Sections 2.3.4 and 3.4 and confirmed that the applicant's LRA does not have any AMR results for the steam and power conversion systems that include stainless steel and copper alloy piping, piping components, and piping elements exposed to raw water. The staff reviewed the applicant's UFSAR and confirmed that no in-scope stainless steel and copper alloy piping, piping components, and piping elements exposed to raw water are present in the steam and power conversion systems; therefore, it finds the applicant's determination acceptable.

LRA Table 3.4.1, item 3.4.1.33 addresses stainless steel heat exchanger components exposed to raw water. The GALL Report recommends use of AMP XI.M20, "Open-Cycle Cooling Water," to manage loss of material due to pitting, crevice, and MIC and fouling for this component group. The applicant stated that this item is not applicable because it has no in-scope stainless steel heat exchanger components exposed to raw water in the condensate, blowdown, or AFW systems. The staff reviewed LRA Sections 2.3.4 and 3.4 and confirmed that the applicant's LRA does not have any AMR results for the steam and power conversion systems that include stainless steel heat exchanger components exposed to raw water. The staff reviewed the applicant's UFSAR and confirmed that no in-scope stainless steel heat exchanger components exposed to raw water are present in the steam and power conversion systems; therefore, it finds the applicant's determination acceptable.

LRA Table 3.4.1, item 3.4.1.34 addresses steel, stainless steel, and copper alloy heat exchanger tubes exposed to raw water. The GALL Report recommends use of AMP XI.M20, "Open-Cycle Cooling Water," to manage reduction of heat transfer due to fouling for this component group. The applicant stated that this item is not applicable because it has no in-scope steel, stainless steel, and copper alloy heat exchanger tubes exposed to raw water in the condensate, blowdown, or AFW systems. The staff reviewed LRA Sections 2.3.4 and 3.4 and confirmed that the applicant's LRA does not have any AMR results for the steam and power conversion systems that include steel, stainless steel, and copper alloy heat exchanger tubes exposed to raw water. The staff reviewed the applicant's UFSAR and confirmed that no in-scope steel, stainless steel, and copper alloy heat exchanger tubes exposed to raw water are present in the steam and power conversion systems; therefore, it finds the applicant's determination acceptable.

LRA Table 3.4.1, item 3.4.1.38 addresses steel bolting and external surfaces exposed to air with borated water leakage. The GALL Report recommends use of AMP XI.M10, "Boric Acid Corrosion," to manage loss of material due to boric acid corrosion for this component group. The applicant stated that this item is not applicable because there are no in-scope steel bolting and external surfaces exposed to borated water leakage in the steam and power conversion systems. The staff reviewed LRA Sections 2.3.4 and 3.4 and confirmed that the applicant's LRA does not have any AMR results in-scope steel bolting and external surfaces exposed to borated water leakage in the steam and power conversion systems. The staff also reviewed the applicant's information in the UFSAR associated with Table 3.4.1, item 3.4.1-38 and confirmed that no steel bolting and external surfaces exposed to air with borated water are present in the in the steam and power conversion systems; therefore, it finds the applicant's determination acceptable.

LRA Table 3.4.1, item 3.4.1.42 addresses steel piping, piping components, and piping elements exposed to air-indoor controlled (external). The applicant stated that this item is not applicable because it has no in-scope steel components exposed to indoor controlled air in the steam and power conversion systems. The staff noted that LRA Table 3.0-1 includes air-indoor controlled within its definition of the plant indoor air environment. The staff reviewed LRA Sections 2.3.4 and 3.4 and noted that there are multiple examples of carbon steel components exposed to plant indoor air in the steam and power conversion systems. Examples include Table 3.4.2-1, accumulator and piping managed for loss of material by the External Surfaces Monitoring Program and closure bolting managed for loss of material by the Bolting Integrity Program. The staff also noted that SRP-LR Table 3.4.1, item 3.4.1.42, GALL Report item SP-1, states that steel piping, piping components, and piping elements exposed to air-indoor controlled (external) have no AERM and no recommended AMP. The staff finds the applicant's determination acceptable because the SRP-LR and GALL Report state that there is no AERM or recommended AMP for this material and environment combination, and the applicant has designated AMPs to manage aging for some steel components exposed to air-indoor controlled (external) beyond the requirements of the GALL Report.

3.4.2.2 Aging Management Review Results Consistent with the Generic Aging Lessons Learned Report for Which Further Evaluation Is Recommended

In LRA Section 3.4.2.2, the applicant further evaluates aging management, as recommended by the GALL Report, for the steam and power conversion systems components and provides information concerning how it will manage the following aging effects:

- cumulative fatigue damage
- loss of material due to general, pitting, and crevice corrosion
- loss of material due to general, pitting, and crevice corrosion; MIC; and fouling
- reduction of heat transfer due to fouling
- loss of material due to general, pitting, crevice, and MIC
- cracking due to SCC
- loss of material due to pitting and crevice corrosion
- loss of material due to pitting, crevice, and MIC
- loss of material due to general, pitting, crevice, and galvanic corrosion
- QA for aging management of nonsafety-related components

For component groups evaluated in the GALL Report, for which the applicant claimed consistency with the report and for which the report recommends further evaluation, the staff reviewed the applicant's evaluation to determine if it adequately addressed the issues further evaluated. In addition, the staff reviewed the applicant's further evaluations against the criteria contained in SRP-LR Section 3.4.2.2. The staff's review of the applicant's further evaluation follows.

3.4.2.2.1 Cumulative Fatigue Damage

In LRA Section 3.4.2.2.1, the applicant stated the evaluation of fatigue is a TLAA as defined in 10 CFR 54.3 and is evaluated in accordance with 10 CFR 54.21(c)(1). PVNGS piping designed to ASME III Class 2, Class 3, and ANSI B31.1 assumes a reduction in the allowable secondary stress range if more than 7,000 full-range thermal cycles are expected in a design lifetime. LRA Section 4.3.5 describes the evaluation of these cyclic-design TLAAs. The applicant further stated the main steam safety valves are ASME III Class 2 components designed with a Class 1 fatigue analysis, and LRA Section 4.3.2.12 describes the evaluation of this TLAA.

The staff reviewed LRA Section 3.4.2.2.1 against the criteria in SRP-LR Section 3.4.2.2.1, which states that Fatigue is a TLAA as defined in 10 CFR 54.3. TLAAs are required to be evaluated in accordance with 10 CFR 54.21(c)(1). Section 4.3, "Metal Fatigue Analysis," of this SRP-LR addresses this TLAA separately. The staff finds the applicant's AMR results are consistent with the recommendations of the GALL Report and SRP-LR, except for in the following areas.

The staff noted that Summary Description, in LRA Section 4.3.5, states that the implicit fatigue analyses discussed in the section are applicable to all ASME Code Class 2 and 3 and ANSI B31.1 piping, piping components, and piping elements. The staff noted that it is not clear if the

Aging Management Review Results

LRA includes all corresponding AMR items for applicable ASME Code Class 2 and 3 or ANSI B31.1 piping, piping components, and piping elements within the scope for license renewal. The staff also noted that this includes those components in the ESF Systems (LRA Section 3.2), Auxiliary Systems (LRA Section 3.3), and the Steam and Power Conversion Systems (LRA Section 3.4). By letter dated July 21, 2010, the staff issued RAI 4.3-13 requesting the applicant clarify if the LRA includes all applicable AMR items with an aging effect of cumulative fatigue damage for those components scoped into license renewal. If not, the staff asked that the applicant to justify why the LRA does not include all corresponding AMR items on cumulative fatigue damage for applicable ASME Code Class 2 and 3 or ANSI B31.1 piping, piping components, and piping elements within the scope for license renewal. The applicant must identify all component types that are within the scope of the implicit fatigue analyses for ASME Code Class 2 and 3 components and B31.1 components in LRA Section 4.3.5 and should, therefore, be within the scope of applicable component-specific AMR items on cumulative fatigue damage. The staff previously identified this as Open Item 4.3-1.

In its response dated August 12, 2010, the applicant stated that for the steam and power conversion systems no additions are required because LRA Tables 3.4.2-1 and 3.4.2-3 include AMR items with an aging effect of cumulative fatigue damage. The staff reviewed LRA Tables 3.4.2-1 and 3.4.2-3 and confirmed that they contain applicable AMR line items associated with the aging effect of cumulative fatigue damage for piping, piping components and piping elements.

Based on its review, the staff finds the applicant's response to RAI 4.3-13, Part 3, related to LRA Section 3.4.2.2.1 acceptable because the applicant confirmed that no additional AMR items associated with ASME Code Class 2 and 3 or ANSI B31.1 piping, piping components, and piping elements were subject to aging management review per 10 CFR Part 54.21(a)(1). In addition, the staff confirmed that the applicant has included AMR line items associated with ASME Code Class 2 and 3 or ANSI B31.1 piping, piping components, and piping elements with an aging effect of cumulative fatigue damage in LRA Tables 3.4.2-1 and 3.4.2-3. The staff's concern described in RAI 4.3-13 is resolved and this part of Open Item 4.3-1 is closed.

Based on its review, the staff concludes that the applicant's proposal to manage cumulative fatigue damage in ASME III Class 2, Class 3, and ANSI B31.1 components meets the SRP-LR Section 3.4.2.2.1 criteria. For those items that apply to LRA Section 3.4.2.2.1, the staff determines that the LRA is consistent with the GALL Report. The applicant has demonstrated that it will adequately manage the effects of aging so that the intended function(s) will be maintained consistent with the CLB during the period of extended operation, as required by 10 CFR 54.21(a)(3). SER Section 4.3 documents the staff's review of the applicant's evaluation of the TLAA for these components.

3.4.2.2.2 Loss of Material Due to General, Pitting, and Crevice Corrosion

The staff reviewed LRA Section 3.4.2.2.2 against the following criteria in SRP-LR Section 3.4.2.2.2:

- LRA Section 3.4.2.2.2.1 refers to Table 3.4.1, items 3.4.1.3 and 3.4.1.4, and addresses steel piping, piping components, piping elements, and heat exchangers exposed to treated water and steel piping, piping components, and piping elements exposed to steam which are managed for the loss of material due to general, pitting, and crevice corrosion by the Water Chemistry and One-Time Inspection Programs. The applicant addressed the further evaluation criteria of the SRP-LR by stating that the above programs will manage the above aging effect for carbon steel and gray cast iron components exposed to secondary water and demineralized water. The applicant noted that the one-time inspection will include selected components at susceptible locations (e.g., stagnant flow locations) where contaminants could accumulate.

Aging Management Review Results

The staff reviewed LRA Section 3.4.2.2.2.1 against the criteria in SRP-LR Section 3.4.2.2.2.1, which states that loss of material due to general, pitting, and crevice corrosion could occur for steel piping, piping components, piping elements, tanks, and heat exchangers exposed to treated water and for steel piping, piping components, and piping elements exposed to steam. The SRP-LR notes that monitoring and control of water chemistry does not preclude this aging effect in locations with stagnant flow conditions. It continues by stating that the applicant should verify the effectiveness of the Water Chemistry Control Program and that a one-time inspection at susceptible locations is an acceptable method.

SER Sections 3.0.3.2.1 and 3.0.3.1.6 document the staff's evaluation of the applicant's Water Chemistry and One-Time Inspection Programs, respectively. In its review of components associate with the items 3.4.1.3 and 3.4.1.4, the staff finds the applicant's proposal to manage aging using the above programs acceptable because the Water Chemistry Program limits the concentrations of chemical species known to cause corrosion and adds chemical species known to inhibit degradation. In addition, the One-Time Inspection Program verifies the effectiveness of the Water Chemistry Program and evaluates aging effects, including loss of material.

Based on the programs identified, the staff concludes that the applicant's programs meet SRP-LR Section 3.4.2.2.2.1 criteria. For those items that apply to LRA Section 3.4.2.2.2.1, the staff determines that the LRA is consistent with the GALL Report. In addition, the applicant has demonstrated that it will adequately manage the effects of aging so that the intended function(s) will be maintained consistent with the CLB during the period of extended operation, as required by 10 CFR 54.21(a)(3).

- LRA Section 3.4.2.2.2.2 refers to LRA Table 3.4.1, item 3.4.1.7 and addresses carbon steel piping, piping components, and piping elements exposed to lubricating oil, which are managed for loss of material due to general, pitting, and crevice corrosion by the Lubricating Oil Analysis and the One-Time Inspection Programs. The applicant addressed the further evaluation criteria of the SRP-LR by stating that it will include a one-time inspection of selected components at susceptible locations where contaminants, such as water, could accumulate.

   The staff reviewed LRA Section 3.4.2.2.2.2 against the criteria in SRP-LR Section 3.4.2.2.2.2 which states that loss of material due to general, pitting, and crevice corrosion could occur for steel piping, piping components, and piping elements exposed to lubricating oil. SRP-LR Section 3.4.2.2.2.2 further states that the existing AMP relies on the periodic sampling and analysis of lubricating oil to maintain contaminants within acceptable limits and, therefore, preserves an environment that is not conducive to corrosion. Further, the applicant verifies the effectiveness of the Lubricating Oil Program through a one-time inspection of selected components at susceptible locations to ensure that corrosion is not occurring and that the component's intended function will be maintained during the period of extended operation. In addition, the GALL Report recommends a further evaluation of programs to manage corrosion by verifying their effectiveness with a one-time inspection of selected components at susceptible locations.

   The staff reviewed the applicant's Lubricating Oil Analysis Program and One-Time Inspection Program in SER Sections 3.0.3.2.17 and 3.0.3.1.6, respectively. In its review of the components associated with the LRA items listed above, the staff finds the applicant's proposal to manage aging effects using the Lubricating Oil Analysis and One-Time Inspection Programs acceptable because: (1) the Lubricating Oil Analysis Program provides for periodic sampling of lubricating oil to maintain contaminants at acceptable limits to preclude loss of material; (2) the applicant will perform one-time inspections of select steel piping, piping components, and piping elements exposed to lubricating oil for loss of material due to general, pitting, and crevice corrosion to verify the effectiveness of the Lubricating Oil Analysis Program in applicable steam and power conversion systems, following the GALL Report recommendation that the "One Time Inspection" is an acceptable AMP to verify the effectiveness of the applicant's Lubricating Oil Analysis Program.

Based on the programs identified, the staff concludes that the applicant's programs meet SRP-LR Section 3.4.2.2.2.2 criteria. For those items that apply to LRA Section 3.4.2.2.2.2, the staff determines that the LRA is consistent with the GALL Report. Further, the applicant has demonstrated that the effects of aging will be adequately managed so that the intended function(s) will be maintained consistent with the CLB during the period of extended operation, as required by 10 CFR 54.21(a)(3).

Based on the programs identified above, the staff concludes that the applicant's programs meet SRP-LR Section 3.4.2.2.2 criteria. For those items that apply to LRA Section 3.4.2.2.2, the staff determines that the LRA is consistent with the GALL Report. The staff also finds that the applicant has demonstrated that the effects of aging will be adequately managed so that the intended function(s) will be maintained consistent with the CLB during the period of extended operation, as required by 10 CFR 54.21(a)(3).

3.4.2.2.3 Loss of Material Due to General, Pitting, Crevice, and Microbiologically-Influenced Corrosion and Fouling

LRA Section 3.4.2.2.3 refers to Table 3.4.1, item 3.4.1.8, and addresses the loss of material due to general, pitting, crevice, and MIC and fouling of steel piping, piping components, and piping elements exposed to raw water. The applicant indicated that these items are not applicable because there are no AFW system components within scope of license renewal exposed to raw water. The staff reviewed LRA Sections 2.3.4 and 3.4, and confirmed that the applicant's LRA does not have any AMR results for the steam and power conversion systems that include steel piping, piping components, and piping elements exposed to raw water. The staff reviewed the applicant's UFSAR and confirmed that no in-scope steel piping, piping components, and piping elements exposed to raw water are present in the AFW system; therefore, it finds the applicant's determination acceptable.

3.4.2.2.4 Reduction of Heat Transfer Due to Fouling

The staff reviewed LRA Section 3.4.2.2.4 against the following criteria in SRP-LR Section 3.4.2.2.4:

- LRA Section 3.4.2.2.4.1 refers to Table 3.4.1, item 3.1.4.9, and addresses the reduction of heat transfer due to fouling of stainless steel and copper alloy heat exchanger tubes exposed to treated water. The applicant stated that this item is not applicable because PVNGS has no in-scope stainless steel or copper alloy heat exchangers with an intended function of heat transfer, exposed to treated water in the condensate, SG blowdown, or AFW systems. The staff reviewed LRA Sections 2.3.4 and 3.4 and confirmed that the applicant's LRA does not have any stainless steel or copper alloy heat exchanger tubes exposed to treated water in the steam and power conversion systems with an intended function of heat transfer. The staff also reviewed the UFSAR to verify the same. Based on its review of the LRA and UFSAR, the staff confirmed that PVNGS does not have any in-scope heat exchangers constructed of stainless steel or copper alloy with an intended function of heat transfer exposed to treated water in the steam and power conversion systems and, therefore, it finds the applicant's determination acceptable.

Aging Management Review Results

- LRA Section 3.4.2.2.4.2 refers to Table 3.4.1, item 3.4.1.10, and addresses the reduction of heat transfer due to fouling for steel, stainless steel, or copper alloy heat exchanger tubes in lubricating oil. The LRA states that the Lubricating Oil Analysis Program and the One-Time Inspection Program will manage this aging effect. The applicant also stated that the one-time inspection will include selected components at susceptible locations where contaminants, such as water, could accumulate.

  The staff reviewed LRA Section 3.4.2.2.4.2 against the criteria in SRP-LR Section 3.4.2.2.4, item 2, which states that monitoring and control of lubricating oil chemistry mitigates the reduction of heat transfer due to fouling, but the control of lubricating oil contaminants may not always have been adequate to preclude corrosion. It continues by stating that the effectiveness of the lubricating oil contaminant control should be verified and that a one-time inspection of susceptible components is an acceptable method to ensure that reduction of heat transfer is not occurring.

  The staff reviewed the applicant's Lubricating Oil Analysis and One-Time Inspection Programs in SER Sections 3.0.3.2.17 and 3.0.3.1.6, respectively. The staff noted that the applicant periodically samples lubricating oil to maintain contaminants within acceptable limits, which will preclude loss of heat transfer due to fouling. The staff also noted that the applicant will conduct one-time inspections for fouling of select stainless steel and copper alloy heat exchanger tubes in applicable steam and power conversion systems exposed to lubricating oil to verify the effectiveness of the Lubricating Oil Analysis Program.

  Based on the programs identified, the staff concludes that the applicant's programs meet SRP-LR Section 3.4.2.2.4.2 criteria. For those items that apply to LRA Section 3.4.2.2.4.2, the staff determines that the LRA is consistent with the GALL Report. The applicant has demonstrated that the effects of aging will be adequately managed so that the intended function(s) will be maintained consistent with the CLB during the period of extended operation, as required by 10 CFR 54.21(a)(3).

Based on the programs identified above, the staff concludes that the applicant's programs meet SRP-LR Section 3.4.2.2.4 criteria. For those items that apply to LRA Section 3.4.2.2.4, the staff determines that the LRA is consistent with the GALL Report and that the applicant has demonstrated that the effects of aging will be adequately managed so that the intended function(s) will be maintained consistent with the CLB during the period of extended operation, as required by 10 CFR 54.21(a)(3).

3.4.2.2.5   Loss of Material Due to General, Pitting, Crevice, and Microbiologically-Influenced Corrosion

The staff reviewed LRA Section 3.4.2.2.5 against the following criteria in SRP-LR Section 3.4.2.2.5:

- LRA Section 3.4.2.2.5.1 refers to Table 3.4.1, item 3.4.1.11, and addresses the loss of material due to general, pitting, crevice, and MIC in steel piping, piping components, piping elements, and tanks exposed to soil. The applicant stated that these items are not applicable because the condensate and AFW systems do not contain steel piping, piping components, piping elements, and tanks exposed to soil. The staff reviewed LRA Sections 2.3.4 and 3.4 and confirmed that the applicant's LRA does not have any AMR results for the condensate and AFW systems that include steel piping, piping components, piping elements, and tanks exposed to soil. The staff reviewed the applicant's UFSAR and confirmed that no in-scope steel piping, piping components, piping elements, and tanks exposed to soil are present in the condensate and AFW systems and, therefore, it finds the applicant's determination acceptable.

Aging Management Review Results

- LRA Section 3.4.2.2.5 refers to LRA Table 3.4.1, item 3.4.1.12 and addresses steel heat exchanger components exposed to lubricating which are managed for loss of material due to general, pitting, crevice, and MIC by the Lubricating Oil Analysis and the One-Time Inspection Programs. The applicant addressed the further evaluation criteria of the SRP-LR by stating that it will include a one-time inspection of selected components at susceptible locations where contaminants, such as water, could accumulate.

    The staff reviewed LRA Section 3.4.2.2.5.2 against the criteria in SRP-LR Section 3.4.2.2.5, item 2, which states that loss of material due to general, pitting, crevice corrosion, and MIC could occur for steel heat exchanger components exposed to lubricating oil. SRP-LR Section 3.4.2.2.5.2 further states that the existing AMP relies on the periodic sampling and analysis of lubricating oil to maintain contaminants within acceptable limits and, therefore, it preserves an environment that is not conducive to corrosion. The applicant verifies the effectiveness of the Lubricating Oil Program through a one-time inspection of selected components at susceptible locations to ensure that corrosion is not occurring and that the component's intended function will be maintained during the period of extended operation. In addition, the GALL Report recommends a further evaluation of programs to manage corrosion by verifying their effectiveness with a one-time inspection of selected components at susceptible locations.

    The staff reviewed the applicant's Lubricating Oil Analysis Program and One-Time Inspection Program in SER Sections 3.0.3.2.17 and 3.0.3.1.6, respectively. In its review of the components associated with the LRA items listed above, the staff finds the applicant's proposal to manage aging effects using the Lubricating Oil Analysis and One-Time Inspection programs acceptable because: (1) the Lubricating Oil Analysis Program provides for periodic sampling of lubricating oil to maintain contaminants at acceptable limits to preclude loss of material; (2) the applicant will conduct one-time inspections of select steel heat exchanger components exposed to lubricating oil for loss of material due to general, pitting, crevice, and MIC to verify the effectiveness of the Lubricating Oil Analysis Program in applicable auxiliary systems, following the GALL Report recommendation that the "One Time Inspection" is an acceptable AMP to verify the effectiveness of the applicant's Lubricating Oil Analysis Program.

    Based on the programs identified, the staff concludes that the applicant's programs meet SRP-LR Section 3.4.2.2.5.2 criteria. For those items that apply to LRA Section 3.4.2.2.5.2, the staff determines that the LRA is consistent with the GALL Report and that the applicant has demonstrated that the effects of aging will be adequately managed so that the intended function(s) will be maintained consistent with the CLB during the period of extended operation, as required by 10 CFR 54.21(a)(3).

- LRA Table 3.4.1, item 3.4.1.2 addresses loss of material due to general, pitting, and crevice corrosion in steel piping, piping components, and piping elements exposed to steam. The applicant stated that this item is not applicable because there are no in-scope steel components exposed to steam in the steam turbine or extraction steam systems. The staff reviewed LRA Sections 2.3.4 and 3.4 as well as the UFSAR and confirmed that the steam turbine and extraction steam functions are not related to maintaining the pressure boundary, but rather to electrical signals associated with an anticipated transient without scram event and de-energization of the heater drain pumps (fire protection function). The staff notes that the applicant chose to address the pressure boundary function of main steam and AFW aging management by items 3.4.1.37 and 3.4.1.29. Item 3.4.1.37 addresses loss of material due to pitting and crevice corrosion in steel, stainless steel, and nickel-based alloy piping, piping

Aging Management Review Results

components, and piping elements exposed to steam by the Water Chemistry Program (consistent with GALL). Item 3.4.1.29 addresses FAC by the Flow-Accelerated Program for both the main steam and AFW systems, which will conduct general corrosion inspections of these lines during the period of extended operation (consistent with GALL). The staff finds the applicant's determination, that AMR item 3.4.1.2 is not applicable, acceptable because the in-scope pressure boundary functions of main steam and AFW are adequately managed for aging by items 3.4.1.37 and 3.4.1.29.

Based on the programs identified above, the staff concludes that the applicant's programs meet SRP-LR Section 3.4.2.2.5 criteria. For those items that apply to LRA Section 3.4.2.2.5, the staff determines that the LRA is consistent with the GALL Report and that the applicant has demonstrated that the effects of aging will be adequately managed so that the intended function(s) will be maintained consistent with the CLB during the period of extended operation, as required by 10 CFR 54.21(a)(3).

3.4.2.2.6 Cracking Due to Stress Corrosion Cracking

LRA Section 3.4.2.2.6 refers to LRA Table 3.4.1, item 3.4.1.14, and addresses stainless steel piping, piping components, piping elements, and heat exchanger components exposed to secondary water (internal) which are managed for cracking due to SCC by the Water Chemistry Program and One-Time Inspection Program. The applicant addressed the further evaluation criteria of the SRP-LR by stating that the Water Chemistry Program and the One-Time Inspection Program will manage cracking due to SCC for stainless steel components exposed to secondary water. The applicant stated the one-time inspection will include selected components at susceptible locations.

The staff reviewed LRA Section 3.4.2.2.6 against the criteria in SRP-LR Section 3.4.2.2.6, which states that cracking due to SCC could occur for stainless steel piping, piping components, piping elements, tanks, and heat exchangers components exposed to treated water greater than 60 degrees C (140 degrees F). The SRP-LR also indicates that cracking due to SCC could occur for stainless steel piping, piping components, and piping elements exposed to steam, and the existing management program relies on the monitoring and control of the water chemistry. The SRP-LR further indicates that a one-time inspection should augment the Water Chemistry Program to verify the absence of cracking.

SER Sections 3.0.3.2.1 and 3.0.3.1.6 document the staff's evaluation of the applicant's Water Chemistry Program and One-Time Inspection Program, respectively. The staff noted that the Water Chemistry Program controls the chemical environment to ensure that the aging effects due to contaminants are limited by managing the primary and secondary water. The staff also noted that this program is effective in creating an environment that is not conducive for cracking to occur in areas of intermediate and high-flow where thorough mixing takes place and the monitoring samples are representative of actual conditions. The staff noted that the applicant's One-Time Inspection Program will conduct NDE inspections of a representative group of components in order to provide verification of the effectiveness of the Water Chemistry Program in low-flow and stagnant areas. The applicant's proposal to manage the SCC of stainless steel components is consistent with GALL Report, item VIII.B1-5. In its review of components associated with item 3.4.1.14, the staff finds the applicant's proposal to manage aging using the Water Chemistry Program and One-Time Inspection Program acceptable because the Water Chemistry Program will create an environment that is not conducive for cracking to occur and the One-Time Inspection Program will verify the effectiveness of the water chemistry and the applicant's use of these programs is consistent with the GALL Report.

Based on the programs identified, the staff concludes that the applicant's programs meet SRP-LR Section 3.3.2.2.6. For those items that apply to LRA Section 3.3.2.2.6, the staff determines that the LRA is consistent with the GALL Report and that the applicant has

Aging Management Review Results

demonstrated that the effects of aging will be adequately managed so that the intended function(s) will be maintained consistent with the CLB during the period of extended operation as required by 10 CFR 54.21(a)(3).

3.4.2.2.7 Loss of Material Due to Pitting and Crevice Corrosion

The staff reviewed LRA Section 3.4.2.2.7 against the following criteria in SRP-LR Section 3.4.2.2.7:

- LRA Section 3.4.2.2.7 refers to LRA Table 3.4.1, items 3.4.1.6, 3.4.1.15, and 3.4.1.16 and addresses steel, stainless steel, aluminum, and copper alloy piping, piping components, and piping elements and stainless steel tanks and heat exchanger components exposed to treated water managed for loss of material due to pitting and crevice corrosion by the Water Chemistry Program, augmented by the One-Time Inspection Program. The applicant addressed the further evaluation criteria of the SRP-LR by stating that it will use the Water Chemistry Program and One-Time Inspection Program to manage loss of material for stainless steel and copper alloy components exposed to secondary water and demineralized water. There were no aluminum AMR items used, so this metal is not addressed. The applicant also stated that the One-Time Inspection Program will include selected components at locations where contaminants could accumulate, such as stagnant flow areas.

The staff reviewed LRA Section 3.4.2.2.7.1 against the criteria in SRP-LR Section 3.4.2.2.7, item 1, which states that loss of material due to pitting and crevice corrosion could occur for stainless steel, aluminum, and copper alloy piping, piping components, and piping elements, and stainless steel tanks and heat exchanger components exposed to treated water. The SRP-LR also states that the GALL Report recommends further evaluation of a plant-specific AMP.

In its review of components associated with item 3.4.1.15, the staff noted that in LRA Table 3.3.2-4, the applicant referred to this item number and GALL Report item VIII.A-5 for copper alloy with greater than 15-percent zinc piping exposed to demineralized water. The staff also noted that LRA Table 3.3.2-4 is for the emergency chilled water system, which is a closed-cycle cooling water system, and that this component is also susceptible to loss of material due to selective leaching. It was unclear to the staff why the applicant referenced item 3.4.1.15 instead of item 3.3.1-84 and GALL Report item VII.C2-7, which recommends the Selective Leaching of Materials Program to manage loss of material due to selective leaching for these components exposed to treated water. The staff issued RAI 3.3.2-1 requesting that the applicant clarify how the Water Chemistry and One-Time Inspection Programs will be used to manage loss of material due to selective leaching.

In its response, dated March 1, 2010, the applicant stated that the copper alloy with greater than 15-percent zinc components exposed to demineralized water are used to provide make-up water to the chilled water system. The applicant further stated that the components are susceptible to loss of material due to selective leaching and that LRA item 3.3.1.84 and GALL Report item VII.C2-7 are referenced in another AMR result for the same components and managed by the Selective Leaching of Materials Program. The staff finds the applicant's response acceptable because the components are being appropriately managed for loss of material due to selective leaching.

SER Sections 3.0.3.2.1 and 3.0.3.1.6 document the staff's evaluation of the applicant's Water Chemistry and One-Time Inspection Programs, respectively. The staff finds the applicant's use of the Water Chemistry and One-Time Inspection Programs acceptable because the Water Chemistry Program monitors and controls the concentration of contaminants in the water in order to minimize corrosion, and the One-Time Inspection

Aging Management Review Results

Program conducts inspections to verify the effectiveness of the Water Chemistry Program.

The staff concludes that the applicant's programs meet SRP-LR Section 3.4.2.2.7, item 1 criteria. For those items that apply to LRA Section 3.4.2.2.7, item 1, the staff determines that the LRA is consistent with the GALL Report and that the applicant has demonstrated that the effects of aging will be adequately managed so that the intended function(s) will be maintained consistent with the CLB during the period of extended operation, as required by 10 CFR 54.21(a)(3).

- LRA Section 3.4.2.2.7 refers to Table 3.4.1, item 3.4.1.17, and addresses the loss of material due to pitting and crevice corrosion in stainless steel piping, piping components, and piping elements exposed to soil. The applicant stated that this item is not applicable because the condensate and AFW systems do not contain stainless steel piping, piping components, and piping elements exposed to soil. The staff reviewed LRA Sections 2.3.4 and 3.4 and confirmed that the applicant's LRA does not have any AMR results for the condensate and AFW systems that include stainless steel piping, piping components, and piping elements exposed to soil. The staff reviewed the applicant's UFSAR and confirmed that no in-scope stainless steel piping, piping components, and piping elements exposed to soil are present in the condensate and AFW systems and, therefore, finds the applicant's determination acceptable.

- LRA Section 3.4.2.2.7 refers to LRA Table 3.4.1, item 3.4.1.18 and addresses loss of material due to pitting and crevice corrosion of copper alloy piping, piping components, and elements exposed to lubricating oil. The GALL Report recommends use of AMPs XI.M39, "Lubricating Oil Analysis," and XI.M32, "One-Time Inspection," to manage loss of material due to pitting and crevice corrosion for this component group. The applicant stated that this item is not applicable because "PVNGS has no in-scope copper alloy components exposed to lube oil in the steam turbine, feedwater, condensate, or AFW systems, so the applicable GALL Report items were not used." The staff reviewed LRA Sections 2.3.4 and 3.4 and confirmed that the applicant's LRA does not have any AMR results for the referenced hardware and systems that include copper alloy components exposed to lube oil. The staff also reviewed the UFSAR to verify the same. Based on its review of the LRA and UFSAR, the staff confirmed that the applicant's plant does not have any in-scope copper alloy piping and components exposed to lubricating oil in the referenced steam and power conversion systems and components, and therefore, it finds the applicant's determination acceptable.

Based on the programs identified above, the staff concludes that the applicant's programs meet SRP-LR Section 3.4.2.2.7 criteria. For those items that apply to LRA Section 3.4.2.2.7, the staff determines that the LRA is consistent with the GALL Report and that the applicant has demonstrated that the effects of aging will be adequately managed so that the intended function(s) will be maintained consistent with the CLB during the period of extended operation, as required by 10 CFR 54.21(a)(3).

3.4.2.2.8 Loss of Material Due to Pitting, Crevice, and Microbiologically-Influenced Corrosion

LRA Section 3.4.2.2.8 refers to LRA Table 3.4.1, item 3.4.1.19 and addresses stainless steel piping, piping components, piping elements, and heat exchanger components exposed to lubricating oil. The GALL Report recommends use of AMPs XI.M39, "Lubricating Oil Analysis," and XI.M32, "One-Time Inspection," to manage for loss of material due to pitting, crevice, and MIC for this component group. The applicant stated that this item is not applicable because "PVNGS has no in-scope stainless steel components exposed to lube oil in the steam turbine, feedwater, condensate, or AFW systems, so the applicable GALL Report items were not used." The staff reviewed LRA Sections 2.3.4 and 3.4 and confirmed that the applicant's LRA does not

Aging Management Review Results

have any AMR results for the referenced hardware that include stainless steel piping, piping components, piping elements, and heat exchanger components exposed to lubricating oil. The staff also reviewed the UFSAR to verify the same. Based on its review of the LRA and UFSAR, the staff confirmed that the applicant's plant does not have any in-scope stainless steel piping, piping components, piping elements, and heat exchanger components exposed to lubricating oil in the referenced steam and power conversion SCs, and therefore, it finds the applicant's determination acceptable.

3.4.2.2.9 Loss of Material Due to General, Pitting, Crevice, and Galvanic Corrosion

LRA Section 3.4.2.2.9 addresses the loss of material due to general, pitting, crevice, and galvanic corrosion for steel heat exchanger components exposed to treated water. The only Table 3.4-1 item is item 3.4.1.5, which is addressed in SER Section 3.4.2.1.1.

3.4.2.2.10 Quality Assurance for Aging Management of Nonsafety-Related Components

SER Section 3.0.4 documents the staff's evaluation of the applicant's QA program.

3.4.2.3 Aging Management Review Results Not Consistent with or Not Addressed in the Generical Aging Lessons Learned Report

In LRA Tables 3.4.2-1 through 3.4.2-3, the staff reviewed additional details of the AMR results for material, environment, AERM, and AMP combinations not consistent with or not addressed in the GALL Report.

In LRA Tables 3.4.2-1 through 3.4.2-3, via Notes F–J, the applicant indicated that combinations of component type, material, environment, and AERM do not correspond to an item in the GALL Report. The applicant provided further information about how it will manage the aging effects. Specifically, Note F indicates that the material for the AMR item component is not evaluated in the GALL Report. Note G indicates that the environment for the AMR item component and material is not evaluated in the GALL Report. Note H indicates that the aging effect for the AMR item component, material, and environment combination is not evaluated in the GALL Report. Note I indicates that the aging effect identified in the GALL Report for the item component, material, and environment combination is not applicable. Note J indicates that neither the component nor the material and environment combination for the item is evaluated in the GALL Report.

For component type, material, and environment combinations not evaluated in the GALL Report, the staff reviewed the applicant's evaluation to determine if the applicant has demonstrated that it will adequately manage the effects of aging so that the intended function(s) will be maintained consistent with the CLB for the period of extended operation. The following sections document the staff's evaluation.

3.4.2.3.1 Main Steam System—Summary of Aging Management Review—License Renewal Application Table 3.4.2-1

The staff reviewed LRA Table 3.4.2-1, which summarizes the results of AMR evaluations for the main steam system component groups.

In LRA Tables 3.3.2-4, 3.3.2-10, 3.3.2.-21, 3.4.2-1, and 3.4.2-3, the applicant stated that calcium silicate and mineral wool insulation exposed to borated water leakage have no AERM and no proposed AMP. The AMR items cite generic Note J. SER Section 3.3.2.3.4 documents the staff's review.

In LRA Tables 3.4.2-1, 3.4.2-2, and 3.4.2-3, the applicant stated that stainless steel closure bolting exposed to plant indoor air and carbon steel closure bolting exposed to atmosphere or weather are managed for loss of preload by the Bolting Integrity Program. The AMR items cite generic Note G, indicating that the environment is not in the GALL Report for this component

Aging Management Review Results

and material. The AMR items also cite plant-specific note 1, indicating that loss of preload is considered to be applicable for all closure bolting.

SER Section 3.0.3.2.3 documents the staff's evaluation of the applicant's Bolting Integrity Program. The staff noted that the mechanisms identified in the GALL Report as causing loss of preload in carbon steel bolts are thermal effects, gasket creep, and self-loosening, which are not all dependent on the bolting material or environment. The staff also noted that activities in the Bolting Integrity Program that control and manage loss of preload are effective for various bolting materials. The staff further noted that the GALL Report, item VIII.H-5 (S-33) recommends using the Bolting Integrity Program to manage the aging effect of loss of preload in carbon steel bolts exposed to air-indoor uncontrolled. Because the GALL Report recommends the Bolting Integrity Program for managing loss of preload in carbon steel bolting exposed to air-indoor uncontrolled, and the Bolting Integrity Program's activities for managing loss of preload are applicable for other bolting materials and environments, the staff finds the applicant's use of the Bolting Integrity Program to manage loss of preload in stainless steel closure bolting exposed to plant indoor air and in carbon steel bolting exposed to atmosphere or weather to be acceptable.

In LRA Table 3.4.2-1, the applicant stated that, for nickel-alloy flexible hoses exposed internally to dry gas, there is no aging effect and no proposed AMP. The AMR items cite generic Note G. The staff reviewed the associated items in the LRA and confirmed that no aging effect is applicable for this component, material, and environment because austenitic materials such as nickel alloys are not subject to loss of material or cracking when exposed to this environment, and these materials are used as corrosion-resistant replacement materials where other materials have degraded. The staff noted that corrosion-resistant materials, such as austenitic and martensitic stainless steels and high-strength nickel-based alloys, offer good protection against loss of material. The staff also noted that the conditions required for cracking due to a variety of mechanisms (SCC, PWSCC, IASCC and IGSCC) to occur, such as being exposed to an aqueous solution (reactor coolant or other corrosive solutions) and high temperatures, do not exist on the surfaces of these components when exposed to dry gas. The staff noted that GALL, AMR item IV.E-1 states that nickel alloy exposed to air-indoor uncontrolled is not subject to an AERM. The staff noted that an air-indoor uncontrolled environment is more aggressive than a dry gas environment because it is possible for condensation to occur in an air-indoor uncontrolled environment. The staff finds the applicant's determination acceptable because nickel alloy is a highly corrosion-resistant material, it is not subject to conditions where cracking is possible, and it is exposed to a less aggressive environment when compared to GALL AMR item IV.E-1, for which there is no AERM.

On the basis of its review, the staff finds that the applicant has appropriately evaluated the AMR results of material, environment, AERM, and AMP combinations not evaluated in the GALL Report. The staff finds that the applicant has demonstrated that the effects of aging will be adequately managed so that the intended function(s) will be maintained consistent with the CLB for the period of extended operation, as required by 10 CFR 54.21(a)(3).

3.4.2.3.2 Condensate Storage and Transfer System—Summary of Aging Management Review—License Renewal Application Table 3.4.2-2

The staff reviewed LRA Table 3.4.2-2, which summarizes the results of AMR evaluations for the condensate storage and transfer system component groups.

SER Section 3.4.2.3.1 documents the staff's evaluation for stainless steel closure bolting exposed to plant indoor air and carbon steel closure bolting exposed to atmosphere or weather managed for loss of preload by the Bolting Integrity Program, citing generic Note G.

Aging Management Review Results

SER Section 3.3.2.3.7 documents the staff's evaluation for stainless steel components exposed to atmosphere or weather conditions, where there are no AERM and no AMP, citing generic Note G.

In LRA Table 3.4.2-2, the applicant stated that the Boiling Integrity Program manages stainless steel closure bolting, exposed to atmosphere or weather, for loss of preload. The AMR items cite generic Note G, indicating that the environment is not in the GALL Report for this component and material. The AMR items also cite plant-specific note 1, indicating that loss of preload is considered to be applicable for all closure bolting. The staff reviewed all AMR results in the GALL Report where the aging effect is loss of preload and confirmed there are no AMR results for stainless steel bolting where the environment is atmosphere or weather.

SER Section 3.0.3.2.3 documents the staff's evaluation of the applicant's Bolting Integrity Program. The staff noted that the mechanisms identified in the GALL Report as causing loss of preload in carbon steel bolts are thermal effects, gasket creep, and self-loosening, which are not all dependent on the bolting material or environment. The staff also noted that activities in the Bolting Integrity Program that control and manage loss of preload are effective for various bolting materials. The staff further noted that the GALL Report item VIII.H-5 recommends using the Bolting Integrity Program to manage the aging effect of loss of preload in carbon steel bolts exposed to air-indoor uncontrolled. Since the GALL Report recommends the Bolting Integrity Program for managing loss of preload in carbon steel bolting exposed to uncontrolled indoor air, and the Bolting Integrity Program's activities for managing loss of preload are applicable for other bolting materials and environments, the staff finds the applicant's use of the Bolting Integrity Program to manage loss of preload in stainless steel closure bolting exposed to atmosphere or weather to be acceptable.

In LRA Table 3.4.2-2, the applicant stated that stainless steel valves exposed externally to atmosphere or weather have no AERM, and it will not implement an AMP for these components. The AMR items cite generic Note G, indicating that the environment is not in the GALL Report for this component and material combination.

The staff reviewed all AMR result lines in the GALL Report for this material and environment combination and noted that the GALL Report recommends that stainless steel components exposed to indoor uncontrolled air or air with borated water leakage have no AERM. The staff also noted that the climate at this location is dry and arid with high average temperatures (68 to 108 degrees F) throughout the year and low average rainfall (8 to 10 inches per year), making the aging effects of the atmosphere or weather at this location similar to those of indoor uncontrolled air. The staff finds the applicant's determination that no AMP is required acceptable because stainless steel components exposed to the atmosphere or weather would not be expected to experience an aging effect at this location.

The staff finds that the applicant has demonstrated that the effects of aging will be adequately managed so that the intended function will be maintained consistent with the CLB for the period of extended operation, as required by 10 CFR 54.21(a)(3).

3.4.2.3.3 Auxiliary Feedwater System—Summary of Aging Management Review—License Renewal Application Table 3.4.2-3

The staff reviewed LRA Table 3.4.2-3, which summarizes the results of AMR evaluations for the AFW system component groups.

In LRA Tables 3.3.2-4, 3.3.2-10, 3.3.2.-21, 3.4.2-1, and 3.4.2-3, the applicant stated that calcium silicate and mineral wool insulation exposed to borated water leakage have no AERM and no proposed AMP. The AMR items cite generic Note J. SER Section 3.3.2.3.4 documents the staff's review.

Aging Management Review Results

In LRA Table 3.4.2-3, the applicant stated that the Water Chemistry and One-Time Inspection Programs manage the carbon steel AFW turbine oil cooler heat exchanger, exposed to secondary water, for reduction of heat transfer. The AMR item cites generic Note G, indicating that for the item, the environment is not in the GALL Report for this component and material.

SER Sections 3.0.3.2.1 and 3.0.3.1.6 document the staff's evaluation of the applicant's Water Chemistry and a One-Time Inspection Programs, respectively.

The staff finds the applicant's proposal to manage aging using the Water Chemistry and One-Time Inspection Programs acceptable because the Water Chemistry Program will minimize the potential development and progress of heat exchanger fouling, and the One-Time Inspection Program will verify the effectiveness of the Water Chemistry Program using inspections with specific attributes related to the reduction of heat transfer.

SER Section 3.4.2.3.1 documents the staff's evaluation for stainless steel closure bolting exposed to plant indoor air and carbon steel closure bolting, exposed to atmosphere or weather, managed for loss of preload by the Bolting Integrity Program, citing generic Note G.

SER Section 3.3.2.3.2.1 documents the staff's evaluation for aluminum heat exchanger (governor oil cooler), valve, and filter exposed to lubricating oil and managed for loss of material by the Lubricating Oil Analysis and One-Time Inspection Programs with generic Note G.

3.4.3 Conclusion

The staff concludes that the applicant has provided sufficient information to demonstrate that the effects of aging for the steam and power conversion systems components within the scope of license renewal and subject to an AMR will be adequately managed so that the intended function(s) will be maintained consistent with the CLB for the period of extended operation, as required by 10 CFR 54.21(a)(3).

3.5 Aging Management of Structures and Component Supports

This section of the SER documents the staff's review of the applicant's AMR results for the SC supports of the following components:

- containment building
- control building
- diesel generator building
- turbine building
- auxiliary building
- radwaste building
- main steam support structure
- SBO generator structures
- fuel building
- spray pond and associated water control structures
- tank foundations and shells
- transformer foundations and electrical structures
- yard structures (in-scope)
- supports

3.5.1 Summary of Technical Information in the Application

LRA Section 3.5 provides AMR results for the SC supports groups. LRA Table 3.5-1, "Summary of Aging Management Evaluations in Chapters II and III of NUREG-1801 for Containments,

Aging Management Review Results

Structures, and Component Supports," is a summary comparison of the applicant's AMRs with those evaluated in the GALL Report for the SC supports groups.

The applicant's AMRs evaluated and incorporated applicable plant-specific and industry operating experience in the determination of AERMs. The plant-specific evaluation included Condition Reports and discussions with appropriate site personnel to identify AERMs. The applicant's review of industry operating experience included a review of the GALL Report and operating experience issues identified since the issuance of the GALL Report.

3.5.2 Staff Evaluation

The staff reviewed LRA Section 3.5 to determine if the applicant provided sufficient information to demonstrate that the effects of aging for the SC supports within the scope of license renewal and subject to an AMR will be adequately managed so that the intended function(s) will be maintained consistent with the CLB for the period of extended operation, as required by 10 CFR 54.21(a)(3).

The staff reviewed AMRs to confirm the applicant's claim that certain AMRs were consistent with the GALL Report. The staff did not repeat its review of the matters described in the GALL Report; however, the staff did verify that the material presented in the LRA was applicable and that the applicant identified the appropriate GALL Report AMPs. SER Section 3.0.3 documents the staff's evaluations of the AMPs, and SER Section 3.5.2.1 provides details of the staff's evaluation.

The staff also reviewed AMRs consistent with the GALL Report and for which further evaluation is recommended. The staff confirmed that the applicant's further evaluations were consistent with the SRP-LR Section 3.5.2.2 acceptance criteria. SER Section 3.5.2.2 documents the staff's evaluations.

The staff also conducted a technical review of the remaining AMRs not consistent with, or not addressed in, the GALL Report.5.2.3.

For SSCs that the applicant claimed were not applicable or required no aging management, the staff reviewed the AMR items and the plant's operating experience to verify the applicant's claims.

Table 3.5-1 summarizes the staff's evaluation of components, aging effects or mechanisms, and AMPs listed in LRA Section 3.5 and addressed in the GALL Report.

Table 3.5-1. Staff Evaluation for Structures and Component Supports Components in the GALL Report

Component Group (GALL Report Item No.)	Aging Effect/ Mechanism	AMP in GALL Report	Further Evaluation in GALL Report	AMP in LRA, Supplements, or Amendments	Staff Evaluation
PWR Concrete (Reinforced and Prestressed) and Steel Containments					
Concrete elements: walls, dome, basemat, ring girder, buttresses, containment (as applicable) (3.5.1-1)	Aging of accessible and inaccessible concrete areas due to aggressive chemical attack, and corrosion of embedded steel	ISI (IWL) and for inaccessible concrete, an examination of representative samples of below-grade concrete, and periodic monitoring of groundwater if environment is non-aggressive. A plant-specific program is to be evaluated if environment is aggressive.	Yes, plant specific, if environment aggressive	ISI (IWL)	Consistent with GALL Report. See SER Section 3.5.2.2.1(1).
Concrete elements; All (3.5.1-2)	Cracks and distortion due to increased stress levels from settlement	Structures Monitoring Program. If a de-watering system is relied upon for control of settlement, then the licensee is to ensure proper functioning of the de-watering system through the period of extended operation	Yes, if not within the scope of the applicant's Structures Monitoring Program or a de-watering system is relied upon.	Structures Monitoring Program	Consistent with GALL Report. See SER Section 3.5.2.2.1(2).
Concrete elements: foundation, sub-foundation (3.5.1-3)	Reduction in foundation strength, cracking, differential settlement due to erosion of porous concrete subfoundation	Structures Monitoring Program. If a de-watering system is relied upon to control erosion of cement from porous concrete subfoundations, then the licensee is to ensure proper functioning of the de-watering system through the period of extended operation	Yes, if not within the scope of the applicant's Structures Monitoring Program or a de-watering system is relied upon.	Not applicable	Not applicable to PVNGS. See SER Section 3.5.2.2.1(2).
Concrete elements: dome, wall, basemat, ring girder, buttresses, containment, concrete fill-in annulus (as applicable) (3.5.1-4)	Reduction of strength and modulus of concrete due to elevated temperature	A plant-specific AMP is to be evaluated.	Yes, plant-specific if temperature limits are exceeded.	Not applicable	Not applicable to PVNGS. See SER Section 3.5.2.2.1(3).

Aging Management Review Results

Component Group (GALL Report Item No.)	Aging Effect/ Mechanism	AMP in GALL Report	Further Evaluation in GALL Report	AMP in LRA, Supplements, or Amendments	Staff Evaluation
Steel elements: drywell; torus; drywell head; embedded shell and sand pocket regions; drywell support skirt; torus ring girder; downcomers; liner plate, ECCS suction header, support skirt, region shielded by diaphragm floor, suppression chamber (as applicable) (3.5.1-5)	Loss of material due to general, pitting and crevice corrosion	ISI (IWE) and 10 CFR Part 50, Appendix J	Yes, if corrosion is significant for inaccessible areas	Not applicable	Not applicable to PWRs. See SER Section 3.5.2.1.1.
Steel elements: steel liner, liner anchors, integral attachments (3.5.1-6)	Loss of material due to general, pitting and crevice corrosion	ISI (IWE) and 10 CFR Part 50, Appendix J	Yes, if corrosion is significant for inaccessible areas	ISI (IWE), and 10 CFR 50, Appendix J	Consistent with GALL Report. See SER Section 3.5.2.2.1(4).
Prestressed containment tendons (3.5.1-7)	Loss of prestress due to relaxation, shrinkage, creep, and elevated temperature	TLAA, evaluated in accordance with 10 CFR 54.21(c)	Yes, TLAA.	TLAA	Consistent with GALL Report. See SER Section 3.5.2.2.1(5).
Steel and stainless steel elements: vent line, vent header, vent line bellows; downcomers (3.5.1-8)	Cumulative fatigue damage (CLB fatigue analysis exists)	TLAA evaluated in accordance with 10 CFR 54.21(c)	Yes, TLAA	Not applicable	Not applicable to PWRs. See SER Section 3.5.2.1.1.
Steel, stainless steel elements, dissimilar metal welds: penetration sleeves, penetration bellows; suppression pool shell, unbraced downcomers (3.5.1-9)	Cumulative fatigue damage (CLB fatigue analysis exists)	TLAA evaluated in accordance with 10 CFR 54.21(c)	Yes, TLAA	TLAA	Consistent with GALL Report. See SER Section 3.5.2.2.1(6).
Stainless steel penetration sleeves, penetration bellows, dissimilar metal welds (3.5.1-10)	Cracking due to SCC	ISI (IWE) and 10 CFR Part 50, Appendix J and additional appropriate examinations & evaluations for bellows assemblies and dissimilar metal welds	Yes, detection of aging is to be evaluated	Not applicable	Not applicable to PVNGS. See SER Section 3.5.2.2.1(7).

Aging Management Review Results

Component Group (GALL Report Item No.)	Aging Effect/ Mechanism	AMP in GALL Report	Further Evaluation in GALL Report	AMP in LRA, Supplements, or Amendments	Staff Evaluation
Stainless steel vent line bellows (3.5.1-11)	Cracking due to SCC	ISI (IWE) and 10 CFR Part 50, Appendix J and additional appropriate examinations & evaluations for bellows assemblies and dissimilar metal welds	Yes, detection of aging is to be evaluated	Not applicable	Not applicable to PWRs. See SER Section 3.5.2.1.1.
Steel, stainless steel elements, dissimilar metal welds: penetration sleeves, penetration bellows; suppression pool shell, unbraced downcomers (3.5.1-12)	Cracking due to cyclic loading	ISI (IWE) and 10 CFR Part 50, Appendix J supplemented to detect fine cracks	Yes, detection of aging is to be evaluated	TLAA	GALL Report items not used. Evaluated as a TLAA. See SER Sections 3.5.2.2.1(8) and 4.6.
Steel, stainless steel elements, dissimilar metal welds: torus; vent line; vent header; vent line bellows; downcomers (3.5.1-13)	Cracking due to cyclic loading	ISI (IWE) and 10 CFR Part 50, Appendix J supplemented to detect fine cracks	Yes, detection of aging is to be evaluated	Not applicable	Not applicable to PWRs. See SER Section 3.5.2.1.1.
Concrete elements: dome, wall, basemat ring girder, buttresses, containment (as applicable) (3.5.1-14)	Loss of material (scaling, cracking, and spalling) due to freeze-thaw	ISI (IWL). Evaluation is needed for plants that are located in moderate to severe weathering conditions (weathering index > 100 day-inch/yr) (NUREG-1557)	Yes, for inaccessible areas of plants located in moderate to severe weathering conditions	ISI (IWL)	Consistent with GALL Report. See SER Section 3.5.2.2.1(9).
Concrete elements: walls, dome, basemat, ring girder, buttresses, containment, concrete fill-in annulus (as applicable) (3.5.1-15)	Cracking due to expansion and reaction with aggregate; increase in porosity, permeability due to leaching of calcium hydroxide	ISI (IWL) for accessible areas. None for inaccessible areas if concrete was constructed in accordance with the recommendations in ACI 201.2R	Yes, if concrete was not constructed as stated in inaccessible areas.	ISI (IWL)	Consistent with GALL Report. See SER Section 3.5.2.2.1(10).
Seals, gaskets, and moisture barriers (3.5.1-16)	Loss of sealing and leakage through containment due to deterioration of joint seals, gaskets, and moisture barriers (caulking, flashing, and other sealants)	ISI (IWE) and 10 CFR Part 50, Appendix J	No	ISI (IWE), and 10 CFR 50, Appendix J	Consistent with GALL Report.

Aging Management Review Results

Component Group (GALL Report Item No.)	Aging Effect/ Mechanism	AMP in GALL Report	Further Evaluation in GALL Report	AMP in LRA, Supplements, or Amendments	Staff Evaluation
Personnel airlock, equipment hatch and CRD hatch locks, hinges, and closure mechanisms (3.5.1-17)	Loss of leak tightness in closed position due to mechanical wear of locks, hinges and closure mechanisms	10 CFR Part 50, Appendix J and Plant TS	No	10 CFR 50, Appendix J and Plant TS	Consistent with GALL Report.
Steel penetration sleeves and dissimilar metal welds; personnel airlock, equipment hatch and CRD hatch (3.5.1-18)	Loss of material due to general, pitting, and crevice corrosion	ISI (IWE) and 10 CFR Part 50, Appendix J	No	ISI (IWE), and 10 CFR 50, Appendix J	Consistent with GALL Report.
Steel elements: stainless steel suppression chamber shell (inner surface) (3.5.1-19)	Cracking due to SCC	ISI (IWE) and 10 CFR Part 50, Appendix J	No	Not applicable	Not applicable to PWRs. See SER Section 3.5.2.1.1.
Steel elements: suppression chamber liner (inner surface) (3.5.1-20)	Loss of material due to general, pitting, and crevice corrosion	ISI (IWE) and 10 CFR Part 50, Appendix J	No	Not applicable	Not applicable to PWRs. See SER Section 3.5.2.1.1.
Steel elements: drywell head and downcomer pipes (3.5.1-21)	Fretting or lock up due to mechanical wear	ISI (IWE)	No	Not applicable	Not applicable to PWRs. See SER Section 3.5.2.1.1.
Prestressed containment: tendons and anchorage components (3.5.1-22)	Loss of material due to corrosion	ISI (IWL)	No	ISI (IWL)	Consistent with GALL Report.
Safety-Related and Other Structures and Component Supports					
All Groups except Group 6: interior and above grade exterior concrete (3.5.1-23)	Cracking, loss of bond, and loss of material (spalling, scaling) due to corrosion of embedded steel	Structures Monitoring Program	Yes, if not within scope of the applicant's structures monitoring program	Structures Monitoring Program	Consistent with GALL Report. See SER Section 3.5.2.2.2(1).
All Groups except Group 6: interior and above grade exterior concrete (3.5.1-24)	Increase in porosity and permeability, cracking, loss of material (spalling, scaling) due to aggressive chemical attack	Structures Monitoring Program	Yes, if not within scope of the applicant's structures monitoring program	Structures Monitoring Program	Consistent with GALL Report. See SER Section 3.5.2.2.2(1).

Aging Management Review Results

Component Group (GALL Report Item No.)	Aging Effect/ Mechanism	AMP in GALL Report	Further Evaluation in GALL Report	AMP in LRA, Supplements, or Amendments	Staff Evaluation
All Groups except Group 6: steel components: all structural steel (3.5.1-25)	Loss of material due to corrosion	Structures Monitoring Program. If protective coatings are relied upon to manage the effects of aging, the structures monitoring program is to include provisions to address protective coating monitoring and maintenance	Yes, if not within the scope of the applicant's Structures Monitoring Program	Structures Monitoring Program	Consistent with GALL Report. See SER Section 3.5.2.2.2(1).
All Groups except Group 6: accessible and inaccessible concrete: foundation (3.5.1-26)	Loss of material (spalling, scaling) and cracking due to freeze-thaw	Structures Monitoring Program. Evaluation is needed for plants that are located in moderate to severe weathering conditions (weathering index > 100 day-inch/yr) (NUREG-1557)	Yes, if not within the scope of the applicant's structures monitoring program or for inaccessible areas of plants located in moderate to severe weathering conditions	Structures Monitoring Program	Consistent with GALL Report. See SER Section 3.5.2.2.2(1).
All Groups except Group 6: accessible and inaccessible interior/exterior concrete (3.5.1-27)	Cracking due to expansion due to reaction with aggregates	Structures Monitoring Program. None for inaccessible areas if concrete was constructed in accordance with the recommendations in ACI 201.2R-77	Yes, if not within the scope of the applicant's structures monitoring program or concrete was not constructed as stated for inaccessible areas.	Structures Monitoring Program	Consistent with GALL Report. See SER Section 3.5.2.2.2(1).
Groups 1-3, 5-9: All (3.5.1-28)	Cracks and distortion due to increased stress levels from settlement	Structures Monitoring Program. If a de-watering system is relied upon for control of settlement, then the licensee is to ensure proper functioning of the de-watering system through the period of extended operation	Yes, if not within the scope of the applicant's structures monitoring program or a de-watering system is relied upon.	Structures Monitoring Program	Consistent with GALL Report. See SER Section 3.5.2.2.2(1).

Aging Management Review Results

Component Group (GALL Report Item No.)	Aging Effect/ Mechanism	AMP in GALL Report	Further Evaluation in GALL Report	AMP in LRA, Supplements, or Amendments	Staff Evaluation
Groups 1-3, 5-9: foundation (3.5.1-29)	Reduction in foundation strength, cracking, differential settlement due to erosion of porous concrete subfoundation	Structures Monitoring Program. If a de-watering system is relied upon for control of settlement, then the licensee is to ensure proper functioning of the de-watering system through the period of extended operation	Yes, if not within the scope of the applicant's Structures Monitoring Program.	Structures Monitoring Program	Not applicable to PVNGS. See SER Section 3.5.2.2.2(1).
Group 4: Radial beam seats in BWR drywell; RPV support shoes for PWR with nozzle supports; SG supports (3.5.1-30)	Lock-up due to wear	ISI (IWF) or Structures Monitoring Program	Yes, if not within the scope of the ISI or structures monitoring	Not applicable	Not applicable to PVNGS. See SER Section 3.5.2.2.2(1).
Groups 1-3, 5, 7-9: below-grade concrete components, such as exterior walls below grade and foundation (3.5.1-31)	Increase in porosity and permeability, cracking, loss of material (spalling, scaling), aggressive chemical attack; cracking, loss of bond, and loss of material (spalling, scaling), corrosion of embedded steel	Structures Monitoring Program. Examination of representative samples of below-grade concrete, and periodic monitoring of groundwater, if the environment is non-aggressive. A plant-specific program is to be evaluated if environment is aggressive.	Yes, plant-specific, If environment is aggressive	Structures Monitoring Program	Consistent with GALL Report. See SER Section 3.5.2.2.2(2).
Groups 1-3, 5, 7-9: exterior above and below grade reinforced concrete foundations (3.5.1-32)	Increase in porosity and permeability, and loss of strength due to leaching of calcium hydroxide	Structures Monitoring Program for accessible areas. None for inaccessible areas if concrete was constructed in accordance with the recommendations in ACI 201.2R-77	Yes, if concrete was not constructed as stated for inaccessible areas	Not applicable	Not applicable to PVNGS. See SER Section 3.5.2.2.2(2).
Groups 1-5: concrete (3.5.1-33)	Reduction of strength and modulus due to elevated temperature	Plant-specific	Yes, plant-specific if temperature limits are exceeded	Not applicable	Not applicable to PVNGS. See SER Section 3.5.2.2.2(3).

Aging Management Review Results

Component Group (GALL Report Item No.)	Aging Effect/ Mechanism	AMP in GALL Report	Further Evaluation in GALL Report	AMP in LRA, Supplements, or Amendments	Staff Evaluation
Group 6: concrete; all (3.5.1-34)	Increase in porosity and permeability, cracking, loss of material due to aggressive chemical attack; cracking, loss of bond, loss of material due to corrosion of embedded steel.	Inspection of Water-Control Structures or FERC/U.S. Army Corps of Engineers dam inspections and maintenance programs and for inaccessible concrete, an examination of representative samples of below-grade concrete, and periodic monitoring of groundwater, if the environment is non-aggressive. A plant-specific program is to be evaluated if environment is aggressive	Yes. Plant-specific if environment is aggressive	Inspection of Water-Control Structures	Consistent with GALL Report. See SER Section 3.5.2.2.2(4).
Group 6: exterior above and below grade concrete foundation (3.5.1-35)	Loss of material (spalling, scaling) and cracking due to freeze-thaw	Inspection of Water-Control Structures or FERC/U.S. Army Corps of Engineers dam inspections and maintenance programs. Evaluation is needed for plants that are located in moderate to severe weathering conditions (weathering index > 100 day-inch/yr) (NUREG-1557)	Yes, for inaccessible areas of plants located in moderate to severe weathering conditions.	Inspection of water-control Structures	Consistent with GALL Report. See SER Section 3.5.2.2.2(4).
Group 6: all accessible and inaccessible reinforced concrete (3.5.1-36)	Cracking due to expansion/re-action with aggregates	Accessible areas: Inspection of Water-Control Structures or FERC/U.S. Army Corps of Engineers dam inspections and maintenance programs. None for inaccessible areas if concrete was constructed in accordance with the recommendations in ACI 201.2R-77	Yes, if concrete was not constructed as stated for inaccessible areas	Inspection of Water-Control Structures	Consistent with GALL Report. SER Section 3.5.2.2.2(4).

Aging Management Review Results

Component Group (GALL Report Item No.)	Aging Effect/ Mechanism	AMP in GALL Report	Further Evaluation in GALL Report	AMP in LRA, Supplements, or Amendments	Staff Evaluation
Group 6: exterior above and below grade reinforced concrete foundation interior slab (3.5.1-37)	Increase in porosity and permeability, loss of strength due to leaching of calcium hydroxide	For accessible areas, Inspection of Water-Control Structures or FERC/U.S. Army Corps of Engineers dam inspections and maintenance programs. None for inaccessible areas if concrete was constructed in accordance with the recommendations in ACI 201.2R-77	Yes, if concrete was not constructed as stated for inaccessible areas	Inspection of Water-Control Structures	Consistent with GALL Report. See SER Section 3.5.2.2.2(4).
Groups 7, 8: Tank Liners (3.5.1-38)	Cracking due to SCC; loss of material due to pitting and crevice corrosion	Plant-specific	Yes	Not applicable	Not applicable to PVNGS. See SER Section 3.5.2.2.2(5).
Support members; welds; bolted connections; support anchorage to building structure (3.5.1-39)	Loss of material due to general and pitting corrosion	Structures Monitoring Program	Yes, if not within the scope of the applicant's Structures Monitoring Program	Structures Monitoring Program	Consistent with GALL Report. See SER Section 3.5.2.2.2(6).
Building concrete at locations of expansion and grouted anchors; grout pads for support base plates (3.5.1-40)	Reduction in concrete anchor capacity due to local concrete degradation/ service induced cracking or other concrete aging mechanisms or other con	Structures Monitoring Program	Yes, if not within the scope of the applicant's Structures Monitoring Program	Structures Monitoring Program	Consistent with GALL Report. See SER Section 3.5.2.2.2(6).
Vibration isolation elements (3.5.1-41)	Reduction or loss of isolation function/radiation hardening, temperature, humidity, sustained vibratory loading	Structures Monitoring Program	Yes, if not within the scope of the applicant's Structures Monitoring Program.	Not applicable	Not applicable to PVNGS. See SER Section 3.5.2.2.2(6).
Groups B1.1, B1.2, and B1.3: support members: anchor bolts, welds (3.5.1-42)	Cumulative fatigue damage (CLB fatigue analysis exists)	TLAA, evaluated in accordance with 10 CFR 54.21(c)	Yes, TLAA	Not applicable	Not applicable to PVNGS. See SER Section 3.5.2.2.2(7) and 4.3.
Groups 1-3, 5, 6: all masonry block walls (3.5.1-43)	Cracking due to restraint shrinkage, creep, and aggressive environment	Masonry Wall Program	No	Masonry Wall Program	Consistent with GALL Report.

Aging Management Review Results

Component Group (GALL Report Item No.)	Aging Effect/ Mechanism	AMP in GALL Report	Further Evaluation in GALL Report	AMP in LRA, Supplements, or Amendments	Staff Evaluation
Group 6 elastomer seals, gaskets, and moisture barriers (3.5.1-44)	Loss of sealing due to deterioration of seals, gaskets, and moisture barriers (caulking, flashing, and other sealants)	Structures Monitoring Program	No	Structures Monitoring Program	Consistent with GALL Report.
Group 6: exterior above and below grade concrete foundation; interior slab (3.5.1-45)	Loss of material due to abrasion, cavitation	Inspection of Water-Control Structures or FERC/U.S. Army Corps of Engineers dam inspections and maintenance	No	Inspection of Water-Control Structures	Consistent with GALL Report.
Group 5: Fuel pool Liners (3.5.1-46)	Cracking due to SCC; loss of material due to pitting and crevice corrosion	Water chemistry and monitoring of spent fuel pool water level and level of fluid in the leak chase channel	No	Water Chemistry and monitoring of spent fuel pool water level in accordance with TS and leakage from the leak chase channels	Consistent with GALL Report.
Group 6: all metal structural members (3.5.1-47)	Loss of material due to general (steel only), pitting, and crevice corrosion	Inspection of Water Control Structures Associated with Nuclear Power Plants. If protective coatings are relied upon to manage aging, protective coating monitoring and maintenance provisions should be included	No	Inspection of Water-Control Structures	Consistent with GALL Report.
Group 6: earthen water control structures - dams, embankments, reservoirs, channels, canals, and ponds (3.5.1-48)	Loss of material, loss of form due to erosion, settlement, sedimentation, frost action, waves, currents, surface runoff, Seepage	Inspection of water-control structures associated with nuclear power plants	No	Not applicable	Not applicable to PVNGS. See SER Section 3.5.2.1.1.
Support members: welds; bolted connections; support anchorage to building structures (3.5.1-49)	Loss of material/ general, pitting, and crevice corrosion	Water chemistry and ISI (IWF)	No	Not applicable	Not applicable to PWRs. See SER Section 3.5.2.1.1.

Aging Management Review Results

Component Group (GALL Report Item No.)	Aging Effect/ Mechanism	AMP in GALL Report	Further Evaluation in GALL Report	AMP in LRA, Supplements, or Amendments	Staff Evaluation
Groups B2, and B4: galvanized steel, aluminum, stainless steel support members; welds; bolted connections; support anchorage to building structure (3.5.1-50)	Loss of material due to pitting and crevice corrosion	Structures Monitoring Program	No	Structures Monitoring Program	Consistent with GALL Report, except for the Class 2 and 3 components under ASME Section XI, Subsection IWF. See SER Section 3.5.2.1.4.
Group B1.1: high strength low-alloy bolts (3.5.1-51)	Cracking due to SCC; loss of material due to general corrosion	Bolting Integrity	No	Bolting Integrity	Consistent with GALL Report.
Groups B2, and B4: sliding support bearings and sliding support surfaces (3.5.1-52)	Loss of mechanical function due to corrosion, distortion, dirt, overload, fatigue due to vibratory and cyclic thermal loads	Structures Monitoring Program	No	Structures Monitoring Program	Consistent with GALL Report.
Groups B1.1, B1.2, and B1.3: support members: welds; bolted connections; support anchorage to building structure (3.5.2-53)	Loss of material due to general and pitting corrosion	ISI (IWF)	No	ISI (IWF)	Consistent with GALL Report.
Groups B1.1, B1.2, and B1.3: Constant and variable load spring hangers; guides; stops (3.5.2-54)	Loss of mechanical function to corrosion, distortion, dirt, overload, fatigue due to vibratory and cyclic thermal loads	ISI (IWF)	No	ISI (IWF)	Consistent with GALL Report.
Steel, galvanized steel, and aluminum support members; welds; bolted connections; support anchorage to building structure (3.5.2-55)	Loss of material due to boric acid corrosion	Boric Acid Corrosion	No	Boric Acid Corrosion	Consistent with GALL Report.
Groups B1.1, B1.2, and B1.3: Sliding surfaces (3.5.2-56)	Loss of mechanical function due to corrosion, distortion, dirt, overload, fatigue due to vibratory and cyclic thermal loads	ISI (IWF)	No	ISI (IWF)	Consistent with GALL Report.

Aging Management Review Results

Component Group (GALL Report Item No.)	Aging Effect/ Mechanism	AMP in GALL Report	Further Evaluation in GALL Report	AMP in LRA, Supplements, or Amendments	Staff Evaluation
Groups B1.1, B1.2, and B1.3: Vibration isolation elements (3.5.2-57)	Reduction or loss of isolation function/ radiation hardening, temperature, humidity, sustained vibratory loading	ISI (IWF)	No	Not applicable	Not applicable to PVNGS. See SER Section 3.5.2.1.1.
Galvanized steel and aluminum support members; welds; bolted connections; support anchorage to building structure exposed to air-indoor uncontrolled (3.5.2-58)	None	None	No	None	Consistent with GALL Report.
Stainless steel support members; welds; bolted connections; support anchorage to building structure (3.5.2-59)	None	None	No	None	Consistent with GALL Report.

The staff's review of the SC supports groups fell into three categories. One category, documented in SER Section 3.5.2.1, reviewed AMR results for components that the applicant indicated are consistent with the GALL Report and require no further evaluation. Another category, documented in SER Section 3.5.2.2, reviewed AMR results for components that the applicant indicated are consistent with the GALL Report and for which further evaluation is recommended. A third category, documented in SER Section 3.5.2.3, reviewed AMR results for components that the applicant indicated are not consistent with, or not addressed in, the GALL Report. Section 3.0.3 documents the staff's review of AMPs credited to manage or monitor aging effects of the SC supports.

3.5.2.1 Aging Management Review Results Consistent with the Generic Aging Lessons Learned Report

LRA Section 3.5.2.1 identifies the materials, environments, AERMs, and the following programs that manage aging effects for the SSCs and their commodity groups:

- 10 CFR Part 50, Appendix J Program
- ASME Section XI, Subsection IWE Program
- ASME Section XI, Subsection IWF Program
- ASME Section XI, Subsection IWL Program
- Bolting Integrity
- Boric Acid Corrosion
- Fire Protection Program
- Masonry Wall Program
- RG 1.127, Inspection of Water-Control Structures Associated with Nuclear Power Plants
- Structures Monitoring Program
- Water Chemistry Program

Aging Management Review Results

LRA Tables 3.5.2-1 through 3.5.2-14 summarize AMRs for the SC supports elements and indicate AMRs claimed to be consistent with the GALL Report.

The staff audited and reviewed the information in the LRA. The staff did not repeat its review of the matters described in the GALL Report; however, the staff did verify that the material presented in the LRA was applicable and that the applicant identified the appropriate GALL Report AMRs. The staff's evaluation follows.

3.5.2.1.1   Aging Management Review Results Identified as Not Applicable

In LRA Table 3.5.1, items 3.5.1.5, 3.5.1.8, 3.5.1.11, 3.5.1.13, 3.5.1.19 through 3.5.1.21, and 3.5.1.49, the applicant states that the corresponding AMR items in the GALL Report are not applicable to PVNGS because the units are a PWR reactor design that incorporates a containment system consisting of a steel-lined prestressed cylindrical concrete structure with a hemispherical dome. The AMR items in the GALL Report are only applicable to steel and stainless steel elements of BWR designs. The staff verified that the stated AMR items in the GALL Report are only applicable to metallic components of BWR designs and are not applicable to the PVNGS LRA. Based on this determination, the staff finds that the applicant has provided an acceptable basis for concluding AMR items 3.5.1.5, 3.5.1.8, 3.5.1.11, 3.5.1.13, 3.5.1.19 through 3.5.1.21, and 3.5.1.49 are not applicable.

LRA Table 3.5.1, item 3.5.1.48 addresses loss of material or form due to erosion, settlement, sedimentation, frost action, waves, currents, surface runoff, and seepage of earthen water-control structures exposed to water. The GALL Report recommends the Inspection of Water-Control Structures AMP to ensure that the aging effect is adequately managed. The applicant stated that this item is not applicable because there are no earthen dams, embankments, reservoirs, channels, canals, or ponds in-scope for license renewal. The staff reviewed LRA Sections 2.4.10 and 3.5 and confirmed that the applicant's LRA does not have any AMR results for earthen water-control structures. The staff also reviewed the UFSAR to verify the same. Based on its review of the LRA and UFSAR, the staff confirmed that the applicant's plant does not have any in-scope earthen water-control structures; therefore, it finds the applicant's determination acceptable.

LRA Table 3.5.1, item 3.5.1.57 addresses reduction or loss of isolation function due to radiation hardening, temperature, humidity, or sustained vibratory loading of vibration isolation elements. The GALL Report recommends the "ASME Section XI, Subsection IWF" AMP to ensure that the aging effect is adequately managed. The applicant stated that this item is not applicable because there are no vibration isolation elements in-scope for license renewal. The staff reviewed LRA Sections 2.4.14 and 3.5 and confirmed that the applicant's LRA does not have any AMR results for vibration isolation elements. The staff also reviewed the UFSAR to verify the same. Based on its review of the LRA and UFSAR, the staff confirmed that the applicant's plant does not have any in-scope vibration isolation elements; therefore, it finds the applicants determination acceptable.

3.5.2.1.2   Cracking Due to Restraint Shrinkage, Creep, and Aggressive Environment

In LRA Tables 3.5.2-2, 3.5.2-4, 3.5.2-5, and 3.5.2-13, which reference item 3.5.1.43 and plant-specific note 1, the applicant credits the Masonry Wall Program and the Fire Protection Program for managing this aging effect and mechanism in a plant indoor air environment. The applicant also included plant-specific note 1 that states, "NUREG-1801 does not provide a line in which Concrete Masonry is inspected per the Fire Protection program."

The staff reviewed the AMR results that referenced Note E and plant-specific note 1. The staff determined, for these items, that the component type, material, environment, and aging effect are consistent with the corresponding line of the GALL Report; however, where the GALL Report recommends AMP XI.S5, "Masonry Wall Program," the applicant has proposed using the

Aging Management Review Results

Masonry Wall Program and the Fire Protection Program. The LRA states that the intended functions related to this item include fire barrier, shelter and protection, and structural support. Appendix B of the LRA states that the Masonry Wall Program is part of the Structures Monitoring Program that implements structures monitoring requirements as specified by 10 CFR 50.65. This program includes cracking of masonry walls and structural steel restraint systems of the masonry walls within scope of license renewal based on guidance provided in IE bulletin 80-11 and NRC IN 87-67. In Appendix B of the LRA, the applicant further states that it uses the Fire Protection Program to manage aging in the form of cracking, spalling, and loss of material by visual inspection, every 18 months, of the fire barrier walls, ceilings, and floors.

Since the applicant uses the Masonry Wall Program, with inspections included as part of the Structures Monitoring Program, with periodic visual inspections of the fire barrier walls, floors, and ceilings also performed under the Fire Protection Program, the staff finds that the applicant addressed the AERM adequately. SER Sections 3.0.3.2.20 and 3.0.3.2.7 document the staff's review of the Structures Monitoring Program and Fire Protection Program, respectively.

3.5.2.1.3 Loss of Material Due to General (Steel Only), Pitting, and Crevice Corrosion

In LRA Table 3.5.2-10, items that reference item 3.5.1.47 and plant-specific note 1, the applicant credits the Structures Monitoring Program for managing this aging effect/mechanism in atmosphere/weather, plant indoor air, and submerged environments. The applicant also included a plant-specific note that states "NUREG 1801, line III.A6-11 specifies RG 1.127 as the program for metal components in water-control structures. RG 1.127 does not address metal components, so the Structures Monitoring Program is used."

The staff reviewed the AMR results lines that referenced Note E and the plant-specific note and determined that, for these items, the component type, material, and aging effect are consistent with the corresponding line of the GALL Report. However, where the GALL Report recommends XI.S7, "Regulatory Guide 1.127, Inspection of Water-Control Structures Associated with Nuclear Power Plants," as the AMP, the applicant has proposed using the Structures Monitoring Program as the AMP for the carbon steel water-control structural components. Appendix B, Section B2.1.33, of the LRA states that the PVNGS Structures Monitoring Program includes, and is consistent with, the recommendations in GALL AMP XI.S7, is in compliance with the requirements of 10 CFR 50.65, and includes inspection and surveillance activities for water-control structures associated with emergency cooling water systems on a frequency of at least once every five years.

Since the applicant has committed to an appropriate AMP with an inspection frequency commensurate with the GALL Report guidance, the staff finds these AMR results to be acceptable. SER Section 3.0.3.2.20 documents the staff's review of the Structures Monitoring Program.

3.5.2.1.4 Loss of Material/Pitting and Crevice Corrosion

In LRA Table 3.5.2-14, the item that references item 3.5.1.50 and plant-specific note 1, the applicant credits the ASME Section XI, Subsection IWF Program for managing this aging effect and mechanism in atmosphere or weather environment. Plant-specific note 1 states "NUREG-1801 does not provide a line to evaluate stainless steel components outdoors under ASME Section XI, Subsection IWF."

The staff reviewed the AMR results lines that referenced Note E and plant-specific note 1 and determined, that the component type, material, and aging effect for these items are consistent with the corresponding line of the GALL Report. However, where the GALL Report recommends XI.S6, "Structures Monitoring Program" as the AMP for stainless steel ASME 2 and 3 supports, the applicant has proposed using ASME Section XI, Subsection IWF as the program to manage loss of material, cracking, and loss of mechanical function that could result

Aging Management Review Results

in loss of intended function for Class 1, 2, and 3 component supports. The IWF Program requires visual inspections at a frequency that meets or exceeds the requirements of the Structures Monitoring Program.

Since the applicant has committed to an appropriate AMP with an inspection frequency and method that meets or exceeds the GALL Report guidance, the staff finds these AMR results to be acceptable. SER Section 3.0.3.1.11 documents the staff's review of the ASME Section XI, Subsection IWF Program.

Conclusion. The staff evaluated the applicant's claim of consistency with the GALL Report. The staff also reviewed information pertaining to the applicant's consideration of recent operating experience and proposals for managing aging effects. On the basis of its review, the staff concludes that the AMR results, which the applicant claimed to be consistent with the GALL Report, are indeed consistent with the corresponding AMRs. Therefore, the staff concludes that the applicant has demonstrated that the effects of aging for these components will be adequately managed so that their intended function(s) will be maintained consistent with the CLB during the period of extended operation, as required by 10 CFR 54.21(a)(3).

3.5.2.2 Aging Management Review Results Consistent with the Generic Aging Lessons Learned Report for Which Further Evaluation Is Recommended

In LRA Section 3.5.2, the applicant further evaluates aging management, as recommended by the GALL Report, for the containments, structures, and component supports and provides information concerning how it will manage aging effects in the following three areas:

(1)   PWR and BWR Containments

- aging of inaccessible concrete areas
- cracks and distortion due to increased stress levels from settlement, reduction of foundation strength, cracking, and differential settlement due to erosion of porous concrete subfoundations if not covered by the Structures Monitoring Program
- reduction of strength and modulus of concrete structures due to elevated temperature
- loss of material due to general, pitting, and crevice corrosion
- loss of prestress due to relaxation, shrinkage, creep, and elevated temperature
- cumulative fatigue damage
- cracking due to SCC
- cracking due to cyclic loading
- loss of material (scaling, cracking, and spalling) due to freeze-thaw
- cracking due to expansion and reaction with aggregate and increase in porosity and permeability due to leaching of calcium hydroxide

(2)   Safety-Related and Other SC Supports

- aging of structures not covered by the Structures Monitoring Program
- aging management of inaccessible areas (below-grade inaccessible concrete areas of Groups 1–5 and 7–9 structures)
- reduction of strength and modulus of concrete structures due to elevated temperature for Group 1–5 structures

Aging Management Review Results

- aging management of inaccessible areas for Group 6 structures (below-grade inaccessible concrete areas)
- cracking due to stress corrosion and loss of material due to pitting and crevice corrosion for Group 7 and 8 stainless steel tank liners
- aging of supports not covered by the Structures Monitoring Program
- cumulative fatigue damage due to cyclic loading

(3) QA for Aging Management of Nonsafety-Related Components

SER Section 3.0.4 documents the staff's evaluation of the applicant's QA program.

3.5.2.2.1 Pressurized Water Reactor and Boiling Water Reactor Containments

The staff reviewed LRA Section 3.5.2.2.1 against the criteria in SRP-LR Section 3.5.2.2.1, which addresses several areas as follows:

- Aging of Inaccessible Concrete Areas. LRA Section 3.5.2.2.1.1 states that reinforced concrete structures were designed, constructed, and inspected in accordance with ACI and ASTM standards to provide good quality, dense, well-cured, and low permeability concrete. Crack control is achieved through proper sizing, spacing, and distribution of reinforcing steel in accordance with ACI 318-71. Concrete structures are not subjected to groundwater for sustained periods. An engineering study was conducted to confirm that groundwater elevations are located below the lowest structures indicating that further evaluation for the effects of aggressive chemical attack and corrosion of embedded steel is not required.

    The staff reviewed LRA Section 3.5.2.2.1.1 against the criteria in SRP-LR Section 3.5.2.2.1.1, which states that increases in porosity and permeability, cracking, loss of material (e.g., spalling, scaling) due to aggressive chemical attack, and cracking, loss of bond, and loss of material (e.g., spalling, scaling) due to corrosion of embedded steel could occur in inaccessible areas of PWR and BWR concrete and steel containments. The GALL report identifies ASME Section XI, Subsection IWL to manage these aging effects and recommends further evaluation of plant-specific programs to manage these aging effects for inaccessible areas if the environment is aggressive.

    The staff confirmed that the ASME Section XI, Subsection IWL Program manages all accessible areas of the concrete containment building for cracking, loss of material, and increase in porosity and permeability. SER Section 3.0.3.1.10 documents the staff's review of the ASME Section XI, Subsection IWL Program. SER Section 3.5.2.2.2 documents the staff's review of the applicant's evaluation of aging management of inaccessible areas not covered here, including the containment-related concrete.

- Cracks and Distortion Due to Increased Stress Levels from Settlement; Reduction of Foundation Strength, Cracking, and Differential Settlement Due to Erosion of Porous Concrete Subfoundations, if not Covered by the Structures Monitoring Program. LRA Section 3.5.2.2.1.2 states that further evaluation of settlement is not required because the concrete components are evaluated under the Structures Monitoring Program and no permanent de-watering system or porous concrete foundations exist at PVNGS.

    The staff reviewed LRA Section 3.5.2.2.1.2 against the criteria in SRP-LR Section 3.5.2.2.1.2, which states that cracks and distortion due to increased stress levels from settlement and reduction in foundation strength, cracking, and differential settlement due to erosion of porous concrete subfoundations could occur. The GALL report identifies the Structures Monitoring Program to manage these aging effects and

Aging Management Review Results

no further evaluation is recommended if this activity is within scope of the Structures Monitoring Program.

The staff confirmed that the Structures Monitoring Program manages the monitoring for settlement of every major structure. SER Section 3.0.3.2.20 documents the staff's review of the Structures Monitoring Program. The staff also confirmed that no permanent de-watering system or porous concrete foundations exist. The staff finds acceptable the applicant's evaluation of this AERM in that it meets the criteria in SRP-LR Section 3.5.2.2.1.2.

- Reduction of Strength and Modulus of Concrete Structures Due to Elevated Temperature. LRA Section 3.5.2.2.1.3 states that the reactor cavity cooling subsystem operates in conjunction with the containment normal cooling units and provides continuous cooling of the primary shield and reactor cavity to limit the concrete temperature to less than the specified limit of 150 degrees F (66 degrees C). The reactor cavity is monitored with four cavity high temperature alarm channels that are annunciated in the control room. Plant TS require that the containment average air temperature not exceed 117 degrees F (47 degrees C). High-temperature piping has been designed to limit the local concrete temperature to 200 degrees F (93 degrees C). In the case of piping carrying hot fluid, the pipe is insulated to prevent excessive concrete temperatures and to prevent excessive heat loss from the fluid.

The staff reviewed LRA Section 3.5.2.2.1.3 against the criteria in SRP-LR Section 3.5.2.2.1.3, which recommends further evaluation of the plant-specific AMP if any portion of the concrete containment components exceeds the specified temperature limits of 150 degrees F (66 degrees C) general and 200 degrees F (93 degrees C) local.

The staff finds the applicant's evaluation acceptable in that this aging effect is not likely to develop because the containment concrete is kept below the allowable temperature limits.

- Loss of Material Due to General, Pitting and Crevice Corrosion. LRA Section 3.5.2.2.1.4 addresses loss of material due to general, pitting, and crevice corrosion for steel elements of accessible and inaccessible areas of containments. The LRA states that reinforced concrete structures were designed, constructed, and inspected in accordance with ACI and ASTM standards that provide good quality, dense, well-cured, and low-permeability concrete. Concrete mixes were designed in accordance with ACI 211.1-74. The applicant uses the ASME Section XI, Subsection IWL Program to identify and manage any cracks in the concrete that could potentially provide a pathway for water to reach inaccessible portions of the steel containment liner. In the LRA, the applicant states that borated water spills are not common and, when detected, the applicant cleans them up promptly. The applicant notes that further evaluation for corrosion in inaccessible areas of the steel liner of the containment is not required.

The staff reviewed LRA Section 3.5.2.2.1.4 against the criteria in SRP-LR Section 3.5.2.2.1.4. The SRP-LR criteria state that loss of material due to general, pitting and crevice corrosion could occur in steel elements of accessible and inaccessible areas for all types of PWR and BWR containments. The existing program relies on ASME Section XI Subsection IWE, and 10 CFR Part 50, Appendix J to manage this aging effect. The GALL Report recommends further evaluation of plant-specific programs to manage this aging effect for inaccessible areas if corrosion is significant. GALL Report, item II.A1-11 states that for inaccessible areas (e.g., embedded steel shell or liner), loss of material due to corrosion is not significant if the following conditions are satisfied:

Aging Management Review Results

- Concrete meeting the specifications of ACI 318 or 349 and the guidance of ACI 201.2R was used for the containment concrete in contact with the embedded containment shell or liner.

- The concrete is monitored to ensure that it is free of penetrating cracks that provide a path for water seepage to the surface of the containment shell or liner.

- The moisture barrier, at the junction where the shell or liner becomes embedded, is subject to aging management activities in accordance with ASME Section XI, Subsection IWE requirements.

- Borated water spills and water ponding on the containment concrete floor is not common and, when detected, is cleaned up promptly.

The staff verified that the containment concrete is monitored for cracks by the IWL AMP and that water ponding is not common on the containment floor. SER Sections 3.0.3.2.20, 3.0.3.2.18, and 3.0.3.1.12 document the staff's review of the applicant's Structures Monitoring Program, ASME Section XI, Subsection IWE Program, and Appendix J Program, respectively. The staff also reviewed the UFSAR and verified that all concrete work was done in accordance with ACI 318; however, the LRA did not discuss the first condition adequately in that it did not provide a comparison of the recommendations in ACI 211.1-74 to the recommendations of ACI 201.2R. Therefore, by letter dated February 19, 2010, the staff issued RAI 3.5.2.2.1-1 requesting the applicant to discuss how the concrete in contact with the embedded steel liner complies with the guidance in ACI 201.2R.

By letter dated March 24, 2010, the applicant responded and explained that the in-scope concrete was designed and constructed in accordance with ACI 211.1-74, which provides procedures for designing concrete mixes that take into consideration requirements for placeability, consistency, strength, and durability. The applicant further explained that the recommendations of ACI 201 are incorporated throughout these procedures and referenced in the ACI 211.1-74 discussions of durability, air-entrainment, and water-cement ratios.

The staff reviewed the applicant's response and found it acceptable because it explains that ACI 211.1-74 includes many of the recommendations included in ACI 201.2R. The staff independently reviewed both standards and noted that both recommend use of a low water-cement ratio and the proper use of air-entrainment. Since the applicant followed the guidance of ACI 318 and ACI 211.1, which references much of the guidance in ACI 201.2R, the staff finds the applicant's response acceptable. The staff's concern in RAI 3.5.2.2.1-1 is resolved. Since the applicant has explained how all four conditions are satisfied, the staff finds that corrosion is not significant for inaccessible areas, the criteria of SRP-LR Section 3.5.2.2.1.4 have been met, and further evaluation is not required.

- <u>Loss of Prestress Due to Relaxation, Shrinkage, Creep, and Elevated Temperature</u>. Loss of prestress due to relaxation, shrinkage, creep, and elevated temperature is a TLAA, as defined in 10 CFR 54.3. TLAAs are required to be evaluated in accordance with 10 CFR 54.21. SER Section 4.5, "Concrete Containment Tendon Prestress Analysis," documents the staff's review of the applicant's evaluation of this TLAA.

- <u>Cumulative Fatigue Damage</u>. LRA Section 3.5.2.2.1.6 states that containment penetrations for the main steam, main feedwater, and recirculation sump suction penetrations are supported by TLAAs. The applicant further stated that there are no penetration bellows within the scope of license renewal. LRA Section 4.6.2 describes the evaluation of the main steam and feedwater penetrations, while Section 4.6.3 describes the evaluation of the recirculation sump suction penetrations.

The staff reviewed LRA Section 3.5.2.2.1.6 against the criteria in SRP-LR Section 3.5.2.2.1.6, which states that fatigue analyses of penetrations are TLAAs, as defined in 10 CFR 54.3. SER Section 4.6 separately addresses the evaluation of this TLAA. The staff also confirmed that there are no containment penetration bellows within the scope of license renewal.

- Cracking Due to Stress Corrosion Cracking. LRA Section 3.5.2.2.1.7 states that PVNGS has no in-scope stainless steel penetration sleeves, penetration bellows, or dissimilar metal welds subject to SCC. The applicant stated that this AERM is not applicable. The staff confirmed that there are no in-scope stainless steel penetration sleeves, penetration bellows, or dissimilar metal welds subject to SCC.

  On the basis of its review, the staff finds the applicant's evaluation of the aging effect "cracking due to stress corrosion cracking" acceptable, because no in-scope stainless steel penetration sleeves, penetration bellows, or dissimilar metal welds subject to SCC exist at PVNGS.

- Cracking Due to Cyclic Loading. In LRA Section 3.5.2.2.1.8, the applicant stated that this section was not applicable because fatigue of metal components is a TLAA, evaluated in accordance with 10 CFR 54.21(c), so it did not use the applicable GALL Report lines.

  The staff reviewed LRA Section 3.5.2.2.1.8 against the criteria in SRP-LR Section 3.5.2.2.1.8, which states that cracking due to cyclic loading of the stainless steel shells (including welded joints) and penetrations (including penetration sleeves, dissimilar metal welds, and penetration bellows) could occur in PWR containments. The existing program relies on ASME Section XI, Subsection IWE and 10 CFR Part 50, Appendix J to manage this aging effect. However, VT-3 visual inspection may not detect fine cracks. The GALL Report recommends further evaluation for detection of this aging effect.

  The staff reviewed the appropriate GALL Report items for this section and confirmed that they only apply if a CLB fatigue analysis does not exist. The staff reviewed the LRA and confirmed that the metal containment components within the scope of license renewal have a fatigue analysis and are reviewed as TLAAs. Therefore, the staff confirms no further evaluation is necessary for this section. SER Section 4.6 documents the staff's review of the containment TLAAs.

- Loss of Material (Scaling, Cracking, and Spalling) Due to Freeze–Thaw. LRA Section 3.5.2.2.1.9 states that loss of material due to freeze-thaw is not applicable. The applicant stated that PVNGS is located in a weathering region classified as negligible according to Figure 1 of ASTM C33-03; therefore, this AERM is not applicable.

  The staff reviewed LRA Section 3.5.2.2.1.9 against the criteria in SRP-LR Section 3.5.2.2.1.9, which recommends further evaluation of loss of material due to freeze-thaw for plants with concrete containments located in moderate to severe weathering conditions. The staff finds acceptable the applicant's evaluation that this aging effect is not applicable because the primary containment structural concrete is located in a weathering region classified as having negligible freeze-thaw effects.

- Cracking Due to Expansion and Reaction with Aggregate, and Increase in Porosity and Permeability, Due to Leaching of Calcium Hydroxide. LRA Section 3.5.2.2.1.10 states that acceptance of aggregate materials was based, in part, on petrographic examination

Aging Management Review Results

in accordance with ASTM C295. Aggregate reactivity was evaluated in accordance with ASTM C289 and C227, and the reinforced concrete structures were designed, constructed, and inspected in accordance with ACI and ASTM standards that provide good quality, dense, well-cured, and low-permeability concrete. The applicant further stated that concrete mixes were designed in compliance with ACI 211.1-74 and that procedural controls were imposed on the concrete throughout the batching, mixing, and placing processes. The applicant also noted that the concrete structures are not subjected to flowing water, and an engineering study shows that the groundwater elevations are below the lowest structures. The applicant states that further evaluation for the effects of reaction with aggregates and leaching of calcium hydroxide are not required.

The staff reviewed LRA Section 3.5.2.2.1.10 against the criteria in SRP-LR Section 3.5.2.2.1.10, which states that cracking due to expansion and reaction with aggregate, and increase in porosity and permeability due to leaching of calcium hydroxide, could occur in concrete elements of concrete and steel containments. The GALL Report recommends further evaluation for concrete not constructed in accordance with the recommendations in ACI 201.2R-77.

The staff confirmed that the applicant uses the ASME Section XI, Subsection IWL Program to manage cracking, loss of material, and increase in porosity and permeability of the concrete containment building, and it evaluated the aggregate materials in accordance with appropriate ASTM standards. SER Section 3.0.3.1.10 documents the staff's review of the applicant's ASME Section XI, Subsection IWL Program. The staff also reviewed the UFSAR and verified that the applicant used ASTM C295 to evaluate the reactivity of the concrete aggregate. In its review, the staff noted that the applicant did not note that the concrete was constructed in accordance with the recommendations in ACI 201.2R. By letter dated February 19, 2010, the staff issued RAI 3.5.2.2.1-1 to ask if concrete was constructed using the recommendations provided in ACI 201.2R.

By letter dated March 24, 2010, the applicant responded and explained that the in-scope concrete was designed and constructed in accordance with ACI 211.1-74, which provides procedures for designing concrete mixes that consider ACI201 recommendations for concrete placeability, consistency, strength, and durability.

The staff reviewed the applicant's response and found it acceptable because it explains that the recommendations of ACI 201 are incorporated into the guidance of ACI 211.1 74. SER Section 3.5.2.2.1(4) provides a more detailed discussion of the staff's review. The staff's concern in RAI 3.5.2.2.1-1 is resolved. Since the reactivity of the aggregates has been evaluated using ASTM C295, and the concrete was constructed in accordance with the guidance in ACI 201.2R, the staff finds that the criteria of SRP-LR Section 3.5.2.2.1.10 have been met, and no further evaluation is required.

Based on the programs and analyses discussed above, the staff concludes that the applicant has met the criteria of SRP-LR Section 3.5.2.2.1. For those items that apply to LRA Section 3.5.2.2.1, the staff determines that the LRA is consistent with the GALL Report and the applicant has demonstrated that it will adequately manage the effects of aging so that the intended functions will be maintained consistent with the CLB during the period of extended operation, as required by 10 CFR 54.21(a)(3).

3.5.2.2.2 Safety-Related and Other Structures and Component Supports

The staff reviewed LRA Section 3.5.2.2.2 against the criteria in SRP-LR Section 3.5.2.2.2, covering several areas as addressed below.

Aging Management Review Results

- Aging of Structures Not Covered by Structures Monitoring Program. LRA Section 3.5.2.2.2.1 states that corrosion of embedded steel, aggressive chemical attack, loss of material due to corrosion, freeze-thaw, reaction with aggregates, and settlement are all aging effects that the Structures Monitoring Program evaluates. The LRA also states that further evaluation of erosion of porous concrete subfoundations is not required because PVNGS does not have porous concrete subfoundations. The LRA states that all in-scope sliding surfaces are evaluated under the Structures Monitoring Program or the ASME Section XI, Subsection IWF Program.

  The staff reviewed LRA Section 3.5.2.2.2.1 against the criteria in SRP-LR Section 3.5.2.2.2.1, which states that the GALL Report recommends further evaluation of certain structure and aging effect combinations if they are not covered by the structures monitoring program. These combinations include those listed below:

  - cracking, loss of bond, and loss of material (spalling, scaling) due to corrosion of embedded steel for Groups 1–5, 7, and 9 structures
  - increase in porosity and permeability, cracking, loss of material (spalling, scaling) due to aggressive chemical attack for Groups 1–5, 7, and 9 structures
  - loss of material due to corrosion for Groups 1–5, 7, and 8 structures
  - loss of material (spalling, scaling) and cracking due to freeze-thaw for Groups 1–3, 5, and 7–9 structures
  - cracking due to expansion and reaction with aggregates for Groups 1–5 and 7–9 structures
  - cracks and distortion due to increased stress levels from settlement for Groups 1–3 and 5–9 structures
  - reduction in foundation strength, cracking, differential settlement due to erosion of porous concrete subfoundation for Groups 1–3 and 5–9 structures

  The LRA further states that lock-up due to wear may occur for Lubrite radial beam seats in BWR drywells, RPV support shoes for PWRs with nozzle supports, SG supports, and other sliding support bearings and sliding support surfaces. The existing program relies on the Structures Monitoring Program to manage this aging effect. The GALL Report recommends further evaluation only for structure-aging effect combinations not within the ISI (IWF) or Structures Monitoring Programs.

  The staff reviewed the LRA and confirmed that the listed aging effects are evaluated within the Structures Monitoring Program. The staff also verified that PVNGS does not have porous concrete subfoundations. Therefore, as noted in the GALL Report, no further evaluation has been conducted.

- Aging Management of Inaccessible Areas (Below-Grade Inaccessible Concrete Areas of Groups 1-5, and 7-9 Structures). The staff reviewed LRA Section 3.5.2.2.2.2 against the following criteria in SRP-LR Section 3.5.2.2.2.2:

  - Loss of material (spalling, scaling) and cracking due to freeze-thaw could occur in below-grade inaccessible concrete areas of Groups 1–3, 5, and 7–9 structures.

    LRA Section 3.5.2.2.2.2.1 states that loss of material due to freeze-thaw is not applicable. The applicant stated that PVNGS is located in a weathering region classified as negligible according to Figure 1 of ASTM C33-03. Therefore, the applicant states that this AERM is not applicable.

Aging Management Review Results

The staff reviewed LRA Section 3.5.2.2.2.1 against the criteria in SRP-LR Section 3.5.2.2.2.2.1, which recommends further evaluation of loss of material due to freeze-thaw for plants with concrete containments located in moderate to severe weathering conditions. The staff finds acceptable the applicant's evaluation that this aging effect is not applicable because the concrete is located in a weathering region classified as negligible.

- Cracking due to expansion and reaction with aggregates could occur in below-grade inaccessible concrete areas for Groups 1–5 and 7–9 structures.

In LRA Section 3.5.2.2.2.2.2, the applicant states that acceptance of aggregate materials was based, in part, on petrographic examination in accordance with ASTM C295 and aggregate reactivity was evaluated in accordance with ASTM C289 and C227. In LRA Section 3.5.2.2.1.10, the applicant stated that the reinforced concrete structures at PVNGS were designed, constructed, and inspected in accordance with ACI and ASTM standards that provide good quality, dense, well-cured, and low-permeability concrete. The applicant further states, in LRA Section 3.5.2.2.1.10, that concrete mixes were designed in compliance with ACI 211.1-74 and those procedural controls were imposed on the concrete throughout the batching, mixing, and placing processes. The applicant states that further evaluation for the effects of reaction with aggregates is not required.

The staff reviewed LRA Section 3.5.2.2.2.2.2 against the criteria in SRP-LR Section 3.5.2.2.2.2.2, which states that the GALL Report recommends further evaluation of inaccessible areas of these groups of structures if the concrete was not constructed in accordance with the recommendations in ACI 201.2R-77. GALL Report item III.A2-2 states that investigations, tests, and petrographic examinations of aggregates performed in accordance with ASTM C295-54 or ASTM C227-50 can demonstrate that the aggregate is not reactive within the reinforced concrete. If either of these conditions is met, the GALL Report notes that aging management is not necessary.

The staff found that the concrete mix design adequately addressed cracking due to expansion and reaction with aggregates. The LRA states that acceptance of aggregate materials was based, in part, on petrographic examination in accordance with ASTM C295, and aggregate reactivity was evaluated in accordance with ASTM C289 and C227. Also, the staff verified in the UFSAR that concrete work was done in accordance with ACI 318.

Based on its review, the staff finds that the aggregates used at PVNGS are nonreactive, and the concrete was constructed in accordance with the recommendations in ACI 318. Therefore, cracking due to expansion and reaction with aggregate in below-grade, inaccessible concrete areas for Groups 1–5 and 7–9 structures are not aging effects for concrete elements, and no additional plant-specific program is required.

- Cracks and distortion due to increased stress levels from settlement and reduction of foundation strength, cracking, and differential settlement due to erosion of porous concrete subfoundations could occur in below-grade inaccessible concrete areas of Groups 1–3, 5, and 7–9 structures.

In LRA Section 3.5.2.2.2.2.3, the applicant states that competent foundation materials are present to establish conservative design and construction criteria for support of the facilities with major structures founded on engineered backfill or undeformed basin sediments, with a minimum thickness in the power block areas of 200 feet. The applicant stated that these sediments are firm, consolidated,

Aging Management Review Results

continuous, and show no evidence of shears, faults, joints, folds, or other tectonic features. The LRA notes that the applicant has not constructed a permanent dewatering system and has monitored the settlement of all major structures at frequent years during the first three years after construction, followed by a monitoring frequency of five years. Results reported in the UFSAR through December 2003 show that the total post-construction recorded settlements are well below the 1.5-inch maximum specified in the UFSAR. The applicant states that further evaluation of settlement is not required because the concrete components are evaluated under the Structures Monitoring Program and no permanent de-watering system or porous concrete foundations exist.

The staff reviewed LRA Section 3.5.2.2.2.2.3 against the criteria in SRP-LR Section 3.5.2.2.2.2.3, which states that the GALL Report recommends verification of the continued functionality of the de-watering system during the period of extended operation if the plant's CLB credits a de-watering system to control settlement. The GALL Report recommends no further evaluation if this activity, and these aging effects are included in the scope of the applicant's Structures Monitoring Program.

The staff confirmed in program basis documents that the Structures Monitoring Program manages the monitoring for settlement of every major structure. SER Section 3.0.3.2.20 documents the staff's review of the Structures Monitoring Program. The staff also confirmed that no permanent de-watering system or porous concrete foundations exist. The staff finds acceptable the applicant's evaluation of this AERM because it meets the criteria in SRP-LR Section 3.5.2.2.2.3.

- Increase in porosity and permeability, cracking and loss of material (spalling, scaling) due to aggressive chemical attack and cracking, loss of bond, and loss of material (spalling, scaling) due to corrosion of embedded steel could occur in below-grade inaccessible concrete areas of Groups 1–3, 5, and 7–9 structures.

LRA Section 3.5.2.2.2.2.4 states that reinforced concrete structures were designed, constructed, and inspected in accordance with ACI and ASTM standards and procedural controls were utilized throughout the batching, mixing, and placement processes to provide for a good quality, dense, well-cured, and low-permeability concrete. The LRA further states that proper sizing, spacing, and distribution of reinforcing steel complied with ACI 318-71 requirements. Concrete structures were noted not to be subjected to groundwater for any sustained period, and an engineering study confirmed that the groundwater elevations are below the lowest structures.

The staff reviewed LRA Section 3.5.2.2.2.2.4 against the criteria in SRP-LR Section 3.5.2.2.2.2.4, which states that the GALL Report recommends further evaluation of plant-specific programs to manage these aging effects and mechanisms in inaccessible areas of these groups of structures if the environment is aggressive. In the GALL Report, it is noted that for inaccessible areas of plants with non-aggressive groundwater or soil (i.e., pH greater than 5.5, chlorides less than 500 parts per million, or sulfates less than 1,500 parts per million), the applicant should consider examinations of the exposed portions of the below-grade concrete, when excavated for any reason as well as periodic monitoring of below-grade water chemistry, including consideration of potential seasonal variations.

Aging Management Review Results

The staff found that the concrete structures were designed, constructed, and inspected following recommended ACI and ASTM standards and procedural controls to provide good quality concrete. The LRA states that the concrete structures are not subjected to groundwater for any sustained periods; however, the applicant does not quantify what it meant by sustained periods. The applicant also failed to demonstrate that the groundwater or soil adjacent to the inaccessible concrete structures is not aggressive, allows for opportunistic inspections of exposed portions of below-grade concrete, or provides for periodic monitoring of below-grade water chemistry, including seasonal variations. By letter dated February 19, 2010, the staff issued RAI 3.5.2.2.2-1 to address this issue.

By letter dated March 24, 2010, the applicant responded and explained that plant-operating experience, including opportunistic inspections of buried structures, has not identified any degradation due to aggressive groundwater or soil. The applicant further explained that the groundwater below the site is a perched aquifer that resulted from irrigation before plant construction. Since cessation of irrigation, the water level has dropped. It is currently 20 feet below the lowest structure and continues to drop. The applicant also stated that an engineering study in 2007 concluded that it is unlikely that the groundwater levels will rise in the future. Furthermore, the applicant stated that the Structures Monitoring Program includes provisions for inspections of inaccessible areas whenever they are made available, as well as actively uncovering below-grade concrete if conditions in adjoining or similar areas indicate that it is necessary.

The staff reviewed the response and found it acceptable because the applicant explained that the below-grade concrete is not exposed to groundwater, and the Structures Monitoring Program examines below-grade concrete when excavated for any reason. In addition, the applicant has not seen any indication of concrete degradation related to aggressive groundwater. The staff's concern in RAI 3.5.2.2.2-1 is resolved.

On the basis of its review, the staff finds that the increase in porosity and permeability, cracking, loss of material (spalling, scaling) due to aggressive chemical attack and the cracking, loss of bond, and loss of material (spalling, scaling) due to corrosion of embedded steel in below-grade inaccessible concrete areas of Groups 1–3, 5 and 7–9 structures require no further evaluation because the concrete is not exposed to aggressive groundwater. In addition, the Structures Monitoring Program will inspect below-grade concrete when exposed for any reason.

- Increase in porosity and permeability, and loss of strength due to leaching of calcium hydroxide, could occur in below-grade inaccessible concrete areas of Groups 1–3, 5, and 7–9 structures.

LRA Section 3.5.2.2.2.2.5 states that reinforced concrete structures were designed, constructed, and inspected in accordance with ACI and ASTM standards and procedural controls were utilized throughout the batching, mixing, and placement processes to provide for a good quality, dense, well-cured, and low-permeability concrete. The LRA also states that concrete mixes were designed in accordance with ACI 211.1-74. In addition, concrete structures are not subjected to flowing water for any sustained periods, and an engineering study confirms that groundwater elevations are below the lowest structures. The applicant stated that further evaluation of this AERM is not required.

Aging Management Review Results

The staff reviewed LRA Section 3.5.2.2.2.2.5 against the criteria in SRP-LR Section 3.5.2.2.2.2.5, which states that the GALL Report recommends further evaluation of this aging effect for inaccessible areas of Groups 1–3, 5 and 7–9 structures if concrete was not constructed in accordance with the recommendations in ACI 201.2R-77.

In their review, the staff noted that the concrete structures were designed, constructed, and inspected in accordance with ACI and ASTM standards and procedural controls were utilized throughout the batching, mixing, and placement processes to provide for a good quality, dense, well-cured, and low permeability concrete. In its review, the staff noted that the applicant did not define what was meant by "not subjected to flowing water for any sustained periods" and did not note that the concrete was constructed in accordance with the recommendations in ACI 201.2R. By letter dated February 19, 2010, the staff issued RAI 3.5.2.2.1-1 to address the compliance of PVNGS concrete to recommendations provided in ACI 201.2R.

By letter dated March 24, 2010, the applicant responded and explained that the in-scope concrete was designed and constructed in accordance with ACI 211.1-74, which provides procedures for designing concrete mixes that take into consideration ACI 201 recommendations for concrete placeability, consistency, strength, and durability. The applicant further explained the two critical environmental conditions to consider when designing durable concrete are groundwater and exposure to freeze-thaw cycles. The applicant stated that an engineering study, conducted in 2007, concluded there was little likelihood of local raising groundwater levels beneath the units in the future. Therefore, site structures are not exposed to groundwater from the perched aquifer. In addition, the weathering index for PVNGS is negligible according to ASTM C33, Figure 1; therefore, the freeze-thaw cycles are not a concern.

The staff reviewed the applicant's response and found it acceptable because it explains that the recommendations of ACI 201 are incorporated into the guidance of ACI 211.1-74. SER Section 3.5.2.2.1(4) supplies a more detailed discussion of the staff's review. The staff's concern in RAI 3.5.2.2.1-1 is resolved. Since the concrete was constructed in accordance with the guidance in ACI 201.2R, the staff finds that the criteria of SRP-LR Section 3.5.2.2.2.2.5 have been met, and no further evaluation is required.

Based on the programs and analyses discussed above, the staff concludes that the applicant has met the criteria of SRP-LR Section 3.5.2.2.2.2. For those items that apply to LRA Section 3.5.2.2.2.2, the staff determines that the LRA is consistent with the GALL Report and the applicant has demonstrated that the effects of aging will be adequately managed so that the intended functions will be maintained consistent with the CLB during the period of extended operation, as required by 10 CFR 54.21(a)(3).

- Reduction of Strength and Modulus of Concrete Structures Due to Elevated Temperature for Groups 1–5 Structures). LRA Section 3.5.2.2.2.3 addresses reduction of concrete strength and modulus due to elevated temperatures that may occur in PWR and BWR Groups 1–5 concrete structures. The applicant stated that the reactor cavity cooling subsystem operates in conjunction with the containment normal cooling units and provides continuous cooling of the primary shield and reactor cavity to limit the concrete temperature to less than the specified limit of 150 degrees F (65 degrees C). The reactor cavity is monitored with four high temperature alarm channels that are annunciated in the control room. Plant TS require that the containment average air temperature not exceed 117 degrees F (47 degrees C). High-temperature piping penetrations have been designed to limit the local concrete temperature to 200 degrees F (93 degrees C). In the case of piping carrying hot fluid, the pipe is insulated to prevent excessive concrete temperatures.

The staff reviewed LRA Section 3.5.2.2.2.3 against the criteria in SRP-LR Section 3.5.2.2.2.3, which states that reduction of strength and modulus of concrete, due to elevated temperatures, may occur in PWR and BWR Groups 1–5 concrete structures. ACI 349-85 specifies the concrete temperature limits for normal operation or any other long-term period and states that general area temperatures shall not exceed 150 degrees F (65 degrees C) except for local areas that are permitted to have temperatures not to exceed 200 degrees F (93 degrees C). The GALL Report recommends further evaluation of a plant-specific program if any portion of the safety-related and other concrete structures exceeds these limits.

The staff reviewed program basis documents and noted that Group 1–5 concrete elements do not exceed temperature limits associated with aging degradation due to elevated temperature. On the basis of its review, the staff finds that reduction in strength and modulus of elasticity due to elevated temperatures in concrete areas of Groups 1–5 structures is not a plausible AERM because concrete temperatures are maintained below limits specified in ACI 349-85. Therefore, the staff finds that this is not an AERM for these components because design and preventive measures preclude occurrence of the elevated temperature condition.

- Aging Management of Inaccessible Areas for Group 6 Structures (Below Grade Inaccessible Concrete Areas). The staff reviewed LRA Section 3.5.2.2.2.4 against the following criteria in SRP-LR Section 3.5.2.2.2.4:

    – Increase in porosity and permeability, cracking, loss of material (spalling, scaling) due aggressive chemical attack and cracking, loss of bond, and loss of material (spalling, scaling) due to corrosion of embedded steel could occur in below-grade inaccessible concrete areas of Group 6 structures.

    LRA Section 3.5.2.2.2.4.1 states that reinforced concrete structures were designed, constructed, and inspected in accordance with ACI and ASTM standards. In addition, it used procedural controls throughout the batching, mixing, and placement processes to provide for a good quality, dense, well-cured, and low-permeability concrete. The LRA further states that proper sizing, spacing, and distribution of reinforcing steel complied with ACI 318-71 requirements. Concrete structures were noted to not be subjected to groundwater for any sustained periods, and an engineering study was performed to confirm that the groundwater elevations are below the lowest structures.

    The staff reviewed LRA Section 3.5.2.2.2.4.1 against the criteria in SRP-LR Section 3.5.2.2.2.4.1, which states that the GALL Report recommends further evaluation of plant-specific programs to manage these aging effects in inaccessible areas of these groups of structures if the environment is aggressive. In the GALL Report, it is noted that for inaccessible areas of plants with non-aggressive groundwater and soil (i.e., pH greater than 5.5, chlorides less than 500 parts per million, or sulfates less than 1,500 parts per million) the applicant should consider examinations of the exposed portions of the below-grade concrete, when excavated for any reason and periodic monitoring of below-grade water chemistry, including consideration of potential seasonal variations.

    The staff noted that inspections of Group 6 structures are performed under the Structures Monitoring Program, which is consistent with and incorporates the elements of RG 1.127, "Inspection of Water-Control Structures Associated with Nuclear Power Plants Program." The staff found that the concrete structures were designed, constructed, and inspected following the recommended ACI and ASTM standards and procedural controls to provide good quality concrete. SER Section 3.0.3.2.20 documents the staff's review of the Structures Monitoring Program. The staff confirmed that Group 6 structures subject to this AMR are in-scope of the Structures Monitoring Program. It was further noted that the concrete structures are not subjected to groundwater for any sustained periods of time; however, the applicant did not quantify what it meant by sustained periods of time. The applicant also failed to demonstrate that the groundwater or soil adjacent to the inaccessible concrete structures is not aggressive, allow for opportunistic inspections of exposed portions of below-grade concrete, or provide for periodic monitoring of below-grade water chemistry, including seasonal variations. By letter dated February 19, 2010, the staff issued RAI 3.5.2.2.2-1 to address these issues.

By letter dated March 24, 2010, the applicant responded and explained that plant operating experience, including opportunistic inspections of buried structures, has not identified any degradation due to aggressive groundwater or soil. The applicant further explained that the groundwater below the site is a perched aquifer that resulted from irrigation before plant construction. Since cessation of irrigation, the water level has dropped. It is currently 20 feet below the lowest structure and continues to drop. The applicant also stated that an engineering study, done in 2007, concluded that there is little likelihood the groundwater levels will rise in the future. Furthermore, the applicant stated that the Structures Monitoring Program includes provisions for inspections of inaccessible areas whenever they are made available, as well as actively uncovering below-grade concrete if conditions in adjoining or similar areas indicate that it is necessary.

The staff reviewed the response and found it acceptable because the applicant explained that the below-grade concrete is not exposed to groundwater, and the Structures Monitoring Program examines below-grade concrete when excavated for any reason. In addition, the applicant has not seen any indication of concrete degradation related to aggressive groundwater. The staff's concern in RAI 3.5.2.2.2-1 is resolved.

On the basis of its review, the staff finds that the increase in porosity and permeability, cracking, loss of material (spalling, scaling) due to aggressive chemical attack and cracking, loss of bond, and loss of material (spalling, scaling) due to corrosion of embedded steel in below-grade inaccessible concrete areas of Group 6 structures requires no further evaluation because the concrete is not exposed to aggressive groundwater, and the Structures Monitoring Program will inspect below-grade concrete when exposed for any reason.

- Loss of material (spalling, scaling) and cracking due to freeze-thaw could occur in below-grade inaccessible concrete areas of Group 6 structures.

LRA Section 3.5.2.2.2.4.2 states that PVNGS is located in a weathering region classified as negligible according to Figure 1 of ASTM C33-03. Therefore, the applicant states that this AERM is not applicable.

The staff reviewed LRA Section 3.5.2.2.2.4.2 against the criteria in SRP-LR Section 3.5.2.2.2.4.2, which recommends further evaluation of loss of material due to freeze-thaw for plants with concrete containments located in moderate to

Aging Management Review Results

severe weathering conditions. The staff finds acceptable the applicant's evaluation that this aging effect is not applicable because the Group 6 structures concrete is located in a weathering region classified as negligible.

- Cracking due to expansion and reaction with aggregates and increase in porosity and permeability, and loss of strength due to leaching of calcium hydroxide could occur in below-grade inaccessible reinforced concrete areas of Group 6 structures.

LRA Section 3.5.2.2.2.4.3 states that concrete in inaccessible areas is evaluated for expansion and cracking due to reaction with aggregate. In LRA Section 3.5.2.2.1.10, the applicant stated that acceptance of aggregate materials was based, in part, on petrographic examination in accordance with ASTM C295. Aggregate reactivity was evaluated in accordance with ASTM C289 and C227, and the reinforced concrete structures were designed, constructed and inspected in accordance with ACI and ASTM standards to provide good quality, dense, well-cured, and low-permeability concrete. The applicant further stated that concrete mixes were designed in compliance with ACI 211.1-74 and that procedural controls were imposed on the concrete throughout the batching, mixing, and placing processes. The applicant also noted that the concrete structures are not subjected to flowing water for sustained periods, and an engineering study shows that the groundwater elevations are below the lowest structures. The applicant stated that further evaluation for the effects of reaction with aggregates and leaching of calcium hydroxide are not required.

The staff reviewed LRA Section 3.5.2.2.2.4.3 against the criteria in SRP-LR Section 3.5.2.2.2.4.3, which states that the GALL Report recommends further evaluation of inaccessible areas if concrete was not constructed in accordance with the recommendations in ACI 201.2R-77.

SER Section 3.5.2.2.2(2)(b) documents the staff's review for cracking due to expansion and reaction with aggregates for inaccessible concrete elements of Groups 1–5 and 7–9 structures. The staff noted that inspections of Group 6 structures are performed under the Structures Monitoring Program, which is consistent with and integrates the elements of RG 1.127, "Inspection of Water-Control Structures Associated with Nuclear Power Plants Program." SER Section 3.0.3.2.20 documents the staff's review of the Structures Monitoring Program. The staff confirmed that accessible portions of Group 6 structures subject to this AMR are in-scope of the Structures Monitoring Program.

SER Section 3.5.2.2.2(2)(e) documents the staff's review for increase in porosity and permeability, and loss of strength due to leaching of calcium hydroxide for inaccessible concrete elements of Groups 1–3, 5, and 7–9 structures. The staff noted that the applicant conducts inspections of Group 6 structures under the Structures Monitoring Program, which is consistent with and integrates the elements of RG 1.127, "Inspection of Water-Control Structures Associated with Nuclear Power Plants Program." SER Section 3.0.3.2.20 documents the staff's review of the Structures Monitoring Program. The staff confirmed that accessible portions of Group 6 structures subject to this AMR are in-scope of the Structures Monitoring Program.

On the basis of its review, the staff finds that cracking due to expansion and reaction with aggregates and increase in porosity and permeability and loss of strength due to leaching of calcium hydroxide in below-grade inaccessible

Aging Management Review Results

concrete areas of Group 6 structures requires no further evaluation because the concrete was constructed in accordance with the guidance in ACI 201.2R.

Based on the programs identified above, the staff concludes that the applicant has met the criteria of SRP-LR Section 3.5.2.2.2.4. For those items that apply to LRA Section 3.5.2.2.2.4, the staff determines that the LRA is consistent with the GALL Report. The applicant has demonstrated that the effects of aging will be adequately managed so that the intended functions will be maintained consistent with the CLB during the period of extended operation, as required by 10 CFR 54.21(a)(3).

- Cracking due to Stress Corrosion Cracking and Loss of Material due to Pitting and Crevice Corrosion for Group 7 and 8 (Stainless Steel Tank Liners). LRA Section 3.5.2.2.2.5 states that the applicant evaluated in-scope tank liners in the CVCS and condensate systems and assigned them to GALL Report Chapters VII and VIII. Therefore, the applicant stated that further evaluation for the effects of cracking due to SCC and loss of material due to pitting and crevice corrosion is not required.

  The staff reviewed LRA Section 3.5.2.2.2.5 against the criteria in SRP-LR Section 3.5.2.2.2.5, which states that cracking due to SCC and loss of material due to pitting and crevice corrosion could occur for Group 7 and 8 stainless steel tank liners exposed to standing water. The GALL Report recommends further evaluation of plant-specific programs to manage these aging effects.

  The staff reviewed the Gall Report, Chapters VII and VIII relative to the CVCS and condensate systems, respectively. In Chapter VII Section E1, which addresses the CVCS, and Chapter VIII Section E, which addresses the condensate system, the staff noted that for stainless steel tanks in a treated borated water environment (temperature greater than 60 degrees C), the GALL Report recommends the Water Chemistry Program as the AMP for cracking due to SCC and loss of material due to pitting and crevice corrosion. The staff confirmed that the applicant evaluated in-scope tank liners in the condensate and CVCS systems and had been assigned to the Gall Report Chapters VII and VIII.

- Aging of Supports Not Covered by Structures Monitoring Program. LRA Section 3.5.2.2.2.6 states that further evaluation of the following components is not required because they will be inspected per the Structures Monitoring Program: building concrete around support anchors, HVAC duct supports, instrument supports, non-ASME mechanical equipment supports, non-ASME supports, and electrical panels and enclosures. The LRA states, in Table 3.5.1, that PVNGS has no in-scope vibration isolation elements.

  The staff reviewed LRA Section 3.5.2.2.2.6 against the criteria in SRP-LR Section 3.5.2.2.2.6, which states that further evaluation of certain component support and aging effect combinations is recommended if not covered by the Structures Monitoring Program. This includes the loss of material due to general and pitting corrosion, for Group B2–B5 supports, the reduction in concrete anchor capacity due to degradation of the surrounding concrete, for Group B1–B5 supports, and the reduction or loss of isolation function due to degradation of vibration isolation elements, for Group B4 supports. Further evaluation is necessary only for structure and aging effect combinations not covered by the Structures Monitoring Program.

  - Loss of Material Due to General and Pitting Corrosion, for Group B2–B5 Supports. The LRA states, in Table 3.5.1, that loss of material due to general and pitting corrosion for Group B2–B5 supports is an aging effect that does not require further evaluation because the components are inspected under the Structures Monitoring Program. Supports identified include the HVAC duct supports, instrument supports, non-ASME mechanical equipment supports, non-ASME supports, and electrical panels and enclosures.

The staff reviewed LRA Section 3.5.2.2.2.6 against the criteria in SRP-LR Section 3.5.2.2.2.6, which states that further evaluation is necessary only for structure and aging effect combinations not covered by the Structures Monitoring Program.

The staff confirmed that the Structures Monitoring Program manages the component support and aging effect combination of loss of material due to general and pitting corrosion for Group B2–B5. SER Section 3.0.3.2.20 documents the staff's review of the PVNGS Structures Monitoring Program. Since the applicant has committed to an appropriate AMP for the period of extended operation, the staff finds these AMR results to be acceptable.

- <u>Reduction in Concrete Anchor Capacity Due to Degradation of the Surrounding Concrete, for Group B1–B5 Supports</u>. The LRA states, in Table 3.5.1, that the Structures Monitoring Program manages the reduction in concrete anchor capacity due to degradation of the surrounding concrete.

  The staff reviewed LRA Section 3.5.2.2.2.6 against the criteria in SRP-LR Section 3.5.2.2.2.6, which states that further evaluation is necessary only for structure and aging effect combinations not covered by the Structures Monitoring Program. The staff confirmed that the Structures Monitoring Program manages the component support and aging effect combination of reduction in concrete anchor capacity due to degradation of the surrounding concrete for Group B2–B5 supports. SER Section 3.0.3.2.20 documents the staff's review of the PVNGS Structures Monitoring Program. Since the applicant has committed to an appropriate AMP for the period of extended operation, the staff finds these AMR results to be acceptable.

- <u>Reduction or Loss of Isolation Function Due to Degradation of Vibration Isolation Elements, for Group B4 Supports</u>. The LRA states, in Table 3.5.1, that PVNGS has no in-scope vibration isolation elements. SER Section 3.5.2.1.1 documents the staff's evaluation.

Based on the programs and analysis identified above, the staff concludes that the applicant has met the criteria of SRP-LR Section 3.5.2.2.2.6. For those items that apply to LRA Section 3.5.2.2.2.6, the staff determines that the LRA is consistent with the GALL Report and the applicant has demonstrated that the effects of aging will be adequately managed so that the intended functions will be maintained consistent with the CLB during the period of extended operation, as required by 10 CFR 54.21(a)(3).

- <u>Cumulative Fatigue Damage due to Cyclic Loading</u>. LRA Section 3.5.2.2.2.7 states that the applicant's review identified no TLAAs supporting the design of these components. SER Section 4.3, "Metal Fatigue Analysis" documents the staff's evaluation of the Class 1 component supports metal fatigue TLAA.

3.5.2.2.3 Quality Assurance for Aging Management of Nonsafety-Related Components

SER Section 3.0.4 documents the staff's evaluation of the applicant's QA program.

3.5.2.3 Aging Management Review Results Not Consistent with or Not Addressed in the Generic Aging Lessons Learned Report

In LRA Tables 3.5.2-1 through 3.5.2-14, the staff reviewed additional details of the AMR results for material, environment, AERM, and AMP combinations not consistent with, or not addressed in, the GALL Report.

Aging Management Review Results

In LRA Tables 3.5.2-1 through 3.5.2-14, the applicant indicated, via notes F–J that the combination of component type, material, environment, and AERM does not correspond to an item in the GALL Report. The applicant provided further information about how it will manage the aging effects. Specifically, Note F indicates that the material for the AMR item component is not evaluated in the GALL Report. Note G indicates that the environment for the AMR item component and material is not evaluated in the GALL Report. Note H indicates that the aging effect for the AMR item component, material, and environment combination is not evaluated in the GALL Report. Note I indicates that the aging effect identified in the GALL Report for the item component, material, and environment combination is not applicable. Note J indicates that neither the component nor the material and environment combination for the item is evaluated in the GALL Report.

For component type, material, and environment combinations not evaluated in the GALL Report, the staff reviewed the applicant's evaluation to determine if the applicant has demonstrated that it will adequately manage the effects of aging so that the intended function(s) will be maintained consistent with the CLB for the period of extended operation. The following sections document the staff's evaluation.

3.5.2.3.1 Containments, Structures, and Component Supports—Summary of Aging Management Evaluation—Containment Building

In LRA Tables 3.5.2-1, 3.5.2-2, and 3.5.2-5, the applicant stated that the Fire Protection Program manages cementitious fire barrier coatings and wraps exposed to air-indoor uncontrolled (external) for loss of material and cracking. The AMR items cite generic Note J, indicating that neither the component nor the material and environment combination is evaluated in the GALL Report. The staff reviewed all AMR result lines in the GALL Report where the component and material is cementitious coating fire barriers or wraps and confirmed that there are no entries for this component or material where the aging effect is loss of material due to cracking.

The staff reviewed the applicant's Fire Protection Program, and SER Section 3.0.3.2.7 documents its evaluation. The staff noted that the applicant's Fire Protection Program conducts visual inspections of fire barriers once every 18 months for detection of cracking and loss of material. The staff also noted that the Fire Protection Program is used for other fire barriers including concrete walls, floors, and ceilings, and that cementitious fire barrier coatings have similar aging effects to concrete. The staff finds the applicant's proposal to manage aging using the Fire Protection Program acceptable because the program performs visual inspections of fire barriers that are capable of detecting loss of material for cementitious fire barrier coatings and wraps.

In LRA Tables 3.5.2-1, 3.5.2-2, and 3.5.2-7 the applicant stated that ceramic fiber and thermo-lag fire barrier seals exposed to air-indoor uncontrolled (external) are being managed for loss of material and cracking by the Fire Protection Program. The AMR items cite generic Note J, indicating that neither the component nor the material and environment combination is evaluated in the GALL Report. The staff reviewed all AMR result lines in the GALL Report where the component and material is ceramic fiber or thermo-lag fire barriers or seals and confirmed that there are no entries for this component and material where the aging effect is loss of material and cracking.

The staff reviewed the applicant's Fire Protection Program, and SER Section 3.0.3.2.7 documents its evaluation. The staff noted that the applicant's Fire Protection Program conducts visual inspections of fire barriers once every 18 months for detection of cracking and loss of material. The staff also noted that thermo-lag and ceramic fibers are used to provide fire barriers for cable trays and conduits. The staff finds the applicant's proposal to manage aging using the Fire Protection Program acceptable because the program performs visual inspections

Aging Management Review Results

of fire barriers that are capable of detecting loss of material and cracking for ceramic and thermo-lag fire barriers and seals.

On the basis of its review, the staff finds that the applicant has appropriately evaluated the AMR results of material, environment, AERM, and AMP combinations not evaluated in the GALL Report. The staff finds that the applicant has demonstrated that the effects of aging for these components will be adequately managed so that their intended function(s) will be maintained consistent with the CLB during the period of extended operation, as required by 10 CFR 54.21(a)(3).

3.5.2.3.2 Containments, Structures, and Component Supports—Summary of Aging Management Evaluation—Control Building

SER Section 3.5.2.3.1 documents the staff's evaluation for cementitious coating fire barrier coatings and wraps exposed to air-indoor uncontrolled (external), with aging effects of loss of material and cracking managed by the Fire Protection Program, with generic Note J.

SER Section 3.5.2.3.1 documents the staff's evaluation for ceramic fiber and thermo-lag fire barrier seals exposed to air-indoor uncontrolled (external), with aging effects of loss of material and cracking managed by the Fire Protection Program, with generic Note J.

In LRA Tables 3.5.2-2, 3.5.2-5, and 3.5.2-13, the applicant stated that gypsum and plaster fire barriers exposed to air-indoor uncontrolled (external) are managed for cracking by the Fire Protection Program. The AMR items cite generic Note J, indicating that the GALL Report does not evaluate either the component or the material and environment combination. The staff reviewed all AMR result lines in the GALL Report where the component and material is gypsum or plaster fire barriers, and it confirmed that there are no entries for this component and material where the aging effect is cracking.

The staff reviewed the applicant's Fire Protection Program, and SER Section 3.0.3.2.7 documents its evaluation. The staff noted that the applicant's Fire Protection Program conducts visual inspections of fire barriers once every 18 months for detection of cracking and loss of material. The staff also noted that gypsum and plaster materials are commonly used to form walls when a fire barrier is required and that the GALL Report recommends the Fire Protection Program to manage aging for fire barrier walls constructed of other materials, such as concrete. The staff finds the applicant's proposal to manage aging using the Fire Protection Program acceptable because the program performs visual inspections of fire barriers that are capable of detecting loss of material and cracking for ceramic and thermo-lag fire barriers and seals.

The staff finds that the applicant has demonstrated that it will adequately manage the effects of aging for these components so that their intended function(s) will be maintained consistent with the CLB during the period of extended operation, as required by 10 CFR 54.21(a)(3).

3.5.2.3.3 Diesel Generator Building—Summary of Aging Management Review—License Renewal Application Table 3.5.2-3

The staff reviewed LRA Table 3.5.2-3, which summarizes the results of AMR evaluations for the diesel generator building component groups. The staff's review did not identify any items with notes F–J, indicating that the combinations of component type, material, environment, and AERM for this system are consistent with the GALL Report.

SER Section 3.0.2.2 documents the staff's evaluation of the items with Notes A–E.

Aging Management Review Results

3.5.2.3.4 Turbine Building—Summary of Aging Management—License Renewal Application Table 3.5.2-4

SER Section 3.5.2.3.1 documents the staff's evaluation for cementitious coating fire barrier coatings and wraps exposed to air-indoor uncontrolled (external) with aging effects of loss of material and cracking managed by the Fire Protection Program, with generic Note J.

3.5.2.3.5 Auxiliary Building—Summary of Aging Management Review—License Renewal Application Table 3.5.2-5

SER Section 3.5.2.3.1 documents the staff's evaluation for cementitious coating fire barrier coatings and wraps exposed to air-indoor uncontrolled (external) with aging effects of loss of material and cracking managed by the Fire Protection Program, with generic note J.

SER Section 3.5.2.3.2 documents the staff's evaluation for gypsum and plaster fire barrier exposed to air-indoor uncontrolled (external) having an aging effect of cracking managed by the Fire Protection Program, with generic note J.

3.5.2.3.6 Radwaste Building—Summary of Aging Management Review—License Renewal Application Table 3.5.2-6

The staff reviewed LRA Table 3.5.2-6, which summarizes the results of AMR evaluations for the radwaste building component groups. The staff's review did not identify any items with notes F through J, indicating that the combinations of component type, material, environment, and AERM for this system are consistent with the GALL Report.

SER Section 3.0.2.2 documents the staff's evaluation of the items with Notes A–E.

3.5.2.3.7 Main Steam Support Structure—Summary of Aging Management Review—License Renewal Application Table 3.5.2-7

In LRA Table 3.5.2-7, the applicant stated that the Fire Protection Program manages the thermo-lag fire barrier seals externally exposed to plant indoor air for loss of material and cracking. SER Section 3.0.3.2.7 documents the staff's evaluation.

3.5.2.3.8 Station Blackout Generator Building—Summary of Aging Management Review—License Renewal Application Table 3.5.2-8

The staff reviewed LRA Table 3.5.2-8, which summarizes the results of AMR evaluations for the SBO generator building component groups. The staff's review did not identify any items with notes F–J, indicating that the combinations of component type, material, environment, and AERM for this system are consistent with the GALL Report.

SER Section 3.0.2.2 documents the staff's evaluation of the items with Notes A–E.

3.5.2.3.9 Fuel Building—Summary of Aging Management Review—License Renewal Application Table 3.5.2-9

In LRA Table 3.5.2-9, for one component type—structural steel—the applicant proposed to assign the component to the Structures Monitoring Program to manage the aging effect of loss of material on carbon steel in a buried environment. This item references Note J, stating the GALL Report does not evaluate either the component or the material and environment combination. The intended function of this component is to provide for thermal expansion or seismic separation.

The applicant stated that the Structures Monitoring Program provides inspection guidelines and walkdown checklists for concrete elements, structural steel, masonry walls, structural features, structural supports, and miscellaneous components such as doors. The applicant notes that the structural component is in a buried environment; however, it is not clear how this component will be inspected under the Structures Monitoring Program to demonstrate that this AERM is being

Aging Management Review Results

effectively managed since the Structures Monitoring Program, in large measure, is visual. Therefore, by letter dated February 19, 2010, the staff issued RAI 3.5.2.3-1 asking the applicant to explain how it will use the Structures Monitoring Program to manage this AERM.

By letter dated March 24, 2010, the applicant explained that the portion of the component that is in a buried environment is normally inaccessible. The Structures Monitoring Program will inspect this portion of the component when it is accessible for any reason, or if conditions in adjoining areas indicate it is necessary to uncover the buried portion for inspection.

The staff reviewed the applicant's response and found it acceptable because it is the Structures Monitoring Program, which is the appropriate AMP for carbon steel in an outdoor environment, per GALL Report Table 5, item 25. In addition, the applicant's Structures Monitoring Program has incorporated GALL Report recommended opportunistic inspections of inaccessible areas. SER Section 3.5.2.2.2, "Aging Management of Inaccessible Areas," provides a more detailed discussion of the staff's review of aging management of inaccessible areas.

3.5.2.3.10 Spray Pond and Associated Water Control Structures—Summary of Aging Management Review—License Renewal Application Table 3.5.2-10

In LRA Table 3.5.2-10, for one component type—screen—the applicant proposed to assign copper alloy to the Structures Monitoring Program to monitor for loss of material in a raw water environment. This item references Note J, "Neither the component nor the material and environment combination is evaluated in NUREG-1801," and plant-specific note 2, which states that the GALL Report does not provide a line in which copper alloy screens are inspected per the Structures Monitoring Program.

The applicant stated that the Structures Monitoring Program provides inspection guidelines and walkdown checklists for concrete elements, structural steel, masonry walls, structural features, structural supports, and miscellaneous components such as doors. The applicant notes that the structural component is used as a filter and will be monitored for loss of material; however, since this material is in a raw water environment and potentially has limited accessibility, it is not clear to the staff how the Structures Monitoring Program will inspect this component to demonstrate that this AERM is being effectively managed. Therefore, by letter dated February 19, 2010, the staff issued RAI 3.5.2.3-2 asking the applicant to explain how it will use the Structures Monitoring Program to manage this AERM.

By letter dated March 24, 2009, the applicant explained that the screens in question are the ESP screens, which are accessible through the exterior deck of the spray pond pump house. The applicant raises the screens above the water level for visual inspection for loss of material.

The staff reviewed the applicant's response and found it acceptable because it explains how the Structures Monitoring Program will be able to complete visual inspections on the component for loss of material. The staff's concern in RAI 3.5.2.3-2 is resolved.

The staff noted that this material, aging effect, and environment combination is very similar to GALL Report item III.A6-11, which credits XI.S7 with managing copper alloys in raw water for loss of material. Since the applicant's Structures Monitoring Program incorporates the requirements of XI.S7 (i.e. visual inspections on a five-year frequency), the staff finds the applicant's use of the Structures Monitoring Program acceptable for this AMR item.

SER Section 3.3.2.3.7 documents the staff's evaluation for stainless steel components exposed to atmosphere or weather conditions, where there are no AERM and no AMP, citing generic note G.

Based on its review, the staff finds that the applicant has appropriately evaluated the AMR results of material, environment, AERM, and AMP combinations not evaluated in the GALL Report. The staff finds that the applicant has demonstrated that the effects of aging will be

Aging Management Review Results

adequately managed so that the intended function(s) will be maintained consistent with the CLB for the period of extended operation, as required by 10 CFR 54.21(a)(3).

3.5.2.3.11 Tank Foundations and Shells—Summary of Aging Management Review—License Renewal Application Table 3.5.2-11

The staff reviewed LRA Table 3.5.2-11, which summarizes the results of AMR evaluations for the tank foundations and shells component groups.

3.5.2.3.12 Transformer Foundations and Electrical Structures—Summary of Aging Management Review—License Renewal Application Table 3.5.2-12

The staff reviewed LRA Table 3.5.2-12, which summarizes the results of AMR evaluations for the transformer foundations and electrical structures component groups. The staff's review did not identify any items with notes F–J, indicating that the combinations of component type, material, environment, and AERM for this system are consistent with the GALL Report.

SER Section 3.0.2.2 documents the staff's evaluation of the items with Notes A–E.

3.5.2.3.13 Yard Structures (In-Scope)—Summary of Aging Management Review—License Renewal Application Table 3.5.2-13

SER Section 3.5.2.3.2 documents the staff's evaluation for gypsum and plaster fire barriers exposed to air-indoor uncontrolled (external) having an aging effect of cracking managed by the Fire Protection Program with generic Note J.

3.5.2.3.14 Supports—Summary of Aging Management Review—License Renewal Application Table 3.5.2-14

In LRA Table 3.5.2-14, for one component type—non-ASME support—the applicant proposed to assign carbon steel and stainless steel having a nonsafety-related structural support function to the Structures Monitoring Program to manage the aging effect of loss of material in a raw water environment. This item references Note G, "Environment not in NUREG-1801 for this component and material." The applicant stated that the PVNGS Structures Monitoring Program provides inspection guidelines and walkdown checklists for concrete elements, structural steel, masonry walls, structural features, structural supports, and miscellaneous components such as doors. The applicant notes that the structural component is used as a nonsafety-related structural support function and that nonsafety-related supports meet the criterion under 10 CFR 54.4(a)(2) when they prevent interaction between safety-related and nonsafety-related components. However, since this material is in a raw water environment and potentially has limited accessibility, it is not clear to the staff how the applicant will inspect this component under the Structures Monitoring Program to demonstrate that it effectively manages this AERM. Therefore, by letter dated February 19, 2010, the staff issued RAI 3.5.2.3-3 asking the applicant to explain how it will use the Structures Monitoring Program to manage this AERM.

In its response, dated March 24, 2010, the applicant explained that the components were included to evaluate the non-code supports for drain pipes located inside the radioactive waste drain sumps. The applicant further explained that it included these supports within the scope of license renewal in error; therefore, the referenced non-code supports have been deleted from LRA Table 3.5.2-14.

The staff verified that removal of the supports from the scope of license renewal was appropriate by comparing with the criteria of 10 CFR 54.4(a)(1), (2), or (3) for SSCs within the scope of license renewal. The staff reviewed the applicant's response and found it acceptable because the supports do not meet the criteria of 10 CFR 54.4(a)(1), (2), or (3).

In LRA Table 3.5.2-14, for component type—Supports ASME 2 and 3—the applicant proposed to credit the IWF Program to manage the aging effect of loss of material for carbon and

Aging Management Review Results

stainless steel in a raw water and a fuel oil environment. This item references Note G, "Environment not in NUREG-1801 for this component and material." Since this material is in a raw water or fuel oil environment and potentially has limited accessibility, it is not clear how this component will be inspected under the IWF Program to demonstrate that this AERM is being effectively managed. Therefore, by letter dated April 28, 2010, the staff issued RAI 3.5.2.3-4 asking the applicant to explain how the IWF Program would manage the effect of aging on carbon and stainless steel components in a raw water or fuel oil environment.

By letter dated May 21, 2010, the applicant explained that the supports in a raw water environment are ASME Class 3 stainless steel supports located in the ESPs. There are no carbon steel supports in the raw water environment of the ESPs. The applicant revised the LRA accordingly. The applicant further explained that it examined the ASME Class 3 supports, according to IWF requirements, using remote cameras. The applicant also explained that the supports in a fuel oil environment are ASME Class 3 carbon steel supports located in the diesel fuel oil storage tank. There are no stainless steel supports in a diesel fuel oil environment. The applicant revised the LRA accordingly. The applicant further explained that these supports are within the scope of the AMSE Section XI, Subsection IWF Program, but are exempt from examination.

The staff reviewed the applicant's response and found the applicant's explanation acceptable regarding the inspection of stainless steel supports in a raw water environment since they are being inspected remotely, according to the IWF requirements. However, the staff needed additional information about the inspection requirements for carbon steel supports. Therefore, the staff requested the basis for which the fuel oil transfer pump supports are exempt from IWF examination requirements.

In a letter dated June 21, 2010, the applicant supplemented its response to RAI 3.5.2.3-4. The applicant stated that a carbon steel Class 3 support assembly, which also supports the diesel fuel oil transfer pump, restrains the three-quarter-inch diameter diesel fuel discharge line in the diesel fuel oil tank. The applicant further stated that diesel fuel oil pump support is exempt from IWF examination based on the size of the pipe.

The staff reviewed the applicant's supplemental information and found it acceptable because the diesel fuel oil tank support also supports the three-quarter-inch diameter diesel fuel discharge line. IWF-1230 exempts supports from examination requirements that are connected to piping that is exempt from volumetric, surface, and VT-1 or VT-3 visual examination in accordance with IWD-1220. According to IWD-1220, Class 3 components and piping segments, 4-inch diameter and smaller, are exempt from VT-1 visual examination.

Based on its review, the staff finds that the applicant has appropriately evaluated the AMR results of material, environment, AERM, and AMP combinations not evaluated in the GALL Report. The staff finds that the applicant has demonstrated that the effects of aging will be adequately managed so that the intended function(s) will be maintained consistent with the CLB for the period of extended operation, as required by 10 CFR 54.21(a)(3).

3.5.3 Conclusion

The staff concludes that the applicant has provided sufficient information to demonstrate that the effects of aging for the structures and component supports within the scope of license renewal and subject to an AMR will be adequately managed so that the intended function(s) will be maintained consistent with the CLB for the period of extended operation, as required by 10 CFR 54.21(a)(3).

Aging Management Review Results

3.6 Aging Management of Electrical and Instrumentation and Controls

The following information documents the staff's review of the applicant's AMR results for the electrical and instrumentation and control (I&C) components and component groups of the following parts:

- connections (metallic parts)
- connectors
- high-voltage insulators
- insulated cables and connections
- metal enclosed bus
- penetrations electrical
- switchyard bus and connections
- terminal block
- transmission conductors and connections

3.6.1 Summary of Technical Information in the Application

LRA Section 3.6 provides AMR results for the electrical and I&C components and component groups. LRA Table 3.6.1, "Summary of Aging Management Evaluations in Chapter VI of NUREG-1801 [the GALL Report] for Electrical Components," is a summary comparison of the applicant's AMRs with those evaluated in the GALL Report for the electrical and I&C components and component groups.

The applicant's AMRs evaluated and incorporated applicable plant-specific and industry-operating experience in the determination of AERMs. The plant-specific evaluation included condition reports and discussions with appropriate site personnel to identify AERMs. The applicant's review of industry-operating experience included a review of the GALL Report and operating experience issues identified since the issuance of the GALL Report.

3.6.2 Staff Evaluation

The staff reviewed LRA Section 3.6 to determine if the applicant provided sufficient information to demonstrate that it would adequately manage the effects of aging for the electrical and I&C components within the scope of license renewal and subject to an AMR will be adequately managed so that the intended functions will be maintained consistent with the CLB for the period of extended operation, as required by 10 CFR 54.21(a)(3).

The staff reviewed AMRs to ensure the applicant's claim that certain AMRs were consistent with the GALL Report. The staff did not repeat its review of the matters described in the GALL Report; however, the staff did verify that the material presented in the LRA was applicable and that the applicant has identified the appropriate GALL Report AMRs. SER Section 3.0.3 documents the staff's evaluations of the AMPs, and SER Section 3.6.2.1 gives details of the staff's evaluation for AMRs that are consistent with the GALL Report.

The staff also reviewed AMRs consistent with the GALL Report and for which further evaluation is recommended. The staff confirmed that the applicant's further evaluations were consistent with the SRP-LR Section 3.6.2.2 acceptance criteria. SER Section 3.6.2.2 documents the staff's evaluations.

The staff also conducted a technical review of the remaining AMRs not consistent with or not addressed in the GALL Report. The technical review evaluated if all plausible aging effects have been identified and if the aging effects listed were appropriate for the material-environment combinations specified. SER Section 3.6.2.3 documents the staff's evaluations.

Aging Management Review Results

For SSCs that the applicant claimed were not applicable or required no aging management, the staff reviewed the AMR items and the plant's operating experience to verify the applicant's claims.

Table 3.6-1 summarizes the staff's evaluation of components, aging effects or mechanisms, and AMPs listed in LRA Section 3.6 and addressed in the GALL Report.

Table 3.6-1. Staff Evaluation for Electrical and Instrumentation and Controls in the GALL Report

Component Group (GALL Report Item No.)	Aging Effect/ Mechanism	AMP in GALL Report	Further Evaluation in GALL Report	AMP in LRA, Supplements, or Amendments	Staff Evaluation
Electrical equipment subject to 10 CFR 50.49 EQ requirements (3.6.1-1)	Degradation due to various aging mechanisms	EQ of Electric Components	Yes	TLAA	Consistent with GALL Report. See Section 3.6.2.2.1.
Electrical cables, connections and fuse holders (insulation) not subject to 10 CFR 50.49 EQ requirements (3.6.1-2)	Reduced insulation resistance and electrical failure due to various physical, thermal, radiolytic, photolytic, and chemical mechanisms	Electrical Cables and Connections Not Subject to 10 CFR 50.49 EQ Requirements	No	Electrical Cables and Collections not Subject to 10 CFR 50.49 EQ Requirements	Consistent with GALL Report.
Conductor insulation for electrical cables and connections used in instrumentation circuits not subject to 10 CFR 50.49 EQ requirements that are sensitive to reduction in conductor insulation resistance (3.6.1-3)	Reduced insulation resistance and electrical failure due to various physical, thermal, radiolytic, photolytic, and chemical mechanisms	Electrical Cables And Connections Used In Instrumentation Circuits Not Subject to 10 CFR 50.49 EQ Requirements	No	Electrical Cables and Connections Not Subject to 10 CFR 50.49 EQ Used in Instrumentation Circuits	Consistent with GALL Report.
Conductor insulation for inaccessible medium voltage (2 kV to 35 kV) cables (e.g., installed in conduit or direct buried) not subject to 10 CFR 50.49 EQ requirements (3.6.1-4)	Localized damage and breakdown of insulation leading to electrical failure due to moisture intrusion, water trees	Inaccessible Medium Voltage Cables Not Subject to 10 CFR 50.49 EQ Requirements	No	Inaccessible Medium-Voltage Cables Not Subject to 10 CFR 50.49 EQ Program	Consistent with GALL Report.
Connector contacts for electrical connectors exposed to borated water leakage (3.6.1-5)	Corrosion of connector contact surface due to intrusion of borated water	Boric Acid Corrosion	No	Boric Acid Corrosion	Consistent with GALL Report.

3-329

Aging Management Review Results

Component Group (GALL Report Item No.)	Aging Effect/ Mechanism	AMP in GALL Report	Further Evaluation in GALL Report	AMP in LRA, Supplements, or Amendments	Staff Evaluation
Fuse Holders (Not Part of a Larger Assembly): Fuse holders - metallic clamp (3.6.1-6)	Fatigue due to ohmic heating, thermal cycling, electrical transients, frequent manipulation, vibration, chemical contamination, corrosion, and oxidation	Fuse Holders	No	Fuse Holder	Consistent with GALL Report.
MEB – bus/connections (3.6.1-7)	Loosening of bolted connections due to thermal cycling and ohmic heating	MEB	No	MEB Program	Consistent with GALL Report.
MEB – insulation/insulators (3.6.1-8)	Reduced insulation resistance and electrical failure due to various physical, thermal, radiolytic, photolytic, and chemical mechanisms	MEB	No	MEB Program	Consistent with GALL Report.
MEB – enclosure assemblies (3.6.1-9)	Loss of material due to general corrosion	Structures Monitoring Program	No	MEB Program	Consistent with GALL Report. See Section 3.6.2.1.1
MEB – enclosure assemblies (3.6.1-10)	Hardening and loss of strength due to elastomers degradation	Structures Monitoring Program	No	MEB Program (B2.1.36)	Consistent with GALL for material, environment, aging effect, but a different AMP is credited (See Section 3.6.2.1.2)
High-voltage insulators (3.6.1-11)	Degradation of insulation quality due to presence of any salt deposits and surface contamination; loss of material caused by mechanical wear due to wind blowing on transmission conductors	A plant-specific AMP is to be evaluated	Yes	None	Further evaluation (See SER Section 3.6.2.2.2)
Transmission conductors and connections; switchyard bus and connections (3.6.1-12)	Loss of material due to wind induced abrasion and fatigue; loss of conductor strength due to corrosion; increased resistance of connection due to oxidation or loss of preload	A plant-specific AMP is to be evaluated	Yes	None	Further evaluation (See SER Section 3.6.2.2.3)

Aging Management Review Results

Component Group (GALL Report Item No.)	Aging Effect/ Mechanism	AMP in GALL Report	Further Evaluation in GALL Report	AMP in LRA, Supplements, or Amendments	Staff Evaluation
Cable Connections - metallic parts (3.6.1-13)	Loosening of bolted connections due to thermal cycling, ohmic heating, electrical transients, vibration, chemical contamination, corrosion, and oxidation	Electrical Cable Connections Not Subject to 10 CFR 50.49 EQ Requirements	No	Electrical Cable Connections Not Subject to 10 CFR 50.49 EQ Requirements Program (B2.1.35)	Consistent with GALL (See Section 3.6.2.1)
Fuse Holders (Not Part of a Larger Assembly) - insulation material (3.6.1-14)	None	None	No	Not applicable	Consistent with GALL. (See Section 3.6.2.1)

The staff's review of the electrical and I&C component groups fell into one of three categories. One category, documented in SER Section 3.6.2.1, reviewed AMR results for components that the applicant indicated are consistent with the GALL Report and require no further evaluation. Another category, documented in SER Section 3.6.2.2, reviewed AMR results for components that the applicant indicated are consistent with the GALL Report and for which further evaluation is recommended. A third category, documented in SER Section 3.6.2.3, reviewed AMR results for components that the applicant indicated are not consistent with or not addressed in the GALL Report. SER Section 3.0.3 documents the staff's review of AMPs credited to manage or monitor aging effects of the electrical and I&C components.

3.6.2.1 Aging Management Review Results Consistent with the Generic Aging Lessons Learned Report

LRA Section 3.6.2.1 identifies the materials, environments, AERMs, and the following programs that manage aging effects for the electrical and I&C components:

- Electrical Cable Connections Not Subject to 10 CFR 50.49 EQ Requirements
- Boric Acid Corrosion
- Electrical Cables and Connections Not Subject to 10 CFR 50.49 EQ Requirements
- Electrical Cables and Connections Not Subject to 10 CFR 50.49 EQ Requirements Used in Instrumentation Circuits
- Inaccessible Medium-Voltage Cables Not Subject to 10 CFR 50.49 EQ Requirements
- Metal Enclosed Bus
- Fuse Holders

In LRA Table 3.6.2-1, the applicant summarized AMRs for the electrical and I&C components and claimed that these AMRs are consistent with the GALL Report.

The staff audited and reviewed the information in the LRA. The staff did not repeat its review of the matters described in the GALL Report; however, the staff did verify that the material presented in the LRA was applicable and that the applicant identified the appropriate GALL Report AMRs.

Aging Management Review Results

The staff reviewed the LRA to confirm that the applicant provided a brief description of the system, components, materials, and environments; stated that the applicable aging effects were reviewed and evaluated in the GALL Report; and identified those aging effects for the electrical and I&C components that are subject to an AMR. On the basis of its review, the staff determines that, for AMRs not requiring further evaluation, as identified in LRA Table 3.6.1, the applicant's references to the GALL Report are acceptable and no further staff review is required.

3.6.2.1.1 Loss of Material Due to General Corrosion

LRA Table 3.6.1, item 3.6.1.9 addresses the loss of material due to general corrosion of MEB enclosure assemblies. The LRA credits the MEB Program. The GALL Report recommends AMP XI.S6, "Structures Monitoring Program," to manage the aging effects for these components. The associated AMR item cites generic Note E, indicating that the LRA AMR is consistent with the GALL Report item for material, environment and aging effect, but credits a different AMP.

For those items with generic Note E, the GALL Report recommends the Structures Monitoring Program, which recommends the use of visual inspections. In its review of components associated with item 3.6.1.9, for which the applicant cited generic Note E, the staff noted that the MEB Program proposes to manage the aging of the MEB enclosure assemblies using visual inspection of these components for evidence of degradation. SER Section 3.0.3.1.14 documents the staff's evaluation of the MEB AMP. The staff finds that using visual inspection as described in the Metal-Enclosed Bus Inspection Program is acceptable to inspect the outside of metal enclosed bus enclosure assemblies for loss of material due to general corrosion.

The staff concludes that the applicant has demonstrated that it will adequately manage the effects of aging of these components so that their intended functions will remain consistent with the CLB during the period of extended operation, as required by 10 CFR 54.21(a)(3).

3.6.2.1.2 Hardening and Loss of Strength Due to Elastomer Degradation

LRA Table 3.6.1, item 3.6.1.10 addresses the effects of hardening and loss of strength of elastomers. The LRA credits the MEB Program. The GALL Report recommends XI.S6, "Structures Monitoring Program," to manage the aging effects for these components. The associated AMR item cites generic Note E, indicating that the LRA AMR is consistent with the GALL Report item for material, environment and aging effect, but credits a different AMP.

For those items with generic Note E, the GALL report recommends the Structures Monitoring Program, which recommends the use of visual inspections. In its review of components associated with item 3.6.1.10, for which the applicant cited generic Note E, the staff noted that the MEB Program proposes to manage the aging of the MEB enclosure assemblies using visual inspection and flexing of the elastomer, as described in MEB Program. SER Section 3.0.3.1.14 documents the staff's evaluation of this AMP. The staff finds that using visual inspection and flexing, as described in the MEB Inspection Program, is acceptable to inspect the MEB elastomer for degradation.

The staff concludes that the applicant has demonstrated that it will adequately manage the effects of aging of these components so that their intended functions will be maintained consistent with the CLB during the period of extended operation as required by 10 CFR 54.21(a)(3).

3.6.2.2 Aging Management Review Results Consistent with the Generic Aging Lessons Learned Report for Which Further Evaluation Is Recommended

In LRA Section 3.6.2.2, the applicant further evaluates aging management, as recommended by the GALL Report, for the electrical and I&C components and provides information concerning how it will manage the following aging effects:

Aging Management Review Results

- electrical equipment subject to EQ

- degradation of insulation quality due to salt deposits or surface contamination and loss of material due to mechanical wear

- loss of material due to wind-induced abrasion and fatigue, loss of conductor strength due to corrosion, and increased resistance of connection due to oxidation or loss of pre-load

- QA for aging management of nonsafety-related components

For component groups evaluated in the GALL Report, for which the applicant claimed consistency with the report and for which the GALL Report recommends further evaluation, the staff reviewed the corresponding AMRs identified in LRA Table 3.6.1 as items 3.6.1.11 and 3.6.1.12. The staff also reviewed the applicant's evaluation to determine if it adequately addressed the issues. In addition, the staff reviewed the applicant's further evaluations against the criteria contained in SRP-LR Section 3.6.2.2. The staff's review of the applicant's further evaluations follows.

3.6.2.2.1 Electrical Equipment Subject to Environmental Qualification

In LRA Section 3.6.2.3, "Environmental Qualification of Electrical and Instrumentation and Control Equipment," the applicant states EQ is a TLAA, as defined by 10 CFR 54.3. TLAAs are required to be evaluated in accordance with 10 CFR 54.21. SER Section 4.4, "10 CFR 50.49 Thermal, Radiation, and Cyclical Aging Analyses" documents the staff's review of the applicant's evaluation of this TLAA.

3.6.2.2.2 Degradation of Insulator Quality Due to Salt Deposits or Surface Contamination and Loss of Material Due to Mechanical Wear

LRA Section 3.6.2.2.2 addresses degradation of insulator quality due to salt deposits or surface contamination and loss of material due to mechanical wear. The applicant states that PVNGS is located in an area where the outdoor environment is not subject to industry air pollution or salt spray. Contamination build-up on the high-voltage insulators is not a problem due to sufficient rainfall in the spring and summer washing the insulators. Additionally, there is no salt spray at the plant since the plant is not located near the ocean. The degradation of insulator quality in the absence of salt deposits and surface contamination is not an AERM at PVNGS. The applicant further stated that the transmission conductors are designed and installed so that they do not swing significantly and cause wear due to wind-induced abrasion and fatigue. The applicant concluded that loss of material due to wind-induced abrasion and fatigue is not an applicable AERM.

The staff reviewed LRA Section 3.6.2.2.2 against SRP-LR Section 3.6.2.2.2, which states that degradation of insulator quality due to salt deposits or surface contamination may occur in high-voltage insulators. The GALL Report recommends further evaluation of plant-specific AMPs for plants at locations of potential salt deposits or surface contamination (e.g., near salt water bodies or industrial pollution). Loss of material due to mechanical wear caused by wind on transmission conductors may occur in high-voltage insulators. The GALL Report recommends further evaluation of a plant-specific AMP to ensure that these aging effects are adequately managed.

The staff noted various airborne materials such as dust, salt, and industrial effluents can contaminate insulator surfaces. However, the buildup of surface contamination is gradual and, in most areas, rain washes away such contamination; the glazed insulator surface aids this contamination removal. Surface contamination can be a problem in areas where the greatest concentration of airborne particles is prevalent, such as near facilities that discharge soot or near the seashore where there is salt spray. Since PVNGS is not located near facilities that discharge soot or near the seashore, the rate of contamination buildup on the insulators is not significant. However, the applicant did not address plant-specific operating experience with high-voltage insulator failures relating to surface contamination.

In a letter dated February 19, 2010, the staff issued RAI 3.6.2.2.2-1 asking the applicant to review plant-specific operating experience to confirm that there have been no failures of high-voltage insulators due to surface contamination. In response to the staff's request, in a letter dated March 24, 2010, the applicant stated that the transmission system owner, Salt River Project, had not identified any documented operating experience failures of high-voltage insulators within the scope of license renewal due to surface contamination. A search of documented operating experience identified the following three high-voltage bushing flashovers that resulted in unit trips:

(1) **July 31, 1988**—B Phase, Unit 3

(2) **November 14, 1991**—A Phase, Unit 3

(3) **March 1, 1996**—C Phase, Unit 1

The applicant also stated that during the initial evaluation of the flashovers, contamination levels were reviewed and found to be minimal such that there was little risk of contamination-induced flashover. The applicant further stated that subsequent additional evaluation of the flashovers concluded that the flashovers were due to the tilt angle of the bushings. Booster sheds were added to channel water away from the bushings during heavy rain; there have not been repeat flashovers.

The staff finds the applicant's response acceptable because the applicant reviewed its plant-specific operating experience and confirmed that there have been no failures of high-voltage insulators due to surface contamination. The staff accepts the applicant's conclusion that the flashover events were not caused by insulator contamination, but due to high conductivity water that channeled between the bushings in heavy rain due to the tilt angle of the bushing. The applicant added booster sheds to channel water away from the surface of the bushings; this design change prevents a water path to the insulators. Based on its review, the staff determined that surface contamination is not a significant AERM for the high-voltage insulators.

The staff noted that mechanical wear is an aging effect for strain and suspension insulators as they are subject to movement. Movement of the insulators can be caused by wind blowing the supported transmission conductor, causing it to swing from side to side. If this swinging is frequent enough, it can cause wear in the metal contact point of the insulator string and between an insulator and supporting hardware. Although this wear is possible, industrial experience has shown that the transmission conductors do not normally swing and that when they do, in a substantial wind, they do not continue to swing for very long once the wind has subsided.

Transmission conductors at PVNGS are designed and installed not to swing significantly and not to wear due to wind-induced abrasion and fatigue. However, the applicant did not address plant-specific operating experience with high-voltage insulator loss of material due to wear. In a letter dated February 19, 2010, the staff issued RAI 3.6.2.2.2-2 asking the applicant to review its operating experience to identify if wear has occurred in high-voltage insulators and transmission conductors. In response to the staff's request, in a letter dated March 24, 2010, the applicant stated that it and the Salt River Project have not identified any documented operating experience of high-voltage insulators and transmission conductors within the scope of license renewal associated with loss of material due to wear that has resulted in a loss of intended function. The staff finds the applicant response acceptable because the applicant has reviewed its documented plant-specific operating experience and confirmed that there has been no failure of high-voltage insulators and transmission conductor due to loss of material due to wear.

Aging Management Review Results

Based on its review, the staff determined that loss of material due to wear is not a significant AERM.

Based on the discussed above, the staff concludes that the applicant has met the SRP-LR Section 3.6.2.2.2 criteria. The staff determines that that the LRA is consistent with the GALL Report and that the applicant has demonstrated that the effects of aging will be adequately managed so that the intended functions will be maintained consistent with the CLB during the period of extended operation, as required by 10 CFR 54.21(a)(3).

3.6.2.2.3 Loss of Material Due to Wind-Induced Abrasion and Fatigue, Loss of Conductor Strength Due to Corrosion, and Increased Resistance of Connection Due to Oxidation or Loss of Pre-Load

LRA Section 3.6.2.2.3 addresses loss of material due to wind-induced abrasion and fatigue, loss of conductor strength due to corrosion, and increased resistance of connection due to oxidation or loss of pre-load.

In LRA Section 3.6.2.2.3, the applicant stated that industry experience has shown that transmission conductors are designed and installed not to swing significantly and cause wear due to wind-induced abrasion and fatigue. Therefore, the applicant concluded that loss of material due to wind-induced abrasion and fatigue is not an applicable AERM for the period of extended operation.

The applicant stated that the most prevalent mechanism contributing to loss of conductor strength of an aluminum conductor steel reinforced (ACSR) transmission conductor is corrosion, which includes corrosion of the steel core and aluminum strand pitting. The applicant further stated the following:

> ACSR conductor degradation begins as a loss of zinc from the galvanized steel core wires. Corrosion rates depend largely on air quality, which involves suspended particles in the air, sulfur dioxide concentration, rain, fog chemistry, and other weather conditions. The PVNGS outdoor environment is not subject to industry air pollution or saline environment that would cause significant corrosion of the transmission conductors.

> The National Electrical Safety Code (NESC) requires that tension on installed conductors be a maximum of 60 percent of the ultimate conductor strength. The NESC also sets the maximum tension a conductor must be designed to withstand heavy load requirements, which includes consideration of ice, wind, and temperature.

> At PVNGS, the ACSR transmission conductors are 2-2156 KCMIL per phase with a core of 19 steel strands having ultimate conductor strength of 60,300 pounds-force. The PVNGS ACSR transmission conductors within the scope of license renewal are installed so that conductor tension does not exceed 18,000 pounds-force at the NESC heavy loading condition (30 percent of the ultimate conductor strength).

The applicant also stated that tests performed by Ontario Hydroelectric on ACSR transmission conductors with a core of 7 steel strands averaging 70–80 years old, showed a 30-percent loss of ultimate conductor strength due to corrosion. Assuming a 30-percent loss of ultimate conductor strength (18,090 pounds-force) due to corrosion over 60 years, the PVNGS ACSR transmission conductors have adequate design margin to offset the loss of strength due to corrosion and still meet the NESC requirement of not exceeding 60 percent of the ultimate conductor strength [(60,300-18,090)x60%=25,326]. The applicant concluded that corrosion is not a credible aging effect that requires management for the period of extended operation.

Aging Management Review Results

The applicant stated that, at the time of installation, it treated transmission conductor and switchyard bus connections with corrosion inhibitors to avoid connection oxidation and torqued to avoid loss of pre-load. Further, the applicant stated the following:

> Based on temperature data in the UFSAR Chapter 2.3, the transmission connections and switchyard bus does not experience thermal cycling. The transmission connections and switchyard bus are subject to average monthly temperatures ranging from 105 °F in July and August to 38 °F in January with minimal ohmic heating. Therefore, increased resistance of connections due to oxidation or loss of pre-load is not an aging effect requiring management for the period of extended operation. These connections are periodically evaluated via thermography as part of the preventive maintenance activities. The periodic thermography will continue into the period of extended operation.

The staff reviewed LRA Section 3.6.2.2.3 against the criteria in SRP-LR Section 3.6.2.2.3, which states that loss of material due to wind-induced abrasion and fatigue, loss of conductor strength due to corrosion, and increased resistance of connection due to oxidation or loss of pre-load could occur in transmission conductors and connections and in switchyard bus and connections. The GALL Report recommends further evaluation of a plant-specific AMP to ensure that this aging effect is adequately managed.

The staff noted that transmission conductors do not normally swing and that when they do, due to a substantial wind, they do not continue to swing for very long once the wind has subsided. Wind loading that can cause a transmission line to vibrate is considered in the design and installation. In addition, the sections of transmission conductors in the scope of license renewal are short spans connecting the switchyard to the startup transformers, and the surface areas exposed to wind loads are not significant. Furthermore, the applicant indicated that it has reviewed the plant-specific operating experience and did not identify issue of loss of material due to wear for transmission conductors. Based on its review, the staff determined that loss of material of transmission conductors due to vibration is not an AERM.

The staff reviewed the testing program performed by Ontario Hydroelectric to determine whether PVNGS transmission conductors have adequate design margin to perform their intended function during the extended period of operation. The study showed about 30-percent loss of conductor strength of an 80-year-old ACSR conductor due to corrosion. The NESC requires that tension on installed conductors be a maximum of 60 percent of the ultimate conductor strength. The NESC also sets the maximum tension; a conductor must be designed to withstand under heavy load requirements, which include consideration of one-half inch of radial ice and 4 pounds per square feet wind.

The staff reviewed the requirements concerning the specific conductors included in the AMR. At PVNGS, the ACSR transmission conductors are 2,156 thousand circular mils. These transmission conductors have a core of 19 steel strands with conductor strength of 60,300 pounds-force. These transmission conductors have 18,000 pounds-force of NESC heavy loading. With the loss of 30 percent conductor strength due to corrosion, the conductor strength would be 42,210 pounds force (60,300x0.7).

The ratio between the heavy loading and the ultimate conductor strength, after losing 30 percent of conductor strength, would be approximately 43 percent. The ratio of heavy loading and the ultimate conductor strength is below the maximum 60 percent NESC requirement. Furthermore, the staff noted that the length of transmission conductors within the scope of license renewal is generally a short span. These transmission conductors connect the switchyard to the startup transformers, providing restoration of offsite power after an SBO event. The loading of these transmission conductors is much less than the calculated heavy loading of a long span transmission line.

Aging Management Review Results

The staff determined that with a 30 percent loss of conductor strength, there is still ample margin between the NESC requirements and the actual conductor strength. Based on this information, the staff determined that loss of conductor strength due to corrosion of transmission conductor is not a significant AERM for the period of extended operation.

SRP-LR Section 3.6.2.2.3 states that loss of pre-load could occur in transmission and switchyard bus connections. EPRI TR 104213 states that an electrical connection must be designed to remain tight and maintain good conductivity through a large temperature range. Meeting this design requirement is difficult if the materials specified for the bolt and the conductor are different and have different rates of thermal expansion. For example, copper and aluminum bus materials expand faster than most bolting materials. If thermal stress is added to stresses inherent at assembly, the joint members or fasteners can yield. If deformation occurs during thermal loading (i.e., heatup), then when the connection cools, the joint will loosen. Increased temperature difference in electrical bolted joints is due to increased current duration. The temperature of an electrical bolted joint will rise, and the stress will increase with increasing current duration. If this temperature increase is not taken into consideration, then loose and failure prone joints will result.

The applicant stated that the transmission connections and switchyard bus does not experience thermal cycling. The applicant also stated that the transmission connections and switchyard bus are subject to average monthly temperatures ranging from 38–105 degrees F with minimal ohmic heating. The applicant concluded that loss of pre-load is not an AERM for the period of extended operation. The thermal expansion, due to ohmic heating and thermal cycling, depends heavily on the load and not the average monthly temperature.

In a letter dated February 19, 2010, the staff issued RAI 3.6.2.2.3-1 asking the applicant to justify why loss of pre-load is not an applicable AERM. In response to the staff's request, in a letter dated March 24, 2010, the applicant stated that loss of pre-load of switchyard bus and connections is not applicable because procedures require that switchyard connections be assembled using a corrosion inhibitor, and connections are also torqued to avoid loss of pre-load. Additionally, switchyard conductor and bus connections are assembled with stainless steel Belleville washers to prevent loss of preload. The applicant further stated that the transmission system owner, Salt River Project, periodically performs infrared scans of switchyard equipment and connections, including before and after scans, to verify connector integrity for equipment undergoing maintenance. The applicant also stated that a search of operating experience identified no evidence of switchyard bus connection or transmission conductor connection loss of pre-load.

The staff finds the applicant response acceptable because the use of Belleville washers on bolted electrical connections of dissimilar metals compensates for temperature changes, maintains the proper torque, and prevents loosening. This method of assembly is consistent with the good bolting practices recommended by industry guidelines (EPRI TR-104213, "Bolted Joint Maintenance & Application Guide"). Furthermore, the applicant reviewed its plant-specific operating experience and did not find any evidence of switchyard bus and transmission conductor connection failures due to loss of pre-load. Based on its review, the staff finds that loss of pre-load of switchyard bus and transmission conductor connections is not an applicable AERM.

Based on the programs identified above, the staff concludes that the applicant has met the SRP-LR Section 3.6.2.2.3 criteria. For those line items that apply to LRA Section 3.6.2.2.3, the staff determines that that the LRA is consistent with the GALL Report and that the applicant has demonstrated that the effects of aging will be adequately managed so that the intended functions will be maintained consistent with the CLB during the period of extended operation, as required by 10 CFR 54.21(a)(3).

Aging Management Review Results

3.6.2.2.4 Quality Assurance for Aging Management of Nonsafety-Related Components

SER Section 3.0.4 documents the staff's evaluation of the applicant's QA program.

3.6.2.3 Aging Management Review Results Not Consistent with or Not Addressed in the Generic Aging Lessons Learned Report

In LRA Table 3.6.2-1, the staff reviewed additional details of the AMR results for material, environment, AERM, and AMP combinations not consistent with, or not addressed in, the GALL Report. The applicant indicated, via Notes F–J that the combination of component type, material, environment, and AERM does not correspond to a line item in the GALL Report. The applicant provided further information about how it will manage the aging effects.

For component type, material, and environment combinations not evaluated in the GALL Report, the staff reviewed the applicant's evaluation to determine if the applicant has demonstrated that the effects of aging will be adequately managed so that the intended functions will be maintained consistent with the CLB for the period of extended operation. The following section documents the staff's evaluation.

Metal Enclosed Bus (Enclosure)—Summary of Aging Management Review—License Renewal Application Table 3.6.2-1. In LRA, Table 3.6.2-1 under MEB (Enclosure), the applicant indicated that it will use the MEB Program to manage the loss of material for aluminum bus enclosures. The applicant included Note J, which states that neither the component nor the material and environment combination is available in the GALL Report. The staff noted that the MEB Program proposes to manage the aging of the MEB enclosure assemblies through visual inspections of these components for evidence of loss of material. SER Section 3.0.3.1.14 documents the staff's evaluation of the MEB AMP. The staff finds that using visual inspections, as described in the MEB Program, is acceptable for inspecting the outside of aluminum bus enclosure assemblies for loss of material due to general corrosion.

3.6.3 Conclusion

The staff concludes that the applicant has provided sufficient information to demonstrate that the effects of aging for the electrical and I&C components within the scope of license renewal and subject to an AMR will be adequately managed so that the intended function(s) will be maintained consistent with the CLB for the period of extended operation, as required by 10 CFR 54.21(a)(3).

3.7 Conclusion for Aging Management Review Results

The staff reviewed the information in LRA Section 3, "Aging Management Review Results," and LRA Appendix B, "Aging Management Programs." On the basis of its review of the AMR results and AMPs, the staff concludes that the applicant has demonstrated that the aging effects will be adequately managed so that the intended functions will be maintained consistent with the CLB for the period of extended operation, as required by 10 CFR 54.21(a)(3). The staff also reviewed the applicable UFSAR supplement program summaries and concludes that the supplement adequately describes the AMPs credited for managing aging, as required by 10 CFR 54.21(d).

With regard to these matters, the staff concludes that there is reasonable assurance that the applicant will continue to conduct the activities authorized by the renewed licenses in accordance with the CLB, and any changes made to the CLB, in order to comply with 10 CFR 54.21(a)(3), are in accordance with the Atomic Energy Act of 1954, as amended, and NRC regulations.

4.0 TIME-LIMITED AGING ANALYSES

4.1 Time-Limited Aging Analyses

4.1.1 Identification of Time-Limited Aging Analyses

This section of the safety evaluation report (SER) addresses the identification of time-limited aging analyses (TLAAs). In license renewal application (LRA) Sections 4.2–4.8, Arizona Public Service Company (APS) (the applicant) addressed the TLAAs for Palo Verde Nuclear Generating Station, Units 1, 2, and 3 (PVNGS). SER Sections 4.2–4.9 document the review of the TLAAs conducted by the U.S. Nuclear Regulatory Commission (NRC) staff (the staff).

TLAAs are certain plant-specific safety analyses that involve time-limited assumptions defined by the current operating term. Pursuant to Title 10 of the *Code of Federal Regulations* (10 CFR), Section 54.21(c)(1), applicants must list TLAAs. The definition of a TLAA is found in 10 CFR Part 54.3, "Definitions."

In addition, pursuant to 10 CFR 54.21(c)(2), applicants must list existing plant-specific exemptions granted in accordance with 10 CFR 50.12, "Specific Exemptions," based on TLAAs. For any such exemptions, the applicant must evaluate and justify the continuation of the exemptions for the period of extended operation.

4.1.2 Summary of Technical Information in the Application

To identify TLAAs, the applicant evaluated calculations against the six criteria specified in 10 CFR 54.3. The applicant said that it identified the calculations that met the six criteria by searching the current licensing basis (CLB). The CLB includes the Updated Final Safety Analysis Report (UFSAR), engineering calculations, technical reports, engineering work requests, licensing correspondence, and applicable vendor reports. In LRA Table 4.1-1, "List of TLAAs" the applicant listed the following applicable TLAAs:

- Reactor vessel neutron embrittlement analysis
- Metal fatigue analysis
- Environmental qualification (EQ) of electrical equipment
- Concrete containment tendon prestress
- Concrete liner plate, equipment hatch and personnel airlocks, penetrations, and polar crane brackets
- Plant-specific TLAA

As required by 10 CFR 54.21(c)(2), the applicant must list all exemptions granted in accordance with 10 CFR 50.12 that are based on TLAAs and evaluated and justified for continuation through the period of extended operation. The LRA states that the applicant reviewed each active exemption to determine whether it was based on a TLAA. The applicant stated that it had no TLAA-based exemptions.

4.1.3 Staff Evaluation

LRA Table 4.1-1 lists the TLAAs the applicant identified as being applicable to PVNGS. The staff reviewed the information to determine whether the applicant had provided sufficient information as required by 10 CFR 54.21(c)(1) and 10 CFR 54.21(c)(2). As defined in 10 CFR 54.3, TLAAs meet the following six criteria:

Time-Limited Aging Analyses

(1) Involve systems, structures, and components within the scope of license renewal, under 10 CFR 54.4(a)

(2) Consider the effects of aging

(3) Involve time-limited assumptions defined by the current operating term (for example, 40 years)

(4) Are determined to be relevant by the applicant in making a safety determination

(5) Involve conclusions, or provide the basis for conclusions, related to the capability of the systems, structures, and components to perform its intended functions, pursuant to 10 CFR 54.4(b)

(6) Are contained or incorporated by reference in the CLB

The staff noted that the applicant assembled the list of potential TLAAs using the following regulatory and industry documents:

- The NUREG-1800, "Standard Review Plan for Review of License Renewal Applications for Nuclear Power Plants," Chapter 4
- Nuclear Energy Institute 95-10, "Industry Guideline for Implementing the Requirements of 10 CFR 54, the License Renewal Rule"
- The 10 CFR Part 54 Final Rule "Statement of Considerations"
- Prior LRAs
- Plant-specific document reviews and interviews with plant personnel

The staff finds the applicant's use of these documents to compile a list of potential TLAAs reasonable since the applicant used all available resources from the staff, Nuclear Energy Institute, past LRAs, and its own plant-specific review.

Using the documents listed above, the applicant reviewed its CLB documents to determine if the design or analysis feature of each potential TLAA, in fact, exists in the licensing basis for the site. The applicant also determined if additional unit-specific TLAAs exists. The applicant reviewed the following documents to formulate the list of potential unit-specific TLAAs:

- UFSAR
- Technical Specifications (TS)
- The SERs for the original operating licenses
- Subsequent Safety Evaluations (SEs)
- APS and NRC docketed licensing correspondence

In accordance with 10 CFR 54.21(c), those potential TLAAs that meet all six criteria defined in 10 CFR 54.3(a) are actual TLAAs and require a disposition. The applicant reviewed the six criteria based on information in the CLB source documents (as listed above), and from other source documents for the potential TLAAs to include the following:

- The *Standard Safety Analysis* Report for CE System 80
- Vendor, NRC-sponsored, and licensee topical reports
- Design calculations
- Code stress reports or code design reports
- Drawings
- Specifications

Time-Limited Aging Analyses

The staff finds the applicant's approach in determining TLAA reasonable because the applicant has performed a comprehensive search through its CLB, based on available staff and industry guidance and experience, and has reviewed these potential TLAAs against the six criteria of a TLAA as defined in 10 CFR 54.3(a).

The applicant provided a list of potential TLAAs from NUREG-1800, "Standard Review Plan for Review of License Renewal Applications for Nuclear Power Plants," (SRP-LR) dated September 2005. The applicant listed those potential TLAAs in LRA Table 4.1-2, "Review of Analyses Listed in NUREG-1800, Table 4.1-2." The applicant further provided a list of its plant-specific TLAAs in LRA Table 4.1-1.

The staff performed an independent search of exemptions in effect during the staff's review of the LRA by reviewing the operating license and conducting a search of the NRC's Agencywide Documents Access and Management System. The staff found five exemptions in effect, with one scheduled to be issued by the staff on July 31, 2010. Three of these exemptions have been granted since submittal of the LRA. The staff confirmed that none of these exemptions are based on a TLAA.

Based on the information provided by the applicant regarding the results of the applicant's search of the CLB to identify these exemptions, and the staff's independent search, the staff has determined, in accordance with 10 CFR 54.21(c)(2), that there are no TLAA-based exemptions which have been justified for continuation through the period of extended operation.

4.1.3.1 Evaluation of the Applicant's Identification of Time-Limited Aging Analyses

4.1.3.1.1   Absence of Time-Limited Aging Analyses for Metal Corrosion Allowances and Corrosion Effects

Summary of Technical Information in the Application. SRP-LR Table 4.1-3 lists examples of potential plant-specific TLAAs. This table includes "metal corrosion allowance" as a possible TLAA. In response to this table entry, the applicant provided LRA paragraph 4.7.2, "Absence of TLAAs for Metal Corrosion Allowances and Corrosion Effects." In this paragraph, the applicant states that, other than the issues described in LRA paragraphs 4.7.4 and 4.7.5, which are addressed in SER Sections 4.7.4 and 4.7.5, it "found no description of time-dependant corrosion allowances, rates, or corrosion-dependent design lives of pressure vessels, system components, piping, or metal containment components" in its review of the CLB. Based on this statement, the applicant implies that TLAAs are not required.

Staff Evaluation and Conclusion. In its review of LRA paragraph 4.7.2, the staff concluded that the applicant intended to state that TLAAs for metal corrosion allowances were not required by 10 CFR 54.3(a)(6), which states that they are not contained or incorporated by reference in the CLB. In evaluating the applicant's assertion, the staff conducted a search of the applicant's UFSAR and TS. The staff also considered additional documents such as NRC general communications and American Society of Mechanical Engineers (ASME) Code requirements, which could incorporate a requirement in the CLB for a corrosion allowance TLAA. Following this review, the staff finds no reason to disagree with the applicant's assertion that calculations related to corrosion allowances, other than those described in paragraphs 4.7.4 and 4.7.5, are not included in the applicant's CLB. The staff concludes that the absence of a TLAA, pursuant to 10 CFR 54.21, for metal corrosion allowances and corrosion effects is acceptable because the requirement of 10 CFR 54.3(a)(6) is not met.

4.1.3.1.2   Absence of a Time-Limited Aging Analysis for Reactor Vessel Underclad Cracking Analyses

Summary of Technical Information in the Application. In LRA Section 4.7.6, the applicant stated that the reactor pressure vessel (RPV) underclad cracking has been addressed by the choice of material and weld cladding processes that are designed to avoid these defects, consistent with

Time-Limited Aging Analyses

regulatory guide (RG) 1.43, "Control of Stainless Steel Weld Cladding of Low-Alloy Steel Components." The applicant stated further that they have not discovered any cracks.

The vessel shell and head plates are constructed of SA-533, Grade B, Class 1 Steel, which is immune to underclad cracking. The RPV nozzles and flange are constructed of susceptible material SA-508, but these components were clad with low-heat-input processes, which are not known to cause underclad cracking. The determination that the RPV material is not susceptible to underclad cracking is not based on time-dependent analyses and, therefore, the applicant states underclad cracking is not a TLAA.

Staff Evaluation and Conclusion. The staff noted that underclad cracks have been reported to exist only in SA-508, Class 2, RPV forgings manufactured with a coarse grain microstructure and clad by high heat input, submerged arc welding processes. Since high heat input, submerged arc welding processes were not used for the fabrication of the cladding over their RPV nozzles and flange, the staff agrees with the applicant that RPV underclad cracking is not an issue and is not a TLAA because the requirement of 10 CFR 54.3(a)(4) is not met.

4.1.3.1.3 Absence of a Time-Limited Aging Analysis for a Reactor Coolant Pump Flywheel Fatigue Crack Growth Analysis

Summary of Technical Information in the Application. In LRA Section 4.7.7, the applicant stated that the reactor coolant pump (RCP) flywheel burst is the subject of RG 1.14, "Reactor Coolant Pump Flywheel Integrity" and its predecessor, Safety Guide 14. The CLB commits to 10-year interval inspections in accordance with Safety Guide 14, Revision 0, Position c.4.b.

PVNGS relies on flywheel design, material, fabrication, and the periodic inspections in accordance with Safety Guide 14, Position c.4.b. No crack growth analysis or time-dependent probabilistic failure assessment has been performed for the flywheels, either to extend the inspection interval for less than the design life or to support a safety determination for the design life. Therefore, no TLAA exists.

Staff Evaluation and Conclusion. The applicant stated that "[t]he current PVNGS licensing basis commits to the 10-year-interval inspections of Safety Guide 14 (Rev. 0) Position c.4.b." It further stated that no crack growth analysis has been performed for the flywheels; therefore, no TLAA exists. For the period of extended operation, RG 1.14, Rev. 1 should be used and referenced unless the RCP inspection in the CLB is independent of the underlying stress and fracture mechanics analyses of the flywheels, which may contain a time-limited analysis such as a fatigue analysis or a fatigue crack growth analysis. Hence, the staff issued RAI 4.7.7-1 for clarification.

The applicant's response to RAI 4.7.7-1 confirmed that, "[t]he stress and fracture mechanics analysis in the RCP flywheel design report does not contain any fatigue or time-dependent fatigue crack evaluations. The design report includes an evaluation of cracks that will permit either a ductile or brittle burst at overspeed." This statement clarified that the applicant's RCP flywheel evaluation in the CLB is not a TLAA. However, if indications from the past RCP flywheel inspections were detected and evaluated for continued operation for a limited time, these flaw evaluations are very likely to be TLAAs. Hence, the staff issued RAI 4.7.7-2.

The applicant's response to RAI 4.7.7-2 confirmed that, "APS has performed ultrasonic test examinations approximately every three years and eddy current test examinations every 10 years in accordance with the Inservice Inspection (ISI) Program on the flywheel of each of the RCPs. No indications of degradation have been found in any of the RCP flywheels, and, therefore, no flaw evaluations have been performed." Hence, the staff concludes that the applicant's RCP flywheel evaluation is not a TLAA because the requirement of 10 CFR 54.3(a)(6) is not met. Also, there is no TLAA on flaw evaluations for detected flaws in the RCP flywheels because no indications of degradation have been found in them.

Time-Limited Aging Analyses

4.1.3.1.4 Absence of Time-Limited Aging Analyses in Fatigue Crack Growth Assessments and Fracture Mechanics Stability Analyses for the Leak-Before-Break Elimination of Dynamic Effects of Primary Loop Piping Failures

Summary of Technical Information in the Application. The staff has approved the Leak-Before-Break (LBB) application for the primary coolant loop piping in the three PVNGS units. The approved LBB application permits elimination of the postulated large breaks in the main reactor coolant loops and their jet and pipe whip effects and the removal of jet barriers and whip restraints. The containment pressurization and equipment qualification analyses retain the large-break assumptions. NUREG-0857, "Safety Evaluation Report Related to the Operation of PVNGS Units 1, 2, and 3," November 1981, with Supplements 1–12, documents NRC's approval of the LBB application. The applicant claims that the fatigue crack growth analysis and the fracture mechanics stability analyses are not TLAAs.

Staff Evaluation. The staff reviewed LRA Section 4.3.2.15 to verify, pursuant to 10 CFR 54.3(a), that the fatigue crack growth assessments and fracture mechanics stability analyses for the LBB elimination of dynamic effects of primary piping failures are not TLAAs.

By letter dated October 11, 1984, the staff approved the LBB analysis for the main reactor coolant loops, including the hot, cold, and crossover legs in each of the PVNGS units.

In general, the two TLAA issues regarding LBB evaluations are the fatigue crack growth analysis and the flaw evaluation of the cast austenitic stainless steel (CASS). The fatigue crack growth analysis postulates several representative flaws in high stress locations and calculates their final crack size using transient cycles to determine their acceptability at the end of the licensed life. The transient cycles are time-dependent; therefore, the fatigue crack growth analysis is usually considered a TLAA. CASS experiences thermal aging embrittlement as it ages, which reduces its fracture toughness. Fracture toughness is a material property that resists crack initiation and propagation; therefore, the use of CASS material usually involves a TLAA.

The applicant stated that the primary coolant loop piping does not contain CASS. Therefore, thermal aging embrittlement of CASS is not a TLAA issue. The staff verified that there is no CASS in the primary coolant loop piping and, therefore, finds this acceptable. The following evaluation provides the staff's determination of the fatigue crack growth analysis.

Fatigue Crack Growth. Section 3.6.3 of the SRP-LR, paragraph III.10 states that, "[t]he reviewer should determine that the candidate piping does not have a history of fatigue cracking or failure. An evaluation to ensure that the potential for pipe rupture due to thermal and mechanical induced fatigue is unlikely should be performed."

In a letter dated January 14, 2010, the staff issued RAI 4.3.2.15-1 asking the applicant to demonstrate that fatigue is not an active degradation mechanism during the period of extended operation and that the original LBB evaluation is valid for the extended period of operation. By letters dated March 1 and May 21, 2010, the applicant responded that within the context of SRP-LR Section 3.6.3, fatigue is not an active degradation mechanism for the following reasons:

- Fatigue crack growth analyses for the proposed licensed operating period are acceptable (in this case, including the period of extended operation). The acceptability of the LBB fatigue crack growth analyses for the licensed operating period including the period of extended operation is addressed below.

- Crack stability analyses are acceptable and, if time-dependent, are acceptable for the proposed licensed operating period (in this case, including the period of extended operation). The LBB crack stability analyses are not time-dependent and are discussed below.

Time-Limited Aging Analyses

- Material fracture toughness parameters, including effects of long-term thermal aging are acceptable. The LBB crack stability analyses include appropriate material toughness parameters, which are not time-dependent and are discussed below.

- There are no indications of pressurized water stress corrosion cracking, erosion, erosion-corrosion, water hammer, creep, other cracking, leakage, or other evidence of actual or incipient fatigue or failures that would indicate that the LBB analysis is invalid within the scope of the piping exempted from crack postulation by the LBB analysis of NRC Mechanical Engineering, Branch Technical Position MEB-3.1, *Postulated Rupture Locations for Fluid System Piping Inside and Outside of Containment*, as part of Standard Review Plan Section 3.6.2 (Note: MEB-3.1 has been renamed Branch Technical Position BTP 3-4, Revision 2 and is a stand-alone document in Chapter 3 of the Standard Review Plan). No such indications or failures have been observed in the scope of PVNGS LBB piping.

The technical basis for the LBB evaluation is provided in the CE topical report entitled "Leak Before Break Evaluation of the Main Coolant Loop Piping of a CE Reactor Coolant System." This report was provided as an attachment to a letter dated June 14, 1983 (also, Revision 1 of the CE topical report was submitted to the staff as an attachment to a letter dated December 23, 1983). Section 3 of the topical report describes fatigue calculations to demonstrate the acceptability of fatigue crack growth for various postulated flaws, which demonstrates that fatigue is not an active degradation mechanism of concern. One of these calculations, for a relatively small flaw of 1 inch in depth and 8 inches in length, demonstrates that the crack will not penetrate through wall for a very large number of cycles, principally heatup and cooldown cycles.

The technical basis report for approval of LBB incorporates fatigue crack growth calculations that have a time basis (40 years) or consider numbers of cycles in the calculations. However, LRA Section 4.3.2.15 states that the LBB fatigue crack analyses are not TLAAs because the postulated fatigue cracks grow slowly and the fatigue evaluation does not depend on the design life. In RAI 4.3.2.15-1, the staff asked the applicant to provide technical basis to support the notion that the LBB evaluation is not a TLAA.

By letter dated March 1, 2010, in response to RAI 4.3.2.15-1, the applicant stated that the acceptability of the LBB evaluation depends on a fracture mechanics crack stability analysis and a fatigue crack growth analysis. By letter dated May 21, 2010, the applicant provided additional clarification to its response to RAI 4.3.2.15-1, as described below.

Crack Stability Analysis - The applicant postulated three semi-elliptical cracks (0.5 x 39 inches, 1.0 x 34 inches, and 0.35 x 45.5 inches) for the crack stability analysis. For the large crack sizes postulated, the applicant calculated that through-wall leaks would occur at 21, 4, and 38 years, respectively; determined that preferential growth would be in the radial direction, and determined that the rate of growth between a through-wall detectable leak and a critical crack size would be acceptable for defects much larger than any anticipated actual initial defect. The staff noted that the LBB analysis is based on existence of a leaking flaw in the pipe and the pipe is allowed to leak. The above crack stability analysis shows when a leak will occur based on postulated flaws as part of the applicant's sensitivity study for the LBB evaluation.

The applicant stated that two crack stability criteria have been used to assess the likelihood that a crack with opening stress intensity K, or a J-integral at the tip would remain stable. The applicant contended that both methods are independent of time. The first involves the use of linear-elastic fracture mechanics fracture toughness K_{Ic} which is a measure of the material's resistance to fracture. A K_I value below which there is no crack extension is K_c. As a practical consideration, K_{Ic} is a measure of the stress intensity at which fracture takes place. Its value has been empirically correlated to the material's Charpy V-Notch test value using the

Rolfe-Novak-Barsom correlation that is independent of time. The second method involves the elastic-plastic crack instability theory using the J-integral crack tip parameter T, the tearing modulus, when the volume of plastically deformed material is appreciable. A $T_{Applied}$ value below which there is no crack extension is $T_{Material}$. $T_{Material}$ is only a function of the J-integral, modulus of elasticity, E, and the material's yield stress, S_y, and not time. The applicant concluded that all of these flaws will be acceptable at the end of 60 years.

The applicant concluded that the fracture mechanics crack stability analysis is not time-dependent and, therefore, remains applicable for the period of extended operation. The staff agrees that the crack stability analysis is not a TLAA because the requirement of 10 CFR 54.3(a)(3) is not met.

Fatigue Crack Growth Analysis - The applicant postulated two cracks (8 x 18 inches and 1 x 8 inches) for the fatigue crack growth calculations. The existing fatigue crack growth evaluation demonstrates that initially postulated cracks larger than those required by the LBB rule will remain within allowable sizes for an order of magnitude longer than the 40-year current licensed operating period. Since the safety determination supported by this evaluation does not depend on the design life, and, therefore, does not meet Criteria 3 and 5 of the 10 CFR 54.3(a) TLAA definition, the applicant did not classify this fatigue crack growth evaluation as a TLAA.

The applicant stated further that since the safety determination supported by the existing fatigue crack growth evaluation is valid for several times the 60-year design life, it is valid for at least the period of extended operation.

In Enclosure 2 of the May 21, 2010 letter, the applicant clarified the fatigue crack growth calculations. The applicant stated that a linear elastic fracture mechanics analysis was performed to determine the crack growth of the various postulated semi-elliptical shaped inner surface cracks in the reactor coolant system (RCS) main loop piping. The method of analysis is based on the ASME, Section XI, Appendix A, subsurface flaw evaluation procedure, where the fatigue crack growth rate, da/dN, of the material is characterized in terms of the range of applied stress intensity factor, ΔK.

This characterization is generally of the form, $da/dN = C_0 (\Delta K_I)^n$, which has been determined experimentally. The material constants for carbon steel fatigue crack growth in a water environment used by CE in their evaluation are as follows:

$n = 3.726$

$C_0 = 3.795 \times 10^{-10}$

The applicant stated that the fatigue crack growth curve used in the evaluation also included upper-bound data to envelope the ASME Code Section XI curve. From this method of stress intensity factor determination, the ΔK_I level is calculated based on the crack size and loading conditions. Using a stepping procedure for the number of cycles of loading in a given time period, depth and length crack growth rates are calculated, and the corresponding change in crack size is determined as well as the time required to penetrate the entire pipe wall and produce a leak. The start-up and shutdown transient was found to be the greatest contributor to the usage factor for the main loop piping. A cyclic stress of 18 ksi, conservatively enveloping the start-up and shut down stress, was applied to the hypothetical flaws (1 x 8 inch and 8 x 18 inches) in the circumferential direction in both the 42-inch diameter hot leg and 30-inch diameter cold leg piping. The applicant stated that the number of start-up and shutdown cycles necessary to cause a 1-inch deep crack from 8 to 18 inches long to grow through the pipe wall and leak is at least 3,000 to 8,000 cycles which is significantly greater than the 40 year design value of 500 cycles and the projected value of 213 actual cycles in 60 years. Thus the applicant stated that the partial through-wall cracks will not propagate through the entire pipe wall for more than 400 years.

Time-Limited Aging Analyses

The basis for the safety determination is, therefore, not that the crack will remain within an acceptable size within a 40-year design lifetime but, (1) that the rate of growth for any anticipated crack size is acceptable, even following wall penetration and detectable leakage, and (2) that initial defects larger than the ASME Code Section III initial inspection criteria will not grow through the pipe wall in several 40-year design lifetimes.

The applicant determined the frequency of load application by assuming a uniform distribution of a typical 40-year set of CE design basis loading events over a 40-year life. LRA Table 4.3-3 demonstrates that the rates of accumulation of transient cycles have remained less than those assumed as the basis for the LBB evaluation, with a few exceptions that have no significant effect on the bases for the LBB fatigue crack growth evaluation. Therefore, the basis of the fatigue crack growth analysis, the basis for the safety determination, and the conclusion of the safety determination will not change with an increase in the licensed operating period. The applicant concluded that the fatigue crack growth analyses, therefore, do not "...involve time-limited assumptions *defined by the current operating term*, for example, 40 years." They are *time-dependent*, but for an indefinite period, and are, therefore, not *time-limited*.

In a letter dated March 1, 2010, in response to RAI 4.3.2.15-1, the applicant stated that "...the Metal Fatigue aging management program is not implemented to monitor the transient cycles to confirm that the transient cycles used in the fatigue crack growth analyses for the LBB piping exceed the actual transient cycles because the existing LBB fatigue crack growth evaluation is valid for the period of extended operation..." The staff needed further clarification of the applicant's determination that the transient cycles used in the fatigue crack growth analyses for the LBB piping do not need to be tracked by the applicant's enhanced metal fatigue AMP. In a May 19, 2010, conference call, the staff requested the applicant to verify and confirm that the transient cycles used in the fatigue crack growth calculation exceed and bound the actual operating transient cycles.

In a letter dated May 21, 2010, the applicant submitted a follow-up response stating that the LBB analysis for the RCS main loop piping consisted of two bounding evaluations that remove the time dependency from the projected crack growth evaluation. In the fatigue crack growth evaluation, cracks larger than allowed by the ASME Code Section III initial inspection acceptance criteria would take 3,000 to 8,000 cycles of the most significant contributor to fatigue usage factor (start-up and shutdown) to propagate through the entire pipe wall. The 40-year design value for this transient is 500 cycles and the projected value for 60 years is 213 cycles. In the crack stability evaluation, cracks significantly larger than the ASME allowable were demonstrated to remain stable after leaking under the most severe loading, which is the safe shutdown earthquake. Therefore, given the results of the above sensitivity evaluations, the staff confirmed that the number of transient cycles accumulated over a 60-year period will not affect the results of the LBB evaluations and finds it acceptable that the metal fatigue program is not implemented to monitor the fatigue crack growth analyses for the LBB piping.

The staff finds the applicant's clarification acceptable because the analysis does not meet 10 CFR 54.3(a)(3). The staff noted the analysis showed that the postulated flaws are acceptable for 8,000 transient cycles while the transient cycles expected through the period of extended operation is 213 cycles. The analysis does not meet 10 CFR 54.3(a)(3) criteria where the time-limited assumptions are defined by the current operating term, for example, 40 years. The staff's concern in RAI 4.3.2.15-1 is resolved.

Alloy 82/182 Dissimilar Metal Welds. Nickel-based Alloy 600/82/182 material in the pressurized water reactor (PWR) environment has been shown to be susceptible to primary water stress corrosion cracking (PWSCC). In RAI 4.3.2.15-3, the staff asked the applicant to identify any Alloy 82/182 weld metal and Alloy 600 components used in the LBB approved piping for both units. The staff also asked the applicant to discuss any measures (such as weld overlays or mechanical stress improvement) that have been or will be implemented to reduce the

Time-Limited Aging Analyses

susceptibility of PWSCC in the LBB piping components if it contains alloy 600/82/182 material. The staff also asked the applicant to discuss the inspection history and future inspection frequency of the Alloy 82/182 dissimilar metal butt welds. By letter dated March 1, 2010, in response to RAI 4.3.2.15-3, the applicant stated that no Alloy 82/182 weld metal or Alloy 600 components remain in the main reactor coolant loops within the scope of the LBB analysis, except the branch, instrument, and resistance temperature detector (RTD) nozzle connections shown in Table 4.1-1 below. The applicant noted that none of the components in the table below have shown any degradation.

Table 4.1-1. Alloy 82/182 Dissimilar Metal Welds Mitigation

RC Loop Nozzle to:	Welds	Inspection Methods	Mitigation Strategy
Shutdown cooling line 1 & 2	Alloy 82/182	100% Volumetric once in the next 5 years, if no additional indications/growth, continue with the existing Code examination program for unflawed condition or approved alternative	Full Structural Overlay: Unit 1 fall 2008 Unit 2 spring 2008 Unit 3 spring 2009
		Bare metal visual examination once every 3 refuel outages (RFOs) when volumetric exams are not performed	
Pressurizer surge line	Alloy 82/182	100% Volumetric once in the next 5 years, if no additional indications/growth, continue with the existing Code examination program for unflawed condition or approved alternative	Full Structural Overlay: Unit 1 spring 2007 Unit 2 spring 2008 Unit 3 fall 2007
		Bare metal visual examination once every 3 RFO when volumetric exams are not performed	
Pressurizer Spray line 1A & 1B	Alloy 82/182	Bare metal visual examination once every 3 RFO	Potential for future structural weld overlay or mechanical stress improvement
Safety injection line	Alloy 82/182	100% Volumetric every 6 yrs & bare visual examination once every 3 RFO when volumetric exams are not performed	None
Drain line 1A, 1B, & 2A	Alloy 82/182	Bare metal visual examination once every 3 RFO	None
Letdown line 2B	Alloy 82/182	Bare metal visual examination once every 3 RFO	None
Charging line 2A	Alloy 82/182	Bare metal visual examination once every 3 RFO	None
Cold leg RTD Nozzles	Alloy 600	Bare metal visuals	None
RCP Instrument Taps	Alloy 600	Bare metal visuals	None

The staff finds that the applicant has clarified that no Alloy 82/182 weld metal or Alloy 600 components remain in the main reactor coolant loops within the scope of the LBB analysis except those shown in the table. The staff reviewed the table to evaluate if the applicant is using appropriate inspection methods to manage the branch, instrument, and RTD nozzle connections adequately. The staff finds that the applicant will manage the Alloy 82/182 and Alloy 600 components in the LBB-approved RCS piping adequately and in accordance with ASME Code requirements as discussed below. The staff's concern in RAI 4.3.2.15-3 is resolved.

In RAI 4.3.2.15-4, the staff asked the applicant to discuss the inspection history and results of the LBB-approved piping. If indications or flaws remain in inservice LBB piping, the staff asked the applicant to discuss monitoring of indications and flaws to the end of the period of extended

Time-Limited Aging Analyses

operation and inspection schedules for each of the LBB pipes (other than existing indications and flaws).

By letter dated March 1, 2010, in response to RAI 4.3.2.15-4, the applicant stated that welds and piping that are part of the LBB piping have been examined under the ASME Code Section XI ISI starting in the first interval and will be examined under the rules of the ASME Code Section XI ISI in future intervals. The piping in question is part of the overall population of welds subject to examination. The applicant has found no rejectable indications, and it will continue to examine piping in accordance with the ASME Code Section XI ISI. The applicant performs surface and volumetric exams on 25 percent of the welds spread out over each 10-year interval and a visual exam every outage.

The staff finds that the applicant will follow the ASME Code Section XI ISI to inspect the LBB-approved primary coolant piping during the period of extended operation. Therefore, the structural integrity of the reactor coolant loop piping will be maintained to the end of 60 years. The staff's concern in RAI 4.3.2.15-4 is resolved.

Effects of Power Uprate and Steam Generator Replacement on the LBB Analysis. The applicant has evaluated the effect of power uprate and SG replacement on the LBB-approved primary coolant piping and found no change to the conclusion of the LBB analysis. In RAI 4.3.2.15-5, the staff asked the applicant to clarify whether the LBB analyses have been reanalyzed to determine the effect of operating conditions due to system modifications such as power uprates or SG modifications on the LBB analyses for the period of extended operation.

By letter dated March 1, 2010, in response to RAI 4.3.2.15-5, the applicant stated it evaluated the effects of power uprate and SG replacement, and the evaluation resulted in no change to the conclusion of the LBB analysis (Supplement 11 to NUREG-0857, "Safety Evaluation Report Related to the Operation of PVNGS, Units 1, 2, and 3"). Since the applicant reviewed the LBB analyses for effects of power uprate and SG replacement with no effect on the conclusion and, since they are not TLAAs, the increase in operating life for the period of extended operation does not affect them. The staff's concern in RAI 4.3.2.15-5 is resolved.

The staff finds that the power uprate and SG replacement will not affect the LBB evaluation during the period of extended operation.

Conclusion. Based on its review, the staff concludes that the applicant has demonstrated that the LBB analysis is not a TLAA because the analysis does not meet 10 CFR 54.3(a)(3) criteria.

4.1.3.1.5 Absence of Supplemental Fatigue Analysis Time-Limited Aging Analyses in Response to Bulletin 88-08 for Intermittent Thermal Cycles due to Thermal-Cycle-Driven Interface Valve Leaks and Similar Cyclic Phenomena

Summary of Technical Information in the Application. In amended LRA Section 4.3.2.8 (dated May 27, 2010), the applicant references Bulletin 88-08, "Thermal Stresses in Piping Connected to Reactor Cooling Systems," which recommends that a high-cycle fatigue analysis be performed for the auxiliary pressurizer spray systems. This section states that a "supplemental bounding thermal gradient stress analysis to determine the effect of low cycle fatigue," was performed and that the analysis did not evaluate the effects of high-cycle fatigue on these lines, as recommended in Bulletin 88-08.

Staff Evaluation. The staff confirmed that the applicant's response to Bulletin 88-08, dated October 3, 1988, did not commit to the performance of a high-cycle fatigue analysis.

LRA Section 4.3.2.7, subsection "Flow Stratification Thermal Gradient in the Auxiliary Spray Line and Tee" states that, "[t]he analysis of the thermal gradient demonstrated that the cumulative fatigue usage factor, including the effects of this thermal gradient, meets ASME Code Section III Subsection NB-3600 for a 40-year plant life." Based on this statement, it

Time-Limited Aging Analyses

appears that this analysis meets the definition of a TLAA in accordance with 10 CFR 54.3(a). By letter dated July 21, 2010, the staff issued RAI 4.3-15 asking that the applicant identify the low-cycle fatigue analysis that is being referred to in LRA Sections 4.3.2.7 and 4.3.2.8 and clarify whether the low-cycle fatigue analysis on the auxiliary pressurizer spray systems included an applicable, implicit fatigue analysis, cycle-based fatigue flaw growth, or cycle-based fracture mechanics analysis. The staff also asked the applicant justify why the low-cycle fatigue analysis would not need to be identified as a TLAA if it is determined that the analysis does include a cycle-dependent analysis. This was previously identified as part of Open Item 4.3-1.

The applicant's response dated August 12, 2010, stated that the low-cycle fatigue analysis referred to in LRA Section 4.3.2.7 and 4.3.2.8 is an analysis that considered deadweight, thermal (including stratification), seismic and LOCA load case to determine the effects on the existing fatigue cycle stress range. The applicant further stated that the calculation did not contain an implicit fatigue analysis, cycle-based fatigue flaw growth or a cycle-based fracture mechanics analysis. The applicant stated that the analysis concluded that the effect of thermal stratification does not negatively impact the auxiliary pressurizer spray systems or the stress ranges of the fatigue analysis. The applicant also stated that the low cycle fatigue referred to in LRA Sections 4.3.2.7 and 4.3.2.8 in response to NRC Bulletin 88-08 does not include cycle-based assumptions.

Based on its review, the staff finds the applicant's response to RAI 4.3-15 acceptable because the applicant clarified that the low cycle fatigue analysis referred to in LRA Sections 4.3.2.7 and 4.3.2.8 is not a TLAA and does not contain an implicit fatigue analysis, cycle-based fatigue flaw growth, or a cycle-based fracture mechanics analysis. This part of Open Item 4.3-1 is closed.

Conclusion. The staff concludes that the applicant has demonstrated that the applicant does not have a fatigue analysis TLAAs in response to Bulletin 88-08 for intermittent thermal cycles due to thermal-cycle-driven interface valve leaks and similar cyclic phenomena as part of its CLB, and therefore does not meet the definition of a TLAA, in accordance with 10 CFR 54.3.

4.1.3.1.6 Absence of Time-Limited Aging Analyses in Evaluations of Effects of Vibration on the Unit 1, Train A Shutdown Cooling System Suction Line Fatigue Analysis and of Vibration Limits Established for its Isolation Valve Actuator

Summary of Technical Information in the Application. LRA Section 4.3.2.13 summarizes the evaluation of the absence of a TLAA for the evaluations of effects of vibration on the Unit 1 Train A shutdown cooling system (SCS) suction line fatigue analysis, and of vibration limits established for its isolation valve actuator.

In March of 2006, the applicant conducted a test to diagnose the causes of high vibration in the Unit 1 Train A SCS suction line. The applicant stated that the correction included moving the Unit 1 UV651 valve inboard, to increase the acoustic response above the line and valve resonance. Unit 1 has since operated at 100 percent power with acceptable vibration levels. The evaluation of the UV651 valve actuator determined that maintaining vibration below the administrative limit would maintain accelerations below the revised vibration limits established for indefinite, continuous operation. These evaluations are, therefore, not time-limited and are therefore not TLAAs.

Staff Evaluation. The staff reviewed LRA Section 4.3.2.13 to evaluate the absence of a TLAA in evaluations of effects of vibration on the Unit 1 Train A SCS suction line fatigue analysis, and of vibration limits established for its isolation valve actuator. The applicant stated that in 2006, a test to diagnose causes of high vibration in the Unit 1 Train A SCS suction line was conducted. The Train A SCS suction line is connected to the Loop 1 hot leg. This test operated both Loop 1 reactor coolant pumps but only one Loop 2 pump. This condition produced high Loop 1 flow, which caused brief excursions of an SCS Train A vibration monitor beyond both the administrative and analytical limits. The applicant stated that at these vibration levels, the time

Time-Limited Aging Analyses

required for operator action to shut down the unit might result in unacceptable fatigue usage and eventual failure of the piping or isolation valve motor operator. The applicant stated that the correction included moving the Unit 1 UV651 valve inboard, to increase the acoustic response above the line and valve resonance. Unit 1 has since operated at 100 percent power with acceptable vibration levels. The applicant has also moved the corresponding valves in Units 2 and 3 to prevent similar problems. The staff noted the applicant's conclusion that maintaining vibration below the administrative limit would maintain alternating stresses below the endurance limit at the most limiting location. The staff also noted that the applicant demonstrated that vibration levels remaining below the administrative limit would maintain accelerations below the revised vibration limits established for the UV651 valve actuator for an indefinite period. The staff finds it reasonable that if the vibrations are maintained below the endurance limit, then the fatigue life can be considered infinite. This is reasonable because the alternating stress is less than the stress that would result in fatigue failure and because the material can endure an extremely large number of cycles (10^7 cycles) without failing. The staff noted that the applicant's evaluation did not qualify the piping or valve for any similar excursions during the remaining life of the plant, and therefore does not meet the definition of a TLAA in accordance with 10 CFR 54.3(a)(3).

Based on its review, the staff finds the applicant's evaluation of high vibration of the Unit 1 Train A SCS suction line and of the actuator of its UV651 motor-operated isolation valve is not a TLAA because the applicant demonstrated that maintaining vibration below the administrative limit would maintain alternating stresses below the endurance limit for the UV651 valve actuator for an indefinite period. Further, it is acceptable because the applicant's evaluation did not qualify the piping or valve for any similar excursions during the remaining life of the plant, and, therefore, it does not meet the definition of a TLAA in accordance with 10 CFR 54.3(a)(3).

Conclusion. Based on its review, the staff concludes that the applicant has demonstrated that the evaluations of the effects of vibration on the Unit 1 Train A SCS suction line fatigue analysis and of vibration limits established for its isolation valve actuator, are not TLAAs because they do not meet 10 CFR 54.3(a)(3) criteria.

4.1.4 Conclusion

On the basis of its review, the staff concludes that the applicant has provided an acceptable list of TLAAs, as required by 10 CFR 54.21(c)(1), and that no exemption has been granted on the basis of a TLAA for which continuation has been justified during the period of extended operation.

4.2 Reactor Vessel Neutron Embrittlement

"Neutron embrittlement" is the term for changes in mechanical properties of RPV materials caused by exposure to a fast neutron flux, energy (E) values greater than 1 mega electron-volt (MeV) ($E > 1\ MeV$), within the vicinity of the reactor core, called the "beltline region." The most pronounced material change is a reduction in fracture toughness. As fracture toughness decreases with cumulative fast neutron exposure, the material's resistance to cleavage and ductile fracture decreases. Fracture toughness also depends on temperature. The reference temperature (RT_{NDT}), above which the material behaves in a ductile manner and below which the material behaves in a brittle manner, increases as the fluence increases and requires higher temperatures for continued ductility. All light-water reactors are required by 10 CFR 50.60 to meet the fracture toughness, pressure-temperature (P-T) limits, and material surveillance program requirements for the reactor coolant pressure boundary in Appendices G and H of 10 CFR Part 50. The RT_{NDT} value, which is evaluated at one-quarter or three-quarters of the RPV wall thickness (¼T or ¾T) for a specified effective full power years (EFPYs), is usually referred to as the "adjusted reference temperature" (ART) in the P-T limit applications. In

Time-Limited Aging Analyses

10 CFR Part 50.61, fracture toughness requirements are supplied for protecting the RPV of a PWR against the consequences due to a pressurized thermal shock (PTS) event—a severe overcooling event concurrent with or followed by significant pressure in the RPV. Neutron fluence, upper shelf energy (USE), PTS, and P-T limits are time-dependent items that must be investigated to evaluate RPV embrittlement or reduction of fracture toughness. The CLB analyses evaluating reduction of fracture toughness of the RPV for 40 years are TLAAs. The following sections address neutron fluence, USE, PTS, and P-T limits for RPV beltline materials for the period of extended operation.

4.2.1 Neutron Fluence, Upper Shelf Energy and Adjusted Reference Temperature

4.2.1.1 Summary of Technical Information in the Application

LRA Section 4.2.1 summarizes the evaluation of neutron fluence, USE, and ART for the period of extended operation.

The applicant analyzed the most-recently-examined 230-degrees, Capsule 5 dosimeters from each of the three Units (reports were submitted to the staff in letters dated April 5, 2005; April 4, 2006; and September 26, 2005) to project the neutron fluence at 54 EFPYs, including effects of the power uprate. The applicant's revised fluence values were determined with transport calculations using the Discrete Ordinates Radiation Transport Code and the Bugle-96 cross-section library, which is derived from the Evaluated Nuclear Data Files, version B-VI. The applicant stated that the neutron transport and dosimetry evaluation methodologies follow the guidance and meet the requirements of RG 1.190, "Calculational and Dosimetry Methods for Determining Pressure Vessel Neutron Fluence," and are consistent with Westinghouse Commercial Atomic Power (WCAP) Report WCAP-14040-NP-A, "Methodology Used to Develop Cold Overpressure Mitigating System Setpoints and RCS Heatup and Cooldown Limit Curves."

The LRA states that the clad-base metal interface fluences at 54 EFPY, projected from measured exposures and lead factors of Capsule 5, are $2.51E+19$, $2.83E+19$, and $2.93E+19$ neutrons per centimeter squared (n/cm^2) for $E > 1$ MeV for Units 1, 2, and 3, respectively. The applicant states that these values are less than the original 32 EFPY projection of $3.15E+19$ n/cm^2 ($E > 1$ MeV) used in the PTS evaluation dated January 17, 1986, or the $3.29E+19$ n/cm^2 ($E > 1$ MeV) used to determine the end-of-license (EOL) ART and USE in the NRC's Reactor Vessel Integrity Database (RVID).

The CLB predictions of USE and ART in PVNGS vessel materials at 32 EFPYs indicated that Unit 1 plate materials will be most limiting for both USE and ART. The recently-measured ΔRT_{NDT} values in LRA Table 4.2-2 confirm that the limiting Unit 1 plate material will remain limiting for ART in the period of extended operation.

For the USE evaluation, the applicant reproduced from the RVID, the copper content, initial USE, and neutron fluence values for RPV beltline materials in LRA Tables 4.2-3 to 4.2-5. The estimated 54 EFPY USE values in these tables are obtained using Position 1.2 of RG 1.99, "Radiation Embrittlement to Reactor Pressure Vessel Materials," Revision 2. They are greater than 50 ft-lbs. The applicant stated that the most recent measured USEs show that the decline in USEs is less than originally predicted by RG 1.99, Revision 2. Hence, the applicant concludes that the USE of the limiting material will remain adequate for the period of extended operation.

In summary, the applicant asserts that the evaluation of the acceptability of neutron fluence, USE, and ART remains valid for the period of extended operation, in accordance with 10 CFR 54.21(c)(1)(i). In addition, neutron fluence, USE, and ART will be managed for the period of extended operation by continuing the Reactor Vessel Surveillance Program in accordance with 10 CFR 54.21(c)(1)(iii). See SER Section 3.0.3.2.10 for the staff's evaluation of the Reactor Vessel Surveillance Program.

Time-Limited Aging Analyses

4.2.1.2 Staff Evaluation

The staff reviewed LRA Section 4.2.1 to verify, pursuant to 10 CFR 54.21(c)(1)(i), that the neutron fluence, USE, and ART analyses remain valid for the period of extended operation or, pursuant to 10 CFR 54.21(c)(1)(iii), that the validity of these parameters and their associated analyses will be adequately managed for the period of extended operation using the Reactor Vessel Surveillance Program.

Neutron fluence is an input for determining the USE and ART. The staff reviewed the referenced surveillance capsule dosimetry reports to determine that anisotropic scattering in the fluence analyses was treated with a P_5 Legendre expansion, and that angular discretization was modeled with an S_{16} order of angular quadrature. Based on the review, the staff concluded that the applicant performed fluence calculations in accordance with RG 1.190 and the results are, therefore, acceptable. This is because the Bugle-96 cross section library is derived from Evaluated Nuclear Data Files, version B-VI-based nuclear data and because the scattering approximations and angular quadrature exceed the minimum values specified in RG 1.190.

The uncertainty specified by the applicant of 13 percent, is within the 20-percent tolerance specified in RG 1.190 for calculational uncertainty, which is acceptable to the staff. The staff also considered the acceptability of fluence projections to the end of the period of extended operation based on 54 EFPYs of exposure. The applicant increased its assumed capacity factors from 80 to 90 percent. The staff finds that the assumption of a 90-percent capacity factor is acceptable because capacity factors in the past five years have averaged less than 90 percent as documented in NUREG-1350, "2009-2010 Information Digest," Volume 21, August 2009.

Based on the above evaluation, the staff finds that the applicant's fluence calculations are acceptable to support the period of extended operation.

For USEs, 10 CFR 50, Appendix G contains screening criteria that establish limits on how far the USE values for a RPV material may be allowed to decrease due to neutron irradiation exposure. The regulation requires the initial USE value be greater than 75 ft-lbs in the unirradiated condition and that the value must be greater than 50 ft-lbs in the fully irradiated condition throughout the licensed life of the plant. USE values of less than 50 ft-lbs may be acceptable to the staff if it can be demonstrated that these lower values will provide margins of safety against brittle fracture equivalent to those required by ASME Code Section XI, Appendix G.

As discussed earlier, the staff accepts the 54 EFPY fluence values used by the applicant. These 54 EFPY fluence values are bounded by the fluence value in the CLB for 32 EFPYs. LRA Tables 4.2-3 to 4.2-5 summarize the 54 EFPY USE. Upon review, the staff found apparent discrepancies in these tables. Hence, the staff issued RAI 4.2.1-1 on November 3, 2009. This RAI also included a finding in LRA Table 4.2-6 for Unit 1, 54 EFPY reference temperature for pressurized thermal shock (RT_{PTS}).

The applicant responded in a letter, dated December 18, 2009, which appropriately corrected LRA Tables 4.2-3 to 4.2-8, including additional corrections to errors not identified by the staff. As a result, the staff was able to verify consistency of RPV material information between LRA Section 4.2 and the staff's RVID. The applicant obtained the 54 EFPY USE values in LRA Tables 4.2-3 to 4.2-5 using Position 1.2 of RG 1.99 (without using surveillance data). All USE values exceed 50 ft-lbs. Surveillance data from three withdrawn capsules were not used because of the applicant's conclusion: "[t]he most recent coupon examination results also show that the decline in USE and increase in RT_{NDT} in plate and weld materials are less than originally predicted by Regulatory Guide 1.99, Revision 2...." This justification of not using the surveillance data is not in accordance with RG 1.99, Revision 2. Hence, the staff issued RAI 4.2.1-2 to determine the applicant's basis for not considering all surveillance data.

Time-Limited Aging Analyses

The applicant responded in a letter, dated December 18, 2009, but did not provide adequate justification for not using surveillance data in predicting 54 EFPY USE drops for RPV materials. However, this information is available in the additional references listed below:

- WCAP-16374-NP, "Analysis of Capsule 230° from Arizona Public Service Company Palo Verde Unit 1 Reactor Vessel Radiation Surveillance Program," February 2005

- WCAP-16524-NP, "Analysis of Capsule 230° from Arizona Public Service Company Palo Verde Unit 2 Reactor Vessel Radiation Surveillance Program," February 2006

- WCAP-16449-NP, "Analysis of Capsule 230° from Arizona Public Service Company Palo Verde Unit 3 Reactor Vessel Radiation Surveillance Program," August 2005.

Acceptance of the applicant's approach depends on the examination of the measured USE drops for all RPV materials having at least two surveillance data.

The staff reviewed the measured USE drops for all RPV surveillance materials having at least two surveillance data in these surveillance reports. The staff found that the measured USE drops are less than the predicted values using Position 1.2 of RG 1.99, Revision 2, except for one Unit 2 surveillance data where the measured USE drop is more than the predicted value by 1 percent. The staff determined that the applicant's 54 EFPY USE values, based on Position 1.2, are adequate because the 1-percent difference in the USE value for the surveillance data is within test and curve fitting uncertainty. Hence, RAI 4.2.1-2 is resolved, and the applicant's USE analysis, with its results summarized in LRA Tables 4.2-3 to 4.2-5, is in accordance with RG 1.99, Revision 2 and remains valid for the period of extended operation.

LRA Section 4.2.1 also discussed the applicant's evaluation of ART values for RPV materials. For RPV materials having specimens in surveillance capsules, RG 1.99, Revision 2 requires that all surveillance data in the surveillance capsule reports be considered in determining their chemistry factors. The staff discovered inconsistent information in the surveillance capsule reports. Hence, the staff issued RAI 4.2.1-3 on November 3, 2009, to confirm that the applicant misidentified surveillance specimens for Unit 1 in the WCAP-15589 report and confirm that Units 2 and 3 did not experience similar misidentifications.

In a letter dated December 18, 2009, the applicant confirmed the misidentification of specimens in Capsule 38-degrees, for Unit 1 in the WCAP-15589 report and stated that WCAP-15589, Revision 1 report corrected this problem. The applicant submitted the WCAP-15589, Revision 1 report on November 13, 2009. Hence, the WCAP-16374 report for Capsule 230-degrees and the USE and ART evaluations in the LRA reflect correct information on the use of surveillance data. To rule out misidentification of surveillance specimens for Units 2 and 3, the applicant stated that unlike Unit 1, which has three capsules containing M-4311-1 base metal material and three containing M-6701-2 base metal material, Units 2 and 3 have only one type of base metal material in their surveillance capsules. Hence, the staff concludes that the misidentification of surveillance specimens that happened to Unit 1 is unlikely to happen to Units 2 and 3, and the staff considers RAI 4.2.1-3 resolved.

No criterion for ARTs is given in 10 CFR 50. However, as the most important parameter for determining the fracture toughness of the RPV material, it affects the P-T limits directly. The staff found that the chemistry factor for the limiting Unit 1 plate material in the CLB is greater than that based on surveillance data, and determined that the applicant can continue to use it to calculate the ART for the limiting material. The use of ART in the P-T limits for the period of extended operation is discussed in Section 4.2.3 of this SER.

4.2.1.3 Updated Final Safety Analysis Report Supplement

The applicant provided an UFSAR supplement summary description of its TLAA evaluation of the neutron fluence, USE, and ART values for RPV materials in LRA Section A3.1.1. The staff

Time-Limited Aging Analyses

reviewed LRA Section A3.1.1 against the acceptance criteria in SRP-LR Section 4.2.2.2. Based on its review of the UFSAR supplement, the staff concludes that the summary description of the applicant's actions to address the neutron fluence, USE, and ART is adequate per 10 CFR 54.21(d).

4.2.1.4 Conclusion

On the basis of its review, as discussed above, the staff concludes that the applicant has demonstrated, pursuant to 10 CFR 54.21(c)(1)(i), that for neutron fluence, USE, and ART, the analyses remain valid for the period of extended operation. The staff also finds it acceptable that the plant's Reactor Vessel Surveillance Program will provide information for further validating or modifying its projected neutron fluence, ART, and USE values during the period of extended operation, pursuant to 10 CFR 54.21(c)(1)(iii). Finally, the staff finds that the UFSAR supplement contains an appropriate summary description of the TLAA evaluations, as required by 10 CFR 54.21(d).

4.2.2 Pressurized Thermal Shock

4.2.2.1 Summary of Technical Information in the Application

LRA Section 4.2.2 summarizes the PTS evaluation of the beltline materials for the period of extended operation against the screening criteria established in accordance with 10 CFR 50.61, "Fracture Toughness Requirements for Protection Against Pressurized Thermal Shock Events." The screening criteria are 270 degrees F for plates, forging, and axial weld materials and 300 degrees F for circumferential weld materials.

The applicant claims that since the 54 EFPY fluence is expected to remain within the values originally predicted for a 32 EFPY life, the 54 EFPY RT_{PTS} is also expected to remain within the values originally predicted for a 32 EFPY life. Hence, the conclusions of the original evaluation are unaffected, and the original evaluation of the PTS screening parameter and the conclusion remains valid for the period of extended operation in accordance with 10 CFR 54.21(c)(1)(i).

4.2.2.2 Staff Evaluation

The staff reviewed LRA Section 4.2.2 to verify, pursuant to 10 CFR 54.21(c)(1)(i), that the analyses remain valid for the period of extended operation.

The 10 CFR 50.61 provides the fracture toughness requirements protecting the RPVs of PWRs against the consequences of PTS. Applicants are required to perform an assessment of the RPV materials' projected RT_{PTS} values through the end of their operating license. The rule requires each applicant to calculate the EOL RT_{PTS} value for each RPV beltline material. The RT_{PTS} value for each beltline material is the sum of the unirradiated RT_{NDT}, a shift in the RT_{NDT} value caused by neutron irradiation of the material (ΔRT_{NDT}), and a margin value to account for uncertainties (M). 10 CFR 50.61 also provides screening criteria, against which the calculated values are to be evaluated.

As stated in LRA Section 4.2.2.1, the screening criteria are 270 degrees F for plates, forging, and axial weld materials and 300 degrees F for circumferential weld materials. 10 CFR 50.61 provides a discussion regarding the calculations of ΔRT_{NDT} and the M value (defined in 10 CFR 50.61(c)(1)(iii)). In 10 CFR 50.61, ΔRT_{NDT} is the product of a chemistry factor and a fluence factor, where the fluence factor is dependent upon the neutron fluence at the clad-to-base metal interface and the chemistry factor is dependent upon information from either the surveillance material or from the tables in 10 CFR 50.61. If the RPV beltline material is not represented by surveillance material, its chemistry factor may be determined using the tables and the methodology documented in 10 CFR 50.61. The chemistry factor determined from the tables in 10 CFR 50.61 depends upon the amount of copper and nickel in the material. If the RPV beltline material is represented by surveillance material, its chemistry factor may be

Time-Limited Aging Analyses

determined from the surveillance data using the methodology documented in 10 CFR 50.61. The methods of determining RT_{PTS} values in 10 CFR 50.61 are equivalent to the methods of determining RT_{NDT} values in RG 1.99, Revision 2.

In LRA Tables 4.2-6 to 4.2-8, the applicant reproduced the information (RPV materials data, neutron fluence, and the projected RT_{PTS} results) for PVNGS from the NRC's RVID to demonstrate that the units comply with 10 CFR 50.61. Instead of using the RVID labeling, "EOL," the applicant labeled the fluence-dependent parameters in these tables as "54 EFPY" to indicate that the PTS evaluation is valid for the period of extended operation. The tabulated 54 EFPY RT_{PTS} values are based on the neutron fluence value of $3.29E+19$ n/cm^2 ($E > 1$ MeV), which bounds the 54 EFPY neutron fluence values of $2.51E+19$ n/cm^2, $2.83E+19$ n/cm^2, and $2.93E+19$ n/cm^2 ($E > 1$ MeV) for the PVNGS units. The staff accepts these values (see Section 4.2.1.2 of this SER). LRA Tables 4.2-6 to 4.2-8 also show that the RT_{PTS} for the limiting RPV beltline material (the intermediate shell plate M-6701-2 and M-6701-3) is 122.5 degrees F, meeting the PTS criteria.

LRA Section 4.2.2 further states that the PVNGS 10 CFR 50.61 PTS submittal, dated January 17, 1986, projected an RT_{PTS} of 132 degrees F for the limiting plate material at the $3.15E+19$ n/cm^2 ($E > 1$ MeV) clad-base metal interface fluence. Based on the copper and nickel values reported in the January 17, 1986, PTS evaluation, the staff believes that the applicant's limiting material then was the Unit 1 intermediate shell plate M-6701-1, not the Unit 1 intermediate shell plates M-6701-2 or M-6701-3 identified in the CLB (or RVID). This discrepancy is not important because the RVID shows that the difference in RT_{PTS} caused by using the different limiting plates identified above is only 0.8 degree F. Considering this, the January 17, 1986, evaluation still bounds the applicant's PTS evaluation at 54 EFPYs for the period of extended operation.

Based on the above discussion, the staff concludes that RPV beltline materials satisfy the PTS requirements of 10 CFR 50.61 through the period of extended operation. The applicant's TLAA for calculating the RT_{PTS} values of the RPV beltline materials at the end of the period of extended operation is acceptable because the calculated values are bound by the existing analysis and meet the requirements of 10 CFR 54.21(c)(1)(i). This ensures that the RPV materials will have adequate RT_{PTS} values and fracture toughness through the period of extended operation.

4.2.2.3 *Updated Final Safety Analysis Report Supplement*

The applicant provided an UFSAR supplement summary description of its TLAA evaluation of PTS in LRA Section A3.1.2. Based on its review of the UFSAR supplement, the staff concludes that the summary description of the applicant's actions to address PTS is adequate.

4.2.2.4 *Conclusion*

On the basis of its review, as discussed above, the staff concludes that the applicant has demonstrated, pursuant to 10 CFR 54.21(c)(1)(i), that for PTS, the analyses remain valid for the period of extended operation. The staff also concludes that the UFSAR supplement contains an appropriate summary description of the TLAA evaluation, as required by 10 CFR 54.21(d).

4.2.3 Pressure-Temperature Limits

4.2.3.1 *Summary of Technical Information in the Application*

LRA Section 4.2.3 summarizes the evaluation of P-T limits for the period of extended operation. The applicant states that the current license includes P-T limit curves calculated for embrittlement effects originally determined to be valid up to 32 EFPYs. However, they were based on projections of EOL ART that depended on an originally-estimated 32 EFPY beltline high-energy neutron fluence of $3.29E+19$ n/cm^2 ($E > 1$ MeV), which exceeds the maximum fluence now expected at 54 EFPYs, $2.93E+19$ n/cm^2 ($E > 1$ MeV). Hence, the present P-T limit curves for 32 EFPYs are still valid for the period of extended operation in accordance with 10 CFR 54.21(c)(1)(i). APS will confirm their basis for 54 EFPYs prior to operation beyond 32 EFPYs and will update documents in accordance with the provisions of 10 CFR 50.59.

4.2.3.2 Staff Evaluation

The staff reviewed LRA Section 4.2.3 to verify that, pursuant to 10 CFR 54.21(c)(1)(i), the P-T limit analyses remain valid for the period of extended operation.

The staff approved the current P-T limits on February 25, 2010, through issuance of Amendment 178 which revised the TS to relocate the P-T limits and the low temperature overpressure protection (LTOP) system enable temperatures from the TS to a licensee-controlled document, the Pressure Temperature Limits Report. The associated request for exemption from 10 CFR 50, Appendix G on P-T limits calculation was approved on February 24, 2010. The current P-T limits are for 32 EFPYs with a neutron fluence of $3.29E+19$ n/cm^2 ($E > 1$ MeV) at the RPV clad-to-base metal interface (the RPV surface). The staff found that the limiting material for the P-T limits, which apply to all units, is the Unit 1 intermediate shell plate M-6701-2 or M-6701-3. The copper and nickel contents, the initial RT_{NDT} values, and the 32 EFPY neutron fluence at the RPV surface for the limiting material are identical to the information in the RVID. The applicant continued to use this material information in its P-T limit evaluation for the period of extended operation.

As evaluated in Section 4.2.1.2 of this SER, the staff accepts the applicant's 54 EFPY neutron fluence values of $2.51E+19$ n/cm^2, $2.83E+19$ n/cm^2, and $2.93E+19$ n/cm^2 ($E > 1$ MeV). These fluence values are bounded by the 32 EFPY neutron fluence value of $3.29E+19$ n/cm^2 ($E > 1$ MeV) for the current P-T limits. Since the copper and nickel contents and the initial RT_{NDT} values for the limiting material of the P-T limits remain unchanged during the period of extended operation, the fact that the neutron fluence for the current P-T limits bounds the 54 EFPY neutron fluence shows that the current P-T limits bound the 54 EFPY P-T limits.

Based on the above discussion, the staff determines that the applicant's P-T limit evaluation of the RPV beltline materials during the period of extended operation is acceptable because the current P-T limits remain valid for the period of extended operation in accordance with 10 CFR 54.21(c)(1)(i). This P-T limit evaluation will ensure that the RPV materials will have adequate fracture toughness and meet the requirements of 10 CFR 50.60 and 10 CFR 50, Appendix G during the period of extended operation.

As mentioned, the applicant was approved on February 25, 2010, to relocate the P-T limits for PVNGS to a licensee-controlled Pressure Temperature Limits Report. As such, as long as the P-T limit methodology stays the same, future changes to the P-T limit curves will be processed through the provisions of 10 CFR 50.59 instead of the license amendment process, as stated by the applicant.

4.2.3.3 Updated Final Safety Analysis Report Supplement

The applicant provided an UFSAR supplement summary description of its TLAA evaluation of P-T limits in LRA Section A3.1.3. Based on its review of the UFSAR supplement, the staff concludes that the summary description of the applicant's actions to address P-T limits is adequate.

4.2.3.4 Conclusion

On the basis of its review, as discussed above, the staff concludes that the applicant has demonstrated, pursuant to 10 CFR 54.21(c)(1)(i), that, for P-T limits, the analyses remain valid for the period of extended operation. The staff also concludes that the UFSAR supplement

Time-Limited Aging Analyses

contains an appropriate summary description of the TLAA evaluation, as required by 10 CFR 54.21(d).

4.2.4 Low Temperature Overpressure Protection

4.2.4.1 Summary of Technical Information in the Application

LRA Section 4.2.4 summarizes the evaluation of LTOP for the period of extended operation. The applicant states that TS Limited Condition for Operation 3.4.13 requires LTOP, which is provided by relief valves in the two suction lines of the SCS or by operating with the RCS depressurized and with an open RCS vent of sufficient size to protect the SCS and RCS. The LTOP enable temperatures (the temperatures below which LTOP must be established), and those analyses that confirm the ability to protect the system's pressure limits, depend on the P-T limit curves and the ART. The LTOP enable temperatures and the supporting design basis calculations are TLAAs. The mass and energy addition transient analyses in the LTOP licensing basis, however, are not time-dependent. The applicant uses the enable temperatures and P-T heatup and cooldown limits as input to determine maximum system temperature at the time of the event and the heatup and cooldown rates with the system aligned.

The applicant states further that the only time-limited analyses upon which the LTOP setpoints are based are those for the P-T curves and ART. These will remain valid for the period of extended operation. Therefore the LTOP licensing and design basis analyses will remain valid for the period of extended operation, in accordance with 10 CFR 54.21 (c)(1)(i).

4.2.4.2 Staff Evaluation

The staff reviewed LRA Section 4.2.4 to verify, pursuant to 10 CFR 54.21(c)(1)(i), that LTOP remains valid for the period of extended operation.

In Sections 4.2.1.2 and 4.2.3.2 of this SER, the staff concludes that the applicant's ART and P-T limit evaluations of the RPV beltline materials during the period of extended operation are acceptable because the current P-T limits remain valid for the period of extended operation in accordance with 10 CFR 54.21(c)(1)(i). Since the ART and the P-T limits, which are the only time-dependent inputs to the LTOP evaluation, remain valid for the period of extended operation, the LTOP evaluation will also remain valid for the period of extended operation, in accordance with 10 CFR 54.21(c)(1)(i).

4.2.4.3 Updated Final Safety Analysis Report Supplement

The applicant provided an UFSAR supplement summary description of its TLAA evaluation of LTOP in LRA Section A3.1.4. Based on its review of the UFSAR supplement, the staff concludes that the summary description of the applicant's actions to address LTOP is adequate.

4.2.4.4 Conclusion

On the basis of its review, as discussed above, the staff concludes that the applicant has demonstrated, pursuant to 10 CFR 54.21(c)(1)(i), that, for LTOP, the analyses remain valid for the period of extended operation. The staff also concludes that the UFSAR supplement contains an appropriate summary description of the TLAA evaluation, as required by 10 CFR 54.21(d).

4.3 Metal Fatigue Analysis

LRA Section 4.3 provides the assessment of metal fatigue analyses in the CLB, which the applicant determined to be TLAAs for license renewal. The applicant divides this section of the LRA into the following subsections:

- LRA Section 4.3.1, "Fatigue Aging Management Program" and its subsections

Time-Limited Aging Analyses

- LRA Section 4.3.2, "ASME III Class 1 Fatigue Analysis of Vessels, Piping, and Components" and its subsections
- LRA Section 4.3.3, "Fatigue and Cycle-Based TLAAs of ASME III Subsection NG Reactor Pressure Vessel Internals"
- LRA Section 4.3.4, "Effects of the Reactor Coolant System Environment on Fatigue Life of Piping and Components [Generic Safety Issue (GSI) 190]"
- LRA Section 4.3.5, "Assumed Thermal Cycle Count for Allowable Secondary Stress Range Reduction Factor in ANSI B31.1 and ASME III Class 2 and 3 Piping"

The applicant identifies that the following metal fatigue analyses constitute TLAAs for the LRA:

- "Reactor Pressure Vessel, Nozzles, Head and Studs" (LRA Section 4.3.2.1)
- "Control Element Drive Mechanism (CEDM) Nozzle Pressure Housings" (LRA Section 4.3.2.2)
- "Reactor Coolant Pump Pressure Boundary Components" (LRA Section 4.3.2.3)
- "Pressurizer and Pressurizer Nozzles" (LRA Section 4.3.2.4)
- "Steam Generator ASME III Class 1, Class 2 Secondary Side, and Feedwater Nozzle Fatigue Analyses" (LRA Section 4.3.2.5)
- "ASME III Class 1 Valves" (LRA Section 4.3.2.6)
- "ASME III Class 1 Piping and Piping Nozzles" (LRA Section 4.3.2.7)
- "Bulletin 88-11 Revised Fatigue Analysis of the Pressurizer Surge Line for Thermal Cycling and Stratification" (LRA Section 4.3.2.9)
- "Class 1 Fatigue Analyses of Class 2 Regenerative and Letdown Heat Exchangers" (LRA Section 4.3.2.10)
- "Class 1 Fatigue Analyses of Class 2 HPSI and LPSI Safety Injection Safeguard Pumps for Design Thermal Cycles" (LRA Section 4.3.2.11)
- "Class 1 Analysis of Class 2 Main Steam Safety Valves" (LRA Section 4.3.2.12)
- "High Energy Line Break Postulation Based on Fatigue Cumulative Usage Factor" (LRA Section 4.3.2.14)

The staff evaluated these TLAAs in the subsections that follow. The applicant also identified the following metal fatigue analyses in LRA Section 4.3.2 that do not comply with the definition of a TLAA, as defined in 10 CFR 54.3:

- "Absence of Supplemental Fatigue Analysis TLAAs in Response to Bulletin 88-08 for Intermittent Thermal Cycles due to Thermal-Cycle-Driven Interface Valve Leaks and Similar Cyclic Phenomena" (LRA Section 4.3.2.8)
- "Absence of TLAAs in Evaluations of Effects of Vibration on the Unit 1 Train A Shutdown Cooling System Suction Line Fatigue Analysis, and of Vibration Limits Established for its Isolation Valve Actuator" (LRA Section 4.3.2.13)
- "Absence of TLAAs in Fatigue Crack Growth Assessments and Fracture Mechanics Stability Analyses for the Leak-Before-Break (LBB) Elimination of Dynamic Effects of Primary Loop Piping Failures" (LRA Section 4.3.2.15)

Time-Limited Aging Analyses

The staff evaluated the applicant's basis for claiming that these analyses are not TLAAs in SER Section 4.1.3.1.

During the acceptance review of the LRA, the staff noted that Table 4.3-9, "Summary of PVNGS Class 1 Valve Fatigue Analyses," did not give the information necessary for the staff's review. The staff, therefore, ended the acceptance review of the application; issued a letter dated February 13, 2009, to the applicant describing the incomplete information; and asked the applicant to provide a plan for resolving the identified deficiency. The applicant provided its response by letter dated February 25, 2009, and stated that it would supplement the LRA before April 15, 2009. By letter dated April 14, 2009, the applicant submitted Supplement 1 to the LRA which provided the missing information. The staff then accepted the LRA (74 FR 22978) and began its review.

The staff noted other discrepancies and inconsistencies during the review of LRA Section 4.3, "Metal Fatigue Analysis." The staff held several conference calls with the applicant concerning metal fatigue analysis issues. The topics of these conference calls are captured in a summary document dated July 14, 2010. Additionally, the staff held a public meeting to discuss metal fatigue issues on May 6, 2010. The public meeting summary can be found in a document dated June 25, 2010.

By letter dated April 28, 2010, the applicant submitted Amendment 14 to the LRA to clarify and correct LRA Section 4.3.1 as discussed with the staff. By letter dated May 27, 2010, the applicant submitted Amendment 16 to the LRA to provide conforming changes to the remaining Section 4.3 subsections and related sections (e.g., Appendix B, Section B3.1 "Metal Fatigue of Reactor Coolant Pressure Boundary"). By letter dated June 29, 2010, the applicant submitted Amendment 18 to the LRA, which further modified appropriate sections and provided responses to the staff's RAIs issued on June 2, 2010 (these RAIs will be discussed later in this section).

Additional metal fatigue amendments were provided by the applicant by letters dated July 7, 2010 (Amendment 19), August 12, 2010 (Amendment 22), October 13, 2010 (Amendment 25) and December 3, 2010 (Amendment 28). These are discussed in Sections 4.3 and 4.7.

4.3.1 Enhanced Fatigue Aging Management Program

In LRA Section 4.3.1, "Enhanced Fatigue Aging Management Program (B3.1)," the applicant provides a general discussion on its use of the Metal Fatigue of Reactor Coolant Pressure Boundary Program. This discussion includes how the applicant will use the program to track the number of occurrences for the plant's design basis transients and their effects on the fatigue analysis for ASME Code Class components.

The LRA describes the enhanced AMP in the following subsections:

- LRA Section 4.3.1.1, "Licensing and Design Basis of the PVNGS Component Cyclic and Transient Limit Program"
- LRA Section 4.3.1.2, "Enhanced PVNGS Fatigue Management Program (B3.1)"
- LRA Section 4.3.1.3, "Seismic History"
- LRA Section 4.3.1.4, "Present and Projected Status of Monitored Locations"
- LRA Section 4.3.1.5, "Program Scope, Action Limits, and Corrective Actions"

In LRA Section 4.3.1, the applicant identifies that the enhanced metal fatigue AMP will apply one of the following fatigue monitoring methodologies for ASME Code Class components:

- Cycle counting (CC)

Time-Limited Aging Analyses

- Cycle-based fatigue per cycle (CBF-C)
- Cycle-based fatigue with partial cycles (CBF-PC)
- Event pairing cycle-based fatigue (CBF-EP)
- Stress-based fatigue (SBF)

In this section, the applicant clarifies when it is appropriate to use these monitoring methods as the basis for accepting the metal fatigue TLAAs in accordance with the TLAA acceptance requirement in 10 CFR 54.21(c)(1)(iii).

The staff's evaluation of LRA Section 4.3.1 and its subsections follows.

4.3.1.1 Licensing and Design Basis of the Palo Verde Nuclear Generating Station Component Cyclic and Transient Limit Program

4.3.1.1.1 Summary of Technical Information in the License Renewal Application

The applicant provides a summary of the licensing and design information for the "Component Cyclic and Transient Limit Program" in LRA Section 4.3.1.1. TS 5.5.5 requires the applicant to include an administrative program that "provides controls to the UFSAR Section 3.9.1.1 cycle and transient occurrences to ensure that components are maintained within the design limits."

The applicant also states that UFSAR Section 3.9.1.1 includes, by reference, information and transient definitions from the following UFSAR sections and tables, listed in LRA Table 4.3-1:

- UFSAR Section 3.7.3.2, "Operating Basis Earthquake (OBE) Cycles"
- UFSAR Table 3.9.1-1, "ASME III Class 1 Components by the NSSS Vendor (CE)"
- UFSAR Table 3.9-1, "ASME III Class 1 Piping Not By the NSSS Vendor (CE)"
- UFSAR Section 3.9.3, "ASME III Class 2 and 3 Components"
- UFSAR Section 5.4.1, "Reactor Coolant Pumps"
- UFSAR Section 5.4.2, "Steam Generators"
- UFSAR Section 5.4.3, "Reactor Coolant Piping"
- UFSAR Section 5.4.10, "Pressurizer"

4.3.1.1.2 Staff Evaluation

The staff reviewed the TS and UFSAR to assess whether the sections referenced by the applicant in LRA Section 4.3.1.1 and in LRA Table 4.3-1 were the applicable CLB and current design basis documents. The staff confirmed that TS 5.5.5 gives the licensing requirements for tracking the occurrences of the design basis transients, and the TS references the transients listed and evaluated in UFSAR Section 3.9.1.1. The staff verified that UFSAR Section 3.9.1.1 refers to those design basis transients that UFSAR Table 3.9.1-1 lists for ASME Code Class 1 Nuclear Steam Supply System (reactor vessel) components, UFSAR Table 3.9-1 for ASME Code Class 1 Non-Nuclear Steam Supply System (NSSS), and the UFSAR sections referenced in the above bulleted list. Based on this verification, the staff determined that the applicant appropriately referenced the appropriate CLB and UFSAR sections for tracking the design basis transients that are applicable to the fatigue assessments.

4.3.1.1.3 Conclusion

Based on this review, the staff concludes that the applicant's Component Cyclic and Transient Limit Program references the appropriate TS requirement and UFSAR sections and that LRA Section 4.3.1.1, as administratively amended in LRA Amendment 14, provides an accurate summary of the TS requirements and UFSAR sections that are applicable to this program.

4.3.1.2 Enhanced Fatigue Aging Management Program

4.3.1.2.1 Summary of Technical Information in the License Renewal Application

In LRA Section 4.3.1.2, the applicant provides a brief description of the general basis for the enhanced metal fatigue AMP that will be implemented during the period of extended operation. LRA Section 4.3.1.2 includes LRA Table 4.3-2, "PVNGS Unit 1, 2, and 3 Licensing Basis Transients," which provides a summary of the design basis transients that are applicable to this TLAA and the design basis limits for these transients.

4.3.1.2.2 Staff Evaluation

The staff reviewed the information in LRA Section 4.3.1 against the SRP-LR Section 4.3.2.1.1.3 for ASME Code Section III, Code Class 1 components and SRP-LR Section 4.3.2.1.2.3 for ASME Code Section III, Code Class 2 and 3 components designed to ANSI B31.1 requirements. Specifically, the staff reviewed the general scope, monitoring method basis, corrective actions, and analytical margin information in LRA Section 4.3.1 to evaluate whether the monitoring method bases were in conformance with those given in the enhanced metal fatigue AMP. The staff also evaluated whether the bases would be adequate for managing the metal fatigue in ASME Code Class components or in piping, piping components, or piping elements designed to American National Standards Institute (ANSI) B31.1 requirements in accordance with 10 CFR 54.21(c)(1)(iii).

The applicant gives their basis for using CBF-C, CBF-PC, and CBF-EP monitoring methods in LRA Section 4.3.1. The staff noted that this is an acceptable basis for how they would apply these methods. The applicant clarified that it would count the number of cycles for transients used in the analysis for the locations monitored by these methods. The staff also noted that these methods would periodically update the cumulative usage factor (CUF) values based on actual cycle count data. The staff finds that this is consistent with the "detection of aging effects" program element recommendation in the Generic Aging Lessons Learned (GALL) Report AMP X.M1, "Metal Fatigue of Reactor Coolant Pressure Boundary." The applicant also clarified the differences between the CBF-C, CBF-PC, and CBF-EP monitoring methods and explained how these methods meet ASME Code Section III requirements for CUF calculations. The staff finds the applicant's basis for using CBF-C, CBF-PC, and CBF-EP methods, as amended, acceptable because the methods comply with ASME Code Section III requirements and are consistent with the recommendations of the GALL Report.

The staff noted there was conflicting information between the scope of the information provided in the original LRA Section 4.3.1 and relevant information in other subsections of LRA Section 4.3 concerning the use of "Global" and SBF monitoring methods.

The staff noted the original LRA Section 4.3.1 states that the "Global" monitoring method will be used to count and track transient event cycles affecting the location to ensure that the numbers of transient events assumed by the design basis calculations will not be exceeded. However, under this monitoring method, the fatigue AMP will not periodically calculate accumulated fatigue usage of the component location being monitored. The staff noted that, in contrast, the "detection of aging effects" program element in the GALL Report Metal Fatigue AMP recommends periodic updates of the CUF calculations. Thus, the staff determined that the basis for applying the "Global" monitoring method is not consistent with the recommendations of the GALL Report Metal Fatigue AMP. The staff also noted that LRA Section 4.3.1.5 states that the use of the "Global" monitoring method would only be applied to component locations with low calculated design basis CUF values. However, the staff noted that in the original LRA Table 4.3-4, the applicant applied the use of the "Global" monitoring method to both components with low and high calculated design basis CUF values. Thus, the staff noted that there was conflicting information in the TLAA on how the "Global" monitoring method would be applied and that the "Global" monitoring method was not consistent with the CUF update

Time-Limited Aging Analyses

recommendation in the GALL Report Metal Fatigue AMP. A metal fatigue conference calls summary document, dated July 14, 2010, summarizes the staff's discussion with the applicant concerning this issue.

With respect to SBF monitoring methods, the staff noted that on page 4.3-3 of the original LRA Section 4.3.1, the applicant said that it intends to apply the SBF monitoring to those component locations with high CUF values for which a more refined approach is necessary to show long-term structural acceptability. The staff noted that the applicant clarified that SBF monitoring updates the CUF calculations for these components using "real time" temperature, pressure, and flow histories for the components. The applicant further stated that the monitoring method depends on "global-to-local" correlation or "transfer" functions, which calculate local transient pressures and temperatures from data collected by the limited number of plant instruments to determine local stresses and fatigue usage. The staff noted the original LRA Table 4.3-4 states that the SBF monitoring basis could be applied on a bounding basis, where the application of SBF monitoring for one component with high valued CUF values would also be used as a SBF monitoring basis for other components with high valued CUFs. The associated sections of the original LRA (Sections 4.3.1 and 4.3.1.5), however, do not provide sufficient justification that the SBF monitoring method could be applied on a bounding SBF monitoring basis. While the applicant did provide its action limits and correction actions for SBF monitoring in LRA Section 4.3.1.5, the applicant's bases do not establish how it would apply corrective actions for the bounding SBF monitoring basis. Specifically, the staff noted that the applicant had not established or justified what type of corrective actions it would apply to unmonitored, highly-valued CUF component locations if a CUF action limit was reached for a monitored location. A metal fatigue conference calls summary document, dated July 14, 2010, summarizes the staff's discussion with the applicant concerning this issue.

The staff reviewed the information in TS 5.5.5 and in the original LRA Sections 4.3.1 and 4.3.1.2 and LRA Table 4.3-2 against relevant design basis information in USAR Sections 3.9.1.1, 3.7.3.2, 3.9.3, 5.4.1, 5.4.2, 5.4.3, and 5.4.10 and UFSAR Tables 3.9.1-1 and 3.9-1 for consistency. The staff also reviewed the information in the original LRA Section 4.3.1.2 and LRA Table 4.3-2 against relevant information in other subsections, including the original LRA Sections 4.3.2, 4.3.3, 4.3.4 and 4.3.5 and LRA Tables 4.3-3 through 4.3-8.

TS 5.5.5, "Component Cyclic or Transient Limits," provides controls to track UFSAR Section 3.9.1.1 cyclic and transient occurrences to ensure the applicant maintains components within the design limits. During its review, the staff noted there were many inconsistencies between the information in LRA Sections 4.3.1, 4.3.1.2, or Table 4.3-2 and design basis information in the UFSAR. The staff also noted inconsistencies between subsections and between subsections and tables contained in LRA Section 4.3. The following items provide examples of the inconsistencies that the staff noted in the original LRA:

- A given transient is listed in LRA Table 4.3-2 (and and in LRA Table 4.3-3) as a normal operating condition, upset condition, or test condition transient but is listed under a different transient category in either UFSAR Table 3.9.1-1 or 3.9-1.

- Normal operating condition, upset condition, or test condition transients that are listed in either UFSAR Table 3.9.1-1 or 3.9-1 are not accounted for in LRA Tables 4.3-2 and 4.3-3.

- Design basis limit information for a given transient in LRA Table 4.3-2 fails to reflect all design basis information or is different from that listed for the corresponding transient in either UFSAR Table 3.9.1-1 or 3.9-1.

Time-Limited Aging Analyses

- LRA Table 4.3-2 states that a given transient in LRA 4.3-2 will be counted under the program's tracking activities, but LRA Table 4.3-3 contradicts this by indicating that the transient will not be counted under the program's monitoring activities.

- Omission of emergency or faulted events in LRA Tables 4.3-2 and 4.3-3 that are within the scope of emergency or faulted design basis transients in UFSAR Table 3.9.1-1 or 3.9-1.

A metal fatigue conference calls summary document, dated July 14, 2010, describes the staff's discussions with the applicant concerning these and other issues.

By letter dated April 28, 2010, the applicant submitted Amendment 14 to the LRA to address several of these issues. In this letter, the applicant provided updated information concerning the enhanced metal fatigue AMP monitoring bases for this TLAA. Specifically, the staff noted that the applicant amended the LRA to make the following changes and clarifications:

- LRA Section 4.3 was amended to use the terminology "cycle counting" monitoring to replace the term "Global" monitoring. The staff finds this change to be acceptable because the change is administrative and does not affect the staff's basis for accepting the monitoring bases for the enhance fatigue AMP.

- LRA Section 4.3.1 was changed to clarify that the scope of the enhanced metal fatigue AMP will include all ASME Code Section III Class 1 components and Class 2 portions of the SGs that have been analyzed to ASME Code requirements for Class 1 components.

- The enhanced metal fatigue AMP was clarified such that it will continue to monitor for plant transients required by TS 5.5.5. In addition, CUFs will be calculated for a subset of ASME Code Class 1 reactor coolant pressure boundary vessel and piping components and ASME Code Class 2 SG locations that were conservatively analyzed using ASME Code Class 1 CUF analysis bases.

- LRA Sections 4.3.1 and 4.3.1.2 were changed to clarify that the enhanced program continues to count transient cycles and will monitor the CUF values for bounding locations, as given in amended LRA Table 4.3-4 of Amendment 14.

- The LRA was amended to clarify the enhanced metal fatigue AMP action limits on tracked cycles and CUF values and establish appropriate corrective actions to be taken before the licensing basis limits on fatigue effects, at any location, are exceeded.

The staff evaluated the monitoring bases for the enhanced metal fatigue AMP, as amended and verified that the applicant is crediting the enhanced metal fatigue AMP to disposition the TLAAs for ASME Code Class 1 components that were designed to ASME Code Section III or for ASME Code Class 2 SG components that were analyzed in accordance with ASME Code Section III CUF design calculations. The staff verified that the applicant has dispositioned its implicit fatigue analyses for safety Class 1 piping designed to ANSI B31.1 requirements or ASME Code Class 2 or 3 components designed to ASME Code Section III requirements in accordance with 10 CFR 54.21(c)(1)(i) or 10 CFR 54.21(c)(1)(ii). Based on this review and verification, the staff finds that the applicant's scope of the enhanced metal fatigue AMP acceptable because it is appropriately being used for those ASME Code Class 1 and 2 components that were analyzed to ASME Code Section III CUF analysis criteria.

The applicant amended LRA Sections 4.3.1 and 4.3.1.2 to clarify the differences between the current fatigue AMP and the enhanced version of the program that it will carry out during the period of extended operation. The applicant clarified that the enhanced metal fatigue AMP will use CC, CBF-C, CBF-PC, CBF-EP, and SBF monitoring bases. The staff verified that the applicant appropriately revised and updated the contents of LRA Table 4.3-4.

Time-Limited Aging Analyses

In amended LRA Table 4.3-4, the applicant credits the following enhanced metal fatigue AMP monitoring bases for ASME Code Class 1 components:

- SBF monitoring as the 10 CFR 54.21(c)(1)(iii) aging management monitoring basis for the pressurizer surge line elbow, which is the limiting environmentally-assisted fatigue (EAF) location (i.e., limiting NUREG/CR-6260 location)

- CBF-PC monitoring as the 10 CFR 54.21(c)(1)(iii) aging management monitoring basis for the pressurizer spray nozzles, which are the limiting non-environmental CUF components for the current fatigue AMP (limiting design basis CUF value of 0.9923)

The staff noted that under the amended basis in LRA Table 4.3-4, as given in Amendment 14, the applicant currently credits SBF monitoring only for 10 CFR 54.21(c)(1)(iii) management of the pressurizer surge line elbow, which according to the LRA is the limiting ASME Code Class 1 location for EAF. For the current fatigue AMP, the pressurizer spray nozzles are the limiting ASME Code Class 1 component (limiting design basis CUF of 0.9923). The updated table does not credit SBF for this limiting component. By letter dated July 21, 2010, the staff issued RAI 4.3-7 asking the applicant to justify its basis for not evaluating the pressurizer spray nozzles for EAF, considering that the pressurizer spray nozzles have a limiting design basis CUF of 0.9923. This issue was previously identified as part of Open Item 4.3-1.

In its response dated August 12, 2010, the applicant stated that the surge line elbow location is an adequate sentinel location for monitoring because the EAF usage factor is a product of the environmental factor (F_{en}) and design basis CUF. Transients with large, sudden temperature shocks that give rise to a high effective strain rate and a lower F_{en} dominate the pressurizer spray nozzle fatigue analysis. By contrast, the applicant stated that the surge line elbow experiences a mix of rapid (e.g., insurge, outsurge) and slow (e.g. heatup or cooldown stratification) transients, thus, it experiences a higher F_{en} compared to the pressurizer spray nozzle. The staff noted that this is reasonable based on the type of transients experienced by each component because components that experience transients with a lower strain rate will have a larger F_{en} value. Furthermore, the applicant also conservatively stated that the surge line elbow analysis includes effects from stratification mechanisms while it is known that the pressurizer spray nozzle does not experience stratification effects.

Based on its review, the staff finds the applicant's response to RAI 4.3-7 acceptable because the applicant clarified that the higher F_{en} value for the surge line elbow will result in a higher EAF usage factor compared to the pressurizer spray nozzle. Further, since the stratification effect of the surge line and the fatigue analysis are only associated with the surge line elbow, this results in the surge line elbow as the bounding component compared to the pressurizer spray nozzle. The staff's concern described in RAI 4.3-7 is resolved and this part of Open Item 4.3-1 is closed.

In Amendment 14, the applicant modified LRA Sections 4.3.1 and 4.3.1.2 to clarify that it will enhance the current fatigue AMP to include additional location-specific CUF calculations and an automated and computerized management software program for CC and CBF monitoring within two years of entering the period of extended operation. The applicant amended the LRA to clarify that the CC monitoring method will track and count transient event cycles to ensure that it will not exceed the number of transient events assumed by the design basis calculations, but this monitoring method will not perform periodic updates of the CUF calculations. The applicant clarified that the automated and computerized software program will automatically track and count the design basis transients for the applicant's facility and that this will supplement the applicant's manual counting of design basis transient occurrences.

The applicant amended the LRA to clarify the differences between the CBF-C, CBF-PC, and CBF-EP monitoring methods and to clarify that the monitoring methods will use both CC and CUF monitoring by periodically updating the CUFs for the appropriate components. The

Time-Limited Aging Analyses

applicant also amended the LRA to clarify that FatiguePro® will be used as the enhanced program's software basis for implementing the CC and CBF monitoring methods.

The staff reviewed the amended CC monitoring basis against SRP-LR Section 4.3.2.1.1.3 and to the "parameters monitored or detected," and "detection of aging effects" program elements in the GALL Report Metal Fatigue AMP. The staff noted that for the amended basis for CC monitoring, the applicant clarified that the monitoring method will only track and count cycles for the design basis transients that are applicable to this TLAA. The staff also noted that, in Amendment 14, the applicant clarified that the CC monitoring methodology will apply corrective actions if an action limit is reached. The staff noted that in LRA Section 4.3.1.5, the applicant states that these corrective actions will include an assessment of the need to perform an updated CUF calculation for a component if an action limit is reached.

Based on this review, the staff finds the applicant's CC monitoring basis to be acceptable because: (1) the amended basis conforms to the staff's recommendation in the "parameters monitored or inspected" program element in the GALL Report Metal Fatigue AMP for cycle tracking; (2) the amended basis will conform to the staff recommendation in the "detection of aging effects" program element in the GALL Report Metal Fatigue AMP for performing periodic CUF updates when a CC action limit is reached and an update of the CUF calculation is determined to be the appropriate corrective action for the applicable component; and (3) this is consistent with the recommendations in SRP-LR Section 4.3.2.1.1.3.

The staff reviewed the amended CBF-C, CBF-PC, and CBF-EP monitoring bases against SRP-LR Section 4.3.2.1.1.3 and the "parameters monitored or detected," and "detection of aging effects" program element recommendations in the GALL Report Metal Fatigue AMP. The staff noted that the FatiguePro® software basis for implementing the CBF-C, CBF-PC and CBF-EP methodologies include both CC applications and periodic CUF update bases. The staff determined that the CBF-C, CBF-PC, and CBF-EP monitoring methodologies are all acceptable ways of performing periodic CUF calculations because they involve an acceptable ASME Code Section III stress calculation methodology.

Based on this review, the staff finds the applicant's basis for applying CBF-C, CBF-PC, and CBF-EP monitoring methods to be acceptable for three reasons: (1) the amended basis conforms to the staff's recommendation in the "parameters monitored or inspected" program element in the GALL Report Metal Fatigue AMP for cycle tracking; (2) the amended basis conforms to the staff's recommendation in the "detection of aging effects" program element in the GALL Report Metal Fatigue AMP for periodic CUF updates; and (3) this conforms to the recommendations in SRP-LR Section 4.3.2.1.1.3 to accept TLAAs on metal fatigue and manage metal fatigue in accordance with 10 CFR 54.21(c)(1)(iii).

In Regulatory Information Summary (RIS) 2008-30, the staff raised technical concerns related to the conservatism of using one-dimensional stress models for the evaluation of EAF in limiting locations. For these locations, the staff recommended that the applicant use three-dimensional stress models conforming to ASME Code Section III requirements to confirm that one-dimensional stress models are conservative.

In Amendment 14, the applicant committed to the use of a software program for SBF monitoring of the pressurizer surge line (hot leg) elbow that incorporates a three-dimensional, six-element stress tensor method to meet the ASME Code Section III NB-3200 requirements. The applicant also committed to the implementation of this software at least two years before entering the period of extended operation. The applicant amended the "Methods" statement in LRA Section 4.3.1 to clarify how it would use SBF monitoring methods relative to ASME Code Section III requirements, based on the results of real stress histories for the components evaluated.

Time-Limited Aging Analyses

The staff reviewed the amended SBF monitoring basis against SRP-LR Section 4.3.2.1.1.3 and the "detection of aging effects" and "monitoring and trending" program elements in the GALL Report Metal Fatigue AMP. The staff also reviewed the amended basis against the staff's requirements for performing stress analyses and CUF calculations in the ASME Code Section III, as invoked by reference in 10 CFR 50.55a, "Codes and Standards," and the staff's recommendations for performing these type of analyses in RIS 2008–30.

In Amendment 14, the applicant modified Commitment No. 39. LRA Sections A2.1, A3.2, and B3.1 noted this change to reflect the use of a fatigue monitoring software program and methods for SBF monitoring that will carry out a three-dimensional, six-element tensor stress analysis method and conform to the requirements of ASME Code Section III Article NB-3200. Thus, the staff noted that this commitment has been placed on both the UFSAR supplement for this TLAA and the UFSAR supplement for the applicant's enhanced metal fatigue AMP for purposes of addressing the technical issues raised and discussed in RIS 2008–30.

The staff also noted that, in Amendment 14, the applicant no longer credits its SBF monitoring on a bounding basis. The staff finds that this change resolves the staff's concern discussed earlier in this section.

Based on this review, the staff finds the applicant's amended basis for using SBF monitoring to be acceptable based on the following criteria and conclusions:

- The applicant no longer credits SBF monitoring on a bounding basis and will apply SBF monitoring methods to each applicable component.

- The amended basis does not credit a version of FatiguePro® which uses a one-dimensional stress-intensity term in lieu of a six-element stress tensor as the software basis for SBF monitoring. Instead, it addresses the need to implement a SBF monitoring software program and methodology that comply with the requirements in the ASME Code Section III and that conform to the technical recommendations in RIS 2008–30.

- The staff has verified that Commitment No. 39 reflects this basis, as updated in LRA Amendment 14, and that the commitment is in the UFSAR supplements for both this TLAA and the enhanced metal fatigue AMP.

- When the enhanced program is implemented for SBF monitoring during the period of extended operation, the software program and methodology will be in compliance with the stress analysis criteria in the ASME Code Section III, Article NB-3200, the requirements in 10 CFR 50.55a, "Codes and Standards," and in conformance with the technical analysis recommendations in NRC RIS 2008–30.

- The amended basis conforms to the recommendations in SRP-LR for using SBF monitoring as a basis for accepting TLAAs in accordance with 10 CFR 54.21(c)(1)(iii) and with the recommendations in the GALL Report Metal Fatigue AMP for performing periodic updates of CUF calculations.

Based on its review, the staff finds the applicant has demonstrated that the enhanced Metal Fatigue Program is acceptable because the applicant has described how it will implement the monitoring methods (CC, CBF-C, CBF-PC, CBF-EP, and SBF) consistent with the recommendations of the GALL Report AMP for those components that require aging management for cumulative fatigue damage.

4.3.1.2.3 Conclusion

On the basis of its review, the staff concludes that the applicant has provided an acceptable demonstration, pursuant to 10 CFR 54.21(c)(1)(iii) that the effects of aging due to fatigue on the

Time-Limited Aging Analyses

intended functions of the components within the scope of the enhanced Metal Fatigue Program will be adequately managed for the period of extended operation.

4.3.1.3 Seismic History

4.3.1.3.1 Summary of Technical Information in the License Renewal Application

In LRA Section 4.3.1.3, the applicant gives a brief description of seismic design basis requirements and seismic transient history. The applicant clarifies that those design analyses that compared seismic loads to allowable component or structure stress allowable loads are not TLAAs. The applicant states, however, that the design of systems, structures, and components may include seismic loads in the fatigue analyses or may assume a stated number of seismic load cycles for the purpose of establishing an allowable stress or stress range (e.g., as would be used in implicit fatigue analyses of ANSI B31.1 components or Code Class 2 or 3 components designed to ASME Code Section III).

The applicant states that for design purposes, the safe-shutdown earthquake (SSE) is based on a 0.20 gravity ground-motion stress, and the OBE is based on a 0.10 gravity ground-motion stress. The applicant states that for the purposes of evaluating actual earthquake events, an SSE is defined as an earthquake that results in a categorization of eight on a Mercalli intensity scale seven (i.e., results in ground-motion ranging stresses ranging from 0.15 gravity to 0.33 gravity). An OBE is defined as an earthquake that results in a categorization of seven on a Mercalli intensity scale (i.e., results in ground-motion stresses ranging from 0.072 gravity to 0.15 gravity). The applicant summarizes that, as of 2008, only seven minor earthquakes have occurred and that the strongest of these earthquakes resulted in ground-motion stresses of only approximately 0.015 gravity. The applicant states that there have not been any recorded SSE or OBE events to date.

4.3.1.3.2 Staff Evaluation

The staff reviewed the SSE and OBE information in LRA Section 4.3.1.3 against the applicant's transient categories for these events in LRA Tables 4.3-2 and 4.3-3. The staff noted that in the original LRA, Tables 4.3-2 and 4.3-3 included two earthquake transient categories: (1) Transient 27, "Operating Basis Earthquake," which represents that transient category for OBE events, and (2) Transient 39, "Seismic Event Up To and Including One-half of the Safe-shutdown Earthquake, at 100% Power," which represents the transient category for non-SSE and non-OBE seismic events.

The staff noted that UFSAR Table 3.9.1-1 lists Transient 27, "Operating Basis Earthquake," as one of the upset condition transients that is applicable to the ASME Code Class 1 NSSS components (the reactor vessel components), and the design basis sets a limit of 200 occurrences for this transient. The staff also noted that UFSAR Table 3.9-1, item I.F.2.a lists Transient 39, "Seismic Event Up To and Including One-half of the Safe-shutdown Earthquake, at 100% Power," as an upset condition transient that is applicable to the Class 1 RCS piping components. The staff verified that LRA Tables 4.3-2 and 4.3-3 appropriately reflected the cycle occurrence design limit of 200 for Transient 27 and the cycle occurrence design limit of 2 for Transient 39. Thus, the staff finds that LRA Tables 4.3-2 and 4.3-3 reflected the appropriate design basis cycle occurrence limit information for Transients 27 and 39. The staff noted that in Amendment 14, the applicant administratively changed the transient number for the "Operating Basis Earthquake" transient from Transient 27 to Transient 32, and the "Seismic Event Up To and Including One-half of the Safe-shutdown Earthquake, at 100% Power," transient from Transient 39 to Transient 44. SER Section 4.3.1.4 supplies the staff's evaluation on the applicant's basis for projecting the number of cycles that will occur for these transients through the expiration of the period of extended operation (60-year cycle projections for these transients).

Time-Limited Aging Analyses

The staff also noted that the design basis in UFSAR Section 3.9.1.1 includes the faulted condition transient "Seismic Event Up To and Including One-half of the Safe-shutdown Earthquake, at 100% Power," which is listed as an UFSAR Table 3.9-1 faulted condition transient I.F.4.a for Class 1 RCS piping components and faulted condition transient II.E4.a for Class 1 portions of the chemical and volume control system. The staff determined that the applicant appropriately accounted for this transient in the amended LRA Tables 4.3-2 and 4.3-3. SER Section 4.3.1.4.2 provides the staff's evaluation as to whether the applicant should track and count this transient under the enhanced metal fatigue AMP.

4.3.1.3.3   Conclusion

Based on its review, the staff concludes that LRA Tables 4.3-2 and 4.3-3 reflect the appropriate design basis limit values for Transient 27, "Operating Basis Earthquake," and Transient 39, "Seismic Event Up To and Including One-half of the Safe-shutdown Earthquake, at 100% Power." SER Section 4.3.1.4 gives the staff's evaluation of the applicant's 60-year cycle projections for Transients 27 and 39 and basis for omitting the "Seismic Event Up To and Including One-half of the Safe-Shutdown Earthquake, at 100% Power," transient from the scope of LRA Tables 4.3-2 and 4.3-3.

4.3.1.4   Present and Projected Status of Monitored Locations

4.3.1.4.1   Summary of Technical Information in the License Renewal Application

The applicant provides its basis for establishing the current transient occurrence values (cycle values) and the transient values that are projected for the period of extended operation (60-year cycle values) in LRA Section 4.3.1.4. In this section, the applicant includes a summary of the methodology used to project the cycle occurrence values for the design basis transients to the expiration of the period of extended operation (i.e., 60-year cycle projection methods). The section also includes LRA Table 4.3-3, which gives the applicant's current cycle and 60-year cycle data based on the applicant's implementation of its 60-year cycle projection methods. LRA Table 4.3-2 summarizes the design basis limit criteria for the design basis transients that are involved with this TLAA derived from transient information in UFSAR Tables 3.9.1-1 or 3.9-1.

In LRA Amendment 14, the applicant revised its basis methodology for projecting the cycle occurrence values for the design basis transients to the expiration of the period of extended operation (i.e., 60-year cycle projection methodology for the design basis transients).

4.3.1.4.2   Staff Evaluation

The staff reviewed the 60-year cycle projection methodology in LRA Section 4.3.1.4 and the 60-year cycle projection data in LRA Table 4.3-3 against SRP-LR Section 4.3.2.1.1.1 to determine whether sufficient information has been provided to demonstrate that "the number of assumed transients would not be exceeded during the period of extended operation" and to ensure that existing CUF or implicit fatigue calculations remain valid for the period of extended operation in accordance with 10 CFR 54.21(c)(1)(i). The staff also reviewed the 60-year projection methodology, the 60-year projection basis, and data in LRA Table 4.3-3 against applicable requirements for design basis transient cycle tracking in TS 5.5.5 and applicable design basis transient information in UFSAR Section 3.9.1.1 and UFSAR Tables 3.9.1-1 and 3.9-1.

The staff noted the applicant's footnotes for LRA Tables 4.3-2 and 4.3-3 stated that only "normal operation condition," "upset condition," and "test condition" transients needed to be tracked under the current version and enhanced version of its fatigue AMP. However, the staff also noted that there were inconsistencies between the design basis transients that the applicant had included in LRA Tables 4.3-2 and 4.3-3 and the design basis transients that were listed in UFSAR Section 3.9.1.1 and Tables 3.9.1-1 and 3.9-1. The staff also noted that there were

Time-Limited Aging Analyses

issues with respect to the applicant's basis for projecting cycles to the expiration of the period of extended operation. The following bulleted list gives examples of the types of inconsistencies that the staff noted in the LRA prior to LRA Amendment 14:

- Upset condition and test condition transients in UFSAR Table 3.9-1 are missing from the scope of LRA Tables 4.3-2 and 4.3-3.

- Faulted and Emergency transient events listed in UFSAR Tables 3.9.1-1 and 3.9-1 are missing from the scope of LRA Tables 4.3-2 and 4.3-3.

- Inconsistencies exist between a given transient definition in LRA Tables 4.3-2 and 4.3-3 from the definition in UFSAR Tables 3.9.1-1 or Table 3.9-1 (i.e., the LRA originally lists Transient 50, "Depressurization by MSSV at 100% Power," as an upset condition and UFSAR Table 3.9-1 item I.C.3.a identifies this transient as an emergency condition transient)

- For a given transient, inconsistencies, such that the information for the given transient in LRA Section 4.3-2, are contradicted by information for the same transient in LRA Table 4.3-3.

- Proposals to track one design basis transient as a basis for tracking a different design basis transient when the UFSAR says that the transients are being applied to a different set of components.

By letter dated April 28, 2010, the applicant submitted LRA Amendment 14 to address these issues. In the amendment, the applicant indicated that it had performed a revised recount of the design basis transients that had occurred before January 1996 in order to reconstitute a best estimate of the transients that occurred from the time of initial operations through 1995. The applicant reviewed the following documents to perform the recount activities and its best estimate transient recount numbers: control room logs, NRC monthly operating reports, and Licensee Event Reports. The staff finds this to be an acceptable basis for performing the recount activities because these documents appropriately record applicable normal operating condition, upset condition, and test condition transients, from which the applicant may reconstitute transient occurrences.

The applicant said that they added the updated design basis transient recount numbers to those transients that were actually tracked and counted in accordance with TS 5.5.5 for the period from January 1996 through the end of December 2005. The staff noted that the totals were used to establish the applicant's best count estimate for the design basis transients from initial unit operation through end of 2005. The staff also noted that, based on these count totals, the applicant used the following linear extrapolation model to project the number of transient occurrences that would occur through to the end of the period of extended operation:

- For most transients, the applicant used Unit 3 time of operation through year 2005 (18 years) as the basis deriving the linear scaling factor for the analysis (i.e., 60/18 years = scaling factor of 3.33).

- For some transients, the applicant applied a scaling factor of 6.66 when the available transient data was only available for a 10-year period (1995–2005).

- To derive the 60-year projection totals for each transient, the applicant used the highest accumulation total from the three unit counting activities to derive the transient total for the TLAA. The applicant multiplied this value by the applicable scaling factor (3.33 if the count totals were based on counts performed over an 18–20 year period and 6.66 if the count totals were based on counts performed over a 10-year period).

Time-Limited Aging Analyses

The staff finds this revised projection basis to be acceptable because it relies on actual transient count data and uses the unit with the least amount of operating time to derive the linear scaling factor for the applicant's 60-year transient projections.

The staff noted that the amendment modified Tables 4.3-2 and 4.3-3 to update the list of transients provided in the original LRA, such that the list and descriptions of transients are consistent with those with the design basis transients. The staff noted that the applicant's amendment to the tables are consistent with UFSAR Section 3.9.1.1 and UFSAR Tables 3.9.1-1 and 3.9-1, which are referenced in TS 5.5.5. The staff noted that in the amended Table 4.3-3, the applicant increased the number of transients listed from 61 to 83 transients. The staff also noted that in amended Table 4.3-3, the applicant provided an update to the cumulative cycle counts for the transients from plant start up through the beginning of 2006. The applicant provided its revised 60-year transient projections, which were based on the applicant's updated recount activities and new 60-year linear scaling model (except for 6 transients in which the applicant still assumed 25-percent of the design basis limit).

In the update of Tables 4.3-2 and 4.3-3, the applicant no longer used bolded and non-bolded text as a basis for designating whether a given design basis transient would be counted or not. This resolved the staff's concern in the original LRA on the inconsistency in counting basis information for Transient 8, "RC Pump Starting"; Transient 9, "RC Pump Stopping"; Transient 10, "Cold Feedwater Following Hot Standby"; Transient 22, "Initiation of Shutdown Cooling"; and Transient 34, "Partial Loss of Condenser Cooling at 100% Power."

In the update of Table 4.3-3, the applicant no longer tied its tracking of upset condition Transient 32 (OBE Condition) as the basis for counting the number of occurrences for the Transient 44, "Seismic Event up to and Including One-Half of the Safe Shutdown Earthquake, at 100% Power." This revision made the design basis, recount, and 60-year projection bases for Transient 32 and Transient 44 consistent with those for the transients in UFSAR Table 3.9.1-1 and Table 3.9-1, respectively, and resolved the staff's concern.

Following modification of LRA Tables 4.3-2 and 4.3-3 in Amendment 14, the staff is unable to determine which of the transients the 25-percent assumed occurrences basis is applied. By letter dated June 2, 2010, the staff issued RAI 4.3-1 asking the applicant to clarify when the 25-percent assumed transient occurrence basis was used in LRA Tables 4.3-2 and 4.3-3 and to justify why this assumption yields a conservative 60-year cycle basis. This was previously identified as part of Open Item 4.3-1.

In its response dated June 29, 2010, the applicant stated that it elected to retain the 25 percent assumed transient accumulation for fourteen transients. The staff noted that the list of these transients is provided in Table RAI 4.3-1 of the applicant's response. The applicant stated that for Transients 13, 26, 27, 57, 59, 60, 80, 82, and 83, a review of logs, Licensee Event Reports, NRC Monthly Operating Reports and test records revealed that either the transients had not occurred between 1985–1995 or that the 25 percent assumption was not exceeded. The staff noted that since these transients were confirmed not to have occurred during 1985–1995 or did not exceed the 25 percent assumption, it is conservative for the applicant to assume that these transients occurred at 25 percent of the design limit. For these transients, the staff finds it acceptable that the applicant continued to use the 25 percent assumption because the applicant conservatively assumed the transient has occurred even though it was confirmed that the transients did not occur or did not exceed this assumption by a review of documentation from 1985-1995.

The applicant also stated in its June response that for Transients 8, 9, and 18, the counted accumulation of events between 1995–2005 were less than 5 percent of the limiting values. The staff noted that for these transients the review of plant records did not confirm an actual count because of inconsistencies or lack of specific details in plant records. Therefore, the

Time-Limited Aging Analyses

applicant compared the actual occurrences of these transients and compared it to the 25 percent assumption used from 1985–1995. The staff noted that this comparison indicated that the number of occurrence in the 25 percent assumption was greater than the number of actual occurrences from 1995–2005. For these transients, the staff finds it acceptable that the applicant continued to use the 25 percent assumption because the applicant conservatively assumed a larger of number of occurrences when compared to actual occurrences from 1995–2005 to account for variations in earlier years of operation.

The applicant also stated in its June response that for Transients 20 and 21, the tests occur at scheduled intervals with occasional tests being performed for post maintenance testing, so the rate of occurrence is constant, which lends itself to a reasonably accurate prediction of accumulation. The staff noted that the LPSI pump runs were assumed to occur at a rate of 15.6 occurrences per year for 1985–1995 compared to a rate of 12.7 occurrences per year between 1995–2005. For these transients, the staff finds it reasonable that the applicant continued to use the 25 percent assumption because of the routine and consistent occurrence of these transients. The applicant's assumption was conservative compared to the actual occurrences from 1995–2005.

The applicant stated the transient totals were projected to the end of the period of extended operation for information only. The applicant also stated there will be specific and targeted action limits to ensure actual fatigue limits are not exceeded. These corrective actions will be triggered by the limit established in the enhanced Metal Fatigue Program.

The staff finds the applicant's response to RAI 4.3-1 acceptable because the applicant justified its use of the 25 percent assumption for the transients as described above to obtain a conservative baseline for transient occurrences. Also, it is acceptable because the applicant will use the enhanced Metal Fatigue Program for continued tracking of these transients. This will ensure that when an action limit is reached, corrective actions are taken to maintain fatigue usage below the design limit of 1.0. The staff's concern described in RAI 4.3-1 is resolved and this part of Open Item 4.3-1 is closed.

In addition, the staff noted that the update of LRA Table 4.3-3 indicated that the applicant would not count Transient 3 (5 percent per minute ramp increase from 15 percent to 100 percent power) and Transient 4 (5 percent per minute ramp decrease from 15 percent to 100 percent power). The staff noted that these transients are in UFSAR Table 3.9-1 and UFSAR Section 3.9.1.1, which is referenced in TS 5.5.5. By letter dated June 2, 2010, the staff issued RAI 4.3-2 asking the applicant to clarify with justification whether these transients are required to be counted per TS 5.5.5 and UFSAR Section 3.9.1.1. If so, the staff asked that the applicant clarify the actions that it will take to resolve the apparent inconsistency if it determines there is a valid technical basis for not counting these transients. The staff also asked the applicant to clarify whether Transient 3 or Transient 4 has ever occurred and to justify its basis for not counting these transients. This was previously identified as part of Open Item 4.3-1.

In its response to RAI 4.3-2 requests 1 and 2, dated June 29, 2010, the applicant stated the program specified in TS.5.5.5 provides controls to track the UFSAR Section 3.9.1.1 cyclic and transient occurrences to ensure that components are maintained within the design limits. The applicant further stated that the controls to track cyclic and transient occurrences are implemented by either counting the occurrences or by accounting for the occurrences such that components are maintained within the design limits. The applicant also stated that a Licensing Document Change Request is being developed to add this clarification to UFSAR Section 3.9.1.1.

The applicant stated that the intent of Transients 3 and 4 was primarily to address the daily changes in grid demand that have been historically observed at other plants. The applicant stated that its design accommodates these types of cyclic load swings as well as the infrequent

Time-Limited Aging Analyses

variations in power required by equipment maintenance, Technical Specifications action statements, or other operational considerations.

The applicant stated that it has followed a base load strategy since initial operation in each of the three units and that using a 90-percent capacity factor and 60 years of operation, one can calculate that 15,000 power change cycles would require one cycle every 31.6 hours. The staff noted that unless a plant operates with a load following strategy, this number is not credible since power changes for maintenance, Technical Specifications action statements and operational considerations are infrequent. The applicant stated that its operating strategy does not include load following, therefore, these transients are accounted for such that components are maintained within the design limits.

The applicant stated that there is no design feature that would prevent it from making power changes to load-follow at the request of the load dispatcher. However, a review performed by the applicant of control room logs for the period of 1985–1995 to reconstruct transient history did not identify any load following power changes as defined in UFSAR Table 3.9.1-1. The staff noted that the applicant also reviewed dispatch procedures, the PVNGS owner-participant agreement, and a recent operating agent filing of annual resource planning. The applicant stated that this review determined that in the event of a grid condition requiring power reduction, the PVNGS units have priority to operate as base load power (not fluctuating), so that fossil-fuel power plants absorb changes in consumer demand. Further, power generation planning models used by the applicant indicate the intent to operate PVNGS as a base load plant.

The applicant stated that its reviews support the conclusion that it has not experienced Transients 3 or 4 due to load following and that the intention for the foreseeable future is to continue to operate its units as base loaded units. The applicant provided a table showing the number of power changes that occurred in each unit during the 24-month period of 2006–2007. The staff noted that the power changes experienced by each unit were the result of refueling outages, maintenance, post reactor trip startups or Technical Specifications action statements.

Based on its review, the staff finds the applicant's response to RAI 4.3-2, requests 1 and 2 acceptable because: (1) the applicant demonstrated that power changes are the result of refueling outages, maintenance, post reactor trip startups or TS action statements and not the result of load following, (2) the applicant's review confirmed that load following power changes have not occurred from 1985–1995, (3) the applicant's intent is to operate the plant as a base load plant, thus accounting for Transients 3 and 4, and (4) the applicant will update its UFSAR to clarify that it can track transient occurrences by counting the occurrences or by accounting for the occurrences. The staff's concern described in RAI 4.3-2, request 1 and 2 is resolved and this part of Open Item 4.3-1 is closed.

In its response to RAI 4.3-2, request 3, dated June 29, 2010, the applicant stated that the analyses for all three units include the same load following cycles (15,000 increase and 15,000 decrease) and the differences are not due to differences in geometry, materials, loading, or transients, but are due to modeling and analysis methods and assumptions. The applicant stated that one difference is that the Unit 1 analysis used a more conservative treatment of vortex shedding. The applicant further stated that some modeling differences resulted in a slightly different limiting location between the three units and *arithmetic* load addition was used instead of *vector* load addition at the limiting Unit 1 location. Furthermore, the vortex shedding difference produced a larger number of assumed vortex shedding load cycles for Unit 1 which was a significant factor in the difference.

The staff was unclear if vortex shedding was accounted for in the fatigue analysis for Units 2 and 3 and why the Unit 1 analysis treats vortex shedding so conservatively. It was also not clear to the staff why the stress ranges were slightly lower for the analyses for Units 2 and 3 as compared to Unit 1. The staff held a teleconference with the applicant on September 22, 2010,

Time-Limited Aging Analyses

for clarification. By letter dated October 13, 2010, the applicant clarified its response to RAI 4.3-2, request 3, by stating that vortex shedding at the instrument nozzle is present at all times while the unit is in operation or the reactor coolant pumps are running, therefore, it is applicable to all transients associated with the reactor vessel instrument nozzle analysis. The applicant stated that CE designed the instrument nozzle such that the natural frequency of the nozzle (approximately 347 cycles per second) was not close to the vortex shedding frequency (254 cycles per second) to avoid a resonance condition, and it accounted for the hydraulic loads imposed on the nozzle and J-weld attachment to the vessel wall.

The applicant stated that the reports for all three units considered vortex shedding in the analyses and the same analyst prepared all three of the reports: Unit 1 in 1978, Unit 2 in 1979, and Unit 3 in 1981. The applicant explained that the main differences between each of the analyses are as follows:

- The reports for Units 2 and 3 utilized a more thorough evaluation in that more cuts were used in the determination of stresses in the critical areas of the instrument nozzle.

- All three reports account for the operating loads, external loads as well as hydraulic loads from vortex shedding, but the reports for Units 2 and 3 implicitly demonstrate that the vortex shedding hydraulic loads and their corresponding alternating stresses are below the endurance limit. As such, it is not required to be superimposed as a separate transient with all the other design transients. Nevertheless, this external load was included with the other loads in the fatigue analysis and it was shown that the CUF was below 1.0.

This is consistent with the ASME NB-3200 fatigue analysis where vibration is not combined with other service loads in the fatigue evaluation. The Unit 1 report performed a more simplified conservative analysis and normalized all of the plant transients to 254 cps so that the vortex shedding load transient (with equivalent 109 cycles) could be superimposed as a separate transient and paired up with other plant design transients. In addition, the Unit 1 report utilized a commercial fatigue curve in lieu of the ASME Figure I-9.2 to calculate a usage factor beyond 10^6 cycles which resulted in the higher CUF factor.

The staff determined that the "thorough evaluation" performed by the applicant in the Units 2 and 3 reports is reasonable since the applicant used more cuts in the determination of stresses which provides a more refined model of stresses in the components. The staff also noted that the applicant demonstrated in the Unit 2 and 3 analyses that the vortex shedding hydraulic loads and their corresponding alternating stresses are below the endurance limit. The staff noted that if the alternating stresses are below the endurance limit the fatigue life can be considered infinite because of the extremely large number of cycles the material can endure (10^7 cycles). The staff noted that all three analyses considered the effects of vortex shedding. Furthermore, for the Unit 1 analyses the applicant used a simplified method by normalizing the vortex shedding transient so it was possible to pair the vortex shedding transient with other plant design transients. The staff finds this to be a conservative approach because the applicant paired the vortex shedding transient with other transients even though the vortex shedding hydraulic loads and its alternating stresses are below the endurance limit. The staff noted that this conservative treatment accounts for the larger CUF value.

Based on its review, the staff finds the applicant's response to RAI 4.3-2, request 3, as amended, acceptable because the applicant accounted for the operating loads, external loads and hydraulic loads from vortex shedding in the analyses for Units 1, 2 and 3, and because the applicant conservatively addressed vortex shedding in the Unit 1 analyses even though the alternating stresses were below the endurance limit. Finally, the response is acceptable

Time-Limited Aging Analyses

because the applicant clarified why the CUF for the RPV instrument nozzle was higher in the Unit 1 analysis when compared to the Unit 2 and 3 analyses. The staff's concern described in RAI 4.3-2, request 3 is resolved and this part of Open Item 4.3-1 is closed.

In the updated LRA Table 4.3-2, the applicant lists Transient 17, "Initiation of Auxiliary Spray," as an applicable normal operating condition transient. In the updated LRA Table 4.3-3, the applicant stated that it will correlate the tracking of Transient 17 to the tracking of pressurizer cooldown events, which is listed in these updated tables as Transient 12, "Pressurizer cooldown from 563°F to 70°F at a rate of ≤ 200°F/hr." It is not clear to the staff whether Transient 17 is referring to an initiation of the pressurizer spray system or an initiation of the containment spray system. It is also not clear to the staff why it is valid to correlate the tracking of Transient 17 to the tracking of Transient 12. By letter dated July 21, 2010, the staff issued RAI 4.3-8 asking the applicant clarify whether Transient 17 is referring to an initiation of the pressurizer spray system or an initiation of the containment spray system. The staff also asked the applicant to provide its basis for correlating Transient 17 to Transient 12. This was previously identified as part of Open Item 4.3-1.

In its response dated August 12, 2010, the applicant stated that Transient 17 in LRA Table 4.3-2 and 4.3-3, "Initiation of auxiliary spray during cooldown," refers to the initiation of auxiliary pressurizer spray during pressurizer cooldown. The applicant also stated that auxiliary pressurizer spray is used to complete pressurizer cooldown when the main pressurizer spray becomes unavailable during plant cooldown. During plant cooldown and depressurization, the reactor coolant pumps must be secured prior to full depressurization; this results in a loss of the reactor coolant pump differential pressure which drives the normal pressure spray. The applicant concluded that the initiation of auxiliary spray during cooldown (Transient 17) is related to the number of pressurizer cooldowns (Transient 12). The staff noted that based on the applicant's operating practices it is reasonable to correlate the tracking of Transient 17 to the tracking of Transient 12. The staff confirmed the applicant's response by verifying the information in the UFSAR Section 5.4.10 that discusses that the auxiliary spray line is provided to allow cooling if the reactor coolant pumps are secured.

Based on its review, the staff finds the applicant's response to RAI 4.3-8 acceptable because the applicant clarified the initiation of the transient and provided the relationship between the two transients. The applicant explained the sequence of events to initiate auxiliary spray during cooldown which provided the basis for the correlation between Transient 17 and Transient 12. The staff's concern described in RAI 4.3-8 is resolved and this part of Open Item 4.3-1 is closed.

The staff noted that the update of LRA Table 4.3-3 provided counting and 60-year projections for Transient 25, "Standby to SI hot leg injection check valve stroke test to standby (using the high-pressure safety injection (HPSI) pump)." The applicant stated that the test is conducted during refueling outages and that the transient is not currently counted because it was recently identified and added to UFSAR Table 3.9-1. The applicant also stated that the transient will be counted when it is added to the scope of the transient CC procedure. The staff noted that the applicant identified 16 occurrences of this transient for Units 1 and 3, and 17 occurrences for Unit 2, through December 31, 2005. The staff also noted this transient is projected to occur 57 times through the end of the period of extended operation. The staff determined there is an inconsistency in the recording of occurrences for this transient between January 1, 2006, and the time when the transient will be accounted in a future revision of the transient CC procedure. By letter dated July 21, 2010, the staff issued RAI 4.3-9, asking the applicant to clarify if the transient CC procedure has been updated to include Transient 25 and, if not, when the procedure will be updated to include this transient. The staff also asked that the applicant to explain how it will ensure that it accounts for all occurrences of Transient 25. This was previously identified as part of Open Item 4.3-1.

Time-Limited Aging Analyses

The applicant's August 12, 2010, response stated that in LRA Amendment 14 (April 28, 2010) the applicant committed (Commitment No. 55) to the following:

> The transient in UFSAR Table 3.9-1, items I.E.1.b and III.A.1.f, "Standby to SI hot leg injection check valve stroke test to standby (using the HPSI pump)" will be added to the CC surveillance procedure 73T-9RC02 by August 25, 2010.

The applicant also stated that LRA Table 4.3-3 was revised in Amendment 14 to include the total number of occurrences for the period 1985–2005 based on plant refueling history.

Based on its review, the staff finds the applicant's response to RAI 4.3-9 acceptable because the CC surveillance procedure was updated to include Transient 25 and the applicant's enhanced Metal Fatigue Program will monitor this transient during the period of extended operation to ensure that it does not exceed the design limit. The staff's concern described in RAI 4.3-9 is resolved. This part of Open Item 4.3-1 is closed.

In the update to Table 4.3-3, the applicant provided its counting and 60-year projections for Transient 79, "Reactor coolant system leak test." For this transient, the applicant stated that its recent recount found that the transient occurred five times for Unit 1, four times for Unit 2, and two times for Unit 3 through end of December 2005. It is not clear to the staff whether this transient represents the system leak test for the RCP boundary, mandated by ASME Code Section XI, Table IWB-2500-1, Examination Category B-P and 10 CFR 50.55a. The staff noted that this ASME Code requirement involves pressurizing the RCP boundary once every refueling outage to normal operating pressure and performing a visual examination of the system's components for evidence of reactor coolant leakage. The staff noted that the applicant has been operating for about 22–24 years of licensed operation. Thus, based on the time from initial operation, the staff estimates that the RCS leak test would have occurred approximately 14–16 times since initial operations of the units. By letter dated July 21, 2010, the staff issued RAI 4.3-10 asking the applicant clarify whether Transient 79 is different from the system leak test that is required by ASME Code Section XI. If the transient and the ASME Code Section XI system leak are different, the staff asked the applicant to clarify how it will track the ASME Code Section XI system leak. Furthermore, if these two are not different, the applicant must justify the number of occurrences stated in the LRA. This was previously identified as part of Open Item 4.3-1.

In its response dated August 12, 2010, the applicant indicated that the Transient 79, "Reactor Coolant System Leak Test," test condition is listed in UFSAR Table 3.9.1-1 and does represent the ASME Code Section XI system leak test. The applicant explained that the associated ASME Code fatigue analyses determined the cumulative fatigue resulting from the specified transients in UFSAR Table 3.9.1-1, including 500 heatup and cooldown cycles in which the pressure and temperature range from 15 psia and 80 degrees F to 2250 psia and 565 degrees F and back to 15 psia and 80 degrees F. Furthermore, the analyses also considers 200 additional transients in which the pressure cycles from 400 psia and 160 degrees F to 2250 psia and 400 degrees F and back to 400 psia and 160 degrees F. The staff noted that for the ASME Code fatigue analyses, the fatigue effects for these two transients were determined as separate events and is additive in the analyses.

In actual operating practice, however, the applicant stated that the ASME Code Section XI leak test is performed at normal operating pressure and temperature (2250 psia and 565 degrees F, respectively) in Mode 3 hot standby as part of the normal plant heat up. The RCS pressure and temperature are not typically reduced as part of or following the leak test. The staff finds it reasonable that the applicant determined that no actual fatigue effects occur as a result of the test. The fatigue effects are due only to the plant heatup because the RCS pressure and temperature are not reduced and cycled as a separate evolution from normal heatup during

Time-Limited Aging Analyses

actual operating practice. The staff also finds it reasonable that the applicant's fatigue monitoring program records the plant evolution only as a plant heatup.

The applicant stated that even if the RCS pressure and temperature were reduced due to the need for repairs, the evolution would be recorded as a heatup and cooldown cycle since the transient profile would be better represented by the heatup and cooldown profile. The staff finds this acceptable because the applicant has accounted for the plant heatup and cooldown evolutions and ASME Code Section XI leak tests and the associated fatigue effects. The staff noted that the Transient 79 events counted to date are RCS leak tests where the units were cycled from cold conditions to normal operating pressure and temperature (2250 psia and 565 degrees F) and back to cold conditions, as part of pre-operational tests.

Based on its review, the staff finds the applicant's response to RAI 4.3-10 acceptable because the applicant clarified that its operating practice is to perform the ASME Code Section XI leak test concurrently with a plant heatup without a separate thermal transient and, therefore, the fatigue effects are appropriately accounted for as a plant heatup transient and not as a separate leak test transient. Further, the applicant clarified that the ASME Code fatigue analyses account for the fatigue effects of plant heatup, cooldown, and ASME Code Section XI leak test as separate transients. The staff's concern described in RAI 4.3-10 is resolved and this part of Open Item 4.3-1 is closed.

Based on its review, the staff finds the applicant has demonstrated it will monitor transients with its enhanced Metal Fatigue Program and take corrective actions prior to the design limit exceeding 1.0 because the applicant has accounted for all actual transient occurrences or provided conservative assumptions, as described above, such that there is a baseline that can be monitored by the enhanced Metal Fatigue of Reactor Coolant Pressure Boundary Program to ensure design limits and design calculations remain valid. The staff also finds the applicant has demonstrated that existing CUF or implicit fatigue calculations remain valid during the period of extended operation acceptable because the applicant has shown that the number of assumed design transients will not be exceeded during the period of extended operation.

4.3.1.4.3   Conclusion

On the basis of its review, the staff concludes that the applicant has an appropriate baseline for all transients and that these transients will be monitored by the enhanced Metal Fatigue Program such that the effects of aging due to fatigue on the intended functions will be adequately managed for the period of extended operation.

4.3.1.5   Enhanced Metal Fatigue Reactor Coolant Pressure Boundary Program Scope, Action Limits, and Corrective Actions

4.3.1.5.1   Summary of Technical Information in the License Renewal Application

In LRA Amendments 14 (dated April 28, 2010) and 16 (dated May 27, 2010), the applicant revised the bases and discussion provided in LRA Section 4.3.1.5. This section provides the scope and basis for defining the action limits for the applicant's enhanced metal fatigue AMP and the corrective action options the applicant will carry out if it reaches an action limit on CC or CUF monitoring.

4.3.1.5.2   Staff Evaluation

Scope and Method. The staff noted that in the update of LRA Section 4.3.1.5, "Scope" and "Method" subsections, the applicant stated that the scope of the enhanced metal fatigue AMP will include all ASME Code Section III Class 1 components and components with a Class 1 fatigue analysis. The staff confirmed that the applicant is applying the enhanced metal fatigue AMP as the basis for dispositioning CUF-based TLAAs, as noted in applicable subsections in LRA Section 4.3.2, 4.3.3, and 4.3.4.

Time-Limited Aging Analyses

The applicant noted that the enhanced metal fatigue AMP uses five monitoring methods, consistent with the updated information in LRA Table 4.3-4: (1) CC monitoring, (2) CBF-C monitoring, (3) CBF-PC monitoring, (4) CBF-EP monitoring, or (5) SBF monitoring. The staff evaluated the appropriateness of these methods in SER Section 4.3.1.2 and found them acceptable.

The applicant stated that transient event cycles required to be monitored by TS 5.5.5 will continue to be tracked to ensure that the numbers of transient events assumed by the design basis calculations will not be exceeded. The staff noted that, in the update of LRA Table 4.3-3, the applicant said it would need to track many of the design basis transients. The staff has evaluated these transients in SER Section 4.3.1.4.2 and found them acceptable.

Action Limits. The staff noted that, in the update of LRA Section 4.3.1.5, the applicant stated that the current metal fatigue AMP is based on CC monitoring with the exception of the pressurizer spray nozzle. This is the limiting Class 1 component for CUF, which is currently monitored using the applicant's CBF-PC method of CUF monitoring. The staff noted that the applicant clarified that the current program sets action limits on CC monitoring at 90 percent of the design basis limit values for the transients and action limits on CUF monitoring of the pressurizer spray nozzle at 0.65.

The staff noted that in the enhanced metal fatigue AMP (amended), corrective action limits will ensure that corrective actions are taken before the design limits are exceeded. These limits ensure that the applicant initiates re-evaluation or other appropriate corrective actions while sufficient margin remains to allow at least one occurrence of the worst case (highest fatigue usage per cycle), low probability transient that is included in the design specifications.

The staff noted that the applicant clarified for NUREG/CR-6260 locations, that the CUF calculations would include application of the appropriate F_{en} environmental factor. This is consistent with the GALL Report Metal Fatigue AMP, which recommends the "acceptance criteria" program element "... involves maintaining the fatigue usage below the design code limit considering environmental fatigue effects as described under the program description."

The staff noted that the applicant's action limit basis in the Action Limit Margins Section is consistent with the staff's recommendation in the GALL Report. Based on this review, the staff finds the basis for the program's action limits to be acceptable because the applicant reflected this basis in the update of LRA enhanced metal fatigue AMP, and it is consistent with the recommendation in the "acceptance criteria" program element in the GALL Report Metal Fatigue AMP.

The applicant stated that, for action limits on CC, it will establish the limits based on the design limit on a specified number of accumulated cycles. The staff noted that an applicant's decision on the degree of conservatism that should be applied to action limits on CC is not mandated by any NRC requirements. As a minimum, the applicant would be required to take appropriate corrective action if the design limit for a design basis transient was reached. Based on Commitment No. 39, the applicant will establish the action limits on CC for the enhanced metal fatigue AMP that will be implemented during the period of extended operation, and this action limit will include an appropriate margin on the design limit. The staff finds this acceptable.

Corrective Actions. The staff noted that the update to LRA Section 4.3.1.5, "Cycle Count Action Limits and Corrective Actions" states:

> Since sufficient margin must be maintained to accommodate any design transient regardless of probability, the enhanced Metal Fatigue of Reactor Coolant Pressure Boundary program (B3.1) corrective actions will be taken before the remaining number of allowable occurrences for any specified transient becomes less than one. Corrective actions will be required when the cycle count for any of

Time-Limited Aging Analyses

the significant contributors to usage factor is projected to reach the action limit defined the enhanced Metal Fatigue of Reactor Coolant Pressure Boundary program (B3.1) before the end of the next fuel cycle.

The staff noted that the applicant will require CC corrective actions only for those design basis transients that the applicant considers significant contributors to fatigue usage. By letter dated July 21, 2010, the staff issued RAI 4.3-11, asking the applicant to clarify the definition of the term "significant contributors to usage factor" and explain how this is associated with the corrective action limits in the enhanced Metal Fatigue Program. This was previously identified as part of Open Item 4.3-1.

In its response dated August 12, 2010, the applicant stated that all of the transients listed in UFSAR Tables 3.9.1-1 and 3.9-1, as shown in LRA Table 4.3-2 are significant contributors to fatigue usage factor. The applicant also stated that each transient in LRA Table 4.3-2 will have appropriate corrective action limits associated with it and these limits will reflect the UFSAR transient limits and assumptions made in the analyses of record.

Based on its review, the staff finds the applicant's response to RAI 4.3-11 acceptable because the applicant clarified that the "significant contributors" to fatigue include all transients listed in the UFSAR tables, and because the CC corrective action limits associated with all transients listed in LRA Table 4.3-2 will be tracked by the enhanced Metal Fatigue Program. This will ensure that the assumptions made in the analyses of record and design limits are not exceeded. The staff's concern described in RAI 4.3-11 is resolved and this part of Open Item 4.3-1 is closed.

The applicant stated that the enhanced metal fatigue AMP will use an automated three-dimensional, six-element stress tensor SBF monitoring management software module for the monitored locations and that "cycle-based CUFs will be calculated periodically." In the update of LRA Section 4.3.1, the applicant said that, of the four monitoring methods that involve cycle counting (CC monitoring, CBF-C monitoring, CBF-PC monitoring, or CBF-EP monitoring), only the three CBF methods would involve both CC and periodic updates of the CUF calculations. The staff noted that in the update of LRA Section 4.3.1.2, the applicant said that the CC monitoring method would not perform periodic updates of a component's CUF values, but if the action limit is reached, then corrective actions are necessary.

The staff noted that the applicant stated that CUF corrective actions will be required when the calculated CUF (from cycle-based or SBF monitoring) is projected to reach a value of 1.0 within the next two or three fuel cycles. The staff also noted that the applicant's basis also factored F_{en} environmental adjustments into the selection of its CUF based action limits. The staff finds this to be a reasonable basis for applying CUF based corrective actions when compared to the "acceptance criteria" program element in the GALL Report Metal Fatigue AMP, which recommends that action limits on CUFs be taken before the design limit is reached and that the acceptance criteria should account for environmental effects.

However, the staff also noted that the applicant said that corrective actions must also be taken while there is still sufficient margin to accommodate at least one occurrence of the worst case (highest fatigue usage per cycle) design transient event, in order to accommodate occurrence of a low probability transient. The staff noted that this basis statement appeared to be more relevant to CC activities and not to CUF monitoring activities. The staff has evaluated the action limits for CC monitoring earlier in this evaluation.

The applicant's updated basis included the following seven possible corrective actions:

(1) Determine whether the scope of the enhance fatigue management program must be enlarged to include additional affected reactor coolant pressure boundary locations

(2) Adjust fatigue monitoring methods to confirm continued conformance to the Code design limit

(3) Repair or modify the component

(4) Replace the component

(5) Perform a more rigorous analysis of the component to demonstrate that the design code limit will not be exceeded

(6) Modify plant operating practices to reduce the fatigue usage accumulation rate

(7) Perform a flaw tolerance evaluation, impose component-specific inspections under ASME Code Section XI Appendices A or C (or their successors), and obtain required approval of the NRC

The staff noted that all of these corrective actions on CUF monitoring are acceptable for addressing a component whose CUF value is approaching a value of 1.0. Specifically, the staff noted that corrective actions 1, 2, 5, and 6 would all ensure that the CUF value for a component would remain within the ASME Code allowable limit of 1.0. The staff also noted that corrective action options 3, 4, and 7 would address those components for which the applicant could not ensure that the CUF value would remain below a value of 1.0.

With regard to corrective action 1, the applicant says it is only applicable to RCP boundary components. However, in its review of LRA Section 4.3.2, the staff confirmed that the TLAA includes the CUF results for some ASME Code Class 2 components that were analyzed to ASME Code Section III CUF requirements for Code Class 1 components. As a result, the staff noted that the action in corrective action 1 might also be applicable to those ASME Code Class 2 components that were analyzed to ASME Code Section III CUF requirements for Code Class 1 components.

By letter dated July 21, 2010, the staff issued RAI 4.3-12 requesting the applicant clarify if the scope of corrective action 1 on CUF monitoring includes all components with ASME Code Section III CUF calculations for Code Class 1 components and ASME Code Class 2 components that were analyzed to ASME Code Section III CUF requirements for Code Class 1 components. If the scope of correction action 1 does not include both sets of components, the staff asked that the applicant justify why they are not within scope. This was previously identified as part of Open Item 4.3-1.

In its response dated August 12, 2010, the applicant stated that the scope of the enhanced Metal Fatigue Program includes the ASME Code Section III Class 1 components and components with Class 1 fatigue analysis, which includes Class 2 components that were analyzed to ASME Code Section III CUF requirements for Class 1 components.

Based on its review, the staff finds the applicant's response to RAI 4.3-12 acceptable because the applicant clarified that the scope of the enhanced Metal Fatigue Program includes all components, including Class 2 and 3 components, with a CUF analysis. Further, the response is acceptable because the applicant's program will ensure that the design limit of 1.0 is not exceeded or corrective actions will be taken to reanalyze, repair, or replace the component before the design limit is exceeded. The staff finds that this approach provides effective corrective actions for these components. The staff's concern described in RAI 4.3-12 is resolved. This part of Open Item 4.3-1 is closed.

The staff noted that the GALL Report states that a program consistent with GALL AMP X.M1 is an acceptable option for managing metal fatigue for the reactor coolant pressure boundary, considering environmental effects and no further evaluation is recommended for license renewal if the applicant selects this option pursuant to 10 CFR 54.21(c)(1)(iii) to evaluate metal fatigue for the reactor coolant pressure boundary.

Based on its review, the staff finds the applicant has demonstrated it will monitor transients with its enhanced Metal Fatigue Program and take corrective actions prior to the design limit exceeding 1.0 because the applicant provided the details of the scope, action limits and corrective actions associated with its enhanced Metal Fatigue Program and the staff has found these consistent with the recommendations in the GALL Report Metal Fatigue AMP.

4.3.1.5.3   Conclusion

On the basis of its review, the staff concludes that the applicant has demonstrated that the effects of aging on the intended functions will be adequately managed for the period of extended operation.

4.3.2   American Society of Mechanical Engineers III Fatigue Analysis of Class 1 Vessels, Piping, and Components

LRA Section 4.3.2 summarizes the evaluation of the CUF analyses that comprise the "ASME III Fatigue Analysis of Class 1 Vessels, Piping and Components," for the period of extended operation. These TLAAs are based on the analyses in the applicant's current design basis CUF calculations for these components. In these TLAAs, the applicant provides its bases for dispositioning the CUF analyses for its ASME Code Class 1 reactor vessel, pressurizer, piping, and SG components and for the Class 2 SG components that were analyzed to ASME Code Section III CUF criteria, in accordance with the TLAA acceptance criteria in 10 CFR 54.21(c)(1)(i), (ii), or (iii).

4.3.2.1   Reactor Pressure Vessel, Nozzles, Head, and Studs

4.3.2.1.1   Summary of Technical Information in the Application

LRA Section 4.3.2.1 and Amendment 16 dated May 27, 2010, summarize the evaluation of the fatigue analyses for the reactor pressure vessels which were designed, built, and analyzed by Combustion Engineering to the standards of ASME Code Section III, Subsection NB (Class 1), 1971 Edition with addenda through winter 1973. The applicant stated that pressure-retaining and support components of the reactor pressure vessels are subject to an ASME Boiler and Pressure Vessel Code, Division 1, Section III, fatigue analysis. Furthermore, these analyses have been updated to incorporate redefinitions of loads and design basis events, operating changes, and a power uprate with SG replacement. The applicant further stated that the currently applicable fatigue analyses of these components are TLAAs.

The amended LRA Section 4.3.2.1 dispositions all of the analyses in this section consistent with 10 CFR 54.21(c)(1)(ii), that the analyses have been projected to the end of the period of extended operation, or 10 CFR 54.21(c)(1)(iii), that the enhanced Metal Fatigue Program will track the transients identified in these analyses to ensure that appropriate reevaluation or other corrective action will be initiated if an action limit is reached.

4.3.2.1.2   Staff Evaluation

The staff reviewed this section using the guidance of SRP-LR Sections 4.3.2.1.1.2 and 4.3.2.1.1.3 to verify that the analyses in this section demonstrate, consistent with 10 CFR 54.21(c)(1)(ii), that the analyses have been projected to the end of the period of extended operation, or consistent with 10 CFR 54.21(c)(1)(iii), that the effects of aging on the intended functions of these components will be adequately managed during the period of extended operation.

LRA Section 4.3.2.1 provides a CUF value of 0.823 for the RPV studs and a CUF value of 0.954 for the RPV bottom head support or shear lugs. LRA Section 4.3.2.1 also states that the RPV studs are the more limiting component because they will experience more severe stresses during each transient event, even though they are limited to a lower design limit on the number of allowable heatup and cooldown events. The staff noted that in the updated LRA Table 4.3-4, the applicant stated that both of these component locations will be monitored using CC monitoring methods. The applicant's CC monitoring methods do not include automatic periodic updates of CUF calculations. Therefore, it is not clear to the staff if CC will be applied only to the RPV studs, even though the RPV lugs have an existing CUF of 0.954.

By letter dated July 21, 2010, the staff issued RAI 4.3-14 requesting the applicant clarify whether it is using CC monitoring methods on the RPV studs. The staff requested the following additional actions if the cycle-based monitoring will be performed only on the RPV studs: (1) summarize transients that were used for the CUF calculations for the RPV studs and RPV bottom head lugs, and (2) clarify the quantitative contribution to fatigue usage for each transient analyzed. This was previously identified as part of Open Item 4.3-1.

In its response dated August 12, 2010, the applicant stated that both the RPV studs and RPV external bottom head shear lugs will be monitored by CC and appropriate corrective action limits will be applied to both.

Based on its review, the staff finds the applicant's response to RAI 4.3-14 acceptable because the applicant clarified that each component, studs and RPV external bottom head shear lugs, will be monitored individually by CC, and action limits for the enhanced Metal Fatigue Program will be established to allow for corrective actions before the design basis number of events is exceeded. The staff's concern described in RAI 4.3-14 is resolved and this part of Open Item 4.3-1 is closed.

Based on its review, the staff finds the applicant has demonstrated pursuant to 10 CFR 54.21(c)(1)(ii), that the analyses for the reactor pressure vessel, nozzles, and head components are acceptable because the applicant's revised fatigue analyses for the reactor pressure vessel, nozzles, and head components demonstrate that the design limit of 1.0 is met for the period of extended operation. The staff also finds that the applicant has demonstrated pursuant to 10 CFR 54.21(c)(1)(iii), that the effects of aging on the intended functions for the reactor vessel studs will be managed during the period of extended operation because the reactor vessel studs will be managed by the applicant's enhanced Metal Fatigue Program which ensures that the number of cycles assumed in the design calculation are not exceeded and the CUF will not exceed the code limit of 1.0.

4.3.2.1.3 Updated Final Safety Analysis Report Supplement

LRA Section A3.2.1.1 provides the UFSAR supplement summarizing the evaluation of the fatigue analyses for the reactor pressure vessel, nozzles, head, and studs. Based on its review of the UFSAR supplement, the staff concludes that the applicant provided an adequate summary description of its actions to address the TLAA for the reactor pressure vessel, nozzles, head, and studs is adequate, as required by 10 CFR 54.21(d).

4.3.2.1.4 Conclusion

On the basis of its review, the staff concludes that the applicant has demonstrated, pursuant to 10 CFR 54.21(c)(1)(ii), that the fatigue analyses for the reactor pressure vessel, nozzle and head components have been projected to the end of the period of extended operation. The staff also concludes that the applicant has demonstrated, pursuant to 10 CFR 54.21(c)(1)(iii) that the effects of aging on the intended functions of the reactor vessel studs will be adequately managed for the period of extended operation. The staff also concludes that the UFSAR supplement contains an appropriate summary description of the TLAA evaluation, as required by 10 CFR 54.21(d).

Time-Limited Aging Analyses

4.3.2.2 Control Element Drive Mechanism Nozzle Pressure Housings

4.3.2.2.1 Summary of Technical Information in the Application

LRA Section 4.3.2.2 and Amendment 16 dated May 27, 2010, summarize the evaluation of the fatigue analyses of the CEDM and reactor vessel level monitoring system pressure housings. The applicant stated that the CEDM and reactor vessel level monitoring system pressure housings will be replaced and the replacements are designed to ASME Code Section III Subsection NB. The applicant also stated that the fatigue analyses of the replacement components are sufficient for a 40-year design life, based on the design basis transient events specified for the original reactor vessel heads. The design life of these replacement heads, nozzles, and CEDM and reactor vessel level monitoring system pressure housings therefore extends beyond the end of the period of extended operation. The applicant provided a disposition of this TLAA in accordance with 10 CFR 54.21(c)(1)(ii), that the analyses have been projected to the end of the period of extended operation.

4.3.2.2.2 Staff Evaluation

The staff reviewed LRA Section 4.3.2.2 and Amendment 16 to verify pursuant to 10 CFR 54.21(c)(1)(ii), that the analyses have been projected to the end of the period of extended operation.

The applicant stated that it is replacing reactor vessel heads for all three units, including their nozzles, CEDM pressure housings, RVLMS pressure housing, and caps for the spare nozzles. As of December 1, 2010, the applicant stated that Units 1, 2, and 3 reactor vessel heads have been replaced.

The applicant stated that the CEDM pressure housings replacements are designed to ASME Code Section III Subsection NB, 1998 Edition with addenda up to and including the 2000 Addenda, for a 40-year operating period. The applicant also stated that the design report for the replacements calculated fatigue usage factors at the two limiting locations: 0.4210 in the motor housing and 0.2240 in the lower end of the upper pressure housing.

The applicant stated that the reactor vessel level monitoring system pressure housings replacements are designed to ASME Code Section III Subsection NB, 1998 Edition with addenda up to and including the 2000 Addenda, as ASME III NCA-1260 Code Class 1 appurtenances; for a 40-year operating period. The applicant also reported a usage factor of 0.654 in the upper flange.

Effects of Power Uprate and SG Replacement on the CEDM and Reactor Vessel Level Monitoring System Pressure Housing Analyses. LRA Section 4.3.2.2 and Amendment 16 state that the revised OBE and faulted loads on the original CEDM nozzle are less than the maximum allowed in the analyses of record. Thus, there is no change to the original design report fatigue usage. Furthermore, the applicant explained that the analyses of the replacement CEDM and reactor vessel level monitoring system pressure housings were based on the original set of design basis transient events, and were also not affected by the power uprate and SG replacement modifications.

Effects of Combustion Engineering Infobulletin 88-09. The applicant stated that the CE Owner's Group review did not identify any effects on the original fatigue analysis of the CEDM. The applicant further explained that the conclusions of the analysis of the replacement CEDM and RVLMS pressure housings, based on the original set of design basis transient events, were also not affected by the Infobulletin 88-09 evaluation.

Based on its review, the staff finds the applicant has demonstrated, pursuant to 10 CFR 54.21(c)(1)(ii), that the analyses for the CEDM and reactor vessel level monitoring system pressure housings have been projected to the end of the period of extended operation

Time-Limited Aging Analyses

because the fatigue analyses of the replacement components are sufficient beyond the period of extended operation. Thus, fatigue usage factors in the housings are below the design limit of 1.0 for the period of extended operation.

4.3.2.2.3 Updated Final Safety Analysis Report Supplement

LRA Section A3.2.1.2 provides the UFSAR supplement summarizing the evaluation of the fatigue analyses for the CEDM and reactor vessel level monitoring system pressure housings. Based on its review of the UFSAR supplement, the staff concludes that the applicant provided an adequate summary description of its actions to address the TLAA for the CEDM and reactor vessel level monitoring system pressure housings as required by 10 CFR 54.21(d).

4.3.2.2.4 Conclusion

On the basis of its review, the staff concludes that the applicant has demonstrated, pursuant to 10 CFR 54.21(c)(1)(ii), that the fatigue analyses for the CEDM and RVLMS pressure housings have been projected to the end of the period of extended operation. The staff also concludes that the UFSAR supplement contains an appropriate summary description of the TLAA evaluation, as required by 10 CFR 54.21(d).

4.3.2.3 Reactor Coolant Pump Pressure Boundary Components

4.3.2.3.1 Summary of Technical Information in the Application

LRA Section 4.3.2.3 and Amendment 16 dated May 27, 2010, summarize the evaluation of the fatigue analyses of the reactor coolant pump pressure boundary components. The applicant stated that these components are designed to ASME Code Section III, Subsection NB. The applicant also stated that the analysis was reexamined for the power uprate and steam generator replacement modifications. The applicant provided a disposition of this TLAA in accordance with 10 CFR 54.21(c)(1)(iii), that the effects of aging will be adequately managed for the period of extended operation.

4.3.2.3.2 Staff Evaluation

The staff reviewed LRA Section 4.3.2.3 and Amendment 16 to verify pursuant to 10 CFR 54.21(c)(1)(iii), that the effects of aging on the intended functions will be adequately managed for the period of extended operation.

The applicant stated that the reactor coolant pump pressure boundary components are designed to ASME Code Section III, Subsection NB, 1974 Edition (no addenda). The applicant also stated that a fatigue analysis, in accordance with Subparagraph NB-3222.4(e), was performed only for pump casing components. The high pressure cooling system and seal housing adapters invoked the fatigue analysis waiver of NB-3222.4(d), or were designed to requirements other than those of Section III Class 1.

The applicant stated that the maximum total CUF for all components is 0.988 for the pump casing closure studs. The analysis of the pump casing closure studs initially resulted in usage factors greater than 1.0. To reduce the usage factor below 1.0, the applicant explained that the number of heatup and cooldown cycles, which are the most significant contributors to usage factor in all pump components, was reduced to 475 events. The applicant stated that the reduced number of heatup and cooldown cycles is incorporated into the fatigue-monitoring program and is under review for addition to UFSAR Section 3.9.1.1.

<u>Effects of Power Uprate and Steam Generator Replacement on the CEDM and RVLMS Pressure Housing Analyses</u>. LRA Section 4.3.2.3 and Amendment 16 state that the original fatigue analyses of record are still valid and the effects of the steam generator replacements and power uprate loads on the analysis of record have been reconciled.

Time-Limited Aging Analyses

Effects of Combustion Engineering Infobulletin 88-09. The applicant stated that the CE Owner's Group review did not identify any effects on the fatigue analysis of the reactor coolant pumps.

Based on its review, the staff finds the applicant has demonstrated, pursuant to 10 CFR 54.21(c)(1)(iii), that the effects of aging on the intended functions of the reactor coolant pump pressure boundary components will be adequately managed for the period of extended operation because action limits of the enhanced Metal Fatigue Program will be established to permit completion of corrective actions before the design basis number of events specified in the Design Specification, UFSAR Table 3.9-1, and the RCP closure studs' more restrictive number of heatup and cooldown events are exceeded. Thus, the fatigue usage factor will not exceed the design limit of 1.0 during the period of extended operation.

4.3.2.3.3 Updated Final Safety Analysis Report Supplement

LRA Section A3.2.1.3 provides the UFSAR supplement summarizing the evaluation of the fatigue analyses for the reactor coolant pump pressure boundary components. Based on its review of the UFSAR supplement, the staff concludes that the applicant provided an adequate summary description of its actions to address the TLAA for the reactor coolant pump pressure boundary components as required by 10 CFR 54.21(d).

4.3.2.3.4 Conclusion

On the basis of its review, the staff concludes that the applicant has demonstrated, pursuant to 10 CFR 54.21(c)(1)(iii), that the effects of aging on the intended functions of the reactor coolant pump pressure boundary components will be adequately managed for the period of extended operation. The staff also concludes that the UFSAR supplement contains an appropriate summary description of the TLAA evaluation, as required by 10 CFR 54.21(d).

4.3.2.4 Pressurizer and Pressurizer Nozzles

4.3.2.4.1 Summary of Technical Information in the Application

The pressurizers are designed to the ASME Code Section III, Subsection NB (Class 1), 1971 Edition with addenda through winter 1973. The pressurizers are welded, vertical cylindrical pressure vessels with hemispherical heads, fabricated of carbon steel with the interior surface clad with stainless steel. A cylindrical support skirt and flange is attached to the lower head shell with a forged knuckle support ring.

The central vertical surge nozzle, two vertical lower level instrument nozzles, and 36 heater sleeves penetrate the lower head. Four shear lugs, welded to the upper shell, stabilize the vessel against seismic and other overturning loads. The central vertical spray nozzle, manway, four horizontal upper instrument nozzles, and four horizontal safety valve nozzles penetrate the upper head.

The surge, spray, and safety valve nozzles attached to the pressurizer contain safe ends for welding to the attached stainless steel piping. Recently, the applicant installed weld overlays on these nozzles, safe ends, and safe end welds to mitigate potential PWSCC. All of the Alloy 600 instrument nozzles have been replaced with Alloy 690 materials, which are less susceptible to PWSCC.

The heater sleeves and heaters have all been replaced with Alloy 690 heater sleeves, which are attached to the lower vessel head by half-nozzle repairs and welded to external reinforcing pads. The heater sheaths are attached to the outer ends of the Alloy 690 heater sleeves by fillet seal welds. The sheaths of the electric heater, the welds between the end plug and sheath, and the fillet seal welds to the heater sleeves, are Class 1 pressure boundary welds. The Unit 1 heater sleeve (B18) and Unit 2 heater sleeves (A06 and B18) have been closed with welded Type 316 stainless steel plugs.

Time-Limited Aging Analyses

The pressurizers have operated since startup with a continuous spray flow to prevent boron concentration stratification and to mitigate spray line and nozzle fatigue. This continuous flow is achieved via regulating bypass valves around each of the two main spray valves.

Table 4.3-7 from LRA Amendment 16 (May 27, 2010) provides a disposition for each pressurizer component analysis. With the design basis set of transients, including power uprate, SG replacement, and other effects described above, the worst-case calculated 40-year fatigue usage factors exceed 0.9 in a few pressurizer components. Other fracture mechanics or fatigue analyses depend on the limiting number of occurrences assumed for a 40-year design life.

Some of the revised time-dependent component evaluations were based on a 60-year extended licensed operating period and, if valid for the period of extended operation, are not TLAAs. Others evaluations were for shorter than 40 years and did not extend to the end of the current licensed operating period, and therefore were not TLAAs. The fatigue analyses for materials adjacent to the surge and spray nozzle overlay repairs extend to the period of extended operation, but were able to meet the 1.0 usage factor acceptance criterion only for a 40-year life and are therefore TLAAs.

For those analyses that are TLAAs, the applicant dispositioned them in accordance with 10 CFR 54.21(c)(1)(iii), that the effects of aging will be adequately managed for the period of extended operation using the enhanced metal fatigue AMP.

4.3.2.4.2 Staff Evaluation

The staff reviewed LRA Section 4.3.2.4 to verify, pursuant to 10 CFR 54.21(c)(1)(iii), that the effects of aging on the intended function(s) of the pressurizer and pressurizer nozzles will be adequately managed for the period of extended operation.

TLAAs of the pressurizer and pressurizer nozzle are focused on the crack growth calculations and CUF calculations that were done as part of the CLB in light of additional 20 years of service beyond the current operating license. The affect of the period of the extended operation on the crack growth calculations and CUF calculations is related to the transient cycles used.

The applicant stated that pressure retaining and support components of the pressurizer are subject to an ASME, Section III, fatigue analysis. These analyses have been updated to incorporate updated definitions of loads, design basis events, operating changes, power uprate, and modifications to include the following:

- Effects of indications in a Unit 2 pressurizer support skirt forging weld
- Effects on the pressurizer of NRC Bulletin 88-11 thermal stratification in the surge line not included in the original analyses
- Effects on the pressurizer of insurge-outsurge transients not included in the original analyses
- Effects on the pressurizer of CE Infobulletin 88-09 "Nonconservative Calculation of Cumulative Fatigue Usage"
- Replacement instrument nozzles
- Crack growth and fracture mechanics stability analyses of postulated defects in original heater sleeve attachment welds remaining in the pressurizer lower heads
- Replacement heaters
- Replacement heater sleeves and their welds to the heaters
- Thermal effects on the Unit 3 pressurizer of incorrectly installed replacement heaters

4-47

Time-Limited Aging Analyses

- Compressive weld overlays of the surge, spray, and safety valve nozzles and safe ends and welds

The staff grouped the above items into the following components and events in this evaluation:

- Pressurizer support skirt forging weld
- Impact of power uprate and SG replacement
- Pressurizer surge line nozzle
- Pressurizer heater sleeves
- Pressurizer overheating event
- Weld overlay of the surge line, spray, and safety valve nozzles

Pressurizer Support Skirt Forging Weld. Section 4.3.2.4 of the LRA (page 4.3–39) states that the 1991 CE Owners Group (CEOG) review of CE Infobulletin 88-09, "Nonconservative Calculation of Cumulative Fatigue Usage," found that the fatigue usage factor in the worst-affected location (bottom head-support skirt) of the PVNGS Unit 1, 2, and 3 pressurizers might increase 32 percent above the design basis calculated value of 0.8895. Therefore, the applicant evaluated these effects further and amended the design reports. The revised worst-location 40-year design basis CUF, including these effects, is 0.7223.

In RAI 4.3.2.4-1, the staff asked the applicant to supply the CUF for the worst location at the pressurizer for the period of extended operation because the above information supplies the CUF for the worst location for the 40-year design life. The staff also asked the applicant to discuss whether the fatigue analysis of the pressurizer bottom head support skirt is a TLAA.

The applicant's March 1, 2010, response to RAI 4.3.2.4-1 stated that the pressurizer bottom head support skirt is the most affected location of the Infobulletin 88-09 reanalysis, not that it is the highest (i.e., the worst) CUF location as determined by the ASME Code analysis of the pressurizer.

The applicant explained that in Units 1 and 2, the highest CUF locations for a fatigue TLAA are the short heater sleeve plugs (LRA Table 4.3-7, line 15), with a CUF of 1.0. In Unit 3, the highest CUF location for a fatigue TLAA is the spray nozzle and safe end with overlay repair (LRA Table 4.3-7, line 20), with a 40-year calculated CUF of 0.9923. These values show that 60-year usage factors, calculated on the same bases, would exceed the ASME Code allowable of 1.0. Disposition of these TLAAs for the period of extended operation depends on 10 CFR 54.21(c)(1)(iii) aging management using the enhanced metal fatigue AMP which the applicant identified in the amended LRA.

The applicant stated that all fatigue analyses of the pressurizer and its subcomponents are TLAAs, except those already extended to a 60-year design life under analyses for the CLB per 10 CFR 54.3(a)(3), and those are not the basis for a safety determination per 10 CFR 54.3(a)(4).

The staff finds that the applicant has clarified that the CUF of 0.7233 from the support skirt is not the worst CUF for the pressurizer. The worst case CUF for Units 1 and 2 is in the short heater sleeve plugs and for Unit 3 is the overlaid spray nozzle. The applicant also noted that the enhanced metal fatigue AMP will be used to monitor the worst case CUFs to ensure that they will not exceed the ASME Code allowable of 1.0. Therefore, the staff's concern in RAI 4.3.2.4-1 is resolved.

In Section 4.3.2.4 of the LRA, the applicant reported that two flaws were detected in the Unit 2 pressurizer support skirt-forging weld and that the fatigue crack growth analysis predicted growth from the as-found size of 0.59 inch to a size of 0.6921 inch over the design life. In RAI 4.3.2.4-2, the staff asked the applicant to discuss how many years were assumed for the design life and were assumed in the fatigue crack growth calculation of the detected flaws. The

Time-Limited Aging Analyses

applicant's March 1, 2010, response stated that the fatigue crack growth analysis has shown that no repair is required for continued operation over the service life of the vessel, which, at the time of the analysis (1993), was understood to be 40 years. The applicant stated that a 60-year analysis is not required under the license renewal rule if it manages the aging effect.

As stated in LRA Section 4.3.2.4, the linear elastic fracture mechanics fatigue crack growth analysis of indications in the Unit 2 pressurizer support skirt-forging weld is valid for up to 500 plant startup and shutdown cycles, 480 plant trips, and 2 million normal and upset cycles. The Metal Fatigue of Reactor Coolant Pressure Boundary Program (enhanced metal fatigue AMP) will track these events, and action limits will ensure that appropriate corrective actions are completed before the design basis number of these events is exceeded. Appropriate corrective actions may include repair, replacement, or reanalysis. Growth of the Unit 2 pressurizer skirt indications will be managed for the period of extended operation in accordance with 10 CFR 54.21(c)(1)(iii). The parameter in the fatigue crack growth calculation that is affected by the time is the transient cycles used. The transient cycles will be different between 40 years and 60 years. The applicant will use the enhanced metal fatigue AMP to monitor the transient cycles to ensure that the transient cycles assumed in the fatigue crack growth calculation for the cracks found in the pressurizer support skirt-forging weld bound the actual transient cycles at the end of 60 years.

The staff held a teleconference with the applicant on September 22, 2010, to clarify how design basis transient cycle tracking and counting activities are accounted for in the CLB for ASME Code Section XI supplemental fatigue flaw growth or cycle dependent fracture mechanics evaluations. The staff also requested the applicant to justify the use of design basis transient cycle tracking and counting activities as the basis to disposition the ASME Code Section XI analyses in LRA Section 4.7.4 in accordance with 10 CFR 54.21(c)(1)(iii), if the scope of the applicant's CLB does not include this activity.

By letter dated October 13, 2010, the applicant described how it will use CC methods to track these supplemental evaluations. The staff reviewed the response and finds it acceptable that the applicant will use its enhanced Metal Fatigue Program to monitor the transient cycles used in the fatigue crack growth calculations. The staff's evaluation of the applicant's clarification is documented in SER Section 4.7.4.2.

The staff finds it acceptable that the applicant will use the enhanced metal fatigue AMP to monitor the transient cycles used in the fatigue crack growth calculation for the detected flaws in the Unit 2 pressurizer support skirt to ensure that the crack growth calculation is valid during the period of extended operation. Therefore, the staff's concern in RAI 4.3.2.4-2 is resolved.

Impact of Power Uprate and Steam Generator Replacement. In LRA, Section 4.3.2.4, the applicant states that power uprate and SG replacement have no effect on the design reports for any of the pressurizers in the three units. In RAI 4.3.2.4-3, the staff asked the applicant to reference the design reports associated with pressurizer and pressurizer nozzles that have been reviewed to determine the affect of power uprate and SG replacement. The staff asked that the applicant describe these reports briefly in the context of Section 4.3.2.4 and clarify whether the loadings on the pressurizer and pressurizer nozzles are affected by the power uprate and SG replacement.

The applicant's March 1, 2010, response to RAI 4.3.2.4-3 provided a list of the power uprate and replacement SG evaluation reports.

The power uprate licensing reports are included in letters from the applicant, dated December 21, 2001, and July 9, 2004. The reports did not explicitly cite or revise the analyses of record. Instead, these reports reviewed the supporting design transients as documented in the code design specifications for nuclear steam supply system SCs.

Time-Limited Aging Analyses

In the reports, the applicant concluded that the original design transients are more limiting than the corresponding limiting calculated transients associated with the power uprate. Hence, the original SC design specifications remain bounding to the new operating conditions associated with the power uprate.

The applicant stated further that the power uprate and replacement SG evaluations demonstrated that the loadings on the pressurizer remain within the conditions assumed for the analysis of record. Loads on the pressurizer and its nozzles remained less than or equal to those used for the analyses of record, and, therefore, no changes to the analyses of record were necessary.

By letter dated May 21, 2010, the applicant submitted Amendment 15, which supplemented the response to RAI 4.3.2.4-3 and further clarified that the Westinghouse design report addendum for the pressurizer confirms that the power uprate and SG replacement modifications have no effect on the pressurizer design reports for any of the three units. This conclusion applies to the severity of the design basis transient events, is unaffected by the number of occurrences of each transient event assumed by the analyses, and is unaffected by the design life. In conclusion, the power uprate and SG replacement modifications have no effect on the pressurizer design reports through the period of extended operation. The staff's concern in RAI 4.3.2.4-3 is resolved.

The staff finds that the applicant has demonstrated that power uprate and SG replacement do not significantly affect the loading on the pressurizer and its nozzles.

Pressurizer Surge Line Nozzle. LRA Section 4.3.2.4 states that the original stress and fatigue analysis of the pressurizer surge line nozzle have been superseded by the reanalysis for a compressive overlay, which included the thermal stratification and insurge-outsurge effects. In RAI 4.3.2.4-4, the staff asked the applicant to describe the analysis input, method, results, and acceptance criteria in detail, demonstrating that the structural integrity of the surge nozzle will be maintained to the end of 60 years.

The applicant's March 1, 2010, response to RAI 4.3.2.4-4 stated that the reanalysis of the overlaid surge nozzle is a worst-case calculation of the projected usage factor for a 60-year lifetime. The calculation for the 60-year cycle usage factor for overlaid surge line nozzle multiplied the 40-year cycle usage factor of 0.960 by 1.5 to obtain a 60-year value of 1.440. However, the pressurizer surge line nozzle will be monitored for fatigue usage, and the fatigue CUF will not exceed the code limit of 1.0, so long as the number of applied load cycles does not exceed the number specified by the design specification for this nozzle and used in the original analysis. The analysis includes effects of thermal stratification and insurge-outsurge.

The staff finds that the CUF for the surge line nozzle is predicted to exceed the allowable of 1.0 at the end of 60 years. However, the applicant will monitor the CUF for the surge line nozzle based on the enhanced metal fatigue AMP to ensure that it will not exceed the CUF allowable of 1.0. Before the CUF exceeds the allowable code limit, the metal fatigue program requires corrective actions. The staff finds that the structural integrity of the pressurizer surge line will be maintained during the period of extended operation because the applicant is using appropriate monitoring methods to ensure that the pressurizer surge line nozzles will not exceed a CUF allowable of 1.0. Therefore, the staff's concern in RAI 4.3.2.4-4 is resolved.

Pressurizer Heater Sleeves. PWR operating experience has shown that Alloy 82/182 welds that join the pressurizer heater sleeves to the pressurizer shell are susceptible to PWSCC. LRA Section 4.3.2.4 discusses the fatigue crack growth analysis in the original pressurizer heater sleeve attachment welds. In RAI 4.3.2.4-5, the staff asked the applicant to complete the following tasks:

- Discuss the postulated initial cracks in the original sleeve-to-inner-wall attachment welds

Time-Limited Aging Analyses

- Discuss the projected final flaw size for the postulated cracks at the end of 60 years
- Discuss the allowable flaw size
- Discuss the results
- Describe the methods used in the "subsequent" and "code design" reports
- Provide the references for these reports

The applicant's March 1, 2010, response to RAI 4.3.2.4-5 stated that the postulated initial crack size in the crack growth analyses was a flaw size of 0.6 inches in the Alloy 82/182 weld. The projected final flaw size at the end of 60 years is 1.16 inches. The fracture mechanics analysis permits an allowable flaw size of 1.2 inch. The fatigue crack growth analysis described in the LRA was performed in support of the temporary mechanical nozzle seal assembly (MNSA) repairs to three Unit 3 pressurizer heater sleeves and was performed for a 60-year period. As a result of the replacement of all heater sleeves with the half-nozzle repair method, the fatigue crack growth analysis of the remnant nozzles, performed in support of the MNSA repairs, has been superseded by the analyses done to support the heater sleeve replacement modification and associated RR 29. This RR asked for relief from the requirements of the ASME Code Section XI, IWA-3300, IWA-4310, IWB-2420, IWB-3242.4, and IWB-3610 for the pressurizer heater sleeve and Alloy 82/182 remnant nozzles.

The applicant performed a fatigue crack growth analysis for the pressurizer heater sleeve half-nozzle repair supporting RR 29. By letter dated November 5, 2004, the staff approved RR 29. The applicant stated that these analyses are superseding and applicable to all three units. A subsequent evaluation of effects of the Unit 3 lower head overheating events determined that the calculated increase in the growth of the postulated defect would be a negligible $4.44E$-5 inch. This evaluation was, therefore, not included in the Unit 3 code design report. The applicant stated that this evaluation applied only to the brief period of time for which the Unit 3 pressurizer lower head was subject to overheating; therefore, it is not a TLAA, since it does not meet the criteria of 10 CFR 54.4(a)(4).

The currently applicable analyses used finite-element models to calculate stress intensity factors for postulated flaws for various operating conditions and calculated the crack growth per cycle for the significant contributors to crack growth, using the methodology of the 1992 edition of the ASME Code Section XI for a water environment. The applicant's analyses then determined the maximum permissible crack size for which the most limiting allowed stress intensity would not be exceeded, based on the assumption that the applied stress intensity factor is proportional to the square root of the crack dimension. The crack growth was calculated as the sum of the products of lifetime design cycles times their respective crack growth increments per cycle.

The applicant stated that these analyses assume a 60-year design life, and are, therefore, not TLAAs, per 10 CFR 54.3(a)(3). The currently applicable analyses and supporting calculations were incorporated by reference, unchanged, in the three pressurizer code design reports.

The applicant stated that LRA Amendment 10 corrected the heading for this discussion, and two paragraphs below it in the LRA, to recognize the RR and associated analysis. It also specified the analysis results discussed above.

The staff finds that the applicant has clarified the analyses of the heater sleeve repairs. The analyses document the technical basis of the heater sleeve repairs, which the staff previously approved. The staff finds that the applicant has responded to RAI 4.3.2.4-5 satisfactorily, and the staff's concern is resolved.

LRA Section 4.3.2.4 states that for the half-nozzle repair method, the original sleeve-to-inner wall attachment welds (i.e., J-groove welds) are analyzed for 60 years and, therefore, the analysis is not a TLAA. The staff notes that other welds and components were used for the

Time-Limited Aging Analyses

half-nozzle repair (e.g., welds were used to join the half nozzle to the weld pad). In RAI 4.3.2.4-6, the staff asked the applicant to discuss why the fatigue analysis of the half nozzles, associated new attachment welds, and weld pads are not discussed as part of the review.

The applicant's March 1, 2010, response to RAI 4.3.2.4-6 stated that the fatigue analyses of the pressurizer heater sleeve half-nozzle repairs for all three units were evaluated for a period of 60 years. These analyses are, therefore, not TLAAs per 10 CFR 54.3(a)(3). The 60-year usage factors calculated by these repair analyses can be found in LRA Table 4.3-7, item 16 for Unit 2 and item 17 for Units 1 and 3. The applicant noted that LRA Table 4.3-7, item 16, is an analysis intended to be applicable to all three units; however, this half nozzle repair was installed only in the Unit 2 pressurizer. Item 16 is, therefore, a design basis analysis for Unit 2 only. Item 17 is the equivalent analysis for Units 1 and 3.

The staff finds that the applicant analyzed the repaired heater sleeve by the half-nozzle repair method for a period of 60 years as shown in LRA Table 4.3-7, items 16 and 17. Therefore, the analyses for the half-nozzle repair do not require a TLAA evaluation. The staff finds this acceptable because the applicant's fatigue analysis has demonstrated that the repaired heater sleeves will maintain its structural integrity at the end of 60 years. The staff's concern in RAI 4.3.2.4-6 is resolved.

LRA Section 4.3.2.4 states, "[t]he analysis of the weld pads does not explicitly supersede the results of the fatigue analysis with the tapped anchor holes. Therefore, both fatigue analysis results apply ...." In RAI 4.3.2.4-7, the staff asked the applicant to discuss whether the anchor holes and weld pads on the pressurizer bottom are a TLAA, clarify the above two statements in the context of TLAA, and clarify whether the CUFs for the anchor bolts and weld pads for 60 years are within the allowable factor of 1.0.

The applicant's March 1, 2010, response to RAI 4.3.2.4-7 stated that the anchor hole analysis was for a 40-year design basis set of cycles and is, therefore, a TLAA as shown in LRA Table 4.3-7, item 12. The fatigue analysis of the weld pads was for a 60-year design basis set of cycles and is, therefore, not a TLAA as shown in LRA Table 4.3-7, item 17. The applicant stated that LRA Table 4.3-7, item 16, is an analysis intended to be applicable to all three units; however, the replacement heater sleeve was installed only in the Unit 2 pressurizer. Item 16 is, therefore, a design basis analysis for Unit 2 only. Item 17 is the equivalent analysis for Units 1 and 3.

The applicant explained that the statement, "[t]he analysis of the weld pads does not explicitly supersede the results of the fatigue analysis with the tapped anchor holes. Therefore, both fatigue analysis results apply ...," means that the fatigue analysis that evaluated the effects of the anchor holes as shown in LRA Table 4.3-7, item 12, was not included in the superseding analysis of the weld pads as shown in LRA Table 4.3-7, item 17, that overlaid them. Therefore, the results of both analyses are applicable to the safety determination of the pressurizer pressure boundary.

The 40-year Unit 3 anchor hole analysis resulted in a maximum CUF of 0.443, which, projected to the end of the period of extended operation, would be 0.6645. The 60-year Units 1 and 3 weld pad analysis resulted in a maximum CUF of 0.551.

The staff finds that the applicant has clarified the CUFs in items 12, 16 and 17 of LRA Table 4.3-7. LRA Table 4.3-7, item 12 is related to the MNSA attachment holes in the Unit 3 lower head. LRA Table 4.3-7, item 16 is related to the heater sleeve half-nozzle repair with a weld pad for all three units, but the repair is only installed in Unit 2. LRA Table 4.3-7, item 17 is related to the heater sleeve MNSA weld pad repair for Units 1 and 3. The staff's concern in RAI 4.3.2.4-7 is resolved.

Time-Limited Aging Analyses

Pressurizer Overheating Event. In 2005, Framatome identified a fabrication error that had installed longer-than-specified replacement heaters in the Unit 3 pressurizer, extending them into the lower region of the heater sleeves. This error subjected local regions of the surrounding pressurizer head base metal to temperatures above those for which design stress intensity values are given in the ASME Code Section III, Appendix I. The applicant stated that all 36 of these Framatome heaters have since been replaced.

The Unit 3 pressurizer is designed to the ASME Code Section III, 1971 edition through winter 1973 addenda, installed to 1974 edition–1975 addenda. The base metal is SA-533, Grade A, Class 1, carbon steel. This ASME Code material has a maximum temperature of 700 degrees F for which design stress intensity values are given; this is also the limiting temperature for application of the ASME Code Section III, NB-1120, Figure I-9.1. The applicant found that the pressurizer base material surrounding the heater sleeves had been subjected to temperatures up to 779 degrees F for up to 3,700 hours. By letter dated June 28, 2005, APS submitted a RR for an alternative to NB-1120 for the Unit 3 pressurizer lower head, including an evaluation of the creep effects for the Unit 3 pressurizer lower head.

The applicant's evaluation applied the elevated temperature rules of the ASME Code Section III, Subsection NH, which permits design to specific Subsection NB-3000 rules if creep and relaxation are negligible. The evaluation demonstrated that creep was negligible for the 3,700-hour exposure period; therefore, the ASME Code Section III, Subsection NB rules could be used, with the adjusted design stress intensity factors. The applicant found no immediate adverse effects on the overheated material, and the staff granted the relief. Although this relief was asked "for the remainder of plant life," as is appropriate for a request supported, in part, by an evaluation of fatigue effects, the supporting evaluation of creep effects was limited to the 3,700-hour exposure to elevated temperature, and the evaluation of the creep effects is therefore not a TLAA. However, overheating did affect the code fatigue analysis.

In RAI 4.3.2.4-8 and RAI 4.3.2.4-9, the staff asked the applicant to provide the technical basis supporting that the evaluation of creep effects is not a TLAA. The applicant's March 1, 2010, response stated that the evaluation of the creep effect is not a TLAA because the effect ended with the replacement of the heaters. The evaluation of the creep effect is, therefore, not a TLAA because its assumptions are not time-limited as defined by the current operating term as specified in 10 CFR 54.3(a)(3). The evaluation and its conclusions will not change, therefore, with an extension to the licensed operating period.

The applicant stated further that the 2005 overheating event did not affect the currently installed heaters or their code fatigue analyses because the affected heaters were replaced, but the event did affect the code fatigue analysis of the Unit 3 pressurizer lower head. The effects of the overheating on the code fatigue analysis of the pressurizer lower head were evaluated for a 60-year design life, as summarized in LRA Table 4.3-7, item 18. Since this addendum to the design report was done for a 60-year design life, PVNGS did not classify it as a TLAA; however, the affected code design report is a TLAA.

The staff's concern was that excessive temperatures beyond the design limit might degrade the material property of the pressurizer shell; however, the applicant demonstrated that the overheating event is not time-dependent, and the creep effect is negligible. The staff finds that because the applicant will be implementing the enhanced metal fatigue AMP to monitor certain components of the pressurizer, the nozzles, and heater sleeves, these components will be inspected per the ASME Code Section XI, and any potential degradation of the pressurizer due to the creep effects should be detected. The staff's concerns in RAIs 4.3.2.4-8 and 4.3.2.4-9 are resolved.

Weld Overlay of the Surge line, Spray and Safety Valve Nozzles. The applicant stated that the weld overlay of the surge line, spray valve, and safety valve nozzles are supported by fracture

Time-Limited Aging Analyses

mechanics analyses and periodic inspections under ASME Code Section XI as the means to address aging in the overlaid welds. The applicant stated that the fracture mechanics and fatigue crack growth analyses of the overlaid nozzles assume 1.5 times the transient cycles used in the 40-year design basis, but they do not support safety determinations for a defined design lifetime. Therefore, they are not TLAAs. However, the revised fatigue analyses of the adjacent materials affected by the overlays are time-dependent and are TLAAs unless successfully projected to the end of the period of extended operation. The revised fatigue analyses include the period from initial operation to overlay installation since the adjacent materials were not replaced.

In RAI 4.3.2.4-10, the staff asked the applicant, (1) to clarify whether the CUF calculation and fatigue crack growth calculation were calculated for the adjacent materials to the end of 60 years, (2) to submit the revised fatigue analyses for the adjacent materials or describe in detail the analysis input, methods, acceptance criteria, and result, (3) to identify all the pressurizer nozzles in all three units that have been weld overlaid and identify "the adjacent materials" affected by the weld overlays, and (4) to discuss the actions that will be taken as a result of the TLAA determination.

The applicant's March 1, 2010, response to RAI 4.3.2.4-10 stated that crack growth analyses per the ASME Code Section XI, IWB-3640 address potential growth of cracks in the susceptible dissimilar weld metal overlay region, but no fatigue crack growth analysis was done for materials *adjacent* to the overlay. The fatigue crack growth analysis pertains to original materials *under* the overlay. In effect, the analyses assume a 60-year design life.

The flaw growth analyses for the weld overlays projected flaw sizes either to the end of the 60-year design life or to the next scheduled ISI; or the analyses determined the maximum time permitted between ISIs to support the safety determination. In none of these cases did the flaw evaluation support a safety determination for a time period defined by the current licensed operating term, and, therefore, the flaw evaluation was not a TLAA in any of these cases since it does not meet the criteria of 10 CFR 54.3(a)(3).

The applicant stated that the ASME Code Section III fatigue analyses were done only for regions adjacent to the overlay not affected by cracks or assumed cracks. CUFs were calculated adjacent to the overlay on a similar (i.e., 60-year) basis. Therefore, CUF analyses that met the ASME Code fatigue usage factor limit of 1.0 for 60 years were determined not to be TLAAs. Calculations for fatigue in the surge and spray nozzles that did not meet the 60-year life, and qualified the location for only 40 years, were determined to be TLAAs. The adjacent materials are those determined by the analyses to have effects on their design stresses due to the overlays. LRA Table 4.3-7, items 19, 20, and 21, show CUFs for surge, spray, and safety nozzles for all three units.

The applicant calculated a very low CUF for the safety valve nozzles for a 60-year life as shown in LRA Table 4.3-7, item 21. This analysis is, therefore, not a TLAA per 10 CFR 54.3(a)(3); however, the CUF calculations in the surge and spray nozzles for a 60-year life were higher. For the surge and spray nozzles, the applicant was able to demonstrate an acceptable fatigue usage for only the design basis number of events assumed for 40 years as shown in LRA Table 4.3-7, items 19 and 20. Fatigue for the surge and spray nozzles are managed, therefore, by the enhanced metal fatigue AMP for the period of extended operation.

The staff finds that the applicant has clarified the flaw growth calculations and CUF calculations for the overlaid surge line, safety valve, and spray valve nozzles. The staff agrees that the flaw growth calculations for the postulated flaw in the overlaid nozzles assumed transient cycles for 60 years and, because the calculations were performed for 60 years, the flaw growth calculations are not TLAAs. The CUF calculations for the surge line and spray valve nozzles are TLAA because the CUFs were calculated for 40 years. The safety valve CUF calculation is

Time-Limited Aging Analyses

not a TLAA because it was performed for 60 years. The applicant will use the enhanced metal fatigue AMP to monitor the CUFs of the surge line and spray valve nozzles, and the staff finds this acceptable. Therefore, the staff's concern in RAI 4.3.2.4-10 is resolved.

LRA Section 4.3.2.4 discusses the fatigue usage factors for the surge and spray nozzles for a 60-year life as being 1.44 and 1.49, respectively. The applicant stated that the surge nozzle is monitored for fatigue usage, and the fatigue usage factor will not exceed the code limit of 1.0 as long as the number of applied load cycles does not exceed the number specified by the design specification used in the analyses. In RAI 4.3.2.4-11, the staff asked the applicant to clarify whether the spray nozzle will be monitored for fatigue usage and discuss why a plastic analysis was not done in accordance with NB-3228 of the ASME Code Section III when the CUFs for surge and spray piping exceed 1.0, as calculated by the elastic analysis of NB-322. LRA Section 4.3.2.9 discusses a plastic analysis done by CE on the surge line that lowered the CUF.

The applicant's March 1, 2010, response to RAI 4.3.2.4-11 stated that, as shown in LRA Table 4.3-4, item 17, the spray nozzle will be monitored by the enhanced metal fatigue AMP, which will maintain the usage factor of the surge and spray nozzles at less than 1.0 for the period of extended operation or ensure that other acceptable actions are taken to maintain the basis of the safety determination. The applicant stated that a plastic analysis was not required since a monitoring method is being used. LRA Section 4.3.2.9 (and Section 4.3.4) provides the plastic analysis of the surge line because it has been performed in support of the safety determination for the current licensed operating term and is, therefore, a TLAA. The staff finds that LRA Table 4.3-4, items 16 and 17, state that the surge line and spray nozzle will be monitored by the enhanced metal fatigue AMP because of high CUFs. The staff finds that because the surge line and spray valve nozzles will be monitored by the enhanced metal fatigue AMP, the structural integrity of these nozzles will be maintained. Therefore, the staff's concern has been addressed and RAI 4.3.2.4-11 is resolved.

LRA Section 4.3.2.4 states that the enhanced metal fatigue AMP will ensure that the fatigue usage factors, based on those transient events, will remain within the ASME Code limit of 1.0 for the period of extended operation, or will ensure that re-evaluation or other corrective actions will be taken before the design basis number of these events is exceeded. In RAI 4.3.2.4-15, the staff asked the applicant to identify the pressurizer subcomponents that will be reanalyzed in the context of Section 4.3.2.4.

The applicant's March 1, 2010, response to RAI 4.3.2.4-15 stated that a subcomponent will be reanalyzed or reanalyzed if it is determined by the enhanced metal fatigue AMP that fatigue in the subcomponent has reached an action limit that requires mitigation of fatigue effects and that re-evaluation or reanalysis is the preferred mitigation action. Other possible corrective actions include repair, replacement, or a fatigue crack growth analysis plus augmented inspection. The pressurizer subcomponents that will be reanalyzed will be determined by the state of fatigue effects in the pressurizer at that time, as tracked by the enhanced metal fatigue AMP, and upon review of the affected Class 1 pressurizer analyses. LRA Section 4.3.1 and Appendix B3.1 provide further description of the enhanced metal fatigue AMP, including locations to be monitored and the bases for action limits. The staff finds that the applicant has satisfactorily addressed the circumstances in which the re-evaluation of the fatigue usage factors will be performed. Therefore, the staff's concern in RAI 4.3.2.4-15 is resolved.

In summary, the staff finds that the applicant has adequately addressed time-dependent analyses (i.e., CUF calculations and crack growth calculations) of the pressurizer and associated components in all three units in terms of TLAAs. Therefore, the staff finds that the pressurizer will maintain its structural integrity during the period of extended operation.

Time-Limited Aging Analyses

4.3.2.4.3 Updated Final Safety Analysis Report Supplement

The applicant provided an UFSAR supplement summary description of its TLAA of the pressurizer and pressurizer nozzles in LRA Section A3.2.1.4. Based on its review of the UFSAR supplement, the staff concludes that the summary description of the applicant's actions to address the TLAA for the pressurizer and pressurizer nozzles is adequate.

4.3.2.4.4 Conclusion

Based on its review, the staff concludes pursuant to 10 CFR 54.21(c)(1)(iii), that the applicant has demonstrated that the effects of aging on the intended function of the pressurizer and pressurizer nozzles will be adequately managed for the period of extended operation. The UFSAR supplement contains an appropriate summary description of the TLAA evaluation of the pressurizer and pressurizer nozzles, as required by 10 CFR 54.21(d).

4.3.2.5 Steam Generator American Society of Mechanical Engineers Code, Section III Class 1, Class 2 Secondary Side, and Feedwater Nozzle Fatigue Analyses

4.3.2.5.1 Summary of Technical Information in the Application

LRA Section 4.3.2.5 and Amendment 16 dated May 27, 2010, summarize the evaluation of the fatigue analyses of ASME Code Section III, Class 1 SG, Class 2 secondary side, and the feedwater nozzles. The applicant stated that the replacement SGs are designed to ASME Code Section III, Subsection NB (Class 1) and NC (Class 2). The LRA identified that the Unit 1 and 3 SGs will be within their 40-year design life at the end of the period of extended operation, and the Unit 2 SGs will be at year 42 of operation at the end of the period of extended operation.

The original LRA dispositioned the Unit 2 SG TLAAs in accordance with 10 CFR 54.21(c)(1)(iii), that the effects of aging on the intended function(s) will be adequately managed for the period of extended operation. The Units 1 and 3 SG TLAAs were dispositioned in accordance with 10 CFR 54.21(c)(1)(i), that the analyses remain valid during the period of extended operation.

4.3.2.5.2 Staff Evaluation

The staff reviewed LRA Section 4.3.2.5 and Amendment 16 to verify, pursuant to 10 CFR 54.21(c)(1)(i), that the TLAA of the ASME Code Section III Class 1 SG, Class 2 secondary side, and feedwater nozzles remain valid during the period of extended operation. Further, the staff's review evaluated, pursuant to 10 CFR 54.21(c)(1)(iii), if the effects of aging on the intended function(s) of the ASME Code Section III Class 1 SG, Class 2 secondary side, and feedwater nozzles will be adequately managed for the period of extended operation.

The applicant stated that pressure-retaining and support components of the primary coolant side of the SGs are subject to an ASME Code Section III fatigue analysis. Although the secondary side is Class 2, all pressure retaining parts of the SGs satisfy the Class 1 criteria, including Section III fatigue analysis.

The staff reviewed LRA Section 4.3.2.5 as related to the replacement recirculating SG tube CUF calculations. The applicant stated that this analysis is not a TLAA because the safety determination does not depend on it and due to periodic SG tube inspection schedules. The staff does not consider the applicant's apparent use of SCC mechanisms and ASME Code examinations as a valid basis for concluding that the CUF calculations would not qualify the tubes for metal fatigue during the remainder of the licensed life of the tubes. By letter July 21, 2010, the staff issued RAI 4.3-13, requesting that the applicant justify its basis for concluding that the CUF calculation for the replacement recirculating SG tubes is not a TLAA.

The applicant responded to RAI 4.3-13 by letter dated August 12, 2010 (Amendment 22). The staff's evaluation of RAI 4.3-13 is documented in SER Section 3.1.2.2.1. The staff noted that the applicant amended LRA Section 4.3.2.5 to identify the SG tube fatigue analysis as a TLAA

Time-Limited Aging Analyses

and to disposition the TLAA for the SG tubes in accordance with 10 CFR 54.21(c)(1)(i). The applicant stated that the cyclic stress range for the SG tubes is less than the endurance limit, and this allows the SG tubes to withstand an infinite number of cycles, so the CUF was determined to be zero. The applicant further explained that since the SG tube CUF is zero, the analysis of record remains valid through the period of extended operation for all three PVNGS units. The staff finds it acceptable that the applicant amended its LRA to identify the SG tube code fatigue analysis as a TLAA, because this analysis meets the definition of a TLAA as defined in 10 CFR 54.3. The staff finds it acceptable that the SG tube code fatigue TLAA is dispositioned in accordance with 10 CFR 54.21(c)(1)(i), because the CUF value for the SG tubes was calculated to be zero for all three units and, therefore, the SG tubes will endure an infinite number of cycles. The staff's concern in RAI 4.3-13 related to the SG tube code fatigue analysis is resolved.

LRA Table 4.3-8 provides the design basis CUF for the replacement SGs (considering the power uprate) for all three units. The applicant explained that although the replacement SG designs are essentially identical, the Unit 2 code analysis was performed first, under separate contract. The calculated CUFs therefore differ in some components. Furthermore, the applicant stated that since the ASME code does not specify all locations which must be analyzed, this left many of the detailed choices to the experience and skill of the analyst. For example, the Unit 2 analyst did not elect to perform a fatigue analysis at the support skirt opening or in the economizer cylinder near the tubesheet while the Unit 1 and 3 analyst did so.

For the high usage factors calculated for the primary manway and secondary handhole studs, the applicant stated that the fatigue analysis determines the replacement interval of those components, and the fatigue analysis is, therefore, not a TLAA for these studs.

The applicant also stated that the worst-case usage factors calculated for the specified set of design basis transients for the replacement SGs (considering the power uprate) exceed 0.9 in several SG components. The applicant explained that fatigue usage factors in the SG components do not depend on effects that are time-dependent at steady-state conditions, but depend only on effects of operational and upset transient events. The applicant chose to apply aging management to all the SGs and will use the enhanced Metal Fatigue Program to track events and ensure that appropriate reevaluation or other corrective action will be initiated if an action limit is reached.

Based on its review, the staff finds the applicant has demonstrated, pursuant to 10 CFR 54.21(c)(1)(i), that the TLAA analyses for the SG tubes remains valid during the period of extended operation because the cyclic stress range for the SG tubes is less than the endurance limit allowing an infinite number of cycles. The staff also finds the applicant has demonstrated, pursuant to 10 CFR 54.21(c)(1)(iii), that the effects of aging on the intended functions of the replacement SG pressure boundaries with Class 1 analyses, with the exception of the SG tubes, will be adequately managed for the period of extended operation because action limits of the enhanced Metal Fatigue Program will ensure completion of corrective actions before the design basis number of events is exceeded. Thus, the fatigue usage factor will not exceed the design limit of 1.0 during the period of extended operation.

4.3.2.5.3 Updated Final Safety Analysis Report Supplement

LRA Section A3.2.1.5 provides the UFSAR supplement summarizing the evaluation of the fatigue analyses for ASME Code Section III Class 1 SG, Class 2 secondary side, and feedwater nozzles. Based on its review of the UFSAR supplement, the staff concludes that the applicant provided an adequate summary description of its actions to address the TLAA for the ASME Code Section III Class 1 SG, Class 2 secondary side, and feedwater nozzles, as required by 10 CFR 54.21(d).

Time-Limited Aging Analyses

4.3.2.5.4   Conclusion

On the basis of its review, the staff concludes that, pursuant to 10 CFR 54.21(c)(1)(i), the applicant has demonstrated that the fatigue analysis for the SG tubes remain valid for the period of extended operation. Pursuant to 10 CFR 54.21(c)(1)(iii), the applicant has demonstrated that the effects of aging on the intended functions of the ASME Code Section III Class 1 SG, Class 2 secondary side, and feedwater nozzles will be adequately managed for the period of extended operation. The UFSAR supplement contains an appropriate summary description of the TLAA evaluation, as required by 10 CFR 54.21(d).

4.3.2.6   American Society of Mechanical Engineers Code, Section III, Class 1 Valves

4.3.2.6.1   Summary of Technical Information in the Application

LRA Section 4.3.2.6 and Amendment 16 dated May 27, 2010, summarize the evaluation of the fatigue analyses of the ASME Code Section III, Class 1 valves. The applicant stated that those valves are designed to ASME Code Section III, Subsection NB. The applicant also stated while ASME Code requires a fatigue analysis only for Class 1 valves with an inlet greater than four inches nominal, some Class 1 valves with an inlet four inches or less also require a fatigue analysis. The applicant dispositioned these TLAAs in accordance with 10 CFR 54.21(c)(1)(iii), that the effects of aging on the intended functions will be adequately managed for the period of extended operation.

4.3.2.6.2   Staff Evaluation

The staff reviewed LRA Section 4.3.2.6 and Amendment 16 to verify pursuant to 10 CFR 54.21(c)(1)(iii), that the effects of aging on the intended functions will be adequately managed for the period of extended operation.

In LRA Table 4.3-9, the applicant listed the Class 1 valves and corresponding design basis CUF values. The table showed that the calculated worst-case usage factors are 0.702 and 0.7656, in the 16-inch shutdown cooling suction isolation valves and 2-inch charging line isolation valves, respectively. The applicant explained that fatigue usage factors in these valves do not depend on effects that are time-dependent at steady-state conditions, but depend only on effects of operational, abnormal, and upset transient events.

Effects of Power Uprate and SG Replacement on the CEDM and RVLMS Pressure Housing Analyses. LRA Section 4.3.2.6 and Amendment 16 state that the original fatigue analyses of record are still valid and the effects of the SG replacements and power uprate loads on the analysis of record have been reconciled.

Effects of Combustion Engineering Infobulletin 88-09. The applicant stated that the CE Owner's Group review did not identify any effects on the fatigue analysis of the ASME III Class 1 valves.

Based on its review, the staff finds the applicant has demonstrated, pursuant to 10 CFR 54.21(c)(1)(iii), that the effects of aging on the intended functions of the ASME III Class 1 valves will be adequately managed for the period of extended operation because the action limits of the enhanced Metal Fatigue Program will ensure completion of corrective actions before the design basis number of events is exceeded. Thus, the fatigue usage factor will not exceed the design limit of 1.0 during the period of extended operation.

4.3.2.6.3   Updated Final Safety Analysis Report Supplement

LRA Section A3.2.1.3 provides the UFSAR supplement summarizing the evaluation of the fatigue analyses for the ASME Code Section III, Class 1 valves. Based on its review of the UFSAR supplement, the staff concludes that the applicant provided an adequate summary description of its actions to address the TLAA for the ASME Code Class 1 valves as required by 10 CFR 54.21(d).

Time-Limited Aging Analyses

4.3.2.6.4 Conclusion

On the basis of its review, the staff concludes that the applicant has demonstrated, pursuant to 10 CFR 54.21(c)(1)(iii), that the effects of aging on the intended functions of the ASME Code Section III, Class 1 valves will be adequately managed for the period of extended operation. The staff also concludes that the UFSAR supplement contains an appropriate summary description of the TLAA evaluation as required by 10 CFR 54.21(d).

4.3.2.7 American Society of Mechanical Engineers III Class 1 Piping and Piping Nozzles

4.3.2.7.1 Summary of Technical Information in the Application

The ASME Code Class 1 reactor coolant main loop piping is designed to the ASME Code Section III, Subsection NB, 1974 edition with addenda through summer 1974. The main loop piping fatigue analysis was performed to the 1974 edition with addenda through summer 1974. The fatigue analyses of piping outside the main loop used the 1974 edition with addenda through winter 1975 or the 1977 edition with addenda through summer 1979. These fatigue analyses have been updated to incorporate redefinitions of loads and design basis events, operating changes, power uprate, SG replacement, and minor modifications. The currently applicable fatigue analyses of the Class 1 piping are TLAAs.

4.3.2.7.2 Staff Evaluation

The staff reviewed LRA Section 4.3.2.7 to verify, pursuant to 10 CFR 54.21(c)(1)(iii), that the effects of aging on the intended function(s) of the ASME Code Section III, Class 1, piping and piping nozzles will be adequately managed for the period of extended operation.

The applicant stated that in the primary coolant piping, the most limiting calculated design basis usage factors occur in the charging nozzles and approach the limit of 1.0 due to transient thermal stresses from normal operating and upset injection events. The applicant stated that, with the exception of the charging line nozzles and possibly the pressurizer surge line discussed in Section 4.3.2.9, fatigue usage factors in these components do not depend on effects that are time-dependent at steady-state conditions but depend only on effects of operational, abnormal, and upset transient events. The applicant stated that the Metal Fatigue AMP will track these transient events, and the design basis fatigue usage factor limit of 1.0 will not be exceeded in these locations without an appropriate evaluation and any necessary mitigating actions.

The applicant stated that original codes of record did not invoke the requirement for cycle-based stress limit for pipe support and, as permitted by code rules, later editions were not invoked for any pipe support reanalysis. The staff finds that the pipe supports were designed without considering cycle-based stresses per the ASME Code Section III requirements; therefore, pipe supports do not have a TLAA.

The staff reviewed the following subsections:

- Effects of Power Uprate and SG Replacement on the Piping Fatigue Analyses
- Charging Lines and Nozzles
- Reduced Wall Thickness in the RCS
- Alloy 600 Hot Leg Small-Bore Nozzle Repairs
- Alloy 600 Hot Leg Small-bore Half Nozzle Repairs
- Effect of Unit 3 MNSA Holes on Reactor Coolant Piping
- Redesigned RCS Thermowells
- Safety Injection Nozzle Thermal Sleeves and Auxiliary Spray Line
- Hot Leg Surge and Shutdown Cooling Nozzle Weld Overlays
- Disposition: Aging Management per 10 CFR 54.21(c)(1)(iii)

Time-Limited Aging Analyses

Effects of Power Uprate and Steam Generator Replacement on the Piping Fatigue Analyses. LRA Section 4.3.2.7 states that RCS piping, nozzles, RTD thermowells, and other Class 1 piping satisfy the CLB design number of transients under power uprate and SG replacement. However, it is not clear to the staff whether the piping components have been analyzed for 60 years. By letter dated January 14, 2010, the staff issued RAI 4.3.2.7-1 asking the applicant to clarify whether the Class 1 piping components satisfy the allowable CUF of 1.0 using a projected number of transients at the end of 60 years.

The applicant's March 1, 2010, response to RAI 4.3.2.7-1 stated that the RCS piping, nozzle, RTD thermowell, and other Class 1 piping component analyses that were not based on a 60-year life, are TLAAs. The applicant stated that the disposition of these TLAAs depends on the enhanced metal fatigue AMP to ensure that the 40-year design numbers of transients will not be exceeded during 60 years of operation without appropriate corrective actions. These TLAAs will, therefore, be managed for the period of extended operation in accordance with 10 CFR 54.21(c)(1)(iii).

The staff finds that because the CUF calculations of Class 1 piping were not performed for 60 years, the applicant will use the enhanced metal fatigue AMP to monitor the transient cycles used in the fatigue analyses to ensure they are bound by the actual cycles during the period of extended operation. The staff finds that the applicant's use of the enhanced metal fatigue AMP to monitor the transient cycles is acceptable and the staff's concern in RAI 4.3.2.7-1 is resolved.

Charging Lines and Nozzles. LRA Section 4.3.2.7 states that the Metal Fatigue of Reactor Coolant Pressure Boundary Program will calculate SBF in the chemical and volume control system charging nozzle. By letter dated January 14, 2010, the staff issued RAI 4.3.2.7-2, asking the applicant to describe the SBF analysis in detail, such as the analysis input, analytical procedures and method, acceptance criteria, and results. By letter dated March 1, 2010, in response to RAI 4.3.2.7-2, the applicant stated that the calculation has not yet been done. The future SBF monitoring program is discussed in the disposition of LRA Sections 4.3.2.7 and 4.3.1, as follows:

> Stress-based fatigue (SBF) monitoring will compute a "real time" stress history for a given component from actual temperature, pressure, and flow histories. SBF is intended for those high-fatigue components where a more refined approach is necessary to show long-term structural acceptability. SBF monitoring depends on "global-to-local" correlation or "transfer" functions which calculate local transient pressures and temperatures from data collected by the limited number of plant instruments, and from them, local stresses and fatigue usage.

The applicant stated that the SBF monitoring method is an enhancement to the metal fatigue AMP. Furthermore, the analysis details, such as the analysis input, analytical procedures and method, and acceptance criteria have not yet been completed. For SBF monitoring, the applicant has committed to the use of a fatigue-monitoring software program that incorporates a three-dimensional, six-element model, meeting the ASME Code Section III, NB-3200 requirements. This will be implemented at least two years before the period of extended operation. By letter dated February 19, 2010, the applicant submitted Amendment 9, which updated Commitment No. 39 in LRA Appendix A, Table A4-1, to include the use of the three-dimensional, six-element fatigue-monitoring model.

By letter dated May 27, 2010, the applicant submitted Amendment 16 which revised the monitoring method for the charging lines and nozzles to use a CBF-EP monitoring method. The staff noted that this method of monitoring is consistent with the recommendations of the "parameters monitored/inspected" program element of GALL AMP X.M1, which states that more detailed local monitoring of the plant transient may be used to compute the actual fatigue usage

Time-Limited Aging Analyses

for each transient. The staff's review of the applicant CBF-EP monitoring method is documented in SER Section 4.3.1.2.2.

The staff finds it acceptable that the applicant will use the CBF-EP monitoring method for the charging lines and nozzles because it monitors the transient events and thermals cycles to ensure that the usage factor in the CVCS charging nozzles is within the ASME Code Section III allowable limit of 1.0, consistent with the recommendations of GALL AMP X.M1. The staff's concern in RAI 4.3.2.7-2 is resolved.

Reduced Wall thickness in the Reactor Coolant System. The applicant reviewed the fatigue analysis for the RCS hot leg and cold leg piping for the SG replacement and power uprate. As a result of the review, the applicant amended two design reports that account for two piping configurations.

The LRA states that the first modification involved the intended design configuration, which assumes full carbon steel field welds. These results continue to remain applicable to the actual pipe runs, but not the field welds. This configuration results in a maximum calculated usage factor well below 1.0 for the hot leg and hot leg elbow. This fatigue analysis assumes the design basis transients for a 40-year plant life; therefore, it is a TLAA that will be managed through the enhanced metal fatigue AMP.

The LRA states that the second configuration assumed reduced piping wall thicknesses at both "postulated" and "acceptable" minimum wall thicknesses. The postulated minimum wall thickness values were bounding values for all three units. The acceptable minimum wall thickness values were based on design condition stress limits. This evaluation also evaluated all design basis transients for a 40-year plant life. All of the fatigue calculations use conservative bending stress intensification factors that are specifically applicable only to the crotch region of elbows. This evaluation also assumed a reduced wall thickness for the entire pipe run, rather than the welds. This evaluation concluded that the acceptable minimum wall thickness values in all field weld locations meet all ASME Code requirements.

The applicant stated that the evaluation for reduced wall thicknesses calculated fatigue usage factors approaching 1.0. Fatigue in RCS piping can be adequately managed during the period of extended operation using the CC method. The applicant also stated that CC monitoring will ensure that re-evaluation or other corrective action is initiated if the CC action limit is reached. The applicant stated that this is adequate to manage the fatigue in the welds because the revised calculated fatigue usage in the welds has the same transient event cycle count basis. Action limits will permit completion of corrective actions before the design basis number of events is exceeded.

By letter dated February 14, 2010, the staff issued RAI 4.3.2.7-3, asking that the applicant do four things: (1) clarify why the fatigue usage factor for the RCS hot and cold leg piping with reduced wall thickness was not calculated for 60 years, (2) list all Alloy 82/182 welds in the RCS hot leg and cold leg piping, (3) discuss whether there are any indications or flaws detected in the Alloy 82/182 welds that remained in service, and (4) if there are flaws in the Alloy 82/182 welds, perform a fatigue crack growth analysis for the 60-year plant life or justify why flaw evaluations are not need to demonstrate the structural integrity of the affected welds at the end of 60 years.

The applicant's March 1, 2010, response to RAI 4.3.2.7-3 stated that fatigue usage factors in the RCS hot and cold leg piping with reduced wall thickness were not projected to 60 years because the disposition of the TLAA depends on the enhanced metal fatigue AMP to ensure that the 40-year design numbers of transients will not be exceeded during 60 years of operation without appropriate corrective actions. These TLAAs will, therefore, be managed for the period of extended operation in accordance with 10 CFR 54.21(c)(1)(iii). The applicant stated further that there are no Alloy 82/182 welds in the main loop RCS hot and cold leg piping.

Time-Limited Aging Analyses

The staff finds that it is acceptable that the applicant will use the enhanced metal fatigue AMP to ensure that the 40-year design basis transients will not be exceeded during the period of extended operation. The staff concern in RAI 4.3.2.7-3 is resolved.

Alloy 600 Hot Leg Small-Bore Nozzle Repairs. The applicant stated that all the Alloy 600 instrumentation nozzles have been replaced in the hot legs and pressurizer for all three units in an effort to reduce the potential for PWSCC. Welded plugs, full nozzle, half nozzle, and three-quarter nozzle repairs have been used for the inservice RTDs. The half nozzle repair applies to the removal of the lower half (axial length) of the original nozzle. The three-quarter nozzle repair applies to the removal of lower three-quarter length of the original nozzle. The full nozzle repair applies to the removal of the entire original nozzle and installation of a complete new nozzle. The methods and new design basis for the repairs used in the RCS hot leg small-bore nozzles are discussed below. The pressurizer nozzle repairs are evaluated in SER Section 4.3.2.4.

LRA Section 4.3.2.7 states that the original RCS hot legs contained 27 Alloy 600 small-bore nozzles in each unit. In 1992, the applicant replaced seven pressure differential transmitter (PDT) nozzles and one sample nozzle in Unit 2 with full nozzles. The applicant stated that the remaining hot leg small-bore nozzles were replaced with the Alloy 690 half-nozzle design. However, it is not clear to the staff whether these remaining nozzles are located in Unit 1, 2, or 3 and it is not clear why small-bore nozzles in Units 1 and 3 are not discussed in this section. By letter dated February 14, 2010, the staff issued RAI 4.3.2.7-4, asking the applicant to complete the following tasks:

- Provide a table, similar to Table 4.3-7, with the following information: list all 27 small-bore nozzles for each unit, identify the type of the nozzle (e.g., RTD or PDT) or systems, identify the repair method for each nozzle, identify whether a fatigue analysis was performed for 60 years for each nozzle, and specify whether a TLAA is needed.

- If a nozzle is not analyzed for 60 years, perform a fatigue analysis for 60 years, or justify why a fatigue analysis is not needed to demonstrate that that small-bore nozzle satisfy the ASME Code Section III allowable usage factor of 1.0 at the end of 60 years.

- Discuss whether cold leg piping contains small-bore Alloy 600 nozzles, whether they were replaced with Alloy 690 nozzles, and whether their fatigue usage factors were analyzed for 60 years.

By letter dated March 1, 2010, in response to RAI 4.3.2.7-4, the applicant provided the following account of the small-bore nozzle repairs for hot leg piping (Table 4.3-1).

Table 4.3-1. Alloy 600 Small-Bore Hot Leg Nozzle Repairs

Unit	Number and Type of Nozzle	Repair Type	Maximum CUF
1	9 Pressure and sampling	Half nozzle repair	Fatigue waiver
	8 Spare RTD	Welded plugs	0.051
	10 Inservice RTD	Three quarter nozzle repair	0.0105
2	8 (7 PDT and 1 sampling)	Full nozzle repair	0.863
	1 Pressure and sampling	Half nozzle repair	Fatigue waiver
	8 Spare RTD	Welded plugs	0.051
	10 In service RTD	Three quarter nozzle repair	0.0105
3	9 Pressure and sampling	Half nozzle repair	Fatigue waiver
	8 Spare RTD	Welded plugs	0.051

4-62

Time-Limited Aging Analyses

Unit	Number and Type of Nozzle	Repair Type	Maximum CUF
	10 Inservice RTD	Three quarter nozzle repair	0.0105

The applicant stated that the TLAAs for the above hot leg small-bore nozzles will be managed by the enhanced metal fatigue AMP for the period of extended operation in accordance with 10 CFR 54.21(c)(1)(iii). The disposition of these TLAAs depends on the enhanced metal fatigue AMP to ensure that the number of transients for the 40-year design will not be exceeded during 60 years of operation without appropriate corrective actions.

The applicant stated that the RCS cold leg piping still contains 12 small-bore Alloy 600 nozzles. The maximum CUF associated with the cold leg RTD nozzles is 0.0591. The fatigue analysis was performed using the 40-year design numbers of transients. These TLAAs will, therefore, be managed by the enhanced metal fatigue AMP for the period of extended operation in accordance with 10 CFR 54.21(c)(1)(iii). The enhanced metal fatigue AMP will ensure that the numbers of transients for the 40-year design will not be exceeded during 60 years of operation without appropriate corrective actions.

The staff finds that the applicant has clarified the number and type of the small-bore nozzles in the hot leg pipe, the types of repairs, and the maximum CUF for specific components in each unit. For the cold leg, there are still 12 Alloy 600 nozzles. The staff finds that the small-bore nozzles in the hot leg and cold leg will be managed by the enhanced metal fatigue AMP during the period of extended operation. Therefore, their structural integrity will be maintained such that they can perform their intended functions. The staff's concern in RAI 4.3.2.7-4 is resolved.

Alloy 600 Hot Leg Small-bore Half Nozzle Repairs. LRA Section 4.3.2.7 discusses that the PDT and sampling half-nozzle repairs do not need a fatigue analysis (fatigue analysis waiver) as required by Section NB-3222.4(d) of the ASME Code Section III. However, the welded plugs for the RTD nozzles repairs were analyzed for fatigue per ASME Code Section III, Article NB-3222.4(e). By letter dated February 14, 2010, the staff issued RAI 4.3.2.7-5, asking the applicant to explain why a fatigue analysis does not need to be performed for the half nozzle repair, but one is required for the weld plug repair.

The applicant's March 1, 2010, response to RAI 4.3.2.7-5 stated that the Class 1 main loop piping fatigue analysis was performed to the ASME Code Section III, Subsection NB, 1974 edition with addenda through summer 1974. The PDT and sampling half-nozzle repairs satisfied the fatigue waiver evaluation of ASME Code Section III, Article NB-3222.4(d). Per Article NB-3222.4(a), "If the specified operation of the component meets all of the conditions of NB-3222.4(d), no analysis for cyclic operation is required..." The NB-3222.4(d) fatigue waiver option for the welded plugs was not pursued.

The staff finds that the applicant followed the requirements of the ASME Code Section III, NB-3222.4(a) and NB-3222.4(d) and agrees that the applicant is not required to do a fatigue analysis for the PDT and sampling nozzles. The staff's concern in RAI 4.3.2.7-5 is resolved.

Effect of Unit 3 MNSA Holes on Reactor Coolant Piping. The applicant stated that a MNSA had been installed at a leaking thermowell in Unit 3. This MNSA was replaced with a three-quarter nozzle repair via the Alloy 600 replacement program. The tapped holes in the hot leg for the MNSA attachment were not repaired after the nozzle replacement. This portion of the Unit 3 hot leg has a higher CUF at the tapped hole location, as identified in the MNSA design report. The CUF was confirmed in the replacement SG and power uprate design report. The applicant also stated that the higher CUFs associated with the MNSA tapped holes will not affect the fatigue monitoring of the RCS piping. The enhanced metal fatigue AMP CC monitoring action limit for the RCS will initiate re-evaluation or other corrective actions to address this Unit 3 location.

Time-Limited Aging Analyses

Action limits will permit completion of corrective actions before the design basis number of events is exceeded.

The MNSA has been replaced with a three-quarter nozzle repair and the tapped hole is a remnant of the MNSA repair. The staff finds that the applicant will use the CC method in the metal fatigue program to monitor the fatigue analysis of the tapped hole location at a replaced thermowell at Unit 3. The staff finds this acceptable because the applicant will monitor the effects of aging with the enhanced metal fatigue AMP such that corrective actions are taken before exceeding the design limit.

Redesigned Reactor Coolant System Thermowells. LRA Section 4.3.2.7, subsection "Redesigned Reactor Coolant System Thermowells," states that the thermowell modifications did not affect the previous conclusion concerning fatigue of the thermowells and that there is no safety determination based on the plant life for these high-cycle loads. Therefore, this is not a TLAA. By letter dated February 14, 2010, the staff issued RAI 4.3.2.7-6, asking the applicant to explain why this issue is not a TLAA since the thermowells experience high-cycle fatigue that is time-dependent. Also the applicant was requested to perform a fatigue analysis of the thermowells for 60 years, or justify why a fatigue usage factor analysis for 60 years is not needed to demonstrate that the new thermowell design satisfy the ASME Code Section III allowable CUF of 1.0.

The applicant's March 1, 2010, response to RAI 4.3.2.7-6 stated that the failure mechanism of the thermowells was high-cycle fatigue caused by the resonance between the thermowell's natural frequency and the vortex shedding frequencies of the coolant inside the pipe. Furthermore, the analysis and testing of the redesigned thermowells determined that the new design was not susceptible to this failure mechanism. This determination did not consider the plant life; therefore, the evaluation for high-cycle fatigue is not a TLAA, in accordance with 10 CFR 54.3(a)(3).

The staff finds that the new redesigned thermowells are not susceptible to high-cycle fatigue degradation mechanism, which was caused by resonance between the old thermowell natural frequency and the vortex shedding frequencies of the coolant inside the pipe. The staff agrees that the evaluation for high-cycle fatigue is not a TLAA and is not required to be considered for the new thermowells. This staff's concern in RAI 4.3.2.7-6 is resolved.

Safety Injection Nozzle Thermal Sleeves and Auxiliary Spray line. LRA Section 4.3.2.7 concludes that the modification on thermal sleeves of the safety injection nozzles did not affect the previous conclusion concerning fatigue of the safety injection nozzles. However, the applicant did not discuss whether the fatigue analysis of the safety injection nozzles or the auxiliary spray line was based on 40-year or 60-year transient cycles. Also, the LRA states that the CUF, including the effects of thermal gradient in the auxiliary spray line, meets the ASME Code Section III, for a 40-year plant life. By letter dated February 14, 2010, the staff issued RAI 4.3.2.7-7 and RAI 4.3.2.7-8, asking the applicant to perform a fatigue analysis for 60 years or justify why a fatigue analysis of the safety injection nozzles and auxiliary spray line and tee for the end of 60 years is not needed to demonstrate that the CUFs of the subject nozzles at the end of plant life satisfies the ASME Code Section III allowable of 1.0. The staff also asked whether this analysis is a TLAA.

The applicant's March 1, 2010, response to RAIs 4.3.2.7-7 and 4.3.2.7-8 stated that the fatigue analysis of the safety injection nozzles was performed using the 40-year design numbers of transient cycles; therefore, it is a TLAA and will be managed for the period of extended operation in accordance with 10 CFR 54.21(c)(1)(iii). The disposition of the fatigue analysis of the safety injection nozzles depends on the enhanced metal fatigue AMP to ensure that the 40-year design numbers of transients will not be exceeded during 60 years of operation without appropriate corrective actions. The applicant responded that the auxiliary spray line and tee

Time-Limited Aging Analyses

fatigue analysis is a TLAA and will be managed for the period of extended operation in accordance with 10 CFR 54.21(c)(1)(iii). The disposition of the fatigue analysis of the auxiliary spray line and tee depends on the enhanced metal fatigue AMP to ensure that the 40-year design number of transients will not be exceeded during 60 years of operation without appropriate corrective actions.

The staff finds that the applicant has clarified that the fatigue analysis of the safety injection nozzles and auxiliary spray line and tee is a TLAA, and the enhanced Metal Fatigue Program will monitor the fatigue analysis of the safety injection nozzles. The staff finds that the enhanced Metal Fatigue Program will maintain the structural integrity of the safety injection nozzles and auxiliary spray line and tee; therefore, the staff's concerns in RAIs 4.3.2.7-7 and 4.3.2.7-8 are resolved.

Hot Leg Surge and Shutdown Cooling Nozzle Weld Overlays. The applicant stated that PWSCC is a degradation mechanism for Alloy 82/182 welds. The applicant further stated that while no flaws have been detected at PVNGS, it will install full structural weld overlays over the pressurizer surge, spray, safety and relief valve nozzles, and the hot leg surge and shutdown cooling nozzle welds to ensure structural integrity of the RCS boundary. The applicant stated that these weld repairs meet the requirements of ASME Code Class 1 components and are supported by fracture mechanics analyses and periodic inspections. The applicant also stated that the fracture mechanics analyses of the materials overlaid by the weld repair are not TLAAs. The staff determined that it needed further information as discussed below.

By letter dated February 14, 2010, the staff issued RAI 4.3.2.7-9, asking the applicant to explain why the fracture mechanics analyses of the hot leg surge and shutdown cooling nozzles overlaid by the weld repair are not TLAAs. By letter dated March 1, 2010, in response to RAI 4.3.2.7-9, the applicant stated that the fatigue crack growth analyses calculate the crack propagation in order to demonstrate that a postulated crack will not exceed the acceptance criterion of the analysis during the inspection interval. The inspection interval is less than the plant life; therefore, the fatigue crack growth analyses are not TLAAs, in accordance with 10 CFR 54.3(a)(3).

The applicant stated further that the fatigue crack growth analyses of the weld overlay of the hot leg surge line and shutdown cooling line nozzles are not TLAAs because the postulated flaw is calculated for the inspection interval. The staff agrees that the fatigue crack growth analyses are not TLAAs when the nozzles are inspected during every 10-year inspection interval or are inspected within the time period before the flaws in the nozzles are projected to reach an unacceptable size, as shown in the fatigue crack growth analyses.

By letter dated May 8, 2008, the applicant submitted for the staff's review and approval, RR 36 for the weld overlay repair of the hot leg surge line and shutdown cooling line nozzles for the third ISI interval for Units 1 and 3. By letter dated November 10, 2008, the staff authorized the use of RR 36.

In these submittals, the applicant required inspection schedules of the weld overlaid nozzles. For example, RR 36 requires that overlaid nozzles be examined during the first or second refueling outage following weld overlay installation. Also, RR 36 requires fatigue crack growth calculations to be updated periodically to meet certain ASME requirements to ensure structural integrity of the subject nozzles.

Because the weld overlay has been installed on the hot leg surge and shutdown cooling nozzles to mitigate potential PWSCC and RR 36 has provided specific periodic inspection requirements, the staff finds that the structural integrity of the subject nozzles will be maintained adequately. For this case, the staff finds that the fatigue crack growth analyses for the subject nozzles are not a TLAA because the fatigue crack growth analyses are performed for every inspection interval (i.e. 10 years). The staff's concern in RAI 4.3.2.7-9 is resolved.

Time-Limited Aging Analyses

Disposition: Aging Management per 10 CFR 54.21(c)(1)(iii). LRA Section 4.3.2.7 states that the enhanced metal fatigue AMP will continue to confirm that usage factors and waivers are not time-dependent or that re-evaluation or other corrective action is initiated if an action limit is reached. By letter dated February 14, 2010, the staff issued RAI 4.3.2.7-10. This RAI asked that the applicant specify the exact piping components and systems that will be monitored under the enhanced metal fatigue AMP in the context of Section 4.3.2.7 and discuss how often the actions will be performed under the enhanced metal fatigue AMP (e.g., monitoring the transient cycles and reviewing the records). LRA Section 4.3.1 discusses that a FatiguePro® computer software program is used to monitor the transient cycles; however, it is not clear how often the applicant performs the monitoring and when corrective actions are taken.

The applicant's March 1, 2010, response to RAI 4.3.2.7-10 stated that the piping components that will be monitored are listed in LRA Table 4.3-4. LRA Section 4.3.1.5 states that the scope of the bounding set of monitored locations is sufficient to ensure that fatigue in any other locations of concern, not included in the set, is within the same system and subject to the same transients, or within a system affected by the same transients. LRA Section 4.3.1.5 states that the current metal fatigue AMP requires this evaluation at least once per fuel cycle. This schedule will apply to the enhanced metal fatigue AMP for the period of extended operation.

The staff finds that the applicant has clarified that the piping components in LRA Table 4.3-4 will be monitored by the enhanced Metal Fatigue Program at least once per fuel cycle. LRA Table 4.3-4 provides a summary of fatigue usages from the Class 1 piping analyses and methods of management by the metal fatigue of RCP boundary program. The staff finds that the Class 1 piping covered in LRA Table 4.3-4 is sufficiently comprehensive, and the monitoring frequency of every fuel cycle is adequate. The staff's concern in RAI 4.3.2.7-10 is resolved. The staff finds that the structural integrity of the Class 1 piping will be maintained during the period of extended operation based on the enhanced metal fatigue AMP.

4.3.2.7.3   Updated Final Safety Analysis Report Supplement

The applicant supplied a UFSAR supplement summary description of its TLAA of the ASME Code Section III, Class 1 Piping and Piping Nozzles in LRA Section A.3.2.1.7. Based on its review of the UFSAR supplement in LRA Section A.3.2.1.7, the staff concludes that the summary description of the applicant's actions to address the TLAA for the ASME Code Section III, Class 1 Piping and Piping Nozzles is adequate.

4.3.2.7.4   Conclusion

On the basis of its review, the staff concludes that, pursuant to 10 CFR 54.21(c)(1)(iii), the applicant has demonstrated that the effects of aging on the intended function of the ASME Code Section III, Class 1 Piping and Piping Nozzles will be adequately managed for the period of extended operation. The UFSAR supplement contains an appropriate summary description of the TLAA evaluation of the ASME Code Section III, Class 1 Piping and Piping Nozzles, as required by 10 CFR 54.21(d).

4.3.2.8   Absence of Supplemental Fatigue Analysis Time-Limited Aging Analyses in Response to Bulletin 88-08 for Intermittent Thermal Cycles due to Thermal-Cycle-Driven Interface Valve Leaks and Similar Cyclic Phenomena

The staff reviewed LRA Section 4.3.2.8 to verify that the supplemental fatigue analysis time-limiting aging analyses in response to Bulletin 88-08 for intermittent thermal cycles due to thermal-cycle-driven interface valve leaks and similar cyclic phenomena are not TLAAs. The staff's evaluation is found in SER Section 4.1.3.1.5.

Time-Limited Aging Analyses

4.3.2.9 Bulletin 88-11 Revised Fatigue Analysis of the Pressurizer Surge Line for Thermal Cycling and Stratification

4.3.2.9.1   Summary of Technical Information in the Application

LRA Section 4.3.2.9 states that the pressurizer surge lines are designed to the ASME Code Section III, Subsection NB, 1977 edition with addenda through summer 1979. The surge line design was reanalyzed in 1991 through the CEOG in response to concerns expressed in Bulletin 88-11, "Pressurizer Surge Line Thermal Stratification." The purpose of Bulletin 88-11 was to request that licensees establish and implement a program to confirm pressurizer surge line integrity in view of the occurrence of thermal stratification and to require licensees to inform the staff of the actions taken to resolve this issue. The applicant dispositions this TLAA in accordance with 10 CFR 54.21(c)(1)(iii), that the effects of aging on the intended functions of the pressurizer surge line will be adequately managed for the period of extended operation.

4.3.2.9.2   Staff Evaluation

The staff reviewed LRA Section 4.3.2.9 to verify, per 10 54.21(c)(1)(iii), that the effects of aging on the intended function(s) of the pressurizer surge line will be adequately managed for the period of extended operation.

Effects of Thermal Stratification on the Surge Line Piping Fatigue Analysis. As a result of thermal stratification occurring in pressurizer surge lines in the 1980s, the staff published Bulletin 88-11 to alert PWR licensees about the issue and to require licensees to mitigate thermal stratification and monitor potential degradation. Bulletin 88-11 requires PWR licensees to visually inspect the surge line, demonstrate that the surge line meets applicable design codes, and update stress and fatigue stress analyses of the surge line.

The applicant stated that CE performed a fatigue evaluation of surge lines in CEOG plants with thermal stratification loading. The analysis assumed the design basis number of 500 heatup transients. The CEOG analysis is based on a limiting set of thermal stratification transients defined from data collected from several CE units, not including PVNGS, but used the PVNGS surge line for the limiting analysis because its geometry produced the most-limiting stresses. Insurge-outsurge and thermal stratification effects doubled the 40-year CUF of the original analysis of record at the limiting location in the surge line elbow at the pressurizer.

The applicant reported that the elastic analysis produced a CUF of 1.65 in the surge line elbow. To decrease the CUF below the ASME fatigue limit of 1.0, CE performed a plastic analysis, resulting in a limiting CUF of 0.937 in the surge line elbow. This CEOG limiting-case analysis is conservative because it did not include any credit for mitigating actions, or the actual severity of transients, experienced during operation. A reanalysis for more realistic transients should, therefore, be able to demonstrate considerable margin. PVNGS collected and reduced their data independently from the other plants; hence, there is no specific thermal transient information from PVNGS within the CEOG report. However, in the absence of any analysis more specific to PVNGS, the applicant confirmed this bounding analysis as the fatigue analysis of record for this component.

By letter dated February 14, 2010, the staff issued RAI 4.3.2.9-1, asking the applicant to describe the analyses in detail including methodology, input, acceptance criteria, and results and to clarify whether the CUF analysis is based on a 40-year period or 60-year period. LRA Section 4.3.2.9 states that the Metal Fatigue of Reactor Coolant Pressure Boundary Aging Management Program will be used to monitor the surge line. The staff asked the applicant to discuss whether the metal fatigue AMP will initiate actions based on the elastic analysis result (CUF of 1.65) or plastic analysis result (CUF of 0.937).

The applicant's March 1, 2010, response to RAI 4.3.2.9-1 stated that calculation 13-MC-ZZ-595 performed fatigue evaluations for the pressurizer surge line, including the Bulletin 88-11

Time-Limited Aging Analyses

additional thermal cycling and stratification effects. This analysis incorporated results of CEOG calculation MISC-ME-C-115 and report CEN-387-P.

The applicant further stated that these analyses found that the highest CUF (i.e., 1.65) in the surge line is in the elbow below the pressurizer. The 1.65 CUF is a result of a preliminary shakedown analysis per the ASME Code Section III, NB-3228.4. As stated in the scope of calculation MISC-ME-C-115, "[t]he fatigue evaluation program developed to analyze the shakedown analysis results is used only to rank each point in the elbow. The output from this program in no way represents the fatigue usage for the actual transients listed...." The same analysis re-analyzed this highest-ranking 1.65 CUF location and calculated an actual CUF of 0.778. As stated in the conclusion of calculation MISC-ME-C-115, "[u]sing the transients analyzed in the shakedown analysis as a method to rank locations in the limiting surge line elbow, the location of greatest usage is analyzed for its actual usage.... The point of highest actual usage is U = 0.778."

In response to Bulletin 88-11, CEOG performed the evaluation reported in CEN-387-P on pressurizer surge line thermal stratification. CEN-387-P reported a plant-specific analysis CUF of 0.937 at the surge line elbow location, also using the methods of NB-3228.4. The acceptance criterion is a calculated 40-year CUF that is less than or equal to 1.0. This plant-specific analysis is the analysis of record for this component at PVNGS. CEN-387-P included effects of insurge-outsurge and thermal stratification at the limiting surge line elbow location.

The applicant stated that if CC were used as the fatigue management method, the metal fatigue AMP would, therefore, initiate actions based on the analysis of record result (CUF of 0.937). However, this is a sample location based on NUREG/CR-6260, *"Application of NUREG/CR-5999 Interim Fatigue Curves to Selected Nuclear Power Plant Components,"* and when a conservative estimate of the multiplier for effects of the reactor coolant environment is used, the calculated CUF becomes several times the code acceptance criterion of 1.0. The applicant stated that fatigue in this location will, therefore, be managed by SBF monitoring, and the action limits for this location will, therefore, depend on calculated actual fatigue usage, not on a 40-year (or 60-year) value determined by the analysis of record. LRA Section 4.3.4 provides additional information on the NUREG/CR-6260 sample locations, and LRA Table 4.3-4, item 24, provides information for monitoring of this location.

The applicant stated that LRA Section 4.3.2.9 misidentified the 1.65 CUF as the result of an elastic analysis. By letter dated March 1, 2010, the applicant submitted Amendment 10, which revised the sentence to read, "[a] preliminary shakedown analysis produced..." The related paragraphs in LRA Section 4.3.4 also misidentify the 1.65 CUF as the result of an elastic analysis and include an unnecessary sentence. In LRA Amendment 10, the applicant revised the last paragraph on LRA page 4.3-64 as follows:

> Pressurizer Surge Line (Hot Leg) Elbow (Location 4): Combustion Engineering (CE) performed a fatigue evaluation of surge lines in various CE Owners Group (CEOG) plants, with thermal stratification loading. The analysis assumed the design basis number of 500 heatup transients. A preliminary shakedown analysis produced a cumulative usage factor of 1.65 in the comparable (and more limiting) surge line pressurizer elbow. To decrease the CUF below the ASME fatigue limit of 1.0, CE then performed a plastic analysis resulting in a limiting CUF of 0.937 in the pressurizer elbow. APS confirmed this bounding analysis as the fatigue analysis of record for this component at PVNGS. See Section 4.3.2.9.
>
> To evaluate effects of the reactor coolant environment, APS re-evaluated the CUF in the pressurizer surge line hot leg elbow using design basis transient

Time-Limited Aging Analyses

cycles and ASME Subsection NB-3200 6-component stress tensors. The simplified elastic-plastic analysis produced a CUF of 1.9396, which is above the ASME code allowable fatigue limit of 1.0. The CE plastic analysis described in Section 4.3.2.9 that calculated a CUF of 0.937 is more precise than the APS reevaluation; however, the APS hot leg elbow reevaluation will....

The staff finds that the applicant has clarified the CUF calculation and the CUF results of the pressurizer surge line. Per NRC Bullet 88-11, the applicant has considered the thermal cycles and stratification in CUF calculation of the surge line. The metal fatigue program will monitor the CUF using the SBF monitoring method. The staff finds that the structural integrity of the pressurizer surge line will be adequately monitored during the period of extended operation. The staff also finds that the applicant satisfies the TLAA requirements of 10 CFR 54.21(c)(1)(iii) because the metal fatigue program will be used to monitor the CUF of the pressurizer surge line. The staff's concern in RAI 4.3.2.9-1 is resolved.

Effect of Bulletin 88-11 on Risk-Informed Inservice Inspection Program. PVNGS augmented its ASME Code Section XI, ISI Program to include inspections of the surge line elbow, which were done to address NRC Bulletin 88-11 concerns. PVNGS subsequently proposed the alternative Risk-Informed ISI (RI-ISI) in RR-32. The RI-ISI application is based on the EPRI RI-ISI Program, which explicitly considered NRC Bulletin 88-11 concerns in its application. Therefore, the PVNGS RI-ISI Program addresses the NRC Bulletin 88-11 concerns.

By letter dated February 14, 2010, the staff issued RAI 4.3.2.9-2, asking the applicant to confirm that the surge line elbow is a component that requires a nondestructive examination to be performed under the ISI program. This RAI also asks that the applicant, discuss how often the surge line elbow will be inspected in each of the 10-year ISI intervals through the sixth interval and discuss the nondestructive examination method that will be used for each inspection.

The applicant's March 1, 2010, response to RAI 4.3.2.9-2 stated that the surge line elbow is subject to the ASME Code Section XI, ISI Program. Subsequent to the initial LRA submittal, no RR has been filed to permit use of a RI-ISI program for the current, third inspection interval. The applicant revised LRA Section 4.3.2.9 as follows:

> PVNGS augmented its ASME Section XI, ISI program to include inspections of the surge line elbow, which were performed to address NRC Bulletin 88-11 concerns. PVNGS subsequently proposed the alternative RI-ISI in RR 32 for the third period of the second ISI interval. The RI-ISI application was based on the EPRI RI-ISI program, which explicitly considered NRC Bulletin 88-11 concerns in its application. The NRC Bulletin 88-11 concerns were therefore addressed by the PVNGS RI-ISI program. However the program was approved only for the third period of the second ISI interval, and no relief request has been filed for the current, third interval.

The applicant stated that the ASME Code Section XI requires that the surge line elbow in each of the PVNGS units be visually (VT-2) examined each refueling outage. The ISI Program is revised to the requirements of 10 CFR 50.55a for each inspection interval.

The staff finds that the applicant will follow the ASME Code Section XI to inspect the pressurizer surge line elbow that has the highest CUF. The applicant will visually inspect the elbow each refueling outage. The staff finds that the structural integrity of the pressurizer surge line elbow will be adequately maintained during the period of extended operation. The staff finds acceptable that the implementation of the RI-ISI program is limited to only the third period of the second ISI interval. The staff's concern described in RAI 4.3.2.9-2 is resolved.

Effects of Power Uprate and Steam Generator Replacement on the Surge Line Piping Fatigue Analysis. The applicant stated that the evaluation of the power uprate and SG replacement found that the resulting changes in temperature ranges have no effect on the surge line fatigue analysis.

The applicant stated that the surge line elbow will be subject to SBF monitoring under the enhanced Metal Fatigue of Reactor Coolant Pressure Boundary Program. The program will maintain a record of the CUF. This record will be reviewed and evaluated at intervals specified by the program, at a frequency sufficient to ensure that appropriate corrective action will be initiated if an action limit is reached. Action limits will be established to permit completion of corrective actions before the code limit is exceeded. The effects of fatigue in the Class 1 surge line will thereby be managed for the period of extended operation, in accordance with 10 CFR 54.21(c)(1)(iii).

In RAI 4.3.2.9-3, the staff asked the applicant to discuss how often they will review and evaluate the record of the worst case CUFs. The applicant's March 1, 2010, response to RAI 4.3.2.9-3 stated that as stated in LRA Section 4.3.1.5, page 4.3-23, "The PVNGS fatigue management program currently ... requires this evaluation at least once per fuel cycle." This schedule will apply to the enhanced metal fatigue AMP for the period of extended operation. The staff finds that the monitoring frequency of once per fuel cycle for the CUF of the surge line elbow is adequate to ensure its structural integrity because it will ensure the applicant takes corrective actions before exceeding the design limit. The staff's concern described in RAI 4.3.2.9-3 is resolved.

4.3.2.9.3 Updated Final Safety Analysis Report Supplement

The applicant provided a UFSAR supplement summary description of its TLAA of the pressurizer surge line for thermal cycling and stratification in LRA Section A3.2.1.8. On the basis of its review of the UFSAR supplement in LRA Section A3.2.1.8, the staff concludes that the summary description of the applicant's actions to address the TLAA for pressurizer surge line for thermal cycling and stratification is adequate.

4.3.2.9.4 Conclusion

On the basis of its review, the staff concludes that, pursuant to 10 CFR 54.21(c)(1)(iii), the applicant has demonstrated that the effects of aging on the intended function of the pressurizer surge line for thermal cycling and stratification will be adequately managed for the period of extended operation. The UFSAR supplement contains an appropriate summary description of the TLAA evaluation of the pressurizer surge line for thermal cycling and stratification, as required by 10 CFR 54.21(d).

4.3.2.10 Class 1 Fatigue Analyses of Class 2 Regenerative and Letdown Heat Exchangers

4.3.2.10.1 Summary of Technical Information in the Application

LRA Section 4.3.2.10 and Amendment 16 dated May 27, 2010, summarize the evaluation of the ASME Code Class 1 fatigue analyses of ASME Code Class 2 regenerative and letdown heat exchangers. The applicant stated that the specifications of those exchangers require an ASME Code Class 1, NB-3222 analysis, including a fatigue evaluation for a specified set of events, each for a specified number of occurrences, for a 40-year design life. The applicant dispositions this TLAA in accordance with 10 CFR 54.21(c)(1)(iii), that the effects of aging on the intended functions of the ASME Code Class 2 regenerative and letdown heat exchangers will be adequately managed for the period of extended operation.

4.3.2.10.2 Staff Evaluation

The staff reviewed LRA Section 4.3.2.10 and Amendment 16 to verify pursuant to 10 CFR 54.21(c)(1)(iii), that the effects of aging on the intended functions will be adequately managed for the period of extended operation.

Time-Limited Aging Analyses

LRA Amendment 16 revised Section 4.3.2.10 and stated that for both types of heat exchangers, the fatigue analyses were performed with transients specified in the CE general specification for System 80 plants. The staff noted that the applicant did not identify which transients were evaluated in the System 80 CUF calculations for these heat exchangers. Furthermore, the LRA stated that the fatigue effects of the heat exchangers are bounded by the fatigue of the charging nozzle. However, the current design basis CUF values for the heat exchangers were not provided. By letter dated July 21, 2010 the staff issued RAI 4.3-16 requesting the applicant to clarify the current design basis CUF values for the regenerative heat exchangers and letdown heat exchangers; identify the transients that were evaluated in the CUF calculations of these heat exchangers; and identify the design basis limits for the transients analyzed in these calculations. This was previously identified as part of Open Item 4.3-1.

The applicant's response dated August 12, 2010, provided the design basis CUFs for the components and identified the charging nozzle as the limiting location. The response also explained the design basis transients analyzed and associated limits for the transients. During the review of the analyses of record for the heat exchangers, the applicant noted that the analysis assumed a higher number of cycles for significant design transients and a lower number of cycles for less significant transients than those stated in the UFSAR for several transients. The applicant stated that none of the transient limits have been challenged by current operating history. The applicant also stated that the inconsistency between the transient assumptions in the UFSAR and those in the analyses is in the corrective action process for evaluation and resolution (tracking number (CRAI) 3494095).

Based on its review, the staff finds the applicant's response to RAI 4.3-16 acceptable because, (1) the CUF values confirm that the charge nozzle is the bounding location; (2) the number of events in the design basis specifications are consistent with or greater than the number of transients that will be used as CC action limits in the enhanced Metal Fatigue Program which ensures that the CC action limits are appropriate for the heat exchangers; and (3) the applicant identified and initiated an action to track and resolve the inconsistency between the transient assumptions in the UFSAR and those in the analysis. The staff's concern described in RAI 4.3-16 is resolved and this part of Open Item 4.3-1 is closed.

Based on its review, the staff finds the applicant has demonstrated, pursuant to 10 CFR 54.21(c)(1)(iii), that the effects of aging on the intended functions of the Class 2 regenerative and letdown heat exchangers will be adequately managed for the period of extended operation because the action limits of the enhanced Metal Fatigue Program will be established to permit completion of corrective actions before the design basis number of events is exceeded. Thus, the fatigue usage factor will not exceed the design limit of 1.0 during the period of extended operation.

4.3.2.10.3 Updated Final Safety Analysis Report Supplement

LRA Section A3.2.1.9 provides the UFSAR supplement summarizing the evaluation of the Class 1 fatigue analyses for Class 2 regenerative and letdown heat exchangers. Based on its review of the UFSAR supplement, the staff concludes that the applicant provided an adequate summary description of its actions to address the TLAA for the Class 2 regenerative and letdown heat exchangers as required by 10 CFR 54.21(d).

4.3.2.10.4 Conclusion

On the basis of its review, the staff concludes that the applicant has demonstrated, pursuant to 10 CFR 54.21(c)(1)(iii), that the effects of aging on the intended functions of the Class 2 regenerative and letdown heat exchangers will be adequately managed for the period of extended operation. The staff also concludes that the UFSAR supplement contains an appropriate summary description of the TLAA evaluation, as required by 10 CFR 54.21(d).

Time-Limited Aging Analyses

4.3.2.11 Class 1 Fatigue Analyses of Class 2 High-Pressure Safety Injection and Low-Pressure Safety Injection Safeguard Pumps for Design Thermal Cycles

4.3.2.11.1 Summary of Technical Information in the Application

LRA Section 4.3.2.11 and Amendment 16, dated May 27, 2010, summarize the evaluation of the Class 1 fatigue analyses of Class 2 HPSI and LPSI pumps. The applicant stated that the design of the Class 2 pumps includes no fatigue analysis. The applicant dispositions this TLAA per 10 CFR 54.21(c)(1)(iii), that the effects of aging on the intended functions of the Class 2 HPSI and LPSI pumps will be adequately managed for the period of extended operation.

4.3.2.11.2 Staff Evaluation

The staff reviewed LRA Section 4.3.2.11 and Amendment 16, to verify per 10 CFR 54.21(c)(1)(iii), that the effects of aging on the intended functions of the Class 2 HPSI and LPSI pumps will be adequately managed for the period of extended operation.

The applicant stated that the design of the Class 2 pumps does not include a fatigue analysis. However, UFSAR Section 3.9.3.5.3.3 describes the design for a stated number of thermal transient cycles. The applicant stated that the structural integrity and operability analyses for both pumps cite the Class 1 methods of ASME Code Section III, Article NB-3222.4 when addressing these thermal transients.

The staff noted that both the HPSI and LPSI pumps are designed for the injection initiation transient temperature change of 40 degrees to 300 degrees F. The applicant stated that using the design temperature of the transient, ASME Code Section III, Appendix I, Figure I-9.2 allows approximately 550 operating cycles and 23,500 operating cycles for the HPSI and LPSI pumps, respectively. Since the design of the pumps assumed 10 cycles, the applicant concluded that there is sufficient margin to support the period of extended operation.

The staff noted that the LPSI pumps are designed for the shutdown cooling initiation transient temperature change of 70 degrees to 350 degrees F. The applicant stated that by using the design temperature of the transient, approximately 18,000 operating cycles are allowed for the LPSI pump. Since the design of the pump assumed 500 cycles, the applicant concluded that there is sufficient margin to support the period of extended operation.

Based on its review, the staff finds the applicant has demonstrated, pursuant to 10 CFR 54.21(c)(1)(iii), that the analyses for the effects of aging on the intended functions of the Class 2 HPCI and LPCI pumps remain valid for the period of extended operation because the enhanced Metal Fatigue Program will track events to ensure that appropriate corrective action will be initiated and completed before the design basis number of events is exceeded such that the fatigue usage factor will not exceed the design limit of 1.0.

4.3.2.11.3 Updated Final Safety Analysis Report Supplement

LRA Section A3.2.1.10 provides the UFSAR supplement summarizing the evaluation of the Class 1 fatigue analyses for the Class 2 HPSI and LPSI pumps. Based on its review of the UFSAR supplement in the LRA and UFSAR Section 3.9.3.5.3.3, the staff concludes that the applicant provided an adequate summary description of its actions to address the TLAA for the Class 2 HPSI and LPSI pumps as required by 10 CFR 54.21(d)

4.3.2.11.4 Conclusion

On the basis of its review, the staff concludes that the applicant has demonstrated, pursuant to 10 CFR 54.21(c)(1)(iii), that the effects of aging on the intended functions of the Class 2 HPSI and LPSI pumps will be adequately managed for the period of extended operation. The staff also concludes that the UFSAR supplement contains an appropriate summary description of the TLAA evaluation, as required by 10 CFR 54.21(d)

Time-Limited Aging Analyses

4.3.2.12 Class 1 Analysis of Class 2 Main Steam Safety Valves

4.3.2.12.1 Summary of Technical Information in the Application

LRA Section 4.3.2.12 and Amendment 16 dated May 27, 2010, summarize the evaluation of the Class 1 analysis of Class 2 main steam safety valves (MSSVs). The applicant stated that the design of the Class 2 MSSVs includes a Class 1 fatigue analysis to ASME Code Section III, Article NB-3550, "Cyclic Loads for Valves." The applicant stated that since the cyclic design basis is described in the UFSAR, the fatigue analysis is, therefore, a TLAA. The applicant dispositions this TLAA in accordance with 10 CFR 54.21(c)(1)(i), that the analyses remain valid for the period of extended operation.

4.3.2.12.2 Staff Evaluation

The staff reviewed LRA Section 4.3.2.12 and Amendment 16, to verify pursuant to 10 CFR 54.21(c)(1)(i), that the analyses remains valid during the period of extended operation.

The applicant stated that the design of the Class 2 MSSVs includes a Class 1 fatigue analysis to ASME Code Section III, Subsubarticle NB-3550, "Cyclic Loads for Valves." The applicant provided the cyclic design basis as described in the UFSAR Section 5.2.2.4.3.2. The applicant further provided usage factors at two critical areas of the valves, the inlet crotch and the disc. The two usage factors are less than one-ninth of the design limit of 1.0. Based on its review, the staff finds the applicant has demonstrated, pursuant to 10 CFR 54.21(c)(1)(i), that the Class 1 fatigue analyses for the Class 2 MSSVs remains valid during the period of extended operation because the valves are suitable to operate adequately for at least nine times the original 40-year design lifetime.

4.3.2.12.3 Updated Final Safety Analysis Report Supplement

LRA Section A3.2.1.11 provides the UFSAR supplement summarizing the evaluation of Class 1 analysis of Class 2 MSSVs. Based on its review of the UFSAR supplement, the staff concludes that the applicant provided an adequate summary description of its actions to address the TLAA for the Class 2 MSSVs as required by 10 CFR 54.21(d).

4.3.2.12.4 Conclusion

On the basis of its review, the staff concludes that the applicant has demonstrated, pursuant to 10 CFR 54.21(c)(1)(i), that the Class 1 fatigue analyses for the Class 2 MSSVs remain valid during the period of extended operation. The staff also concludes that the UFSAR supplement contains an appropriate summary description of the TLAA evaluation, as required by 10 CFR 54.21(d).

4.3.2.13 Absence of Time-Limited Aging Analyses in Evaluations of Effects of Vibration on the Unit 1, Train A Shutdown Cooling System Suction Line Fatigue Analysis and of Vibration Limits Established for its Isolation Valve Actuator

The staff reviewed LRA Section 4.3.2.13 to verify that the evaluations of the effects of vibration on the Unit 1, Train A SCS system suction line fatigue analysis and vibration limits established for its isolation valve actuator are not TLAAs. The staff's evaluation is found in SER Section 4.1.3.1.6.

4.3.2.14 High Energy Line Break Postulation Based on Fatigue Cumulative Usage Factor

4.3.2.14.1 Summary of Technical Information in the Application

LRA Section 4.3.2.14 describes the applicant's TLAA for high energy line break (HELB) postulation. Break locations are determined in accordance with Standard Review Plan for License Renewal (SRP-LR) Branch Technical Position (BTP) MEB 3-1. Breaks in piping with ASME Code Section III, Class 1 fatigue analyses are identified based on the CUF values (with

Time-Limited Aging Analyses

the stated exception of the reactor coolant system primary loops), and these determinations, therefore, are TLAAs. The RCS primary loop piping is eliminated from consideration by the leak before break (LBB) analyses. The applicant dispositions this TLAA in accordance with 10 CFR 54.21(c)(1)(iii), that the effects of aging on the HELB postulation will be adequately managed for the period of extended operation.

4.3.2.14.2 Staff Evaluation

The staff reviewed LRA Section 4.3.2.14, to verify, pursuant to 10 CFR 54.21(c)(1)(iii), that the effects of aging on the HELB postulation will be adequately managed for the period of extended operation.

The LRA states that breaks in piping with ASME Code Class 1 fatigue analyses are identified based on CUF (with the exception of the RCS primary loops, as stated above, which are eliminated by the LBB analyses). Break location postulations, which depend on usage factor, will remain valid as long as the calculated usage factors are not exceeded. The applicant also stated that the enhanced Metal Fatigue Program will ensure that appropriate reevaluation or other corrective actions are initiated if an action limit is reached. Action limits for the HELB design basis permit completion of corrective actions before the calculated design basis usage factors in Class 1 lines (outside the reactor coolant system primary loops) are exceeded. The staff determined that the applicant appropriately accounted for the HELB postulations because the break locations remain valid as long as the cumulative usage factors are less than 0.1.

Based on its review, the staff finds the applicant has demonstrated, pursuant to 10 CFR 54.21(c)(1)(iii), that the effects of aging on the HELB postulation will be adequately managed for the period of extended operation because the enhanced Metal Fatigue Program will ensure that appropriate corrective actions will be initiated if an action limit is reached. Action limits for the HELB design basis will be established to permit completion of corrective actions before the calculated design basis usage factors in Class 1 lines (outside the reactor coolant system primary loops) is exceeded.

4.3.2.14.3 Updated Final Safety Analysis Report Supplement

LRA Section A3.2.1.12 provides the UFSAR supplement summary description for the HELB postulation of TLAAs that were evaluated in LRA Section 4.3.2.14. Based on its review of the UFSAR supplement, the staff concludes that the applicant provided an adequate summary description of its actions to address the HELB postulation.

4.3.2.14.4 Conclusion

On the basis of its review, the staff concludes that the applicant has demonstrated, pursuant to 10 CFR 54.21(c)(1)(iii), that the effects of aging on the HELB postulation will be adequately managed for the period of extended operation. The staff also concludes that the UFSAR supplement contains an appropriate summary description of the TLAA evaluation, as required by 10 CFR 54.21(d).

4.3.2.15 Absence of Time-Limited Aging Analyses in Fatigue Crack Growth Assessments and Fracture Mechanics Stability Analyses for the Leak-Before-Break Elimination of Dynamic Effects of Primary Loop Piping Failures

The staff reviewed LRA Section 4.3.2.15 to verify that the LBB (elimination of dynamic effects of primary loop piping failures) fatigue crack growth and fracture mechanics stability analyses are not TLAAs. The staff's evaluation is found in SER Section 4.1.3.1.4.

Time-Limited Aging Analyses

4.3.3 Fatigue and Cycle-Based Time-Limited Aging Analyses of American Society of Mechanical Engineers III, Subsection NG, Reactor Pressure Vessel Internals

4.3.3.1 Summary of Technical Information in the License Renewal Application

LRA Section 4.3.3 summarizes the evaluation of the CUF analyses that comprise the "Fatigue and Cycle-Based TLAAs of ASME III Subsection NG Reactor Pressure Vessel Internals," for the period of extended operation. These TLAAs are based on the CUF analyses in the applicant's current design for the applicant reactor vessel internal (RVI) components. The applicant dispositions these TLAAs per 10 CFR 54.21(c)(1)(iii), that the effects of aging on the intended functions of the reactor vessel internals will be adequately managed for the period of extended operation.

4.3.3.2 Staff Evaluation

On May 27, 2010, the applicant submitted LRA Amendment 16. In this LRA Amendment, the applicant submitted its conforming changes to LRA Section 4.3.3 to address staff concerns that were discussed with the applicant in a public meeting on May 6, 2010.

In LRA Amendment 16, the applicant amended Section 4.3.3 to note that some of the reactor vessel internal (RVI) components were designed to the 1974 Edition of the ASME Code Section III, Subsection NG, or to more recent endorsed versions of the ASME Code Section III. LRA Section 4.3.3 identifies that the design codes require CUF calculations for these ASME Code Section III, Subsection NG components and that these analyses are TLAAs. The staff noted that the Materials Reliability Program Report (MRP-227) identifies the following CE RVI components as ASME Code Class 1 components:

- Guide lugs and guide lug inserts and bolts
- Fuel alignment pins
- RVI components in the upper flange assembly

The staff noted that the assessment in LRA Section 4.3.3 does not identify which of the RVI components were designed to these ASME Code requirements and were required to have a CUF calculation. By letter dated July 21, 2010, the staff issued RAI 4.3-17 asking that the applicant identify which RVI components are designed to Subsection NG requirements and are required to have a CUF design calculation. The staff also asked that the applicant identify the design basis CUFs for those components and design basis limits for those transients analyzed. The staff also requested the applicant justify the use of cycle-based monitoring if the existing design basis CUF value for any RVI component with a high CUF value (e.g. greater than or equal to 0.9). This was previously identified as part of Open Item 4.3-1.

The applicant's response dated August 12, 2010, provided a list of the RVI components as described in UFSAR Section 3.9.5. In addition to tabulating the fatigue usage factors of the components of the RVI, the applicant also provided the design basis transients and associated limiting number of events. The applicant stated that since the RVI fatigue usage factors depend on the effects of transient events, the increase in operating life to 60 years will not have a significant effect on the fatigue usage factors if the numbers of design transient cycles remain within the numbers assumed by the original 40-year analysis. The applicant explained that monitoring the transient counts to remain less than their 40-year value will ensure that the CUF remain less than design basis CUFs. Furthermore, since any design basis CUF less than 1.0 is an acceptable result, the applicant stated that no additional action is required to be taken for components with CUFs close to, but less than 1.0.

Based on its review, the staff finds the applicant's response to RAI 4.3-17 acceptable because the applicant identified the ASME Code Section III, Subsection NG RVI components and clarified the design basis CUFs and transients for those components. Further, the applicant will

4-75

Time-Limited Aging Analyses

use CC in its enhanced Metal Fatigue Program to track these transients to ensure that when action limits are reached, that corrective actions are taken to maintain fatigue usage below the design limit of 1.0. The staff's concern described in RAI 4.3-17 is resolved and this part of Open Item 4.3-1 is closed.

The applicant stated that the fatigue usage factors for the RVI components do not depend on flow-induced vibration or other high-cycle effects that are time-dependent at steady-state conditions. The staff reviewed UFSAR Section 3.9.2.5 and noted there is a discussion regarding flow-induced vibrations and confirmed that there are no TLAAs associated with the evaluation of flow-induced vibration or other high-cycle effects. The staff further noted that for those RVI components that have a CUF analysis, the usage factor is below the design limit of 1.0.

Based on its review, the staff finds the applicant has demonstrated pursuant to 10 CFR 54.21(c)(1)(iii), that the effects of aging will be adequately managed for the period of extended operation because the applicant will use its enhanced Metal Fatigue Program to monitor the number of transient cycles to ensure that corrective actions are taken if an action limit is reached, to ensure that the assumption made in the design calculations remain valid.

4.3.3.3 Updated Final Safety Analysis Report Supplement

LRA Section A3.2.2 provides the UFSAR supplement summarizing the TLAA for the ASME Code Section III, Subsection NG RVI components. The staff reviewed LRA Section A3.2.2 against the acceptance criteria in SRP-LR Sections 4.3.2.3 and 4.3.3.3. Based on its review, the UFSAR supplement is consistent with SRP-LR Sections 4.3.2.3 and 4.3.3.3. The staff determines that the applicant provided an adequate summary description of its actions to address the ASME Code Section III, Subsection NG RVI components, as required by 10 CFR 54.21(d).

4.3.3.4 Conclusion

On the basis of its review, the staff concludes that the applicant has provided an acceptable demonstration, pursuant to 10 CFR 54.21(c)(1)(iii), that the effects of aging on the intended functions of the RVI components will be adequately managed for the period of extended operation. The staff also concludes that the UFSAR supplement contains an appropriate summary description of the TLAA evaluation, as required by 10 CFR 54.21(d).

4.3.4 Effects of the Reactor Coolant System Environment on Fatigue Life of Piping and Components (Generic Safety Issue 190)

4.3.4.1 Summary of Technical Information in the License Renewal Application

LRA Section 4.3.4 summarizes the evaluation of the CUF analyses that comprise the "Effects of the Reactor Coolant System Environment on Fatigue Life of Piping and Components (Generic Safety Issue 190)," for the period of extended operation. These EAF analyses are not mandated by applicant's CLB for ASME Code Class 1 components. Instead, the applicant identifies that, although these types of analyses are not part of the existing design basis, they were included in order to conform with acceptance criteria and review procedure recommendations in the SRP-LR, Sections 4.3.2.2 and 4.3.3.2, respectively.

The applicant stated that the EAF analyses for its ASME Code Class 1 components were done in order to resolve the concerns identified in GSI-190 and in accordance with the staff's recommendations in NUREG/CR-6260 "Application of NUREG/CR-5999 Interim Fatigue Curves to Selected Nuclear Power Plant Components." The applicant stated that NUREG/CR-6260 recommended that the following CE component locations be analyzed for EAF:

- Reactor vessel (RV) shell and lower head

- RV inlet nozzles
- RV outlet nozzles
- Surge line
- Charging system nozzle
- Safety injection system nozzle
- Shutdown cooling line

In LRA Table 4.3-11, the applicant notes that the following component locations were selected as the limiting Class 1 locations that correspond to the EAF assessment locations recommended for CE facilities in NUREG/CR-6260:

- RV shell and lower head
- RV inlet nozzle
- RV outlet nozzle
- Surge line (hot leg) elbow
- Charging system nozzle (safe end location)
- Safety injection nozzle (forging knuckle)
- Safety injection nozzle (safe end)
- Shutdown cooling line (long radius elbow)
- Pressurizer heater locations (not identified as locations in NUREG/CR-6260)

LRA Table 4.3-11 also includes the applicant's EAF factors (F_{en} factors) that it used to adjust the CUF calculations of these components and the EAF usage factor results (F_{en} adjusted CUF values) that the applicant had calculated for these components at the end of the period of extended operation.

The applicant stated that the EAF analyses for these components are projected to the end of the period of extended operation in accordance with the criterion in 10 CFR 54.21(c)(1)(ii). Otherwise, the affect of environmentally-assisted, fatigue-induced cracking on the intended pressure boundary functions of the components will be managed for the period of extended operation in accordance with the TLAA acceptance criterion in 10 CFR 54.21(c)(1)(iii).

4.3.4.2 Staff Evaluation

The staff noted that the applicant conservatively addressed the effects of the reactor coolant environment on component fatigue life consistent with the guidance in the SRP-LR and the staff's recommendations for resolving Generic Safety Issue No. 190 (GSI-190), dated December 26, 1999. The staff also noted that, consistent with Commission Order No. CLI-10-17, dated July 8, 2010, the evaluations associated with the effects of the reactor coolant environment on component fatigue life do not fall within the definition of a TLAA in 10 CFR 54.3(a) because these evaluations are not in the applicant's CLB. Based on Commission Order No. CLI-10-17, the staff finds the applicant's evaluation of the effects of the reactor coolant environment on component fatigue life is conservative and is an acceptable practice consistent with the staff's recommendations in the SRP-LR and the closure of GSI-190.

On May 27, 2010, the applicant submitted LRA Amendment 16 which provided conforming changes to LRA Section 4.3.4 to address staff concerns that were discussed in a public meeting on May 6, 2010.

The staff reviewed the applicant's bases for dispositioning each of the EAF analyses in order to confirm whether or not the applicant had provided a valid basis for demonstrating that each of the CUF analyses would be acceptable in accordance with 10 CFR 54.21(c)(1)(ii), that the analysis has been projected to the end of the period or extended operation, or 10 CFR 54.21(c)(1)(iii), that the effects of aging on the intended functions will be adequately managed for the period of extended operation.

Time-Limited Aging Analyses

LRA Section 4.3.4 states that the maximum applicable environmental factors (F_{en}) for low alloy steel was used for RPV shell and lower head, RPV inlet and outlet nozzles, and safety injection nozzle (forging knuckle). These factors were determined following NUREG/CR-6583, "Effects of LWR Coolant Environments on Fatigue Design Curves of Carbon and Low-Alloy Steels." However, the staff noted that LRA Section 4.3.4 does not give sufficient information to confirm this statement. By letter dated June 2, 2010 the staff issued RAI 4.3-4 asking that the applicant demonstrate that the F_{en} factor used for assessment of the reactor coolant environmental affect on the RPV shell and lower head, RPV inlet and outlet nozzles, and safety injection nozzle (forging knuckle) are the maximum applicable for a given material. The staff also asked that the applicant provide a basis and justification for any assumptions that were made for the parameters in the assessment, such as strain rate, dissolved oxygen, temperature, and sulfur content. This was previously identified as part of Open Item 4.3-1.

The applicant's response dated June 29, 2010, stated that the "maximum applicable" F_{en} factors for the low alloy steel RPV shell and lower head, RPV inlet and outlet nozzles, and safety injection nozzle (forging knuckle) were all computed using NUREG/CR-6583. The applicant further stated that in each case, a constant bounding F_{en} value was computed, using the following assumptions:

- Low concentration of dissolved oxygen (DO < 0.05 ppm) for times when water temperature was above 150 degrees Celsius (302 degrees F)

- Most conservative value of T* (transformed temperature) (= 200 for LAS)

- Most conservative value for ε* (transformed total strain rate) (= ln(0.001))

- Most conservative value of S* (transformed sulphur content) (= 0.015 for LAS)

The applicant stated that the dissolved oxygen value was selected based on industry experience and confirmed by the PVNGS chemistry staff. The applicant noted only a few instances when dissolved oxygen exceeded 0.05 ppm for a relatively short period of time. These occurred following the startup of a third RCP while in hot standby after refueling. The applicant stated that these infrequent exceptions do not impact the validity of the assumed dissolved oxygen level.

The staff finds the applicant's operation with a dissolved oxygen level of less than 0.05 ppm is reasonable. The applicant stated that it confirmed this level had been maintained with the only exceptions occurring during the startup of a third reactor coolant pump in hot standby after refueling. The staff noted that this time duration is insignificant when compared to the amount of time the plant is operated with dissolved oxygen levels less than 0.05 ppm and, therefore, its impact is negligible.

The staff noted the use of NUREG/CR-6583 is consistent with the GALL Report Metal Fatigue Program. The staff confirmed that the assumptions used by the applicant from this report were the most conservative for calculating the F_{en} value for low-alloy steel components and that the resultant F_{en} value is 2.455. The staff noted that the applicant's use of the F_{en} value of 2.455 for low-alloy steel components is acceptable and appropriate, as described above.

Based on its review, the staff finds the applicant's response to RAI 4.3-4 acceptable because the applicant justified its use of a dissolved oxygen level of less than 0.05 ppm and used the most conservative assumptions from NUREG/CR-6583 to calculate the F_{en} value for low-alloy steel components. Finally, the applicant used a maximum F_{en} value of 2.455 based on the acceptable assumptions described above. The staff's concern described in RAI 4.3-4 is resolved and this part of Open Item 4.3-1 is closed.

The GALL Report metal fatigue AMP states that the impact of the reactor coolant environment on a sample of critical components should include the locations identified in NUREG/CR-6260

Time-Limited Aging Analyses

as a minimum, and that additional locations may be needed. In LRA Table 4.3-11, there are eight plant-specific locations listed, based on the seven generic locations identified in NUREG/CR-6260, and one additional location (pressurizer heater penetrations). The applicant discussed in the response to RAI 4.3-7 (August 12, 2010) that the pressurizer surge line elbow is the bounding location for the pressurizer surge line. During its review, the staff was unclear whether the applicant verified that the plant-specific components listed in the LRA Table 4.3-11 per NUREG/CR-6260 were bounding for the generic NUREG/CR-6260 locations. Furthermore, the staff noted that the applicant's plant-specific configuration may contain locations that should be analyzed for the effects of the reactor coolant environment in addition to the generic locations identified in NUREG/CR-6260.

The staff requested the applicant confirm and justify that the plant-specific components or locations listed in LRA Table 4.3-11 (except the pressurizer surge line pressurizer elbow) are bounding for the generic NUREG/CR-6260 locations and the additional location (pressurizer heater penetrations). The staff also requested the applicant to confirm and justify that the LRA Table 4.3-11 locations selected for EAF analyses consists of the most limiting locations for the plant. If these locations are not bounding, clarify the locations that require an EAF analysis and the actions that will be taken for these additional locations.

By letter dated December 3, 2010, the applicant provided additional information to address the staff's concern. The applicant committed (Commitment No. 63) to complete the following:

a) No later than two years prior to the period of extended operation, APS will confirm that the plant-specific components listed in LRA Table 4.3-11 (except the pressurizer surge line pressurizer elbow) are bounding for the generic NUREG/CR-6260 locations and the additional location (pressurizer heater penetrations). If locations are found that are not bounded by the Table 4.3-11 components, APS will perform new analyses as necessary to bound such locations, and

b) No later than two years prior to the period of extended operation, APS will confirm that the LRA Table 4.3-11 locations selected for environmentally assisted fatigue analyses consist of the most limiting cumulative usage factor (CUF) locations for the plant (beyond the generic EAF locations identified in the NUREG/CR- 6260 guidance). If the Table 4.3-11 locations are not bounding, APS will perform an environmentally assisted fatigue analysis for the additional CUF locations not bounded by the Table 4.3-11 locations. If the component with the most limiting CUF is composed of nickel alloy, the methodology used to perform the environmentally-assisted fatigue calculation for nickel alloy will be consistent with NUREG/CR-6909.

Based on its review, the staff finds the applicant's response to draft RAI 4.3.4-1 and Commitment No. 63 acceptable because, (1) the applicant committed to confirm that plant specific components or locations evaluated for environmental fatigue are the limiting locations to ensure that additional locations do not require an EAF analysis, (2) EAF analyses for the additional CUF locations not bounded by LRA Table 4.3-11 locations will be performed, (3) NUREG/CR-6909 will be used for determining a conservative F_{en} factor for any new nickel alloy components that require EAF analysis, and (4) Commitment No. 63 is consistent with the recommendations in SRP-LR Sections 4.3.2.2 and 4.3.3.2, and the GALL Report metal fatigue AMP, to consider environmental effects for the NUREG/CR 6260 locations.

Notes 7 and 9 of LRA Table 4.3-11 state the applicant's reanalysis computed F_{en} values for load set pairs with a significant fatigue contribution for the charging system nozzle (safe end) and the safety injection nozzle (safe end), respectively. LRA Section 4.3.4 does not contain sufficient information on the assumptions used for the environmental F_{en} factor calculations. By letter

Time-Limited Aging Analyses

dated June 2, 2010, the staff issued RAI 4.3-5 asking the applicant to describe the methodology used for the environmental F_{en} factor calculation of the charging system nozzle and the safety injection nozzle. The staff also asked that the applicant provide a basis for any assumptions that were made for the parameters, such as strain rate, dissolved oxygen, and temperature, in the assessment of a computed F_{en} value for the load set pairs with a significant fatigue contribution. Lastly, the staff asked the applicant to confirm the value of the maximum F_{en} factor used for all remaining load set pairs. This was previously identified as part of Open Item 4.3-1.

The applicant's June 29, 2010, response stated that the F_{en} analyses for these locations are documented in detail in plant calculations, and the F_{en} values were determined for each load-set pair using NUREG/CR-5704 for stainless steel components. The applicant further stated that the detailed F_{en} values were computed for load-set pairs that contributed more than 0.001 to the CUF for the given location. The applicant further stated that the load-set pairs that contributed less usage were conservatively assigned an F_{en} value of 15.35. The staff noted that the applicant provided a table with this information for the charging nozzle and safety injection nozzle.

The staff noted that for the detailed F_{en} value, the applicant used the maximum temperature for each time slice, which is conservative. The applicant stated that a dissolved oxygen concentration of less than 0.05 ppm was assumed for stainless steel. The staff noted that NUREG/CR-5704 provides guidance to calculate the F_{en} value for stainless steel and confirmed that the assumption of dissolved oxygen less than 0.05 ppm is conservative since it maximizes the F_{en} value. The applicant stated that both the strain rate and water temperature were calculated from the design transient specifications and corresponding stress analyses, and thus, no assumptions were made for these parameters.

The staff finds the applicant's approach reasonable since for load-set pairs that contributed more than 0.001 to the CUF, the applicant assumed dissolved oxygen levels that would maximize the F_{en} value for stainless steel. The staff also finds the applicant's approach reasonable because it used the strain rate and water temperature that were calculated from the design transient specifications and stress analyses, which allows for a more accurately calculated F_{en} value. Finally, the approach is reasonable because the applicant conservatively assumed a maximum F_{en} value of 15.35 for all remaining load-set pairs.

Based on its review, the staff finds the applicant's response to RAI 4.3-5 acceptable because the applicant provided the details of the methodology used to calculate more accurate F_{en} values for the charging nozzle and safety injection nozzle. The staff's concern described in RAI 4.3-4 is resolved and this part of Open Item 4.3-1 is closed.

The staff noted that for the other stainless steel components that required an EAF analyses, which include the surge line (hot leg) elbow and shutdown cooling line (long radius elbow), the applicant used an F_{en} value of 15.35, which is conservative and acceptable because it is the maximum that can be calculated consistent with NUREG.CR-5704.

LRA Section 4.3.4 states that a bounding F_{en} factor of 1.49 was used for the Alloy 600 pressurizer heater penetrations as determined from NUREG/CR-6335, "Fatigue Strain - Life Behavior of Carbon and Low-Alloy Steels, Austenitic Stainless Steels, and Alloy 600 in LWR Environments." This report provides the statistical characterizations used to derive this F_{en} factor of 1.49 for Alloy 600 and also states that the fatigue S-N database (fatigue per load cycle curves) for Alloy 600 is extremely limited and does not cover an adequate range of material and loading variables that might influence fatigue life. It further states that the data were obtained from relatively few heats of material and is inadequate to establish the effect of strain rate on fatigue life in air or of temperature in a water environment.

The staff noted that NUREG/CR-6909, "Effect of LWR Coolant Environments on the Fatigue Life of Reactor Materials," incorporates more recent fatigue data using a larger database for determining the F_{en} factor of nickel alloys. The staff noted that the applicant's value for F_{en} factor of 1.49 for nickel alloys may be non-conservative. NUREG/CR-6909 states that F_{en} for nickel alloys, varies based on temperature, strain rate, and dissolved oxygen. The staff further noted that, based on actual plant operating conditions, the F_{en} factor can vary from 1.0 to 4.52 based on this methodology. Therefore, the CUF value for the pressurizer heater penetrations may be as high as 2.86 using the CUF presented in the LRA and the maximum F_{en} derived from NUREG/CR-6909, which would exceed the design limit of 1.0 when considering environmental effects of reactor coolant during the period of extended operation.

By letter dated June 2, 2010, the staff issued RAI 4.3-6 asking the applicant justify using a value of 1.49 for the F_{en} factor for this nickel alloy component. The staff further asked that the applicant describe the current or future planned actions to update the CUF calculation with F_{en} factor for the Alloy 600 component only, consistent with the methodology in NUREG/CR-6909. If there are no current or future planned actions to update the CUF calculation with F_{en} factor for the Alloy 600 component consistent with the methodology in NUREG/CR-6909, the applicant must provide a justification for not performing the update. This was previously identified as part of Open Item 4.3-1.

The applicant's June 29, 2010, response to RAI 4.3-6 included a commitment (Commitment No. 57) to confirm the conservatism of the F_{en} value of 1.49 using the methods specified in NUREG/CR-6909 and to use the new F_{en} value if it is more conservative than the 1.49 value. It will complete this commitment no later than two years prior to the period of extended operation. The applicant also committed (Commitment No. 58) to perform a reanalysis of the pressurizer heater penetrations to consider EAF effects using the methodology given in NUREG/CR-6909.

The staff finds the applicant's response to RAI 4.3-6 acceptable because the applicant committed (Commitment No. 57) to confirm the conservatism of its use of the F_{en} value of 1.49 or perform a reanalysis of the pressurizer heater penetrations using a F_{en} value calculated using the methodology in NUREG/CR-6909. The staff's concern described in RAI 4.3-4 is resolved and this part of Open Item 4.3-1 is closed.

Based on its review, the staff finds the applicant's disposition of 10 CFR 54.21(c)(1)(ii) acceptable for the reactor vessel inlet and outlet nozzles and RPV shell and lower head locations because the applicant has demonstrated that, when considering environmental effects of reactor water, the CUF is projected to remain below the design limit of 1.0 for the period of extended operation. Further, the applicant will continue to monitor these locations with its enhanced Metal Fatigue Program during the period of extended operation.

Based on its review, the staff finds the applicant's disposition of 10 CFR 54.21(c)(1)(iii) acceptable for the surge line (hot leg) elbow, charging system nozzle (safe end), safety injection nozzle (forging knuckle and safe end), shutdown cooling line (long radius elbow) and pressurizer heater penetrations. It is acceptable because the applicant will continue to manage the effects of EAF for these components with its enhanced Metal Fatigue Program to ensure that the design limit of 1.0 is not exceeded, or it will take corrective actions to reanalyze, repair, or replace the affected component. Finally, it is acceptable because the applicant committed to reanalyze the nickel alloy pressurizer heater penetrations

4.3.4.3 Updated Final Safety Analysis Report Supplement

LRA Section A3.2.3 provides the UFSAR supplement summarizing the evaluation of the effects of the RCS environment on fatigue life of piping and components (Generic Safety Issue 190). The staff reviewed LRA Section A3.2.3 against the acceptance criteria in SRP-LR Section 4.3.2.3. Based on its review of the UFSAR supplement, consistent with SRP-LR Sections 4.3.2.3 and 4.3.3.3, the staff determines that the applicant provided an adequate summary description of its actions to address effects of the reactor coolant system environment on fatigue life of piping and components, as required by 10 CFR 54.21(d).

4.3.4.4 Conclusion

On the basis of its review, the staff concludes that the applicant's evaluations on the effects of the reactor coolant environment on component fatigue life is not a TLAA as defined by 10 CFR 54.3(a) and is consistent with Commission Order No. CLI-10-17 (July 8, 2010). The staff also concludes that the applicant has provided an acceptable demonstration, pursuant to 10 CFR 54.21(c)(1)(ii), that the effects of fatigue, including environmental effects of reactor coolant water on the intended functions of the reactor vessel inlet and outlet nozzles, and RPV shell and lower head locations, have been projected to the end of the period of extended operation. The staff also concludes that the applicant has demonstrated, pursuant to 10 CFR 54.21(c)(1)(iii), that the effects of fatigue, including environmental effects of reactor coolant water on the intended functions of the surge line (hot leg) elbow, charging system nozzle (safe end), safety injection nozzle (forging knuckle and safe end), shutdown cooling line (long radius elbow), and pressurizer heater penetrations, will be adequately managed for the period of extended operation. The staff finally concludes that the UFSAR supplement contains an appropriate summary description of the TLAA evaluation, as required by 10 CFR 54.21(d).

4.3.5 Assumed Thermal Cycle Count for Allowable Secondary Stress Range Reduction Factor in American National Standards Institute B31.1 and American Society of Mechanical Engineers III Class 2 and 3 Piping

4.3.5.1 Summary of Technical Information in the License Renewal Application

LRA Section 4.3.5 summarizes the evaluation of "Assumed Thermal Cycle Count for Allowable Secondary Stress Range Reduction Factor in ANSI B31.1 and ASME III Class 2 and 3 Piping" TLAAs for the period of extended operation. These TLAAs are based on the criteria for performing implicit fatigue analyses for ANSI B31.1 piping components, as given in the ANSI B31.1 design code, and for ASME Code Class 2 and 3 components, as specified in ASME Code Section III, Article NC-3000 for components designed to ASME Code Section III Class 2 component requirements, and Article ND-3000 for components designed to ASME Code Section Class3 component requirements.

In this TLAA, the applicant noted that, with the exception of the implicit fatigue analyses for the reactor coolant hot leg sampling lines and the SG downcomer and feedwater recirculation lines, all of the implicit fatigue analyses for the ANSI B31.1 piping components and for the ASME Code Class 2 and 3 components remain valid for the period of extended operation in accordance with 10 CFR 54.21(c)(1)(i). For these analyses, the applicant states that the total number of occurrences for the full thermal transients that are applicable to these components is projected to be less than 7,000 through the end of the period of extended operation.

For the implicit fatigue analyses for the reactor coolant hot leg sampling lines and the SG downcomer and feedwater recirculation lines, the applicant states that the analyses have been projected through the end of the period of extended in accordance with 10 CFR 54.21(c)(1)(II). For these implicit fatigue analyses, the applicant states that the total number of full thermal range transients that are applicable to the lines are projected to be in excess of 7,000 cycle occurrences. For these components, applicable stress reduction factors were applied to maximum allowable stress limit criteria for the analyses in order to demonstrate that the existing stress loadings on the components would still be acceptable for the period of extended operation even under the reduced acceptance limit criteria for the analyses.

4.3.5.2 Staff Evaluation

On May 27, 2010, the applicant submitted LRA Amendment 16 which included conforming changes to LRA Section 4.3.5 to address staff concerns discussed with the applicant in a public meeting on May 6, 2010.

Time-Limited Aging Analyses

The staff reviewed the applicant's bases for dispositioning each of the applicant's implicit fatigue analyses for ANSI B31.1 components and ASME Code Class 2 and 3 components. The staff confirmed that the applicant had provided a valid basis for demonstrating that each of the CUF analyses would be acceptable in accordance with 10 CFR 54.21(c)(1)(i), that the analysis remains acceptable for the period of extended operation, or 10 CFR 54.21(c)(1)(ii) that the analysis has been projected to the end of the period or extended operation period of extended operation.

The staff noted that LRA Section 4.3.5 states that the calculated stresses in limiting locations were less than the allowable in the revised design analyses for the reactor coolant hot leg sample lines piping and the SG downcomer and feedwater recirculation lines piping. However, the staff noted that LRA Section 4.3.5 does not give sufficient information for the staff to confirm these assertions. By letter dated June 2, 2010, the staff issued RAI 4.3-3 asking the applicant supply the code allowable stress limits and the stress ranges obtained in the revised design analyses for the reactor coolant hot leg sample line piping and the SG downcomer and feedwater recirculation line piping. The staff also asked the applicant to provide the ASME Code edition and specific subsection used for the revised design analyses for these piping components. This was previously identified as part of Open Item 4.3-1.

The applicant's June 29, 2010, response provided information related to the reactor coolant hot leg sample line piping and the SG downcomer and feedwater recirculation line piping and the code allowable stress limits and stress range reduction factors. The staff noted that for the reactor coolant hot leg sample line, the applicant used a stress range reduction factor (SRRF) of 0.9 because it expected that this component would exceed the original 7,000 cycle limit (SRRF = 1.0), with an estimated 8,273 cycles. The staff noted that this is consistent with ASME Code Section III and SRP-LR, Table 4.3-1. The staff noted that for the SG downcomer and feedwater recirculation line piping, the applicant used an SRRF of 0.8 because it expected that this component would exceed the original 10,224 cycle limit (SRRF = 0.9), with an estimated 15,336 cycles. The staff noted that this is also consistent with ASME Code Section III and SRP-LR Table 4.3-1. The staff noted that for both reactor coolant hot leg sample line piping and the SG downcomer and feedwater recirculation line piping, the revised allowed stress is less than the code allowable limit.

The applicant identified that the revised design analyses for these piping components was performed to the requirements of ASME Code, Section III, 1974 up to and including winter 1975 Addenda, and the SRRF was obtained from Table NC-3611.2(e)-1. The applicant further stated that the comparison of the calculated stress range versus the allowable stress limit was performed per the requirements of paragraph NC-3652.3.

Based on its review, the staff finds the applicant's response to RAI 4.3-3 acceptable because the applicant used the appropriate SRRF of 0.9 and 0.8 for the reactor coolant hot leg sample line piping and the SG downcomer and feedwater recirculation line piping, respectively, which is consistent with the ASME Code Section III and SRP Table 4.3-1. Further, the response is acceptable because the revised allowed stress for these components is less than the ASME Code allowable limit. The staff's concern described in RAI 4.3-3 is resolved and this part of Open Item 4.3-1 is closed.

The amended LRA Section 4.3.5 identified that all implicit fatigue analyses for ANSI B31.1 and ASME Class 2 and 3 piping components will be remain valid for the period of extended operation except for the implicit fatigue analysis of RCS hot leg sampling lines and the recirculating SG downcomer and feedwater recirculation lines. The staff noted that the implicit fatigue analysis table provided for the RCS hot leg sampling lines includes a column "Max. Calculated Stress per Eq. (11) (psi)." However, the column does not identify the source document for the referenced equation 11. Similarly, the implicit fatigue analysis table provided for the RSG DC and FW recirculation lines includes a column "Max. Calculated Stress Range

Time-Limited Aging Analyses

per Eq. (10) (psi)." However, the column does not note the source document for the referenced equation 10. The staff also noted that in the assessment of the recirculating SG downcomer and feedwater recirculation lines, the applicant discussed two different analyses; the original implicit fatigue analysis and an updated pipe break analysis. LRA Section 4.3.5 does not clarify whether the pipe break analysis has a relationship to the original implicit fatigue analysis for these lines. It is also not clear whether both analyses are relied upon for the CLB or whether the pipe-break analysis is a replacement for the original implicit fatigue analysis. It is not clear to the staff which of the analyses is the current analysis of record for the CLB and thus needs to be assessed as a TLAA for these lines. By letter dated July 21, 2010, the staff issued RAI 4.3-18 asking the applicant to identify the source documents for the stated equation references. The staff also requested the applicant to clarify which of the implicit fatigue analyses discussed in LRA Section 4.3.5 for the recirculating SG downcomer and feedwater recirculation lines is the analysis of record for these lines (i.e., the original analysis, the pipe break analysis, or both analyses). This was previously identified as part of Open Item 4.3-1.

The applicant's August 12, 2010, response clarified that equations 10 and 11 in LRA Section 4.3.5 are those listed in ASME Code Section III, Subsection NC-3600, paragraph NC 3652.3, for Class 2 piping and Subsection ND-3600, paragraph ND 3652.3, for Class 3 piping. The applicant also stated that implicit fatigue analyses, discussed in LRA Section 4.3.5 for the recirculating SG downcomer and feedwater recirculation lines, refer to methodology prescribed in subsection NC and ND of ASME Code Section III. The analyses determine that if the number of full-range thermal cycle is expected to be 7,000 or more, then the ANSI B31.1 and ASME Code Section III, Subsection NC and ND for Class 2 and 3 piping require the application of a stress range reduction factor to the allowable stress range for expansion stress. The applicant stated that these analyses are TLAAs.

Based on its review, the staff finds the applicant's response to RAI 4.3-18 acceptable because the applicant clarified the 7000-thermal cycles fatigue analysis is the analysis of record for the recirculating SG downcomer and feedwater recirculation lines and has been identified as a TLAA consistent with 10 CFR 54.21(c)(1). The staff's concern described in RAI 4.3-18 is resolved and this part of Open Item 4.3-1 is closed.

Based on its review, the staff finds the applicant has demonstrated that, with the exception of the reactor coolant hot leg sampling lines and the SG downcomer and feedwater recirculation lines, all analyses for the ANSI B31.1 and ASME III Class 2 and 3 piping remain valid for the period of extended operation pursuant to 10 CFR 54.21(c)(1)(i). The staff finds it acceptable because the number of projected cycles excepted to occur during the period of extended operation is significantly lower than the component design life of 7,000 cycles. The staff's evaluation of the applicant's projection of transients for the period of extended operation is documented in SER Section 4.3.1.4.2. The staff also finds the applicant has demonstrated pursuant to 10 CFR 54.21(c)(1)(ii), that the analyses for the reactor coolant hot leg sampling lines and the SG downcomer and feedwater recirculation lines have been projected through the period of extended operation and are acceptable because the applicant applied the applicable stress range reduction factor, consistent with ANSI B31.1 and ASME Code Section III Subsection NC and ND and the SRP-LR.

4.3.5.3 *Updated Final Safety Analysis Report Supplement*

LRA Section A3.2.4 provides the UFSAR supplement summarizing the TLAAs for ANSI B31.1 and ASME III Class 2 and 3 piping. The staff reviewed LRA Section A3.2.4 against the acceptance criteria in SRP-LR Section 4.3.2.3. Based on its review of the UFSAR supplement, consistent with SRP-LR Sections 4.3.2.3 and 4.3.3.3, the staff determines that the applicant provided an adequate summary description of its TLAAs for ANSI B31.1 and ASME III Class 2 and 3 piping, as required by 10 CFR 54.21(d).

4.3.5.4 Conclusion

On the basis of its review, the staff concludes that the applicant has provided an acceptable demonstration, pursuant to 10 CFR 54.21(c)(1)(i), that the analyses for the ANSI B31.1 and ASME III Class 2 and 3 piping, with the exception of the reactor coolant hot leg sampling lines and the SG downcomer and feedwater recirculation lines, remain valid during the period of extended operation. Further, the staff concludes that the applicant has provided an acceptable demonstration, pursuant to 10 CFR 54.21(c)(1)(ii), that the analyses for the reactor coolant hot leg sampling lines and the SG downcomer and feedwater recirculation lines have been projected to the end of the period of extended operation. The staff also concludes that the UFSAR supplement contains an appropriate summary description of the TLAA evaluation, as required by 10 CFR 54.21(d).

4.4 Environmental Qualification of Electrical Equipment

The EQ requirements established by 10 CFR Part 50, Appendix A, Criterion 4, and 10 CFR 50.49 specifically require each applicant to establish a program to qualify electrical equipment so that such equipment, in its end of life condition, will meet its performance specifications during and following design basis accidents. The 10 CFR 50.49 EQ program is a TLAA for purposes of license renewal. Electrical equipment with a qualified life equal to or greater than the duration of the current operating term is covered by TLAAs. The TLAA of the EQ of electrical components includes certain electrical and instrumentation and control (I&C) components that are important to safety and are located in a harsh environment. The harsh environment includes those areas subject to environmental effects caused by a LOCA, high-energy line break, and a post-LOCA environment.

4.4.1 Summary of Technical Information in the Application

LRA Section 4.4, "Environmental Qualification (EQ) of Electrical Equipment," summarizes the applicant's evaluation of EQ of plant electrical and I&C equipment for the period of extended operation. The EQ Program is an existing program established to meet commitments for 10 CFR 50.49. The applicant also stated that the EQ Program manages applicable component thermal, radiation, and cyclical aging effects based on 10 CFR 50.49 for the current operating license, using methods of demonstrating qualification for aging and accident conditions established by 10 CFR 50.49(f). The applicant selected 10 CFR 54.21(c)(1)(iii) as the means to demonstrate the adequacy of the EQ Aging Management Program for the period of extended operation. The applicant stated that maintaining qualification through the extended license renewal period requires existing EQ evaluations (Electrical Equipment Qualification Data Files) to be reanalyzed. The applicant stated that the effects of power uprate and SG replacement have been evaluated and equipment re-qualified as required. The applicant further stated that the important attributes of reanalysis include analytical methods, data collection and reduction methods, underlying assumptions, acceptance criteria, and corrective actions (if acceptance criteria are not met). The applicant further stated that, if qualification cannot be extended by reanalysis, the component is refurbished or replaced before exceeding the period for which current qualification remains valid.

The applicant concluded that continuing the existing EQ Program ensures that the aging effects will be managed and that the EQ components will continue to perform their intended functions for the period of extended operation. The applicant also concluded that aging effects addressed by the EQ program will thereby be managed for the period of extended operation, in accordance with 10 CFR 54.21(c)(1)(iii).

Time-Limited Aging Analyses

4.4.2 Staff Evaluation

The staff reviewed LRA Section 4.4; Appendix B, Section B3.2; program basis documents; and information supplied to the staff during the audit and interviewed plant personnel to determine if the applicant's EQ Program meets the requirements of 10 CFR 54.21(c)(1). The applicant's EQ TLAA Program is implemented per the requirements of 10 CFR 54.21(c)(1)(iii), which requires that the effects of aging on the intended functions will be adequately managed for the period of extended operation. The staff confirmed the applicant's EQ Program conforms to the requirements of 10 CFR 50.49, including the management of aging effects, to confirm that electric equipment requiring EQ will continue to operate consistent with the CLB during the period of extended operation. Per the GALL Report, plant EQ Programs that meet the requirements of 10 CFR 50.49 are considered acceptable AMPs under license renewal in accordance with 10 CFR 54.21(c)(1)(iii). GALL AMP X.E1, "Environmental Qualification (EQ) of Electric Components," provides a means to meet the requirements of 10 CFR 54.21(c)(1)(iii).

Based on the staff's review of LRA Section 4.4 and Appendix B, Section B3.2, including the audit results, the staff concludes that the applicant's EQ of Electric Equipment TLAA is carried out per the requirements of 10 CFR 54.21(c)(1)(iii).

Therefore, the staff finds that the applicant's EQ Program demonstrates, pursuant to 10 CFR 54.21(c)(1)(iii), that the effect of aging on the intended functions will be adequately managed for the period of extended operation. The applicant's EQ Program is capable of managing the qualified life of components within the scope of license renewal, and the continued implementation of the EQ Program provides assurance that the aging effects will be managed and that electric equipment will continue to perform their intended functions for the period of extended operation.

4.4.3 Updated Final Safety Analysis Report Supplement

In LRA Appendix A, Section A3.3 provides the UFSAR supplement for the EQ of Electrical Equipment Program. The staff reviewed the UFSAR supplement description of the program against the recommended description for this type of program as described in SRP-LR Tables 4.4-1 and 4.4-2 and noted that it did not include reanalysis attributes consistent with the description of the TLAA in LRA Section 4.4 or the AMP in LRA Section B3.2. GALL AMP EQ of Electric Components, states that reanalysis of an aging evaluation is normally done to extend the qualification by reducing excess conservatism incorporated in the prior evaluation. Important attributes of a reanalysis include analytical methods, data collection and reduction methods, underlying assumptions, acceptance criteria, and corrective actions (if acceptance criteria are not met).

By letter dated December 29, 2009, the staff issued RAI 4.4-1 to ask the applicant to provide justification for not including reanalysis attributes in accordance with 10 CFR 54.21(c)(1)(iii) in the UFSAR supplement. The applicant responded by letter dated February 19, 2010, and stated LRA Sections A2.2 and A3.3 have been revised to include the following: "[r]eanalysis of aging evaluations to extend the qualifications of components is performed on a routine basis as part of the EQ Program. Important attributes for the reanalysis of aging evaluations include analytical methods, data collection and reduction methods, underlying assumptions, acceptance criteria and corrective actions (if acceptance criteria are not met)."

With the information provided by the applicant's RAI response, the staff finds the UFSAR supplement acceptable because the applicant revised LRA Sections A2.2 and A3.3 to be consistent with the guidance of SRP Table 4.4.2. The staff considers RAI 4.4-1 resolved.

The staff determines that the information in the UFSAR supplement is an adequate summary description of the program, as required by 10 CFR 54.21(d).

4.4.4 Conclusion

On the basis of its review of the applicant's EQ of Electric Equipment TLAA and RAI response, the staff concludes that the applicant has demonstrated that the effects of aging will be adequately managed so that the intended functions will be maintained, pursuant to 10 CFR 54.21(c)(1)(iii), for the period of extended operation. The staff also reviewed the UFSAR supplement and concludes that it provides an adequate summary description of the program, as required by 10 CFR 54.21(d).

4.5 Concrete Containment Tendon Prestress Analyses

4.5.1 Summary of Technical Information in the Application

LRA Section 4.5 summarizes the evaluation of concrete containment tendons prestress for the period of extended operation. The LRA states that the containment is a prestressed concrete, hemispherical, dome-on-a-cylinder structure with a steel membrane liner. Post-tensioned tendons compress the concrete and permit the structure to withstand design-basis accident internal pressures. The steel tendons, in tension, relax with time and the concrete structure, which the tendons hold in compression, both creeps and shrinks with time. Therefore, the applicant stated that to ensure the integrity of the containment pressure boundary under design-basis accident loads, an inspection program confirms whether the tendon prestress remains within design limits throughout the life of the plant. The applicant further stated that the original design predictions of loss of prestress and the regression analyses of surveillance data that predict the future performance of the post-tensioning system to the end of design life are TLAAs and it dispositioned them in accordance with 10 CFR 54.21(c)(ii).

The LRA describes the post-tensioning system of each unit as consisting of vertical, inverted-U-shaped tendons and horizontal circumferential tendons. The applicant described the vertical, inverted-U tendons as anchored through the bottom of the conventionally-reinforced concrete basemat. The LRA further states that the horizontal hoop tendons are anchored at three exterior buttresses, which are 120 degrees apart. Each hoop tendon extends 240 degrees around the containment building, passing under an intervening buttress. The applicant also stated that the tendons are not bonded to the concrete but inserted in tendon ducts after concrete cure and tensioned in the prescribed sequence. Each tendon consists of up to 186 one-quarter-inch diameter high-strength steel wires with cold-formed button heads on each end bearing on a stressing anchorhead. The total tendon load is carried by shim stack to steel bearing plates embedded in the structure.

LRA, Appendix B, Section B3.3 summarizes the TLAA AMP, "Concrete Containment Tendon Prestress Program." The applicant stated that before September 1996, RG 1.35 governed the tendon examinations. Additionally, the applicant stated that the tendon lift-off surveillances were done for Units 1 and 3, at one, three, and five years post-structural-integrity test, and at each succeeding five-year interval. Unit 2 tendons were examined visually and in other ways, but their lift-off test surveillances were encompassed within the Unit 1 tests, under rules then applicable to 2-unit plants with virtually identical containments. Beginning with License Amendment 151, the program was governed by ASME Code Section XI, Subsection IWL, 1992 Edition with 1992 Addenda and supplemental requirements in 10 CFR 50.55a. A licensing change under approved RR L4 imposed the surveillance of Unit 2 prestress tendon lift-off forces, beginning with its 20th year, and extended the surveillance interval to 10 years for all 3 units. The applicant states that the assessment of the results of the tendon prestressing force measurements and acceptance criteria are in accordance with the edition and addenda of ASME Code Section XI, Subsection IWL referenced above, as incorporated in 10 CFR 50.55a.

Time-Limited Aging Analyses

The applicant stated that the condition of the containment prestressing system meets the criteria for revision for the period of extended operation as described in NUREG-1800, Section 4.5.3.1.2. The applicant discussed these criteria as follows:

- The lift-off trend lines were calculated by regression of individual tendon lift-off data, including the results of the 2005, Unit 2, 20-year surveillance. Therefore, these calculations are consistent with NRC Information Notice 99-10, "Degradation of Prestressing Tendon Systems in Prestressed Concrete containments," Attachment 3.

- The regression analysis of surveillance lift-off data extends the trend lines for both the vertical and horizontal cylinder tendons to 60 years.

- The trend line for all tendon groups remain above the minimum required values (MRVs) for the period of extended operation.

4.5.2 Staff Evaluation

The staff reviewed LRA Section 4.5 to verify using SRP-LR Section 4.5.3.1.2, that the trend of prestressing forces in each tendon group was projected to the end of the period of extended operation and the projected prestressing forces were above their respective MRVs per 10 CFR 54.21(c)(1)(ii). Also, the staff reviewed LRA Section 4.5 to verify that the trend lines for each tendon group, presented in LRA Figures 4.5-1–4.5-6, are based on individual tendon lift-off forces as specified in Information Notice 99-10. The figures show that the projected prestressing forces trend lines for the vertical and horizontal tendons remain above their respective MRVs through the period of extended operation. The trend lines also remain above RG 1.35.1 predicted prestress forces through the period of extended operation; except for Unit 3 horizontal tendon prestress (Figure 4.5-3), indicating that the loss of prestress is less than originally predicted. As stated by the applicant, the trend lines do not include the Unit 1, 25-year tendons lift-off surveillance results. The applicant explained in LRA Section 4.5 that the surveillance was not completed in time to be included in the LRA. By letter dated March 2, 2010, the staff issued RAI 4.5-1 to ask the applicant to provide the tendon regression analyses that include the results of 25-year containment tendon prestressing surveillance for Unit 1.

The applicant's April 1, 2010, response to RAI 4.5-1 stated that it is revising the regression analysis to incorporate the Unit 1, 25-year tendon surveillance data. The applicant committed to submitting the revised LRA Figures 4.5-1, 4.5-2, 4.5-5, and 4.5-6 in LRA amendment by May 28, 2010. The applicant also explained that the evaluation of the Unit 1, 25-year surveillance data shows that the recalculated regression lines for horizontal and vertical tendons will remain well above their respective MRVs through the period of extended operation. By letter dated May 21, 2010, the applicant submitted the revised LRA Figures 4.5-1, 4.5-2, 4.5-5, and 4.5-6. The staff's review of the figures confirmed that the revised regression analysis incorporates the Unit 1, 25-year tendon surveillance data and that the prestress, for each tendon group, will remain above their respective MRVs through the period of extended operation. The staff finds the applicant's response acceptable because the projected prestressing forces trend lines will remain above their respective MRVs through the period of extended operation. The staff's concern in RAI 4.5-1 is resolved.

In LRA Section 4.5, the applicant credits, per 10 CFR 54.21(c)(1)(iii), the "Concrete Containment Tendon Prestress" AMP, described in LRA Section B3.3, for managing loss of prestress in the tendons during the period of extended operation. The staff reviewed LRA Section B3.3 in accordance with the guidance in SRP-LR Section 4.5.3.1.3, to verify the applicant identified the appropriate program as described and evaluated in the GALL Report. The applicant stated that the AMP is an existing program that, following enhancement, will be consistent with NUREG-1801, Section X.S1, "Concrete Containment Tendon Prestress." The

Time-Limited Aging Analyses

staff noted that the applicant referenced RR L4, which permits the 10-year interval between tendon prestress surveillance for the three PVNGS units during the current 40-year term. By letter dated March 2, 2010, the staff issued RAI 4.5-2 to ask the applicant to provide information on how the aging of the containment tendons will be managed during the period of extended operation.

The applicant's April 1, 2010, response to RAI 4.5-2 stated that aging of the tendons will be managed through inspections as described in the applicable edition and addenda of the ASME Code Section XI, Subsection IWL, in accordance with 10 CFR 50.55(a), including any NRC approved RRs. The staff finds the applicant's response acceptable because the tendon prestress surveillance interval will be in accordance with the applicable edition and addenda of ASME Code Section XI, Subsection IWL incorporated in 10 CFR 50.55(a) during the period of extended operation. The staff's concern in RAI 4.5-2 is resolved.

In LRA Section 4.5, the applicant provided Table 4.5-1, Tendon Regression Analysis Input Data for PVNGS Units 1, 2, and 3. The staff's review of the tabulated tendon lift-off data observed that only the "shop end" force is provided for tendons H21-04, V07, and V015. Also, the lift-off force for tendon H21-04 was measured in the third year surveillance and again in the fifth year and the Unit 3 dome horizontal tendon lift-off average forces are greater than the wall horizontal lift-off average forces, in some cases by nearly 100 kips. By letter dated, March 2, 2010, the staff issued RAI 4.5-3 to ask the applicant to explain the anomalies and confirm they have no affect on its conclusion that regression analysis trend lines show that tendon prestress will remain above their respective MRVs through the end of the period of extended operation.

By letter dated May 21, 2010, the applicant responded to RAI 4.5-3 and addressed each anomaly identified by the staff as well as other self-identified anomalies. In its response, the applicant provided technical and licensing bases for its conclusions to show that the anomalies have no significant effect on the regression analysis results. The staff reviewed the applicant's response and found the applicant has adequately addressed the issue because it explained the anomalies and confirmed they have no affect on the regression analysis conclusion. The staff's concern in RAI 4.5-3 is resolved.

The staff reviewed the applicant's operating experience related to the containment tendon prestress force. The results of the review are documented in the staff evaluation of the applicant's Concrete Containment Tendon Prestress Program in SER Section B3.3. The results show that the applicant's program has adequately considered plant-specific operating experience.

Based on this review, the staff concludes the applicant has noted the appropriate program and has stated the GALL Report is applicable to its plant with respect to its program that assesses the concrete containment tendon prestressing forces.

4.5.3 Updated Final Safety Analysis Report Supplement

The applicant provided a UFSAR supplement summary description of its TLAA evaluation of concrete containment tendon prestress in LRA Section A3.4. Based on its review of the UFSAR supplement, the staff concludes that the summary description of the applicant's actions to address concrete containment tendon prestress is adequate.

4.5.4 Conclusion

On the basis of its review, the staff concludes that the applicant has demonstrated, pursuant to 10 CFR 54.21 (c)(1)(ii), that the effects of aging on the concrete containment prestressing tendons have been projected to the end of the period of extended operation. The staff also concludes that the applicant has appropriately credited, pursuant to 10 CFR 54.21(c)(1)(iii), the Concrete Containment Tendon Prestress Program for managing loss of tendon prestress during

Time-Limited Aging Analyses

the period of extended operation. Additionally, the staff determined that the UFSAR supplement contains an appropriate summary description of the TLAA on containment tendon loss of prestress analysis for the period of extended operation, as required by 10 CFR 54.21(d).

4.6 Containment Liner Plate, Equipment Hatch and Personnel Air Locks, Penetrations, and Polar Crane Brackets

4.6.1 Absence of a Time-Limited Aging Analysis for Containment Liner Plate, Polar Crane Brackets, Equipment Hatch and Personnel Air Locks, and Containment Penetrations (Except Main Steam, Main Feedwater, and Recirculation Sump Suction Penetrations)

4.6.1.1 Summary of Technical Information in the Application

The post-tensioned concrete containments were designed in accordance with ASME Boiler and Pressure Vessel Code, Section III, Division 2, Article CC-3000, supplemented by the design methods and criteria of Bechtel Topical Reports BC-TOP-1, "Containment Building Liner Plate Design Report," Revision 1, and BC-TOP-5-A, "Prestressed Concrete Nuclear Reactor Containment Structures," Revision 3. The interior of the containments is lined with steel membrane liners designed to BC-TOP-1 Revision 1. No credit is taken for the liner for the pressure design of the containment, but the liner and penetrations ensure the containments are leak-tight, and their electrical, process, personnel airlock, and equipment hatch penetrations are part of the containment pressure boundary.

LRA Section 4.6.1 summarizes the evaluation of absence of a TLAA for containment liner plate, polar crane bracket, equipment hatch, air lock, and containment penetration design (except main stream, main feedwater, and recirculation sump suction penetrations) for the period of extended operation. The liner plate provides a leak-tight barrier to prevent uncontrolled release of fission products from the containment during normal plant operation and in the unlikely event of an accident. SRP-LR Section 4.6.1 notes that in some designs, "fatigue of the liner plates or metal containments may be considered in the design based on an assumed number of loading cycles for the current operating term." The cyclic loads include reactor building interior temperature variation during the heatup and cooldown of the RCS, a LOCA, annual outdoor temperature variations, thermal loads due to the high energy containment penetration piping lines (such as steam and feedwater lines), seismic loads, and pressurization due to periodic Type A integrated leak rate tests. The applicant states that its examination of the controlling reports BC-TOP-1, BC-TOP-5A, the design specification, and design report found no evidence that heatup and cooldown or seasonal temperature variations were considered cyclic loads on the containment building and liner. The applicant further states that the UFSAR contains no description of cyclic loads or design cycles for the entire containment building; but UFSAR Section 3.8.1.5.4.B describes design cycles that are to be included in the design of the liner plate and penetrations. However, the review of the design specification, design report, and design calculations found time-dependent aspects of some penetration designs but none for liner plate design; therefore, the liner plate design is not supported by a TLAA.

The applicant also explained that the polar crane is supported on a system of girders, which are supported by a series of brackets that are attached to the containment shell. BC-TOP-1 Revision 1 reviews design of the polar crane brackets. The report does not include or specify the requirement for a fatigue analysis, or any other design for a stated number of crane lifts, cyclic loads, or other cyclic events. Therefore, design of the polar crane brackets for a finite number of loads is not supported by a TLAA.

For the equipment hatch and personnel air locks, the applicant said that the components were designed to ASME Code Section III, Division 1, Subsection NE - Class MC Components,

1974 W74. Subparagraph NE-3222.4 gives rules for a fatigue analysis of MC components for cyclic loads, if specified. However, the review of licensing basis documents, specifications, and the design report identified no time-dependent analyses. Designs of the equipment hatch and personnel air locks are, therefore, not supported by TLAAs.

For containment penetrations, the applicant stated that a search of the licensing basis and the review of the design documents found no evidence of any TLAAs applicable to containment penetrations; except for the main stream, main feedwater penetration design in BC-TOP-1 Part II, supporting design calculations described in Section 4.6.2 below, and the recirculation sump suction penetration design described in Section 4.6.3 below. The containment penetrations include no bellows or expansion joints whose design is supported by a TLAA.

4.6.1.2 Staff Evaluation

The staff reviewed LRA Section 4.6.1 to evaluate the absence of a TLAA for the containment liner plate, polar crane brackets, equipment hatch and personnel air locks, and containment penetrations (except main stream, main feedwater, and recirculation sump suction penetrations). The containments were designed in accordance with ASME Boiler and Pressure Vessel Code Section III, Division 2, Article CC-3000, supplemented by the design methods and criteria of BC-TOP-1 and BC-TOP-5-A. The staff noted that BC-TOP-1 and BC-TOP-5-A do not include a fatigue analysis or require evaluation of these components for cyclic loading; except by a reference to ASME Code Section III, Division 2, Article CC-3760, which requires the designer to ensure suitability of the liner plate for cyclic loads established in the design specification. The staff noted that UFSAR Section 3.8.1.5.4.B says that 500 thermal cycles due to variation in the interior temperature of the containment during the heatup and cooldown of the reactor, 40 cycles due to annual outdoor temperature variation, and 1 LOCA cycle are considered in the liner design for its 40-year life. As a result, the staff issued RAI 4.6-1 asking the applicant to evaluate the liner plate system for cyclic loading during the period of extended operation, consistent with UFSAR 3.8.1.5.4.B, or give additional technical basis to demonstrate that this evaluation is not required.

By letter dated April 1, 2010, the applicant stated that the containment liner plate system was evaluated for cyclic loading during the period of extend operation consistent with UFSAR Section 3.8.1.5.4.B requirements. The applicant concluded that the design basis analyses are conservative and remain valid for the period of extended operation, pursuant to 10 CFR 54.4(c)(1)(i).

In its review of the analyses, the staff noted that the applicant did not provide the actual thermal cycles for the current term and the projected thermal cycles through the period of extended operation. However, the applicant stated that the assumed 500 containment interior operational heatup and cooldown cycles correspond to 8⅓ cycles per year for 60-year plant life, which is conservative. The staff agrees that 8⅓ thermal cycles per year are conservative and finds the applicant's response acceptable because the containment liner plate system is evaluated for cyclic loading, consistent with UFSAR Section 3.8.1.5.4.B. The applicant noted that Palo Verde Action Request 3451141 was initiated to clarify UFSAR Section 3.8.1.5.4.B. The staff's concern in RAI 4.6-1 is resolved.

For polar crane brackets, equipment hatch and personnel air locks, and containment penetrations (except main stream, main feedwater, and recirculation sump suction penetrations), the staff agrees with the applicant that neither the Bechtel Topical Reports nor the ASME Code editions and addenda invoked by them impose a fatigue analysis or evaluation for cyclic loading.

Time-Limited Aging Analyses

4.6.1.3 Updated Final Safety Analysis Report Supplement

The applicant provided a UFSAR supplement summary description of its TLAA evauation in LRA Section A3.5, as provided in Amendment 17 (June 21, 2010). Based on its review of the UFSAR supplement, the staff concludes that the summary description of the applicant's actions to address design cycles for the containment liner plate is adequate.

4.6.1.4 Conclusion

On the basis of its review, the staff concludes that the applicant has demonstrated, pursuant to 10 CFR 54.21(c)(1)(i), that the evaluation for the containment liner plate system remains valid during the period of extended operation. Additionally, the absence of a TLAA for the polar crane brackets, equipment hatch and personnel air locks, and containment penetrations (except main steam, main feedwater, and recirculation sump suction penetrations) for the period of extended operation is adequate. The staff also concludes that the UFSAR supplement contains an appropriate summary description of the TLAA evaluation, as required by 10 CFR 54.21(d).

4.6.2 Design Cycles for the Main Steam and Main Feedwater Penetrations

4.6.2.1 Summary of Technical Information in the Application

LRA Section 4.6.2 summarizes the evaluation of design cycles for the main steam and main feedwater penetrations for the period of extended operation. The applicant states that the design of main steam line penetrations includes 100 lifetime steady state operating thermal gradient plus normal operating cyclic loads (i.e., Loading Condition V) and 10 steady state operating thermal gradient plus steam pipe rupture cyclic loads (i.e., Loading Condition IV), as specified in BC-TOP-1. The BC-TOP-1 analysis of effects of Loading Condition IV and V cyclic loads does not calculate a usage factor but uses a simplified ASME Code Section III, Subparagraph NB-3228.3, elastic-plastic analysis to compare the maximum allowed alternating stress range to the calculated maximum alternating stress intensity. The applicant noted that neither BC-TOP-1 nor the main steam penetration design calculation explicitly include the main feedwater penetrations; but the design calculation for "remaining penetrations" refers to the main steam penetration design calculation for both main steam and main feedwater penetrations. The applicant concluded the assumed cyclic loads for the main steam penetrations and the elastic-plastic evaluation of BC-TOP-1 is applicable to the main feedwater penetrations.

The LRA states that the original design basis 100 operating thermal cycles (Load Condition V) assumed for the main stream penetrations (also applicable to main feedwater penetrations) will be exceeded during the period of extended operation. The applicant stated that based on its plant operating experience, 250 full-range thermal cycles (BC-TOP-1 Part II "Condition V" events) could be expected in 60 years. The applicant then used 250 cycles in its evaluation of the TLAA, in addition to the 10 Loading Condition IV events, and concluded that the design analyses of the main steam and main feedwater penetrations remain valid for the period of extended operation.

4.6.2.2 Staff Evaluation

The staff reviewed LRA Section 4.6.2 to verify, pursuant to 10 CFR 54.21(c)(1)(i), that the analyses remain valid for the period of extended operation. The main steam penetrations analyses, also applicable to the main feedwater penetrations, are based on BC-TOP-1 Loading Condition V thermal cycle events, which are directly dependent on startup-shutdown cycles, and Loading Condition IV events, which do not change with the licensed plant life. Based on plant operating experience, the applicant projected 250 Load Condition V thermal cycles for 60 years. The design basis equivalent usage factor for the 10 assumed condition IV events is 0.270, and the design basis equivalent usage factor for the original assumed 100 Condition V events is

0.028. Using the projected 250 Condition V events, multiplied by a factor of 10 for a total of 2500 cycles, the applicant calculated the equivalent usage factor:

$$0.270 + (2500/100) \times 0.028 = 0.970 < 1.0$$

The staff finds the calculated usage factor of 0.97 near the acceptable limit of 1.0. However, the use of 2500 cycles in the analyses is conservative, and additional margin is available in the design. The staff finds that the calculations will remain valid during the period of extended operation in accordance with 10 CFR 54.21(c)(1)(i).

4.6.2.3   Updated Final Safety Analysis Report Supplement

The applicant provided a UFSAR supplement summary description of its TLAA evaluation of design cycles for the main steam and feedwater line penetrations in LRA Section A3.5. Based on its review of the UFSAR supplement, the staff concludes that the summary description of the applicant's actions to address design cycles for the main steam and main feedwater line penetrations is adequate.

4.6.2.4   Conclusion

Based on its review, the staff concludes that the applicant has demonstrated, pursuant to 10 CFR 54.21(c)(1)(i), that for design cycles for the main steam and main feedwater line penetrations, the analyses remain valid for the period of extended operation. The staff also concludes that the UFSAR supplement contains an appropriate summary description of the TLAA evaluation, as required by 10 CFR 54.21(d).

4.6.3   Design Cycles for the Recirculation Sump Suction Line Penetrations

4.6.3.1   Summary of Technical Information in the Application

LRA Section 4.6.3 summarizes the evaluation of design cycles for the recirculation sump suction line penetrations for the period of extended operation. The applicant stated that recirculation suction line penetrations were evaluated for ASME Code Section III, NE-3222.4(d) "Vessels Not Requiring Analysis for Cyclic Operation" exemption from fatigue analysis. The exemption criteria depend on the number of cycles for which loads are applied, therefore, the exemption is a TLAA. In this TLAA, the applicant noted that the analysis for the design cycles for the recirculation sump suction line penetrations remain valid for the period of extended operation in accordance with 10 CFR 54.21(c)(1)(i).

4.6.3.2   Staff Evaluation

The staff reviewed LRA Section 4.6.3 to verify pursuant to 10 CFR 54.21(c)(1)(i), that the analysis remains valid for the period of extended operation. The applicant stated that the analysis of the recirculation sump suction line penetrations was based on the alternating stress range for pressure cycles. The analysis demonstrated that the allowable number of cycles is 1E+4, which is far greater than the number of cycles expected for the period of extended operation. The staff also reviewed NE-3222.4(d), "Vessels Not Requiring Analysis for Cyclic Operation," to confirm that the fatigue analysis for theses penetrations is not required. The staff confirmed that a fatigue analysis is not required since the applicant met the requirements of NE-3222.4(d). The staff finds that the analyses will remain valid during the period of extended operation in accordance with 10 CFR 54.21(c)(1)(i).

4.6.3.3   UFSAR Supplement

The applicant provided a UFSAR supplement summary description of its TLAA evaluation of design cycles for the recirculation sump suction line penetrations in LRA Section A3.5.2. On the basis of its review of the UFSAR supplement, the staff concludes that the summary description

Time-Limited Aging Analyses

of the applicant's actions to address design cycles for the recirculation sump suction line penetrations are adequate.

4.6.3.4 Conclusion

On the basis of its review, the staff concludes that the applicant has demonstrated, pursuant to 10 CFR 54.21(c)(1)(i), for design cycles for the recirculation sump suction line penetrations, that the existing analysis remains valid for the period of extended operation. The staff also concludes that the UFSAR supplement contains an appropriate summary description of the TLAA evaluation, as required by 10 CFR 54.21(d).

4.7 Other Plant-Specific Time-Limited Aging Analyses

4.7.1 Load Cycle Limits of Cranes, Lifts, and Fuel Handling Equipment Designed to Crane Manufacturers Association of America Standard-70

4.7.1.1 Summary of Technical Information in the Application

In the LRA Section 4.7.1, the applicant provided a list of lifting machines to Crane Manufacturers Association of America (CMMA) standard 70 as follows.

4.7.1.1.1 Cranes

Containment Building Polar Crane. The applicant stated that the polar crane is designed to CMAA-70, Class A, with 225-ton main and 35-ton auxiliary hoists. The applicant also stated that the crane has three operational requirements: SG construction, plant operation, and SG removal.

Cask Handling Crane. The applicant stated that the cask handling crane is an indoor electrical overhead traveling bridge crane with a single-failure-proof trolley. The main hoist is rated at 150 tons, and the auxiliary hoist is rated at 15 tons. The applicant also stated that the cask-handling crane currently meets CMAA-70, service level A standards.

SAFLIFT™ Strongback Canister Hoist. The applicant stated that the SAFLIFT™ strongback canister hoist is a combined 125-ton lift beam plus 50-ton single-failure-proof canister hoist. The applicant also stated that the SAFLIFT™ strongback canister hoist is designed to CMAA-70 class C (2000), NUREG-0554 (1979), and NUREG-0612 Appendix C (1980) standards.

New Fuel Handling Crane. The applicant stated that the new fuel handling crane is a CMAA-70 service level C, 10-ton bridge crane. It is also used to perform activities associated with spent fuel reconstitution and re-caging.

4.7.1.1.2 Fuel and Control Element Assembly Handling Machines

Spent Fuel Handling Machine. The applicant stated that the spent fuel handling machine transfers fuel between the new fuel elevator, the transfer system, the spent fuel storage racks, and the spent fuel storage canister in the cask-loading pit. The applicant also stated that the specification requires design for 60,000 cycles of full speed hoist operation and 30,000 cycles of bridge and trolley operation. The hook capacity is 2,000 pounds.

Refueling Machine. The applicant stated that the refueling machine moves fuel assemblies in and out of the core and between the core and the transfer system. The applicant also stated that the specification requires design for 60,000 cycles of full speed hoist operation and 30,000 cycles of bridge and trolley operation. The hook load is limited to 2,600 pounds over "fuel only regions" and 1,600 pounds over "fuel plus hoist-box regions."

Control Element Assembly (CEA) Change Platform. The applicant stated that the CEA change platform is used to move the CEAs within the upper guide structure or between the upper guide structure and the CEA elevator. The applicant also stated that the specification requires design

Time-Limited Aging Analyses

for 30,000 cycles of full speed operation. The hook capacity is 2,000 pounds. The CEA change platform is not expected to do any over-capacity lifts during its lifetime.

Fuel Transfer System (i.e., Upenders, Trolley). The applicant stated that the fuel transfer system moves the fuel between the containment building and the fuel building through the transfer tube. The applicant also stated that the specification requires design for 10,000 cycles of operation, where one cycle consists of the transport and handling operations associated with the exchange of fuel assemblies between the fuel handling and containment buildings. The fuel transfer components are not expected to do any over-capacity transfers during their lifetime.

New Fuel Elevator. The applicant stated that the new fuel elevator is used to introduce new fuel into the spent fuel pool so that it can be moved to the transfer system by the spent fuel-handling machine. The applicant also stated that the specification requires design for 20,000 cycles of operation, where one cycle is defined as one complete up and down movement of the elevator. The capacity is 2,000 pounds. The new fuel elevator is not expected to do any over-capacity lifts during its lifetime.

CEA Elevator. The applicant stated that the CEA elevator is used to introduce new CEAs into the refueling pool and may be used to hold the spent CEAs while they are being disassembled for disposal. The applicant also stated that the specification requires design for 10,000 cycles of operation. The capacity is 2,000 pounds. The CEA elevator is not expected to do any over-capacity lifts during its lifetime.

4.7.1.2 Staff Evaluation

The staff reviewed LRA Section 4.7.1 to verify, pursuant to 10 CFR 54.21(c)(1)(i), that the analyses remain valid for the period of extended operation as follows.

Polar Crane. The overhead crane in the containment (225-ton/35-ton) for reactor servicing operations is of the polar configuration and is seated on a girder bracketed off the containment wall. The polar crane is designed to CMAA-70, class A requirement. The crane, therefore, was designed to 100,000 maximum-rated load cycles for a 40-year life.

The number of maximum rated load cycles for the polar crane originally projected for 40 years was 243. The number of maximum rated cycles for a 60-year life, based on 40 refueling outages, is 390. This is fewer than the 100,000 permissible cycles and, therefore, is acceptable.

Cask Handling Crane. The cask handling crane is an indoor electrical overhead traveling bridge crane with a single failure proof trolley. The cask handling crane currently meets CMAA-70, service level A requirements. The crane, therefore, was designed to 100,000 maximum-rated load cycles for a 40-year life.

The number of maximum rated load cycles for the cast handling crane originally projected for 40 years was 864. The number of maximum rated cycles for a 60-year life, based on 40 refueling outages, is 1,296. This is fewer than the 100,000 permissible cycles and, therefore, is acceptable.

SAFLIFT™ Strongback Canister Hoist. The SAFLIFT™ strongback canister hoist is a combined 125-ton lift beam plus 50-ton single-failure-proof canister hoist. The SAFLIFT™ strongback canister hoist is designed to CMAA-70 class C (2000), NUREG-0554 (1979), and NUREG-0612 Appendix C (1980) standards. The SAFLIFT™ strongback canister hoist currently meets CMAA-70, service level C requirement. The hoist, therefore, was designed to 500,000 maximum-rated load cycles for a 40-year life.

The number of maximum rated load cycles for the hoist originally projected for 40 years was 2,565. The number of maximum rated cycles for a 60-year life, based on 40 refueling outages, is 3,848. This is fewer than the 500,000 permissible cycles and, therefore, is acceptable.

Time-Limited Aging Analyses

New Fuel Handling Crane. The new fuel handling crane (10-ton) currently meets CMAA-70, service level C requirement. The new fuel handling crane, therefore, was designed to 500,000 maximum-rated load cycles for a 40-year life.

The number of maximum rated load cycles for the new fuel handling crane originally projected for 40 years was 13,770. The number of maximum rated cycles for a 60-year life, based on 40 refueling outages, is 20,655. This is fewer than the 500,000 permissible cycles and, therefore, is acceptable.

Spent Fuel Handling Machine. The spent fuel handling machine (2,000 pounds) currently meets CMAA-70, service level A requirements. The spent fuel handling machine, therefore, was designed to 100,000 maximum-rated load cycles for a 40-year life.

The number of maximum rated load cycles for the spent fuel handling machine originally projected for 40 years was 43,389. The number of maximum rated cycles for a 60-year life, based on 40 refueling outages, is 65,084. This is fewer than the 100,000 permissible cycles and, therefore, is acceptable.

Refueling Machine. The refueling machine hook load is limited to 2,600 pounds over "fuel only regions" and 1,600 pounds over "fuel plus hoist-box regions," and it currently meets CMAA-70, service level A requirements. The refueling machine, therefore, was designed to 100,000 maximum-rated load cycles for a 40-year life.

The number of maximum rated load cycles for the refueling machine originally projected for 40 years was 21,546. The number of maximum rated cycles for a 60-year life, based on 40 refueling outages, is 32,319. This is fewer than the 100,000 permissible cycles and, therefore, is acceptable.

CEA Change Platform. The hook of the CEA change platform capacity is 2,000 pounds and currently meets CMAA-70, service level A requirements. The refueling machine, therefore, was designed to 100,000 maximum-rated load cycles for a 40-year life.

The number of maximum rated load cycles for the CEA change platform originally projected for 40 years was 1,458. The number of maximum rated cycles for a 60-year life, based on 40 refueling outages, is 2,187. This is fewer than the 100,000 permissible cycles and, therefore, is acceptable.

Fuel Transfer System (i.e., Upenders, Trolley). The fuel transfer system moves the fuel between the containment building and the fuel building through the transfer tube. The fuel transfer system currently meets CMAA-70, service level A requirements. The fuel transfer system, therefore, was designed to 100,000 maximum-rated load cycles for a 40-year life.

The number of maximum rated load cycles for fuel transfer system originally projected for 40 years was 19,521. The number of maximum rated cycles for a 60-year life, based on 40 refueling outages, is 29,282. This is fewer than the 100,000 permissible cycles and, therefore, is acceptable.

New Fuel Elevator. The new fuel elevator (2,000 pounds) currently meets CMAA-70, service level A requirements. The new fuel elevator, therefore, was designed to 100,000 maximum-rated load cycles for a 40-year life.

The number of maximum rated load cycles for new fuel elevator originally projected for 40 years was 4,374. The number of maximum rated cycles for a 60-year life, based on 40 refueling outages, is 6,562. This is fewer than the 100,000 permissible cycles and, therefore, is acceptable.

Time-Limited Aging Analyses

CEA Elevator. The CEA elevator (2,000 pounds) currently meets CMAA-70, service level A requirements. The CEA elevator, therefore, was designed to 100,000 maximum-rated load cycles for a 40-year life.

The number of maximum rated load cycles for CEA elevator originally projected for 40 years was 729. The number of maximum rated cycles for a 60-year life, based on 40 refueling outages, is 1,094. This is fewer than the 100,000 permissible cycles and, therefore, is acceptable.

4.7.1.3 Updated Final Safety Analysis Report Supplement

The applicant supplied an UFSAR supplement summary description of its TLAA evaluation of load cycle limits of cranes, lifts, and fuel handling equipment to CMAA-70 in LRA Section A3.6.1. Based on its review of the UFSAR supplement, the staff concludes that the summary description of the applicant's actions to address crane load cycles is adequate.

4.7.1.4 Conclusion

Based on its review, as discussed above, the staff concludes that the applicant has demonstrated, pursuant to 10 CFR 54.21(c)(1)(i), that for load-cycle limits of cranes, lifts, and fuel handling equipment to CMAA-70, the analyses remain valid for the period of extended operation. The staff also concludes that the UFSAR supplement contains an appropriate summary description of the TLAA evaluation, as required by 10 CFR 54.21(d).

4.7.2 Absence of Time-Limited Aging Analyses for Metal Corrosion Allowances and Corrosion Effects

The staff reviewed LRA Section 4.7.2 to verify that the metal corrosion allowances analysis is not a TLAA. The staff's evaluation is found in SER Section 4.1.3.1.1.

4.7.3 Inservice Flaw Growth Analyses that Demonstrate Structural Stability for 40 Years

4.7.3.1 Summary of Technical Information in the Application

Defects discovered by ISI or component failures may be repaired or replaced to restore the basis of the original design analysis, may be repaired or replaced to a different configuration, or may be analyzed to confirm that the as-found condition is acceptable. For ASME components, ASME Code Section XI controls these activities. A flaw analysis of a Class 1 component usually requires a fatigue crack growth analysis, which is a TLAA if it qualifies the component for the plant design life. A thorough review of the PVNGS licensing basis, supported by interviews with plant staff familiar with the history of Class 1 components, found the following TLAA evaluations of indications discovered during ISIs:

- A linear elastic fracture mechanics fatigue crack growth analysis of indications in a Unit 2 pressurizer support skirt-forging weld as discussed in Section 4.3.2.4 and similar evaluations of postulated (rather than actual) initial defects

- Crack growth and fracture mechanics stability analyses of postulated defects in original heater sleeve attachment welds remaining in the pressurizer lower heads following heater sleeve replacements as discussed in Section 4.3.2.4

- Fatigue crack growth and fracture mechanics stability analyses in support of pressurizer nozzle overlays as discussed in Section 4.3.2.4

- Fatigue crack growth and fracture mechanics stability analyses in support of hot leg surge and shutdown cooling nozzle weld overlays as discussed in Section 4.3.2.7

Time-Limited Aging Analyses

- Fatigue crack growth assessments and fracture mechanics stability analyses in support of the LBB evaluation, but no TLAAs (Section 4.3.2.15)
- Fatigue crack growth and fracture mechanics stability analyses of half-nozzle repairs to alloy 600 materials in reactor coolant hot legs as discussed in Section 4.7.4

4.7.3.2 Staff Evaluation

The staff's evaluation for Sections 4.3.2.4, 4.3.2.7, 4.3.2.15, and 4.7.4 can be found in the corresponding sections of this SE.

4.7.3.3 Updated Final Safety Analysis Report Supplement

The UFSAR supplement for Sections 4.3.2.4, 4.3.2.7, and 4.7.4 can be found in the corresponding sections in the staff's SE.

4.7.3.4 Conclusion

The staff's conclusions on these fatigue analyses are found in SER Sections 4.3.2.4, 4.3.2.7, 4.3.2.15 and 4.7.4.

4.7.4 Fatigue Crack Growth and Fracture Mechanics Stability Analyses of Half-Nozzle Repairs to Alloy 600 Material in Reactor Coolant Hot Legs and Supporting Corrosion Analyses

4.7.4.1 Summary of Technical Information in the Application

The applicant stated that for the half-nozzle repair of the Alloy 600 nozzles in the hot leg, the staff authorized a RR for the flaw removal and successive inspection requirements of the 1992 edition and addenda of the ASME Code Section XI, IWA-3300 and IWB-2420 for the alternative half-nozzle method used to repair Alloy 600 small-bore nozzles in the hot leg.

LRA Section 4.7.4 states that as part of the RR, the applicant was required to perform a fatigue crack growth calculation, flaw stability analysis, and corrosion analysis. The applicant recognized, however, that the RR permitting these repairs was granted only through the fourth 10-year inspection interval, and must, therefore, be extended for the period of extended operation. The safety determination supporting this ASME Code exemption is also supported by a commitment to track time at cold shutdown conditions, which must also be continued for the period of extended operation.

The corrosion analysis in the hot leg piping walls exposed by the repairs, depends on time at cold shutdown. The original LRA Section 4.7.4 concluded that the corrosion analysis is not a TLAA since the analysis was extended beyond 60 years. As evaluated below, LRA Amendment 19 changed the evaluation to a TLAA, to be dispositioned per 10 CFR 54.21(c)(1)(i), that the analysis remains valid for the period of extended operation.

Fatigue crack growth and flaw stability analyses of nozzle remnants and welds left in the hot leg depend on the number of heatup, cooldown, and OBE cycles assumed for a 40-year life. These analyses are, therefore, TLAAs. The LRA dispositions these TLAAs per 10 CFR 54.21(c)(1)(iii), that the effects of aging on the intended function of the hot leg half-nozzle repairs will be managed during the period of extended operation

4.7.4.2 Staff Evaluation

The staff reviewed LRA Section 4.7.4 to verify, per 10 CFR 54.21(c)(1)(i) that, for the fatigue crack growth and flaw stability analyses of half-nozzle repairs to Alloy 600 material in reactor coolant hot legs, the effects of aging on the intended function will be adequately managed for the period of extended operation. In addition, the staff reviewed LRA Section 4.7.4 to verify that the corrosion analyses in support of half-nozzle repairs to the hot legs are not TLAAs.

Time-Limited Aging Analyses

During the staff's review, the staff noted that the applicant dispositioned ASME Code Section XI supplemental fatigue flaw growth or cycle-dependent fracture mechanics evaluations in accordance with 10 CFR 54.21(c)(1)(iii). The applicant proposed to use the CC activities from its enhanced Metal Fatigue Program to manage the effects of aging and verify the continued validity of these ASME Code Section XI analyses during the period of extended operation.

The staff noted that the applicant's proposal to use CC activities to verify the continued validity of these ASME Code Section XI analyses may be beyond the applicant's CLB. The staff noted that TS 5.5.5 and UFSAR Section 3.9.1.1 discuss cycle tracking and counting against design limits and design calculations, but does not appear to discuss design transient tracking and counting for ASME Code Section XI supplement fatigue flaw growth or cycle dependent fracture mechanics evaluations. Per TS 5.5.5 and UFSAR Section 3.9.1.1, cyclic and transient occurrences are tracked to ensure that components are maintained within the design limits. However, the applicant's CC procedure does not discuss the types of analyses this requirement is applicable to or the action limits and corrective actions that may be taken for these fatigue related or fracture mechanics evaluations. The staff noted that these corrective actions should be specified in the applicant's procedures and the action limits and corrective actions should be associated with the specific type of analysis.

The staff held a teleconference with the applicant on September 22, 2010, for clarification on how design basis transient cycle tracking and counting activities are accounted for in the CLB for ASME Code Section XI supplemental fatigue flaw growth or cycle dependent fracture mechanics evaluations. The staff also discussed the applicant's justification of the use of design basis transient cycle tracking and counting activities as the basis to disposition the ASME Code Section XI analyses in LRA Section 4.7.4 in accordance with 10 CFR 54.21(c)(1)(iii).

By letter dated October 13, 2010, the applicant stated that although the cycle tracking and counting activities of design basis transients for these evaluations are not explicitly described in the CLB today, the applicant recognizes the benefit of enhancing the UFSAR and the plant design transient tracking procedure to provide this guidance. The applicant recognized the importance of providing explicit procedures to assist an analyst if a design transient assumption or CUF limit is approached. Consequently, by letter dated October 13, 2010, the applicant committed (Commitment No. 60) to complete the following by November 30, 2010:

> The reactor coolant system transient and cycle tracking procedure 73ST-9RC02 and UFSAR Section 3.9.1 will be enhanced to discuss corrective actions that need to be taken prior to ASME Code Section III fatigue design limits being exceeded and to state that corrective actions may be required for other fatigue-related analyses, such as certain ASME Code Section XI supplemental fatigue flaw growth or cycle-dependent fracture mechanics evaluations that are dependent on the number of occurrences of design transients.

Based on its review, the staff finds the applicant's clarification and Commitment No. 60 acceptable because the applicant recognized that its CLB does not currently account for design basis transient cycle tracking activities for ASME Code Section XI supplemental fatigue flaw growth or cycle-dependent fracture mechanics evaluations and has committed to update its UFSAR and cycle tracking procedure to account for these evaluations and associated corrective actions. Further, the staff concludes that the applicant has demonstrated that the effects of aging will be managed for the fatigue crack growth calculation, flaw stability analysis, and corrosion analysis of half-nozzle repairs to Alloy 600 material in reactor coolant hot legs.

Time-Limited Aging Analyses

4.7.4.2.1 Absence of Time-Limited Aging Analysis in Corrosion Analyses for Hot Leg Half-Nozzle Repairs

LRA Section 4.7.4 concluded that the corrosion analyses in support of hot leg half-nozzle repairs are not TLAAs. By letter dated July 7, 2010, the applicant submitted LRA Amendment 19 which identified the corrosion analyses for hot leg half nozzle repairs as TLAAs and dispositioned them in accordance with 10 CFR 54.21(c)(1)(i).

The applicant stated that in March 2004, Westinghouse Electric released a topical report approved by staff as WCAP-15973-P-A, "Low-Alloy Steel Component Corrosion Analysis Supporting Small-Diameter Alloy 600/690 Nozzle Repair/Replacement Programs, and calculation CN-CI-02-71, Summary of Fatigue Crack Growth Evaluation Associated with Small Diameter Nozzles in CEOG Plants." These documents support half nozzle and MNSA repairs in CE plants.

On March 25, 2005, the applicant submitted Relief Request RR-31 for the repair of RCS hot leg small-bore nozzle repair for NRC review and approval. By letter dated May 5, 2005, the staff authorized RR-31 (ADAMS Accession No. ML051290123). This relief request uses CN-CI-02-71 and WCAP-15973-P-A in support of a request for exemption from the flaw removal and successive inspection requirements of ASME Code Section XI (1992), Sections IWA-3300 and IWB-2420, for the alternative half-nozzle method used for the ten Unit 2 small-bore, hot leg nozzles to be repaired during the spring 2005 refueling outage. WCAP-15973-P-A calculated corrosion rates of 1.53 mils-per-year for Alloy 600 nozzles. In response to the conditions of the final SE for the Westinghouse topical report, the applicant calculated that a limiting corrosion rate of 1.377 mils-per-year for Unit 3 would not exceed the allowable diameter until 2058, 60 years after the repair and 10 years after the end of the period of extended operation. The LRA stated that this calculation is not a TLAA since it does not meet the criterion of 10 CFR 54.3(a)(3) and is valid for the period of extended operation. The applicant stated in the LRA that the corrosion rate for Unit 3 is limiting for all three units and bounds the corrosion rates of Units 1 and 2.

In the relief request submittal, the applicant made an ongoing commitment to track the time at cold shutdown conditions:

> APS commits to continue to track the time at cold shutdown conditions against the assumptions made in the corrosion analysis to assure that the allowable bore diameter is not exceeded over the life of the plant. If the analysis assumptions are exceeded, APS shall provide a revised analysis to the staff and provide a discussion on whether volumetric inspection of the area is required.

This commitment was made because the corrosion rate at cold shutdown conditions is significantly higher than at operating conditions. This request was authorized by the staff, consistent with the APS commitment, and is valid for the second, third, and fourth 10-year inspection intervals. The applicant states in LRA Section 4.7.4 that an extension of this authorization will be required for continued relief from the ASME Code sections.

The provisions of 10 CFR 54.3(a)(3) state that TLAAs "[i]nvolve time-limited assumptions defined by the current operating term, for example, 40 years." The estimated life of 60 years after the repair and 10 years after the end of the extended period operation for the hot leg nozzles is a calculated result using the WCAP-15973-P-A corrosion methodology, not an assumption adopted in the methodology. Further, the WCAP states, "[t]he following assumptions were used in developing an overall corrosion rate for carbon and low alloy steels in a crevice environment and an estimate of the total corrosion for the remaining plant lifetimes..." This statement strongly suggested that the corrosion rate is based on test data considering operating experience and assumed valid for 40 years. The applicant's July 7, 2010, response to RAI 4.7.5-1 amended LRA Section 4.7.4 to identify the evaluation of the corrosion analysis for

Time-Limited Aging Analyses

the hot leg half-nozzle repairs as a TLAA, with a disposition in accordance with 10 CFR 54.21(c)(1)(i), that the corrosion analysis is valid for the period of extended operation. The applicant supported this disposition by citing bounding calculations that demonstrate that the analyses are valid beyond the period of extended operation.

LRA Section 4.7.4 states that the applicant committed to monitoring the cold shutdown conditions against the assumptions made in the corrosion analysis to assure that the allowable bore diameter is not exceeded over the life of the plant for the second, third, and fourth 10-year inspection intervals. By letter dated February 14, 2010, the staff issued RAI 4.7.4-1, asking the applicant to discuss whether the same commitment will be implemented in the fifth and sixth inspection intervals.

The applicant's March 1, 2010, response to RAI 4.7.4-1 stated that LRA Appendix A, Table A4-1, Commitment No. 46, documents the applicant's commitment to continue to monitor the cold shutdown conditions via the current tracking method for the period of extended operation, that is, for the fifth and sixth inspection intervals. The staff verified that LRA Appendix A, Table A4-1, Commitment No. 46 provides the requirement to monitor the cold shutdown conditions for the period of extended operation. Therefore, the staff finds that the structural integrity of the repaired small-bore nozzles will be maintained during the period of extended operation. The staff's concern in RAI 4.7.4-1 is resolved.

Because the estimated repair lifetime for the hot leg nozzle repairs exceeds 60 years and the operational assumptions will be monitored by Commitment No. 46, the staff concludes that the applicant's analysis is acceptable per 10 CFR 54.21(c)(1)(i). The staff also concludes that the bore diameter of the repaired nozzles will have adequate dimensions, and the bore will not be degraded by corrosion through the period of extended operation.

4.7.4.2.2 Fatigue Crack Growth and Stability Analysis for Hot Leg Half-Nozzle Repairs

Westinghouse calculation CN-CI-02-71 found that postulated defects left in remnants of the hot leg nozzles would not grow beyond an acceptable size, assuming 500 heatup and cooldown cycles and 200 OBE cycles, which are the design basis limiting cycles for a 40-year life. The CN-CI-02-71 fatigue crack growth and stability analysis is, therefore, identified in the LRA as a TLAA and dispositioned per 10 CFR 54.21(c)(1)(iii), that the effects of aging on the intended function of the hot leg half-nozzle repairs will be managed during the period of extended operation. To manage the effects of aging for this TLAA, the applicant will carry out the enhanced metal fatigue program to monitor the transient cycles in the analysis during the period of extended operation to ensure that appropriate re-evaluation or other corrective action is initiated if an action limit is reached. The staff finds that the TLAA is acceptable to manage the crack growth and stability analysis of the half nozzle repairs of small-bore nozzles per 10 CFR 54.21(c)(1)(iii).

4.7.4.2.3 Extension to All Hot Leg Small-Bore Nozzles

After reconciling the WCAP 15973-P-A topical report with the non-Westinghouse documentation that it had originally used as a basis, the applicant submitted Revision 1 to RR 31 in a letter dated August 16, 2005. RR 31, Revision 1 added the 63 previously repaired small-bore hot leg nozzles in all three units to those already covered in the initial RR. By letter dated September 12, 2006, the staff approved RR 31, Revision 1. All of the small diameter hot leg nozzles have been replaced. PVNGS has 27 small diameter hot leg penetrations per unit, as described in Section IV of RR 31, Revision 1.

LRA Section 4.7.4, Subsection "Extension to All Hot Leg Small-Bore Nozzles," states that the 63 previously repaired small-bore hot leg nozzles in all three units were added to RR 31. The applicant also stated there are 27 small-bore hot leg penetrations per unit. If there are 27 small-bore nozzles in each unit, the total number of small-bore nozzles in all three units

Time-Limited Aging Analyses

should be 81. It is not clear whether the exact number of small-bore nozzles on the hot leg piping is 63 or 81. By letter dated February 14, 2010, the staff issued RAI 4.7.4-2, asking the applicant to provide the exact number of small-bore nozzles in the hot leg piping in each unit, the number of small-bore nozzles that have been repaired in each unit, and the number of small-bore nozzles that have not been repaired. The applicant was also asked to discuss whether any small-bore nozzles in hot leg piping that are not covered under RR 31 and to confirm that the small-bore nozzles in hot leg piping that are not covered under RR 31 were analyzed for TLAA criteria.

The applicant's March 1, 2010, response to RAI 4.7.4-2 clarified and itemized the number of small-diameter hot leg penetrations in each unit with respect to Revisions 0 and 1 of RR 31 as shown below:

8	Unit 2 nozzles repaired in 1991 via a full nozzle repair (not covered under RR 31)
10	Unit 2 nozzles repaired in 2005 under RR 31, Revision 0
9	Unit 2 nozzles repaired under RR 31, Revision 1
27	Unit 1 nozzles repaired under RR 31, Revision 1
27	Unit 3 nozzles repaired under RR 31, Revision 1
81	Total

All 81 nozzles in all three units have been either repaired or replaced. As shown above, some of the nozzle repairs were covered under RR 31, Revision 0, and some of them were covered under RR 31, Revision 1. RR 31, Revision 0, addressed 10 nozzles replaced in Unit 2 during the spring of 2005. RR 31, Revision 1, added the following previously repaired Alloy 600 small-bore hot leg nozzles as follows:

Unit 1	27 nozzles
Unit 2	9 nozzles
Unit 3	27 nozzles
Total	63 nozzles

The last 63 nozzles covered by RR 31, Revision 1 were initially repaired under the Alloy 600 replacement program, from approximately October 1999 to April 2003. The staff notes that the 63 nozzles are part of the total 81 nozzles in all three units.

RR 31 does not cover the eight Unit 2 nozzles repaired in 1991 by a full nozzle repair. They were repaired in accordance with the ASME Code. The fatigue crack growth analysis and corrosion analysis are not applicable for these eight nozzles because they were repaired with a new, full-length nozzle. LRA Section 4.3.2.7 addresses the TLAAs associated with the design of the eight Unit 2 full nozzle repairs. In addition, the detailed list of the nozzle and the repair methods are provided in the response to RAI 4.3.2.7-4 as discussed in Section 4.3.2.7.2 of this SE.

The staff finds that the applicant has clarified the number of small-bore nozzles in the RCS primary loop piping that have been repaired. Also, the applicant has committed in Commitment No. 46 to monitor the bore diameter of the small-bore nozzle repairs in the RCS primary loop piping and to monitor cold shutdown conditions. Therefore, the repaired 81 nozzles in all three units will be monitored as a part of TLAA per 10 CFR 54.21(c)(1)(iii). The staff's concerns in RAI 4.7.4-2 are resolved.

4.7.4.3 Updated Final Safety Analysis Report Supplement

The applicant provided an amended UFSAR supplement summary description of its TLAA of the fatigue crack growth and fracture mechanics stability analyses of half-nozzle repairs to Alloy 600 material in reactor coolant hot legs in LRA Section A3.6.2. Based on its review of the UFSAR

Time-Limited Aging Analyses

supplement in LRA Section A3.6.2, the staff concludes that the summary description of the applicant's actions is adequate.

to address the TLAA for the half-nozzle repairs to Alloy 600 material in reactor coolant hot legs is adequate.

4.7.4.4 Conclusion

Based on its review, the staff concludes that, pursuant to 10 CFR 54.21(c)(1)(i), the applicant has demonstrated that the TLAA associated with the corrosion analyses of hot leg half-nozzle repairs remain valid for the period of extended operation. Further, the staff concludes that, pursuant to 10 CFR 54.21(c)(1)(iii), the applicant has demonstrated that the effects of aging on the intended function of the hot leg half-nozzle repairs will be adequately managed for the period of extended operation. The UFSAR supplement contains an appropriate summary description of the TLAA evaluations as required by 10 CFR 54.21(d).

4.7.5 Corrosion Analyses of Pressurizer Ferritic Materials Exposed to Reactor Coolant by Half-Nozzle Repairs of Pressurizer Heater Sleeve Alloy 600 Nozzles

4.7.5.1 Summary of Technical Information in the Application

LRA Section 4.7.5 summarizes the evaluation of the general corrosion of pressurizer ferritic materials exposed to reactor coolant as a result of half-nozzle repairs for all heater sleeves. The applicant stated that the bounding case for general corrosion in pressurizer heater sleeves in the WCAP-15973-P-A report, "Low Alloy Steel Component Analysis Supporting Small Diameter Alloy 600/690 Nozzle Repair/Replacement Program," provided an estimated repair life of 194 years; therefore, the applicant stated that the analysis is valid for the period of extended operation. The applicant further concluded that the corrosion analysis is a TLAA.

4.7.5.2 Staff Evaluation

The staff's evaluation of the general corrosion analysis supporting half-nozzle repairs of small-diameter Alloy 600/690 nozzles was documented in an SE, dated January 12, 2005, for the WCAP-15973-P-A report. The staff's evaluation of the plant-specific application of the WCAP-15973-P-A report, including its corrosion analysis, to pressurizer heater sleeves was documented in an SE dated November 5, 2004, for Relief Request RR-29. The plant-specific application was approved for the second 10-year ISI interval. However, since the time period accepted and approved by the staff for the corrosion analysis was the second 10-year ISI interval, applicability of the WCAP-15973-P-A report to the extended period of operation has not been established. Further, the original version of LRA Section 4.7.5 concluded that the corrosion analysis is not a TLAA. The staff issued RAI 4.7.5-1, asking the applicant to identify the plant-specific submittal addressing general corrosion in support of the half-nozzle repairs installed in the pressurizer heater sleeves.

The applicant's response to RAI 4.7.5-1, dated March 1 and May 21, 2010, provided plant-specific operating data and calculations, demonstrating that the WCAP-15973-P-A corrosion results are applicable for the period of extended operation. To ensure that the operation at the period of extended operation will be consistent with the current operating data, the applicant committed (Commitment No. 46) to continue "the cold shutdown time monitoring program [using the current tracking method]" for the period of extended operation. Since the estimated repair lifetimes for the pressurizer heater sleeve nozzles is bounded by the generic 194 years by a significant margin and the operation assumptions will be monitored by Commitment No. 46, the staff concludes that the applicant's analysis is acceptable. Therefore, the analysis meets the requirements of 10 CFR 54.21(c)(1)(i) and will ensure that the repaired nozzles will have adequate dimensions through the period of extended operation.

Time-Limited Aging Analyses

The staff considers the WCAP-15973-P-A corrosion analysis a TLAA because assumptions were used in developing the overall corrosion rate for carbon and low alloy steels in a crevice environment to estimate the total corrosion for the remaining 40-year plant lifetime. The July 7, 2010, response provided an LRA revision concluding that the corrosion analysis is a TLAA. The staff's concerns in RAI 4.7.5-1 are resolved.

4.7.5.3 Updated Final Safety Analysis Report Supplement

The applicant provided a revised UFSAR supplement summary description of its TLAA evaluation of the corrosion analyses of pressurizer ferritic materials exposed to reactor coolant by half-nozzle repairs of pressurizer heater sleeve Alloy 600 nozzles in LRA Section A.3.6.4. Based on its review of the UFSAR supplement, the staff concludes that the summary description of the applicant's actions to address the subject is adequate.

4.7.5.4 Conclusion

On the basis of its review, as discussed above, the staff concludes that the applicant has demonstrated, pursuant to 10 CFR 54.21(c)(1)(i), that, for the general corrosion of ferritic materials exposed to reactor coolant as a result of half-nozzle repairs to pressurizer heater sleeves, the analyses remain valid for the period of extended operation. The staff also concludes that the UFSAR supplement contains an appropriate summary description of the TLAA evaluation, as required by 10 CFR 54.21(d).

4.7.6 Absence of a Time-Limited Aging Analysis for Reactor Vessel Underclad Cracking Analyses

The staff reviewed LRA Section 4.7.6 to verify that the reactor vessel underclad cracking analysis is not a TLAA. The staff's evaluation is found in SER Section 4.1.3.1.2.

4.7.7 Absence of a Time-Limited Aging Analysis for a Reactor Coolant Pump Flywheel Fatigue Crack Growth Analysis

The staff reviewed LRA Section 4.7.7 to verify that the RCP flywheel fatigue crack growth analysis is not a TLAA. The staff's evaluation is found in SER Section 4.1.3.1.3.

4.7.8 Building Absolute or Differential Heave or Settlement, Including Possible Effects of Changes in Perched Groundwater Lens

4.7.8.1 Summary of Technical Information in the Application

LRA Section 4.7.8 summarizes building absolute or differential heave or settlement, including possible effects of changes in the perched groundwater lens. The applicant described the perched groundwater lens as a locally elevated region of groundwater above an impermeable layer, charged by irrigation before construction. An increase in the water level of this lens above the foundation elevations could affect stability, and a decline in the water level could conceivably result in building settlement that exceeds expectations. The applicant stated that groundwater monitoring data shows no potential for settlement due to changes in groundwater level. The original projections of increases in groundwater levels described in UFSAR Section 2.4.13.2.4.D were very conservative, and the conclusion of foundation stability remains valid through the period of extended operation.

UFSAR Section 2.5.4.10.2 documents the applicant's general analysis and evaluation of building heave and settlement. Settlement of major structures is monitored during the current license term on a frequency of five years and will continue through the life of the plant. The settlement surveillance is done as a part of the Structures Monitoring Program described in LRA Section B2.1.32. The staff's evaluation of the program is found in SER Section 3.0.3.2.20. This program provides the requirements to measure settlement of each individual structure and differential settlement at a common point between any two adjacent structures having critical connections as well as containment tilt angle. The post-construction settlement acceptance criteria for each individual structure are as follows:

- The post-construction settlement is less than 1.5 inches.

- Post-construction differential settlement at a common point between any two adjacent structures having critical connections is less than 0.5 inch.

- The post-construction containment tilt angle is less than 0.057 degrees.

The applicant stated that the first action limit is 90 percent of each acceptance criterion and requires an increase in survey frequency and, if necessary, a remedial action.

The applicant stated that the largest short-term, post-construction differential settlement (as of 1984) measured between any two category I structures was 0.3 inches. LRA Table 4.7-2 summarizes the results of 2003 settlement monitoring inspections. The results show that post-construction differential settlement between the Unit 2 auxiliary and radwaste buildings exceeds the maximum allowable 0.5 inches. The table shows that the measured post-construction differential settlement at Units 1 and 3 is nearly 90 percent of the maximum allowable value of 0.5 inches. The applicant said that the increased monitoring frequency verified that a significant trend of settlement or differential settlement was not occurring. The frequency has since been reduced to the normal frequency of five years.

4.7.8.2 Staff Evaluation

The staff reviewed LRA Section 4.7.8 to verify, pursuant to 10 CFR 54.21(c)(1)(i), that the original projection of changes in groundwater level remain valid through the period of extended operation. The staff also reviewed LRA Section B2.1.32 to confirm that, pursuant to 10 CFR 54.21(c)(1)(iii), settlement of structures in the scope of license renewal will be adequately managed during the period of extended operation to ensure their intended functions are maintained consistent with the CLB. In its review of LRA Sections 4.7.8, A3.6.3, and A1.32 and UFSAR Sections 2.4.13 and 2.5.4, the staff determined that additional information, described below, is needed to complete its review.

LRA Section 4.7.8 describes the affect of perched groundwater level increase or decrease on foundation stability and settlement. The applicant stated that the perched groundwater levels will not exceed the levels assumed for building foundation designs and will, therefore, not affect building stability. The applicant explained that the only potential sources of significant recharge of the perched groundwater lens near the units are the 85-acre and 45-acre reservoirs and noted that the reservoirs were lined in 2006 with a double liner system that should prevent any future recharge of the shallow aquifer from the reservoirs. The applicant also noted that the wells near the reservoirs have not yet shown any effects of increased leakage to the groundwater.

In reviewing the above information, the staff was unclear whether the applicant credits monitoring of the perched groundwater level and the double liner system for evaluation of the TLAA in addition to settlement monitoring activities conducted in accordance with its Structures Monitoring Program.

In a conference call dated February 24, 2010, the applicant clarified that the discussion on the perched groundwater lens level and the double liner system was provided as additional information. The applicant's monitoring of the groundwater lens and the reservoirs' double liner system are not credited for settlement. Only the settlement monitoring activities conducted in accordance with the Structures Monitoring Program are credited for managing aging of structural settlement during the period of extended operation. The staff found the applicant's clarification acceptable because the implementation of settlement monitoring activities

Time-Limited Aging Analyses

described in the Structures Monitoring Program provides reasonable assurance that the affect of groundwater level increase or decrease on settlement of structures will be detected before a loss of an intended function.

LRA Section 4.7.8 states that the Structures Monitoring Program monitors on 5-year intervals the foundation responses and ground movement of "major structures." A review of the applicant's Structures Monitoring Program showed that similar wording is included in the program description. The applicant did not specify, however, which structures would be monitored during the period of extended operation and whether the inspection frequency will be adjusted, as described in UFSAR Section 2.5.4.13, if post-construction settlement reaches 90 percent of the design criteria values. As a result, the staff issued RAI 4.7.8-1 asking the applicant to supply a list of structures within the scope of license renewal that will be monitored for the effects of settlement during the period of extended operation and justify excluding any structure from settlement monitoring that is within the scope of license renewal. The RAI also asked that the applicant supply a list of structures, included in the scope of 10 CFR 54.4, that will be monitored on a different frequency or using different instrumentation than specified in the UFSAR, Section 2.5.4.13 and Table 2.5-19.

By letter dated April 1, 2010, the applicant responded to RAI 4.7.8-1 stating that all structures that are within the scope of license renewal in accordance with 10 CFR 54.4(a)(1) or (2), are listed in LRA Table 2.2-1. The applicant further stated that the structures will be monitored for settlement consistent with the frequency (every five years) and instrumentation specified in UFSAR Section 2.5.4.13 and Table 2.5-19 during the period of extended operation. The five-year frequency will be maintained during the period of extended operation unless a more frequent monitoring periodicity is required based on inspection results.

The applicant went on to state that the fire water pump house, the transformer foundations, and the station blackout (SBO) generator structures, which are in scope of license renewal pursuant to 10 CFR 54.4(a)(3), are not in the scope of the settlement program described in UFSAR Section 2.5.4.13. These structures will be visually monitored for aging effects due to settlement on a 10-year frequency during the period of extended operation, in accordance with the Structures Monitoring Program.

The staff found the applicant's response acceptable for the structures within the scope of license renewal in accordance with 10 CFR 54.4(a)(1) or (a)(2) because settlement will be monitored in accordance with the UFSAR requirements during the period of extended operation. For structures within the scope license renewal, pursuant to 10 CFR 54.4(a)(3), the staff questioned how visual inspections performed in accordance with the Structures Monitoring Program will provide the data necessary to trend settlement and differential settlement of the fire water pump house, the transformer foundations, and the SBO generator structures. In a conference call dated May 14, 2010, the applicant was asked to give more details on whether settlement of the fire water pump house, the transformer foundations, and the SBO generator structures is within the scope of this TLAA. If it is, the applicant was asked to explain how visual inspections conducted on a 10-year frequency in accordance with the Structures Monitoring Program will effectively manage the effects of settlement as required by 10 CFR 54.21(c)(1)(iii).

By letter dated May 21, 2010, the applicant supplemented its response to RAI 4.7.8-1(b) and stated that the fire water pump house, the transformer foundations, and the SBO generator structures are not within the scope of the TLAA for settlement monitoring described in UFSAR Section 2.5.4.13 because the structures are not incorporated by reference in UFSAR Section 2.5.4.13. Therefore, they do not meet the criterion of 10 CFR 54.3(a)(6). The staff finds the applicant's response acceptable because settlement of these structures does not meet TLAA criterion 10 CFR 54.3(a)(6). The staff's concerns in RAI 4.7.8-1 are resolved.

4-106

Time-Limited Aging Analyses

LRA Section 4.7.8, Table 4.7-2 provides the 2003 summary results of the settlement monitoring program. The results show that the measured containment building tilt angle and the measured post-construction settlement are less than the maximum allowable values; however, the measured post-construction differential settlement of 0.8748 inches between the Unit 2 auxiliary and the radwaste buildings exceeds the maximum allowable value of 0.5 inches. In addition, the post construction differential settlement of Units 1 and 3 between the auxiliary building and the radwaste building is about 90 percent of the maximum allowable value of 0.5 inches. By letter dated March 2, 2010, the staff issued RAI 4.7.8-2 asking the applicant to provide 1998 and 2008 settlement data for locations provided in the Table 4.7-2 with the 2003 results. In addition, the staff asked the applicant to describe the corrective actions taken to address the affect of exceeding the maximum allowable post-construction differential settlement on Unit 2 structures and critical piping.

By letter dated April 1, 2010, the applicant provided 1997 and 2010 post-construction settlement and post-construction differential settlement summary results in response to RAI 4.7.8-2. The applicant stated the 1998 settlement measurements were taken in 1997, and the 2008 measurements were taken in 2010 as permitted by an established 25-percent grace period for the five-year frequency.

The staff reviewed the tabulated 1997, 2003, and 2010 settlement measurement data and observed that Units 1, 2, and 3 post-construction settlement continues to show a slightly increasing trend. However, the measured post-construction total settlement remains below the maximum allowable limit of 1.5 inches. The 2010 Unit 1 measured post-construction settlement of 1.3524 inches exceeds the action limit of 1.35 inches established in UFSAR Section 2.5.4.13. As a result, the applicant increased the inspection frequency to one-month intervals, as specified in UFSAR Section 2.5.4.13. The staff's review of the data noted no significant changes in total post-construction differential settlement between 1997 and 2010. The measured differential settlement remains below the maximum allowable limit of 0.5 inches, except between the Unit 2 auxiliary and radwaste buildings. The applicant stated that corrective actions were taken as required by UFSAR Section 2.5.4.13 as a result of exceeding the maximum allowable limit. The staff finds the applicant's response acceptable because the 1997, 2003, and 2010 settlement monitoring data support the conclusion that the ongoing settlement and differential settlement show no significant trend and that most of the differential settlement between the auxiliary and the radwaste buildings occurred before 1997. The staff's concern in RAI 4.7.8-2 is resolved.

Based on its review of LRA Section 4.7.8, the staff finds the analyses for groundwater affect on post-construction heave, settlement, and differential settlement, described in the UFSAR Section 2.5.4.10.2, remain valid for the period of extended operation in accordance with 10 CFR 54.21(c)(i). The staff also finds that the applicant's use of the Structure Monitoring AMP provides assurance that the provisions of UFSAR Section 2.5.4.10.2, 2.5.4.11, and 2.5.4.13.1 will be extended through the period of extended operation. The staff concludes that the applicant's commitment to continue monitoring the effects of heave and settlement on structures and differential settlement between structures in the vicinity of critical connections will provide assurance that the affects of heave and settlement and differential settlement will be adequately managed for the period of extended operation in accordance with 10 CFR 54.21(c)(iii).

4.7.8.3 Updated Final Safety Analysis Report Supplement

The applicant provided an UFSAR supplement summary description of its TLAA of building absolute or differential heave or settlement, including possible effects of changes in perched groundwater lens in LRA Section A3.6.3. LRA Section A1.32 provides the summary of the settlement monitoring AMP. Based on its review of the UFSAR supplement, the staff concludes that the summary description of the applicant's actions to address building absolute or

Time-Limited Aging Analyses

differential heave or settlement, including possible effects in perched groundwater lens is adequate.

4.7.8.4 Conclusion

Based on its review, the staff concludes, pursuant to 10 CFR 54.21(c)(1)(i), that the analyses for groundwater affect on post-construction heave, settlement, and differential settlement will remain valid for the period of extended operation. Further, the staff concludes, pursuant to 10 CFR 54.21(c)(1)(iii), the applicant will adequately manage the effects of heave and settlement and differential settlement for the period of extended operation. The staff also concludes that the UFSAR supplement contains an appropriate summary description of this TLAA evaluation for the period of extended operation, in accordance with the requirements of 10 CFR 54.21(d).

4.8 Absence of Time-Limited Aging Analyses Supporting Title 10, Part 50.12, Exemptions, of the Code of Federal Regulations

As required by 10 CFR 54.21(c)(2), the applicant must list all exemptions granted in accordance with 10 CFR 50.12 based on TLAAs. The staff's evaluation is found in SER Section 4.1.3.

4.9 Conclusion for Time-Limited Aging Analyses

The staff reviewed the information in LRA Section 4, "Time-Limited Aging Analyses." On the basis of its review, the staff concludes that the applicant has provided a sufficient list of TLAAs, as defined in 10 CFR 54.3. The applicant has demonstrated that: (1) the TLAAs will remain valid for the period of extended operation, as required by 10 CFR 54.21(c)(1)(i); (2) the TLAAs have been projected to the end of the period of extended operation, as required by 10 CFR 54.21(c)(1)(ii); or (3) that the effects of aging on intended function(s) will be adequately managed for the period of extended operation, as required by 10 CFR 54.21(c)(1)(iii). The staff also reviewed the UFSAR supplement for the TLAAs and finds that the supplement contains descriptions of the TLAAs sufficient to satisfy the requirements of 10 CFR 54.21(d). In addition, the staff concludes, as required by 10 CFR 54.21(c)(2), that no plant-specific, TLAA-based exemptions are in effect.

With regard to these matters, the staff concludes that there is reasonable assurance that the activities authorized by the renewed licenses will continue to be carried out in accordance with the CLB. In addition, any changes made to the CLB, in order to comply with 10 CFR 54.29(a), will be in accordance with the Atomic Energy Act of 1954, as amended, and NRC regulations.

5.0 REVIEW BY THE ADVISORY COMMITTEE ON REACTOR SAFEGUARDS

The NRC staff issued its safety evaluation report (SER) with open items related to the renewal of the operating license for Palo Verde Nuclear Generating Station, Units 1, 2, and 3 (PVNGS) on August 8, 2010. On September 8, 2010, the applicant presented its license renewal application, and the staff presented its review findings to the ACRS Plant License Renewal Subcommittee. The staff reviewed the applicant's comments on the SER and completed its review of the license renewal application. The staff's evaluation is documented in an SER that was issued by letter dated January 11, 2011.

During the 580th meeting of the ACRS, February 10-12, 2011, the ACRS completed its review of the PVNGS license renewal application and the NRC staff's SER. The ACRS documented its findings in a letter to the Commission dated March 1, 2011. A copy of this letter is provided on the following pages of this SER Section.

UNITED STATES
NUCLEAR REGULATORY COMMISSION
ADVISORY COMMITTEE ON REACTOR SAFEGUARDS
WASHINGTON, DC 20555 - 0001

March 1, 2011

The Honorable Gregory B. Jaczko
Chairman
U.S. Nuclear Regulatory Commission
Washington, D.C. 20555-0001

SUBJECT: REPORT ON THE SAFETY ASPECTS OF THE LICENSE RENEWAL APPLICATION FOR THE PALO VERDE NUCLEAR GENERATING STATION

Dear Chairman Jaczko:

During the 580th meeting of the Advisory Committee on Reactor Safeguards (ACRS), February 10-12, 2011, we completed our review of the license renewal application (LRA) for the Palo Verde Nuclear Generating Station, Units 1, 2, and 3 (PVNGS) and the final Safety Evaluation Report (SER) prepared by the NRC staff. Our Plant License Renewal subcommittee also reviewed this matter during its meeting on September 8, 2010. During these reviews, we met with representatives of the NRC staff, Arizona Public Service Company (APS or the applicant), and a member of the public. We also had the benefit of the documents referenced. This report fulfills the requirement of 10 CFR 54.25 that the ACRS review and report on all license renewal applications.

CONCLUSION AND RECOMMENDATION

1. The programs established and committed to by the applicant to manage age-related degradation provide reasonable assurance that the PVNGS units can be operated in accordance with their licensing bases for the period of extended operation without undue risk to the health and safety of the public.

2. The application for the renewal of the operating licenses of the PVNGS units should be approved.

BACKGROUND AND DISCUSSION

PVNGS is located approximately 26 miles west of Phoenix, Arizona. The site consists of three pressurized water reactors of Combustion Engineering (CE) design with dry ambient containments. Each of the PVNGS units utilizes a System 80 nuclear steam supply system provided by CE. Each unit operates at a licensed power output of 3,990 megawatt-thermal. APS requested renewal of the PVNGS licenses for 20 years beyond the current license terms, which expire on June 1, 2025 (Unit 1), April 24, 2026 (Unit 2), and November 25, 2027 (Unit 3). In the final SER, the staff documented their review of the license renewal application and other information submitted by the applicant or obtained during three staff audits and a two-week inspection conducted at the plant site. The staff reviewed the completeness of the applicant's identification of structures, systems, and components (SSCs) that are within the scope of license renewal; the integrated plant assessment process; the applicant's identification of the plausible aging mechanisms associated with passive, long-lived components; the adequacy of the applicant's Aging Management Programs (AMPs); and the identification and assessment of time-limited aging analyses (TLAAs) requiring review.

The applicant identified the SSCs that fall within the scope of license renewal and performed an aging management review for these SSCs. The applicant will implement 40 AMPs for license renewal. These include 29 existing programs and 11 new programs. Of the existing programs, nine AMPs are consistent with guidance in Revision 1 of the Generic Aging Lessons Learned (GALL) Report, five are consistent with exceptions, ten are consistent with enhancements, four are consistent with both enhancements and exceptions, and one is plant-specific. We reviewed the plant-specific programs and the AMP exceptions to the GALL Report.

The applicant identified the systems and components requiring TLAAs and reevaluated them for the period of extended operation. The staff concluded that the applicant has provided an acceptable list of TLAAs, as defined in 10 CFR 54.3. Furthermore, the staff concluded that in all cases the applicant has met the requirements for TLAAs specified in 10 CFR 54.3. We concur with the staff's conclusions that the TLAAs have been properly identified and that the required criteria will be met for the period of extended operation.

The staff conducted three audits and one inspection at PVNGS. The audits verified the appropriateness of the aging management review, scoping and screening methodology, and associated AMPs. The inspection examined the scoping and screening of non-safety related SSCs and verified the adequacy of the guidance, documentation, and implementation of selected AMPs. The audit and inspection teams also performed independent examinations of PVNGS condition reports to confirm that plant-specific operating experience was addressed during the AMP development and implementation processes. Based on the audits and inspections, the staff concluded in the final SER that the proposed activities will adequately manage the aging of SSCs identified in the application and that the intended functions of these SSCs will be maintained during the period of extended operation. We agree with these conclusions.

Following issuance of the draft SER with open items, the applicant submitted additional commitments that expand the scope and/or the means to detect aging effects in several license renewal programs. Among the most significant are those summarized below.

In response to issues identified during the staff's review of the Enhanced Fatigue Aging Management Program, the applicant improved the originally proposed program to track the number of occurrences for comparison to the plant's design basis transients.

The applicant provided a summary of licensing and design information in the LRA and modified its Technical Specifications (TS) to include an administrative program that provides controls to assure that components are maintained within design limits. The staff concluded that the applicant's Component Cycle and Transient Limit Program will be properly described in the applicable Updated Final Safety Analysis Report and TS sections.

The staff has identified industry operating experience which indicates that power cables energized to 480V and higher can experience failures where extended exposure to moisture is a contributing factor. The Inaccessible Medium Voltage Cable Program described in Revision 1 of the GALL Report does not recommend testing for inaccessible cables energized to less than 2kV and does not require testing of inaccessible cables that are not normally energized. The applicant has addressed the staff's concerns by expanding the scope of the Medium Voltage Power Cable Program to include all inaccessible 480V to 2kV power cables, whether energized or not. This expanded scope of cable monitoring is consistent with Revision 2 of the GALL Report.

The staff has concluded that external visual inspections do not provide adequate assurance that cracks are not present at the internal radius of socket welds in Class 1 small bore piping systems. There are currently no approved industry standard methods or qualified techniques to perform volumetric examinations of these welds. The applicant has experienced cracking in two small bore socket welds. In addition to visual inspections, the applicant will enhance the One-Time Inspection Program by committing to perform volumetric examinations of 10 percent of the Class 1 socket welds, up to a maximum of 25 welds for each unit (75 total), prior to the start of the period of extended operation. The applicant will use ultrasonic testing techniques.

The staff has noted a number of recent industry events involving leakage from buried and underground piping and tanks within the scope of license renewal. Buried steel piping is coated, and recent inspections of excavated fire protection and diesel generator fuel oil piping demonstrate that coatings are in very good condition, with appropriate backfill. The applicant has committed to continue to periodically inspect components in soil. The applicant will maintain the availability of cathodic protection of the buried portions of the in-scope buried piping at least 90 percent of the time. Surveys of cathodic protection, consistent with guidance from the National Association of Corrosion Engineers, will be conducted at least annually during the period of extended operation. Visual inspections of in scope piping in a soil environment will be performed each 10 year period starting 10 years prior to the period of extended operation. At least two inspections of stainless steel piping will be conducted at each unit for piping that is not cathodically protected. The staff has concluded that with these enhancements, the proposed programs will adequately monitor and manage the aging of buried piping and tanks.

We agree with the staff that there are no issues related to the matters described in 10 CFR 54.29(a)(1) and (a)(2) that preclude renewal of the operating licenses for PVNGS Units 1, 2, and 3. The programs established and committed to by the applicant provide reasonable assurance that PVNGS can be operated in accordance with their current licensing bases for the period of extended operation without undue risk to the health and safety of the public. The APS application for renewal of the operating licenses for the PVNGS units should be approved.

Harold B. Ray did not participate in the Committee's deliberations regarding this matter.

Sincerely,

/RA/

Said Abdel-Khalik
Chairman

REFERENCES

1. U.S. Nuclear Regulatory Commission, "Safety Evaluation Report Related to the License Renewal of Palo Verde Nuclear Generating Station Units 1, 2, and 3," dated August 2010 (ML102210072)

2. Palo Verde Nuclear Generating Station, Units 1, 2 and 3 - License Renewal Application, Part 1 of 3, dated December 11, 2008 (ML083510612)

3. Palo Verde Nuclear Generating Station, Units 1, 2 and 3 - License Renewal Application, Part 2 of 3, dated December 11, 2008 (ML083510614)

4. Palo Verde Nuclear Generating Station, Units 1, 2 and 3 - License Renewal Application, Part 3 of 3, dated December 11, 2008 (ML083510615)

5. NRC Letter, "Audit Report Regarding the Palo Verde Nuclear Generating Station, Units 1, 2, and 3 License Renewal Application (TAC NOS. ME0254, ME0255, ME0256), dated April 7, 2010 (ML100221296)

6. NRC Letter, "Palo Verde Nuclear Generating Station-NRC License Renewal Inspection Report 05000528/2010007; 05000529/2010007; 05000530/2010007," dated April 29, 2010 (ML101190585)

7. NRC Letter, "Scoping and Screening Audit Report for the Palo Verde Nuclear Generating Station, Units 1, 2, and 3, License Renewal Application (TAC NOS. ME0254, ME0255, and ME0256)," dated July 13, 2010 (ML101740217)

8. U.S. Nuclear Regulatory Commission, "Safety Evaluation Report Related to the License Renewal of Palo Verde Nuclear Generating Station Units 1, 2, and 3," dated January 11, 2011 (ML110110411)

9. Comments from a Member of the Public, dated February 7, 2011 (ML110480377)

6.0 CONCLUSION

The U.S. Nuclear Regulatory Commission staff (the staff) reviewed the license renewal application (LRA) for Palo Verde Nuclear Generating Station in accordance with NRC regulations and NUREG-1800, Revision 1, "Standard Review Plan for Review of License Renewal Applications for Nuclear Power Plants," dated September 2005. Title 10, Section 54.29 of the *Code of Federal Regulations* (10 CFR 54.29) sets the standards for issuance of a renewed license.

On the basis of its review of the LRA, the staff determines that the requirements of 10 CFR 54.29(a) have been met.

The staff notes that the requirements of 10 CFR 51, Subpart A, will be documented in a plant specific supplement to NUREG-1437, "Generic Environmental Impact Statement for License Renewal of Nuclear Plants (GEIS)."

APPENDIX A

Palo Verde Nuclear Generating Station Units 1, 2, and 3 License Renewal Commitments

During the review of the Palo Verde Nuclear Generating Station, Units 1, 2, and 3 (PVNGS), license renewal application (LRA) by the U.S. Nuclear Regulatory Commission (NRC) staff (the staff), Arizona Public Service Company (the applicant) made commitments related to aging management programs to manage aging effects for structures, systems and components. The following table lists these commitments along with the implementation schedules and sources for each commitment.

Table A-1. Palo Verde Nuclear Generating Station License Renewal Commitments

Item Number	Commitment	License Renewal Application Section	Implementation Schedule
1	The summary descriptions of aging management programs, time-limited aging analyses, and license renewal commitments contained in LRA Appendix A, "Updated Final Safety Analysis Supplement," as required by 10 CFR 54.21(d), will be incorporated in the Updated Final Safety Analysis Report for PVNGS Units 1, 2, and 3 in the next update required by 10 CFR 50.71(e) following the issuance of the renewed operating licenses.	A0	The next 10 CFR 50.71(e) Updated Final Safety Analysis Report update, following issuance of the renewed operating licenses
2	Existing Quality Assurance Program is credited for license renewal.	A1 B1.3 Summary Descriptions of Aging Management	Ongoing
3	Existing ASME Section XI Inservice Inspection, Subsections IWB, IWC, and IWD Program is credited for license renewal.	A1.1 B2.1.1 ASME Section XI Inservice Inspection, Subsections IWB, IWC, and IWD	Ongoing
4	Existing Water Chemistry Program is credited for license renewal.	A1.2 B2.1.2 Water Chemistry	Ongoing
5	Existing Reactor Head Closure Studs Program is credited for license renewal.	A1.3 B2.1.3 Reactor Head Closure Studs	Ongoing
6	Existing Boric Acid Corrosion Program is credited for license renewal.	A1.4 B2.1.4 Boric Acid Corrosion	Ongoing
7	Existing Nickel-Alloy Penetration Nozzles Welded to the Upper Reactor Vessel Closure Heads of Pressurized Water Reactors Program is credited for license renewal.	A1.5 B2.1.5 Nickel-Alloy Penetration Nozzles Welded to the Upper Reactor Vessel Closure Heads of Pressurized Water Reactors	Ongoing
8	Existing Flow-Accelerated Corrosion Program is credited for license renewal.	A1.6 B2.1.6 Flow-	Ongoing

Appendix A

Item Number	Commitment	License Renewal Application Section	Implementation Schedule
		Accelerated Corrosion	
9	Existing Bolting Integrity Program is credited for license renewal.	A1.7 B2.1.7 Bolting Integrity	Ongoing
10	Existing Steam Generator Tube Integrity Program is credited for license renewal.	A1.8 B2.1.8 Steam Generator Tube Integrity	Ongoing
11	Existing Open-Cycle Cooling Water System Program is credited for license renewal, AND Prior to the period of extended operation, the program will be enhanced to clarify guidance in the conduct of piping inspections using NDE techniques and related acceptance criteria.	A1.9 B2.1.9 Open-Cycle Cooling Water System	Prior to the period of extended operation[1]
12	Existing Closed-Cycle Cooling Water System Program is credited for license renewal, AND Prior to the period of extended operation, procedures will be enhanced to incorporate the guidance of EPRI TR-107396 with respect to water chemistry control for frequency of sampling and analysis, normal operating limits, action level concentrations, and times for implementing corrective actions upon attainment of action levels.	A1.10 B2.1.10 Closed-Cycle Cooling Water System	Prior to the period of extended operation[1].
13	Existing Inspection of Overhead Heavy Load and Light Load (Related to Refueling) Handling Systems Program is credited for license renewal, AND Prior to the period of extended operation, procedures will be enhanced to inspect for loss of material due to corrosion or rail wear.	A1.11 B2.1.11 Inspection Of Overhead Heavy Load and Light Load (Related to Refueling) Handling Systems	Prior to the period of extended operation[1].
14	Existing Fire Protection Program is credited for license renewal, AND Prior to the period of extended operation procedures will be enhanced to perform the testing of the electro-thermal links and functional testing of the halon and CO_2 dampers every 18 months or at the frequency specified in the current licensing basis in effect upon entry into the period of extended operation.	A1.12 B2.1.12 Fire Protection	Prior to the period of extended operation[1].
15	Existing Fire Water System Program is credited for license renewal, AND Prior to the period of extended operation, the following enhancements will be implemented: • Specific procedures will be enhanced to include review and approval requirements under the Nuclear Administrative Technical Manual (NATM). • Procedures will be enhanced to be consistent with the current code of record or NFPA 25, 2002 Edition. • Procedures will be enhanced to field service test a representative sample or replace sprinklers prior to 50 years in service and test thereafter every 10 years to ensure that signs of degradation are detected in a timely manner. • Procedures will be enhanced to be consistent with NFPA 25, Sections 7.3.2.1, 7.3.2.2, 7.3.2.3, and 7.3.2.4.	A1.13 B2.1.13 Fire Water System	Prior to the period of extended operation[1].

Appendix A

Item Number	Commitment	License Renewal Application Section	Implementation Schedule
16	Existing Fuel Oil Chemistry Program is credited for license renewal, AND Prior to the period of extended operation: • Procedures will be enhanced to extend the scope of the program to include the station blackout generator (SBOG) fuel oil storage tank and SBOG skid fuel tanks. • Procedures will be enhanced to include ten-year periodic draining, cleaning, and inspections on the diesel-driven fire pump day tanks, the SBOG fuel oil storage tank, and SBOG skid fuel tanks. • Ultrasonic testing (UT) or pulsed eddy current (PEC) thickness examination will be conducted to detect corrosion-related wall thinning if degradation is found during the visual inspections and once on the tank bottoms for the EDG fuel oil storage tanks, EDG fuel oil day tanks, diesel-driven fire pump day tanks, SBOG fuel oil storage tank, and SBOG skid fuel tanks. The onetime UT or PEC examination on the tank bottoms will be performed before the period of extended operation.	A1.14 B2.1.14 Fuel Oil Chemistry	Prior to the period of extended operation[1].
17	Existing Reactor Vessel Surveillance Program is credited for license renewal, AND Prior to the period of extended operation: • The schedule will be revised to withdraw the next capsule at the equivalent clad-base metal exposure of approximately 54 effective full-power year (EFPY) expected for the 60-year period of operation, and to withdraw remaining standby capsules at equivalent clad-base metal exposures not exceeding the 72 EFPY expected for a possible 80-year second period of extended operation. This withdrawal schedule is in accordance with NUREG-1801, Section XI.M31, item 6, and with the ASTM E 185-82 criterion which states that capsules may be removed when the capsule neutron fluence is between one and two times the limiting fluence calculated for the vessel at the end of expected life. This schedule change must be approved by the NRC, as required by 10 CFR 50, Appendix H. • If left in the reactor beyond the presently-scheduled withdrawal, the next scheduled surveillance capsule in each unit will reach a clad-base metal 54 EFPY equivalent at about 40 actual operating EFPY (40, 39, and 42 actual EFPY in Units 1, 2, and 3, respectively). • Procedures will be enhanced to identify the withdrawal of the remaining standby capsules at 72 EFPY, at about 50 to 54 actual operating EFPY, near the end of the extended licensed operating period. The need to monitor vessel fluence following removal of the remaining standby capsules, and ex-vessel or in-vessel methods, will be addressed prior to removing the remaining capsules.	A1.15 B2.1.15 Reactor Vessel Surveillance	Prior to the period of extended operation[1].
18	The One-Time Inspection Program conducts one-time inspections of plant system piping and components to verify the effectiveness of the Water Chemistry Program (A1.2), Fuel Oil Chemistry Program (A1.14), and Lubricating Oil Analysis Program (A1.23). The aging effects to be evaluated by the One-Time Inspection	A1.16 B2.1.16 One-Time Inspection	Within the ten year period prior to the period of extended operation[1].

A-3

Appendix A

Item Number	Commitment	License Renewal Application Section	Implementation Schedule
	Program are loss of material, cracking, and reduction of heat transfer.		
19	The Selective Leaching of Materials Program is a new program that will be implemented prior to the period of extended operation. Industry and plant-specific operating experience will be evaluated in the development and implementation of this program.	A1.17 B2.1.17 Selective Leaching of Materials	Within the ten year period prior to the period of extended operation[1].
	The Selective Leaching of Materials Program includes a one-time inspection (visual and/or mechanical methods) of a selected sample of components' internal surfaces to determine whether loss of material due to selective leaching is occurring. A sample size of 20 percent of the population, up to a maximum of 25 component inspections, will be established for each of the system material and environment combinations at the PVNGS site. If indications of selective leaching are confirmed, follow-up examinations or evaluations will be performed.		
20	The Buried Piping and Tanks Inspection Program is a new program that will be implemented prior to the period of extended operation. Within the ten year period prior to entering the period of extended operation an opportunistic or planned inspection of buried tanks at the PVNGS site will be performed. The visual inspections noted below of piping in a soil environment within the scope of license renewal will be conducted within the ten-year period prior to entering the period of extended operation, and during each ten year period after entering the period of extended operation, except the initial diesel generator fuel oil piping inspection will be performed between January 1, 2012, and December 31, 2015. Each inspection will: • select accessible locations where degradation is expected to be high; • excavate and visually inspect the circumference of the pipe • examine at least ten feet of pipe a. Metallic Piping not Cathodically-Protected At least two excavations and visual inspections of stainless steel piping will be conducted in each unit. Stainless steel piping within the scope of license renewal exists in the following systems: • Chemical and Volume Control (CH) • Condensate Transfer and Storage (CT) • Fire Protection (FP) b. Steel Piping Cathodically-Protected At least two excavations and visual inspections of cathodically-protected steel piping will be conducted in each unit. In one of the units, at least one of these inspections will be performed on diesel generator fuel oil piping. c. Steel Piping with Potentially Degraded Cathodic	A1.18 B2.1.18 Buried Piping and Tanks Inspection	Perform the buried piping and tanks inspections within the ten year period prior to the period of extended operation[1], except the initial diesel generator fuel oil piping inspection will be performed between 1/1/12 and 12/31/15. AND Perform the buried piping inspections during each ten year period after entering the period of extended operation. AND Implement the additional enhancements to the Buried Piping and Tanks Inspection Program prior to the period of operation[1].

Appendix A

Item Number	Commitment	License Renewal Application Section	Implementation Schedule
	Protection		
	At least three excavations and visual inspections of fire protection steel piping with potentially degraded bonding straps will be conducted at the PVNGS site.		
	Prior to the period of extended operation, the Buried Piping and Tanks Inspection Program will include provisions to: (1) ensure electrical power is maintained to the cathodic protection system for in-scope buried piping at least 90 percent of the time (e.g., monthly verification that the power supply circuit breakers are closed or other verification that power is being provided to the system), and (2) ensure that the National Association of Corrosion Engineers cathodic protection system surveys are performed at least annually.		
21	The One-Time Inspection of ASME Code Class 1 Small-Bore Piping Program is a new program that will be implemented prior to the period of extended operation. Industry and plant-specific operating experience will be evaluated in the development and implementation of this program.	A1.19 B2.1.19 One-Time Inspection of ASME Code Class 1 Small-Bore Piping	Within the six year period prior to the period of extended operation[1].
	For ASME Code Class 1 small-bore piping, volumetric examinations on selected butt weld locations will be performed to detect cracking. Butt weld volumetric examinations will be conducted in accordance with ASME Section XI with acceptance criteria from Paragraph IWB-3000 and IWB-2430. Weld locations subject to volumetric examination will be selected based on the guidelines provided in EPRI TR-112657. Socket welds that fall within the weld examination sample will be examined following ASME Section XI Code requirements. At least 10 percent of the socket welds in ASME Code Class 1 piping that is less than four inches nominal pipe size and greater than or equal to one inch nominal pipe size will be selected per unit for ultrasonic testing examination, up to a maximum of 25 weld examinations. The sample will be selected based on risk insights and those welds with the potential for aging degradation.		
22	The External Surfaces Monitoring Program is a new program that will be implemented prior to the period of extended operation. Industry and plant-specific operating experience will be evaluated in the development and implementation of this program.	A1.20 B2.1.20 External Surfaces Monitoring Program	Prior to the period of extended operation[1].
23	The applicant will complete the tasks described: a. Reactor Coolant System Nickel Alloy Pressure Boundary Components Implement applicable (1) NRC Orders, Bulletins and Generic Letters associated with nickel alloys and (2) staff-accepted industry guidelines, (3) participate in the industry initiatives, such as owners group programs and the EPRI Materials Reliability Program, for managing aging effects associated with nickel alloys, (4) upon completion of these programs, but not less than 24 months before entering the period of extended operation, APS will submit an inspection plan for reactor coolant system nickel alloy pressure boundary components to the NRC for review and approval, and	A1.21 B2.1.21 Reactor Coolant System Supplement 3.1.2.2.16.2 Pressurizer Spray Head Cracking	Not less than 24 months prior to the period of extended operation[1].

Appendix A

Item Number	Commitment	License Renewal Application Section	Implementation Schedule
	b. Reactor Vessel Internals		
	(1) Participate in the industry programs for investigating and managing aging effects on reactor internals; (2) evaluate and implement the results of the industry programs as applicable to the reactor internals; and (3) upon completion of these programs, but not less than 24 months before entering the period of extended operation, APS will submit an inspection plan for reactor internals to the NRC for review and approval.		
	c. Pressurizer Spray Heads		
	Comply with applicable NRC Orders and implement applicable (1) Bulletins and Generic Letters, and (2) staff-accepted industry guidelines.		
24	The Inspection of Internal Surfaces in Miscellaneous Piping and Ducting Components Program is a new program that will be implemented prior to the period of extended operation. Industry and plant-specific operating experience will be evaluated in the development and implementation of this program.	A1.22 B2.1.22 Inspection of Internal Surfaces in Miscellaneous Piping and Ducting Components	Prior to the period of extended operation[1].
25	Existing Lubricating Oil Analysis Program is credited for license renewal.	A1.23 B2.1.23 Lubricating Oil Analysis	Ongoing
26	The Electrical Cables and Connections Not Subject to 10 CFR 50.49 Environmental Qualification Requirements Program is a new program that will be implemented prior to the period of extended operation. Industry and plant-specific operating experience will be evaluated in the development and implementation of this program.	A1.24 B2.1.24 Electrical Cables and Connections Not Subject to 10 CFR 50.49 Environmental Qualification Requirements	Prior to the period of extended operation[1].
27	Existing Electrical Cables And Connections Not Subject To 10 CFR 50.49 Environmental Qualification Requirements Used In Instrumentation Circuits Program is credited for license renewal , AND Prior to the period of extended operation: • Procedures will be enhanced to identify license renewal scope, require cable testing of ex-core neutron monitoring cables, require an evaluation of the calibration results for non-EQ area radiation monitors, and require acceptance criteria for cable testing be established based on the type of cable and type of test performed.	A1.25 B2.1.25 Electrical Cables and Connections Not Subject to 10 CFR 50.49 Environmental Qualification Requirements Used in Instrumentation Circuits	Prior to the period of extended operation[1].
28	The Inaccessible Medium Voltage Cables Not Subject to 10 CFR 50.49 EQ Requirements Program is credited for license renewal, AND Prior to the period of extended operation procedures will be enhanced to: • Extend the scope of the program to include low voltage (480V and above) non-EQ inaccessible power cables and associated manholes. • Perform the cable inspections on at least an annual frequency and perform the cable testing on a six year	A1.26 B2.1.26 Inaccessible Medium Voltage Cables Not Subject to 10 CFR 50.49 Environmental Qualification Requirements	Prior to the period of extended operation[1].

Appendix A

Item Number	Commitment	License Renewal Application Section	Implementation Schedule
	frequency.		
29	Existing ASME Section XI, Subsection IWE Program is credited for license renewal.	A1.27 B2.1.27 ASME Section XI, Subsection IWE	Ongoing
30	Existing ASME Section XI, Subsection IWL Program is credited for license renewal.	A1.28 B2.1.28 ASME Section XI, Subsection IWL	Ongoing
31	Existing ASME Section XI, Subsection IWF Program is credited for license renewal.	A1.29 B2.1.29 ASME Section XI, Subsection IWF	Ongoing
32	Existing 10 CFR 50, Appendix J Program is credited for license renewal.	A1.30 B2.1.30 10 CFR 50, Appendix J	Ongoing
33	Existing Masonry Wall Program is credited for license renewal, AND Prior to the period of extended operation, procedures will be enhanced to specify ACI 349.3R-96 as the reference for qualification of personnel to inspect structures under the Masonry Wall Program, which is part of the Structures Monitoring Program.	A1.31 B2.1.31 Masonry Wall Program	Prior to the period of extended operation[1].

34	Existing Structures Monitoring Program is credited for license renewal, AND
	Prior to the period of extended operation:
	
	• The Structures Monitoring Program will be enhanced to specify ACI 349.3R-96 as the reference for qualification of personnel to inspect structures under the Structures Monitoring Program.
	
	• For structures within the scope of license renewal, the Structures Monitoring Program will be enhanced to establish the frequency of inspection for each unit at a 5 year interval, with the exception of exterior surfaces of the following nonsafety-related structures, below-grade structures, and structures within a controlled interior environment, which will be inspected at an interval of 10 years:
	
	   – Fire Pump House (Yard Structures)
	   – Radwaste Building
	   – Station Blackout Generator Structures
	   – Turbine Building
	   – Non-Safety Related Tank Foundations and Shells
	   – Non-Safety Related Transformer Foundations and Electrical Structures
	
	• The Structures Monitoring Program will be enhanced to quantify the acceptance criteria and critical parameters for monitoring degradation, and to provide guidance for identifying unacceptable conditions requiring further technical evaluation or corrective action. Procedures will also be enhanced to incorporate applicable industry codes, standards and guidelines (e.g., ACI 349.3R-96,
		A1.32
		
		B2.1.32 Structures Monitoring Program
			Prior to the period of extended operation[1].

Appendix A

Item Number	Commitment	License Renewal Application Section	Implementation Schedule
	ANSI/ASCE 11-90, etc.) for acceptance criteria.		
35	Existing Regulatory Guide 1.127, Inspection Of Water-Control Structures Associated With Nuclear Power Plants Program is credited for license renewal, AND Prior to the period of extended operation, procedures will be enhanced to specify that the essential spray ponds inspections include concrete below the water level.	A1.33 B2.1.33 RG 1.127, Inspection of Water-Control Structures Associated with Nuclear Power Plants	Prior to the period of extended operation[1].
36	Existing Nickel Alloy Aging Management Program is credited for license renewal.	A1.34 B2.1.34 Nickel Alloy Aging Management Program	Ongoing
37	The Electrical Cable Connections Not Subject to 10 CFR 50.49 Environmental Qualification Requirements Program is a new program that will be implemented prior to the period of extended operation. Industry and plant-specific operating experience will be evaluated in the development and implementation of this program.	A1.35 B2.1.35 Electrical Cable Connections Not Subject to 10 CFR 50.49 Environmental Qualification Requirements	Prior to the period of extended operation[1].
38	The Metal Enclosed Bus Program is a new program and will be completed before the period of extended operation and once every 10 years thereafter. Industry and plant-specific operating experience will be evaluated in the development and implementation of this program.	A1.36 B2.1.36 Metal Enclosed Bus	Prior to the period of extended operation and once every ten years thereafter.
39	No later than two years prior to the period of extended operation, the following enhancements will be implemented • Cumulative usage factor (CUF) tracking will be implemented for NUREG/CR-6260 locations not monitored by cycle counting (CC) (the reactor vessel shell and lower head (juncture) location will be monitored by CC). For PVNGS locations identified in NUREG/CR-6260 and monitored by CUF, fatigue usage factor action limits will be required for including effects of the reactor coolant environment. • The Metal Fatigue of Reactor Coolant Pressure Boundary Program will be enhanced to include a computerized program to track and manage both CC and fatigue usage factor. FatiguePro® will be used for CC and cycle-based fatigue monitoring methods. FatiguePro® is an EPRI-licensed product. • The enhanced Metal Fatigue of Reactor Coolant Pressure Boundary Program will monitor plant transients as required by PVNGS Technical Specification 5.5.5. CUFs will be calculated for a subset of ASME III Class 1 reactor coolant pressure boundary vessel and piping locations and component locations with Class 1 analyses. The following methods will be used: – The Metal Fatigue of Reactor Coolant Pressure Boundary Program will be enhanced to use cycle-based fatigue and stress-based fatigue CUF calculations to monitor fatigue. FatiguePro® will be used for CC and cycle-based fatigue monitoring	4.3.1 Fatigue Aging Management Program A2.1 B3.1 Metal Fatigue of Reactor Coolant Pressure Boundary	No later than two years prior to the period of extended operation[1].

Appendix A

Item Number	Commitment	License Renewal Application Section	Implementation Schedule
	methods. FatiguePro® is an EPRI-licensed product. – The stress-based fatigue method will use a fatigue monitoring software program that incorporates a three-dimensional, six-component stress tensor method meeting ASME III NB-3200 requirements. • The enhanced Metal Fatigue of Reactor Coolant Pressure Boundary Program will provide action limits on cycles and on CUF that will initiate corrective actions before the licensing basis limits on fatigue effects at any location are exceeded. – In order to ensure sufficient cycle count margin to accommodate occurrence of a low-probability transient, corrective actions must be taken before the remaining number of allowable occurrences for any specified transient becomes less than 1.0. – CUF action limits will be established to require corrective action when the calculated CUF (from cycle-based or stress-based monitoring) for any monitored location is projected to reach 1.0 within the next two or three operating cycles. In order to ensure sufficient margin to accommodate occurrence of a low-probability transient, corrective actions will be taken while there is still sufficient margin to accommodate at least one occurrence of the worst-case design transient event (i.e., with the highest fatigue usage per event cycle).		
40	Existing Environmental Qualification Program is credited for license renewal, AND Maintaining qualification through the extended license renewal period requires that existing EQ evaluations be re-evaluated.	A2.2 B3.2 Environmental Qualification (EQ) of Electrical Components	Prior to the period of extended operation[1].
41	Existing Concrete Containment Tendon Prestress Program is credited for license renewal, AND • The program will be enhanced to continue to compare regression analysis trend lines of the individual lift-off values of tendons surveyed to date, in each of the vertical and hoop tendon groups, with the minimum required value (MRV) and predicted lower limit (PLL) for each tendon group, to the end of the licensed operating period, and to take appropriate corrective actions if future values indicated by the regression analysis trend line drop below the PLL or MRV. The regression analyses will be updated for tendons of the affected unit and for a combined data set of all three units following each inspection of an individual unit. • Prior to the period of extended operation, procedures will be enhanced to require an update of the regression analysis for each tendon group of each unit, and of the joint regression of data from all three units, after every tendon surveillance. The documents will invoke and describe regression analysis methods used to construct the lift-off trend lines, including the use of individual tendon data in accordance with Information Notice (IN) 99-10, "Degradation of Prestressing Tendon Systems in Prestressed Concrete Containments."	A2.3 B3.3 Concrete Containment Tendon Prestress 4.5 Concrete Containment Tendon Prestress	Prior to the period of extended operation[1].

Appendix A

Item Number	Commitment	License Renewal Application Section	Implementation Schedule
	• The Tendon Integrity test procedure will be revised to extend the list of surveillance tendons to include random samples for the year 45 and 55 surveillances.		
42	The applicant will confirm the reactor coolant system pressure-temperature limits basis for 54 EFPY prior to operation beyond 32 EFPY and will update documents in accordance with the provisions of 10 CFR 50.59. (RCTSAI 3246939)	A3.1.3 Pressure-Temperature Limits	Prior to operation beyond 32 EFPY[1].
43	Completed		
44	Completed		
45	See Item No. 46		
46	An extension of In-Service Inspection Relief Request 31, Revision 1 authorization will be requested for the period of extended operation, supported by a continuation of the cold shutdown time monitoring program.	4.7.4 Fatigue Crack Growth and Fracture Mechanics Stability Analyses of Half-Nozzle Repairs to Alloy 600 Material in Reactor Coolant Hot Legs; Absence of a TLAA for Supporting Corrosion Analyses	Prior to the period of extended operation[1].
47	Deleted (Staff note: this was in the PVNGS Environmental Report)		
48	Deleted (Staff note: this was in the PVNGS Environmental Report)		
49	Deleted (Staff note: this was in the PVNGS Environmental Report)		
50	The Fuse Holder Program is a new program that will be implemented prior to the period of extended operation and once every 10 years thereafter. Industry and plant-specific operating experience will be evaluated in the development and implementation of this program.	A1.37 B2.1.37 Fuse Holder	Prior to the period of extended operation and once every 10 years thereafter.
51	Completed		
52	Deleted (Staff note: this was in the PVNGS Environmental Report)		
53	Completed		
54	Completed		
55	Completed		
56	The spray pond wall rework/repair methods are currently being determined, and the rework/repair is planned to begin in 2011. As Unit 1 spray ponds have the most degradation, work is planned to start there, followed by Units 2 and 3. It is expected that the work will be completed in all three units in 2015.	PVNGS letter dated June 21, 2010	12/31/2015
57	No later than two years prior to the period of extended operation, APS will confirm the conservatism of the F_{en} value of 1.49 using the methods specified in NUREG/CR-6909, and will use the F_{en} calculated using the NUREG/CR-6909 methods if it is more conservative than the 1.49 value.	PVNGS letter dated June 29, 2010	No later than two years prior to the period of extended operation[1].

Appendix A

Item Number	Commitment	License Renewal Application Section	Implementation Schedule
58	No later than two years prior to the period of extended operation, APS will perform a reanalysis of the pressurizer heater penetrations to consider EAF effects using the formulas and methodology given in NUREG/CR-6909.	PVNGS letter dated June 29, 2010	No later than two years prior to the period of extended operation[1].
59	As documented in CRAI 3337611, Engineering Study 13-MS-B089, "Cavitation in Safety Injection System," APS identified 26 components and associated piping in each PVNGS unit potentially susceptible to cavitation under design basis maximum flow conditions. One location in each unit, the HPSI recirculation piping downstream of throttle valve JSIBUV0667, has been confirmed to be susceptible to cavitation erosion, and a 7.5-year time-based replacement schedule described below has been established. All of the remaining 25 locations identified as potentially susceptible to cavitation in Unit 2, 20 of the locations in Unit 1, and 15 of the locations in Unit 3 have been inspected by ultrasonic testing (UT) and demonstrated no degradation. The remaining five locations in Unit 1 are scheduled to be inspected in the Unit 1 fall 2011 refueling outage. Of the remaining ten locations in Unit 3, five will be inspected in the Unit 3 fall 2010 outage and five will be inspected in the Unit 3 spring 2012 outage. Therefore, the inspections in all three units will be completed no later than June 30, 2012. If any of the remaining components and associated piping is found to be susceptible to cavitation or a form of flow-related degradation, it will be incorporated into a replacement plan similar to that for the HPSI recirculation piping downstream of throttle valve JSIBUV0667.	PVNGS letter dated July 30, 2010	6/30/2012
60	The reactor coolant system transient and cycle tracking procedure 73ST-9RC02 and UFSAR Section 3.9.1 will be enhanced to discuss corrective actions that need to be taken prior to ASME Section III fatigue design limits being exceeded and to state that corrective actions may be required for other fatigue-related analyses, such as certain ASME Section XI supplemental fatigue flaw growth or cycle-dependent fracture mechanics evaluations that are dependent on the number of occurrences of design transients.	PVNGS letter dated October 13, 2010	11/30/2010
61	The applicant will perform one of the following three resolution options: 1. Perform an inspection of each steam generator at PVNGS to assess the condition of the divider plate bar welds in all units and the divider plate bars in Unit 2. The examination technique(s) will be capable of detecting PWSCC in the divider plate bar welds in all units, and in the accessible surfaces of the divider plate bars in Unit 2. OR 2. Perform an analytical evaluation of the steam generator divider plate bar welds in all units, and the divider plate bars in Unit 2, in order to establish a technical basis which concludes that the SG reactor coolant system pressure boundary is adequately maintained with the presence of steam generator divider plate bar weld cracking. OR 3. If results of industry and NRC studies and operating	PVNGS letter dated November 23, 2010 as modified by letters dated February 25, 2011 and March 17, 2011	If Option (1) is selected, it will be completed for each SG in each unit during an SG tube eddy-current inspection outage between 20 and 25 calendar years of SG operation. If Option (2) or Option (3) is selected, it will be completed prior to 9/1/2023.

A-11

Appendix A

Item Number	Commitment	License Renewal Application Section	Implementation Schedule
	experience document that potential failure of the SG reactor coolant system pressure boundary due to PWSCC cracking of SG divider plate bar welds and the divider plate bars in Unit 2 is not a credible concern, this commitment will be revised to reflect that conclusion.		
62	The applicant will perform one of the following two resolution options: 1. Perform a one-time inspection of a representative number of tube-to-tubesheet welds in each steam generator to determine if PWSCC cracking is present. If weld cracking is identified: The condition will be resolved through repair or engineering evaluation to justify continued service, as appropriate. An ongoing monitoring program will be established to perform routine tube-to-tubesheet weld inspections for the remaining life of the steam generators. OR 2. Perform an analytical evaluation of the steam generator tube-to-tubesheet welds in order to: Establish a technical basis which concludes that the structural integrity of the steam generator tube-to-tubesheet interface is adequately maintained with the presence of tube-to-tubesheet weld cracking. Establish a technical basis which concludes that the steam generator tube-to-tubesheet welds are not required to perform a reactor coolant pressure boundary function.	PVNGS letter dated November 23, 2010	If Option (1) is selected, it will be completed for each SG in each unit during an SG tube eddy-current inspection outage between 20 and 25 calendar years of SG operation. If Option (2) is selected, it will be completed prior to 9/1/2023.
63	No later than two years prior to the period of extended operation, the applicant will confirm that: The plant-specific components listed in LRA Table 4.3-11 (except the pressurizer surge line pressurizer elbow) are bounding for the generic NUREG/CR-6260 locations and the additional location (pressurizer heater penetrations). If locations are found that are not bounded by the Table 4.3-11 components, APS will perform new analyses as necessary to bound such locations. AND The LRA Table 4.3-11 locations selected for environmentally assisted fatigue analyses consist of the most limiting CUF locations for the plant (beyond the generic EAF locations identified in the NUREG/CR-6260 guidance). If the Table 4.3-11 locations are not bounding, APS will perform an environmentally assisted fatigue analysis for the additional CUF locations not bounded by the Table 4.3-11 locations. If the component with the most limiting CUF is composed of nickel alloy, the methodology used to perform the environmentally-assisted fatigue calculation for nickel alloy will be consistent with NUREG/CR-6909.	PVNGS letter dated December 3, 2010	No later than two years prior to the period of extended operation.

[1] "Prior to period of extended operation," "prior to operation beyond 32 EFPY," and "prior to the end of the current licensed operating period," is prior to the following PVNGS Operating License expiration dates: Unit 1: June 1, 2025; Unit 2: April 24, 2026; Unit 3: November 25, 2027.

APPENDIX B

Chronology

This appendix lists chronologically the routine licensing correspondence between the U.S. Nuclear Regulatory Commission (NRC) staff (the staff) and Arizona Public Service Company (APS) (the applicant). This appendix also lists other correspondence on the staff's review of the Palo Verde Nuclear Generating Station (PVNGS), Units 1, 2, and 3 License Renewal Application (LRA) under Docket Nos. 50-528, 50-529, and 50-530. The date shown in the date column is the date the correspondence was issued and may differ from the actual date of the event.

Date	Subject
12/11/2008	Palo Verde Nuclear Generating Station, Units 1, 2 and 3 - License Renewal Application - Cover Letter (ML083510611)
12/11/2008	Palo Verde Nuclear Generating Station, Units 1, 2 and 3 - License Renewal Application, Part 1 of 3 (ML083510612)
12/11/2008	Palo Verde Nuclear Generating Station, Units 1, 2 and 3 - License Renewal Application, Part 2 of 3 (ML083510614)
12/11/2008	Palo Verde Nuclear Generating Station, Units 1, 2 and 3 - License Renewal Application, Part 3 of 3 (ML083510615)
12/23/2008	Letter from NRC to APS, 12/18/2008, Summary of Public Meeting with Arizona Public Service Company (ML083580225)
12/24/2008	Press Release-08-234, "NRC Announces Availability of License Renewal Application For Palo Verde Nuclear Generating Station" (ML083590238)
1/12/2009	Letter from NRC to APS, Notice of Receipt and Availability of the License Renewal Application for the Palo Verde Nuclear Generating Station Units 1,2, and 3 (ML083530426)
2/13/2009	Letter from NRC to APS, Review Status Of The License Renewal Application For The Palo Verde Nuclear Generating Station (ML090360279)
2/25/2009	Letter from APS to NRC, PVNGS, Units 1, 2, and 3, Letter Regarding Plan to Resolve Deficiency in the PVNGS License Renewal Application (ML090750614)
4/14/2009	PVNGS, Units 1, 2, and 3, Supplement 1 to LRA (ML091130221)
4/15/2009	Summary of telephone conference call held between NRC and APS, 1/30/2009, pertaining to the PVNGS, Units 1, 2, and 3, LRA (ML090830031)
5/8/2009	PVNGS, Units 1, 2, and 3, Errata to Supplement 1 to LRA (ML092600288)
5/11/2009	Letter from NRC to APS, Determination of Acceptability and Sufficiency for Docketing, Proposed Review Schedule, and Opportunity for a Hearing Regarding the Application from Arizona Public Service Co for Renewal of the OL for the PVNGS, Units 1–3 (ML091130106)
5/11/2009	Federal Register Notice, NRC Notice of Acceptance for Docketing of the Application and Notice of Opportunity for Hearing Regarding Renewal of Facility Operating License No's. NPF-41, NPF-51, and NPF-74 for an Additional 20-Year Period Arizona Public Service Co. PVNGS, Units 1,2, 3 (ML091130187)
5/20/2009	Press Release-09-088, "NRC Announces Opportunity to Request Hearing on License Renewal Application for Palo Verde Nuclear Generating Station" (ML091400201)
6/17/2009	Summary of telephone conference call between NRC and APS, 6/9/2009, pertaining to the PVNGS, Units 1, 2, and 3, LRA (ML091620276)
6/26/2009	Letter from APS to NRC, PVNGS, Units 1, 2, and 3, License Renewal Application Online Reference Portal (ML091880425)

Appendix B

Date	Subject
7/23/2009	Letter from Greater Phoenix Chamber of Commerce to Chairman Jaczko, NRC, LTR-09-0380, Ltr. Todd Sanders re: NRC's License Renewal Process for the Palo Verde Nuclear Generating Station (ML092110612)
7/30/2009	Letter from Strategic Teaming & Resource Sharing (STARS) to NRC, Strategic Teaming and Resource Sharing Schedule for STARS License Renewal Applications (ML092120185)
8/11/2009	Letter from NRC to APS, Request for Additional Information for the Review of Palo Verde Nuclear Generating Station, Units 1, 2, and 3, License Renewal Application (ML092180443)
8/14/2009	Summary of telephone conference call held between NRC and APS, 8/3/2009, pertaining to the PVNGS, Units 1, 2, and 3 LRA (ML092230427)
8/25/2009	Letter from NRC to APS, Request for Additional Information for the Review of Palo Verde Nuclear Generating Station, Units 1, 2, and 3, License Renewal Application (ML092380318)
9/10/2009	Letter from APS to NRC, PVNGS, Units 1, 2, and 3, Response to Request for Additional Information Regarding License Renewal Application (ML092610068)
9/16/2009	Letter from NRC to APS, Safety Project Manager Change for the License Renewal Project for Palo Verde Nuclear Generating Station, Units 1, 2, and 3 (TAC Nos. ME0261-ME0266) (ML092430084)
9/30/2009	Summary of telephone conference call between NRC and APS, 9/3/2009, to discuss a draft request for additional information (RAI) for PVNGS, Units 1, 2, and 3, LRA (ML092470463)
10/14/2009	PVNGS, Units 1, 2, and 3, Amendment No. 1 to LRA, Revised Environmental Report, Figure 3-2, and Table 4-2, for the Hassayampa No. 3 Transmission Line (ML092950484)
11/2/2009	Letter from NRC to APS, Request for Additional Information for the Review of Palo Verde Nuclear Generating Station, Units 1, 2, and 3, License Renewal Application (ML092790315)
11/2/2009	Summary of telephone conference call held between NRC and APS, 10/1/2009, concerning a draft RAI pertaining to the PVNGS (ML092800015)
11/3/2009	Summary of telephone conference call held between NRC and APS, 9/16/2009, pertaining to the PVNGS, Units 1, 2, and 3, LRA (ML092750130)
11/3/2009	Letter from NRC to APS, Request for Additional Information for the Review of Palo Verde Nuclear Generating Station, Units 1, 2, and 3, License Renewal Application (ML092750237)
11/12/2009	Summary of telephone conference call held between NRC and APS, 10/13/2009, concerning draft RAIs pertaining to the PVNGS, Units 1, 2, and 3, LRA (ML092890006)
11/12/2009	Letter from NRC to APS, Request for Additional Information for the Review of Palo Verde Nuclear Generating Station, Units 1, 2, and 3, License Renewal Application (ML092890011)
11/13/2009	Letter from NRC to APS, Request for Additional Information for the Review of Palo Verde Nuclear Generating Station, Units 1, 2, and 3, License Renewal Application (ML093090057)
11/18/2009	Letter from NRC to APS, Plan for the Aging Management Program Audit Regarding the Palo Verde Nuclear Generating Station, Units 1, 2, & 3 License Renewal Application Review (TAC Nos. ME0254, ME0255, ME0256) (ML093090246)
12/3/2009	Letter from NRC to APS, Request for Additional Information for the Review of Palo Verde Nuclear Generating Station, Units 1, 2, and 3, License Renewal Application (ML093080557)
12/3/2009	Summary of telephone conference call held between NRC and APS, 10/22/2009, concerning a draft RAI pertaining to the PVNGS, Units 1, 2, and 3, LRA (ML093170428)
12/4/2009	Letter from NRC to APS, Request for Additional Information for the Review of Palo Verde Nuclear Generating Station, Units 1, 2, and 3, License Renewal Application (ML093200413)
12/7/2009	PVNGS, Units 1, 2, and 3, Submittal of annual update to the LRA and LRA, Amendment No. 3 (ML093500101)
12/11/2009	Letter from APS to NRC, PVNGS, Units 1, 2, and 3, Response to Request for Additional Information Regarding License Renewal Application (ML093631562)
12/17/2009	Letter from APS to NRC, PVNGS, Units 1, 2, and 3, Response to Request for Additional Information Regarding License Renewal Application (ML093631139)

Appendix B

Date	Subject
12/18/2009	Letter from APS to NRC, PVNGS, Units 1, 2, and 3, Response to Request for Additional Information Regarding License Renewal Application (ML093640043)
12/21/2009	Letter from APS to NRC, PVNGS, Units 1, 2, and 3, Response to Request for Additional Information Regarding License Renewal Application (ML100040067)
12/23/2009	Letter from NRC to APS, Request for Additional Information for the Review of Palo Verde Nuclear Generating Station, Units 1, 2, and 3, License Renewal Application (ML093380051)
12/23/2009	Letter from NRC to APS, Request for Additional Information for the Review of Palo Verde Nuclear Generating Station, Units 1, 2, and 3, License Renewal Application (ML093650157)
12/29/2009	Letter from NRC to APS, Request for Additional Information for the Review of Palo Verde Nuclear Generating Station, Units 1, 2, and 3, License Renewal Application (ML093490830)
1/14/2010	Letter from NRC to APS, Request for Additional Information for the Review of Palo Verde Nuclear Generating Station, Units 1, 2, and 3, License Renewal Application (ML093290298)
1/18/2010	Letter from APS to NRC, PVNGS, Units 1, 2, and 3, Response to Request for Additional Information Regarding License Renewal Application (ML100260951)
1/22/2010	Summary of telephone conference call held between NRC and APS, 1/14/2010, concerning draft RAIs pertaining to the PVNGS, Units 1, 2, and 3 LRA (ML100141784)
1/28/2010	Letter from NRC to APS, Request for Additional Information for the Review of Palo Verde Nuclear Generating Station, Units 1, 2, and 3, License Renewal Application (ML100150378)
2/5/2010	Letter from APS to NRC, PVNGS, Units 1, 2, and 3, Response to Request for Additional Information Regarding License Renewal Application (ML100490056)
2/12/2010	Summary of telephone conference call held between NRC and APS, 1/13/2010, concerning draft RAIs pertaining to the PVNGS, Units 1, 2, and 3, LRA (ML100140898)
2/16/2010	Summary of telephone conference call between NRC and APS, 1/12/2010, concerning draft RAIs pertaining to the PVNGS, Units 1 and 2, and 3, LRA (ML100131215)
2/19/2010	Letter from NRC to APS, Request for Additional Information for the Review of Palo Verde Nuclear Generating Station, Units 1, 2, and 3, License Renewal Application (ML100270069)
2/19/2010	Summary of telephone conference call held between NRC and APS, 1/28/2010, concerning draft RAIs pertaining to the PVNGS, Units 1, 2, and 3, LRA (ML100320041)
2/19/2010	Letter from APS to NRC, PVNGS, Units 1, 2, and 3, Response to Request for Additional Information Regarding License Renewal Application (ML100610604)
2/19/2010	Letter from NRC to APS, Palo Verde, Unit 1, Notification of Inspection (IR 05000528-10-002) and Request for Information (ML100502137)
3/1/2010	Letter from APS to NRC, PVNGS, Units 1, 2, and 3, Response to Request for Additional Information Regarding License Renewal Application (ML100680517)
3/1/2010	Letter from APS to NRC, PVNGS, Units 1, 2, and 3, Response to Request for Additional Information Regarding License Renewal Application (ML100680518)
3/2/2010	Letter from NRC to APS, Request for Additional Information for the Review of Palo Verde Nuclear Generating Station, Units 1, 2, and 3, License Renewal Application (ML100360296)
3/12/2010	Summary of telephone conference call held between NRC and APS, 1/8/2010, concerning containment coatings pertaining to the PVNGS, Units 1, 2, and 3, LRA (ML100150618)
3/12/2010	Summary of telephone conference call held between NRC and APS, 2/24/2010, concerning draft RAIs pertaining to the PVNGS, Units 1, 2, and 3, LRA (ML100610216)
3/15/2010	Summary of telephone conference call held between NRC and APS, 3/10/2010, concerning the PVNGS, Units 1, 2, and 3 LRA (ML100700618)
3/24/2010	Letter from APS to NRC, PVNGS, Units 1, 2, and 3, Response to Request for Additional Information Regarding License Renewal Application (ML100920055)
4/1/2010	Letter from APS to NRC, PVNGS, Units 1, 2, and 3, Response to Request for Additional Information Regarding License Renewal Application (ML101050045)

Appendix B

Date	Subject
4/2/2010	Letter from APS to NRC, PVNGS, Units 1, 2, and 3, Response to Request for Additional Information Regarding License Renewal Application (ML101050015)
4/7/2010	Letter NRC to APS, Audit Report Regarding the Palo Verde Nuclear Generating Station, Units 1, 2, and 3, License Renewal Application (TAC Nos. ME0254, ME0255, ME0256) (ML100221296)
4/8/2010	Letter from NRC to APS, Request for Additional Information for the Review of Palo Verde Nuclear Generating Station, Units 1, 2, and 3, License Renewal Application (ML100960367)
4/20/2010	Notice of public working meeting with APS, 5/6/2010, to discuss issues related to the PVNGS LRA (ML100980482)
4/20/2010	Letter from NRC to APS, 3/30/2010, Summary of Meeting With Arizona Public Service Company to Discuss Inspection Results for License Renewal Inspection of Nonsafety-related Scoping & Selected Aging Management Programs Conducted Onsite in February 2010 & Documented in IR-10-007 (ML101110595)
4/27/2010	Summary of telephone conference call between NRC and APS, 4/9/2010, concerning draft RAIs pertaining to the PVNGS, Units 1, 2, and 3, LRA (ML101060640)
4/28/2010	Letter from NRC to APS, Request for Additional Information for the Review of Palo Verde Nuclear Generating Station, Units 1, 2, and 3, License Renewal Application (ML101160357)
4/28/2010	Letter from APS to NRC, PVNGS, Units 1, 2, and 3, Supplemental Response to Request for Additional Information Regarding License Renewal Application (ML101320262)
4/29/2010	NRC Inspection Reports; IR 05000528-10-007, 05000529-10-007, 05000530-10-007; 2/1/2010–2/26/2010; PVNGS; Scoping of Nonsafety-Related Affecting Safety-Related Systems and Review of License Renewal Aging Management Programs (ML101190585)
4/29/2010	Letter APS to NRC, PVNGS, Units 1, 2, and 3, Response to Request for Additional Information Regarding License Renewal Application (ML101310227)
5/7/10	Letter from APS to NRC, PVNGS, Units 1, 2, and 3, Commitment to Incorporate in the Updated Final Safety Analysis Report a Requirement to Periodically Assess the Containment Building Interior Coating System (ML101390211)
5/21/2010	Letter from APS to NRC, PVNGS, Units 1, 2, and 3, Response to Request for Additional Information Regarding License Renewal Application (ML101540063)
5/28/2010	Letter from APS to NRC, PVNGS, Units 1, 2, and 3, License Renewal Application Amendment No. 16 (ML101600451)
6/2/2010	Summary of telephone conference call held between NRC and APS, 4/13/2010, concerning draft RAIs pertaining to the PVNGS, Units 1, 2, and 3, LRA (ML101330252)
6/2/2010	Summary of telephone conference call held between NRC and APS, 4/23/2010, concerning the PVNGS, Units 1, 2, and 3, LRA (ML101330286)
6/2/2010	Letter from NRC to APS, Request for Additional Information for the Review of Palo Verde Nuclear Generating Station, Units 1, 2, and 3, License Renewal Application (ML101340100)
6/2/2010	Summary of telephone conference call held between NRC and APS, 4/12/2010, concerning the PVNGS, Units 1, 2, and 3, LRA (ML101340666)
6/2/2010	Summary of telephone conference call held between NRC and APS, 5/14/2010, concerning the PVNGS, Units 1, 2, and 3, LRA (ML101340704)
6/2/2010	Summary of telephone conference call held between NRC and APS, 4/14/2010, concerning draft RAIs pertaining to the PVNGS, Units 1, 2, and 3, LRA (ML101340788)
6/21/2010	PVNGS, Units 1, 2, and 3, Responses to follow-up RAIs regarding buried piping, elastomers, compressed air, containment liner, spray ponds, and supports for the review of the PVNGS (ML101820185)
6/25/2010	Summary of telephone conference calls held between NRC and APS, 5/20/2010 and 6/9/2010, concerning draft RAIs pertaining to the PVNGS, Units 1, 2, and 3, LRA (ML101600547)
6/25/2010	Summary of public meeting held between NRC and APS, 5/6/2010, concerning the metal fatigue review pertaining to the PVNGS, Units 1, 2, and 3, LRA (ML101340802)

Appendix B

Date	Subject
6/29/2010	PVNGS; Units 1, 2, and 3; Docket Nos. STN 50-528, 50-529, and 50-530; 6/2/2010; Response to Request for Additional Information Regarding Metal Fatigue for the Review of the PVNGS License Renewal Application, and License Renewal Application Amendment No. 18 (ML101880278)
7/6/2010	Summary of telephone conference calls held between NRC and APS concerning a draft RAI associated with aging management of the compressed air system related to the PVNGS, Units 1, 2, and 3, LRA (ML101370289)
7/7/2010	PVNGS; Units 1, 2, and 3; Docket Nos. STN 50-528, 50-529, and 50-530; Supplemental Response to Request for Additional Information Regarding Time Limited Aging Analysis for the Review of the PVNGS License Renewal Application, and License Renewal Application Amendment No. 19 (ML101970058)
7/13/2010	Scoping and Screening Audit Report for the PVNGS, Units 1, 2, and 3, LRA (ML101740217)
7/14/2010	Summary of telephone conference calls held between NRC and APS concerning a draft RAI pertaining to the PVNGS, Units 1, 2, and 3, LRA (ML101340829)
7/19/2010	Summary of telephone conference calls held between NRC and APS concerning a draft RAI pertaining to the PVNGS, Units 1, 2, and 3, LRA (ML101760027)
7/20/2010	Summary of telephone conference call held between NRC and APS, 7/8/2010, concerning a draft RAI pertaining to the PVNGS, Units 1, 2, and 3, LRA (ML101900058)
7/21/2010	Palo Verde Nuclear Generating Station (PVNGS), Units 1,2, and 3, Docket Nos. STN 50-528, 50-529 and 50-530, Updated License Renewal Application Commitment List, and License Renewal Application Amendment No. 20 (ML102100096)
7/21/2010	RAI for the review of PVNGS, Units 1, 2, and 3, LRA (ML101890891)
7/23/2010	Summary of telephone conference call held between NRC and APS, 6/17/2010, concerning a draft RAI pertaining to the PVNGS, Units 1, 2, and 3, LRA (ML101730331)
7/30/2010	Palo Verde, Units 1, 2 & 3 - Responses to Follow-Up Request for Additional Information Regarding Small Bore Piping Socket Welds and Other Items for the Review of the PVNGS License Renewal Application, and License Renewal Application Amendment No. 21 (ML102240166)
8/5/2010	PVNGS - Follow-up of License Renewal Unresolved Item - Inspection Report 05000528, 05000529, 05000530/2010010 (ML102190239)
8/6/2010	PVNGS License Renewal - Safety Evaluation Report with Open Items (ML102210072)
8/6/2010	PVNGS - Transmittal letter to Arizona Public Service Company with Safety Evaluation Report with Open Items (ML102150416)
8/9/2010	Transmittal Letter to Advisory Committee on Reactor Safeguards Review of the PVNGS, Units 1, 2, and 3, LRA - SER with Open Items (ML102130350)
8/12/2010	PVNGS, Units 1, 2, and 3, Response to the July 21, 2010, Request for Additional Information Regarding Metal Fatigue for the Review of the PVNGS License Renewal Application, and License Renewal Application Amendment No. 22 (ML102360335)
8/27/2010	PVNGS, Units 1, 2, and 3, Revised Commitment Date to Drain the Spray Chemical Addition Tanks, and License Renewal Application Amendment No. 23 (ML102510187)
9/3/2010	PVNGS, Units 1, 2, and 3, Supplemental Responses to Request for Additional Information Regarding Small Bore Piping Socket Welds and Cavitation Erosion Related to the PVNGS License Renewal Application (ML102571399)
9/8/2010	Transcript of Advisory Committee on Reactor Safeguards Plant License Renewal Subcommittee, PVNGS, Units 1, 2, and 3, on September 8, 2010 in Rockville, MD, pages 1-156 (ML102590478)
9/15/2010	PVNGS, Units 1, 2 and 3 - Supplemental Responses to Request for Additional Information Regarding Small Bore Piping Socket Welds and Cavitation Erosion License Renewal Application (ML102571399)
9/17/2010	PVNGS, Units 1, 2 and 3 - Annual Update to License Renewal Application, and LRA Amendment No. 24 (ML102730057)
9/27/2010	RAI for the review of PVNGS, Units 1, 2, and 3, LRA (ML102560022)

Appendix B

Date	Subject
9/29/2010	PVNGS, Units 1, 2 and 3 - Comments on the Safety Evaluation Report with Open Items Related to License Renewal (ML102810502)
10/5/2010	Summary of telephone conference call held between NRC and APS, 8/31/2010, concerning a draft RAI pertaining to the PVNGS, Units 1, 2, and 3, LRA (ML102430338)
10/7/2010	Summary of telephone conference call held between NRC and APS, 9/22/2010, concerning a draft RAI pertaining to the PVNGS, Units 1, 2, and 3, LRA (ML102660130)
10/13/2010	PVNGS, Units 1, 2 and 3 - Responses to Requests for Additional Information for the Review of the PVNGS License Renewal Application (LRA), and LRA Amendment No. 25 (ML102930032)
10/15/2010	Summary of telephone conference call held between NRC and APS, 7/8/2010, concerning a draft RAI pertaining to the PVNGS, Units 1, 2, and 3, LRA (ML101940142)
10/15/2010	Summary of telephone conference call held between NRC and APS, 7/21/2010, concerning a draft RAI pertaining to the PVNGS, Units 1, 2, and 3, LRA (ML102030500)
10/15/2010	Summary of telephone conference call held between NRC and APS, 9/22/2010, concerning a draft RAI pertaining to the PVNGS, Units 1, 2, and 3, LRA (ML102660067)
10/18/2010	Summary of telephone conference call held between NRC and APS, 6/16/2010, concerning a draft RAI pertaining to the PVNGS, Units 1, 2, and 3, LRA (ML101740110)
11/10/2010	PVNGS, Units 1, 2, and 3, Response to Draft Request for Additional Information for the Review of License Renewal Application and LRA Amendment No. 26 (ML103280058)
11/16/2010	Summary of telephone conference calls held between NRC and APS, 10/22/2010 and 11/3/2010, concerning a Draft RAI pertaining to the PVNGS, Units 1, 2, and 3, LRA (ML102990530)
11/16/2010	Summary of telephone conference call held between NRC and APS, 10/28/2010, concerning a Draft Follow-up RAI pertaining to the PVNGS, Units 1, 2, and 3, LRA (ML103010523)
11/23/2010	PVNGS, Units 1, 2, and 3, Response to Draft Request for Additional Information for the Review of License Renewal Application and LRA Amendment No. 27 (ML103420101)
12/3/2010	PVNGS, Units 1, 2, and 3, Response to Draft Request for Additional Information for the Review of License Renewal Application and LRA Amendment No. 28 (ML103490138)
12/6/2010	Summary of telephone conference call held between NRC and APS, 11/19/2010, concerning a Draft Follow-up RAI pertaining to the PVNGS, Units 1, 2, and 3, LRA (ML103280174)
12/16/2010	PVNGS, Units 1, 2, and 3, Correction to License Renewal Application Section B2.1.10, Closed-Cycle Cooling Water System (ML103630422)
1/7/2011	PVNGS, Units 1, 2, and 3, Region IV Administrator License Renewal Recommendation Letter (ML110100400)
1/11/2011	Summary of telephone conference call held between NRC and APS, 1/11/2011, concerning an RAI response pertaining to the PVNGS, Units 1, 2, and 3, LRA (ML110070256)
1/26/2011	Update to License Renewal Application (LRA) Section A3/2/1/5, LRA Amendment No. 29 (ML110350035)
2/25/2011	PVNGS, Units 1, 2, and 3, Correction to Response to Request for Additional Information for the Review of the PVNGS License Renewal Application (LRA), and LRA Amendment No. 30 (ML110670168)
3/1/2011	Letter to the Honorable Gregory B. Jaczko, Chairman, from Said Abdel-Khalik, ACRS Chairman, dated March 1, 2011, Subject: Report on the Safety Aspects of the License Renewal Application for the Palo Verde Nuclear Generating Station (ML110700690)
3/4/2011	PVNGS, Units 1, 2, and 3, Correction of Typographical Error in Safety Evaluation Report Realted to License Renewal (TAC Nos. ME0254, ME0255, and ME0256) (ML110490540)
3/17/2011	PVNGS, Units 1, 2, and 3, Clarification to Response to Request for Additional Information for the Review of the PVNGS License Renewal Application (LRA), and LRA Amendment No. 31 (ML110880072)

APPENDIX C

Principal Contributors

This appendix lists the principal contributors for the development of this safety evaluation report and their areas of responsibility.

Name	Responsibility
A. Hiser	Management Oversight
A. Klein	Management Oversight
A. Johnson	Reviewer—Reactor Systems and Mechanical
A. Paulson	Reviewer—Mechanical
A. Sheikh	Reviewer—Structural
A. Ulses	Management Oversight
A. Wong	Reviewer—Mechanical
B. Fu	Reviewer—Reactor Systems
B. Holian	Management Oversight
B. Lehman	Reviewer—Structural
B. Parks	Reviewer—Reactor Systems
B. Rogers	Reviewer—Scoping and Screening Methodology
C. Ng	Reviewer—Reactor Systems
C. Doutt	Reviewer—Electrical
D. Alley	Reviewer—Mechanical
D. Hoang	Reviewer—Structural
D. Nguyen	Reviewer—Aging Management Programs and Electrical
D. Pelton	Management Oversight
D. Wrona	Management Oversight
E. Smith	Reviewer—Scoping and Screening Methodology
E. Wong	Reviewer—Chemical
G. Casto	Management Oversight
G. Cranston	Management Oversight
G. Shukla	Management Oversight
H. Ashar	Reviewer—Structural
J. Bettle	Reviewer—Mechanical
J. Collins	Reviewer—Reactor Systems
J. Davis	Reviewer—Aging Management Programs
J. Dozier	Management Oversight
J. Gavula	Reviewer—Mechanical
J. Medoff	Reviewer—Reactor Systems
J. Robinson	Reviewer—Mechanical
J. Rowley	Project Management

Appendix C

Name	Responsibility
J. Shea	Reviewer—Scoping and Screening Methodology
J. Tsao	Reviewer—Reactor Systems
L. Banic	Reviewer—Reactor Systems
L. Regner	Senior Project Management
M. Kichline	Reviewer—Mechanical
M. Mitchell	Management Oversight
N. Iqbal	Reviewer—Fire Protection
O. Yee	Reviewer—Mechanical
R. Auluck	Management Oversight
R. Dennig	Management Oversight
R. Sun	Reviewer—Mechanical
S. Lee	Management Oversight
S. Min	Reviewer—Reactor Systems
S. Sheng	Reviewer—Reactor Systems
W. Smith	Reviewer—Mechanical
W. Holston	Reviewer—Mechanical
Contractors	
Advanced Technologies and Laboratories International, Inc.	
J. Davis	Reviewer—Reactor Systems
W. Jackson	Reviewer—Reactor Systems
E. Patel	Reviewer—Reactor Systems
W. Pavanich	Reviewer—Reactor Systems
Center for Nuclear Waste Regulatory Analysis	
K. Axler	Reviewer—Mechanical
K. Chiang	Reviewer—Reactor Systems
T. Mintz	Reviewer—Mechanical
R. Kazban	Reviewer—Reactor Systems
E. Trillo	Reviewer—Mechanical
Oak Ridge National Laboratory	
D. Naus	Reviewer—Structural

APPENDIX D

References

This appendix lists the references used throughout this safety evaluation report for review of the license renewal application for the Palo Verde Nuclear Generating Station, Units 1, 2, and 3.

References

10 CFR Part 50, "Domestic Licensing of Production and Utilization Facilities"

10 CFR Part 54, "Requirements for Renewal of Operating Licenses for Nuclear Power Plants"

10 CFR Part 100, "Reactor Site Criteria"

American Concrete Institute (ACI), 201, "Durability of Concrete"

ACI, 201.2R, "Guide to Durable Concrete"

ACI, 211.1, "Standard Practice for Selecting Proportions for Normal, Heavyweight, and Mass Concrete"

ACI, 318, "Building Code Requirements for Structural Concrete"

ACI, 349, "Code Requirements for Nuclear Safety-Related Concrete Structures"

American National Standards Institute (ANSI), B31.1, "Power Piping"

American Society of Mechanical Engineers (ASME) Boiler and Pressure Vessel Code, Section III, "Rules for Construction of Nuclear Power Plant Components"

ASME Boiler and Pressure Vessel Code, Section XI, "Rules for Inservice Inspection of Nuclear Power Plant Components"

ASME Code, Section XI, Appendix G, "Fracture Toughness Criteria for Protection Against Failure"

ASME Code, Case N-512, "Assessment of Reactor Vessels with Low Upper-Shelf Charpy Impact Energy Levels"

American Society for Metals (ASM), "Metals Handbook® Desk Edition," 1985

American Society for Testing Materials (ASTM), Special Technical Publication 1005, "Distillate Fuel: Contamination, Storage and Handling"

ASTM, C 33, "Standard Specification for Concrete Aggregates"

ASTM, D 96, "Standard Test Method for Water and Sediment in Crude Oil by Centrifuge Method (Field Procedure)"

ASTM, E 185, "Standard Practice for Design of Surveillance Programs for Light-Water Moderated Nuclear Power Reactor Vessels"

ASTM, C 227, "Standard Test Method for Potential Alkali Reactivity of Cement-Aggregate Combinations (Mortar-Bar Method)"

ASTM, C 289, "Standard Test Method for Potential Alkali-Silica Reactivity of Aggregates (Chemical Method)"

ASTM, C 295, "Standard Guide for Petrographic Examination of Aggregates for Concrete"

ASTM, D 445, "Standard Test Method for Kinematic Viscosity of Transparent and Opaque Liquids"

Appendix D

ASTM, D 975, "Standard Practice for Manual Sampling of Petroleum and Petroleum Products"

ASTM, D 1796, "Standard Practice for Manual Sampling of Petroleum and Petroleum Products"

ASTM, D 2276, "Standard Test Method for Particulate Contaminant in Aviation Fuel by Line Sampling"

ASTM, D 2709, "Standard Practice for Manual Sampling of Petroleum and Petroleum Products"

ASTM, D 4057, "Standard Practice for Manual Sampling of Petroleum and Petroleum Products"

ASTM, D 4378, "Standard Practice for In-Service Monitoring of Mineral Turbine Oils for Steam and Gas Turbines"

ASTM, D 4951, "Standard Test Method for Determination of Additive Elements in Lubricating Oils by Inductively Coupled Plasma Atomic Emission Spectrometry"

ASTM, D 6217, "Standard Test Method for Particulate Contamination in Middle Distillate Fuels by Laboratory Filtration"

ASTM, D 6224, "Standard Practice for In-Service Monitoring of Lubricating Oil for Auxiliary Power Plant Equipment"

ASTM, D 6595, "Determination of Wear Metals and Contaminants in Used Lubricating Oils or Used Hydraulic Fluids by Rotating Disc Electrode Atomic Emissions Spectroscopy"

Bauer, Scott A., Arizona Public Service Company (APS), letter to U.S. Nuclear Regulatory Commission (NRC), "Palo Verde Nuclear Generating Station Unit 2, Reactor Vessel Material Surveillance Capsule at 230°," 102-05457-SAB/TNW/RJR, Docket 50-529, April 4, 2006 (Agencywide Document Access and Management System (ADAMS) Accession Number ML061040590)

Booser, R.E., "CRC Handbook of Lubrication," CRC Press, Inc., Volumes 1 and 2, 1983

Combustion Engineering (CE), Letter LD-83-053, A. E. Scherer to Darrell G. Eisenhut, NRC Docket No. STN 50-470F, "Basis for Design of Plant Without Pipe Whip Restraints," June 14, 1983 (ADAMS Public Legacy Library Accession No. 8306200254).

CE, Letter LD-83-108, A. E. Scherer to Darrell G. Eisenhut, NRC Docket No. STN 50-470F, "Basis for Design of Plant Without Pipe Whip Restraints," December 23, 1983 (ADAMS Public Legacy Library Accession No. 8312300117) With enclosed, *Leak Before Break Evaluation of the Main Loop Piping of a CE Reactor Coolant System*, Revision 1, November 1983

CE, Letter LD-83-053, A. E. Scherer to Darrell G. Eisenhut, NRC Docket No. STN 50-470F, "Basis for Design of Plant Without Pipe Whip Restraints," June 14, 1983 (ADAMS Public Legacy Library Accession No. 8306200254)

CE, Letter to A. E. Scherer, "CESSAR System 80 Safety Evaluation Report (NUREG-0852) Removal of Pipe Whip Restraints," October 11, 1984 (ADAMS Public Legacy Library Accession No. 8410240182)

"Cooper-Bessemer Model KSV Emergency Diesel Generator Lubricating Oil and Jacket Water Analysis Guidelines," Revision 1, 1993

Electric Power Research Institute (EPRI), EPRI No. CS-4555, "Guidelines for Maintaining Steam Turbine Lubrication Systems," July 1986

EPRI, NP-4916, "Lubrication Guide," January 1987

EPRI, NP-5067, "Good Bolting Practices, A Reference Manual for Nuclear Power Plant Maintenance Personnel," Volume 1: "Large Bolt Manual," 1987 and Volume 2: "Small Bolts and Threaded Fasteners," 1990

Appendix D

EPRI, NP-5769, "Degradation and Failure of Bolting in Nuclear Power Plants," Volumes 1 and 2, April 1988

EPRI, Technical Report (TR)-104213 "Bolted Joint Maintenance and Application Guide," December 1995

EPRI, TR-107396, "Closed Cooling water chemistry Guideline," October 31, 1997

EPRI, TR-112657, "Revised Risk-Informed Inservice Inspection Evaluation Procedure," Revision B-A, January 12, 2000

EPRI, TR-1000701, "Interim Thermal Fatigue Management Guideline" (MRP-24), January 16, 2001

EPRI TR-1008224, "PWR Secondary Water Chemistry Guidelines," Revision 6, December 13, 2004

EPRI, TR-1016456, "Recommendations for an Effective Program to Control the Degradation of Buried Pipe," December 2008

"Evaluation of Reactor Pressure Vessels with Charpy Upper-Shelf Energy Less than 50 Ft-Lb," Draft Guide (DG)-1023, September 1993

Generic Letter (GL) 86-10, "Implementation of Fire Protection Requirements," April 24, 1986

GL 88-05, "Boric Acid Corrosion of Carbon Steel Reactor Pressure Boundary Components in PWR plants," March 17, 1988

GL 89-08, "Erosion/Corrosion-Induced Pipe Wall Thinning," May 2, 1989

GL 89-13, "Service Water System Problems Affecting Safety-Related Equipment," July 18, 1989

GL 92-01, "Reactor Vessel Structural Integrity," February 28, 1992

GL 2004-02, "Potential Impact of Debris Blockage on Emergency Recirculation during Design Basis Accidents at Pressurized-Water Reactors," September 13, 2004

GL 2007-01, "Inaccessible or Underground Power Cable Failures that Disable Accident Mitigation Systems or Cause Plant Transients," February 7, 2007

International Organization for Standardization (ISO), ISO 4406, "Hydraulic Fluid Power - Fluids - Method for Coding the Level of Contamination by Solid Particles - Second Edition"

Letter from Christopher I. Grimes, NRC, to Douglas J. Walters, NEI, License Renewal Issue No. 98-0030, "Thermal Aging Embrittlement Of Cast Austenitic Stainless Steel Components," May 19, 2000 (ML003717179)

Letter from David B. Matthews, NRC, to Messrs. Alan Nelson and David Lochbaum, "Staff Guidance on Scoping of Equipment Relied on to Meet the Requirements of the Station Blackout Rule (10 CFR 50.63) for License Renewal (10 CFR 54.4(a)(3))," April 1, 2002

Letter from Pao-Tsin Kuo, NRC, to Messrs. Alan Nelson and David Lochbaum, "Standardized Format for License Renewal Applications," April 7, 2003

License Renewal Interim Staff Guidance (LR-ISG)-2007-02, "Changes to Generic Aging Lesson Learned (GALL) Report Aging Management Program (AMP) XI.E6, 'Electrical Cable Connections Not Subject to 10 CFR 50.49 Environmental Qualification Requirements," issued for public comment by letter dated August 29, 2007 (ML072420437)

Metals Handbook, Ninth Edition, Vol. 13, Metals Park, Ohio, 1987

National Academy of Sciences (NAS), 1638, "Cleanliness Requirements of Parts Used in Hydraulic Systems," August 2001

Appendix D

National Fire Protection Association (NFPA), "NFPA 25, Standard for the Inspection, Testing, and Maintenance of Water-Based Fire Protection Systems," 1998 and 2002 editions

"Notice of availability of the final ISG LR ISG-2007-02," *Federal Register*, December 23, 2009 (74 FR 68287)

Nuclear Administrative Technical Manual (NATM)

Nuclear Energy Institute (NEI), "NUMARC 93-01, Rev. 2, Industry Guideline for Monitoring the Effectiveness of Maintenance of Nuclear Power Plants," April 1996

NEI, TR 95-10, Revision 6, "Industry Guideline for Implementing the Requirements of 10 CFR Part 54 - The License Renewal Rule," June 2005.

NEI, TR 97-06, "Steam Generator Program Guidelines"

NEI, TR 07-07, "Industry Ground Water Protection Initiative, Final Guidance Document," August 2007

Nuclear Management and Resources Council, Letter to W.H. Rasin, "Safety Assessment of WCAP-13587, Revision 1, 'Reactor Vessel Upper Shelf Energy Bounding Evaluation for Westinghouse Pressurized Water Reactors,' September 1993," April 21, 1994 (ADAMS Public Legacy Library Accession No. 9405060287)

NRC, Branch Technical Position (BTP) APCSB 9.5-1, "Guidelines for Fire Protection for Nuclear Power Plants"

NRC, BTP IQMB-1, "Quality Assurance for Aging Management Programs"

NRC, BTP RLSB-1, "Aging Management Review – Generic"

NRC, Bulletin 02-01, "Reactor Pressure Vessel Head Degradation and Reactor Coolant Pressure Boundary Integrity," March 18, 2002

NRC, Bulletin 02-02, "Reactor Pressure Vessel Head and Vessel Head Penetration Nozzle Inspection Programs," August 9, 2002

NRC, Bulletin 03-02, "Leakage from Reactor Pressure Vessel Lower Head Penetrations and Reactor Coolant Pressure Boundary Integrity," August 21, 2003

NRC, Bulletin 80-11, "Masonry Wall Design," May 8, 1980

NRC, Bulletin 88-02, "Rapidly Propagating Fatigue Cracks in Steam Generator Tubes," February 5, 1988

NRC, Bulletin 88-08, "Thermal Stresses in Piping Connected to Reactor Cooling Systems," June 22, 1988

NRC, Bulletin 88-09, "Thermal Stresses in Piping Connected to Reactor Cooling Systems," July 26, 1988

NRC, Bulletin 88-11, "Pressurizer Surge Line Thermal Stratification," December 20, 1988

NRC, Information Notice (IN) 00-14, "Non-Vital Bus Fault Leads to Fire and Loss of Offsite Power," September 27, 2000

NRC, IN 02-12, "Submerged Safety-Related Electrical Cables," April 21, 2002

NRC, IN 09-04, "Age-Related Constant Support Degradation", February 18, 2004

NRC, IN 87-67, "Lessons Learned from Regional Inspections of Licensee Actions in Response to IE Bulletin 80-11," December 31, 1987

NRC, IN 89-53, "Rupture of Extraction Steam Line on High Pressure Turbine," June 13, 1989

Appendix D

NRC, IN 89-64: "Electrical Bus Bar Failures," September 7, 1989

NRC, IN 90-04, "Cracking of the Upper Shell-to-Transition Cone Girth Welds in Steam Generators," January 26, 1990

NRC, IN 91-19, "Steam Generator Feedwater Distribution Piping Damage," March 12, 1991

NRC, IN 94-59, "Accelerated Dealloying of Cast Aluminum-Bronze Valves Caused by Microbiologically Induced Corrosion," August 17, 1994

NRC, IN 94-63, "Boric Acid Corrosion of Charging Pump Casing Caused by Cladding Cracks," August 30, 1994

NRC, IN 97-11, "Cement Erosion From Containment Subfoundations at Nuclear Power Plants," March 21, 1997

NRC, IN 97-84, "Rupture in Extraction Steam Piping as a Result of Flow-Accelerated Corrosion," December 11, 1997

NRC, IN 99-10, "Degradation of Prestressing Tendon Systems in Prestressed Concrete Containments," April 13, 1999

Nuclear Safety Analysis Center (NSAC)-202L-R2, "Recommendations for an Effective Flow-Accelerated Corrosion Program," April 1999

NSAC-202L-R3, "Recommendations for an Effective Flow Accelerated Corrosion Program," December 2006

NUREG/CR-6260, "Application of NUREG/CR-5999 Interim Fatigue Curves to Selected Nuclear Power Plant Components," 1995

NUREG/CR-6583, "Effects of LWR Coolant Environments on Fatigue Design Curves of Carbon and Low-Alloy Steels," 1998

NUREG-0138, "Staff Discussion of Fifteen Technical Issues Listed in Attachment to November 3, 1976 Memorandum from Director, NRR to NRR Staff," November 1976.

NUREG-0313, "Technical Report on Material Selection & Processing GL For BWR Coolant Press Boundary Piping," July 1980

NUREG-0588, "Interim Staff Position on Environmental Qualification of Safety-Related Electrical Equipment," November 1979

NUREG-0800, "Standard Review Plan for the Review of Safety Analysis Reports for Nuclear Power Plants" September 2005

NUREG-0857, "Safety Evaluation Report related to the Operation of Palo Verde Nuclear Generating Station Units 1, 2 and 3, Docket Nos. STN 50-528, STN 50-529, and STN 50-530, Arizona Public Service Company," with Supplements 1 through 12, November 1981

NUREG-1339,"Resolution of Generic Safety Issue 29: Bolting Degradation or Failure in Nuclear Power Plants," June 1990

NUREG-1350, "2009-2010 Information Digest," Volume 21, August 2009 (ML092370512)

NUREG-1557, "Summary of Technical Information and Agreements from Nuclear Management and Resources Council Industry Reports Addressing License Renewal"

NUREG-1785, "Safety Evaluation Report Related to the License Renewal of H.B. Robinson Steam Electric Plant, Unit 2," March 2004

NUREG-1786, "Safety Evaluation Report Related to the License Renewal of R.E. Ginna Nuclear Power Plant," May 2004

Appendix D

NUREG-1800, "Standard Review Plan for Review of License Renewal Applications for Nuclear Power Plants," Revision 1, September 2005

NUREG-1801, "Generic Aging Lessons Learned (GALL) Report," Revision 1, September 2005

NUREG-1833, "Technical Bases for Revision to the License Renewal Guidance Documents" October 2005

NUREG-1929, "Safety Evaluation Report, Related to the License Renewal of Beaver Valley Power Station, Units 1 and 2, FirstEnergy Nuclear Operating Company," Volume 2, October 2009

Office of the Federal Register, Title 10 to the *Code of Federal Regulations*, Part 20 (10 CFR 20), "Standards for Protection Against Radiation"

Regulatory Guide (RG) 1.35, "Inservice Inspection of Ungrouted Tendons in Prestressed Concrete Containments," July 11, 1990

RG 1.65, "Materials and Inspections for Reactor Vessel Closure Studs," October 1973

RG 1.99, "Radiation Embrittlement of Reactor Vessel Materials," Revision 2, May 1988 (ML003740284)

RG 1.121, "Bases for Plugging Degraded PWR Steam Generator Tubes (for Comment)," August 1976

RG 1.127, "Inspection of Water-Control Structures Associated with Nuclear Power Plants," Revision 1, March 1978

RG 1.137, "Fuel-Oil Systems for Standby Diesel Generators," Revision 1, October 1979

RG 1.147, "Inservice Inspection Code Case Acceptability, ASME Section XI, Division 1," Revision 15, October 2007

RG 1.155, "Station Blackout," Revision 0, August 1988

RG 1.160, "Monitoring the Effectiveness of Maintenance at Nuclear Power Plants," Revision 2, March 1997

RG 1.161, "Evaluation of Reactor Pressure Vessels with Charpy Upper-Shelf Energy Less Than 50 Ft-Lb," June 1995 (ML003740038)

RG 1.163, "Performance-Based Containment Leak-Test Program," September 1995

RG 1.188, "Standard Format and Content for Applications to Renew Nuclear Power Plant Operating Licenses," Revision 1, September 2005

RG 1.190, "Calculational and Dosimetry Methods for Determining Pressure Vessel Neutron Fluence," March 31, 2001 (ML010890301)

Roff, W.J., *Fibres, Plastics, and Rubbers: A Handbook of Common Polymers*, Academic Press Inc., New York, 1956

Shell Oil Company, "Lubricants Guide," 1981

Westinghouse Electric Company, WCAP-14040-A, "Methodology Used to Develop Cold Overpressure Mitigating System Setpoints and RCS Heatup and Cooldown Limit Curves," Revision 4, May 2004 (ML050120209)

WCAP-16374-NP, "Analysis of Capsule 230° from Arizona Public Service Company Palo Verde Unit 1 Reactor Vessel Radiation Surveillance Program," February 2005

Appendix D

WCAP-16449-NP, "Analysis of Capsule 230° from Arizona Public Service Company Palo Verde Unit 3 Reactor Vessel Radiation Surveillance Program," August 2005

WCAP-16524-NP, "Analysis of Capsule 230° from Arizona Public Service Company Palo Verde Unit 2 Reactor Vessel Radiation Surveillance Program," February 2006

NRC FORM 335 (12-2010) NRCMD 3.7	U.S. NUCLEAR REGULATORY COMMISSION	1. REPORT NUMBER (Assigned by NRC, Add Vol., Supp., Rev., and Addendum Numbers, if any.)
BIBLIOGRAPHIC DATA SHEET *(See instructions on the reverse)*		NUREG-1961

2. TITLE AND SUBTITLE	3. DATE REPORT PUBLISHED	
Safety Evaluation Report Related to the License Renewal of the Palo Verde Nuclear Generating Station, Units 1, 2, and 3	MONTH	YEAR
	April	2011
	4. FIN OR GRANT NUMBER	

5. AUTHOR(S)	6. TYPE OF REPORT
Lisa M. Regner	
	7. PERIOD COVERED *(Inclusive Dates)*

8. PERFORMING ORGANIZATION - NAME AND ADDRESS *(If NRC, provide Division, Office or Region, U.S. Nuclear Regulatory Commission, and mailing address; if contractor, provide name and mailing address.)*

Division of License Renewal
Office of Nuclear Reactor Regulation
U.S. Nuclear Regulatory Commission
Washington, DC 20555-0001

9. SPONSORING ORGANIZATION - NAME AND ADDRESS *(If NRC, type "Same as above"; if contractor, provide NRC Division, Office or Region, U.S. Nuclear Regulatory Commission, and mailing address.)*

Same as above

10. SUPPLEMENTARY NOTES

11. ABSTRACT *(200 words or less)*

This safety evaluation report (SER) documents the technical review of the Palo Verde Nuclear Generating Station, Units 1, 2, and 3 (PVNGS), license renewal application (LRA) by the U.S. Nuclear Regulatory Commission (NRC) staff. By letter dated December 11, 2008, as supplemented by letter dated April 14, 2009, Arizona Public Service Company (the applicant) submitted the LRA in accordance with Title 10 of the Code of Federal Regulations, Part 54 "Requirements for the Renewal of Operating Licenses for Nuclear Power Plants." The applicant requested renewal of the PVNGS operating licenses NPF-41 (Unit 1), NPF-51 (Unit 2), and NPF-74 (Unit 3), for a period of 20 years beyond the current expiration dates of midnight on June 1, 2025, April 24, 2026, and November 25, 2027, respectively.

On August 6, 2010, the NRC staff issued an SER with Open Item Related to the License Renewal of Palo Verde Nuclear Generating Station, Units 1, 2, and 3, in which one open item and five confirmatory items were identified as needing further review. This SER presents the status of the staff's review of information submitted through March 17, 2011, the cutoff date for consideration in the SER. The open and confirmatory items identified in the SER with Open Item were resolved before the staff made its final determination. SER Sections 1.5 and 1.6 summarize these open and confirmatory items. SER Section 6.0 provides the staff's final conclusion of the LRA review.

12. KEY WORDS/DESCRIPTORS *(List words or phrases that will assist researchers in locating the report.)*	13. AVAILABILITY STATEMENT
Palo Verde Nuclear Generating Station, Units 1, 2, and 3 Arizona Public Service Company License Renewal Nuclear Power Plant 10 CFR Part 54 Docket Nos. 50-528, 50-529, 50-530 Aging Management Scoping and Screening Time-Limited Aging Analysis	unlimited
	14. SECURITY CLASSIFICATION
	(This Page) unclassified
	(This Report) unclassified
	15. NUMBER OF PAGES
	16. PRICE

www.ingramcontent.com/pod-product-compliance
Lightning Source LLC
Chambersburg PA
CBHW081101170526
45165CB00008B/2286